APPLIED PHYSICS

Fourth Edition

APPLIED PHYSICS

Fourth Edition

Prof. T. Bhima Sankaram
Retd. Professor in Physics,
Osmania University,
Hyderabad.

Dr. G. Prasad
Professor in Physics,
Osmania University,
Hyderabad.

BSP BS Publications
A unit of **BSP Books Pvt., Ltd.**

4-4-309, Giriraj Lane, Sultan Bazar,
Hyderabad - 500 095 - A.P.
Phone : 040 - 23445605, 23445688

Fourth edition, 2014
Third revised edition, 2009
Third edition, 2008
Second edition, 2006
First edition, 2002

Published by :

BSP BS Publications
A unit of **BSP Books Pvt., Ltd.**

4-4-309, Giriraj Lane, Sultan Bazar,
Hyderabad - 500 095 - A.P.
Phone : 040 - 23445605, 23445688
e-mail : info@bspbooks.net
www.bspbooks.net

Printed at

Aditya Offset Process (I) Pvt. Ltd.
Hyderabad.

Price : Rs. 275.00
ISBN : 978-81-7800-314-6

Dedicated

to

In loving Memory of

Late Prof. K.G.Bansigir

and

Late Prof.K.V.Rao

Preface to Fourth Edition

It gives us a great pleasure in releasing a fourth edition of our book "Applied Physics". The revision has been done to reach greater number of readers in the country. Our object has always been to present the subject in a lucid manner so that the readers understand and appreciate the content. The object of any book is to make the student remember the book and go back to the book whenever they feel like referring to certain topic they studied in this book. We are glad that we are sticking to that concept. In this fourth edition book, we have added topics on interference, diffraction and polarisation that come under physical optics. It is very essential for all engineering students to learn the concepts regarding the topics as many applications in devices in their lives involve optical phenomenon. In fact future technology in many fields is likely to be optical based - systems. We also a brief information about the origin of Maxwell's equations that are so important in entire communication world that involve generation, transmission and receiving the signals like optical fiber systems. We tried to be concise as our object is to give just concepts over which the students will develop their concepts during their careers.

Teachers, especially the teachers who motivate us to develop an urge to learn and cultivate an original thinking in our lives are the real guides. We were lucky we had two teachers whom we consider as the two inspirers who motivated us to go for teaching jobs and research pursuits. One was late Prof. K.G. Bansigir (Physics head and dean in Jiwaji university, Gwalior, teacher and research guide of Prof. T. Bhima Sankarm) and late Prof. K.V. Rao (professor, dept. of Physics, IIT, Kharagpur and research guide of Prof. G. Prasad.)

We dedicate this revised to the memory of those two teachers whom we always hold in our hearts with great respect.

-Authors

Contents

Chapter - 3 X-Ray Diffraction

Chapter - 4 Defects in Crystals

CHAPTER - 5 | Elements of Statistical Mechanics

Chapter - 7 | Band Theory of Solids

Chapter - 8 | Semiconductor Physics

CHAPTER - 9 | Physics of Semiconductor Devices

CHAPTER - 10 | Dielectric Properties

CHAPTER - 11 Magnetic properties

CHAPTER - 12 | Lasers

CHAPTER - 15 | Nanotechnology

Chapter - 16 | Physical Optics for Instruments

CHAPTER - 17 | Photoconductors and Solar Cells

CHAPTER - 18 | Superconductivity

Chapter 1

Bonding in Solids

1.1 Introduction

Solids are formed when the atoms (molecules) in a liquid state or gaseous state are brought closer into a condensed state and each atom is in close proximity of its neighbours. The condensation into a particular structure of atomic arrangement in solids is determined by the nature and magnitude of the interaction forces between atoms, temperature and pressure. In the process of solidification, the interaction between atoms is such that it reduces the free energy of the system. A system, like a solid, possesses lower energy because of the fact that the electrons occupy lower energy levels than the electrons of typical isolated atom. The state of atoms at an equilibrium separation, thus, corresponds to the condition of stable equilibrium. The interaction forces will be such that the atoms will be at an equilibrium separation to form a crystal of an ordered internal structure.

In general, the chemical bonds that occur in crystals are classified as 1. primary bonds, and 2. secondary bonds. The primary atomic bonds are formed when the interaction forces are relatively strong. The primary bonds are classified as 1. Ionic bonds, 2. Covalent bonds, and 3. Metallic bonds.

The ionic bond is formed between two atoms when one atom transfers an electron to another acceptor atom to give ions of opposite charge. The bonds are formed as a result of strong Coulomb forces. The covalent bond is formed between atoms which share electrons present in the outer most shells. The metallic bond is formed between two metallic atoms which share electron in a delocalized manner. The bond is strong and non-directional.

The secondary bonds are classified into two groups. 1. Permanent dipole bonds: These bonds are formed between molecules that possess permanent dipoles. 2. Fluctuating dipole bonds : These bonds take place among atoms as a result of an asymmetrical distribution of electron densities around their nuclei. The bonding is very weak and is termed fluctuating since the electron density continuously changes with time.

Interatomic Forces

In a gaseous state of matter, the distance of separation between molecules is large. The interatomic force between the atoms or molecules is very weak. The molecules or atoms exist in a continuous motion colliding with each other and the walls of the container in which they are present. When they are brought closer and closer, they first condense into liquid in which the translational motion of the molecules is totally restricted. The energy is then lowered compared to that of free atoms. If the liquid is then condensed to form a solid, the atomic motion is restricted. The atoms will have only vibrational motion, and the rotatory motion, and translational motion are totally absent. The atoms will be held by the bonding forces. The energy of the solid is much lower than that of the liquid state or gaseous state. The atoms in a solid are held by an energy called the bond energy.

The liquid, like the melt of a solid, consists of a short range order that is not stable. There is a continuous agitation of molecules or atoms. The short range order is broken and restored again. This process takes place continuously with a relaxation time. When a solid is formed out of such a state, there is a heat of crystallization that is evolved. The atoms are brought closer to an equilibrium distance at a temperature that is the freezing temperature. In solids, the forces that operate between the atoms are the attractive forces and repulsive forces, and these forces determine the potential energy associated with the atoms in the solid.

Let us examine how does the resultant force between the atoms changes when the atoms are brought closer to each other. At relatively large distances, the attractive force between the atoms dominates. It grows as the interatomic distance diminishes. This variation is shown in Fig. 1.1 as curve 1. When the atoms come sufficiently closer, the repulsive forces come into play ($F_{repul} \propto 1/r^n$) and they grow faster than attractive forces. When the atoms are brought closer, their charge distributions overlap. This overlapping causes electrostatic energy to change. The overlap energy is repulsive at close distances. The repulsion occurs because Pauli exclusion principle prevents multiple occupancy. When atoms reach a distance of separation $r = r_0$ the attractive forces just balance the repulsive forces and the resultant force becomes zero (curve 3). The distance of separation, r_0 is called 'the equilibrium distance'.

The interaction energy U also changes as $F = -\partial U / \partial r$. The interaction energy at $r = r_0$ is called binding energy of pair of atoms. The state defined by $(U)_{r = r_0}$ is a stable equilibrium state. What is said above is valid for all types of primary bonds and the general character of the forces is the same for all types of solids.

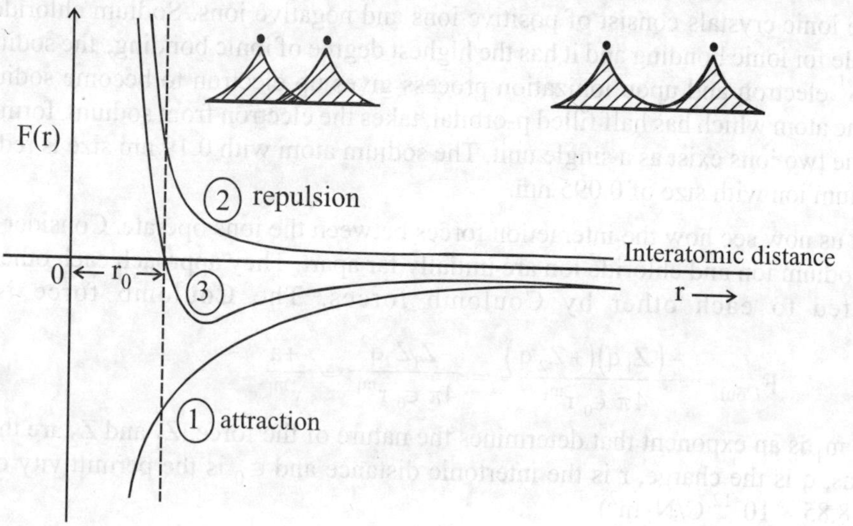

Fig. 1.1 The variation of force between atoms as they are brought closer to each other curve ① attractive force ② repulsive force ③ the resultant force between the atoms. The solid circles and the shaded peaks represent charge distribution overlap.

1.2 Types of Bonding

We shall now see the different types of bonding in crystals in the following sections.

1.2.1 Ionic Bond

Ionic bond is formed when the atoms with small number of electrons on the outer shells, have a tendency to give up electrons to the atoms with an almost filled outer shell. Ionic bond is the simplest type of interatomic bond. By loosing electrons, metallic atoms like alkali-metals, alkaline-earth metals etc., go into the nearest noble gas configuration, while the atoms like halogens take electrons to gain the inert gas configuration. Alkali halides are the simplest type of ionic bonded crystals. The formation of Na C*l* by ionic bonding is shown in Fig. 1.2.

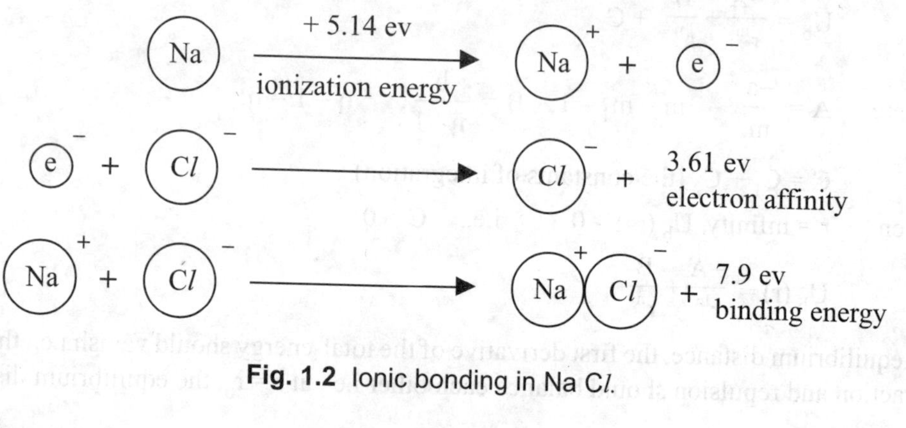

Fig. 1.2 Ionic bonding in Na C*l*.

The ionic crystals consist of positive ions and negative ions. Sodium chloride is the best example for ionic bonding and it has the highest degree of ionic bonding, the sodium atom has one $3S^1$ electron and upon ionization process gives up electron to become sodium ion. The chlorine atom which has half filled p-orbital, takes the electron from sodium, forming chloride ion. The two ions exist as a single unit. The sodium atom with 0.19 nm size is reduced in size to sodium ion with size of 0.095 nm.

Let us now see how the interaction forces between the ions operate. Consider that the two ions, sodium ion and chloride ion are initially far apart. They approach each other as they are attracted to each other by Coulomb forces. The Coulomb force is given as

$$F_{Coul} = \frac{-(Z_1 q)(+Z_2 q)}{4\pi \in_0 r^{m_1}} = \frac{-Z_1 Z_2 q^2}{4\pi \in_0 r^{m_1}} = \frac{+a}{\gamma^{m_1}} \quad .....(1.1)$$

where m_1 is an exponent that determines the nature of the force. Z_1 and Z_2 are the charges of the ions, q is the charge, r is the intertonic distance and $\in_0$ is the permittivity of free space ($\in_0 = 8.85 \times 10^{-12}$ C/N–m^2).

The repulsive force becomes significant when the ions are brought very close to each other, this repulsive force is expressed as

$$F_{rep} = \frac{b}{r^\eta} \quad .....(1.2)$$

where η and b are constants.

The change of energy as the ions come closer to each other can be computed easily.

The potential energy associated with attraction (U_{attr}) and the repulsion (U_{rep}) are given as

$$U_{attr} = -\int \frac{a}{r^{m_1}} dr = \frac{a}{m_1 - 1} \frac{1}{r^{\eta_1 - 1}} + C_1 \quad .....(1.3)$$

$$U_{rep} = -\int \frac{b}{r^\eta} dr = \frac{b}{\eta - 1}\left(\frac{1}{r^{\eta-1}}\right) + C_2 \quad .....(1.4)$$

∴ The binding energy of the ions is

$$U_b = \frac{-A}{r^m} + \frac{B}{r^\eta} + C \quad .....(1.5)$$

where $A = \frac{-a}{m}$; $m = m_1 - 1$; $B = \frac{b}{\eta - 1}$; $\eta - 1 = \eta$;

$C = C_1 + C_2$ (the constants of integration)

when r = infinity, $U_b(\infty) = 0$ i.e., $C = 0$

$$\therefore \quad U_b(r) = \frac{-A}{r^m} + \frac{B}{r^n} \quad .....(1.6)$$

At equilibrium distance, the first derivative of the total-energy should vanish i.e., the forces of attraction and repulsion should balance each other i.e., at $r = r_o$, the equilibrium distance

$$\left(\frac{dU_b}{dr}\right)_{r=r_0} = 0.$$

$$\left[+\frac{A_m}{r_0^{m+1}} - \frac{B_n}{r_0^{n+1}}\right] = 0$$

$$\therefore \qquad \frac{r_0^{m+1}}{r_0^{n+1}} = +\left(\frac{A}{B}\right)\frac{m}{n}$$

i.e., $$r_0^{m-n} = +\left(\frac{A}{B}\right)\left(\frac{m}{n}\right) \qquad .....(1.7)$$

Then $$U_{b(r=r_0)} = -\frac{A}{r_0^m} + A\left(\frac{m}{n}\right)\frac{1}{r_0^m}$$

i.e., $$[U_b]_{r=r_0} = -\frac{A}{r_0^m}\left(1-\frac{m}{n}\right) \qquad .....(1.8)$$

In the case of ionic crystals, $m_1 = 2$; $m = 1$

$$\therefore \qquad [U_b]_{r=r_0} = \frac{-A}{r_0}\left(1-\frac{1}{n}\right) = -\frac{Z_1Z_2q^2}{4\pi \epsilon_0 r_0}\left(1-\frac{1}{n}\right) \qquad .....(1.9)$$

1.2.2 Lattice Energy in Ionic Solids

The above condition that $(du_b/dr)_{r=r_0} = 0$ is the criterion for thermal equilibrium. The stable configuration is possible only when the energy is minimum (curve 3 of Fig. 1.3). Let us see the variation of forces with the distance of separation. This is shown in Fig. 1.1. The net force is zero at the equilbrium position.

When like charges approach each other, they require energy and when unlike charges approach each other they release energy. When the ions are brought closer and closer, there exists a repulsion due to electrons of the adjacent atoms. The Pauli exclusion principle also comes into operation and the electron-repulsion energy becomes significant as the atoms approach closer and closer. The energy (curve 3 in Fig. 1.3), possesses a sharp minimum as the inter atomic distance becomes r_0. This is a stable inter atomic distance. Energy is required to reduce or increase this distance.

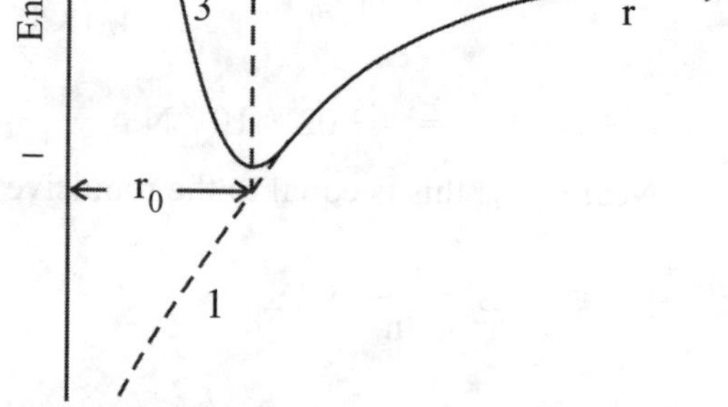

Fig. 1.3 Potential energy variation. The solid line (3) gives the energy of the solid, the resultant of the two energies. The dotted lines (2 and 1) correspond to the repulsive and attractive energies.

Suppose the lattice is built of N_p ion pairs, then the energy of the lattice is obtained as follows.

$$\frac{nB}{r_0^{n+1}} = \frac{-A}{r_0^2}$$

i.e.,
$$B = \frac{Z_1 Z_2 q^2 r_0^{n-1}}{4\pi \epsilon_0 n}$$

for the U_b eq. (1.9) gives

$$U_b = -\left(\frac{q^2 Z_1 Z_2}{4\pi \epsilon_0 r_0}\right)\left(1 - \frac{1}{n}\right)$$

As the lattice is built up of 'N_p' ion pairs, the energy of the lattice is

$$U_{Lat} = -\frac{N_p Z_1 Z_2 q^2}{4\pi \epsilon_0 r_0}\left(1 - \frac{1}{n}\right) = \frac{-AN_p}{V_0^{1/3}}\left(1 - \frac{1}{n}\right) \quad .....(1.10)$$

where V_0 represents the volume.

The energy trough will be very narrow for strongly bonded atoms. As a result, the slope, dF/dr, of the force curve at $r = r_0$ in Fig. 1.3 becomes very steep. This produces an interatomic distance which is constant within 0.001 Å at a given temperature. In such a case, we can take the ions as hard balls with definite radii.

We can determine the value of B in the expression in eq. (1.2) for a crystal like Na *Cl*. The interionic distance is 2.76×10^{-10}m.

The attractive force is

$$F_{att} = \frac{-Z_1 Z_1 e^2}{4\pi \epsilon_0 r^2} = -\frac{\left(1.6 \times 10^{-19}\,C\right)^2}{4\pi\left(8.55 \times 10^{-12}\,C^2/N.m^2\right)\left(2.76 \times 10^{-10}\,m\right)^2}$$

$$= -3.02 \times 10^{-9}\,N.$$

Near $r = r_0$, this is equal to the repulsive force. Then, from eq. (1.6), we get

$$B = \frac{r_0^{n+1}}{n} \times F_{rep}$$

i.e.,
$$B = \frac{3.01 \times 10^{-9} \times \left(2.76 \times 10^{-10}\right)^{n+1}}{n}$$

n = 9 for Na *Cl*.

$$\therefore \qquad B = \frac{3.02\times10^{-9}\times\left(2.76\times10^{-10}\right)^{10}}{9} = 8.6 \times 10^{-106}\ N\,.\,m^{10}$$

Then, the binding energy for Na^+-Cl^- pair is

$$U_b = \frac{-q^2}{4p \in_0 r_0} + \frac{8.6 \times 10^{-106}\ N.m^{10}}{\left(2.76 \times 10^{-10}\ m\right)^9}$$

$$= \frac{-\left(1.6\times10^{-19}\,C\right)^2}{4\pi\left(8.85\times10^{-12}\right)\left(2.76\times10^{-10}\right)} + \frac{8.6\times10^{-106}}{\left(2.76\times10^{-10}\right)^9}$$

$$\sim -8.34 \times 10^{-19}\ J + 0.92 \times 10^{-19}\ J$$

$$= -7.42 \times 10^{-19}\ J.$$

The bond energy due to attraction is – 5.23 eV and the contribution of the repulsive force to the energy is 0.575 eV.

1.2.3 Ionic Radii and Coordination Number

In discussing the ionic bond, we have just considered two ions. A crystal has a certain number of ions per unit volume that are distributed in three dimensions. They can be considered as having spherical shape with a characteristic value for radius in crystals. There are a large number of ionic crystals, namely, alkali halides, alkaline earth halides, oxides, transition metal oxides etc. The ionic radius of ions in these crystals depends on the coordination number.

Since ions can be treated as hard spheres, carrying a charge of normal parent atom of the crystal. They interact through central electrostatic forces. The crystal structures of ionic compounds to a large extent, are determined by packing of ions together as closely as possible under the strong electrostatic forces. The ratio of the ionic radii in the crystal structure is very important aspect of ionic solids.

A crystal of sodium chloride consists of positive ions (cation, Na^+) and negative ions (anion, Cl^-). The energy of attraction between the two ions spaced at a distance r_0 apart

$$U_0 = -\frac{Z_1 Z_2 q^2}{4\pi \in_0 r_0} = \frac{-A}{V_0^{1/3}}$$

where q is the charge of the ion. The reduction of energy on attachment of an electron to an anion, is higher than the increase of energy on detachment of an electron from a cation. The energy of an aggregate composed of positive and negative ions is reduced further due to the fact that each ion of one sign has ions of opposite sign as its nearest neighbours. This does occur as a result of coulomb attraction of the oppositely charged ions. The number of neighbours surrounding an ion and the type of crystal lattice are determined by the ratio of atomic radii of cation and anion. If this is R, then

$$R = r_c / r_a \qquad .....(1.11)$$

where r_c and r_a are the radii of the cation and anion respectively. The lattice of a crystal is stable only when the cores of the ions of one sign are in direct contact with the cores of opposite sign.

The coordination number z for the ionic lattices depends on the value of R. The following table gives the R values and the corresponding Z values.

R	0.73 – 1.00	0.41 – 0.73	0.22 – 0.41	0.15 – 0.22
Z	8	6	4	3

The example for Z = 8 is cesium chloride (Cs*Cl*) Fig. 1.4(c), for Z = 6, it is Na*Cl* Fig. 1.4(b), and Z = 4, it is CaF_2 or LiF.

The Fig. 1.4(a) gives the coordination polyhedron of an ionic bond that has Z = 6. When this polyhedron is repeated to fill space, the arrangement of atoms in a crystal is shown in Fig. 1.4(b). The alkali metal halides and hydrides, the alkaline earth oxides and silver halides (except AgI) belong to this type of ionic bonding.

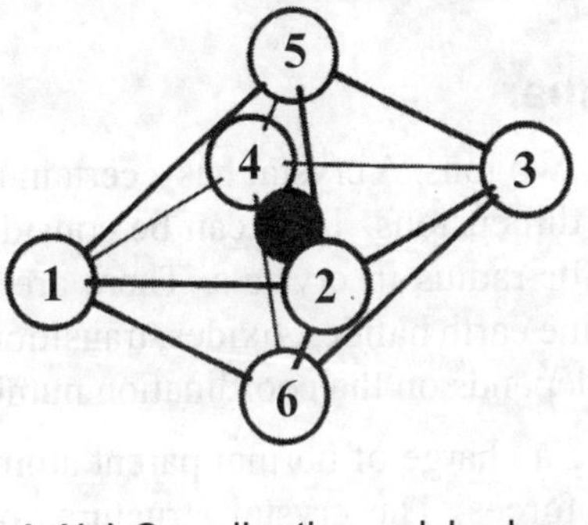

Fig. 1.4(a) Coordination polyhedron for coordination number 6. ● cation ○ anion. The anions 1, 2, 3, 4 are at the apices of the polyhedron.

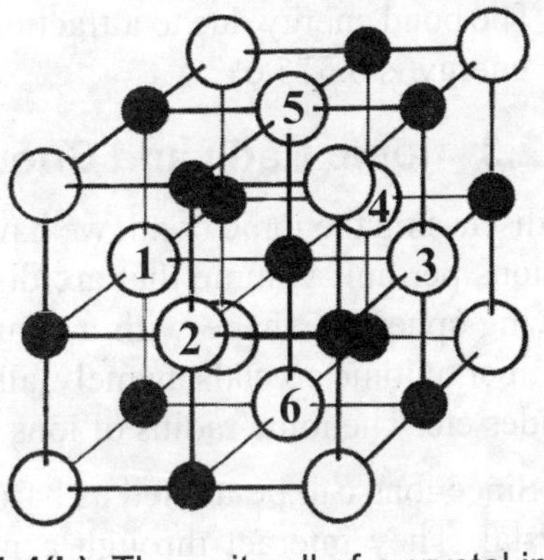

Fig. 1.4(b) The unit cell of a crystal in which Z = 6. The structure is known as rocksalt structure.

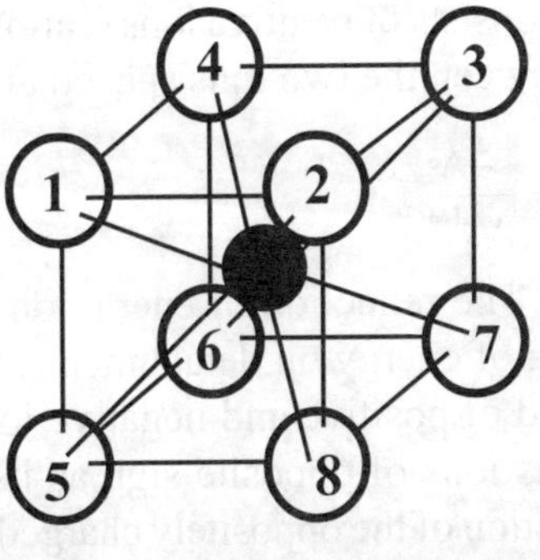

Fig. 1.4 (c) Cs*Cl*, with coordination number 8.

The structure called the rocksalt structure becomes unstable when the anions come into contact along the coordination polyhedron edges. Then , if a further decrease in r_c/r_a is caused, the coordination number falls to 4. In a fourfold structure, tetrahedral structure, the anions are at the apices of a tetrahedron. Decreasing r_c/r_a produces an instability on the hard sphere model of f.c.c when the anions touch along the edges of the tetrahedron. From geometrical considerations, the tetrahedron coordination is possible only for $0.414 > r_c / r_a > 0.225$.

Tetrahedral coordination polyhedra may be arranged or stacked to fill space in two ways. They are 1) the structure showing cubic structure and known as zinc blende 2) wurtzite. ZnS is the example and it exists in both the forms with a coordination number '4'.

1.2.4 Properties of Ionic Crystals

A large class crystals like alkali halides, alkaline earth halides, metal oxides, sulphides, selenides etc., belong to the group of ionic crystals. Alkali halides are the most basic crystals for study. We shall briefly give the general properties of the ionic crystals.

The ionic bond is a strong bond. The properties of crystals depend on the lattice energy. Some basic properties like heat of sublimation, melting point, boiling point, solubility etc., show a regular dependence on the inter ionic distance. For example if we take MF, MC*l*, MBr, MI, the halogenides, of alkali metals (Li, Na, K, Rb, Cs), the boiling points and melting points decrease as one goes from fluorides to iodies. In general, the ionic solids possess high melting points.

The ionic compounds exhibit large band gaps and therefore come under the class of insulators. The resistivities of alkali halides will be of the order of 10^{12} Ω–cm or more at room temperature. The oxides exhibit much higher resistivities. For example, the resistivity of Al_2O_3 is of the order of 10^{16} Ω –cm at room temperature.

The resistivity of all ionic solids decreases with the rise of temperature. That is, they show negative temperature coefficient of resistivity. The resistivity is a strong function of temperature in ionic crystals.

The conduction is basically electrolytic in nature. The ions move through the lattice via the point defects, when electric field is applied to a crystal. The study of electrical conductivity helps to determine the characteristics of defects in ionic crystals.

The defects like point defects and line defects are charged defects. The addition of impurities strongly affects the conductivity of ionic crystals. This is made use of, in making the electrolytic solid state battery cells. Super ionic conductors are useful materials for solid state batteries.

Many ionic crystals are transparent to radiation over a wide wavelength range. Normally the alkalihalides have absorption bands in the infrared region. Therefore, they are useful as infrared filters and windows. When crystals like alkali halides are subjected to high energy radiation like X - rays or γ rays, they get colored i.e., they absorb light in visible radiation at certain wave lengths. The absorption is due to 'Colour centres'. The formation of color centers is very important in energy storage utility. Many oxides, sulphides and alkali halides exhibit luminescence properties if they are subjected to external stimuli like high energy radiation or electric field.

Ionic crystals exhibit high strength. They are also brittle. The fracture takes place along the cleavage planes. The ionic solids are hard. Ionic compounds dissolve in polar solvents.

1.3 Covalent Bond and Covalent Crystals

1.3.1 Covalent Bonds in Crystals

It is known that the word 'covalent' describes the sharing of valence electrons by adjacent atoms. Hydrogen molecule is a good example of covalent bonding. The two protons share two electrons of opposite spins. The electrons lie between the hydrogen atoms for major part of the time. To a first approximation, the covalent bond in a hydrogen molecule can be considered to arise from the attraction of positive ions to the intervening pair of electrons with opposite spins as shown in Fig. 1.5.

$$H_{\bullet} + H^{\bullet} \rightarrow H : H \text{ (Hydrogen molecule)}$$

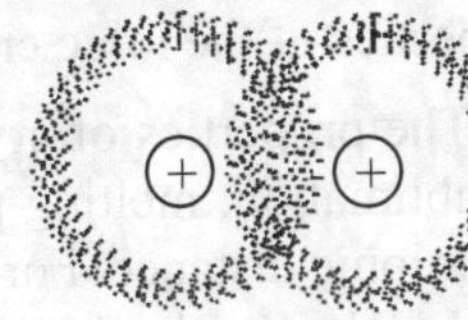

electron pair covalent bond and interaction

Fig. 1.5 Covalent bonding.

In the bonding process of forming the hydrogen molecule, the potential energy of the hydrogen atom is lowered as shown in Fig 1.3, curve 3.

The molecules of fluorine, chlorine, oxygen, nitrogen etc., (the diatomic molecules) are formed on the basis of electron-pair covalent bonds. In these cases, two p-electrons are shared between the atoms. The above examples have single bonds i.e., their coordination number is one. But many elements form multiple bonds, Carbon is the best example. Carbon can form four covalent bonds even though it should form only two-covalent bonds. Sulphur, selenium, tellurium etc., form two bonds; Phosphorous, nitrogen, arsenic, antimony can form three bonds.

The covalent bond is pictured as an overlap of the electron orbitals of the two participating atoms. Pauli's exclusion principle explains why only two electrons can take part in the bond formation. According to it, only two electrons that have the same quantum numbers but opposite spins can pair up. If a third electron has to take part in the bonding, it would have its spin vector in the same sense as either of the earlier electrons. Therefore, it will be repelled. Hund's rule states that, for equivalent orbitals, electrons tend not to share the same orbitals and the electrons occupying different but equivalent orbitals tend to have their spins parallel.

Nitrogen (outer valence electrons $2S^2\ 2p^3$) has three unpaired p-electrons. It can attain the neon noble gas electron configuration ($2S^2\ 2p^6$) by sharing three 2p-electrons with another nitrogen atom to form the diatomic molecule.

The most important covalent bonded material is carbon. The carbon is the basic element for many organic materials like polymeric materials. The carbon atom has the ground state electron configuration $1s^2\ 2s^2\ 2p^2$. Therefore, it should form two covalent bonds with its two half filled 2p-orbitals as per Hund's rule.

However, carbon shows a covalency of four and form four covalent bonds of equal strength. To explain this, let us consider the electron configuration of carbon. There are 2s states and 2p states. The 2s states and 2p states are closely spaced and the energy difference between them is small. Therefore electrons can redistribute themselves as follows (Fig. 1.6).

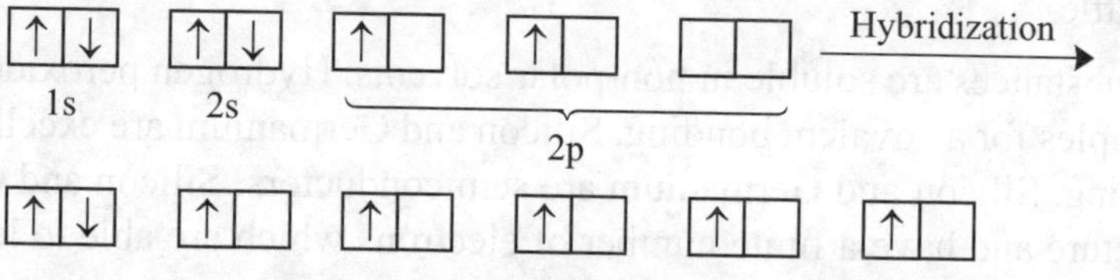

Fig. 1.6 Hybridization of carbon.

The carbon promotes one of its 2s electrons to a 2p orbital, the 2s and three 2p orbitals merge to form sp^3 hybrid orbitals. These are elliptical orbitals which are disposed toward the four corners of a tetra hadron. Each orbital contains one electron. This type of arrangement is known as hybridization. The process of hybridization enables carbon to form four equivalent covalent bonds instead of only two. This process leads to the formation of very stable compounds, or elemental compounds like diamond.

The carbon in the form of diamond exhibits the covalent bonding. The four sp^3 hybrid orbitals are directed symmetrically toward the corners of a regular tetrahedron. This is shown in Fig. 1.7. Each carbon is covalently bonded to four adjacent atoms. Diamond is extremely hard. It has high bond strength and melting point (3550 ^{0}C). Its bond energy is about 711 kJ/mol.

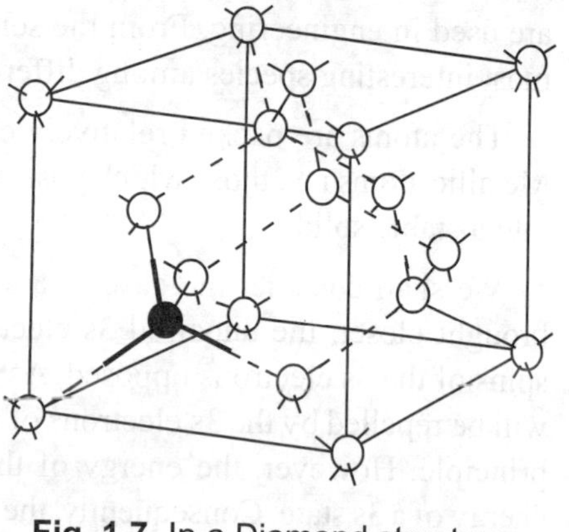

Fig. 1.7 In a Diamond structure, each carbon atom is covalently bonded to four adjacent atoms. The bond angle is 109.5 ^{0}C.

The covalent bonds are highly directional i.e., stereospecific. The coordination numbers for covalently bonded atoms are mostly controlled by radii ratios.

The covalent bonds follow the following characteristics:

1. Multiple bonds have shorter bond lengths.
2. Double bonds are stronger than single bonds. This aspect is very important in polymers.

1.3.2 Properties of Covalent Crystals

The covalent crystals because of their directionality in bonding have different properties compared to ionic crystals. The co-ordination numbers for covalentely bonded crystals are not controlled by radii ratios. There is always a strong covalent attraction only between first neighbours. As with ionic bonds, the lattice energy is associated with covalent crystals. Shortest bond lengths exist only in covalent crystals.

The melting and boiling points of covalent crystals are usually low as compared to those of ionic crystals.

These compound solids exhibit smaller energy band gaps. Pure covalent solids are good insulators at lower temperatures. However, as temperature increases, the conductivity increases. The conduction in most of the covalent solids is electronic in nature. Covalent solids are very hard, for example, diamond is hardest substances known. These solids have high yield strength and are not brittle.

Covalent substances are soluble in non-polar solvents. Hydrogen peroxide, propylene etc., are other examples for a covalent bonding. Silicon and Germanium are excellent examples for covalent bonding. Silicon and Germanium are semiconductors. Silicon and Germanium have the same structure and have a finite number of electrons which are able to leave the covalent bond locations.

1.4 Metallic Bonding

1.4.1 Nature of Metallic Bonding

This is the third type of chemical bonding that is important in materials. It is somewhat intermediate between covalent and ionic bonding. Metals are the major class of materials that are used in engineering. From the scientific point of view also, they place themselves as the most interesting species among different materials.

The atoms are packed relatively close together in a systematic pattern in metallic solids. Metallic atoms i.e., those which possess loosely held valence electrons may be bonded together into a stable solid.

We shall consider the case of a simple metal like sodium. When two sodium atoms are brought closer, the unpaired 3s electrons of the atoms form an electron-pair bond with the spins of the 3s electrons opposed. As a third sodium atom approaches this pair, its 3s electron will be repelled by the 3s electrons of the other two atoms. This is because of Pauli's exclusion principle. However, the energy of the unoccupied 3p states is very nearly the same as the energy of a 3s state. Consequently, the third electron can go over into this state without violating the Pauli exclusion principle. In a similar fashion, a large number of sodium atoms can thus surround any sodium atom. The first sodium atom or the central atom that has only one unpaired electron takes turns forming electron-pair bonds with each of the surrounding atoms. Therefore, the sodium forms less than a whole electron-pair bond with each neighbour. Solid sodium solidifies into a body-centered cubic structure in which each sodium atom is surrounded by eight other atoms. Therefore, each sodium atom can form one-eighth of an electron-pair bond, or one-quarter of an electron bond, with each of its nearest neighbour.

Similarly, copper atoms in crystalline copper are so close together that their outer valence electrons are attracted to the nuclei of their numerous neighbours. Each atom is surrounded by twelve neighbours (closest). The valence electrons are associated with all nuclei and are spread out among all atoms. Now, electrons exist as 'electron gas' in which ion cores are present.

The metallic bond is essentially a covalent bond without saturation. The metal atoms are held together by a mutual sharing of valence electrons. The density of electrons between the atoms is much lower than that is allowed by Pauli exclusion principle. Therefore, the electrons move freely as gas molecules without expending energy. The metallic bond is nondirectional. The coordination number is very high. The metallic lattice is shown in Fig. 1.8.

Fig. 1.8 Metallic lattice.

A large number of metals like Al, Cl Mg, Ti, Ni, Zn, etc., have 12 neighbors for each atom. Other metals like Li, Na, K, V, Mn, W and Fe (at room temperature) have eight neighbours for each atom. The binding energy for potassium is about 89.6 KJ/mol, for Nickel it is 423 KJ/mol, Zinc, it is 131 KJ/mol.

1.4.2 Properties of Metallic Bonding

Metallic bonding is weaker than ionic and covalent bonding, but stronger than Vanderwaals' bonding.

Metals possess high thermal and electrical conductivities. This is because of the free electron movement in the lattice. Metals exhibit ductility. They are brittle below a certain temperature. Metals exhibit brittle to ductile fracture transition at a certain temperature. Many metals like tungsten are tough. But some metal like pure silver, aluminium, gold are soft and malleable. Their melting points vary from 64 ^{0}C (Potassium) to 2000 ^{0}C [Titanium (1812 ^{0}C), Molydenum, Tungsten]. Metallic alloys are extremely useful for all engineering applications. Their high conducting values (gold, platinum, silver, copper etc.,) make them useful as electrodes in scientific and technical devices. Metals are opaque to light and they reflect light energy very efficiently.

1.5 Hydrogen Bonding

A chemical bond, in which a hydrogen atom of one molecule is attracted to an electronegative atom, of the same or another molecule is called 'Hydrogen bond'. Hydrogen forms bonding with electronegative nitrogen, oxygen, or fluorine atom, usually of another molecule. The interaction which occurs when a hydrogen atom, covalently bonded to an electronegative atom (as in AH), interacts with another atom Y to form the aggregate AH ··· Y.

The shortest and strongest bond is indicated as AH, while the secondary and weaker interaction is written as H ··· Y. Thus, AH is a proton donor, while (Y) is a proton acceptor which often contains lone pair electrons and can act as a base.

Hydrogen atom acts as a bridge between two atoms, holding one atom by a covalent bond and the other atom by a hydrogen bond. The hydrogen bond is represented by dotted line (.....), while covalent bond is represented by the solid line(_____).

..... $H^{\delta+} - A^{\delta-}$ .... $H^{\delta+} - A^{\delta-}$ .... $H^{\delta+} - A^{\delta-}$

Examples:

1. Hydrogen fluoride (HF)—The hydrogen atom while remaining bonded with fluorine atom, forms another weak bond with fluorine atom of adjacent bonding molecule. Hence, the formation of a hydrogen bonding results in the formation of a cluster of hydrogen fluoride (HF) molecules as $(HF)_n$, as shown below.

   ......HF......HF.....HF....HF.....HF...

   Hydrogen fluoride can be solidified at low temperatures. In solid form ,it contains long zig-zag chains of molecules as shown in Fig. 1.9.

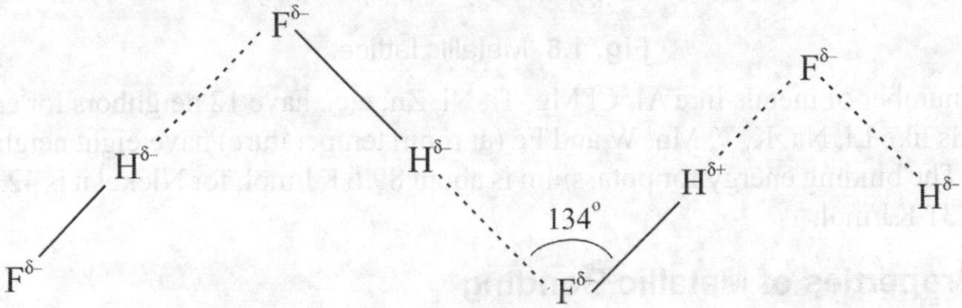

Fig. 1.9 Hydrogen bonding $(HF)_n$ in HF solid state.

Covelent bonding is a strong bond. Compared to it, hydrogen bond is much weaker. The strength of a hydrogen bond is of the order of 10^{-40} KJ-mol^{-1} as compared to the normal covalent bond that would have 40 KJ mol^{-1}. The bond length is longer in hydrogen bonding than that in covalent bonding. In the case of HF molecule, the covalent bond length is 109 pm while the bond length between F and H is 155 pm.

It should be noted that strong electronegativity of the other atom or molecule is a must for the formation of hydrogen bonding. Two good examples we can cite here are: ammonia (NH_3) and water (H_2O).

In the case of ammonia, (NH_3), the electronegative nitrogen atom forms a bond with three hydrogen atoms. As nitrogen is more electronegative and hydrogen acquires a positive charge, the bond takes place with good strength. Because of the above reason, ie the existence of positive and negative sites, the molecules of ammonia is associated through hydrogen bonding in long chains. The bonding is shown in the Fig. 1.10.

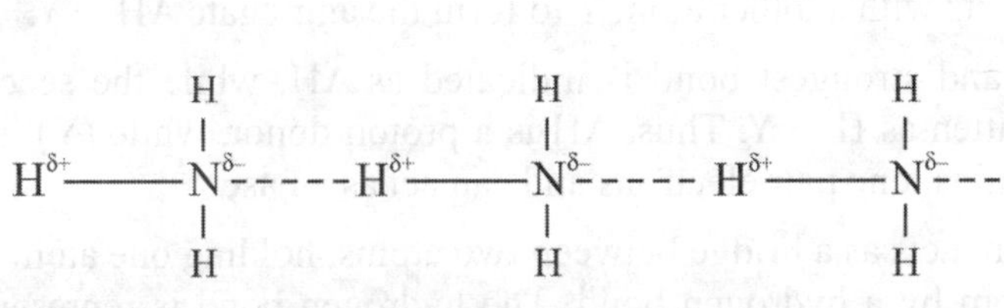

Fig. 1.10 Ammonia bonding, NH_3.

One of the conditions for hydrogen bonding is the size of electronegative atoms or molecule should be small. For example, in the case of NH_3 and HC*l*, only NH_3 shows hydrogen bonding and not HC*l*. This is because of the fact that N is smaller than C*l*.

Water is a good example of polar molecule that exhibits hydrogen bonding. Apart from two (O-H) bonds, water molecule has oxygen atoms forming two hydrogen bonds as shown in the Fig. 1.11.

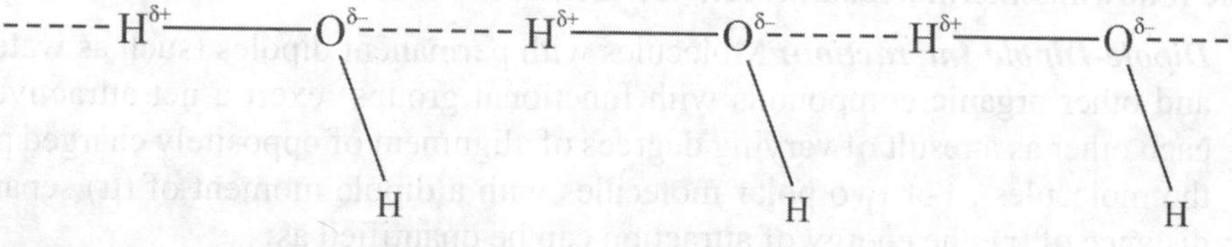

Fig. 1.11 Hydrogen bonding in H_2O.

In H_2O, four hydrogen atoms, two by covalent bonding and two hydrogen bonds tetrahedrally surround each oxygen atom. The molecular formula is $(H_2O)_n$. In the solid form, ice is a very interesting crystal. The crystal lattice consists of linked hexagonal rings of water molecules with considerable open space in the center of the ring. All naturally occurring ice crystals are hexagonal in shape and all snowflakes reflect this basic six-sided **crystal** habit.

Good examples of crystals in which hydrogen bonding plays very important role in determining their properties are Potassium dihydrogen phosphate ($KH_2\ PO_4$, KDP) and Ammonium di hydrogen phosphate ($NH_4H_2PO_4$, ADP). KDP is a ferroelectric below 120K and piezoelectric even at room temperature. KDP exhibits electro-optic effect. ADP is anti-ferroelectric and piezoelectric.

The strongest hydrogen bonds are formed between the most electronegative (A) atoms such as fluorine, nitrogen, and oxygen which interact with (Y) atoms having electronegativity greater than that of hydrogen (C, N, O, S, Se, F, C*l*, Br, I).

The weaker the hydrogen bond, the shorter the lifetime of the complex it forms. An important aspect of weak hydrogen bond formation is that the different molecular aggregates which do form can be easily and reversibly transformed. Thus the small energy changes resulting in the rapid making and breaking of hydrogen bonds in biological systems are of great importance; for example, hydrogen bonding determines the configuration of the famous α-helix of DNA, and the structures of most proteins, thereby serving an important function in determining the nature of all living things.

1.6 Secondary (Van der Waals) Bonding

Primary bonding (ionic, covalent and metallic) is strong and the energies involved range from about 100 to 1000 kJ/mole. In contrast, secondary bonding is weak, involving energies ranging from about 0.1 to 10 kJ/mole. While this type of bonding, also referred to as "residual", is weak, it is essential in the functioning of our environment.

The energy difference between the liquid and vapor states of a given system is given by the heat of vaporization, i.e. the heat required to convert a given liquid into a vapor (normally) at the boiling point temperature at 1 atm pressure. The energy difference is due to intermolecular attraction between molecules at close distance of separation. This phenomenon of attraction through secondary bonding can best be considered between a single pair of molecules, but recognizing that the forces are of longer range.

The following intermolecular forces are identified.

1. ***Dipole-Dipole Interaction:*** Molecules with permanent dipoles (such as water, alcohol and other organic compounds with functional groups) exert a net attractive force on each other as a result of varying degrees of alignment of oppositely charged portions of the molecules . For two polar molecules with a dipole moment of (μ) separated by a distance of (r), the energy of attraction can be quantified as:

$$\mu = L.\partial q \qquad .....(1.12)$$

where L is the distance of separation of the charges. In molecules, the dipole moment is given by the vector sum of electronic charges of the polar bonds. Thus, because of molecular geometry (symmetry), molecules such as CCl_4 exhibit no dipole-moment. The Dipole-Dipole interaction is given by

$$E_{\text{Dipole-Dipole}} = -\frac{2}{3}\frac{\mu^4}{Dr^6} \times \frac{1}{kT} \qquad .....(1.13)$$

where μ is the dipole moment, r is the distance of approach of the oppositely charged molecular portions, k is the Boltzmann constant and T is the absolute temperature in K. The molar energies of attraction associated with dipole-dipole interaction range from 0 to about 10 kJ/mole. These forces are primarily responsible for the liquid state (at room temperature) of most polar organic molecules; they are a contributing factor for H_2O to be liquid at room temperature, and are responsible for alcohol being a liquid.

2. ***Dipole-Induced Dipole Interaction:*** A dipole in one molecule can interact with and polarize the electrons of a neighboring non-polar molecule, thus generating an induced dipole which will experience an attractive force with the polarizing, polar dipole. Debye showed that in a molecule with a "polarizability" of (α) the attractive potential arising from dipole-induced dipole interaction is given as:

$$E_{\text{Dipole-Induced dipole}} = -\frac{2\alpha\mu^2}{Dr^6} \qquad .....(1.14)$$

Induced dipole interaction is important in aqueous solutions and very effective during adsorption of inert molecules on active solid substrates.

3. ***London Dispersion Forces:*** It is a well-known fact that all substances, including rare gases and hydrogen, assume liquid state at finite temperature, an indication of the existence of attractive interatomic and intermolecular forces, even in the absence of permanent dipole systems. The origin of this force has been proposed by F. London in 1930. Accordingly, orbiting electrons will at any instant generate a "temporary dipole", the

configuration of which changes as the electrons move. Since all atoms of a given system similarly experience temporary instantaneous dipole moments, their effect is expected to be cancelled because of the statistically random orientation of dipoles. It is evident that, should the dipoles be synchronized in a given assembly of atoms, then, a net attractive force would result. But, such an attractive force constitutes a lowering of the energy of a given system, synchronization can and will take place because all systems will attempt to assume minimum energy configuration. The London dispersion force can be formulated as:

$$E_{London} = -K\frac{\alpha^2}{r^6} \qquad .....(1.15)$$

where k and α are constants, r is the distance between molecules.

The attractive London forces are small, as manifested by the very low boiling points of some of the rare gases, hydrogen and nitrogen. The dispersion forces are much weaker then covalent bonds within molecules. Dispersion forces occur in addition to dipole-dipole interactions. The molecules is liquids, which have temporary fluctuating dipoles, have low boiling points.

1.7 Cohesive Energy

The energy required to bind the atoms or ions in a molecule is called cohesive energy. It is also the energy necessary to dissociate a molecule into atoms.

While discussing about ionic bond, we have considered two ions in the eq. (1.9). We just have extended it to 'N' ion pairs in eq. (1.10). Let us take up now the same in a general case.

Madelung constant 'A' represents the effect of the interaction of all the ions on any ion in the ionic crystal. All the ions in the lattice will be contributing to the energy of cohesion. We shall discuss a simple way of evaluating the Madelung constant for a crystal like Na*Cl*. In ionic crystal, individual atoms do not exist. Every anion is surrounded by a certain number of cations and every cation is surrounded by as many anions. The coordination number is determined by the ratio of cation and anion radii.

Consider a crystal as arrays of ions (Na*Cl*) as shown in Fig. 1.12., where a two dimensional array is shown. Let u_{ij} be the interaction energy between any two ions i and j. The interactions are pair-wise additive, the total interaction energy between i^{th}

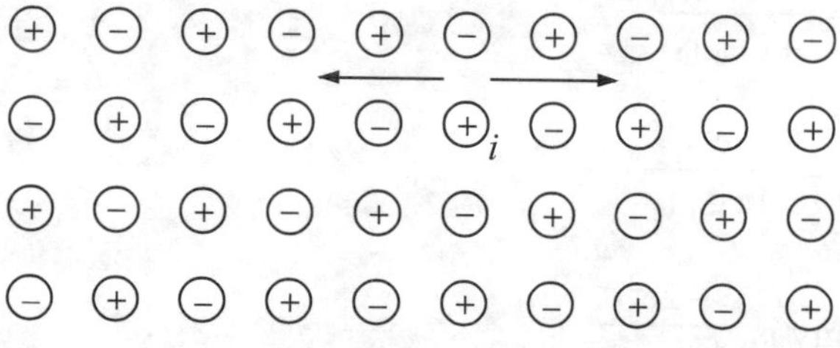

Fig. 1.12 Array of ion pairs in an ionic crystal.

ion taken as a reference ion, and all the other ions in the crystals is given as a summation over all pairs of ions as given below.

$$u_i = {\sum_j}' u_{ij}(r_{ij}) \qquad \text{.........(1.16)}$$

where ${\sum_j}'$ indicates that the summation is taken to include all ions except j = i (i.e., the same ion). This interaction energy does not depend whether the reference ion 'i' is positive or negative or the position of the ion in the array or crystal.

The term u_{ij} consists of first the long-range coulombic interaction. It is a central force as has already been stated, the second force is the short range interaction. We have defined the coulomb energy as

$$u_{ij} = -\frac{Z_i Z_j e^2}{4\pi \epsilon_0 r_{ij}} \qquad \text{.........(1.17)}$$

Suppose $r_{ij} = p_{ij}\, r_0$ and r_0 is the shortest distance between unlike ions in the crystal. Then the total electrostatic energy between the ith ion and other ions in the crystal is given as

$$ {\sum_j}' + \frac{Z_i Z_j e^2}{p_{ij} 4\pi \epsilon_0 r_{ij}} \qquad \text{.........(1.18)}$$

$$= -\frac{Z_i Z_j A}{4\pi \epsilon_0 r_0} \qquad \text{.........(1.19)}$$

where $A = {\sum_j}' \mp \frac{1}{p_{ij}}$ ....(1.20)

The constant A called Madelung constant is introduced in the eqn. (1.6) already.

Madelung constant is the property of a particular crystal structure. It can be calculated.

We consider one row of ions in Fig. 1.4 for Na*Cl*. Consider the ions on either side of the reference ion 'i'.

$$u_{i+1} = \frac{e^2}{4\pi \epsilon_0 r_0}$$

$$u_{i,\, i-1} = \frac{e^2}{4\pi \epsilon_0 r_0}$$

$\therefore$ The net energy is $\frac{-2e^2}{4\pi \epsilon_0 r_0}$.

The next two ions on either side contribute

$$\frac{2e^2}{4\pi \in_0 2r_0}$$

The next-two ions on the either side contribute

$$\frac{-2e^2}{4\pi \in_0 .3r_0}$$

This can be done upto the end of the chain on either side which can be taken as $-\propto$ to $+\propto$. Then

$$\frac{A}{r_0} = 2\left[\frac{1}{r_0} - \frac{1}{2r_0} + \frac{1}{3r_0} - \frac{1}{4r_0} + ...\right] \quad ....(1.21)$$

Thus, the factor two comes in, because there are two ions, one left side and the other on the right side at equal distance.

or $$A = 2\left[1 - \frac{1}{2} + \frac{1}{3} - \frac{1}{4} + ...\right] \quad ....(1.22)$$

i.e., $$A = 2\ ln\ 2 \quad ....(1.23)$$

Thus, the Madelung constant for one dimensional chain is $A = 2\ ln\ 2$.

The cohesive energy decreases monotonously in alkali halides with the generic formula, $M^+ X^-$, where M^+ can be Li^+, Na^+, K^+ or Rb^+ and X^- can be F^-, Cl^-, Br^-, and I^- for same M^+ ion. Thus, the following Table 1.2 gives the cohesive energies for a series of compounds. 'A' depends only on the type of the lattice and is independent of the lattice dimensions and the kind of ions.

The attractive potential energy U of the lattice energy is proportional to $\frac{1}{r_0}$.

The calculated Madelung constants for a few crystals are given below.

Na*Cl*	1.74756
Cs*Cl*	1.76267
$M^{++}X_2^-$	5.03878
Rutile	4.816

For isolated Na^+Cl^- molecule, the Madelung constant has a value equal to one. But in a crystal of Na*Cl*, it is 1.74756 i.e., about 75% more.

If there are 'N' ion pairs, 'Z' is the ion valence, e the elementary charge, 'r_0' is the distance of separation.

$$\text{Then} \quad U_{coul} = -A \frac{NZ^2e^2}{4\pi \epsilon_0} \times \frac{1}{r_0} \qquad ....(1.24)$$

If $Z_{cation} \neq Z$ anion, and Z is taken as the largest common factor.

1.8 Born Repulsive Exponent

The repulsive energy is given as $\frac{B}{r^n}$. B is evaluated in the eq. (1.6). 'n' is called the Born exponent. It is evaluated on the basis of compressibility studies in ionic crystals. During the compression of a crystal, the ions are pushed to closer distances and hence their electron outer shells interact vigorously. Therefore, the study of compressibility gives the information about repulsive forces. The study of compressibility helps is in the estimation of born exponent. If N_p is the number of ion pairs per unit volume, x is the compressibility of the crystal, then,

$$\frac{1}{\chi} = \frac{n-1}{9} A \frac{Z^2e^2}{4\pi \epsilon_0} \cdot \frac{N_p}{r_0^4} \qquad .....(1.25)$$

The value of n is nearly 10 for many ionic crystals. The value of n and the lattice energies for a number of alkali halides is given in the Table 1.2.

Table 1.2 The values of Born's exponent and lattice energies calculated and measured. The energies mentioned are in KCal/mole.

	LiF	LiCl	LiBr	LiI	NaF	NaCl	NaBr	NaICsF	CsF	CsCl	CsBr	CsI
n	6.2	7.3	7.7	7.0	6.4	8.4	8.3	8.0	10.2	10.6	10.5	11.1
U(calculated) KCal/mole	240	194	183	165.7	210	180.9	170	157	176	152	146	138.6
measured in KCal/mole	242	199	188.8	177.7	214	182	173.6	163	172	155.4	149	142.4

The compressibility and repulsion exponent are inversely proportional to one another. Many cases, the value of n is high confirming the fact that the ions can be regarded as hard spheres ($n \rightarrow \propto$) which are arranged in the lattice in such a way that the nearest neighbours touch each other. The repulsion force is principally due to Pauli exclusion principle Knowing the exponent value allows us to calculate the lattice energy, the main part of which is due to coulomb energy. The values of Madelung constants for various crystals are

NaCl – 1.748 ; CsCl - 1.763 ; CaF_2 – 5.039 ; ZnS – 1.641.

Questions

1. Explain the types of interatomic forces.
2. Mention various types of bondings that can exist between atoms.
3. Explain how the force of attraction between two atoms or ions vary as they are brought closer.
4. Explain what is 'ionic bonding'.
5. When two ions are brought closer, what is the nature of energy components between them?
6. What is Coulomb energy? How does it vary with the distance between ions.
7. What do you mean by 'repulsive force' between ions? How does it arise?
8. Derive the expression for the lattice energy of an ionic crystal like $NaCl$.
9. What is Madelung constant?
10. What is a coordination number? Give examples.
11. Explain covalent bonding? What is Hybridization?
12. Explain the bonding in diamond. Give some examples for covalent bonding. In what way covalent bonding is different from ionic bonding?
13. What is a metal? Explain the meaning of electron gas. How is it formed?
14. Give the general properties of ionic solids.
15. Show the minimum cation/anion radius ratio for a coordination number of 6 is 0.414.
16. Explain the formation of Hydrogen bonding. Compare it with covalent bond.
17. Explain how vander Waals bonds are formed.
18. What is cohesive energy. Obtain an expression for cohesive energy of ionic solids.
19. What are ionic solids ? Give examples. Explain formation of an ionic solid.
20. Explain what is cohesive energy ? How do you get the cohesive energy of a molecule like NaCl ?
21. Write short notes on

    1. Binding energy of NaCl molecule 2. Metallic bonding
22. Explain the formation of Van der Waals bonding. Give suitable examples.
23. Write a short notes on

    (a) Dispersive forces (b) Formation of NH_3
24. Discuss the formation of hydrogen bonding with suitable examples.
25. What is the meaning of equilibrium distance between two atoms when they are brought closer. Does it change with temperature ?

26. Discuss with suitable examples, the formation of covalent, metallic and Vander Waals bonding.
27. The ionisation of energy of K is 4.1 ev and the electron affinity of Cl is 3.6 ev. What is the energy required to form a K^+ and Cl^- ion pair ?

Objective Questions

I. Choose the correct choice in the following objective questions.

1. Choose the correct statement
   (a) Primary and secondary bonds are formed by interatomic forces.
   (b) An ionic bond is highly directional.
   (c) Secondary bonds are formed by the electrostatic attraction of electric dipoles or molecules.
   (d) Secondary bonds are formed as a result of coulomb forces between two ions.
2. Ionic solids
   (a) are soluble in polar solvents and they are insulators
   (b) are good conductors and possess low melting points
   (c) are soluble in non polar solvents
3. An ionic bond is
   (a) a homo polar bond
   (b) a heteropolar bond
   (c) an unidirectional bond
   (d) formed by sharing of electrons by two atoms.
4. The repulsive forces occur when two atoms are brought closer because
   (a) the cohesive forces become zero
   (b) the nuclei come too close
   (c) Pauli exclusion principle prevents multiple occupancy of the orbits by electrons
   (d) the electrons are scattered away
5. Secondary bonds are formed
   (a) as a result of an asymmetrical distribution of electron densities around the atoms' nuclei
   (b) as a result of strong coulomb interaction
   (c) as a result of repulsive forces only
   (d) as a result of sharing of electrons in the outer most shells.
6. The attractive force between a pair of K^+ and F^- ions is (the ionic radius of K^+ ion is 1.33 angstroms and the ionic radius of F^- ions is 1.36 angstroms)
   (a) 3.18×10^{-9} N
   (b) 3180×10^{-9} N
   (c) 3.18×10^{-12} N
   (d) 3.18×10^{-9} N.
7. The force of repulsion between a pair of K^+ and F^- ions is 3.18×10^{-9} N. Then the value of 'b' in the repulsive force $nb / r_0 n + 1$ is (the radius of K^+ ion is 1.33 angstrom and the radius of F^- ions is 1.36 Å, the value of n = 8)
   (a) 1.09×10^{-86} N m^8
   (b) 2.93×10^{-96} N m^9
   (c) 3.18×10^{-9} N m
   (d) 3.18×10^{-1} N m.

8. The value of 'b' is 2.93 × 10-96 N.m^9. The ionic distance between a pair of ions is 2.69 Å, then the repulsive energy is
(a) $+1.07 \times 10^{-19}$ J
(b) -1.07×10^{-19} J
(c) 1.07×10^{19} J
(d) 1.07 e V.

9. The Coulomb energy and the repulsive energy between a pair of ions are E_1 and E_2. respectively Then, the net potential energy V for the ionic pair is
(a) $E_1 + E_2$ (b) $E_1 - E_2 = -4$
(c) $E_1 + E_2 = 4$ (d) $-E_1 + E_2 = -4$

10. For CsC*l* crystal, the cation-to-anion radius ratio is 0,934. Then
(a) Twelve Cl^- ions pack around the Cs^+ ion
(b) Six Cl^- ions pack around the Cs^+ ion
(c) Six Cs^+ ions pack around the Cl^- ions
(d) eight Cl^+ ions pack around the Cs^+ ion

11. For NaC*l* crystal, the ratio of the cation-to-anion ratio is 0.525. Then,
(a) only six Cl^- ions pack around one Na^+ ions
(b) only four Cl^- ions pack around one Na^+ ion
(c) twelve Cl^- ions pack around one Na^+ ion
(d) eight Na^+ ions pack around one Cl^- ion

12. (A) As the ratio of cation-to-anion radius decreases, fewer anions can surround a central cation in an ionic crystal.
(B) The ionic bond is directional in character.
Then, the validity of the statements (A) and (B) is,
(a) Both A and B are wrong
(b) both A and B are correct
(c) A is correct, B is wrong
(d) A is wrong , B is correct

13. As the size of the ion in a group in the periodic table (like Li, Na, K, Rb) increases, the lattice energy of the ionic solids (LiC*l*, NaC*l*, KC*l*, RbC*l*) formed with the same type of anion
(a) decreases
(b) increases
(c) remains unaltered
(d) does not follow any rule.

14. The bonding energy between two atoms corresponds to
(a) the maximum of the repulsive energy
(b) the minimum point in the energy vs interatomic distance curve
(c) the maximum of the energy vs interatomic distance curve
(d) the minimum of repulsive force.

15. Secondary bonds are formed
(a) as a result of an asymmetrical distribution of electron densities around the atoms' nuclei
(b) as a result of strong coulomb interaction
(c) as a result of repulsive forces becoming large
(d) as a result of sharing of electrons in the outer most shells.

16. One of the following is an ionic crystal
(a) Silicon (b) gallium arsenide
(c) zinc oxide (d) iron

17. Choose the correct statement
(a) Ionic bonds are formed by electrostatic forces and are highly directional
(b) Covalent bonds can be formed only in diamond
(c) Covalent bonds are highly directional
(d) Polymeric material are formed by ionic bonding

18. The amorphous substances possess
(a) long - range order
(b) short - range order
(c) ionic bonds only
(d) covalent bonds only.

19. An amorphous substance is represented
    (a) by a square lattice
    (b) by a strict periodic lattice
    (c) grains and grain boundaries
    (d) by not a well-defined coordination number of atoms

20. The number of Cl^- ions surrounding each Na^+ ion in NaCl crystal is
    (a) 5 (b) 12
    (c) 8 (d) 6

21. Choose the incorrect statement
    (a) the coordination of each type of ion in CsCl crystal is 8
    (b) a metal that crystallizes in bcc structure becomes a bad conductor
    (c) A unit cell of an ionic crystal shares some of its ions with other unit cells
    (d) The ionic crystals are hard, insulating and possess normally - high melting points

22. Which of the following does not apply to bonding in metals
    (a) mobility of valency electrons
    (b) delocalization of electrons
    (c) overlapping orbitals
    (d) highly directed bonds

23. The crystal structure of CsCl is
    (a) bcc
    (b) simple cube
    (c) hexagonal close packing
    (d) face centered tetragonal

24. Among the three compounds NaCl , KCl, CsCl, LiCl, the one that has the greatest ionicity is
    (a) NaCl (b) CsCl
    (c) KCl (d) LiCl

25. In the compounds LiCl, NaCl, KCl, RbCl, CsCl, the one that has the highest melting point is
    (a) NaCl (b) RbCl
    (c) CsCl (d) KCl

26. In the compounds NaF, NaCl, NaBr, NaI, the one that has the least melting point is
    (a) NaI (b) NaF
    (c) NaBr (d) NaCl

27. The total number of sodium ions per unit cell of NaCl is
    (a) 6 (b) 2
    (c) 8 (d) 4

28. Most of the ionic substances
    (a) are soluble in polar solvents like water
    (b) are good conductors in solid state
    (c) possess too many Voids in the solid state
    (d) contain - highly directional bonds

29. The favorable condition for metallic bond formation is
    (a) there should be orbitals completely filled with electrons
    (b) the nuclear charge of the atom should be low
    (c) the element should have low ionization potential
    (d) the number of valence electrons should be low

II. Fill in the Blanks

1. The primary bonds are (i) __________ (ii) __________ (iii) __________ bonds.
2. A melt of a solid consists of __________ order that is not stable.
3. The repulsive forces between atoms become significant only if atoms are __________ from each other.

4. Born's exponent is evaluated from the __________ studies in ionic crystals.

5. The compressibility of a crystal is __________ proportional to the repulsive exponent.

6. The units of the lattice energy is __________ .

7. The alkali halides show absorption bands in __________ region of the electromagnetic spectrum.

8. The alkali halides can be used as __________ filters and windows.

9. The carbon in the from of diamond exhibits __________ bonding.

10. Multiple bonds in covalent bonding have __________ bond length.

11. Metals basically have __________ structures.

12. The metallic bond is essentially a __________ bond without saturation.

13. The __________ bond is non directional.

14. The metals possess high thermal and electrical conductivities. This is because of __________ movement in the lattice.

15. For hydrogen bonding, the size of electron negative atoms should be __________

16. In the case of NH_3, the nitrogen atom is electro __________ and hydrogen acquires a __________ charge.

17. In H_2O, four hydrogen atoms, two by __________ bonding, and two hydrogen bonds tetrahedrally surround each __________ atom.

18. London-<u>dispers</u> ion forces play the vital role in __________ bonding.

19. Rare gases can condense and exist in liquid form because of __________ bonding.

III. True or False

1. The secondary bonds are formed between the molecules that possess permanent dipole moment

[True/False]

2. The energy of attraction between two ions of opposite charges is given by the expression $Z_1 Z_2 \frac{q^2}{r^2}$

[True/False]

3. Madelung constant is dependent on the lattice dimensions and the nature of ions forming ionic crystal

[True/False]

4. For an isolated NaCl molecule the Madelung constant is equal to 1.747

[True/False]

5. The bonding in an oxide Fe_2O_3 is ionic

[True/False]

6. Ionic crystals dissolve in polar solvents

[True/False]

7. The bonding between two nitrogen atoms in N_2 molecule is ionic

[True/False]

8. The carbon in the form of a diamond exhibits Vander waal bonding

[True/False]

9. The covalent bonds are highly directional or stereo specific

[True/False]

10. Covalent crystals are highly soluble in polar solvents

[True/False]

11. Compared to ionic solids, covalent solids have very long bond lengths

[True/False]

12. Pure sodium exists as a liquid only

[True/False]

13. A metallic bond is essentially a covalent bond without saturation

[True/False]

14. Al_2O_3 is an example of covalent bonding

[True/False]

15. Nitrogen molecule (N_2) is formed due to dipole interaction

[True/False]

16. In carbon, hybridisation between S and P orbitals takes place

[True/False]

17. The weakest hydrogen bonds are formed between the most electronegative

[True/False]

18. HC*l* shows hydrogen bonding

[True/False]

19. All ionic crystals are semiconductors

[True/False]

20. Hydrogen molecule is an example of hydrogen bonding

[True/False]

CHAPTER 2

Crystallography and Crystal Srtuctures

2.1 Introduction

In nature, matter exists in three forms, viz, solids, liquids and gasses. When gasses are cooled, they become liquids (condensation) and on further cooling, solidify into a solid (solidification). As this transformation is taking place, the interaction between the atoms/ions increases and starts forming different type of chemical bonds among them, resulting in a particular arrangement. This general phenomenon takes place at different temperatures for different materials. However, Helium is an exception to this, which on cooling even to 0 K does not become a solid under normal pressure, and requires special technique to make it into a solid. When these materials / elements form into a solid, due to thermodynamical energy considerations, (free energy to be minimum), they either solidify into a crystalline form or into an amorphous form. Most of the materials tend to attain crystalline form, as this state leads to minimum energy compared to amorphous form. In crystalline form, the atoms/ions are arranged in a periodically repeating pattern, where as no such regularity of arrangement is observed in amorphous materials. Glasses, polymers, metallic glasses come under the category of amorphous materials. In solid form, these materials have higher energy state compared to crystalline state. Crystals, in general, melt (transform to liquid) at a fixed temperature, depending upon the pressure, where as glasses or polymers do not show such tendency and take liquid form gradually over a range of temperatures. They first become soft and then start becoming a liquid. This does not happen at a fixed temperature but over a temperature range. They may show some orderly (periodic) arrangement of atoms/ions, but this exists over very small distances, (few Angstroms) and hence amorphous materials possess short range ordering. Unlike amorphous materials, atoms in crystals show orderly or periodic arrangement over large distances.

A particular crystal is formed by cooling the melts of the corresponding solid. This process is called crystallization. The molten state is a liquid state. This liquid state lacks the long-range order but exhibits short range order. In crystals, the radial distribution of atoms is regular. The density of atoms that are present around a reference atom is fixed as per bonding and energy considerations. This is repeated with regular periodicity in three dimensions. This radial distribution of atoms is indicated in Fig. 2.1(a). The distribution is discrete. No atom will be present in between.

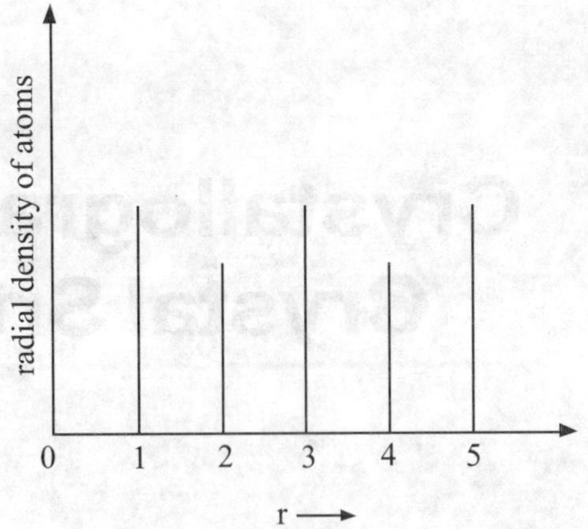

Fig. 2.1(a) Radial distribution of atoms in a crystal over a distance from a reference atom.

Amorphous substances are super cooled liquids. A liquid state consists of molecules that rotate about their axis of rotation. A liquid can also flow showing viscosity. Consider the atom of a liquid like melt of a substance. Let $\rho(r)$ be density of atoms in a small layer (of thickness Δr) of the melt, and ρ_0 be the density averaged over the entire melt. Then the ratio $\rho(r)/\rho_0$ represents the probability of the occurrence of an atom at a given distance r from a reference point or an atom. $\rho(r)/\rho_0$ varies radially with r, as one goes from the reference point as shown in Fig. 2.1(b). The radial density of atoms varies as shown in Fig. 2.1(b).

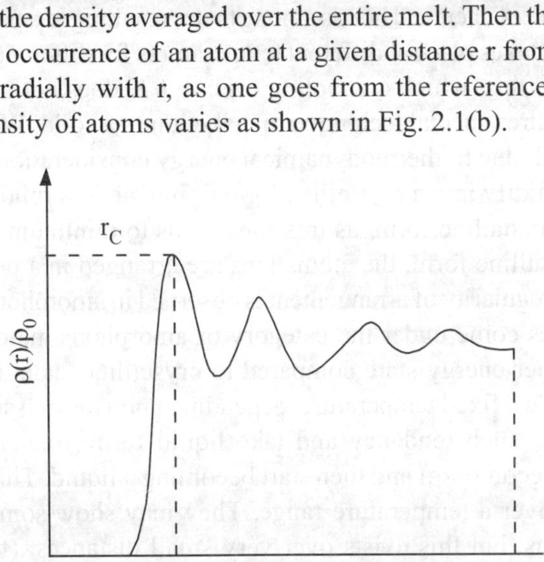

Fig. 2.1(b) Radial distribution of atoms in a melt (liquid).

The radial density of atoms has maxima and minima. It extends to a short distance (r_0) and fades off. This pattern of radial distribution of atoms is called short-range order. It is spread over a few interatomic distances and is lost after that.

Because of different types of periodic arrangement of atoms/ions in crystals, they exhibit physical properties which are direction dependent. This phenomenon is called anisotropy. Most of the amorphous materials show isotropic properties, but deviations from this rule are often observed, because of preparation techniques and materials used. In the following section, let us try to analyse more closely the atomic/ionic arrangement in crystals, as they show very interesting properties, symmetry and orderliness. Even materials which generally solidify into crystalline form, such as metals (Aluminium and other alloys) can be made as glasses and are known as metallic glasses. Their properties vary drastically compared to those in the crystalline form (for the same material). These metallic glasses find many applications in the modern world. Large number of materials like ceramics and polymers can be made in crystalline form as well as in amorphous form.

A crystalline solid can either be a single crystal, where a solid consists of only one crystal (Diamond, Ruby, Quartz etc.), or an aggregate of many small crystals, less than 1mm, separated by well defined boundaries (grain boundaries). In the later form, a solid is said to be polycrystalline (e.g.: aluminium wire, iron rod) and hence the surfaces of these can vary in shape.

2.2 Space Lattice

A crystalline solid, as mentioned above, shows periodicity (regularity) in the arrangement of atoms/ions. However, this arrangement can be of different types depending upon the bonding between them such as ionic, covalent, metallic etc. Covalent bond requires fixed number of neighbours (bonds) because of its directionality and hence results in restriction on the arrangement. The ionic and metallic bonded materials do not have such restrictions.

Careful crystallographic studies (study of crystal, through their appearance/morphology) and X-ray crystal structure studies (study of interior arrangement of atoms) have revealed that there are only few ways in which the atoms can be arranged in space (three dimensions). The atoms, in the form of groups, are arranged at a particular site in space depending upon the bonding and this gets repeated with a certain periodicity in different ways. This arrangement in space is governed by the space lattice which the crystal adopts to. Hence, we shall first study about different space lattices.

"A space lattice is defined as an infinite array of points in three dimensional space, in which every point has an identical environment to all others." The lattice is defined by three fundamental translation vectors **a**, **b**, **c**. These three vectors are called lattice parameters. The atomic arrangement looks identical in every respect when observed from any point **r** as viewed from the point.

$$\mathbf{r}' = \mathbf{r} + l\mathbf{a} + m\mathbf{b} + n\mathbf{c}$$

For easy understanding, let us first consider a two-dimensional array of points arranged to form squares when consecutive points are joined in both directions (X and Y) as shown in Fig. 2.2(a) or to form a hexagon as shown in Fig. 2.2(b).

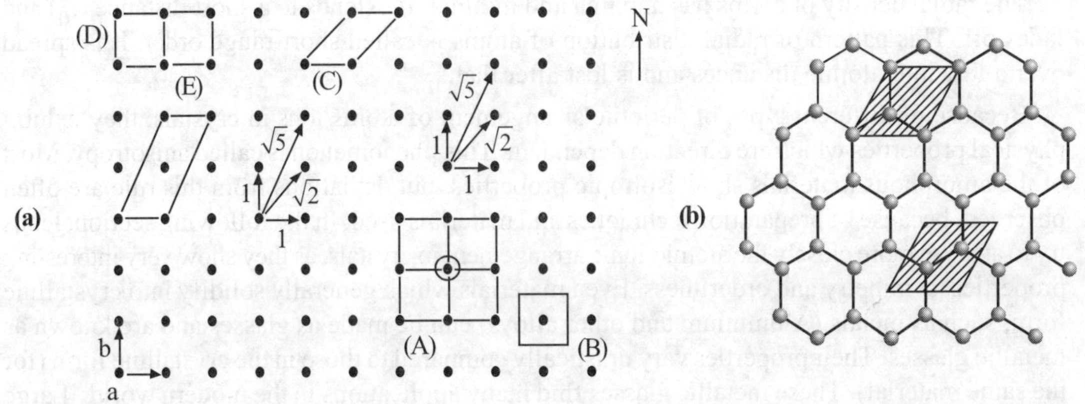

Fig. 2.2 A two dimensional array points **(a)** square lattice **(b)** hexagonal network.

2.3 Two-dimensional Lattice

By repeated translation of the vectors, **a** (along X) and **b** (along Y) in the plane of the paper, we can generate a square array (if **a** = **b** with an angle between them 90^0) and other types of arrays depending upon the values of **a** and **b** and the angular relationship between them as shown in Fig. 2.2 (a) and (b). In Fig 2.2, we can see different types of geometries forming in the array of points, depending upon how the lines are drawn. Some of them include a point in them [(B), (E)], some are joined with different angular relationships but joined from a point. We can also observe in (B) in Fig. 2.2 (a), that the square drawn does not contain any point at the corners but has only one point at the centre. Each of these shapes formed are denoted by different lattice names. In a square lattice (or any other shaped lattice) at any given point in the plane of the paper, the arrangement of points is the same, irrespective of the position in the plane. If we consider any point and look up or right from this point, we see another point at a distance of 1 unit (of **a** and **b**). Along the diagonals of a square, we see the nearest point at a distance of $\sqrt{2}$ units and along the diagonals of a rectangle, we see, the nearest point at a distance of $\sqrt{5}$ units. All these form different set of unit cells in a given lattice, when atoms are arranged in the lattice as shown in Fig. 2.2(a).

Definition of Unit Cell

Unit cell is the smallest unit of a crystal. It is similar to a brick in a brick wall. The characteristic of a crystal is represented by the unit cell. Hence we can define unit cell as :

"The smallest component of the crystal, which when stacked together with pure translation repetition reproduces the whole crystal". It is a small representative of the big crystal. The translation, T is given as

$$T = l\bar{a} + m\bar{b}$$

Definition Primitive unit cell : "Those unit cells which contain only a single lattice point" as shown in Fig. 2.1(a) (A), (B) or (C).

In the case of (A) and (C), even though they contain four lattice points at the corners, the contribution to the unit cell is $1/4^{th}$ from each point as each point is shared by four adjacent unit cells, as shown in Fig 2.1(a), (A). with a small circle around a corner point.

Similarly in a three-dimensional lattice, joining of such consecutive (very next) points would lead to a cube. Each corner of a cube is shared by 8 such cubes in a three dimensional array of points with $\mathbf{a} = \mathbf{b} = \mathbf{c}$, Each point at the corner of the cube would contribute $1/8^{th}$ to the corresponding unit cell.

From Fig. 2.2(b), we can observe that atoms are arranged in a hexagonal fashion (two-dimensions) forming a periodic arrangement with a fixed distance between any two atoms. However, we see in the figure that the same arrangement can result in two types of unit cells. First one has four atoms at the corners, placed at (0, 0), (0, b), (a, 0) and (b, b) and one atom is placed at $(2/3 \times a, 1/3 \times b)$. In this unit cell :

the contributions from four corners are $= 4 \times \frac{1}{4} = 1$

and contribution from one atom inside the cell is $= 1 = 1$

Total $\quad 2$

The two atoms in the unit cell form a basis or motif. Hence a structure of a crystal is built based on the type of lattice and the basis as shown in Fig. 2.3 and 2.4. Here there are three primitive lattices (P) and two non-primitive lattices (NP). Each of these lattices can form different structures, when different basis such as one atom, two atoms, three atoms etc. (as a group with different orientations among themselves) occupy a particular lattice either P or NP. Different structures can also result when a single basis is associated with different lattices indicated in the figure.

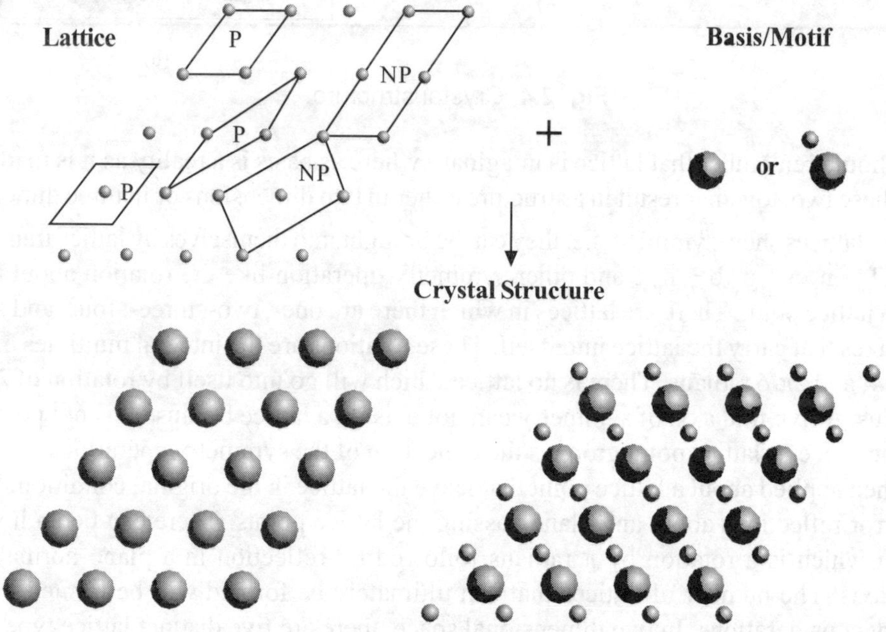

Fig. 2.3 Lattice + Basis = Crystal Structure.

The above can also be understood in a very simple way as follows : Let us assume a class room where chairs are placed with some regularity across the hall and also along the hall. The gap between any two chairs in the two directions can be equivalent to the two translatory vectors (**a**, **b**) mentioned earlier. Hence, the hall with chairs in like a lattice with points. These chairs can be arranged in different ways with different regularity (or periodicity). This is equivalent to different primitive or non-primitive latices. In the hall, if the students occupy the chairs each one sitting on a chair, then each student is like a single atom basis. In this case the centre of the atom coincides with the point. However, this need not be the case always. Let us assume that there are double the number of students compared to the chairs available. Then two students would occupy each chair. If the way they sit is similar in all the chairs, then we are talking about a basis equal to 'two' occupying the same lattice, but leads to different structure, compared to the earlier. The structure is defined as

$$\text{Lattice} + \text{Basis} \rightarrow \text{Structure}$$

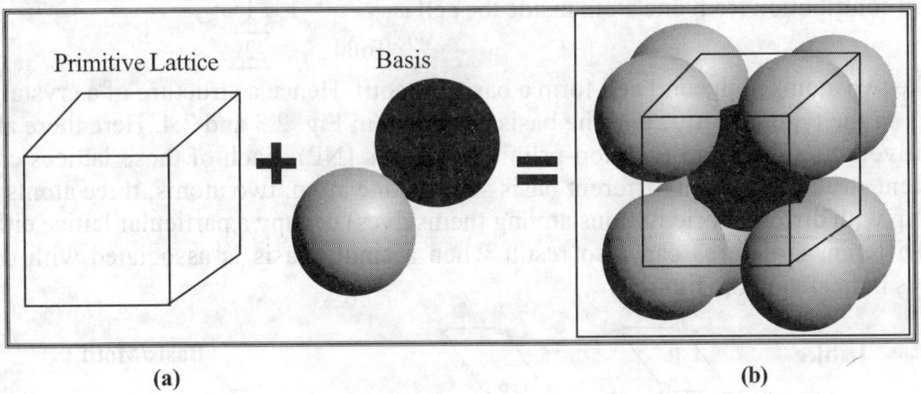

Fig. 2.4 Crystal structure.

One should remember that lattice is imaginary where as basis is a reality as it is made up of atoms. These two, together result in a structure either in two dimensions or in three dimensions.

Crystal lattices show symmetry i.e. they can be brought into themselves by lattice translations T where $T = n_x\, a + n_y\, b + n_z\, c$ and other symmetry operation like the rotation about an axis through a lattice point. There are lattices in which there are one-, two-, three-, four- and six fold rotation axes that carry the lattice into itself. These rotations are the integral multiples of 2π, π, $2\pi/3$, $2\pi/4$, and $2\pi/6$ radians. There is no lattice which will go into itself by rotation of $2\pi/7$, or $2\pi/5$ radius. A five fold axis of symmetry can not exist in a lattice because it is not possible to fill out the space. A lattice point group is the collection of the symmetry operations mentioned above when applied about a lattice point that leave the lattice in the original condition. We can have mirror reflections about any plane passing the lattice points. There can be an inversion operation which is a rotation by π radians followed by reflection in a plane normal to the rotation axis. The number of lattices that can ultimately be formed will be limited by these symmetry considerations. In two dimensional space, there are five distinct lattice types.

2.4 Crystal Systems and Bravais Lattices

The point symmetry groups in three dimensional space select diffferent structures with specific basis or motif. A three-dimensional space lattice is generated by repeated translations of three non-coplanar vectors **a**, **b**, and **c**. The translation 'T' is given as T = la + mb + nc. As in the case of two-dimensions, one can generate many lattices in three dimensions with different primitive and non-primitive cells. One would expect many such lattices in three-dimensions. However, Bravais, showed that the maximum number of independent space lattices are 14 only, the figure fourteen is arrived at on the basis of point symmetry groups. These 14 space lattices are known as Bravais lattices, named after its originator. These 14 lattices are shown in Fig. 2.5. Note that spheres shown in the figures are not atoms but are the lattice points. These 14 lattices are further grouped on the basis of common symmetry among them, based on the dimensions of the unit vectors along the axis and the angular relationship between them. Each of this is called a system and they are seven systems in number. These seven systems along with the correspondings lattices are shown in Fig. 2.5. Notations of lattice parameters **a**, **b**, **c**, α, β, γ are shown in Fig. 2.6. **a**, **b**, and **c** are the unit cell edge lengths and α, β, γ are angles between the axes as shown.

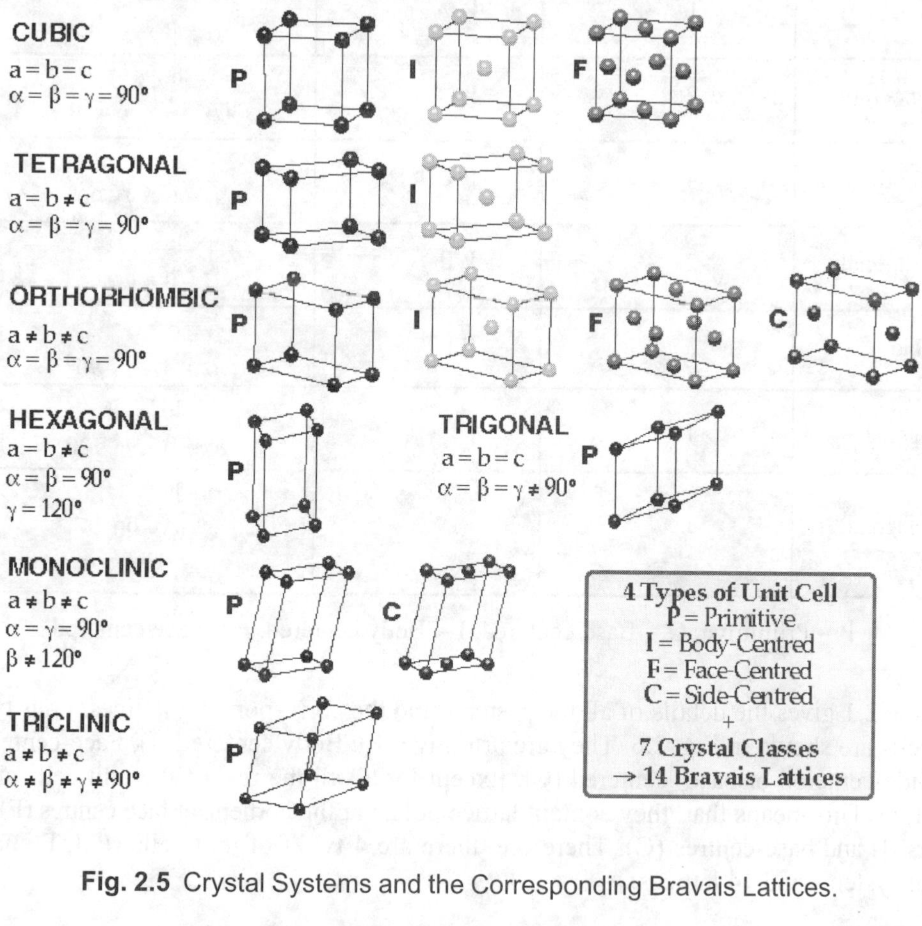

Fig. 2.5 Crystal Systems and the Corresponding Bravais Lattices.

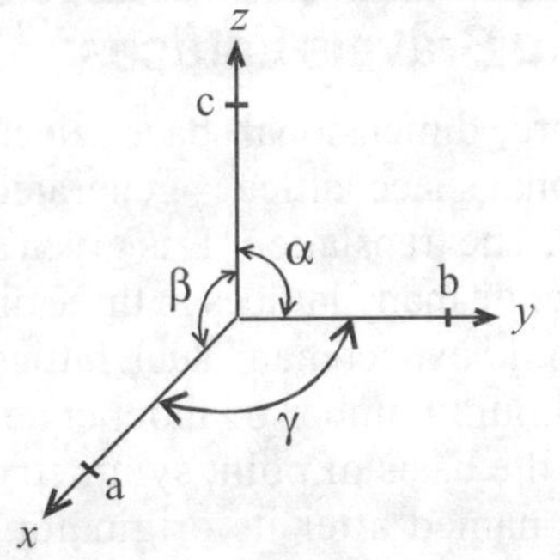

Fig. 2.6 Notation of unit cell parameters.

Table 2.1 The 14 Lattice Types in Three Dimensions.

System	Number of Lattices	Lattice symbols	Restrictions on Convetional Cell Axes and Angels
Triclinic	1	P	$a \neq b \neq c$ $\alpha \neq \beta \neq \gamma, \neq 90^0$
Monoclinic	2	P, C	$a \neq b \neq c$ $\alpha = \gamma = 90^0 \neq \beta$
Orthorhombic	4	P, C, I, F	$a \neq b \neq c$ $\alpha = \beta = \gamma = 90^0$
Tetragonal	2	P, I	$a = b \neq c$ $\alpha = \beta = \gamma = 90^0$
Cubic	3	P, I, F	$a = b = c$ $\alpha = \beta = \gamma = 90^0$
Trigonal	1	P	$a = b = c$ $\alpha = \beta = \gamma < 120^0, \neq 90$
Hexagonal	1	P	$a = b \neq c$ $\alpha = \beta = 90^0$ $\gamma = 120^0$

P = Primittive, C = Base-centered, I = Body centered, F = Face centered

Table 2.1 gives the details of all the systems and the corresponding lattices. Four types of unit cells are shown in Fig. 2.6. They are primitive (P), Body centered (I), Face-centered (F) and Side-centered or base -centered (C). Except for 'P' all the rest of the unit cells are non-primitive. This means that, they contain lattice points at the corners, at face centres (F), body-centers (I) and base-centres (C). Therefore, there are 4 types of unit cells (P, I, F and C), 7 crystal systems and 14 lattices.

2.5 Crystal Planes and Miller Indices

There are several planes possible in a crystal. There are certain rules to be established for determining the coordinate system for the planes. The position of the lattice points and the crystal axes are used for the purpose of indexing the plane systems.

A plane in the three dimensions is described from the analytical geometry as

$$\frac{x}{u}+\frac{y}{v}+\frac{z}{w}=1 \qquad .....(2.1)$$

The plane in three directions in space is intercepting the x–, y– and z– coordinate axis at u, v and w. We may write eqn. (2.1) as

$$hx + ky + lz = 1 \qquad .....(2.2)$$

Where $h = \frac{1}{u}$, $k = \frac{1}{v}$ and $l = \frac{1}{w}$, The *h*, *k* and *l* represent the reciprocals of the corresponding axial intercepts. Therefore, we can conveniently use *h*, *k*, and *l* to represent the planes or identify the planes. A crystal can be imagined as a composite of a large number of planes oriented in different directions. The atomic density on these planes is different. The Fig. 2.7 shows the various planes that can exist in a crystal lattice.

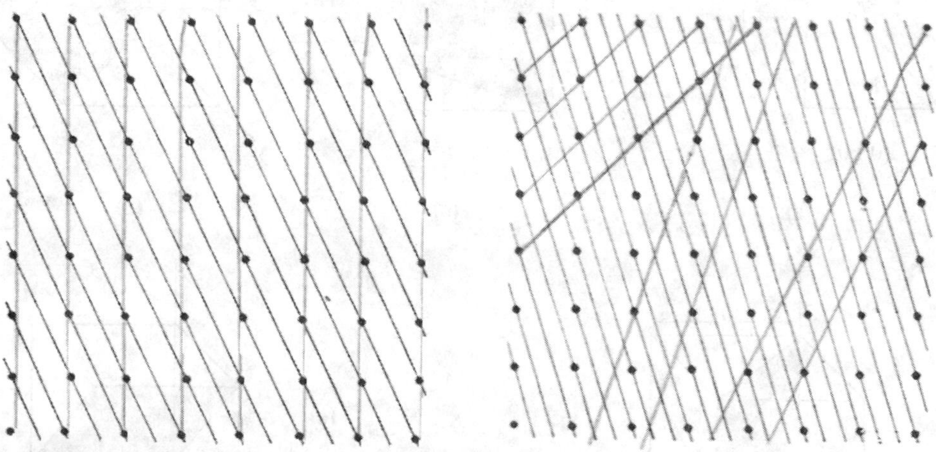

Fig. 2.7 Some lattice planes are shown. A crystal can be imagined to consist of a large number of planes.

The following procedure, for determining the Miller indices for a cubic crystal plane, is followed.

1. First, choose the origin such that the plane does not pass through origin.
2. Find the intercept of the plane in terms of the crystallographic a, b and c or x, y and z-axis for a unit cell. These intercepts may be fractions. Let these intercepts be u, v and w.
3. Take the reciprocal of the intercepts (1/u, 1/v and 1/w).

4. Clear the fractions with the LCM of the denominator integers. Then determine the smallest set of whole numbers which are in the same ratio as the intercepts. These whole numbers are Miller indices. They are written enclosed in parenthesis as $(h\,k\,l)$.

(a) The Fig. 2.8(a) shows some of crystallographic planes of cubic crystal structure. Consider the planes shaded in the Fig. 2.8 (a).

The intercepts of (1 0 0) plane are 1, ∞, ∞. By taking reciprocals of the intercepts, we get 1/1, 0, 0.

the Miller indices are (1 0 0).

This plane that cuts the x-axis at one unit distance i.e., the lattice parameter. The other plane (0 0 1) and (0 1 0) are also shown in the Fig. 2.8(a).

Consider the planes given in the figure in Fig. 2.8. (b), (c), (d), (e).

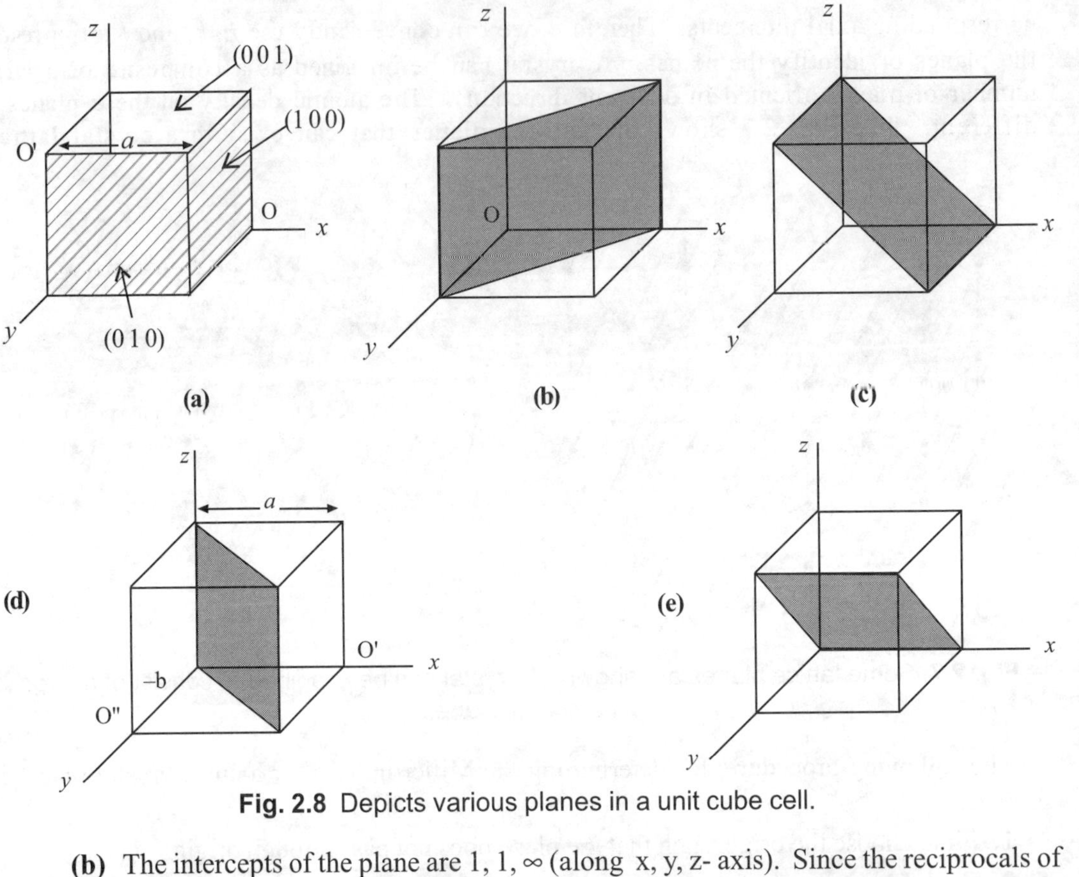

Fig. 2.8 Depicts various planes in a unit cube cell.

(b) The intercepts of the plane are 1, 1, ∞ (along x, y, z- axis). Since the reciprocals of these numbers are 1, 1, 0 which do not involve fractions. The Miller indices of this plane are (1 1 0).

(c) The plane has intercepts 1, ∞, 1; reciprocals 1, 0, 1.

Hence, the plane is (1 0 1)

(d) To find the intercepts, first move the origin to the lower-right back side of the cube (O')

The intercepts are (–1, 1, ∞)

reciprocals –1, 1, 0

The plane is (1 0).

Find what will be the Miller indices if the origin is shifted to O".

(e) The intercepts are (∞, -1, 1)

reciprocals 0, –1, 1

∴ Miller indices $(0\ \bar{1}\ 1)$.

Consider the planes given in Fig. 2.9 (a) in the following.

(a) The intercepts are 1, 1, 1.

reciprocals 1, 1, 1.

∴ Miller indices (1 1 1)

the plane is (1 1 1) plane.

(b) Consider the plane in Fig.2.9 (b).

The intercepts are 1/2, 1/2, 1

Reciprocals 2 2 1

∴ Miller indices (2 2 1)

the plane is (2 2 1)

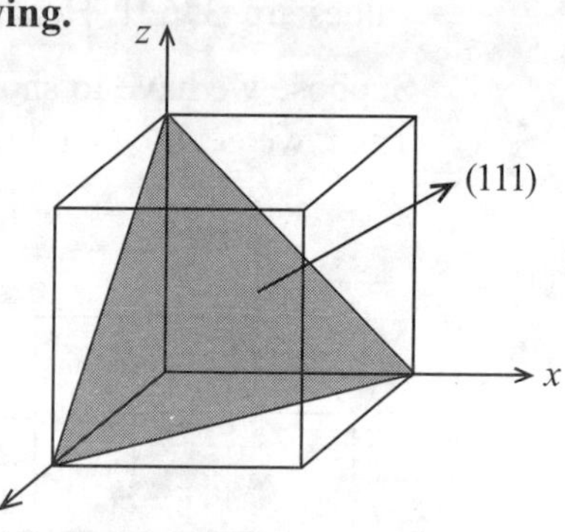

Fig. 2.9(a) (111) plane

(c) Consider the plane shown in Fig. 2.9(c). For this plane, we have to shift the origin to O'.

The intercepts are 1, 1, 1/2
Reciprocals 1 1 2
The plane is (1 1 2).

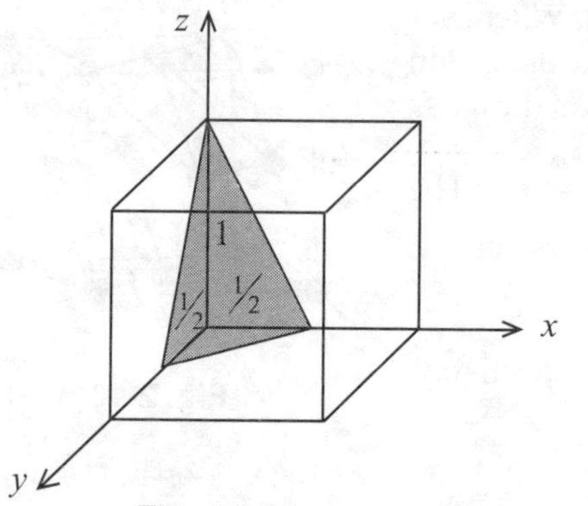

Fig. 2.9(b)

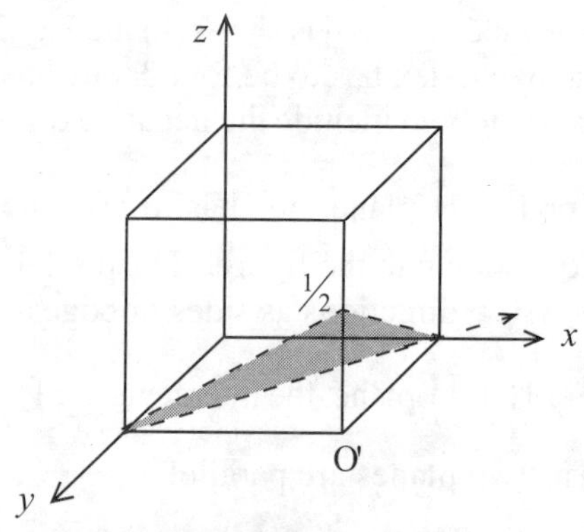

Fig. 2.9(c)

(d) Let us find out the Miller indices of the plane that is marked in Fig. 2.9 (d).

The intercepts are 1/2, 1/2, 1/2
Reciprocals 2, 2, 2
∴ The plane is (2 2 2)
This is parallel to (1 1 1) plane.

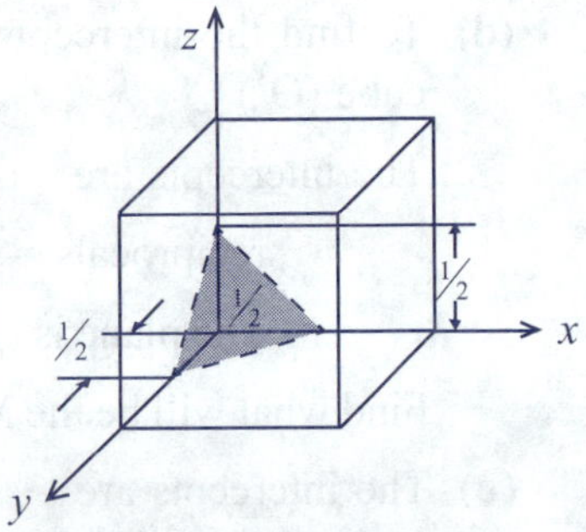

Fig. 2.9(d)

(e) In the Fig. 2.9 (e) the plane gives the intercepts at 2/3, 1/3 and 1 on x–, y– and z– axis respectively.

The reciprocals are : 3/2, 3/1, 1

Multiplying by common LCM of the denominator, we get 3, 6, 1. Then the h, k, l values are 3, 6, 1. Hence, the plane is (3 6 1) plane.

Suppose we have to show 7, 1, 1 plane one has to take the intercepts 1, 1/7, 1/7. Then, we get the (7 1 1) plane (Fig. 2.9f)

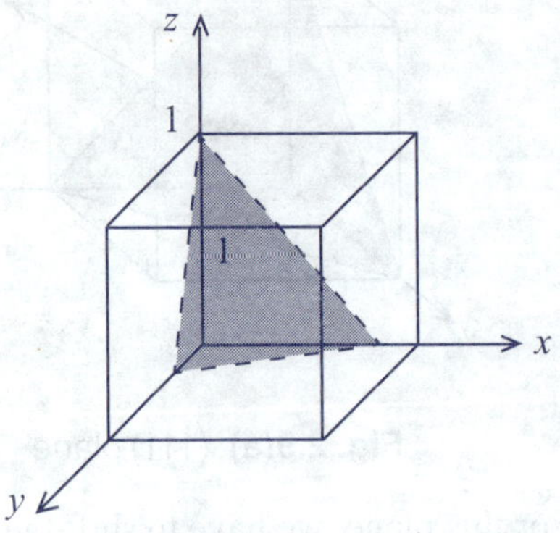

Fig. 2.9(e)

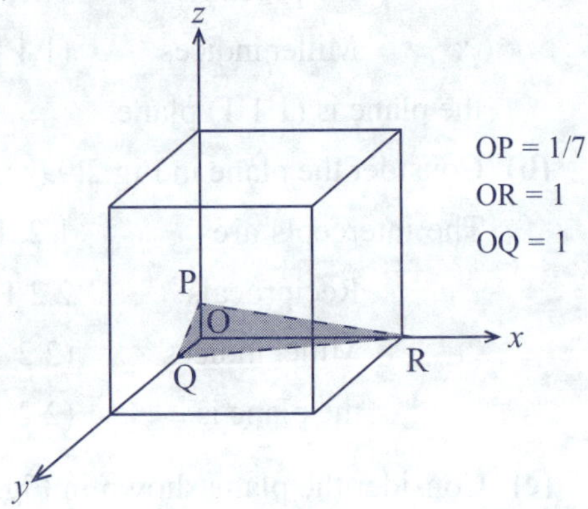

Fig. 2.9(f)

Exercise 1 :

Take a face centered cubic lattice, mark $(\bar{1}\,1\,1)$ and $(1\,\bar{1}\,1)$

The f.c.c. unit cell is shown in the Fig. 2.10(a). Whenever a negative index has to be found, one has to shift the origin appropriately to include the negative directions on the axis.

For $(\bar{1}\,1\,1)$ plane, one has to take the origin at 0'. The plane is shown in the figure. The p { 1 1 1 } planes contain the < 110 > directions as sides or edges.

For $(1\,\bar{1}\,\bar{1})$ plane, the intercepts are 1, –1, –1 respectively.

The two planes are parallel.

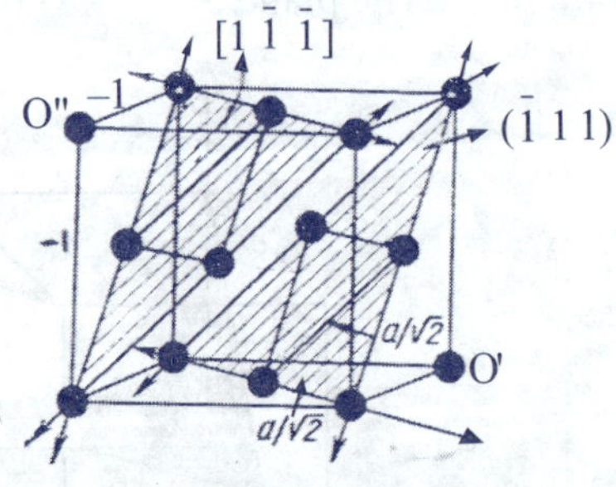

Fig. 2.10(a)

Exercise 2 :

Observe the unit cells in the Fig. 2.10 (b) and mark the (1 1 1) and planes. What is the common edge between them?

The planes are marked in the Fig. 2.10 (b).

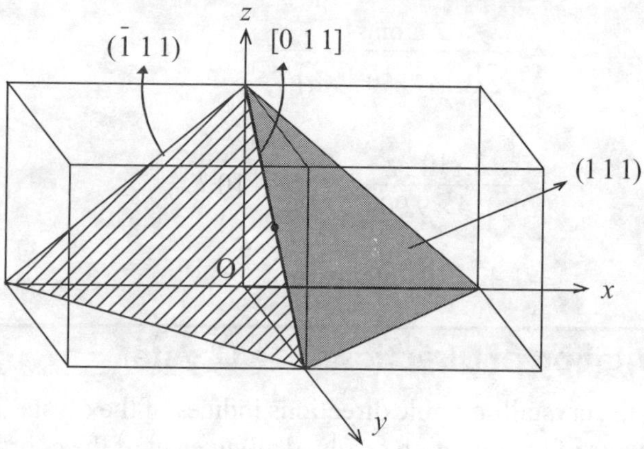

Fig. 2.10(b)

The common edge is [0 1 1].

The two planes intersect at [0 1 1].

They are the twin planes in the crystal.

As an excercise draw (1 1 1) and $\left(1\ 1\ \bar{1}\right)$ planes in the cube.

The {1 1 1} planes in a face centered cubic crystal are densely packed.

Sometimes, it is important to determine the atomic densities on various crystal planes. The planar atomic density is determined by the following relation.

Planar atomic density :

$$P_p = \frac{\text{Equivalent number of atoms whose centres are intercepted by the area of the plane}}{\text{The area}}$$

For example, the copper has a fcc structure.

We have seen in the chapter 1 that the equivalent number of atoms intersected by the (110) plane in terms of the surface area inside fcc cell is 2 atoms.

The area of the plane is $\sqrt{2}\ a^2$.

Then, the number of atoms in (110) planes is

$$\rho_{110} = \frac{2 \text{ atoms}}{\sqrt{2}\left(3.61\times10^{-10}\text{ m}\right)^2} = 1.09\times10^{-19}\text{ atoms/m}^2$$

$$= 1.1 \times 10^{12}/\text{mm}^2.$$

Exercise 3 :

Calculate the planar atomic density on the (110) plane of the bcc molybdenum that has a lattice constant 3.147A^0.

Planar atomic density :

$$= \frac{2 \text{ atoms}}{\sqrt{2}\left(3.147 \times 10^{-10} \text{ m}\right)^2}$$

$$= \frac{2 \times 10^{20}}{1.414 \times 9.904} = \frac{2}{14} \times 10^{20}$$

$$= 1.42 \times 10^{19} \text{ atoms / m}^2$$

2.6 Representation of Directions in a Crystal

For cubic crystals, the crystallographic directions indices of the crystallographic directions are the vector components of the direction resolved along each of the coordinate axes and reduced to the smallest integers. Let us see how the directions in a crystal can be arrived at.

A lattice vector is given as $\mathbf{r} = u\mathbf{a} + v\mathbf{b} + w\mathbf{c}$ and this vector connects a lattice point to the origin. The extension of this vector by a repetitive distance will find another identical point at its end. Thc atoms in the crystal have fixed points.

Consider a cubic cell that is placed on crystallographic axes as shown in Fig. 2.11. Let '0' be the origin. The point P is at the position (a, 0, 0). 'a' represents one unit distance. Therefore, it is at the position (1, 0, 0). The direction of 0 P is given as [1 0 0], and that of P 0 is given as [$\bar{1}$ 0 0]. A number with a bar indicates a negative direction and index.

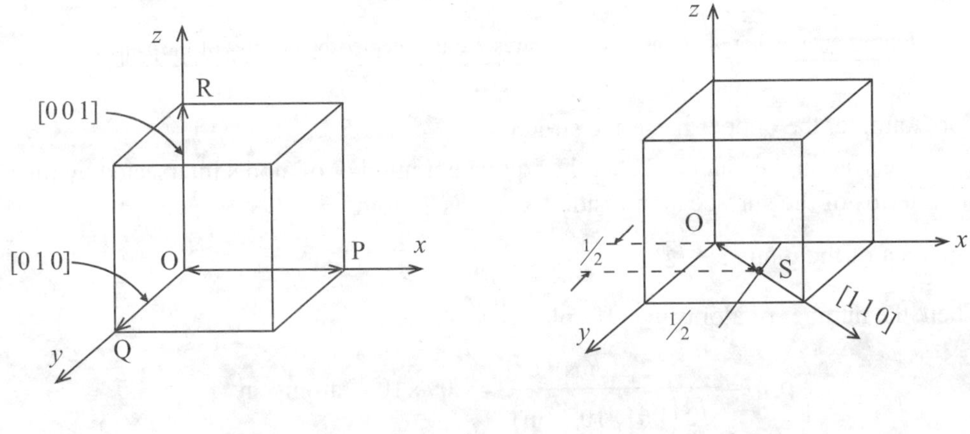

Fig. 2.11 Some directions in a cubic unit cell.

Similarly, points Q and R are in the direction [0 1 0] and [0 0 1] respectively. They are basic crystallographic directions in a cubic unit cell.

Now, consider the direction joining O and S as shown by the arrow. The position coordinates of S are (½, ½, 0). The direction of 0S is (2 × ½, 2 × ½, 0) i.e., [1 1 0]. Suppose the direction $[\bar{1}\ \bar{1}\ 0]$ has to be shown. The direction of the arrow on 0S is reversed as in Fig. 2.11 i.e., SO.

Consider a point T shown in Fig. 2.12(a). The position coordinates of T are (½, ½, ½). The point T is the body centre of the cell. The direction of T O is [1 1 1]. It is extended to end of the line to meet the corner of the cubic cell. [1 1 1] is the direction of the cube diagonal. The direction in the opposite sense gives $(\bar{1}\ \bar{1}\ 1)$ direction i.e., TO.

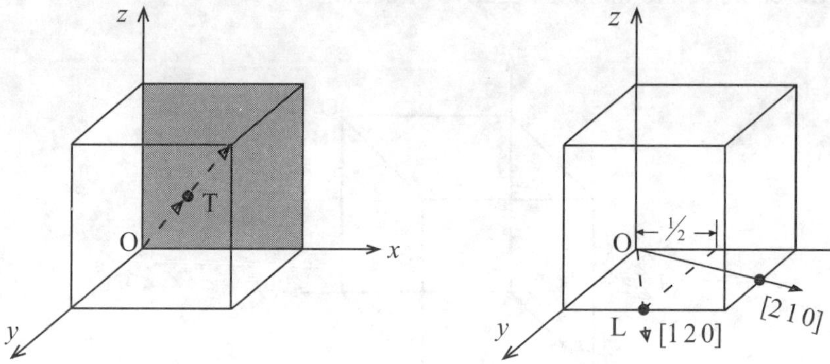

(a) (b)

Fig. 2.12

Consider the position of a point shown in Fig. 2.12(b).

The coordinates of L are (½, 1, 0). The direction of OL is [1 2 0]. To get the direction, you have to multiply all integers by LCM of the denominator in position coordinates. The direction [2 1 0] is also shown in the figure.

Let us find out what are the points that are connected in a cubic cell to show the direction [1 1 2], $[1\ 1\ \bar{2}]$, [3 2 1], $[\bar{3}\ 2\ \bar{1}]$. Consider the Figures in 2.13(a), (b) and (c).

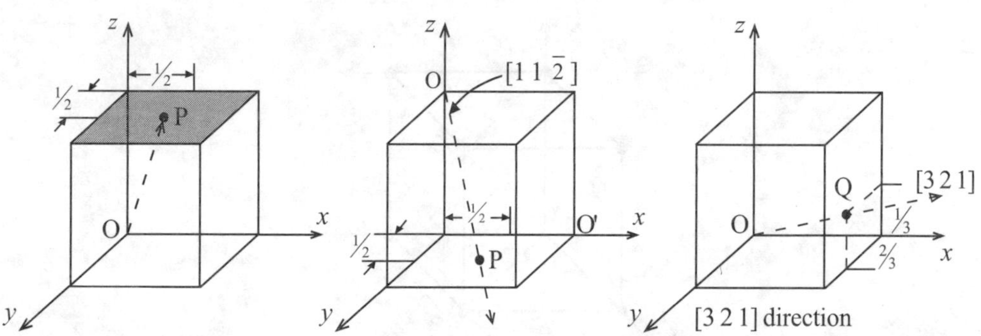

(a) (b) (c)

Fig. 2.13 (a), (b) and (c)

The positional coordinates are obtained by dividing all indices by the largest number. The positions of the first two points are (½, ½, 1); (1/3, 2/3, 1/3) respectively. Let O be the origin. Then the point (½, ½, 1) is at P. The OP is [1 1 2] as shown in Fig. 2.13(a). Similarly, the $[1\ 1\ \bar{2}]$ is shown in the Fig. 2.13(b). The family of <1 1 2> directions pass through the face centers of the cube with the origin at one of the corners of the opposite face.

The point (1, 2/3, 1/3) is Q in Fig. 2.13(c). The direction is [3 2 1]. To show $[\bar{3}\ 2\ \bar{1}]$ direction, we have to consider the origin or shift the origin to O'. The point Q now is at (–1, 2/3, –1/3). Multiply by the common factor in the denominator, we get (–3, 2 –1) i.e., [2] direction along O' Q shown in Fig. 2.13(d). It should be remembered that the origin has to be shifted to another corner of the cubic cell to include negative directions.

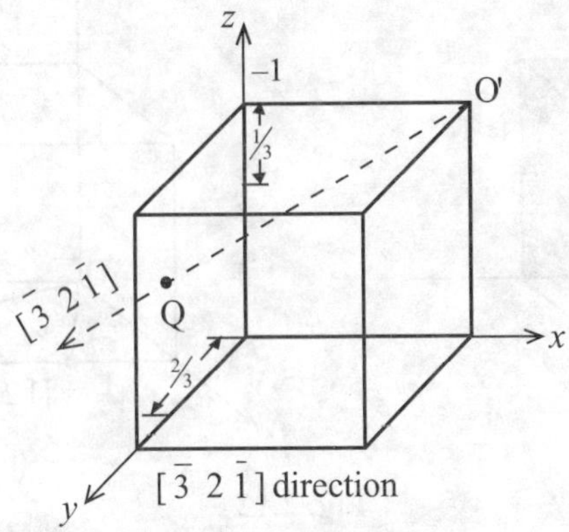

Fig. 2.13(d)

The letters u, v, w are used to indicate the general sense of the direction indices along a, b, c (or x, y, z directions) respectively. The directions are written as [u v w].

All parallel directions or direction vectors have the same direction indices. The figure below shows the parallel direction of [1 1 0] (Fig. 2.14). Similarly,, [0 1 0] or [1 0 0], [0 0 1] have parallel directions. The parallel directions are equivalent.

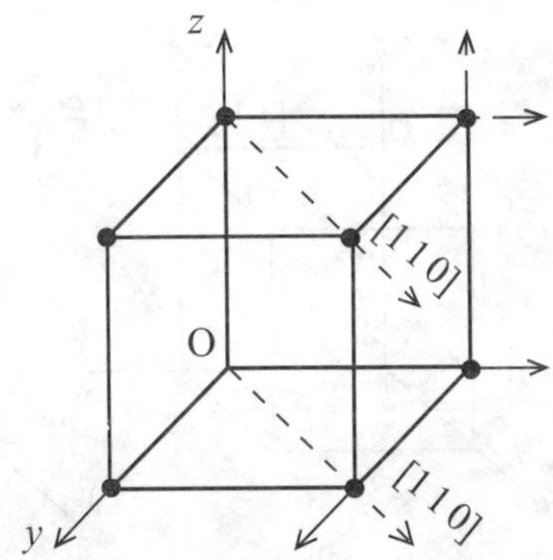

Fig. 2.14 Parallel directions in a unit cell of a cube.

The family of directions are [0 1 0], [0 0 1], [1 0 0], [0 $\bar{1}$ 0], [0 0 $\bar{1}$], [$\bar{1}$ 0 0]. All of them are grouped as $\langle 1\,0\,0 \rangle$. $\langle u\ v\ w \rangle$ represents the family of directions. Similarly $\langle 1\,1\,0 \rangle$ directions, $\langle 1\,1\,1 \rangle$ directions represent the family of [1 1 0] and [1 1 1] directions. They are all present in the cubic cell.

Let us see how a direction like [$\bar{7}$ 3 $\bar{1}$] can be shown. Consider the cubic cell given in Fig. 2.15. There are two negative directions. The origin has to be selected so as to include these negative directions.

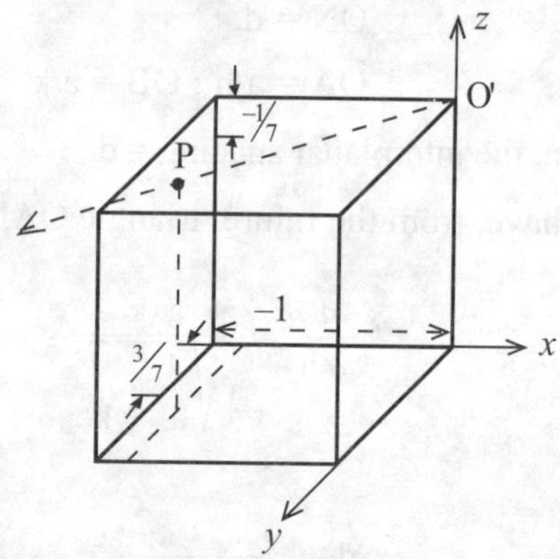

Fig. 2.15 The direction OP is –1/7 below the upper plane and 3/7 before the backside (0 1 0) plane.

The position coordinates of P are (–1, 3/7, –1/7).

We have selected a-axis (i.e., x-axis), and c-axis to include the negative directions. Therefore, the origin is at O'. The direction is shown in the figure. The direction OP is [$\bar{7}$ 3 $\bar{1}$].

2.7 Interplanar Spacing

2.7.1 Separation Between Successive Planes

Consider first a two dimensional lattice as shown in Fig. 2.16. Let the two planes under consideration be A B and C D. One of them C D is passing through origin. Let the intercepts be a/h and a/k along x and y-axes respectively. Drop a perpendicular on to the plane A B.

From the Fig. 2.16,

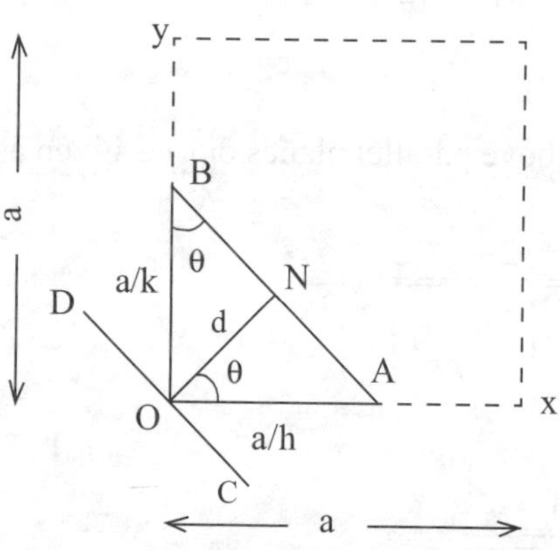

Fig. 2.16 One unit cell of a square lattice.

ON = d.

OA = a/h ; OB = a/k.

Then, the interplanar spacing = d.

We have, from the figure, triangle OAN and OAB,

$$\text{or} \qquad \frac{d}{a/h} = \frac{a/k}{a\sqrt{\dfrac{1}{h^2}+\dfrac{1}{k^2}}}$$

$$\text{or} \qquad d_{hk} = \frac{a}{\sqrt{h^2+k^2}} \qquad .....(2.3)$$

This can be extended to three dimensions. The two planes in the three dimensions can be represented in terms of Miller indices $\{h\,k\,l\}$.

Suppose a/h, a/k, a/l are the intercepts on x, y, z - axis of a cubic crystal.

Then, the equation of the plane is

$$\frac{x}{a/h} + \frac{y}{a/k} + \frac{z}{a/l} = 1$$

$$\text{or} \qquad hx + ky + lz = a$$

The next parallel plane for this plane will have intercepts 2(a/h), 2(a/k), 2(a/l) on x, y, z axis in the crystal.

Then, the equation of that plane is

$$\frac{x}{2(a/h)} + \frac{y}{2(a/k)} + \frac{z}{2(a/l)} = 1$$

$$\text{i.e.,} \qquad hx + ky + lz = 2a.$$

This distance between the above parallel planes d_{hkl} is given as

$$d_{hkl} = \frac{2-1}{\sqrt{\left(\dfrac{h}{a}\right)^2 + \left(\dfrac{k}{a}\right)^2 + \left(\dfrac{l}{a}\right)^2}}$$

$$= \frac{a}{\sqrt{h^2+k^2+l^2}} \qquad .....(2.4)$$

2.8 Cubic Crystal Systems

Fourteen Bravais lattices are grouped into 7 crystal systems as mentioned earlier but are not divided equally among the systems. For instance, a cubic system has three lattices as given below.

(i) ***Simple cubic (Primitive) :*** Lattice points are present only at the corners (Fig. 2.17). The volume of the cell is a^3. The nearest neighbor distance is a.

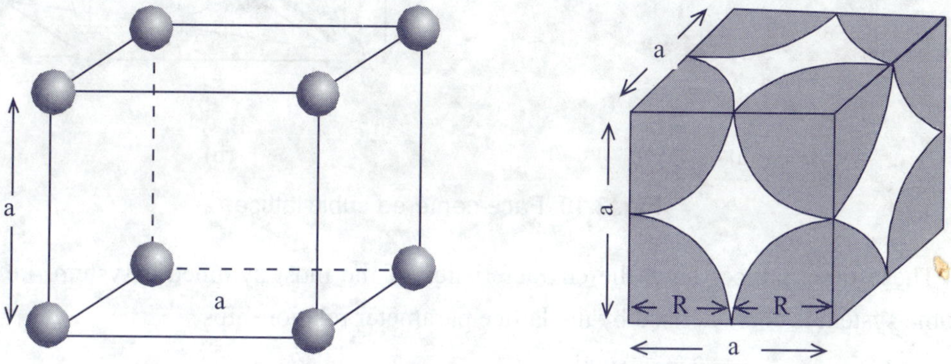

Fig. 2.17 Simple cube. The atoms are at the corners.

(ii) ***Body-centered cubic (BCC) (non-primitive) :*** Lattice Points are present at corners and also at the centre of the cell (Fig. 2.18). The volume of the cell is a^3. The nearest-neighbor distance is $\sqrt{3}\ a/2$.

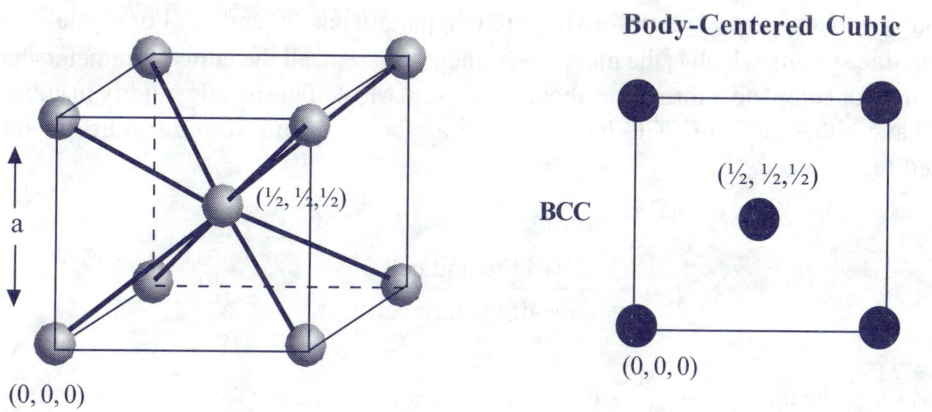

Fig. 2.18 Body-Centered Cubic structure.

(iii) ***Face-centered cubic (FCC):*** Lattice points present at the corners and also at the centres of the faces (Fig. 2.19). The volume of the cell is a^3. The nearest-neighbor distance as $a/\sqrt{2}$.

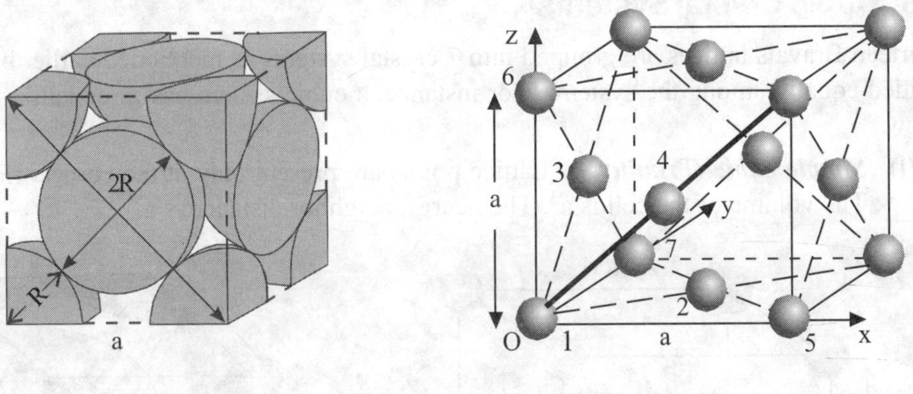

(a) (b)

Fig. 2.19 Face centered cubic lattice.

These three lattices have the characteristics of the most symmetric system, i.e., cubic. Cubic system is characterised by the lattice parameter relationships

$a = b = c$ and $\alpha = \beta = \gamma = 90^0$

That is, any crystal which belongs to cubic system has only one independent characteristic that is its lattice parameter 'a', since $a = b = c$ and the angular relationships are already known. Hence for cubic crystals only 'a' has to be obtained. However in the case of other systems and the respective lattices, other parameters should also be specified. For instance, for crystals belonging to tetragonal system, we require two parameters 'a' and 'c'. For crystals belonging to triclinic system, which is the most unsymmetric system, all the lattice parameters have to be obtained for complete information about the crystal. Most of the metals solidify in cubic system, and have either BCC or FCC lattice. There are some metals which solidify in hexagonal structure.

The density of the crystal is

$$\rho = \frac{\text{mass of the unit cell}}{\text{volume of the unit cell}}$$

$$\text{Mass of the unit cell} = \frac{M}{N_A} \times n$$

N = atomic mass, n = the number of atoms per unit cell, N_A is the Avagadros number.

$$\therefore \qquad \rho = \frac{nM}{N_A a^3}$$

2.9 Atomic Packing Fraction in Crystal

2.9.1 Simple Cubic Structure

This structure is the simplest among all the structures. This is formed when single similar atoms occupy the eight lattice points at the eight corners of the cube as shown in Fig. 2.8. This structure is based on the primitive cell of the cubic system.

In this structure, the distance between any two nearest neighbours is the lattice constant 'a', which is the edge length of the cube. As in the case of other structures, we can obtain the nearest neighbour distance, coordination number, number of lattice points (atoms) per cell and APF. They are as follows :

(i) Lattice points (atoms) per unit cell

$$= \frac{1}{8} \times 8 = 1$$

(ii) Relation between the lattice parameter 'a' and radius of the atom 'R' is

$$a = 2R,$$

as the atoms along the edge touch each other Fig.(2.8(b)). Hence the nearest neighbour distance is 'a' = 2R

(iii) Number of nearest-neighbour : 6

For any given atom at the corner, there are six nearest neighbours, each one along $\pm$X, $\pm$Y and $\pm$Z directions at a distance of 'a', the edge length.

(iv) $$\text{APF} = \frac{V_S}{V_C} = \frac{\frac{4\pi}{3}R^3}{(2R)^3} = \frac{\pi}{6} = 0.52.$$

Hence we observe that simple cubic structure has the least packing fraction among the three structure FCC, BCC and SC.

Metals which crystallize under this structure/lattice are very few with one atom as basis. However, there are many materials which have SC lattice with 2 or more atoms as basis. Cs*Cl* is one such example. But this structure is known as Cs*Cl* structure and not as SC. We will discuss about it later.

Let us first take up the cubic diamond structure which is based purely on covalent bonding between the carbon atoms.

2.9.2 BCC Structure

BCC structure as shown in Fig.2.20, has 9 points, eight at corners and one at the center. The 2-dimensional projection of the structure is also shown in Fig 2.18 where, at the centre, a point is marked (½). This point has coordinates of (½, ½, ½) when measured from one of the corners. The marking (½) indicates, that this ion is displaced by ½ from the bottom along 'z' direction. The projection that we see is that when the cell is observed from the top. Then we see only '4' corners, which have four more points below them, and one at the centre. Hence, this unit cell can be generated with the repetition of coordinates (0, 0, 0) and (½, ½, ½).

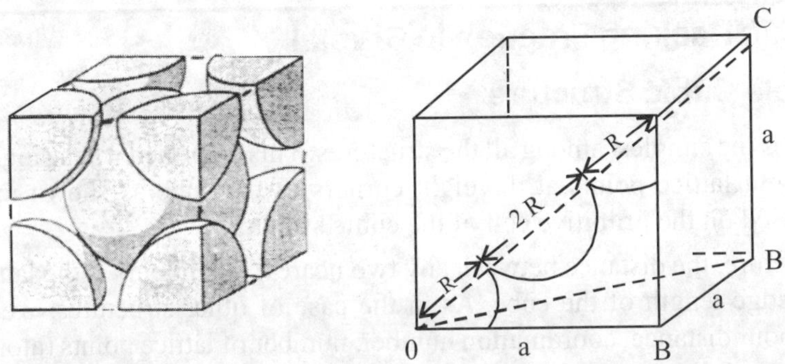

(a) (b)

Fig. 2.20 For the body-centered cubic crystal structure, (a) a hard sphere unit cell representation, (b) diagonal distance.

This structure is commonly observed in metals such as Cr, Fe, W etc., when the atoms are assumed to be hard spheres, it appears like the one shown in Fig.2.20(a). As can be seen from Fig.2.20(a), the centre and corner atoms touch one another along the cube diagonals, and the unit cell length is 'a'. The atomic radius is assumed to be 'R'. From the Fig.2.20(b), we can get the relation between 'a' and 'R' as

$$OB^2 + BC^2 = OC^2$$

$$2a^2 + a^2 = (4R)^2$$

i.e.,
$$a = \frac{4R}{\sqrt{3}}$$

Effective number of points per unit cell :

Number of points or atoms per unit cell in this case is '2' and can be obtained as follows :

No. of atoms at corners : 8

No. of atoms at the centre : 1

Each corner of atom is shared by 8 cubes.

$\therefore$ Contribution of corners to the unit cell $= 8 \times \frac{1}{8} = 1$

Contribution of the atom at the centre of the cell $= 1 = 1$

Total $= 2$

Number of nearest neighbours (coordination number) in this cell is '8'. For instance the atom at centre of the cell has 8 neighbours, which are at the 8 corners and all of them touch the surface of the atom at the centre. Any atom in the lattice can be chosen to be at the centre and the situation described above would be repeated.

Since the radius of each atom is 'R', the nearest neighbour distance would be 2R, which is the distance between the center of the atom situated at the corner and centre of the cell. But R is related to 'a' through $R = \frac{a\sqrt{3}}{4}$, hence the nearest neighbour distance in BCC $= \frac{a\sqrt{3}}{2} =$ 0.866 a.

This indicates that the ions at the corners of this cell do not touch each other.

Atomic Packing Factor (APF) :

It is given as

$$APF = \frac{\text{Volume of atoms in unit cell}}{\text{Volume of unit cell}} = \frac{V_S}{V_C}$$

$$V_S = 2 \times \frac{4\pi}{3} R^3 \; ; V_C = a^3 = \left(\frac{4R}{\sqrt{3}}\right)^3 = \frac{64}{3\sqrt{3}} R^3$$

$$\therefore \qquad APF = \frac{\frac{8\pi}{3} R^3}{\frac{64}{3\sqrt{3}} R^3} = \frac{\pi\sqrt{3}}{8} = 0.68$$

The above result indicates that the atoms are less efficiently packed in BCC compared to FCC, which has APF as 0.74. Reduction of APF indicates that the volume of the unit cell is not completely utilized by atoms.

2.9.3 FCC Structure

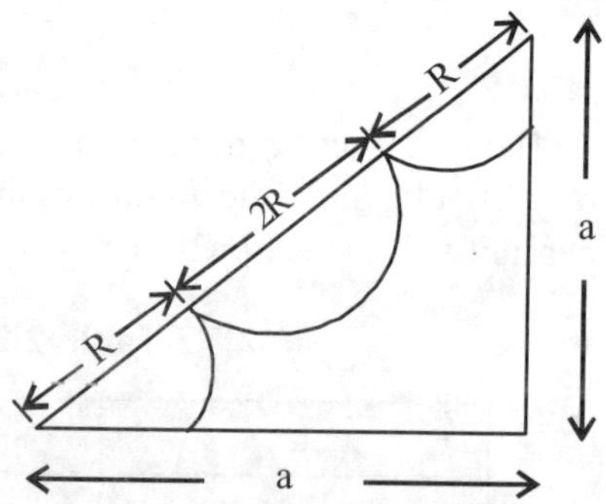

In this structure, atoms are located at each of the corner and centres of all the cube faces, and hence called face-centered cubic (FCC) structure. Fig. 2.19(a) shows a hard sphere model for FCC unit cell, where as in Fig. 2.19(b) the atom centres are represented by small circles as representative of the atoms to give a clear picture of the structure with respect to atomic positions in the unit cell.

As in Fig. 2.19(a), the atoms touch one another along a face-diagonal. The length of the diagonal is 4R, where R is the radius of each atom. The unit cell has a volume a^3. 'a' is the cell edge length. From the Fig. 2.19(a), we see that the total length from one corner to its opposite corner is 4R and hence

$$a^2 + a^2 = (4R)^2$$

$$\text{or} \qquad a = 2\sqrt{2}\, R$$

This is the relationship between the edge length 'a' and the radius of the atom 'R'.

The volume (V_C) of the cell in terms of 'R' can be given as

$$V_C = a^3 = \left(2\sqrt{2}\right)^3 (R)^3 = 16\sqrt{2}\, R^3$$

Similarly we can also obtain the nearest neighbour distance. "Nearest neighbour is that which is closest to a given atom and most probably would be in contact with it." In the case of simple cubic structure, that is the structure formed from primitive cell of cubic, the atom which is nearest are those which are placed at the corners. Hence the nearest neighbour distance is 'a' as given in Table 2.2. However, in the case of FCC, the atoms which are in contact are those which are along the face diagonal of the cube and not the corner atoms. Therefore the nearest distance between the center of the atoms (nearest-neighbours) is 2R or when expressed in terms of lattice parameter 'a' (edge length in cubic) is $a/\sqrt{2}$ i.e., 0.707 a.

Effective Number of Atoms in FCC Crystal :

In FCC, these are 8 atoms at the corners and 6 atoms at the face centres of the cube. As mentioned earlier, all the atoms in a unit cell do not completely belong to that cell but are shared by the adjacent cells in a solid, as the unit cell is not isolated but is a part of the solid. Hence the contributions of the atoms to a given cell can be calculated as follows :

No. of atoms at the corners : 8

No. of atoms at the face-center : 6

Contribution from each corner to the cell under consideration is : 1/8

Each face centred atom is shared by two cubes, hence contribution from each face centre to the cell is : 1/2

∴ Effective total number of atoms belonging to the FCC unit cell is

$$= 8 \times \frac{1}{8} + 6 \times \frac{1}{2} = 1 + 3 = 4$$

Hence effective number of atoms or points belonging to FCC structure or lattice is 4 respectively. Since the volume of the cell is 'a^3', hence lattice points or atoms in a FCC structure per unit volume is ($4/a^3$) as shown in Table 2.2.

Table 2.2 Characteristics of Cubic Lattices

	Simple	Body-centered	Face-centered
Volume, conventional cell	a^3	a^3	a^3
Lattice points per cell	1	2	4
Volume, primitive cell	a^3	$1/2\ a^3$	$1/4\ a^3$
Lattice points per cell	$1/a^3$	$2/a^3$	$4/a^3$
Number of nearest neighbours	6	8	12
Nearest-neighbour distance	a	$\frac{a\sqrt{3}}{2} = 0.866\ a$	$\frac{a}{\sqrt{2}} = 0.707\ a$
Packing fraction*	$(1/6)\pi = 0.524$	$(1/8)\pi\pi\sqrt{3} = 0.680$	$(1/6)\pi\sqrt{2} = 0.740$

Atomic Packing Fraction in FCC Crystal :

In the case of FCC, the coordination number (number of nearest neighbour, and the atomic packing fraction (APF), i.e., the ratio of the volume actually occupied by the atoms to the volume of the unit cell (i.e., volume available to be occupied by the atoms). In the case of metals, with single species of atoms, the coordination number is 12. Let us try to calculate the APF for FCC.

$$\text{APF} = \frac{\text{Volume of the atoms in the unit cell}}{\text{Total volume of the cell}} = \frac{V_U}{V_C}$$

V_U is the filled volume of the cell by spheres occupying it, as it is assumed that the atoms are hard spheres with radius R.

$$\therefore \quad V_U = 4 \times \left(\frac{4}{3}\right) \pi R^3 = \frac{16}{3} \pi R^3$$

we know the value of V_C is $16\sqrt{2}\ R^3$.

$$\therefore \quad \text{APF} = \frac{V_U}{V_C} = \frac{\left(\frac{16}{3}\right)\pi R^3}{16\sqrt{2}R^3} = 0.74$$

For the FCC structure, the atomic packing fraction is 0.74 or 74% of the space in the unit cell is occupied. This is the highest packing fraction that can be achieved where spheres (atoms) of same diameter are packed. For all other structures APF is less, as can be seen from Table 2.2. Hence 'FCC' structure is also known by the name "cubic close packed structure" (CCP) .

2.9.4 Diamond Structure

In crystalline form 'diamond', one of the important precious stones, is available in two structures. It crystallizes in cubic system as well as hexagonal system. We shall discuss about the diamond crystal which has cubic symmetry. Its structure is commonly known as Diamond Cubic (DC). As mentioned earlier, carbon atoms occupy the sites in FCC lattice with basis as '2'. That is, there would be 8 effective atoms in each of the diamond unit cell. This is observed in Fig. 2.21(a), where the covalent bonding among the atoms is shown. There are 14 carbon atoms situated at the corners (8) and face centres (6). Apart from those occupying the FCC lattice points, there are 4 carbon atoms situated along the 2 body diagonals at a height of 1/4 of lattice parameter from the bottom and 3/4 of lattice parameter again from the bottom on the other two body diagonal which are opposite to each other. The basic vectors of the cell are those of FCC (0, 0, 0), (0, ½, ½), (½, 0, ½), (½, ½, 0) and those of second interpenetrating FCC, (¼, ¼, ¼), (¼, ¾, ¾), (¾, ¼, ¾), (¾, ¾, ¼). Each of the carbon atom has coordination number (nearest-neighbour) as '4', unlike in the case of FCC structure in metals where the coordination was '12'.

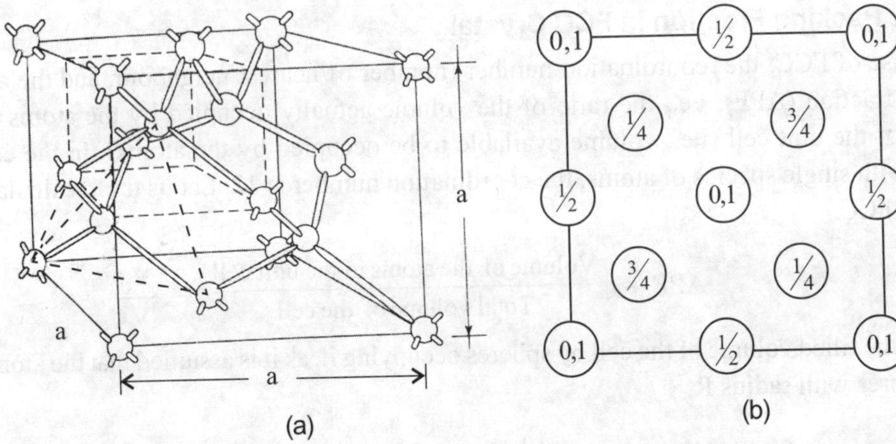

Fig. 2.21 (a) Unit cell of Diamond cubic structure with the tetrahedron formed by the carbon atoms. (b) A view of the structure when seeing from the top (along z-axis) of the unit cell, with positions of the ions measured from the bottom.

Let us calculate APF in the diamond structure.

(a) Effective number of atoms in the Unit cell

(i) Corners $\rightarrow 8 \times \frac{1}{8} = 1$

(ii) Face centers $\rightarrow 6 \times \frac{1}{2} = 3$

(iii) Atoms fully inside the cell $\rightarrow 4 \times 1 = 4$

Total = 8

(b) Volume of each atom $= \frac{4\pi}{3}R^3 = \frac{4\pi}{3}\left(\frac{a\sqrt{3}}{8}\right)^3$,

Since the nearest neighbour to any atom is situated at a distance $\frac{a\sqrt{3}}{4}$, which is the diameter of the atom. It we take any atom at the corner, say P, then the atom 'E' is situated on the body diagonal passing through 'P', at a distance $\frac{1}{4}$th of the distance along the body diagonal. Length of body diagonal is $a\sqrt{3}$ (a – lattice parameter) and $\frac{1}{4}$ of it is $\frac{a\sqrt{3}}{4}$. Hence, only those atoms at a distance of $\frac{a\sqrt{3}}{4}$ (distance between the centres of the atoms) touch each other and all others are at distances greater than this.

$$\therefore \quad APF = \frac{V_S}{V_C} = \frac{8 \times \frac{4\pi}{3}\left(\frac{a\sqrt{3}}{8}\right)^3}{a^3} = 0.34.$$

Hence we see that among the cubic structures, packing efficiency is least in this case, because the number of nearest neighbours is dictated by the number of bonds and not by their size. This indicates that most of the volume of the cell remains unfilled. The tetrahedron formed is shown in Fig. 2.21(a) with PQR as base and 'H' as vertex. Same is also marked in the form of a imaginary cell ABCDEFGH, where the vertex of the tetrahedron is at 'D' and BFG forms the base. Fig. 2.21(b) shows the structure as seen from the top of the cell.

Other fourth column elements in the periodic table such as Si, Ge and grey tin also have diamond cubic structure. As the size of the ion increases with increase of atomic number for Si, Ge and grey tin, the lattice parameter for these crystals also increases. Finally we can conclude that FCC lattice with a basis of 2 carbon atoms results in diamond structure. However, instead of two carbon atoms (similar atom) as the basis, if we choose two different types of atoms, one each, then the same FCC lattice results into another type of structure called zinc blende (ZnS).

2.9.4 Hexagonal Close Packed Crystal Structure

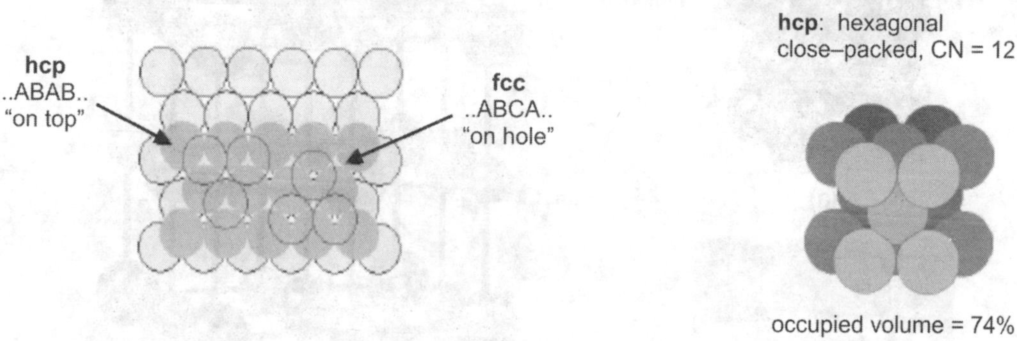

Fig. 2.22 HCP structure.

HCP structure close packed atoms placed one over the other. The stacking sequence of HCP structure is ABAB. Hexagonal close packed structure has a coordination number is 12. This is same as FCC crystal structure. Atomic packing fraction (APF) is given by

$$APF = \frac{2\frac{4}{3}\pi R^3}{\frac{1}{2}\sqrt{3a^2c}} = 0.74$$

Hence occupied volume in the HCP lattice is 74%. Metals do not crystallize in simple HCP structure since APF is too low. The ratio of height c of the prism of the HCP crystal structure to its basal side a is called c/a ratio. The c/a ratio for an ideal HCP crystal structure consisting of uniform spheres packed as tightly as possible is 1.63.

2.10 Sodium Chloride (Rock Salt) Structure

Sodium chloride (Na*Cl*), that is mineral Rock salt, crystallizes in cubic lattice and it has two ions as basis, one forming a fcc structure with Cl^{-1} at (0, 0, 0) as origin and another fcc structure forming with Na^{+} as origin at (1/2, 0, 0). It has 4 molecules of Na*Cl* in one unit cell. The coordination number is 6 for each of the ions (cation and anion). Cations and anions are topologically identical, that is the structure remains identical when a cation or an anion is replaced with an anion or cation respectively at all lattice points.

The fcc nature of lattice can be observed clearly by looking at Fig. 2.23(b), where two types of ions (big and small) are shown. If we consider the big ions as *Cl*, then they occupy all corners and also the face centres of the cube as in the case of fcc structure (lattice). The small ions are of Na^{+} and they occupy the positions exactly half-way between any two Cl^{-1} ions. This is necessary to keep the electrostatic repulsion minimum. Each of cations (Na^{+}) is surrounded by 6 anions (Cl^{-1}) and vice versa. Therefore the coordination number is '6' for each of the ions. There are exactly four units of NaCl molecule in the unit cell. The ions are in the positions described by,

Cl^{-} : 0 0 0 ; ½ ½ 0 ; ½ 0 ½ ; 0 ½ ½ ;

Na^{+} : ½ ½ ½ ; 0 0 ½, 0 ½ 0 ; ½ 0 0

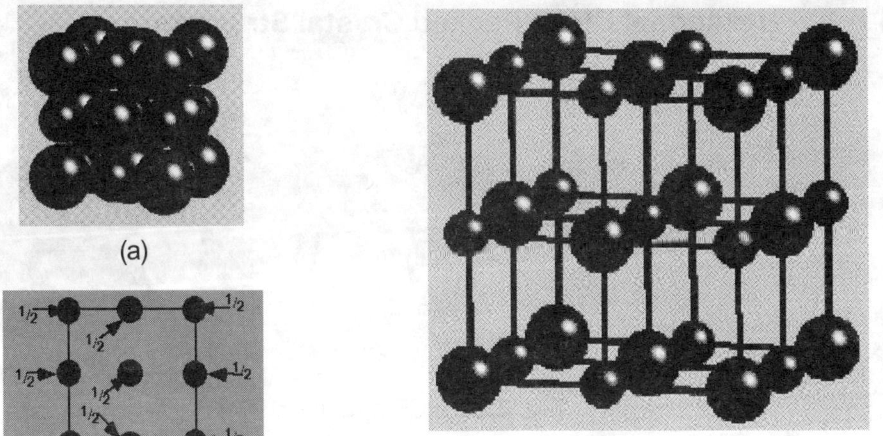

Fig. 2.23 (a) actual structure of NaCl (b) reduced ion-shell model of NaCl (c) two-dimensional projection of reduced ion-shell model.

Looking at Fig. 2.23(b), we may come to the following conclusion about the structure

(a) The structure has two fcc lattices inter-penetrating into each and each lattice displaced by half the lattice constant along the edge of the cube.

or

(b) The structure has a fcc lattice with basis or motif as two ions, i.e., one 'Na' and another '*Cl*' ion. Each of these (cation and anion) is displaced from each other by half the lattice constant along the edge.

2.11 Zinc Sulphide Structure

The cubic zinc sulphide which has two types of atoms, crystallizes into FCC lattice with basis as two. These two atoms are 'Zn' and 'S'. Hence the structure would appear to be different from that of Diamond. The resultant structure, known as zinc blende structure is shown in Fig. 2.24(a). In this, we observe, that, one type of atoms (14) occupy the 14 FCC lattice points and the other type of atoms (4) occupy positions inside the cell along the body diagonal at the position as that occupied by carbon in diamond cubic structure. Hence two-dimensional projection Fig.2.24b shows one type of atoms occupying the corners, edge centres and the centre of the cell, where as the other type of atoms occupy places along the face diagonals of the square. If we imagine that sulphur atoms occupy the FCC positions, then Zn atoms would occupy the sites on the body diagonal. The coordinates of sulphur atoms are (0, 0, 0), (0, ½, ½), (½, 0, ½), (½, ½, 0) and the coordinates of zinc atoms are (¼, ¼, ¼), (¼, ¾, ¾), (¾, ¼, ¾), (¾, ¾, ¼). We can observe that each atom (Zn or S) on the body-diagonal forms four covalent bonds with the other type of atom. That is each Zn atom is bonded to four S atoms, and vice versa. Most often we observe the bonding to be highly covalent in crystals exhibiting this structure. The other materials which show this structure are ZnTe, SiC, CdS, CdTe, GaAs, InSb to mention a few. Because of the covalent nature of the bonds, the materials which crystallize with these (diamond or ZnS) structures have very high hardness and are used as abrasives. Particular examples of this are diamond and SiC. The coordination number, the APF and other parameters in this structure would be same as in the case of diamond structure.

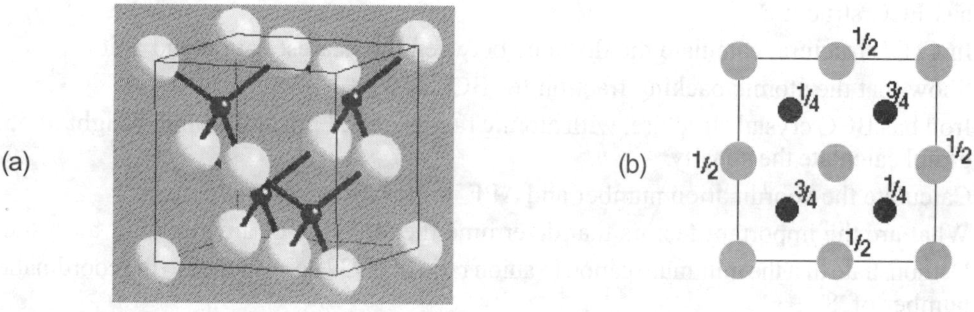

Fig. 2.24 'ZnS' structure.

2.12 CsC*l* Structure

Another structure commonly observed among many ionic solids is that of CsC*l* structure. The structure appears very simple, with C*l* (anions) ions occupying the corner of cube (unit cell) and at the centre we observe Cs ion (cations) as shown in Fig. 2.25. Each of the Cs^+ ions is surrounded by 8 Cl^- ions, leading to a coordination number of '8'. We can also construct an unit cell with Cs^+ ions at the '8' corners of the cell and one and Cl^- ion at the centre of it, again making the coordination number as '8'. The basis of the unit cell has one CS^+ ion at 0 0 0 and one Cl^- ion at ½ ½ ½ positions of the simple cubic lattice. Therefore, we can look at this structure as two inter penetrating simple cubic structure (lattice) displaced by a $\sqrt{3}/2$ distance along the body diagonal of the cube.

One generally gets confused by looking at CsCl structure and assumes it to be of BCC type. This is not correct. In a BCC structure all the lattice points, i.e., 8 corners and the one at centre are occupied by the SAME atom". In the above structure, only corners are occupied by Cl^- and the centre is occupied by the Cs^+ ions or vice versa. Hence CsCl belongs to a simple cubic lattice with basis as '2' (two atoms one Cl^- and one Cs^+). The crystals that show CsCl structure are CsI. CsCl structure is not a close packed structure like fcc structure.

TlBr, NH_4Cl, β-brass, AlNi, CuPd alloys crystallise in this structure.

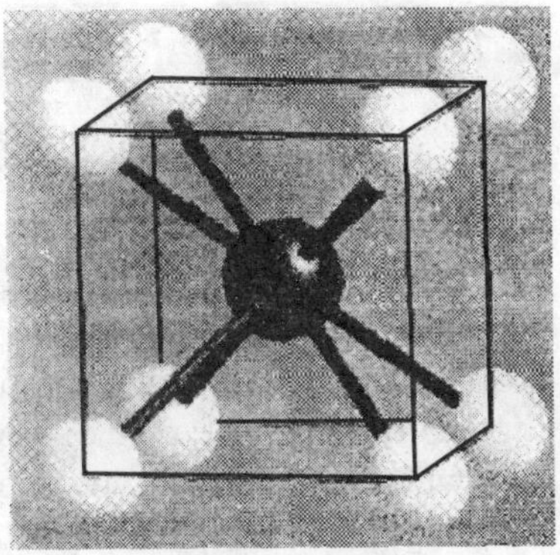

Fig. 2.25 CsCl structure.

Questions

1. How many tetrahedral and octohedral sites are there in a unit cell of cubic close pack structure.
2. Draw the zinc blende structure. Compare this structure with that of diamond structure.
3. What is the difference between the space lattice and a structure.
4. Describe the important features of the three lattices which belong to the cubic system.
5. If the atomic radius of aluminium is 143 pm, calculate the volume of its unit cell. Aluminium has FCC structure.
6. In CsCl structure, calculate the distance between the nearest cation and anion.
7. Show that the atomic packing fraction for BCC is 0.68.
8. Iron has BCC crystal structure, with atomic radius of 124 pm and atomic weight of 55.9 g/mol calculate the density.
9. Calculate the coordination number and APF in the case of ZnS structure.
10. What are the important factors that determine the structure of a material or a crystal?
11. Demonstrate that the minimum cation to anion ratio of 0.732 corresponds to the coordination number of '8'.
12. Compute Atomic packing fraction for CsCl structure.
13. What are the Bravais lattices of NaCl and CsCl?
14. How many lattice parameters are required to completely specify a crystal which belongs to monoclinic system ?
15. What are Bravais Lattice ? Describe them in combination with crystal systems.
16. Calculate the atomic packing fraction of a diamond structure.
17. Calculate the atomic packing fraction in a simple cube.
18. Silicon has the diamond structure. The unit cell has edge a = 0.542 nm. What is the nearest interatomic separation in it? (refer to section 2.9)
19. Write the characteristics of seven crystal systems.
20. What are different space lattices in the cubic system.
21. Explain the structure of hcp lattice. Obtain its APF value.

Objective Questions

I. Choose the correct choice in the following objective questions.

1. According to Bravais, there exists
   (a) fourteen simple primitive lattices
   (b) three hexagonal primitive lattices
   (c) five monoclinic lattices
   (d) seven simple primitive and seven multiple primitive lattices

2. Sodium chloride is
   (a) a body centered crystal
   (b) an inter penetrating fcc lattices of Cl^- ions and Na^+ ions
   (c) simple cubic crystal
   (d) a cubic crystal with Na^+ and Cl^- alternatively at the cube corners

3. The number of atoms per unit cell in diamond structure
   (a) 8 (b) 6
   (c) 4 (d) 12

4. In CsCl structure, Cs^+ is at (0, 0, 0) position in unit cell, then Cl^- is at
   (a) 0, 0, 1/2 (b) 0, 1/2, 0
   (c) 1/2, 1/2, 1/2 (d) 1/4, 1/4, 1/4

5. The unit cell of zinc blende structure
   (a) has two Zn ions and two S ions
   (b) has four Zn ions and two S ions
   (c) has 8 Zinc ions and four Sulphur ions
   (d) has four Zinc ions and four Sulphur ions

6. The amorphous Substances possess
   (a) Long-range order
   (b) Short- range order
   (c) ionic bonds only
   (d) covalent bonds only

7. In an unit cell of NaCl, the distance between nearest Na^+ and Cl^- ions is (let 'a' be the lattice parameter)
   (a) a/4 (b) $a/\sqrt{2}$
   (c) a/2 (d) $\sqrt{2}$ a

8. In a unit cell of NaCl, the distance between the nearest like ions is
   (a) a (b) a/2
   (c) $a/\sqrt{3}$ (d) $a/\sqrt{2}$

9. Copper crystal belongs to
   (a) bcc structure
   (b) a closed packed fcc structure
   (c) a close packed hexagonal structure
   (d) face centered tetragonal structure

10. In Bravais lattices, we have
    (a) the cubic system with four primitive cells
    (b) three tetragonal primitive lattices
    (c) one hexagonal primitive lattices
    (d) three rhombohedral primitive lattices

11. Calcium carbonate belongs to
    (a) bcc structure
    (b) orthorhombic structure
    (c) tetragonal structure
    (d) rhombo hedral structure

12. In NaCl structure, the basis of the cubic lattice with lattice constant 'a'
    (a) Comprises of six lattice points
    (b) Comprises of eight lattice points
    (c) Comprises of twelve lattice points
    (d) Comprises of fourteen lattice pints

13. The atomic packing fraction (APF) in the case of diamond structure is
(a) 0.34 (b) 0.68
(c) 0.73 (d) 0.52

14. In the following, the one that is not a close packed structure is
(a) copper
(b) sodium chloride
(c) cessium chloride
(d) magnesium oxide

15. In a fluorite structure, cations have a coordination number as
(a) 6 (b) 4
(c) 12 (d) 8

16. The cation and anion ratio in the structure of 'Fe O' are 0.077 and 0.140 nanometers . Then the coordination number is
(a) 8 (b) 6
(c) 4 (d) 12

17. The atomic packing traction in FCC crystal is
(a) 0.74 (b) 0.86
(c) 0.34 (d) 0.52

18. The one that has zinc blende structure in the following
(a) FeO (b) MgO
(c) Iron (d) SiC

19. A crystal that exhibits the negative temperature coefficient of resistivity is
(a) copper (b) aluminium
(c) sodium chloride (d) barium

20. The crystal structure of diamond is
(a) fcc with two atom basis of (0 0 0) and a/4 (**i** + **j** + **k**)
(b) simple cubic with two atom basis of (0 0 0) and a/2 (**i** + **j** + **k**)
(c) fcc with two atom basis of (0 0 0) and a/2 (**i** + **j** + **k**)
(d) fcc with one atom basis.

II. Fill in the Blanks

1. Amorphous substances are formed by __________ cooling of the melts of inorganic compounds.
2. A glass is __________ liquid.
3. A substance in amorphous state possesses __________ viscosity.
4. The number of bravais lattices is equal to __________ .
5. The radius of each atom in an fcc structure is 'R', then the volume of the unit cell in terms of R is __________ .
6. The nearest neighbour distance in fcc unit cell is __________ a. (a, the lattice constant)
7. The atoms in bcc structures are __________ efficiently packed compared to fcc.
8. The crystal structures formed based on covalency coordination are __________ and __________ .
9. Silicon carbide and gallium arsenide belong to__________ structure.
10. There are__________number of molecules in NaCl unit cell.
11. Each cation in NaCl structure is surrounded by __________ number of anions.
12. The smallest unit of a crystal is called __________
13. The number of Bravais lattices for orthorhombic system is __________

14. In a monoclinic system, the unit cell lengths are __________ and d and β are __________

15. Tetragonal system has only __________ Bravais lattices.

16. The plane $(1\ \bar{1}\ \bar{1})$ has intercepts __________ on crystal axes.

III. True / False

1. Amorphous substance possesses viscosity.

[True/False]

2. Amorphous substance possesses short range order.

[True/False]

3. Amorphous substance is not in absolute thermal equilibrium although it is stable over years.

[True/False]

4. A crystalline solid possesses a short-range order and slight viscosity

[True/False]

5. The relation $\gamma' = \gamma + la + mb + nc$ where a, b, c are translational vectors representing atomic arrangement in a crystal

[True/False]

6. The intersection of three non-coplanar faces of the unit cell determine the crystallographic axes.

[True/False]

7. Orthorhombic crystal system is represented by $a = b = c$; $\alpha = \beta \neq 90^o$ [True/False]

8. Triclinic crystal system is represented by $a \neq b \neq c$, $\alpha \neq \beta \neq \gamma \neq 90$

[True/False]

9. Packing factor of face centered cubic structure is 74 %

[True/False]

10. The number of molecules is a unit cell of NaCl is one.

[True/False]

11. The bravais lattice of Zns is face centered cubic lattice.

[True/False]

12. The number of atoms per unit cell of BCC structure is two.

[True/False]

13. The diamond unit cell is a simple cube.

[True/False]

14. A crystal plane is given as $(1\ \bar{1}\ 1)$. Then it intersects the y-axis on negative side.

[True/False]

15. Two planes (1 1 1) and $(3\ \bar{3}\ 3)$ are parallel planes.

[True/False]

16. The effective number of atoms in fcc crystals is 2.

[True/False]

17. CsC*l* belongs to body centered cubic structure.

[True/False]

18. A cubic Zinc sulphide has two types of atoms and crystallises into bcc structure with basis as two

[True/False]

19. Silicon carbide belong zinc blende structure.

[True/False]

20. NaCl has two fcc lattices inter penetrating into each and each lattice displaced by half the lattice constant along the edge of the cube.

[True/False]

21. MgO crystallises with NaCl structure.

[True/False]

22. A crystal plane is represented by (3, 6, 1). Then, the intercepts of are $\frac{2}{3}, \frac{1}{3}$ and 1 on x– and y– and z–axes respectively.

[True/False]

CHAPTER 3

X-Ray Diffraction

3.1 Basic Principles

X-rays are the electromagnetic radiations with a wave length of the order of an angstrom. Laue considered that the crystal consists of arrays of atoms in three dimensions. The crystals serve as gratings to the x-rays and produce the diffraction of x-rays. W. H. Bragg and his son W. L. Bragg while studying the properties of X-rays studied the X-ray diffraction pattern of a few crystals. This discovery was the first step towards understanding the crystal structures which subsequently helped in the fields of metallurgy, chemistry, solid state physics etc.

3.2 Bragg Law

It is the periodicity of the crystal structure that makes X-ray diffraction possible. Let us consider the diffraction of X-rays by the atomic arrays shown in Fig. 3.1(a). Let us consider that the atomic array Fig. 3.1(b) is a stacked parallel planes passing through the centres of atoms. Each atom is a centre for X-ray scattering. The monochromatic X-ray beams fall on the parallel planes. They are reflected or scattered by the atoms.

Let the incoming X-rays be MO, M'N. Let the scattered rays be OS', NS". These scattered beams of X-ray form an angle q with the planes of atoms. We represent the rows of atoms as a single horizontal line (shown in Fig. 3.1(c)). The plane of atoms is like a mirror.

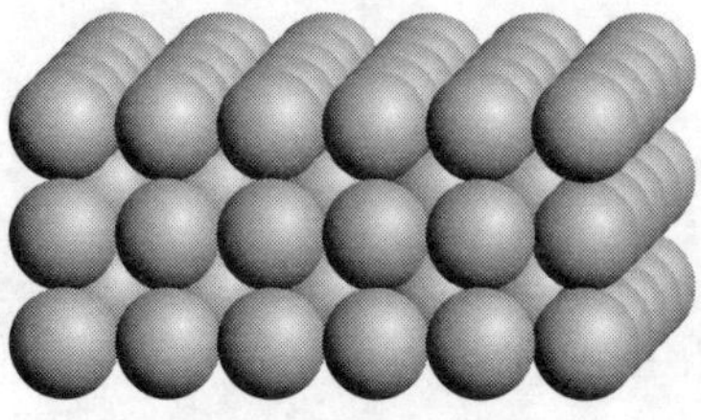

(a) atomic planes

(b) X-ray scattering atom planes

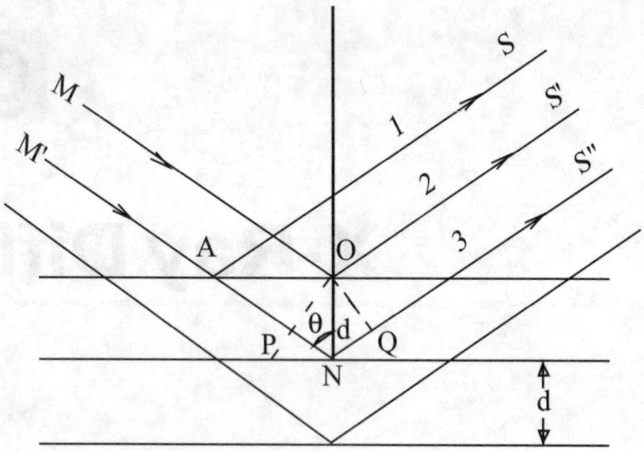

(c) the planes taken as horizontal line

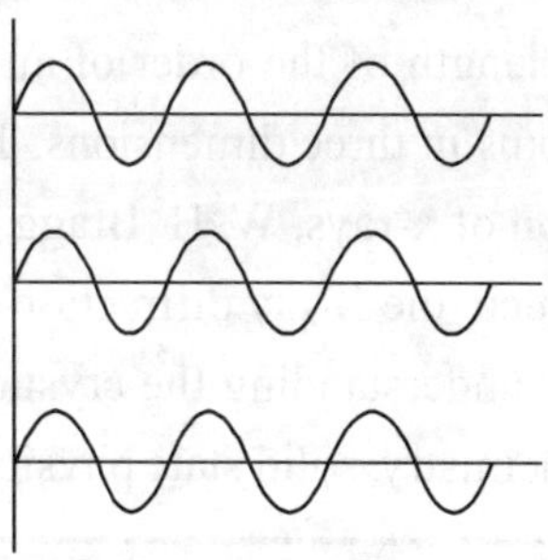

(d) in phase waves

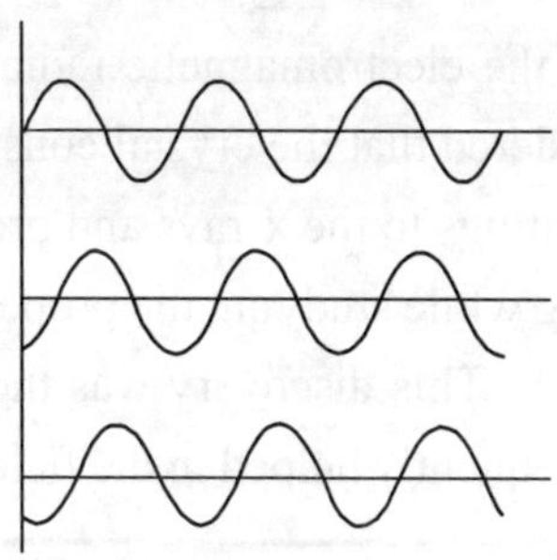

(e) out of phase waves

Fig. 3.1

The scattered rays AS, OS' form an angle θ with the plane (h k l). The total path lengths of the rays M'AS, MOS' are the same. The rays are said to be in phase with each other; that means, the waves of the individual rays arriving at SS' again form a common wave front.

Now, consider the incoming ray M' N and the scattered ray NS". The total path of the x-ray beam is M'N + NS". This is greater than the path lengths of M' A + AS and MO + OS'.

From the Fig. 3.11(c), the path difference is PN + NQ = 2PN.

But $\quad PN = d \sin \theta$

the path difference $\quad \delta = 2d \sin \theta \qquad$ .....(3.1)

If the ray M' N S" arrives at S S' S" in phase with rays M O S' and M' A S or the two planes { h k l } scatter x-rays in phase, then total path difference must be equal to an integral number of wavelengths i.e.

$$2d \sin\theta = n\lambda \qquad .....(3.2)$$

Where n = 0, 1, 2, 3.............. etc.

This is the condition for diffraction to produce an intense beam (d is d_{hkl})

The meaning of above equation is that the diffraction intensities will build up only at certain values of θ, corresponding to specific values of λ and d. 'n' is called the order of the diffraction. All the rays that are scattered by atoms in a set of parallel planes will form a common wave front at an angle θ. Here both λ and θ are important. Under the condition of diffraction given in eq. (3.1), the amplitude of all the waves reinforce to give a resultant wave having the maximum amplitude. This condition is shown in Fig. 3.1(d).

On the other hand, at some other angles, the scattered waves coming from atoms may not be in phase with each other as shown in Fig. 3.1(e). The amplitude of the resultant is zero.

The relation given in eq.(3.2) is known as Bragg law. The reflection angle θ is given as

$$\theta = \sin^{-1}\frac{n\lambda}{2d_{hkl}} \qquad .....(3.3)$$

θ depends on d_{hkl} for a fixed λ and d_{hkl} is the interplanar distance.

In principle, when we keep a crystal in the path of monochromatic X-rays, the diffraction reinforced waves arrive at different angles (θ values). Each diffraction line then corresponds to a diffraction of a set of wavelets scattered from a set of parallel planes. The interplanar distances can be determined in a crystal by determining sin θ values. This is the most important utility of X-ray diffraction. A crystal structure consists of atoms. The atoms in turn consists of electron clouds around the nuclei. It is the scattering of X-rays by the electrons that fixes the angles and the intensity of reflection. The electron cloud as a whole scatters X-rays. X-rays do not penetrate the electron cloud.

The number of electrons composing the cloud around the nucleus decides the intensity of scattering. It should be noted by the students that the word 'reflection' ray is used for 'diffracted ray' in x-ray physics.

3.3 Laue Method

We will confine ourselves to a method which employs a stationary single crystal. Laue method was the first diffraction method ever used. In this method, a beam of white radiation is collinated through a collimeter (c). The 'white' X-rays fall on the stationary single crystal as shown in Fig. 3.2. The Bragg angle θ is therefore fixed for every set of planes in the crystal. Now, each set of planes selects and diffracts a particular wavelength that satisfies the Bragg's law for particular values of d_{hkl} and θ involved. Therefore, we get a number of diffracted beams each corresponding to one set of planes for a particular wavelength. In Laue method λ is variable.

The diffracted pattern is caught on a photographic film. Laue pattern can be obtained either in transmission Laue method (Fig. 3.2(a)) or back-reflection method (Fig. 3.2(b)).

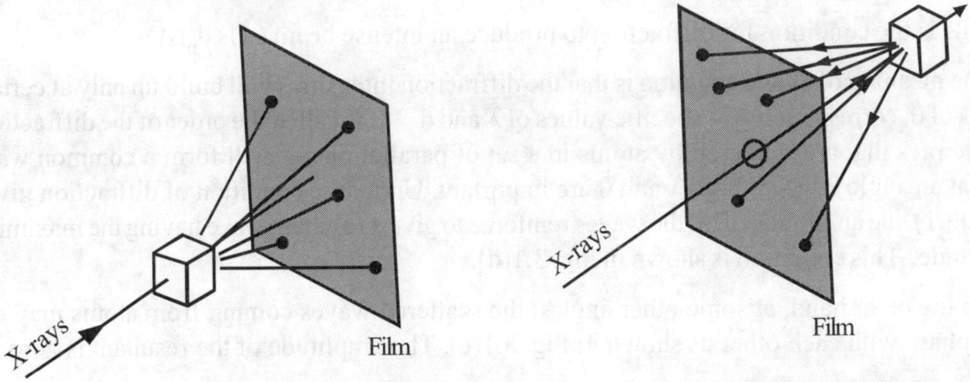

Fig. 3.2 Laue method (a) Transmission (b) Back-reflection.

In both the ways, the diffracted beams give a series of spots on the film as shown in Fig. 3.3.

The spots lie on certain curves. These curves are either ellipses or hyperbolas for transmission Laue pattern, and hyperbolas for back reflection patterns.

Fig. 3.3 Laue Pattern of NaCl crystal.

Although, the experimental method of Laue is very simple, the interpretation of Laue pattern or the diffracted spots is difficult. We can not determine the lattice parameters of a crystal using this method. Laue method is useful to study the symmetry of the crystals and in the determination of orientation of crystals. This is possible because the sample is stationary during the exposure to X-rays. The crystals used should be thin for transmission Laue method. Otherwise, the X-rays are absorbed by the crystals. Then the density of spots on the photographic film will be less.

This interpretation of Laue photograph requires the knowledge of the concept of reciprocal lattice; the topic is outside the syllabus.

The Bragg angle for any transmission Laue spot is given from the relation

$$\tan 2\theta = \frac{r_1}{D} \quad .....(3.4(a))$$

where r_1 is the distance between the centre of the film and the diffraction spot. D is the distance between the specimen and the film (usually 5 cm). Laue spots, near the center of the film or close to the directly transmitted X-ray beam, correspond to the first-order reflections from planes inclined at very small angles to the incident beam. In this case, only lower wavelengths can satisfy the Bragg condition. One can notice that the region close to the centre of the pattern is devoid of any diffraction spots and the size of this region as the intensity of incoming X-ray beam is reduced.

In the case of back reflection pattern, the Bragg angle for any spot may be obtained from

$$\tan(180 - 2\theta) = \frac{r_2}{D} \qquad .....(3.4(b))$$

where r_2 is equal to the distance of a spot from the center of the film and D is the specimen-to-film distance (~3 cm). Normally, the back reflection pattern may have spots close to the centre of the pattern. Such spots are due to high-order-overlapping reflections from (hk*l*) almost perpendicular to the incident beam.

The most essential thing in obtaining the Laue spots is to align the camera (simply a flat plate) correctly oriented with respect to the X-ray beam. That is the reason we require collimeter so that the direct X-ray beam is made to pass straight into the crystal and the centre of the film. Back reflections normally give weak spots. Hence, the exposure of the film should be longer.

As one moves towards the centre of the Laue pattern, the corresponding values of θ decrease. At the same time the indices of planes (hk*l*) increase. We have $\lambda = 2\, d_{hkl} \sin\theta$. When we move towards the centre of the Laue photograph, the corresponding value of d_{hkl} must decrease. Both terms on right hand side of the Bragg relation decrease as we approach the centre of the film. A limit is set by the minimum value of the wavelength in the X-ray spectrum. Hence, the spots can not occur near the centre. If we consider a plane (h, 0 0), then

$$\sin\theta = \frac{h}{\sqrt{h^2 + k^2 + l^2}}$$

$$\text{Then} \quad \theta = \frac{2a}{\sqrt{h^2 + k^2 + l^2}} \times \frac{h}{\sqrt{h^2 + k^2 + l^2}} = \frac{2ah}{(h^2 + k^2 + l^2)} \qquad .....(3.5)$$

Near the centre of the every photograph, h in every spot is equal to unity. Therefore, the spots nearest to the centre are the most likely to contain first-order reflections.

3.4 Powder Method

Many materials that are used in industry are polycrystalline materials. They are used in powder form. Therefore, the X-ray diffraction pattern of such materials is obtained in the form of powder pattern. The powder patterns are fairly simple and are widely used for the characterization of crystals or materials by X-rays diffraction.

Laue photographs give spots. The powder pattern is obtained in the form of lines. The Bragg scattering of X-rays consists of the scattering from all the crystallites in the specimen corresponding to a particular angle of 2θ. The diffracted beams must lie on a cone of apex of which is at the specimen. The specimen is kept on the axis of the X-ray beam.

The axis of the beam coincides with the axis of the cone. The cone has a semi-vertical angle that is equal to 2θ. The formation of such a cone is shown in Fig. 3.4.

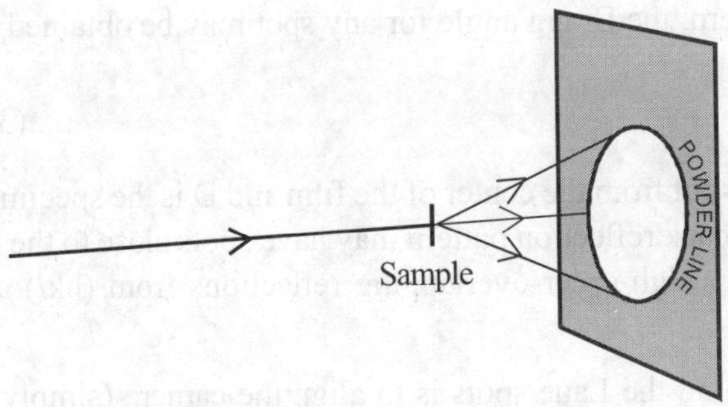

Fig. 3.4 A Circular ring is formed on the photographic film.

A number of lines will be produced for each value of the angle. The flat cameras are not used in general. The specimen is surrounded by a narrow strip of film. This strip intercepts a (Fig. 3.6) large number of cones of X-rays reflections. As the film is of the width of about 1 inch, a part of each cone is recorded. All the reflections for the angles $0 < \theta < 90$ are recorded.

The powder should be extremely fine and the grain size in the powder should be of the order of 1 –3μm. If the grain size is greater than this, the powder pattern will have 'spotiness'. The camera is a cylindrical camera as shown in Fig. 3.5 (a) and (b). The camera is called Hull/ Debye-Scherrer camera.

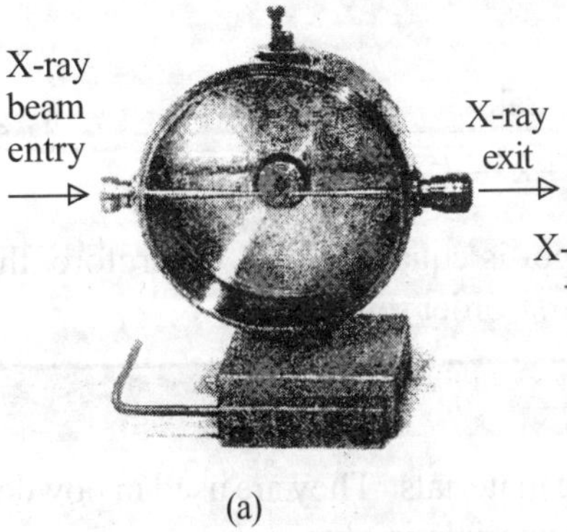

(a)

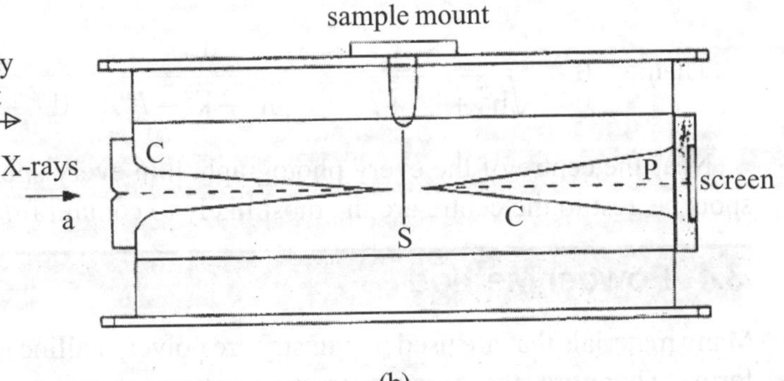

(b)

Fig. 3.5 (a) A cylindrical camera. (b) The top view of the camera. X-rays from left pass through the aperture a and through a collimeter C. They fall on the sample S. The direct beam goes to P.

X-rays are made to pass through a narrow aperture on the left of the camera. The direct beam after passing through the specimen goes to the other end through a narrow aperture and falls on a X-ray luminescent screen. A luminous spot appears on the screen. By adjusting the spot to be exactly at the centre of the screen, the X-rays can be made to go through the sample exactly. The side view of the camera is shown in Fig. 3.5(b).

The sample is stacked in a capillary tube of diameter 0.3 – 0.5 mm. It is kept on the sample holder (M) vertical to the mount.

The geometry of the powder camera is a section of a hollow cylinder. The film strip is enclosed on the inner wall of the camera with an yellow paper sheet on it.

Only monochromatic X-rays are used. Usually all the other radiations from X-ray tube are cut out by keeping a filter before the aperture. The copper target gives two X-rays wavelengths K_{α_1} and K_{α_2}. Therefore, the high angle lines (close to 90°) contain two lines which are due to $K\alpha_1$ and K_{α_2} X-ray beams. This is shown in Fig.3.6(a).

The figure shows the film mounted in the camera and the X-ray powder pattern obtained.

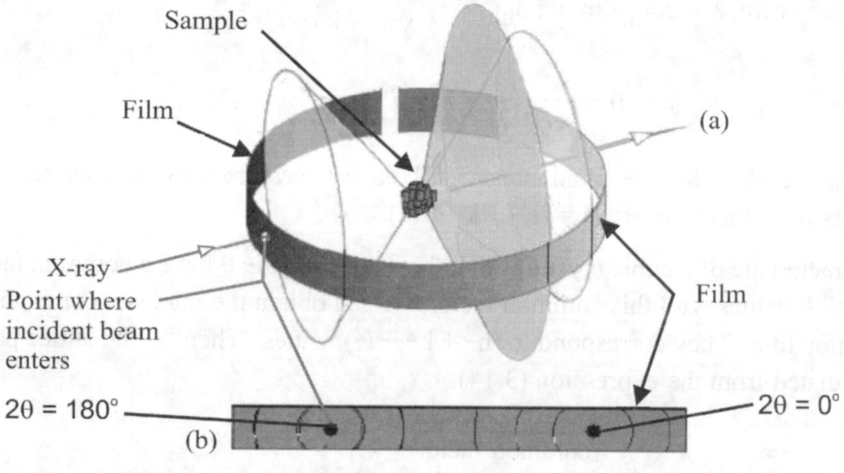

Fig. 3.6 (a) The figure shows the arrangement of the film in the camera. The stretched film shows the diffraction lines. The High angle lines will be due to $K_{\alpha 1}$ and $K_{\alpha 2}$ wavelengths. (b) Powder pattern of KCl-KBr mixed crystal taken with Debye-Scherrer camera.

In the case of the cylindrical camera, θ is proportional to S, where the S is the diameter of the ring that is obtained for an angle θ. Then

$$\theta = \frac{S}{4R}$$

where R represents the radius of the camera.

After taking the photograph, the film is made flat and the distance S is measured with the help of a comparator. A comparator is just a magnifying glass with a cross wire (the cross wire and the film are kept in the same plane) and the glass is fitted to a metal scale. The magnification of the lens is less than four times the diameter of the rings. On the left side of the film, (i.e., the right side of the camera), we get low angle reflections. On the back side i.e., around the point where the X-rays enter the camera, we get the high reflection rings. They are resolved into $K_{\alpha 1}$ and $K_{\alpha 2}$ lines.

Suppose the distance between the centres of holes of the film (they represent the entry of X-rays and exit of the X-ray beams) is D. Then the diffraction angle θ is

$$\theta = \left(90 - 45\frac{S}{D}\right)^0 \quad .....(3.6)$$

The X-rays are scattered by the air molecules in the camera. The direct beam is not allowed to travel in the space. Narrow conic tubes or collimeters are provided so that the tips of the collimeters are close to the sample.

Interpretation of the powder pattern :

In a cubic system, $\lambda = 2d_{hkl} \sin\theta$, $d_{hkl} = a/\sqrt{h^2 + k^2 + l^2}$

This gives $$\sin^2\theta = \frac{\lambda^2}{4a^2}\left(h^2 + k^2 + l^2\right) \quad .....(3.14)$$

The parameter $(h^2 + k^2 + l^2)$ is an integer. All that is necessary is to calculate $\sin^2\theta$ values and N values from them, where $N = (h^2 + k^2 + l^2)$.

The characteristic of a cubic crystal is that the values of $\sin^2\theta$ have a common factor. By dividing $\sin^2\theta$ values with this common factor, we can obtain the integers corresponding to each reflection line. They correspond to $(h^2 + k^2 + l^2)$ values. Then 'a' the lattice parameter can be calculated from the expression (3.14).

$$a = \frac{\lambda}{2}\sqrt{\text{common factor}}$$

The first two lines give the indication that they are from a cubic crystal. The high angles give the most accurate value of the lattice parameter.

The characteristic of a cubic system is that the values of $\sin^2\theta$ have a common factor. This factor can be found by taking the differences of $\sin^2\theta$ values for the successive lines. The least value of difference is the common factor. Then, we divide the $\sin^2\theta$ values with this least difference $(\Delta \sin^2\theta)$ min. We get integer values which correspond to $h^2 + k^2 + l^2$ values. The Miller indices can be found. If the Miller indices come out to be all odd or even, then the cubic system is an f.c.c. lattice. If $h + k + l$ = all even, then the crystal is a b.c.c. crystal.

Let us take an example. An oxide of iron gives the following θ values. Let us find out the Miller indices. It is done in the following table.

S.No.	θ(0)	Sinθ	Sin²θ	ΔSin²θ	$\frac{\text{Sin}^2\theta}{(\Delta\text{Sin}^2\theta)\text{min}}$	N	hkl
1	10.67	0.1852	0.03476	-	3.0	3	(111)
2	17.64	0.3035	0.0929	0.0553	8.0	8	(220)
3	20.80	0.3535	0.12561	0.0328	10.9	11	(311)
4	21.72	0.3709	0.1376	0.0115	11.9	12	(222)
5	25.34	0.4279	0.1831	0.047	16.0	16	(400)
6	31.63	0.5246	0.2753	0.1377	23.9	24	(422)
7	33.814	0.5565	0.3097	0.0344	26.9	27	(333)

The data is taken with X-radiation of wavelength 1.79A

The lattice parameter can be calculated as

$$a = \frac{1.79 \times \sqrt{0.0115}}{2} \times 10^{-8} = 8.3\,\overset{0}{\text{A}}$$

The above cited example is a fcc cubic crystal because all hkl values are either odd or even. For a bcc crystal structure, the diffraction occurs only from those planes whose Miller indices when added together give a total that is even number. i.e., $h + k + l$ = even. The principal diffraction planes in bcc structure are {100}, {200}, {211}, {220}, {310}. Those reflections, for which $h + k + l$ = odd, will be absent i.e., {311}, {100}, {111} will be absent. Higher orders actually are useful for accurate determination of the cell parameters.

Example :

Determine at which (111), (200), (220), (311) planes of a nickel crystal will diffract X-rays ($\lambda = 1.54$). The crystal is a f.c.c. crystal with a lattice parameter a = 3.52 . What are the values of inter planar distances ?

Solution :

h k l	$\sqrt{h^2+k^2+l^2}$	$d = \frac{a}{\sqrt{h^2+k^2+l^2}}\,\overset{o}{\text{A}}$	Sin θ	θ	Sin² θ
(111)	$\sqrt{3}$	$\frac{3.52}{\sqrt{3}}$	0.1263	7.256	0.0159
(200)	2	$\frac{3.52}{2}$	0.1100	6.315	0.0121
(220)	$2\sqrt{2}$	$\frac{3.52}{2\sqrt{2}}$	0.1361	7.82	0.0185
(311)	$\sqrt{11}$	$\frac{3.52}{\sqrt{11}}$	0.0659	3.779	0.0043

3.5 Difference in the Diffraction Pattern of fcc and bcc Crystals

In the case of bcc crystal,

$$h + k + l = \text{Even}$$

For fcc crystal, we get reflections from those planes for which the Miller indices are all odd or even. The planes that reflect X-rays in the case fcc crystal are (111),(200), (220), (311), (222), (400), (331), (420), (422), (511), (333), (440), (531), (442), (600), (620), (533), (622), (444), (551), (711).

We have already mentioned about the difference in the powder pattern of bcc and fcc crystals. Suppose one takes the x-ray diffractogram of a crystal, we have to know whether it is from a bcc crystal or a fcc crystal. First, we have to calculate the Sin θ and $\text{Sin}^2\theta$ values. Since, the wavelength of x-rays and the lattice parameters are fixed for the crystal, the ratio of the $\text{Sin}^2\theta$ values for the first two lines is

$$\frac{\sin^2\theta_1}{\sin^2\theta_2} = \frac{h_1^2 + k_1^2 + l_1^2}{h_2^2 + k_2^2 + l_2^2} \qquad \text{.......(3.7)}$$

Where θ_1 and θ_2 are two different angles associated with the principal diffracting planes $\{h_1, k_1, l_1\}$ and $\{h_2, k_2, l_2\}$, respectively.

The first two sets of planes for bcc are (110) and (200).

$$\frac{\sin^2\theta_1}{\sin^2\theta_2} = \frac{1^2 + 1^2 + 0}{2^2 + 0^2 + 0^2} = 0.5$$

Thus, if the crystal structure of the unknown cubic metal is bcc, the ratio of the $\text{Sin}^2\theta$ values which correspond to the first two principal diffracting planes will be 0.5.

For fcc crystal structure, the first two sets of principal planes are {111} and {200}. Substitution of (h k *l*) values for the planes in the relation for $\text{Sin}^2\theta$, we get

$$\frac{\sin^2\theta_1}{\sin^2\theta_2} = \frac{1^2 + 1^2 + 1^2}{2^2 + 0^2 + 0^2} = 0.75$$

Thus, for a crystal of unknown cubic structure, if the ratio of the first two $\text{Sin}^2\theta$ values comes out as 0.75, it is fcc. The remaining (h k *l*) values can be assigned from the set given.

The following data is given for NaC*l* crystal.

θ	2θ	D Value	sin²θ × 1000	h	k	l	Intensity
13.69	27.38	3.2574	56.01	1	1	1	86.8
15.86	31.75	2.8210	74.68	2	0	0	1000.0
22.73	45.47	1.9947	149.36	2	2	0	663.7
26.95	53.89	1.7011	205.36	3	1	1	23.0
28.25	56.50	1.6287	224.03	2	2	2	215.5
33.13	66.26	1.4105	298.71	4	0	0	95.1
36.55	73.11	1.2944	354.72	3	3	1	11.6
37.67	75.33	1.2616	373.39	4	2	0	255.0
42.02	84.04	1.1517	448.07	4	2	2	190.1
45.23	90.47	1.0858	504.07	5	1	1	9.9
45.23	90.47	1.0858	504.07	3	3	3	3.3

In the observations, $\frac{\text{Sin}^2\theta_1}{\text{Sin}^2\theta_2} = 0.75.$

Therefore, the pattern obtained belongs fcc.

The indices can be ascribed from the table given. The Miller indices should be either all odd or all even.

Calculating for 422 plane, we get the lattice parameter 'a' as

$$a = .077\,\text{nm}\sqrt{\frac{24}{0.4481}} = 0.5635 \text{ nm}$$

3.6 X-ray Diffractometer

In the powder method, we have said that the powder pattern can be recorded on a photographic film. But, these days, the automation has come and the diffracted beam can be recorded by means of a photodetector. The schematic diffractometer is given in Fig. 3.7. The method is same as Debye-scherrer camera method except a movable detector replaces the film. It moves on the circumference of a circle with the sample as its centre.

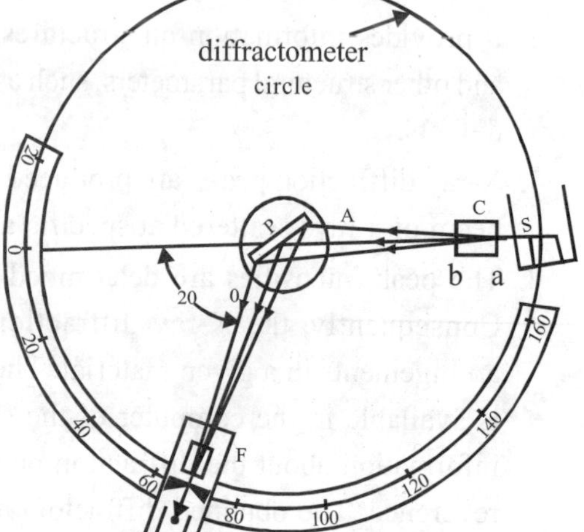

Fig. 3.7 A Schematic diagram of sample and photodetector in x-ray diffractometer.

X-ray from S through a collimeter C fall on the sample at the centre P. The diffractional beam are focussed into the window (F) of the detector (D). The entire set up is placed on a vibration free unit. A carriage (E) supporting the detector and the window rotates about an axis O. The angular motion is accurate to the level of O, 0.1^0 to 0.01^0. Now a days, we have automated systems. The data on the intensities, angular positions of the diffracted beams are collected by a computer through software. One such data collected is given in the Fig. 3.8 for NaC*l*.

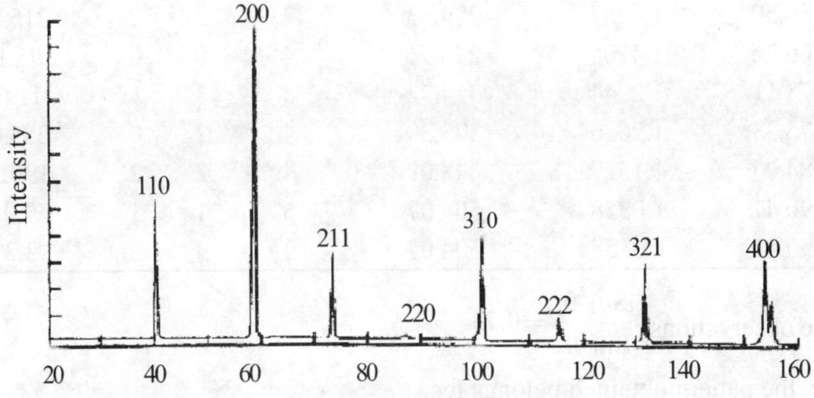

Fig. 3.8 X-ray Diffractogram of a NaC*l* crystal

3.7 Applications of X-ray Diffraction (XRD)

1. X-ray Diffraction (XRD) is a powerful nondestructive technique for characterizing crystalline materials.
2. It provides information on structures, phases, preferred crystal orientations (texture), and other structural parameters, such as average grain size, crystallinity, strain, and crystal defects.
3. X-ray diffraction peaks are produced by constructive interference of a monochromatic beam of x-rays scattered at specific angles from each set of lattice planes in a sample.
4. The peak intensities are determined by the atomic density within the lattice planes. Consequently, the X-ray diffraction pattern is the fingerprint of periodic atomic arrangements in a given material. The advantage with modern XRD is that a data base is available in the computer of the system. The data base consists of the data or the information about the diffraction patterns of thousands of studied materials for ready reverence. The obtained diffractogram by the worker can be compared with standard crystal and it is possible to correctly calculate the lattice parameters, and the impurity peaks. An online search of a standard database for X-ray powder diffraction patterns enables quick phase identification for a large variety of crystalline samples in the modern XRD equipment.

Ideal Uses for XRD Analysis

- Phase identification for a large variety of bulk and thin-film samples.
- Determination of lattice parameters (10^{-4}), and interplanar distances.
- Determination of grain size in polycrystalline blocks.
- Detecting crystalline minority phase compositions (at concentrations greater than ~1%). The example is like magnesium of 1% in aluminum .
- Determination of lattice parameter changes with temperature (thermal expansion) and pressure.
- Determining crystallite size for polycrystalline films and materials.
- Determination of thickness of thin films and multilayers.
- Determination of percentage of material in crystalline form versus amorphous in quenched alloy systems..

Relevant Industries for XRD Analysis

- Measuring sub-milligram loose powder or dried solution samples for phase identification. This includes the identification of crystal lattice type by comparing with readily fed data of structures.
- Analyzing films as thin as 50 angstroms for texture and phase behaviors.
- Measuring residual stress in bulk metals and ceramics.

XRD Analysis: plus points and limitations

- Nondestructive analysis.
- Quantitative measurement of phase contents and texture orientation.
- Minimal or no sample preparation requirements.
- Ambient conditions for all analysis.

Limitations of XRD Analysis

- Cannot identify amorphous materials
- No depth profile information
- Minimum spot size of ~50um

Questions

1. Derive Braggs Law.
2. What is the order of reflection?
3. Calculate the inter planar spacing of atoms for (100), and (110) planes of bcc crystal?
4. Describe the Laue method.
5. Why should we use white radiation of X-rays in Laue method?
6. Where is Laue method useful?
7. Why should you use monochromatic x-ray radiation in the case of powder method?
8. For bcc iron, compute (i) the inter planar spacing (ii) the diffracting angle for the (220) set of planes. The lattice parameter of iron is 0.2889 nm. Assume that a monochromatic radiation having a wavelength of 0.1541 nm is used and the order of reflection is one.
9. Calculate the inter planar spacing for the (110) and (221) sets of planes in aluminium?
10. Rubidium is a bcc crystal. The angle of diffraction for (321) set of planes is at 27.00^0 for the first order reflection. The wavelength of x-rays used is 0.0711 nm. Calculate the interplanar distance of the planes. What is the atomic radius of Rubidium atom?
11. How do you distinguish between bcc and fcc structures from powder pattern?
12. What are X-rays ? What should be the order of wavelengths that should be used for X-ray diffraction of crystals ?
13. Obtain the expression for the interplaner distance d_{hkl} for a cubic lattice
14. Nickel shows a diffraction maximum at $2\theta = 101.502^0$. The wavelength of X-rays is equal to 1.541^0A. Calculate the lattice parameter of nickel cell if the above reflection corresponds to (310) plane (Assume n = 1)
15. A crystal shows the diffraction maxima at the angles $2\theta = 38.184^0$, 44.392^0, 64.576^0 and 72.547^0. If the wavelength of x-rays used is 1.541A^0. Find out whether the crystal used is b.c.c. or f.c.c. What is the lattice constant of the crystal.
16. Describe Laue method of X-ray diffraction. What does the each intense point in X-ray diffraction represent in Laue pattern.
17. Describe the powder method of X-ray diffraction.
18. Discuss how powder photograph is taken.
19. Discuss how powder pattern of a cubic crystal can be analysed ?
20. When a monochromatic X-ray beam of X-rays of wavelength 0.1542 nm is used, the first order reflection from (113) set of planes occurs at θ. What is the value of θ.
21. What is XRD ? What are the applications of XRD ?

Objective Questions

I. Choose the correct choice among the four alternatives.

1. Choose the correct statement
   (a) Mono chromatic X-rays are used in Laue method
   (b) Laue method requires long wave length X-rays only
   (c) 'White' X-radiation is used in laue method
   (d) The sample of crystal should be polycrystalline for Laue method.

2. Read the statements and answer
   (A) Bragg reflections corresponding to (1 1 1), $(1\ 1\ \bar{1})$, $(1\ \bar{1}\ 1)$, in the cubic system have the same Bragg angle for diffraction and same d_{hkl} spacings.
   (B) In the powder diffraction, Bragg reflections from the planes with N = 7,15 are forbidden.
   (a) Both (A) and (B) are correct
   (b) A is wrong, B is wrong
   (c) A is wrong, B is correct
   (d) A is correct, B is wrong

3. The N value for a Bragg reflection is 16, Then, the planes of reflection should be
   (a) (2 2 2) (b) (4 2 0)
   (c) (4 0 0) (d) (2 3 2)

4. In a powder pattern for a cubic crystal, the first two Bragg lines should have indices
   (a) (1 1 1), (2 0 0) (b) (1 1 0), (2 0 0)
   (c) (1 1 1), (1 1 0) (d) (1 1 0), (2 0 0)

5. The ratio of d_{110} and d_{111} is in a cubic crystal is
   (a) $\sqrt{2/3}$ (b) $\sqrt{3/2}$
   (c) $3/2$ (d) $\sqrt{2/3}$

6. The powder pattern of a cubic crystal shows that the $N = \sqrt{h^2 + k^2 + l^2}$ values are 2, 4, 6, 8, 10, 12. Then, the crystal is
   (a) bcc (b) fcc
   (c) hcp (d) simple cube

7. For a body centered cubic lattice, the powder pattern is such that
   (a) (h, k, l) are all odd or even
   (b) $(h^2 + k^2 + l^2)$ should be odd only
   (c) {h, k, l} should be all even only
   (d) (h + k + l) should be even

8. Choose the correct answer about the two statements given
   (A) Bragg reflections from (211) and (110) planes exist for fcc crystal
   (B) The $\sin^2\theta$ values of the first, and the second Bragg reflections have a ratio equal to 0.5 for a bcc crystal.
   (a) (A) is wrong, (B) is correct,
   (b) (A) is correct , B is wrong
   (c) A and B are wrong
   (d) A and B are correct

9. The possible values for $N = (h^2 + k^2 + l^2)$ in the case of a face centered cubic crystal are
   (a) 1, 2, 3, 4, 5..... (b) 3, 4, 5, 6, 8.....
   (c) 3, 4, 8, 11, 12, 16 (d) 2, 4, 6, 8, 10

10. The possible values for $N = (h^2 + k^2 + l^2)$ in the case of a body centered cubic lattice are
    (a) 2, 4, 6, 8, 10 (b) 3, 4, 8, 11, 12
    (c) 1, 2, 3, 4, 5..... (d) 3, 5, 7, 9

11. In the case of a cubic crystal having a lattice parameter a, the inter planar distance d_{hkl} is given as

(a) $a\sqrt{h^2+k^2+l^2}$ (b) $\frac{a}{\sqrt{h^2+k^2+l^2}}$

(c) $\frac{a}{\sqrt{h^2+k^2+l^2}}$ (d) $\frac{a}{h+k+l}$

12. An analysis of X-ray dittractogram of a crystal gives a Bragg reflection at an angle of 13^0. If the wave length of X-rays used is 0.58 Å, The value of inter planar distance is

(a) 2.65×10^{-10}m
(b) 5.0×10^{-10}m
(c) 1.32×10^{10}m
(d) 1.87×10^{-10}m

13. A crystal gives the first two diffraction lines at $\sin^2\theta$ values of 0.1380 and 0.1834. Then the crystal is

(a) bcc (b) hcp
(c) simple cube (d) fcc crystal

14. The first two lines of powder pattern of a fcc crystal are such that the ratio of $\frac{\text{Sin}^2\theta_1}{\text{Sin}^2\theta_2}$ is

(a) 0.67 (b) 0.41
(c) 0.75 (d) 0.25

15. The Laue spots close to the centre of Laue photograph correspond to

(a) low index planes
(b) the low Bragg angles
(c) the high Bragg angles
(d) high d_{hkl} values

16. The camera used to take powder diffraction pattern has a

(a) cylindrical shape (b) spherical shape
(c) flat shape (d) conic shape

17. The diameter of the ring in the diffraction pattern is S, R is the radius of the camera, then angle at which the diffraction ring is obtained on the film is

(a) $\frac{4R}{S}$ (b) $\frac{2R}{S}$

(c) $\frac{S}{R}$ (d) $\frac{S}{4R}$

18. The powder diffraction pattern of a bcc crystal does not contain reflections from

(a) (3 1 1), (1 0 0), (1 1 1)
(b) (2 0 0), (2 1 1)
(c) (3 7 0), (1 1 0)
(d) (2 2 2), (2 2 0)

19. A crystal gives powder diffraction pattern such that the first three lines have $\sin^2\theta$ values are 0.1176, 0.1568, and 0.3136 respectively, then the first principal line and the crystal structure are

(a) (1 1 1), fcc (b) (1 0 0), bcc
(c) (1 1 0), fcc (d) (2 0 0), bcc

20. For recording powder pattern, we should use X-rays that are

(a) 'white' radiation
(b) monochromatic
(c) of very low wavelength
(d) of very-high wavelength

21. In the powder pattern of X-ray diffraction of a cubic crystal,

(a) the first order reflections of (2 0 0) planes and second order reflection of (1 0 0) planes coincide
(b) first order reflection lines of (3 0 0) planes and first order of (1 1 0) planes coincide
(c) The second order reflections do not occur
(d) The third order reflections do not occur

22. If α, β, γ represent the direction cosines, h, k, *l* are the Miller indices, θ is the diffraction angle, d, the inter planar distance, Then

(a) $2d \cos \theta = \frac{\lambda}{a} h$

(b) $2dh = \frac{\lambda}{a} \alpha$

(c) $\frac{2d}{\lambda} \theta = \frac{2\beta}{\lambda} \theta = \frac{2\gamma}{\lambda} \theta \cong d$

(d) $\frac{a}{h} \alpha = \frac{a}{k} \beta = \frac{a}{L} \gamma$

23. For a certain fcc crystal, the cell edge a = 0.563 nm. The smallest angle at which Bragg reflection can occur corresponds to a set of planes whose indices are

(a) 100 (b) 110
(c) 111 (d) 200

24. Silicon has the diamond unit cell edge 'a' equal to 0.542 nm. The nearest interatomic separation in the unit cell is

(a) 0.112 nm (b) 0.234 nm
(c) 0.383 (d) 0.542

(Hint : The nearest atomic distance is along the body diagonal $\gamma = \alpha\sqrt{3}/4$)

25. If d_{100} and d_{111} are in the ratio $\sqrt{3}:1$, then, the corresponding values of sin of Bragg angles will be in the ratio

(a) $\sqrt{3}:1$ (b) $1:\sqrt{3}$
(c) 1 : 3 (d) 3 : 1

II. True or False

1. The defects like point defects or line defects are charged defects in ionic crystals [True/False]

2. For orthorhombic crystal, the Bragg condition is $\sin^2 q_{nkl} \frac{\lambda^2}{4a^2(h^2+k^2)} + \frac{\lambda^2}{4c^2} l2$ [True/False]

3. In a Laue method, mono chromatic X-rays are used for X-ray diffraction [True/False]

4. Laue method deals with a single crystal diffraction. [True/False]

5. In a powder method, monochromatic X-rays are used [True/False]

6. In a powder pattern of X-ray diffraction for fcc crystal, $\sin^2 \theta$ values are proportional to $(h^2 + k^2 + l^2)$ [True/False]

7. An oxide of iron or MgO gives X-ray diffraction powder pattern that is representative of fcc crystal [True/False]

8. For a bcc crystal, the X-ray diffraction pattern from the first two principal diffraction planes has $\sin^2\theta$ values in the ratio 0.75 [True/False]

9. For any fcc crystal, the principle diffraction line in the powder pattern is from (1 1 1) planes. [True/False]

10. A powder pattern of fcc crystal possesses diffraction lines only from planes (1 1 1), (3 3 1), (5 3 1) etc. [True/False]

11. A nickel crystal gives X-ray diffraction pattern that is characteristic of bcc [True/False]

12. XRD of amorphus substance like glass possesses sharp and light intense peaks. [True/False]

CHAPTER 4

Defects in Crystals

4.1 Introduction

We have treated crystals in the earlier chapters as perfect periodic structures. A 'perfect crystal' is an ideal concept, but real crystals that are grown and used contain time-averaged deviations from perfect structure. The defects in crystals are such deviations. The deviations from a perfect three dimensional periodicity of a crystal are known as lattice defects or imperfections. The lattice imperfections are of many types.

A crystal at absolute zero possesses atoms that are frozen in their motion. Above the absolute temperature, the atoms in the crystal vibrate about their mean positions. That means, there is a disturbance or perturbation in the strict periodic structure which can vanish again when the temperature is taken back to the absolute zero. The vibration of atoms is a mechanical vibration. The quanta of mechanical vibrational energy is called 'phonon'. At temperatures above absolute zero, the vibrations will increase i.e., the number of phonons will also increase. A phonon is a dynamic defect in crystals. Its existence depends upon the external excitation, if the excitation has decreased, they disappear. There are many such transient defects such as electronic excitations, exitons, photons, crowdions, magnons etc. They are created by external excitation. These imperfections are called the imperfections of first kind. In the present chapter, we are not interested in such transient defects. We are interested in the imperfections of second kind which are geometrical defects and arise as a result of thermodynamic reasons and mechanical reasons. These imperfections are extremely important as they have a definite role to play in most of the bulk properties of the crystals in some temperature range or other. They are called structural imperfections. They are classified as 1. Point defects (atomic imperfections), vacancies, interstitial, and impurity atoms. They are called zero dimensional defects. 2. Linear defects such as dislocations, 3. Two-dimensional or planar defects such as stacking faults and internal grain boundaries. 4. Volume defects or three dimensional defects, such as voids or pores, impurity precipitates, foreign inclusions, cracks etc.

4.2 Point Defects

The point defects are classified as 1. Vacancies 2. Interstitials 3. Disorder due to impurity atoms.

4.2.1 Vacancies

A vacancy is the simplest defect in a crystal. A vacancy is an atom-site from which the atom is missing. It is known as Schottky defect. Consider a crystal lattice as shown in the Fig. 4.1. A Schottky defect is created by the removal of an atom from an atomic site (like position V_1 or V_2) in the crystal and transferring it to the surface.

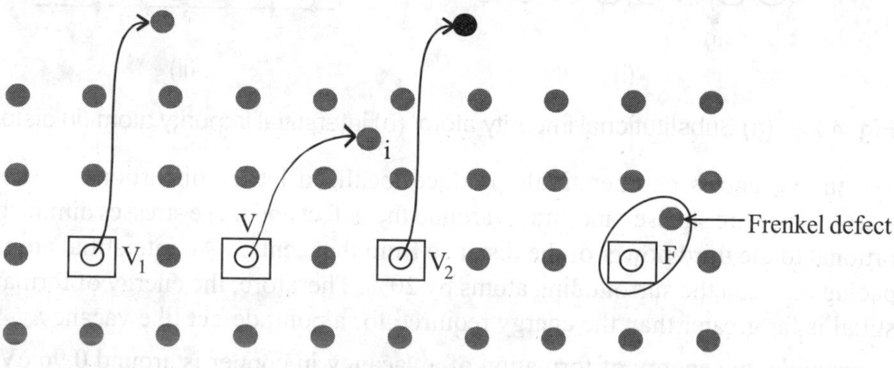

Fig. 4.1 A plane of a metal crystal. V_1 and V_2 are the vacancies created. F is the Frenkel Defect.

If a crystal is given an energy in the form of heat, this energy is distributed among the atoms of the crystals. However, the energy distribution is not uniform. At any temperature, some atoms which have higher energy than their neighbours will be able to move considerable distances from one equilibrium position to another and ultimately go over to the surface. It appears as if the atoms are able to 'vaporise' from their sites and condense elsewhere. The site left by the atom is a vacancy, called Schottky defect. The number of vacancies per unit volume of the crystal depends upon the temperature.

The atom, that has left its place, can go over to an interstitial site '*i*' (called interstice) or to a place which can be far away from it. When it goes over to the next or the nearest interstice, a vacancy-interstitial pair is formed. This vacancy-interstitial pair is called Frenkel defect shown as F in Fig. 4.1. The question comes; how does the atom that is present inside can go to the surface. At any temperature, the surface of a crystal is a partially evaporated. An atom from the surface evaporates and it leaves the surface in vapour state. Then, an interior atom will move into site on the surface leaving behind a vacancy forming a Schottky defect.

4.2.2 Impurities: Substitutional and Interstitial

A crystal can never be hundred percent pure. There are always some unavoidable impurities present. They are regarded as point defects. The impurities can occupy two different sites in the lattice as illustrated in Fig. 4.2. The positions or sites are: (a) The substitutional sites, in which the atom of the parent lattice site is replaced by an impurity atom. They can also replace the existing vacancies. (b) The interstitial positions are those in which the impurity atom is at non-atomic site.

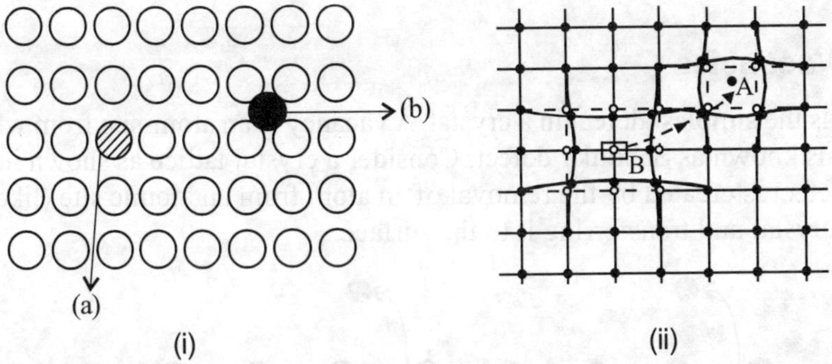

Fig. 4.2 (i) (a) Substitutional impurity atom, (b) interstitial impurity atom (ii) distortion.

Even the vacancies or interstitials produce localized lattice distortions as shown in the Fig. 4.2(ii). There are stresses and strains around the defect and these stresses diminish inversely proportional to the third power of the distance from that centre. An interstitial atom increases the spacing between the surrounding atoms by 20%. Therefore, the energy of formation for an interstitial is far greater than the energy required for a point defect like vacancy.

For example, the energy of formation of a vacancy in copper is around 0.96 eV, where as 2.5 - 3.5 eV energy is required for the formation of an interstitial atom. The parent atom can occupy the interstitial positions, but this is not a very favourable defect.

Impurities can be added intentionally to a crystal to alter its properties or to get required properties for use. The addition of impurities to the crystal and its acceptance to the lattice is based on Hume-Rothery rules. The addition of impurity to the crystal results in substitutional solid solutions like copper-nickel, copper-gold alloys. These alloys are called substitutional solid solutions. Many alloy systems belong to this category. It is necessary as per Hume-Rothery rules, that the impurity atoms and the parent atoms should not differ in size by more than 15%. The crystal structures like the copper and silver alloy, should have the same crystal structure. Preferably the atoms of the alloy elements should be either electropositive or electronegative. The addition of impurity can be done to the extent to which the solubility of the impurity in the crystal is allowed at equilibrium temperature. Interstitial solid solutions are formed between two elements whose atoms differ in size by large amount. The larger atoms go into interstitial positions in the solvent crystal. Some metals can accommodate carbon atoms interstitially to form carbides (iron carbide). The impurities alter the properties of the crystals. The alloys of metals possess enhanced strengths, resistivities, or give different properties as compared to their individual components. For example, one atomic percent of copper dissolved in silver increases the electrical resistivity by 7%. Silicon and germanium doped with acceptor or donor impurities give extrinsic semiconductivity. Similarly mechanical properties of crystals can also be changed by the incorporation of impurities in lattice. Steel is an alloy in which the change of impurities (carbon, manganese, chromium etc.,) or the concentration of the impurities gives various grades of steels.

4.2.3 Frenkel Defects

We can calculate the equilibrium concentration of Frenkel defects under the same assumptions as above. Let the energy necessary for the formation of a Frenkel defect be E_F. A Frenkel defect consists of a vacancy and an interstitial atom, the change in entropy consists of two components.

1. n vacancies are created in the crystal out of N atoms. The expressions of entropy is same as in eqn. (4.2)
2. The n atoms from vacancies can be distributed in W_i different ways over the N_i interstitial sites or positions in the crystal. The corresponding contribution to entropy change is $ln\ W_i$.

∴ the total increase in entropy due to the formation of Frenkel defects will be then

$$S_F = k\left[ln\frac{N!}{(N-n)!n!} + ln\frac{N_i!}{(N_i-n)!n!}\right] \qquad .....(4.1)$$

As in the earlier case, we can apply Stirling approximation and obtain the condition for minimum free energy.

$$\left[ln\frac{N!}{(N-n)!n!} + ln\frac{N_i!}{(N_i-n)!n!}\right]$$

$$\simeq N\,ln\,N + Ni\,ln\,N_i - (N-n)\,ln\,(N-n)$$

$$-(N_i - n)\,ln\,(N_i - n) - 2n\,ln\,n$$

Substituting the above in the relation for free energy, we get

$$F = nE_p - kT\,[N\,ln\,N + N_i\,ln\,N_i - (N-n)\,ln\,(N-n) - (N_i - n)\,ln\,(N_i - n) - 2n\,ln\,n$$

$$\left(\frac{\partial F}{\partial n}\right)_T \simeq EF - kT\,ln\,\frac{(N-n)(N_i-n)}{n^2} \qquad .....(4.2)$$

At equilibrium

$$\left(\frac{\partial F}{\partial n}\right)_T = 0$$

and $N >> n$ and $N_i >> n$

then $$\frac{E_F}{kT} = ln\,\frac{NN_i}{n^2} \qquad .....(4.3)$$

i.e., $$\frac{E_F}{kT} = 2\ ln \frac{\sqrt{NN_i}}{n} \quad .....(4.4)$$

or $$n = \sqrt{NN_i}\ \exp\ (-E_F/2kT) \quad .....(4.5)$$

It is possible that some atoms can go to interstitial sites. An atom in an interstitial site is a defect called 'interstitial defect'. These interstitial atoms can be from lattice sites or from the surfaces. But many metals are close packed structures. The interstitial sites have small volumes. Then, any atom coming into the interstitial produces lot of strains. The order of energy required for the formation of an interstitial is much higher than that required for vacancy formation. For example, copper has a fractional concentration of vacancies at 1000 °C of about 10^{-6} while the fractional concentration of interstitial atoms in copper at 1000 °C is quite low, about 10^{-39}!

The vacancies migrate in metals. The energy required for this migration is about 1 eV for in copper while the interstitial require 0.16 eV for migration i.e., the interstitials would be more mobile than vacancies in metals. But, the vacancy concentration in a particular metal is far higher than that of the interstitials (formed due to dislocated atoms). The diffusion process of atoms (like impurity atoms) in metals is strongly dependent on vacancy concentration.

In metals, both vacancies and Frenkel defects are formed, even though, the energy favours the vacancy formation. The interstitials and atoms can combine and form clusters. Similarly, vacancies can coalesce or combine to form voids. A void is a three dimensional defect. They would form minute spherical structures in crystals like the bubble structure in a glass paperweight.

4.2.4 Equilibrium Concentration of Schottky Defects in Metals

We shall calculate the density of Schottky defects or Vacancies as a function of temperature. We shall have the following assumptions:

1. The crystal consists of N identical atoms. The volume of the crystal is independent of temperature.
2. There exists an energy of formation for a vacancy (E_f). It is independent of temperature.
3. The formation energy is the energy difference between the energy needed to remove an atom from the interior of the crystal and take it to infinity and the energy gained when the atom is placed on the surface of the crystal.
4. The vacancies are treated as indistinguishable particles.
5. The lattice vibrations i.e., the vibrations of atoms in the crystals are not affected by the presence of defects.

Let 'n' be the number of Schottky defects at any temperature T. The necessary equilibrium in the crystal at any temperature T is reached thermodynamically when the free energy F is minimum. The incorporation of a defect in a crystal increases the entropy of the crystal. It also increases the internal energy of the crystal. There are N atomic sites in the crystal. The N sites can be filled by N atoms in N! ways. Out of the N atomic sites, 'n' sites have to be vacancies. The 'n' vacancies can be arranged among themselves in n! ways. The remaining (N-n) sites are occupied by atoms. The atoms can be arranged among themselves in (N-n)! ways. But, they are all indistinguishable. Effectively, the number of ways in which we can create n vacancies in the crystal out of the N atoms will be equal to ${}^{N}C_n$.

Then
$$W = {}^{N}C_n = \frac{N!}{(N-n)!\,n!} \qquad .....(4.6)$$

'W' also represents the probability of a distribution in which n vacancies can be created. The change in entropy as a result of the creation of vacancies is given as

$$S = k \log_e W \qquad .....(4.7)$$

where 'S' is the entropy, k is the Boltzmann constant. The above relation is called the Boltzmann relation between entropy and disorder. The disorder stems from the number of ways in which 'n' vacancies can be created. (ln refers to $\log_e$)

$$\therefore \quad S = k\, ln \frac{N!}{(N-n)!\,n!} \qquad .....(4.8)$$

But, according to Stirling approximation

$$ln\, N! \approx N\, ln\, N - N \qquad .....(4.9)$$

Then eqn 4.3 can be written as

$$S = k[N\, ln\, N - N - (N-n)\, ln\,(N-n) + N - n - n\, ln\, n + n)]$$
$$= k[N\, ln\, N - (N-n)\, ln\,(N-n) - n\, ln\, n) \qquad .....(4.10)$$

In the formation of vacancies, a change in free energy occurs as a result of change in internal energy and a change in entropy. The internal energy is given as

$$U_f = n\, E_f \qquad .....(4.11)$$

where E_f is the energy required for the formation of a vacancy. Then

$$F = n\, E_f - TS \qquad .....(4.12)$$

Then,
$$F = n\, E_f - kT[N\, ln\, N - (N-n)\, ln\,(N-n) - n\, ln\, n)]$$

At equilibrium, the free energy has to be minimum. This is the law of the nature. Therefore, the first derivative with respect n has to be zero.

Then
$$\delta F = E_f\,\delta n - kT[-(N-n)\frac{1}{(N-n)} \times (-\delta n)$$
$$- ln\,(N-n)\,(-\delta n) - \frac{n}{n}\,\delta n - ln\, n\,(\delta n)]$$
$$= \delta n[E_f - kT\, ln\,(N-n) - ln\, n] \qquad .....(4.13)$$

At equilibrium, $\frac{\delta f}{\delta n} = 0$, then

$$\frac{E_f}{kT} = ln\,\frac{(N-n)}{n} \qquad .....(4.14)$$

$N >> n$, N will be in the order of 10^{23} and n will be in the order of 10^7 per unit volume. Then, we can have

$$n \simeq N \exp(-E_f / kT) \qquad .....(4.15)$$

'n' is equal to the number of Schottky defects per unit volume. The concentration of vacancies in metal depends on temperature. As the temperature of the crystal is raised, the concentration increases rapidly. The formation energy of a vacancy depends on the nature of the metal. It is the characteristic of a metal. For example, in the case of copper, the factor $n/N \approx 10^{-19}$ at 300 K. But at 700 K, and 1350 K, it is 10^{-8} and 10^{-4} respectively.

The formation of vacancies at a temperature is such that there is a thermodynamic equilibrium at that temperature. One can increase the concentration of vacancies in a metal at room temperature by heating the crystal of the metal to a high temperature and then suddenly cooling it (this process of sudden cooling is called quenching) to room temperature. Then the crystal would possess the vacancy density that would be there at high temperature. Because of sudden cooling, the vacancies are frozen in the crystal.

4.3 Study of Existence of Points Defects in crystals

The experimental proof for the existence of defects comes from the study of properties with variation in temperature of the crystals. The formation of Schottky defects at constant temperature results in an increase of the crystal volume. As a result of the increase in volume, the density of the crystal is altered. The difference between the measured density and the X-ray density is a measure of the Schottky imperfection's concentration. The X-ray density is given by the atomic mass per unit volume, the number of the molecules in the cell and the volume of the cell. Frenkel defects do not change the density of crystals. The expansion of the solid with increasing temperature is partly due to vacancy creation.

The variation of thermal expansion with temperature or specific heat with temperature yields information about the point defects.

Atomic imperfections are responsible for the diffusion of the impurity atoms or parent atoms in the crystals. Point defects possess mobility and they migrate from one point to another point resulting in the transport of atoms or ions in the crystal. Therefore, the study of diffusion of impurities gives information about point defects in crystals.

4.4 Line Defects

4.4.1 Theoretical Shear Stress

All materials possess mechanical properties. The crystals undergo elastic deformation under low stresses and the plastic deformation under high stresses. They possess mechanical properties like stress-strain relationship, plasticity, yield-stress, flow-stress, fracture, fatigue etc. All these mechanical properties are closely linked with a very important type of imperfections called 'dislocations'. Dislocations are line- defects in crystals and are one dimensional defects. We shall first see how the concept of dislocation was introduced in crystal physics. This will enable us to understand why dislocations are necessary in a crystal.

Imagine a perfect crystal that has atoms arranged in regular order schematically shown in Fig. 4.3. Let 'a' be the distance of separation between atoms. Suppose, a shear stress is applied along x - direction. Let the stress applied be τ.

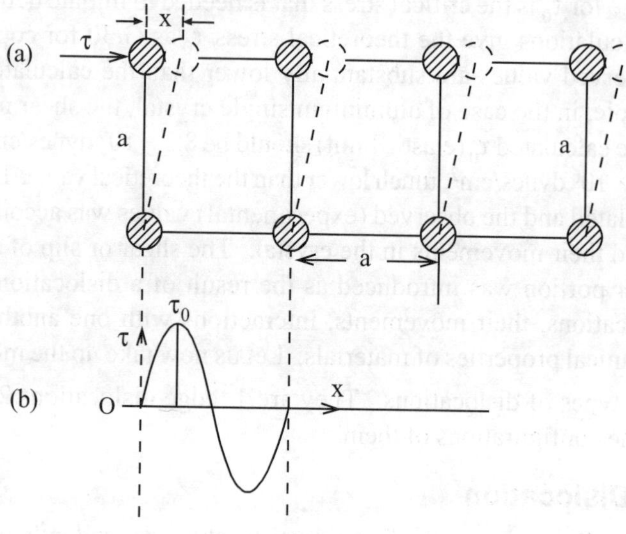

Fig. 4.3 (a) Displacement of atoms when a shear stress t is appled (b) the stress variation in the crystal.

The atoms in the upper row are sheared with respect to the lower adjacent plane. Let this displacement be 'x'. A crystal possesses periodicity. Due to the symmetry of the lattice, we have

$$\tau - 0 \quad \text{at} \quad x = Ka/2, \qquad .....(4.16)$$

where $K = 0,1,2,3, ...$

The Fig. 4.5(b) shows the variation of stress in the crystal due to the periodicity of the crystal.

At points where $\tau > 0$, the lattice offers resistance to the applied stress i.e., at $0 < x < a/2$.

$$\tau < 0 \quad \text{at} \quad a/2 < x < a. \qquad .....(4.17)$$

Then, the stress in the crystal can be expressed as

$$\tau = \tau_0 \sin\left(\frac{2\pi x}{a}\right)$$

Suppose the displacement 'x' is very small which is the case in crystals.

Then, $$\tau \cong \tau_0 \frac{2\pi x}{a} \qquad .....(4.18)$$

For small displacements, Hooke's law is valid.

$$\therefore \quad \text{Stress } \tau = G \times \left(\frac{x}{a}\right) \qquad .....(4.19)$$

Where 'G' is the shear modulus. In such a case, we get

$$\tau_0 = \frac{G}{2\pi} = 0.16\ G \quad .....(4.20)$$

The above value for τ_0 is the critical stress that is needed to initiate deformation of a crystal. More rigorous calculations give the theoretical stress τ_0 as G/30 for copper, silver, gold etc. However, the observed values are substantially lower than the calculated values by several orders. For example, in the case of aluminium single crystal, the shear modulus is 2.5×10^{11} dyne/cm^2, then, the calculated τ_0 (elastic limit) should be 8.3×10^9 dynes/cm^2. But, the observed value was about 4×10^6 dynes/cm^2, much lower than the theoretical value. The striking difference between the calculated and the observed (experimental) values was accounted by the presence of dislocations and their movements in the crystal. The shear or slip of a portion of a crystal relative to another portion was introduced as the result of a dislocation in the crystal. The presence of dislocations, their movements, interactions with one another and other defects explain the mechanical properties of materials. Let us now take up the models of dislocations.

There are two types of dislocations. They are 1. edge dislocations 2. screw dislocations. We will discuss the configurations of them.

4.4.2 Edge Dislocation

Taylor, Orowan and Polanyi were the first scientists who independently in 1934 suggested that the slip in a crystal propagates by the motion of dislocations. We shall now describe edge dislocation from the geometrical point of view.

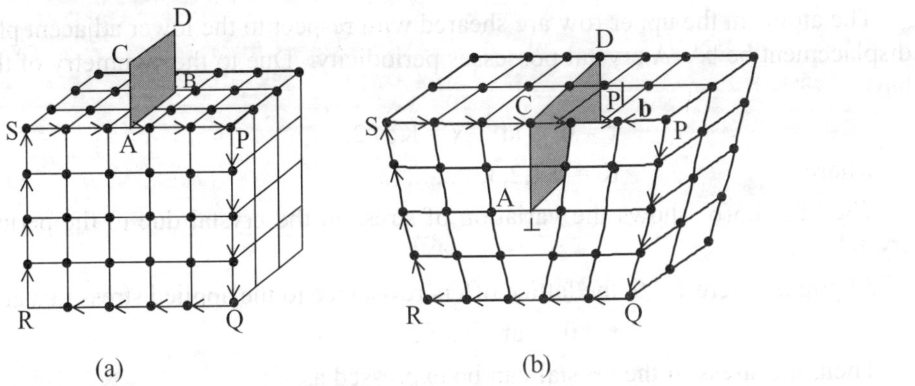

Fig. 4.4 (a) Perfect crystal (b) A linear defect occurs in region just above '⊥'.

Let us take a simple cubic crystal. The atoms in the crystal can be imagined to be held by flexible springs. The Fig. 4.4(a) describes the perfect crystal. Make a cut along CA. Push an extra plane inside as shown in Fig. 4.4(b).

Then one would notice that an extra plane is placed in the upper half of the crystal. Let this plane be ABCD. As a result the atoms in the neighbourhood of the plane ABCD get displaced from their positions as shown in the Fig.4.4(b). The edge of the extra plane AB is called 'edge dislocation' indicated by ⊥. Looking into the paper, the region around the edge of the plane is distorted. The line at the edge of the extra plane is an edge dislocation. It is contained in the plane which is a boundary between slipped and unslipped (lower half) portions of the crystal. As the extra plane is in the upper half of the crystal, the dislocation is called the positive edge dislocation. The point at which the dislocation would lie perpendicular to the plane of the figure is better shown in Fig. 4.5(a) where only the front surface of the crystal is shown. A negative edge dislocation is formed if the extra plane is in the lower half. (shown in Fig. 4.5(b)). When a positive and negative dislocation are on the same plane, they annihilate each other when they meet as a result of stress. If they are on different planes, they may create vacancies or interstitials when they meet. A vacancy creation is shown in Fig. 4.6.

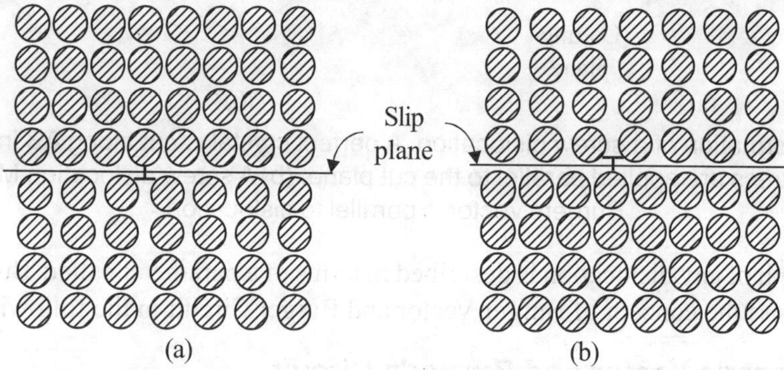

Fig. 4.5 (a) Postive edge dislocation (^) (b) Negative edge dislocation.

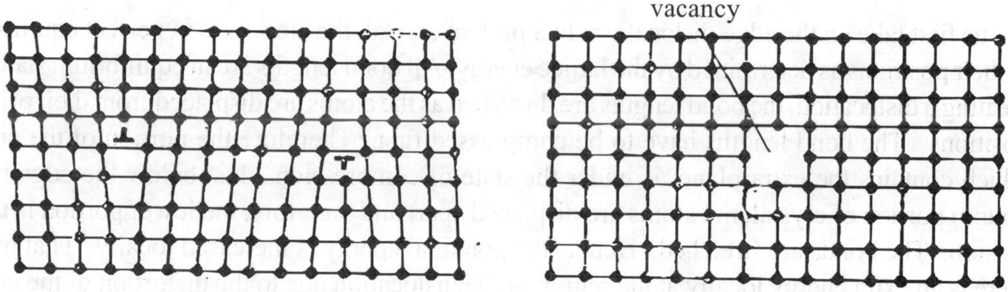

Fig. 4.6 When negative and positive dislocations on different slip planes move under stress and meet, (a), vacancy is created (b)

4.4.3 Screw Dislocation

Consider the perfect crystal once again. Make a cut at AB with the help of a cutting plane. Now, apply a shear stress parallel to the cutting plane such that the right (front) portion of the crystal moves down with respect to the left portion of the crystal. The resulting arrangement is shown in Fig. 4.7(b). A screw dislocation is formed along MN at the point O. The shear stress applied causes a distortion of the crystal lattice in the form of a spiral ramp of distorted atoms around the screw dislocation. The portion or region of distorted crystal is at least a few atoms in diameter.

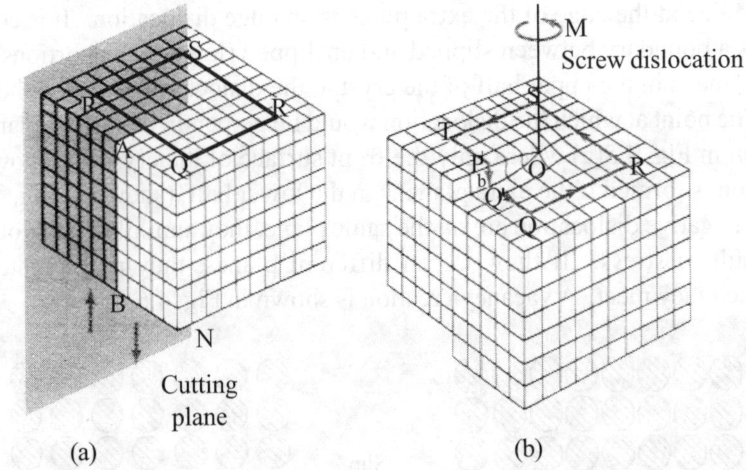

Fig 4.7 (a) Formation of a screw dislocation. A perfect crystal is cut along AB into the crystal and sheer stress are applied parallel to the cut plane. (b) A screw dislocation MN is formed, Burgers vector is parallel to dislocation.

The strength of any dislocation is described in terms of a parameter called 'Burger's Vector'. We shall discuss the concept of Burger Vector and Burger Circuit in the following section.

4.4.4 Burger's Vector and Burger's Circuit

Burger's Vector for an Edge Dislocation

Let us first take up the edge dislocation. In a perfect crystal, the atoms are in perfect equilibrium in their positions as determined by the lattice energy. All bond lengths are in equilibrium state. By creating a dislocation, the bond lengths are disturbed as the atoms are displaced from their original positions. The bond lengths have to be compressed first. Therefore the portion of the crystal which contains the extra plane is under the state of compression. Just below the edge in the bottom portion of crystal, the atoms are displaced apart and therefore, the lower portion is under tension. The bonds are stretched. Hence, the potential energy is increased locally. That means there is an extra energy locally at the region of the dislocation due to the distortion of the lattice. The region of distortion spreads a few lattice distances around the dislocation. The maximum distortion is in the immediate region surrounding the edge of the extra plane.

The distortion represents actually the edge dislocation. The plane in which the dislocation is present is called slip plane. Over this slip plane, we have the slipped portion of the crystal. Below the plane, it is unslipped portion. Therefore, the dislocation forms a boundary between slipped and unslipped portions. This is true for a screw dislocation also.

We consider the perfect crystal as given in Fig. 4.6(a). Start from the point P. We go down by n steps (n = 4) then take m steps to the left (m = 5), then n steps up and finally m steps to the right. We reach the point where we started. This movement around a centre is called Burger circuit. We reach the starting point in a perfect crystal. The Burger's circuit is closed perfectly in a perfect crystal.

Now consider the Burger circuit in a crystal (shown in Fig. 4.6(b)) that contains edge dislocation, once again starting from P. When we reach P', we need an extra step b to reach the original starting point P. This extra step PP' = b is called Burger vector. The Burger vector of an edge dislocation is perpendicular to the dislocation. Actually, the Burger's vector is directed along PP'.

Burger's vector for screw dislocation

In a Burger's circuit, the right hand screw convention is followed (see Fig. 4.9(b)). As one rotates a screw clockwise it goes into the crystal when placed parallel to the dislocation. Consider the Burger's circuit for screw dislocation. Start from Q'. Move a few steps to Q. The turn left and take a few steps to R. Turn left go to S. Then take steps exactly (PS = QR) equal to QR. From P, proceed towards Q. When you reach P'. The loop is not complete. You have to move a distance P' Q' to complete the circuit. Then P' Q' = b. The Burger's circuit is Q' Q R S P P'Q'. In a crystal without dislocation, it is Q R S P Q. The Burgers vector is needed to close Burgers circuit. Thus in the case of screw dislocation, the Burger's vector is parallel to the dislocation. The direction of Burger's vector is indicated by Q'P = b.

Energy of a dislocation

It has already been stated that the region around a dislocation has an extra energy. As the atoms are displaced, the region is strained. Therefore, the dislocation causes a strain around it and hence possesses a stress field around it. A dislocation thus has an elastic energy associated with it. The energy of the dislocation i.e. the elastic strain energy E_L, is proportional to the square of the Burger vector in a crystal and the length of the dislocation. Thus, Burger vector represents the strength of the dislocation. For an edge dislocation, the elastic strain energy of the dislocation is given as

$$E_{edge} = \frac{G\ b^2}{2\pi(1-\nu)}\ l\ ln\frac{R}{r_0} \quad .....(4.21)$$

where, ν is called Poisson ratio of the metal or crystal (which is about 0.2 to 0.4 in metals), l is the length of the dislocation. R represents the radius of the region around the dislocation to which the distortion is taken into consideration, r_o is the radius of the dislocation core.

The energy of a screw dislocation of length 'L' is given as

$$E_s = \frac{Gb^2}{4\pi} L\ ln\frac{R}{r_0} \quad .....(4.22)$$

Normally 'r_o' is comparable with the magnitude of the Burger vector or to the lattice constant. 'R' can not exceed the dimensions of the crystal.

Thus, the energy of the dislocation per unit length is proportional to the shear modulus and the square of the Burgers vector. The energies of edge and screw dislocations are of the same magnitude, but, for the same power, the energy of an edge dislocation is slightly higher than that of a screw dislocation.

A general dislocation is a mixed dislocation. It is a combination of an edge dislocation and screw dislocation. We cannot have straight dislocations like pure edge and pure screw dislocations.

The curved dislocation is a mixed dislocation as shown in Fig. 4.7 (c).

Schematically, we can show how a mixed dislocation is created. Take a crystal. Make a cut along ON push the crystal inside on the upper portion of the crystal. Then we create an extra plane above B. Therefore, at B, we can have an edge dislocation. But at A, the dislocation created is a screw dislocation. In between A and B the dislocation is a mixed dislocation. It means that the curved dislocation is created starting at A and ending at B, At any point on the curve in the middle at a point H, the dislocation has a screw component and an edge component. The curved dislocation is called the Mixed dislocation . It is in the horizontal plane whose edge is N N'

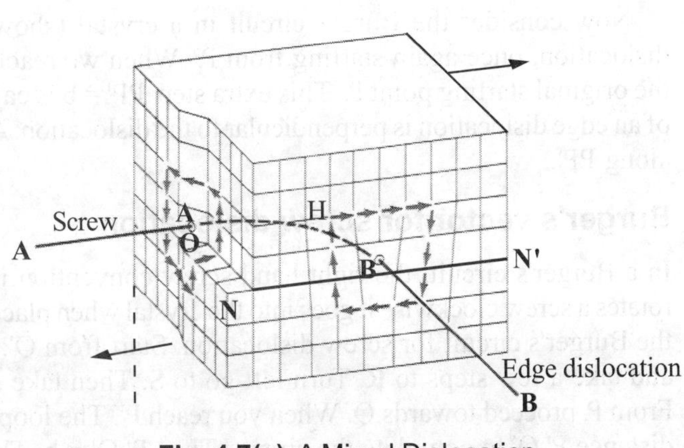

Fig. 4.7(c) A Mixed Dislocation.

At any point H, the Burger vector will have an edge component and a screw component.

In general, Burgers vector can be resolved into basal vectors of the lattice i.e., **b** can be given as $\mathbf{b} = l\,\mathbf{i} + m\,\mathbf{j} + n\mathbf{k}$. There can be an arbitrary angle between a dislocation line and the Burger vector. The basal vectors are the crystallographic directions of the unit cell. If l, m, n = 0, 1, 2, 3.........., then the dislocation is a perfect dislocation. If l, m, n are not equal to the integers, the dislocations are partial dislocations. The magnitude of Burgers vector is

$$b = |\mathbf{b}| = \sqrt{b_x^2 + b_y^2 + b_z^2}$$

where $\mathbf{b}_x = l\mathbf{i}$, $\mathbf{b}_y = m\mathbf{j}$, $\mathbf{b}_z = n\mathbf{k}$.

If the magnitude of Burgers vector is equal to the interatomic spacing, then the dislocation is called unit dislocation.

For a dislocation having a length of one atomic plane, the energy comes out to be a high value. At $G = 3.2 \times 10^{10}\ N/m^2$ and $b = 3.2 \times 10^{-10}$ m and the length of the dislocation equal to **b**, the energy comes out to be 3 eV. The length of the dislocation is normally much greater than b, then, the energy will be much greater. Dislocations can possess energies greater than 10 eV. It is much higher than the thermal energy (at 300 K, $E_{ther} = 0.025$ eV). A large number of dislocations exist in a crystal and in fact, a crystal would have a forest of dislocations. Therefore, a lot of energy is stored in a crystal in the form of the elastic energy of dislocations.

When a stress τ is applied to a crystals, the dislocation is acted upon by a force τb. It moves under the action of the stress resulting in the deformation of the crystal. When it moves to the surface of the crystal, it leaves a step on the surface of the crystal.

4.5 Two dimensional Defects or Surface Defects

The two dimensional defects are the planar defects that separate two regions of the crystal or phases in the crystal. These defects are: the external surfaces, grain boundaries, twin boundaries, stacking faults and phase boundaries. The several different types of planar imperfections, in solids can be grouped into the following categories:

1. Interfaces between solids and gases, which are called free surfaces;
2. Interfaces between regions where there is a change in the electronic structure, but no change in the periodicity of atom arrangement, known as domain boundaries;
3. Interfaces between two crystals or grains of the same phase where there is an orientation difference in the atom arrangement across the interface; these interfaces are called grain boundaries;
4. Interfaces between different phases, called phase boundaries, where there is generally a change of chemical composition and atom arrangement across the interface.

Grain boundaries are peculiar to crystalline solids, while free surfaces, domain boundaries and phase boundaries are found in both crystalline and amorphous solids.

4.5.1 Surfaces

Any crystal that we use is terminated on free surfaces. The surface of a crystal itself is a defect. The atoms on the surface are not well bound. The existence of a surface is based on the surface energy. The bonds of the atoms on the surface are not complete. A surface can have absorbed gas molecules or water molecules. The surfaces contain steps which would be formed during the growth of the crystal. They may contain layers that are not fully grown over the entire surface. Such steps or the edges of the layers attract the molecules or atoms from the environment.

4.5.2 Grain Boundaries

The most important defect in crystalline samples is the grain boundary. When the melts of materials cool, two types of solids can be formed. If the melt is allowed to cool very slowly, the atoms are arranged in a regular fashion and we get a single crystal. If the melt is allowed to cool fast, we get a poly crystalline sample. Any metal wire, or rod or fused materials like quartz rods are in the poly crystalline form.

The solidification of a material is done by cooling its melt. During solidification (at the freezing point), small crystallites (infinitesimally small) are formed first at some places in the melt. They are called 'nuclei'. As the solidification proceeds, these nuclei grow in such as small blocks of the solidified material join them. They are randomly oriented. The crystallites with particular orientation tend to join those with the some orientation.

Large Angle Grain Boundaries

All those crystallites that have the same orientation form a bigger block called 'grain'. the melt now, as it solidifies would consists of a number of grains attached to each other. After complete solidification, the resulting material is a polycrystalline sample. In poly crystalline samples, the grains combine or coalexe and in between the grains, there exists a boundary called 'grain boundary'. This is shown in Fig. 4.8. These grain boundaries which exist between two grains of different orientations are called 'Large angle grain boundaries'. They angle is of the order of a few degrees. They have a width, and are associated with an energy and surface tension. They are defective sites. They are the sources and sinks of point defects like vacancies, interstitials and impurities. Across the grain, the atom mismatch is retained. The Fig. 4.9 gives the scanning electron micrograph of a poly crystalline sample of a ceramic. In the photograph, the grains and the grain boundaries are seen. The black portions represent the voids which are three dimensional defects.

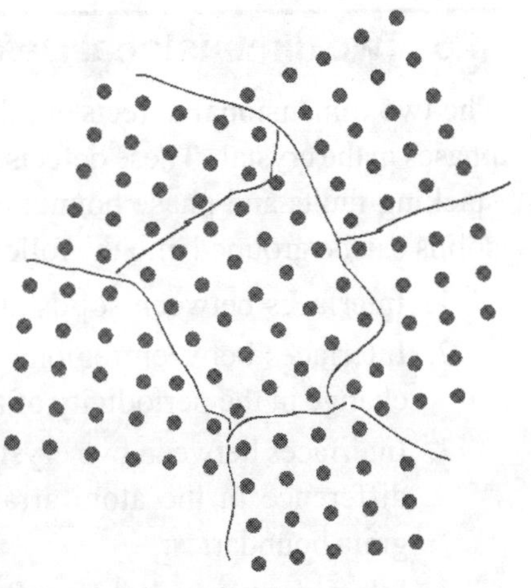

Fig. 4.8 Large angle grain boundaries.

Small Angle Grain Boundaries

Even in single crystals, the grain boundaries exist because of mechanical disturbances caused during solidification. The grain boundaries in single crystals are low angle grain boundaries. They are called sub-grain boundaries. Such sub-grain boundaries are shown in Fig. 4.10. The sample is a chemically etched sodium chloride single crystal. The lines represent the grain boundaries. The small pits randomly distributed are the isolated dislocation end points on the surface of the crystal.

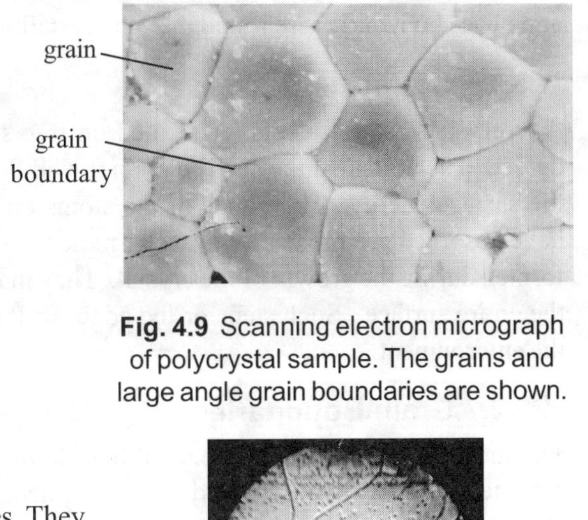

Fig. 4.9 Scanning electron micrograph of polycrystal sample. The grains and large angle grain boundaries are shown.

The sub-grain boundaries are of two types. They are 1. Tilt boundaries and 2. Twist boundaries. The grain boundaries are the arrays or cross grid of a number of dislocations. An array of edge dislocations is called Tilt boundary. The schematic representation is shown in Fig. 4.11. Two perfect regions on either side of the array of edge dislocations are slightly tilted with respect to one another by an angle θ. This angle θ is in most cases far less than 5-7°. In the diagram given in Fig. 4.11, $\frac{b}{h} = \tan\theta \approx \theta$ where h is the vertical spacing between two edge dislocations.

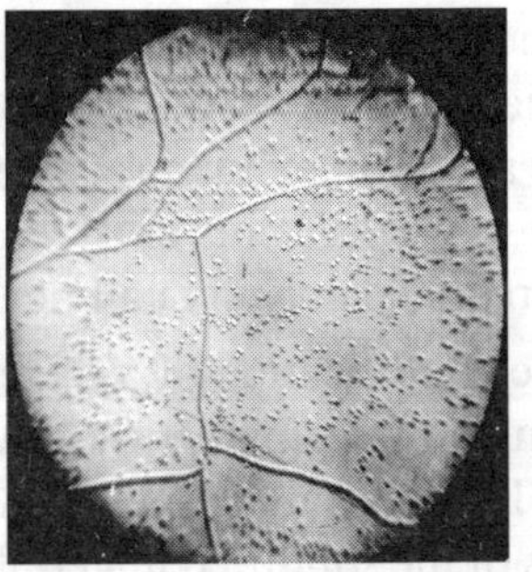

Fig. 4.10 The photograph shows the chemically etched surface of NaC*l* crystal. The pits on the photograph indicate the dislocation end points. The lines indicate the sub grain boundaries.

A cross grid of screw dislocations is called Twist boundary.

The grain boundaries in poly crystalline sample play an important role in the properties of materials like alloys and metals. The affect the creep character, toughness and strain resistance. Yield stress of metals and alloys is dislocation and grain boundary sensitive. The grain boundaries are the sinks of point defects i.e. vacancies and solute atoms. This process of absorption of vacancies or solute atoms gives rise to the formation of voids and precipitates.

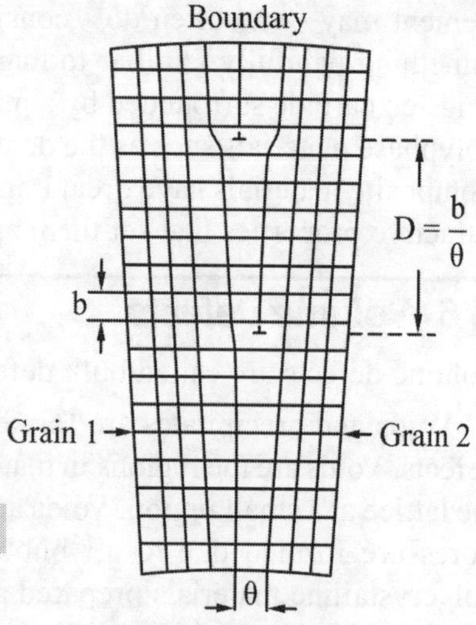

Fig. 4.11 Schematic representation of tilt boundary formed as a vertical array of edge dislocations.

4.5.3 Twin Boundaries

A Twin Boundary happens when the crystals on either side of a plane are mirror images of each other. Two single crystal sections are miss oriented but joined together such that one plane is a mirror image of the other. The boundary common to the two planes is a twin boundary. They are formed during the growth of a crystal.

The boundary between the twinned crystals will be a single plane of atoms (Fig. 4.12).

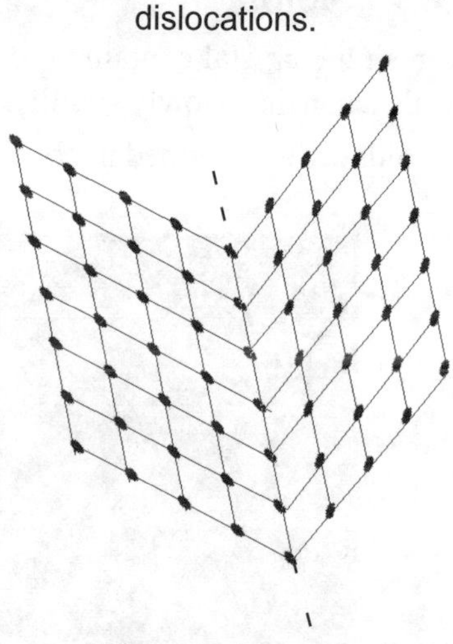

Fig. 4.12 Twin boundary.

4.5.4 Microcracks

A Microcrack occurs where internal broken bonds create new surfaces (Fig. 4.13).

They are about *10 μm* in size and there is a tendency to form on the surface of a solid rather than in the bulk. They also form at grain boundaries and other regions of disorder.

The region across which the bonds are broken is known as the *separation plane*. Microcracks are formed when there is abrasion (or impacts) with dust particles. They are important in determining how, and where, a solid may fracture.

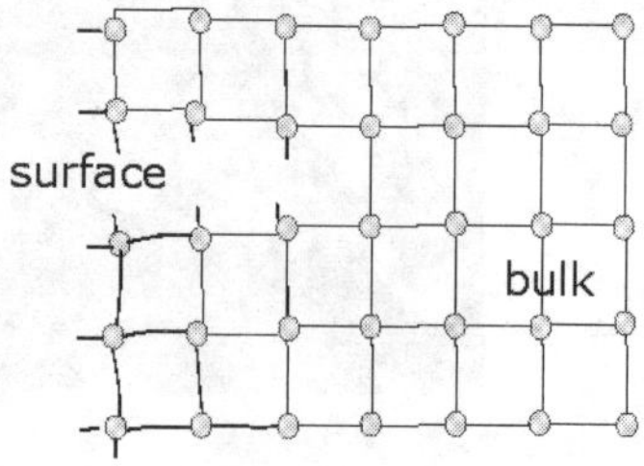

Fig. 4.13

4.5.5 Phase Boundaries

A phase is defined as a homogeneous, physically distinct and mechanically separable portion of the material with a given chemical composition and structure. Phases may be substitutional or interstitial solid solutions, ordered alloys or compounds, amorphous substances or even pure elements; a crystalline phase in the solid state may be either polycrystalline or exist as a single crystal. Solids composed of more than one element may - and often do - consist of a number of phases. For example, a dentist's drill, something painfully familiar to many of us, consists of a mixture of small single crystals of tungsten carbide surrounded by a matrix of cobalt. Here, the cobalt forms a continuous phase. Polyphase materials such as the dentist's drill are generally referred to as composite materials. Composite materials have great importance in the engineering world because they have many attractive properties that set them apart from single-phase matter

4.6 Volume Defects

Volume defects are called bulk defects. They are 1.Voids, 2. Precipitates, 3. Cracks.

Voids and precipitates are formed due to clustering of extrinsic as well as intrinsic point defects. Voids are the regions in materials or crystals where a large number of atoms miss from the lattice at a small region. Voids are formed for a number of reasons. They occur as 1. Pores. Pores are formed due to air bubbles becoming trapped when a material solidifies. The polycrystalline materials prepared as large bars, or even small discs contain pores. They are formed at junctions of the grain boundaries as shown in Fig. 4.14. (a) & (b). 2. Cavities: cavities occur due to shrinkage of metallic samples or alloys due to faster cooling from high temperatures.

4.6.1 Voids

Even single crystal contain voids due to coalescence of vacancies. It is a common feature in metals when they slowly solidify.

Voids are also formed in crystals when the vacancies in the crystal climb to edge dislocation

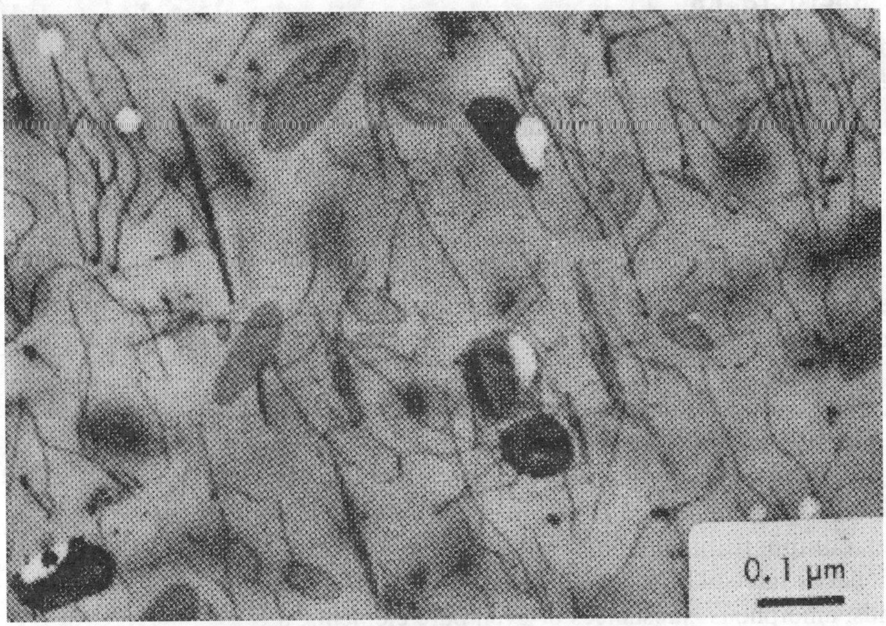

(a)

Fig. 4.14(a) White spots correspond to voids and dark regions the precipitates.

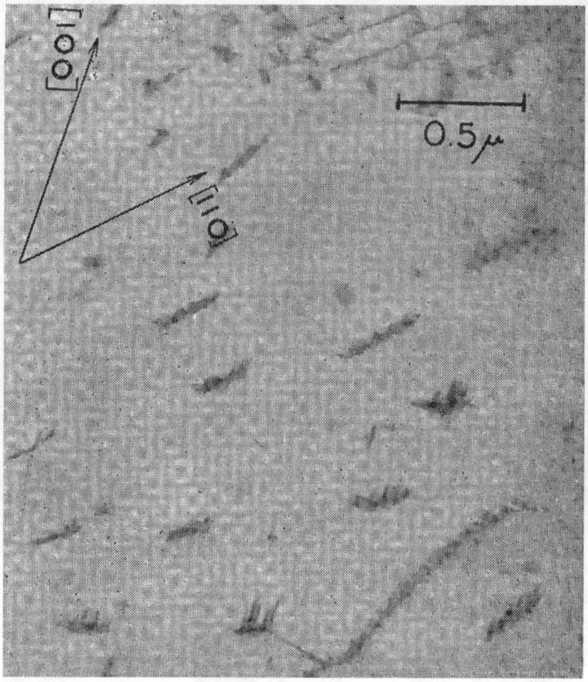

(b)

Fig. 4.14(b) Precipitates formed along dislocations in Iron (b).

at high temperature and do not come back after cooling.

The absence of a number of atoms to form internal surfaces in the crystal leads to formation of voids. They have similar properties as microcracks.

Precipitates: When impurity atoms exceed certain concentration in a crystal, they cluster together to form small regions of a different phase in the crystal. They occupy the space other wise occupied by homogenous host phase. Precipitation of impurities also takes place at the junction of grain boundaries. Impurities tend to cluster at the grain boundaries to relieve the stresses developed at the boundaries. The interstitial defects, if they are in excess at some regions in the crystal, also aggregate to form precipitates. Precipitation is caused also in alloy system to increase the yield strength, the process called precipitation hardening. The precipitates function as barriers to the moving dislocation under stress. At high stresses, dislocation loops are formed around the precipitates. Precipitates or voids tend to form on the dislocations in the crystal. The voids and precipitates are shown in the electron micrographs in Fig. 4.14.

Examples :

(a) Calculate the equilibrium number of vacancies per unit volume at a temperature of 1000^0C. The energy for the formation of vacancy in copper is 0.90 eV. What is the vacancy fraction at 500^0C. (Boltzmann constant $k = 8.62 \times 10^{-5}$ eV/K)

The concentration per unit volume of vacancies (n) is given as

$$n = N\,e^{-E/kT}$$

Now, the value of N, the number of atomic sites in copper is given as

$$N = \frac{N_A \times \rho}{M}$$

where N_A is the Avagadro's number, ρ is the density and M = the atomic weight.

Then $$N = 6.023 \times 10^{23} \times \frac{1}{63.5} \times \frac{8.96 \times 10^{6}}{m^{3}}$$

$$= 8.5 \times 10^{28} \text{ atoms/m}^2$$

$$n = \left(8.5 \times 10^{28}\right) \times \exp\left[-\frac{0.9}{\left(8.62 \times 10^{-5}\right)(1273)}\right]$$

$$= 8.5 \times 10^{28} \times \exp\left[-8.2\right]$$

$$= 8.5 \times 10^{28}\ 10^{\frac{-8.2}{2.303}}$$

$$= 8.5 \times 10^{28} \times 2.75 \times 10^{-4}$$

$$= 2.3 \times 10^{25} \text{ vacancies/m}^3$$

(b) The vacancy fraction at 500^0C

$$\frac{n}{N} = \exp\left[\frac{-0.9}{\left(8.62 \times 10^{-5}\right)(773)}\right] = e^{-13.5} = 10^{-\frac{13.5}{2.303}} = 1.61 \times 10^{-6}$$

The atoms in the grain boundary will not be in perfect crystalline arrangement. Grain boundaries are usually the result of uneven growth when the solid is crystallising. Grain sizes vary from *1 μm to 1 mm.*

Questions

1. Explain how a vacancy is formed in a metal crystal ? Does the vacancy affect the bonding of atoms around ?
2. Illustrate the following type of defects in metallic lattice (a)Vacancy (b) Frenkel defect (c) Interstitial.
3. Derive the expression for the vacancy density in a metal crystal.
4. Explain the formation energy term ? On what factors does it depend ?
5. Derive the expression for the density of Frenkel defects in a metallic crystal ? Which are

the type of defects that predominate in a metallic crystal?

6. The formation energy for a vacancy in a metal crystal is 1eV. What is the proportion of vacancies present in the crystal at 300K and 1000K ? Compare them.
7. Pure magnesium has the formation energy for vacancy as 0.9 eV. Calculate the equilibrium vacancy fraction at 827^{0}C and 527^{0}C.
8. Two metals have the formation energies as 0.7 ev and 0.9 eV. What will the ratio of their vacancy fractions?
9. Write the equation for vacancy concentration in a metal at equilibrium temperature T. Two solids have atoms that have atomic masses in the ratio 1 : 1.3. They have formation energies in the ratio 1:1.3.Calculate the ratio of the vacancy concentration in them at temperature T.
10. The formation energy for a vacancy in pure gold is 0.98 ev. Calculate the equilibrium concentration of vacancies per cubic meter at 827^{0}C and 27^{0}C. Atomic weight of gold is 196.97 M/g mol^{-1} the density = 19320 kgm^{-3}.
11. Explain why vacancies occur in pairs in ionic crystals ? What are the effective charges of the vacancies in ionic crystals?
12. Derive the expression for the vacancy - concentration in an ionic crystal
13. Can a cation vacancy interact with an anion vacancy ? If so, what will be the resultant ?
14. Which is the migration species in an ionic crystal when an electric field is applied across it?
15. Derive the expression for 1. vacancy concentration and 2. Frenkel defect concentration in an ionic crystal
16. What happens when a metal from high temperature is quenched to room temperature from the point of view of the number of vacancies in the crystal ?
17. Write a note on 'impurities' in crystals.
18. In the case of expansion of a solid, what will be the fractional contribution of vacancies at 1000 K in a metal whose linear expansion coefficient is $16.2 \times 10^{-6}/^{\circ}$C and the solid expands linearly by 1.9%.
19. Write a short note on 'Point defects?
20. Give the model of 'edge dislocation' in a metallic crystal. Draw the Burger's circuit.
21. Give the model of 'Screw Dislocation' in a crystal. Draw the Burger's circuit.
22. Give the expression for energy of (a) an edge dislocation (b) screw dislocation.
23. What is 'Burger's vector' ? In what directions do the Burger's vector lie w.r.t (a) an edge dislocation, (b) screw dislocation.
24. Why do you get the deformation of a crystal when a stress is applied to it ?
25. What are surface defects ?

26. Explain different types of volume defects.
27. What are grain boundaries ?
28. What are different types of grain boundaries ?
29. Distinguish between grain boundaries and phase boundaries ?
30. Write a short note on small angle grain boundaries in crystals.
31. How are large angle grain boundaries formed ?
32. Show how a twin boundary is formed.
33. What are voids ?
34. What are precipitates in a crystal ?

Objective Questions

I. Choose the correct choice among the four alternatives.

1. Point defects are
   (a) vacancies, dislocations, impurities, phonons
   (b) grain boundaries, voids
   (c) stacking faults, impurities
   (d) vacancies, impurities, Frenkel defects, interstitial

2. The energy of formation of a schottky defect
   (a) depends on temperature
   (b) is independent of temperature
   (c) is an inverse function of temperature
   (d) is a function of the logarithm of defect concentration

3. In thermal equilibrium in a crystal
   (a) the number of vacancies is exponentially growing function of temperature
   (b) the number of vacancies is exponentially growing function of activation energy
   (c) the number of vacancies is equal to the number of Frenkel defects
   (d) the number of vacancies is equal to the number of interstitials.

4. Any metal crystal in thermal equilibrium contains
   (a) Schotlky and Frenkel defects, unavoidable impurities, interstitial
   (b) only Schotlky defects and Frenkel defects
   (c) only vacancies
   (d) only Frenkel defects and impurities

5. The defects that arise in a crystal as a requirement of thermodynamic equilibrium are
   (a) phonons (b) excitons
   (c) point defects (d) dislocations

6. Interstitial atoms in a metal structure
   (a) cause movement of vacancies in the lattice
   (b) cause a decrease in the lattice parameter of the crystal
   (c) cause a localized strains in the lattice
   (d) can never exist at low temperatures

7. The formation energy of a Frenkel defect is
   (a) the energy needed to bring atom to the surface of the crystal
   (b) the energy needed to bring the defect to the surface of the crystal
   (c) energy needed for taking two atoms and place them on the surface
   (d) the energy needed to bring an atom to an interstial position

8. In close packed structures
   (a) Frenkel defects are the dominating defects
   (b) dislocations are the dominant defects
   (c) Schotlky defects are the dominant defects
   (d) Schotlky defect concentration will be far less than the Frenkel defect concentration

9. Suppose there are N atoms in the crystal. The number of ways in which we can create n vacancies in the crystal is given by
   (a) $\frac{N!}{n!}$ (b) $\left(\frac{N!}{(N-n)!\,n!}\right)$
   (c) (N-n)! (d) $\frac{n!}{N!}$

10. The formation energy of a vacancy in a metallic crystal depends on
    (a) the nature of the metal
    (b) temperature
    (c) the concentration of vacancies
    (d) the total number of atoms in the crystal

11. A copper slab is kept at room temperature. Another slab identical in all respects is heated to 1000 ^{0}C and suddenly cooled to room temperature. Then
    (a) both the slabs will contain the same concentration of vacancies
    (b) the suddenly cooled slab contains more number of vacancies compared to the first slab
    (c) the suddenly cooled slab contains less number of vacancies compared to the first slab
    (d) the second slab does not contain any vacancies as all of them escape away

12. In alkali halides
    (a) Frenkal defects are the dominating defects
    (b) Schotlky defects are the dominating defects
    (c) dislocations are the dominating defects
    (d) vacancy pairs and voids are the dominating defects

13. In silver halides
    (a) no defects can be present
    (b) Schotly defects are dominating defects
    (c) Interstials are the dominating defects
    (d) Frenkel defects are the dominating defects

14. In an ionic crystal
    (a) a vacancy left by a positive ion behaves like a negative charge
    (b) a vacancy left by a positive ion behaves like a positive charge
    (c) a positive ion vacancy is attracted towards negative ions
    (d) a negative ion vacancy gets trapped by a positive ion

15. In an ionic crystal, it is expected that
    (a) the positive ion defects will be a higher than the negative ion vacancies in concentration
    (b) the positive ion vacancies are the only defects that are present
    (c) an equal number of positive ion and negative ion vacancies exist
    (d) the Frenkel defects are dominant defects in all crystals

16. A Schotlky defect in an ionic crystal is
    (a) a pair of negative ion vacancies
    (b) a pair of positive ion vacancies
    (c) a pair of a negative ion vacancy and a positive ion vacancy
    (d) a pair of a vacancy and an ion in the lattice

17. A dislocation is
    (a) always under shear
    (b) a point defect
    (c) a two dimensional defect
    (d) under compression

18. If there are no dislocations in a crystal, the shear stress for deformation is roughly (G = shear modulus of the crystal)
    (a) 0.16G (b) 10^6G
    (c) 10^2G (d) $\frac{G}{100}$

19. The plastic deformation of a crystal is possible due to the presence of
    (a) point defects in the crystal
    (b) impurities in the grain boundaries
    (c) vacancy movement
    (d) the motion of dislocations in the crystal
20. A grain boundary is
    (a) a pair of edge dislocations of opposite nature
    (b) an irregular sequence of layers of atoms
    (c) a series of closely spaced dislocations
    (d) a three dimensional defect
21. The ionic crystals are good insulators at room temperature. They become conducting at higher temperatures because of
    (a) the electrons going to the conduction bond
    (b) the migration of holes in the valence bond
    (c) the movement of a large number of charged vacancies
    (d) the generation of phonons in the crystal
22. The concentration of point defects in a crystal can not be increased by
    (a) elastic deformation
    (b) quenching the crystal from high temperature
    (c) plastic deformation
    (d) increasing the temperature of the crystal

II. Fill in the blanks :

1. A Schotlky defect is also called ________ .
2. The zero dimensional defects are called ________ .
3. The defects in crystals that arise as a result minimum free energy requirements are ________ .
4. If there N atoms in a crystal, the number of ways in which 'n' vacancies can be created is equal to ________ .
5. If E_f is the formation energy of a vacancy, then concentration of vacancies as a function of temperature T is given by the relation ________ where N represents to the total number of atoms.
6. The formation energy E_F of a Frenkel defect in metal is ________ temperature.
7. The concentration of vacancies at 1000 K in a metal (energy of formation of a vacancy at = 0.9eV. The Boltzmann constant k = 8.62×10^{-5} eV|K) will be higher than the concentration of vacancies at 300K by a factor ________
8. Impurities added deliberately to a metal crystal alter the ________ of the crystal.
9. The Frenkel defects do not change the ________ of a metal crystal.
10. The study of thermal expansion at high temperatures yields information about defects like ________.
11. The examples for transient or dynamic defects are ________
12. Vacancies in ionic crystal are formed in ________ .
13. In ionic crystals, the defects are ________ defects.
14. The ________ in metals are more mobile than vacancies.
15. Two metals have the formation energies as 0.7 and 0.9 e V respectively. Then, the ratio of the vacancy fractions at room temperature is ________ .
16. The clustering of vacancies results in the formation of three dimensional defects called ________ in metals.
17. ________ defects are predominant defects in alkali halides.
18. The predominant mobile defects in alkali halides are ________ .
19. A positive ion vacancy possesses effective ________ charge.
20. The impurity atoms in metals can occupy either ________ positions or ________ positions.

21. The energy of formation for an interstitial in metallic lattice is __________ than the energy required for a point defect like vacancy.
22. In substitutional solid solutions like copper-gold alloy, copper occupies __________ position.
23. According to Hume-Rothery rules, the impurity atom and the host atom in a solid solution should not differ by more than __________ for stable solid solution formation.
24. The alloys of metals possess __________ strengths compared to the pure metal.
25. A linear or line defect is called __________.
26. The Burgers vector of an edge dislocation is __________ to it.
27. The Burgers vector of a screw dislocation is __________ to it.
28. A dislocation free crystal can not under go __________ deformation.
29. The energy of a dislocation is proportional to the __________ of the magnitude of Burgers vector.
30. The plastic deformation in any crystal is due to the __________ of dislocations.
31. The mechanical properties of crystals can only be explained on the basis of __________ of dislocations.
32. The striking difference between experimental and theoretical shear stress is explained on the basis of __________ and their __________.
33. The energy of the dislocation per unit length is __________ to the shear modulus and the __________ of the Burger's vector.
34. A curved dislocation is known as a __________.
35. When a dislocation in a crystal moves to the surface of the crystal, it leaves a __________ on the surface.
36. The strength of a dislocation is indicated by the magnitude of __________.
37. A metallic crystal, when deformed, hardens as a result of __________ interactions.
38. The edge dislocations in a crystal, under stress, glide or move in __________ planes.
39. A tilt boundary in a crystal is an array of __________ dislocations.
40. Grain boundaries are __________ dimensional defects.
41. Large angle grain boundaries exist in __________ crystalline samples.
42. Voids can be thought as a cluster of __________ in crystals
43. An aggregate of interstitial defects can give rise to __________ in a crystal.
44. Precipitates in a crystal act as obstacles to moving __________ when a crystal is strained.
45. Grain boundaries are sources and sinks of __________.

III. True or False

1. As the temperature of a crystal is increased, the concentration of vacancies also increases.
2. The formation energy of a vacancy in a metal is independent of temperature.
3. The formation energy of a vacancy is characteristic of a crystal.
4. Line defects are grain boundaries.
5. The Burgers vector of edged location is parallel to the dislocation.
6. Burgers vector of all dislocations meeting at a point sum over to zero.
7. Grain boundaries are sources and sinks of point defects.

CHAPTER 5

Elements of Statistical Mechanics

5.1 Introduction

Thermodynamics deals with the bulk properties of material bodies like solids, liquids and gaseous systems. Such properties like the exchange of heat, working of a heat engine, the expansion of gases, change of state or phase transitions, absorption or evolution of heat by a body etc., are explained in thermodynamics on the basis of excellent mathematical formulations. The relations in thermodynamics explain the bulk behaviour of a physical body to a large extent. It is based on three variables, pressure (P), Volume (V), and Temperature (T). The subject of thermodynamics is based on the three laws of thermodynamics.

The second law of thermodynamics tells about the way in which the heat flows. The heat flows from a reservoir at high temperature to a system at low temperature. This is a natural process. If the reverse is needed, external work has to be done on the system.. The second law also introduces another concept and quantity called 'Entropy". In a physical process in which heat is transferred from one system to another, then, the entropy change involved in the process is given as

$$dS = dQ/T \qquad .....(5.1)$$

where dS is the change in entropy and dQ is the change in heat (heat absorbed or heat emitted.).

In any reversible process, the change in entropy is zero. But, it is difficult to find reversible processes in true sense; therefore, the entropy change in physical or chemical processes is always positive. That means, every time a chemical or physical process takes place, the entropy of the universe would be increasing. Boltzmann while attempting to correlate the thermodynamical parameters to microscopic behaviour of the systems proposed a hypothesis. The matter is made up of small particle, called molecules. The particles are in continuous motion and they would be changing their positions and momenta continuously.

The observed behavior of a system at any temperature depends the way in which the constituent particles of matter would be distributed among different momenta values and positional configuration. This distribution or the permutations that can be possible decides the entropy of the system. Therefore, he proposed a law that the entropy of a system is proportional to the logarithm of the most probable distribution of particles at that temperature. Mathematically, this can be stated as

$$S = k\ln W \qquad .....(5.2)$$

where W represents the probability, S represents the entropy, and k is a proportionality constant, called Boltzmann constant (a universal constant). The value of k is 1.381×10^{-23} Joules / K. This relation is the foundation for both kinetic theory as well as statistical mechanics.

A number of phenomena are explained on the basis of statistical laws. The subject of statistical mechanics is based on the statistical laws. The behaviour of bulk properties of matter is related to the behaviour of microscopic particles of that matter.

5.2 Statistical Mechanics

5.2.1 What is statistical mechanics?

Statistical mechanics supplies or provides the necessary mathematical language for understanding the nature of material systems and formulates the necessary equations for its behaviour in terms of particle parameters in a statistical way and ultimately arrive at the bulk property estimation. A system is an assembly of particles associated with physical and chemical properties. A given macroscopic behaviour of a system is only based on the behaviour of micro systems of that system. These Microsystems are the molecules in a gas or liquid, or atoms in a solid, electrons or holes in crystals, an assembly of photons or phonons etc. They form microsystems of a macrosystem like a gas enclosed in an enclosure.

In order to connect the molecular behaviour to the macroscopic behaviour, one has to understand the motion of molecules or atoms. Because of the large number, it is impossible to set up equation of motion for each and every molecule, study the motions with time and average them over all molecules. The microscopic behaviour of molecules is ultimately connected to the properties of bulk material.

To understand how molecules move in a gaseous system, observe the motion of small flying insects that group around a street lamp. Thousands of insects would be moving around in all directions just around the lamp. It appears as if the lamp is the source of energy and it keeps them flying. One can not hold or concentrate on one insect and observe its motion. If we want to set up equations of motion, we have to set up thousands of equations and solve them. This is an impossible task. Let us see one more thing. Place a number of street lamps in a row distanced a few meters, and see that no other source of light is present in the vicinity. First, swith on only one lamp, say, the first lamp in the row. The insects will be around it. Now, switch of the first lamp and put on the second lamp. We see all the insects moving to second and flying. One can repeat this by following the above process in a sequence and see that the insects reach the last lamp in the row. We can not see the individual insect's movement, but ultimately the entire group of insects would be in motion. The bulk observation that we make finally is the movement of insects from the first lamp to the last. The same situation can be imagined in the case of molecules or particles in a material system. We consider the entire group of particles as a point of observation. This group of particles will be undergoing changes inside at every instant of time. The changes can be mentally picturised as different points in a special space.

Boltzmann laid the foundation for statistical mechanics in 19 th century and Gibbs developed the formulism to understand the motion of the particles in a thermodynamical system at any temperature. This formulism later by application of quantum conditions has helped in understanding the physical phenomenon that occur or materials properties and their variations with external parameters. Two states are used to describe a system. One is the 'microstates' and the other is the 'macrostate'.

Thermo dynamical relations are based on empirical laws derived from the observations based on experimentation. The properties observed should ultimately relate to the constituents like molecules or atoms or ions, or electrons etc. The statistical mechanics actually does that job.

The statistical mechanics considers 'system' which interacts with other systems surrounding it and reach an equilibrium state. The 'system' can be an assembly of molecules of a gas or liquid, electrons, holes, photons, phonons, magnons, polarons, muons, ,or any particles that have a nature to move and exchange the energy or diffuse or react chemically.

5.2.2 The characteristics or features of a 'system'

A system like a gas in a container is specified by the volume, temperature and number of particles. The parameters volume, energy, pressure, temperature describe the macrostate of the system. The macrostate of the system corresponds to the one which exhibits an average behavior that represents the experimentally observed state. However, the macrosystem is composed of a large number of microstates. These microstates are the states of the constituents i.e., the particles of the system at any instant of time.

The systems under consideration in statistical mechanics are associated with features that are given below:

1. It contains very large number of particles in a unit volume. To start with, the particles are identical particles. The particles are in continuous motion randomly and make collisions with each other or the walls of the container.
2. The system can exchange energy with any other system that is in contact or exchange particles as well as energy. Normally, no system can remain isolated unless artificially forced. The state of any system can be specified in terms of energy, temperature, volume, or by number, energy, and temperature.
3. The particles in the system will be in continuous motion. The state of particles in the system is specified by the positional coordinates, and momentum coordinates.

Therefore, each particle in a system is associated with six coordinates, three positional coordinates, x.y.z and momenta coordinates p_x, p_y, and p_z, along the three axes.

4. Any system under observation will be reaching an equilibrium state when it comes in contact with some other system (example: a liquid in a vessel, coming into contact with a reservoir of heat like a heater or another system at higher temperature, exchanges energy with reservoir). This equilibrium can be in terms of statistical equilibrium, thermal equilibrium, mechanical equilibrium or chemical equilibrium.

Macrostate and microstates:

The system is associated with a macrostate and it passes through a number of microstates before reaching an equilibrium macrostate. The macrostate is specified by the properties of he system or the variables: the pressure, volume, and temperature. The movement of the particles continuously in the system gives a number of microstates. The distribution of particles goes on changing at every instant of time. That means that the system goes through a number of microstates as stated earlier. This aspect is shown in Fig.5.1.

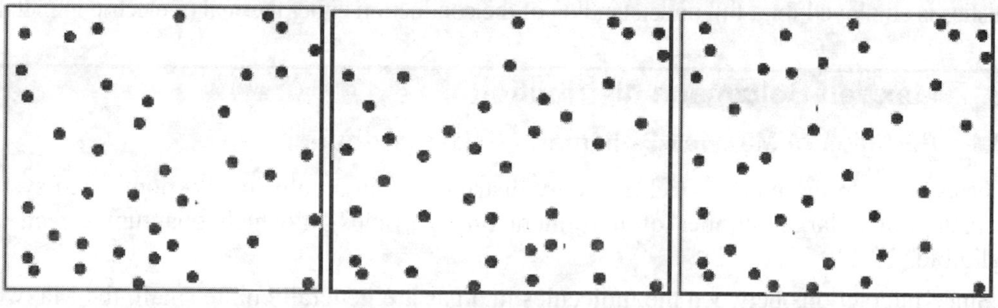

Fig. 5.1 The three pictures shown in here represent three micro states the molecular system undergoes. The molecules here are represented in full circles and smaller number is chosen for brevity. If we take the picture of the molecular motion, at different instants, we can see the microstates that the system undergoes. It is a mental picture. This is only to indicate that the system of particles goes through a number of microstates and the macrostate is unaltered.

The particle distribution in the three figures represents the microstates, as if they are obtained by photographing at high speed. They are the mental picturisations the microstates. All the microstates are equally accessible by the particles. The total number of accessible microtates is related to the distribution function. The microstates do not effect the macrostate of a system. The macrostate of a system is the equilibrium state.

Statistical mechanics usually approaches systems from the point of view of microscopic, or atomic-scale. In its modern form, statistical mechanics recognizes three broad types of systems:

They are: 1. Those that obey Maxwell-Boltzmann statistics, 2. Those that obey Bose-Einstein statistics, and 3. Those that obey Fermi-Dirac statistics.

Maxwell-Boltzmann statistics apply to systems of classical particles, such as the atmosphere, in which considerations from the quantum theory are small enough that they may be ignored.

The other two types of statistics namely Bose-Einstein and Fermi-Dirac statistics are concerned about the systems in which quantum-mechanical effects cannot be ignored.

Bose-Einstein statistics apply to systems of particles that have integral values of the quantum mechanical property called spin; these particles are called 'Bosons" named after Styandra Bose, an Indian physicist from Bengal and worked with Einstein. An unlimited number of bosons can be placed in the same state. Photons, for instance, are bosons, and so the study of electromagnetic radiation, such as the radiation of a black body involves the use of Bose-Einstein statistics.

Fermi-Dirac statistics apply to systems of particles that have half-integral values of spin; Pauli's exclusion principle is applicable. The particles are called 'Fermions'. No two fermions can exist in the same energy state. Electrons are fermions, and so Fermi-Dirac statistics is employed for a full understanding of the conduction either thermal or electric in metals.

5.3 Maxwell-Boltzmann distribution

5.3.1 Particles of Maxwell-Boltzmann (MB) distribution

It corresponds to the most probable energy distribution, in a collisionally-dominated system consisting of a large number of non-interacting particles in which quantum effects are negligible.

Since interactions between the molecules in a gas are generally quite small, the Maxwell-Boltzmann distribution provides a very good approximation of the conditions in the case of an "ideal gas" consisting of non-interacting molecules in the ground state, all energy is in the form of kinetic energy.

The relationship between kinetic energy and momentum for massive particles is $E = p^2/2m$, where p^2 is the square of the momentum vector $p = [p_x, p_y, p_z]$.

Boltzmann statistics is based on the following characteristics of the particle system. The particle system is a non interacting particle system like the molecules in an ideal gas.

1. The system consists of very large number of constituent particles per unit volume. The particles are identical.
2. The particles are in continuous motion and the state of each particle is defined by its positional coordinates(x,y,z) and momentum coordinates (p_x , p_y ,p_z). The particles collide with each other and the collisions are purely elastic.
3. The particles are *distinguishable* particles.
4. The particles are distributed among the energy levels that are spread from zero to infinity on the energy scale.
5. There is no restriction on a particle from occupying any energy state available at the temperature at which the system is kept. It means that all energy states are equally accessible for the particles.

5.3.2 MB Statistics

Consider a system of particles like the molecules of an ideal gas enclosed in a volume, V. These particles are distributed among various energy levels that are distributed continuously. The distribution is as per the following pattern:

n_1 particles in ε_1 level; n_2 particles in ε_2; (5.2A)
n_3 particles in ε_3; n_i particles in ε_i energy level etc.

The conditions on the system for the above configuration are:

1. The total number (N) of particles in the system under consideration is constant i.e.,

$$\sum_i n_i = N \tag{5.2B}$$

2. The total energy (E) of the system under consideration is constant. The system of particles undergoes a number of microstates while reaching the equilibrium state.

$$\sum_i \varepsilon_i n_i = E \tag{5.2C}$$

Let us see, in how many ways the configuration given in (5.2A) can be achieved. That is related to the probability of the distribution and ultimately to the bulk behaviour of particles as a system.

Let us first consider that n_1 particles should be selected from N particles to occupy ε_1 level. They have to be chosen from N particles. The number of ways in which this can be done is

$$N(N-1)(N-2)\ldots\ldots(N-n_1+1) = N!$$

But, it should be noted that the permutations of n_1 particles in the energy level ε_1 does not give rise to a new distribution. Imagine the students in a class. The sitting of the students in which ever bench that is available does not alter the class structure. There is no restriction on any student or a condition that he/she should occupy a particular seat. All seats are equally accessible to all students. Similarly, all energy levels are equally accessible to all the particles. There is no restriction on the number of particles in any energy. The number of possible permutations for n_1 particles to be arranged among themselves is $n_1!$. In a similar way, we have $(N-n_1)!$ permutations that also do not give a new distribution.

The number of ways in which n_1 particles occupy energy level ε_1 is

$$w(n_1) = \frac{N!}{(N-n_1)!n_1!.} \quad \ldots..(5.3)$$

Similarly for energy level ε_2, the number of ways in which the selection can be made out of $(N-n_1)$ particles. Therefore for the second level the number possible permutations will be

$$w(n_2) = \frac{N!}{(N-n_1-n_2)!n_2!.} \quad \ldots..(5.4)$$

Similarly we can get expressions for the possible number of ways of filling the energy levels ε_3,

E_4.... and so on. The total number of possible ways in which a distribution mentioned in the configuration (5.2A) can be achieved is given by the expression

Continuing this type of estimate, we arrive at the total number of ways in which the distribution given in (5.2A) can be achieved as

$$w(n_1)w(n_2).....w(n_i) = \frac{N!}{n_1!n_2!n_3!...} \quad .....(5.5)$$

The probability of getting the above distribution is proportional to the expression given in (5.7). Hence,

$$P = \frac{1}{C}\frac{N!}{n_1!n_2!n_3!...} \quad .....(5.6)$$

where, P, represents the probability and C is a constant. Taking logarithm of the above expression and differentiating it with respect to n_i and then equating it zero for maximum probability, we get the expression for the number of microstates in the system under consideration as

$$n_i = \frac{Ne^{-\beta\varepsilon_i}}{\sum_i e^{-\beta\varepsilon_i}} \quad .....(5.7)$$

5.3.3 M.B Distribution Function

The probability that the system under consideration can be in the state defined by (5.2A) is given as

$$P_i = \frac{n_i}{N} = \frac{e^{-\beta\varepsilon_i}}{\sum_i e^{-\beta\varepsilon_i}} \quad .....(5.8)$$

The derivation of the above equation is given below for those who wish to go into details.

Taking logarithm on both sides, we obtain

$$\ln P = \ln N! - \ln \sum_i n_i! - \ln C \quad .....(5.9)$$

but, $ln\, N! \approx NlnN - N$ (Stirling's approximation)

$$\therefore \quad \ln P = N\ln N - N - \sum_i n_i \ln n_i + \sum_i n_i + \ln C$$

For maximum probability that represents the equilibrium condition, we have to differentiate the above eqn. and set it to zero for maximum probability.

Then, we get

$$\partial \sum n_i \, ln n_i = \sum (\partial n_i + ln n_i \partial n_i) = 0 = \sum_i ln n_i \partial n_i \qquad \text{.....(5.10)}$$

Differentiating the expressions (5.2B) and (5.2C), and multiplying the first expression with α and the second expression with β respectively, and adding, we get

$$\sum_i (\ln n_i + \alpha + \beta) \partial n_i = 0 \qquad \text{.....(5.11)}$$

The constants α and β are called Lagrange's multipliers. Simplifying the above expression we get

$$n_i = e^{-\alpha} e^{-\beta \varepsilon_i} \qquad \text{.....(5.12)}$$

Substituting eq. (5.12) in (A), we get

$$n_i = \frac{N e^{-\beta \varepsilon_i}}{\sum_i e^{-\beta \varepsilon_i}} \qquad \text{.....(5.13)}$$

The probability that the system under consideration can be in the state defined by (5.6) is given as

$$P_i = \frac{n_i}{N} = \frac{e^{-\beta \varepsilon_i}}{\sum_i e^{-\beta \varepsilon_i}} \qquad \text{.....(5.14)}$$

$\sum e^{\frac{-C_i}{kT}}$ is a parameter aoosicated, with a system of particles and is taken as a constant of the system and can be represented by the distribution function in general is written as

$$P(MB) = Ae^{-\beta \varepsilon_1} = Ae^{-\varepsilon_1/kT}$$

The expression given above represents the particle distribution in a system at a temperature, T and that reacts with a neighbouring system thermally only. The above one is a classical statistical distribution in which there is no restriction on the particle regarding the energy or its occupation.

The expression $\sum_i e^{-\beta \varepsilon_i}$ is represented by the symbol Z. It is called the partition function of the system. The constant β is a function of temperature and is the inverse of thermal energy (kT) and is given as, $\beta = \frac{1}{kT}$. The partition function is an important parameter of a system and represents the state in which the system is present at an equilibrium condition.

The knowledge of partition function enables us to estimate all the thermal properties or magnetic or electrical properties of a system like a gaseous system or a liquid system which contains molecules in motion, or a crystal in which vibrations of atoms are present. It is applied.

5.3.4 Maxwell – Boltzmann distribution of energy

The energy distribution of particles like the molecules of an ideal gas follows the distribution function given as

$$n(\varepsilon)d\varepsilon = \frac{2\pi N}{(\pi kT)^{3/2}}\sqrt{\varepsilon}\, e^{-\varepsilon/kT}\, d\varepsilon \qquad \text{.....(5.15)}$$

The distribution of particles against the energy for a gaseous system is given in Fig.5.2. The curve spreads from a very low energy to very large values of energy.

The curves exhibit peaks indicating that maximum number of particles possess that energy. The area under the curve gives the total number of particles in the system.

Very few particles exist at very high energies. As the temperature of the gas increases, the curve tends to become flat and the shifts to higher energy side and the particles tend to spread among energy levels more uniformly.

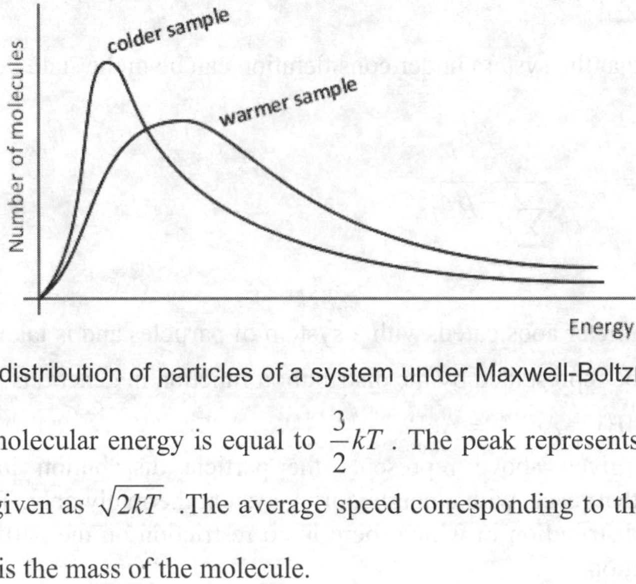

Fig. 5.2 The distribution of particles of a system under Maxwell-Boltzmann statistics.

The average molecular energy is equal to $\frac{3}{2}kT$. The peak represents the most probable energy which is given as $\sqrt{2kT}$ .The average speed corresponding to the average energy is $\sqrt{\frac{8kT}{m}}$, where m is the mass of the molecule.

5.4 Bose-Einstein (BE) Statistics

5.4.1 Introduction

Maxwell-Boltzmann statistics considers that the particles in a system are distinguishable particles. The energies of the particles can be from zero to infinity in a continuous manner from zero to infinity. The MB statistics can effectively be applied to molecular system in a gaseous system where the considerations from quantum theory are small enough to be neglected. Satyandra Bose (in 1920s) while attempting to solve the radiation field problem came across the idea that, if particles like photons are indistinguishable, then the counting can not be done as in Boltzmannn statistics. He simply showed that the number of permutations would come down for indistinguishable particles. Quantum theory has proved that each energy level, ε_i contains g_i distinct sublevels, all of which have the same energy, and are distinguishable energy states. For example, two particles may have different momenta, in which case they are distinguishable from each other, yet they can still have the same energy. The value associated with g_i is called degeneracy.

If we consider the generalization of classical picture, we should also associate the sublevels i.e., a degeneracy with the energy levels. Let the number of sublevels be equal to g_i. Let the number of particles be n_i and they should be adjusted among the energy states. The first particle can choose any of the cells or sublevels in g_i ways, the second particle in g_i ways, the third particle in g_i ways, and so on. That means, n_i particles can be placed in g_i sublevels in $(g_i)^{n_i}$ ways. The permutations of particles n_i among themselves does not give a new distribution. Therefore ,the number of microstates or the number of ways in which n_i particles can be distributed in g_i levels is equal to $< n_i > = \frac{n_i!(g_i)^{n_i}}{n_i!}$.

5.4.2 A Case Study of MB and BE Statistics

Let us take an example: Consider that there are three particles, indexed as a, b, and c, and the particles are distinguishable. Suppose that there are three energy states available. The particles have to be distributed in these energy states. They can be distributed in three boxes (represent three energy states) as per the Maxwell- Boltzmann statistics as follows:

(A) Maxwell-Boltzmann statistics:

-	a	bc
a	-	bc
a	b	c
b	a	c
c	b	a
b	c	a
a	c	b
c	a	B
Abc	-	-
-	abc	-
-	-	abc
ac	b	-
ac	-	b
b	ac	-
-	ac	b
-	b	ac
b	-	ac
c	ab	-
-	ab	c

-	c	ab
c	-	ab
ab	c	-
ab	-	c
a	bc	-
-	bc	a
bc	a	-
bc	-	a

In the example given above, the total number of ways in M-B statistics is equal to $w_i = (g_i)^{n_i} = 27$. These arrangements are shown in the table given above.

According to Bose-Einstein statistics, the three particles are indistinguishable and can be represented by a, a, a. Then, they can be distributed in three boxes as follows:

(B) Bose-Einstein statistics:

For the same particle and cell system, the particles are indistinguishable. Particles = 3; the number of cells = 3.

The arrangement is given as

a	a	a
a	aa	-
-	aa	a
-	a	aa
-	a	aa

aa	a	-
aa	-	-
aaa	-	-
-	aaa	-
-	-	aaa

In the above distribution, we can see that there are 10 ways of arrangements. There fore, the number of microstates is reduced as compared to that of classical distribution when particles are indistinguishable particles.

On the basis of the above concept, Bose-Einstein and the distribution function were worked out. We shall see the details.

5.4.2 The characteristic features or postulates of Bose-Einstein Statistics are as follows

1. The system under consideration consists of very large number of non-interacting and identical particles.
2. The particles are quantum particles and the energy of the particle is quantized. The particles possess discrete eigen (energy) values. The particles have intrinsic angular momenta in integral units of ***h*/(2p)** ,where **p** is the momentum of the particles , that is, $p = \frac{2\pi\hbar}{L} n$.

3. Each particle is associated with a wave function. The wave functions are symmetric. That means, if there are two particles and if they are interchanged, their wave functions are unaltered or remain the same. This is typical characteristic of Bose-Einstein (BE) statistics. The particles which follow BE statistics are called 'Bosons" named after Bose.
4. The state of the system is specified by specifying occupation numbers n_i such that n_i particles are in ε_i (energy level) that has the energy states (or sublevels), g_i.
5. The total number (N) of particles is constant in the system such that

$$\sum_i n_i = N \tag{A}$$

6. The total energy (E) of the system is Constant.

$$\sum_i \varepsilon_i n_i = E \tag{B}$$

7. Bosons have integral values of the quantum mechanical property called 'spin'. The spin number of the particles can be 0, 1 .....
8. An unlimited number of bosons can be placed in the same energy state. There is no restriction on the particle to occupy any state of energy.
9. All those particles that mediate the fundamental forces of nature are bosons.

Bosons are said to be the particles that transmit interactions (force carriers), or the constituents of radiation

The examples for particles that follow Bose-Einstein statistics are photons, phonons (packets of mechanical vibrational energy packets), pions, the W and Z particles, gluons. The nucleus of a carbon-12 atom contains 6 protons and 6 neutrons (all fermions) and is a boson. The atom helium-4 (^{4}He), is made of 2 protons, 2 neutrons and 2 electrons, and is a boson.

We can now obtain a general expression for the average occupation number of particles in available energy states. Then, we mention the Bose-Einstein distribution function and apply it to the case of photons.

5.4.3 Bose-Einstein Statistics

(a) ***Number of accessible microstates:*** Consider a system of particles that are enclosed, and held at a temperature, T. Suppose we have a number of energy levels, ε_1, ε_2, ε_3,....ε_i labeled by i = 1,2,3,...... The first energy level having energy ε_1 contains a total of n_1 particles. Suppose each level is subdivided into g_1 distinct sublevels (eigen states), all of which have the same energy.

An energy level is said to be degenerate if it contains two or more different states. The number of different states for a particular energy level is called that level's degeneracy.

Any number of bosons can occupy any of the sublevels in g_1. That means, there is no restriction on the number of particles that can occupy an energy state. A microstate is a full specification of all degrees of freedom of a system. A system may be conveniently defined as having *N* degrees of freedom confined to a volume *V*. In general, microscopic degrees of freedom are *quantum numbers.*

Let us consider first, the occupation of sublevels .We imagine them to be like cells that can hold the particles as shown in Fig. 5.3. The sequence of the cells is 1,2,3,4,.......

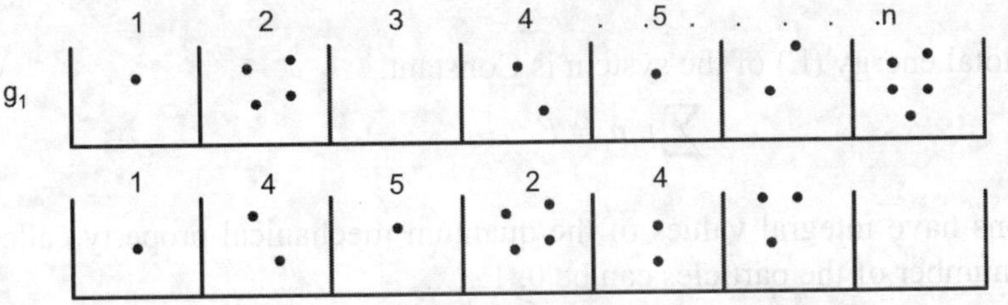

Fig. 5.3 The cells and the particles selected in sequence.

If we have to form a sequence of cells, the head of the sequence can be selected in g_1 ways. The remaining sections are $(g_1 - 1)$. These cells and the particles n_1 can be arranged in $(n_1 + g_1 - 1)!$ ways. Then, the total number of possible sequences of g_1 and n_1 particles is given as

$$g_1(n_1+g_1-1)! \qquad \text{...(5.16)}$$

The sequence of cells can be changed as shown in Fig. 5.3. But, the macrostate of the arrangement is not changed. Therefore, there are g_1 groups in the sequence, one for each physical state, the number of different sequences of g_1 groups is $g_1!$ Similarly, the permutations by which n_1 particles among themselves will not give new any new arrangement. Hence, the above expression should be divided by $n_1!$. The number of distributions of n_1 particles and g_1 states in the first group of eigen states is

$$< n_1 > = \frac{g_1(n_1+g_1-1)!}{g_1!n_1!} = \frac{(n_1+g_1-1)!}{(g_1-1)!n_1!} \qquad \text{.....(5.17)}$$

5.4.4 Bose-Einstein Distribution Function

Considering all the groups of energy states associated with energy levels ε_1, ε_2, ε_3,....ε_i, we get similar expression for each group.

The total number of ways in which n_1 particles are in g_1 states, n_2 particles are in g_2 states, n_3 particles in g_3 states ....n_i particles in g_i states as given by the expression

$$w_i = \frac{(n_1+g_1-1)!}{n_1!(g_1-1)!} \times \frac{(n_2+g_2-1)!}{n_2!(g_2-1)!} \times ...... \times \frac{(n_i+g_i-1)!}{n_i!(g_i-1)!}$$

$$= \prod_i \frac{(n_i+g_i-1)!}{n_i!(g_i-1)!} \quad .....(5.18)$$

The values of g and n in each group far exceed 1, hence 1 can be neglected in the above expression, the , the total number of microstates can be written as

$$<n_i> = \prod_i \frac{(n_i+g_i)!}{n_i!(g_i)!} \quad .....(5.19)$$

The probability of distribution is proportional to the total number of microstates.

$$P = \prod_i \frac{(n_i+g_i)!}{n_i!(g_i)!} \times C \quad .....(5.20)$$

where C is a constant. Following the same procedure as followed in MB distribution, we get the expression for Bose-Einstein distribution of the particles among various energy states of a system of indistinguishable particles. The distribution function is given as func

$$<n_i>_{BE} = \frac{g_i}{e^{(\alpha+\beta\varepsilon_i)}-1} \quad .....(5.21)$$

where $\alpha = -\frac{\mu}{kT}$, and $\beta = \frac{1}{kT}$ .....(5.22)

μ is called the chemical potential of the system.

The Bose-Einstein distribution describes the statistical behavior of integer spin particles (bosons). At low temperatures, bosons can assemble in unlimited number in the same energy state, a phenomenon called "condensation".

Many bosons can occupy a single quantum state. This allows them to behave collectively and is responsible for the behavior of lasers and super fluid helium.

photons, ^{4}He atoms, gluons

The derivation of the distribution function is given below for the benefit for better understanding for interested readers.

We can get the Bose-Einstein distribution by simply taking the logarithm on both sides and differentiating with respect to n_i.

Taking logarithm on both sides of the expression for the probability, we get

$$\ln P = \sum_i [\ln(n_i+g_i)! - \ln n_i! - \ln g_i! + \ln C \quad .....(5.23)$$

Using Stirling's approximation, we obtain

$$\ln P = \sum_i [(n_i + g_i)\ln(n_i + g_i) - n_i \ln n_i - g_i \ln g_i + \ln C$$

Differentiating the above expression with variable n and equating the result to zero for maximum probability ,we get

$$\partial \ln P = \sum_i [\ln(n_i + g_i) - \ln n_i]\partial n_i = 0 \qquad .....(5.24)$$

$$= \sum_i [\ln \frac{(n_i + g_i)}{n_i}]\partial n_i = 0$$

we have $\sum_i \partial n_i = 0 \qquad \sum_i \varepsilon_i \partial n_i = 0$

multiplying the first eqn by α and the second eqn. by β and adding those terms to eqn.(5.24) we obtain,

$$\sum_i [\ln \frac{(n_i + g_i)}{n_i} + \alpha + \beta\varepsilon_i]\partial n_i = 0 \qquad .....(5.25)$$

Simplifying, we obtain

$$1 + \frac{g_i}{n_i} = e^{(\alpha + \beta\varepsilon_i)} \qquad .....(5.26)$$

Then, we have

$$n_i = \frac{g_i}{e^{(\alpha + \beta\varepsilon_i)} - 1} \qquad .....(5.27)$$

The above expression actually represents the most probable distribution of microstates (or particles) in the various energy states (eigen values) of a system of particles that obey BE statistics.

5.5 Fermi-Dirac (FD) Statistics

5.5.1 The characteristic features or postulates Fermi-Dirac statistics:

1. The system under consideration consists of very large number of non-interacting and identical particles.
2. The particles are quantum particles and the energy of the particle is quantized. The particles possess discrete eigen (energy) values. The particles have intrinsic angular momenta in integral units of ***h/(2p)***,where p is the momentum of the particles, that is, $p = \frac{2\pi\hbar}{L} n$.

3. The particles have half-integral values of the quantum mechanical property called spin. (Spin numbers would be $\frac{1}{2}, \frac{3}{2}, \frac{5}{2}$ .....)
4. The particles, which come under FD statistics follow, Pauli- exclusion principle.
5. Each particle is associated with a wave function. The wave functions are anti symmetric. That means, if there are two particles and if they are interchanged, the sign of wave functions is altered. This is typical characteristic under Fermi-Dirac (FD) statistics. The particles which follow FD statistics are called 'Fermions'.
6. The state of the system is specified by specifying occupation numbers n_i such that n_i particles are in εi (energy level) that has the energy states, g_i.
7. They are "antisocial" in the sense that two fermions cannot exist in the same state i.e. each energy state may contain either 0(no particle) or one particle.
8. The summation of all energies of all particles in different energy states (eigen states) is the total energy of the system.

   This is expressed as

$$\sum \epsilon_i n_i = E \qquad \text{(A)}$$

9. The total number of particles in the system is constant. This is expressed as

$$\sum_i n_i = N \qquad \text{(B)}$$

10. Protons, neutrons, electrons, and many other elementary particles are fermions.

FD Statistics

Consider a system of particles having a total number of particles equal to N.

Let the energy states the $g_1, g_2, g_3, \ldots . g_i$., and let the particles be grouped as $n_1, n_2, n_3, \ldots \ldots n_i$..

Fermi-Dirac distribution considers the distribution of particle and the eigen states or energy states together.

The number of ways in which g_1 states can be arranged among themselves is $g_1!$ corresponding to the permutations of g_1 states.

Pauli-exclusion principle insists that no energy state or cell can take more than one particle. In the g_1 states of the first group, only n_1 states are occupied and $(g_1 - n_1)$ states are empty. The possible permutations of g_1 cells or energy states is equal to $g_1!$.

The permutations the empty $(g_1 - n_1)$ states is given as $(g_1 - n_1)!$. This distribution of the empty cells among themselves does not give any new distribution.

Similarly, the particles are indistinghishable and they occupy n_1 energy states or cells the permutation of particles also does give a new distribution. The number of permutations of n_1 occupied cells is equal to $n_1!$.

Therefore, the number of distinguishable arrangements of n_1 particles and g_1 energy states (cells) among themselves is

$$\frac{g_1!}{n_1!(g_1-n_1)!} \quad .....(5.28)$$

Similarly, n_2 particles are arranged in g_2 states, and n_3 particles are arranged in g_3 states, ... and n_i particles are arranged in g_i states.

The total number of ways in which all the permutations may take place or the total number of the microstates is given as

$$w_i = \prod_i \frac{g_i!}{n_i!(g_i-n_i)!} \quad .....(5.29)$$

The above expression represents the total number of microstates for the a system of particles that obey FD statistics.

All energy states are equally probable for the particles and the probability of distribution is given as

$$P_i = \prod_i \frac{g_i!}{n_i!(g_i-n_i)!} \times constant \quad .....(5.30)$$

Suppose we have to arrange fermions represented by a, a, a, among three energy states. This can be done in only one way as shown below. There is no other way of arranging two particles in one box as Pauli's exclusion principle does not permit it. Then, the Fermi –Dirac distribution for the given system is given as

a	a	a

We can compare this with M.B. and B.E distribution given in 5.5.1. The number of micro states is drastically decreased.

Fermi-Dirac distribution function: Can the basis of the above discussion, the Fermi-Dirac distribution function is given below.

Total number of microstates for the distribution discussed above

$$<n_i> = \frac{g_i}{e^{\alpha+\beta\varepsilon_i}+1} \quad .....(5.31)$$

and F.D. distribution function is

$$f_{FD} = \frac{1}{e^{\alpha+\beta\varepsilon_i}+1} \quad .....(5.32)$$

The following text gives the mathematical derivation of the F.D. distribution function. Those who are interested in the derivation may go through it.

The Fermi distribution law can be obtained in the following way.It needs to find the most probable distribution.

Taking logarithms on both sides of the above expression, we get

$$\ln P_i = \sum [\ln g_i! - \ln n_i! - \ln(g_i - n_i)!] + \ln(\text{constant})$$

The number of particles varies continuously, and differentiating the above eqn, and equating it to zero for maximum probability, we get

$$\sum_i \frac{n_i}{(g_i - n_i)} \partial n_i = 0 \qquad \text{.....(5.33)}$$

Take the equations (A) and (B), multiply the first eqn. by α and the second by β and partially differentiate with respect to n_i, then, we get

$$\alpha \sum \partial n_i = 0 \qquad \text{and} \qquad \beta \sum_i \varepsilon_i \partial n_i = 0$$

and

$$\sum_i [\ln (\frac{n_i}{(g_i - n_i)}) + \alpha + \beta \varepsilon_i] \partial n_i = 0 \qquad \text{.....(5.34)}$$

Simplifying, we obtain

$$\frac{n_i}{(g_i - n_i)} = \exp(\alpha + \beta \varepsilon_i) \qquad \text{.....(5.35)}$$

Further simplifying, we get

$$< n_i > = \frac{g_i}{e^{\alpha + \beta \varepsilon_i} + 1} \qquad \text{.....(5.36)}$$

The above expression gives the most probable distribution of the particles among various energy states in a system that follows the Fermi-Dirac statistics. The above expression is also called Fermi-Dirac distribution law.

In the expression given above, we have,

$$\alpha = -\frac{\mu}{kT}, \quad and \ \beta = \frac{1}{kT} \qquad \text{.....(5.37)}$$

μ is called the chemical potential of the particle, and T is the absolute temperature, and k is the Boltzmann constant.

The particles which follow FD statistics are called fermions. The assembly of the indistinguishable particles that follow Pauli exclusion principle is called Fermi-Dirac gas.. An assembly of free electrons such the one in metals is one such gas and that group of electrons is called 'electron gas.' This will be dealt separately.

5.6 Black Body Radiation Law

5.6.1 Early Models

When a body is heated to a high temperature, it emits radiation that is spectrally continuous. The energy of radiance is distributed over a wide wavelength region spreading from infrared into ultraviolet region. The energy (spectral) radiancy is defined as the radiant energy flux (power per unit area) emitted by the surface of a glowing body per unit wavelength interval. For every material, there exists a family of spectral radiancy curves like the one shown in Fig. 5.4. The radiation is spread over from short wavelength to long wavelength (infrared). There exists a peak. The radiation is spread over from short wavelength to long wavelength (infrared). There exists a peak at a particular wavelength, λ_m.

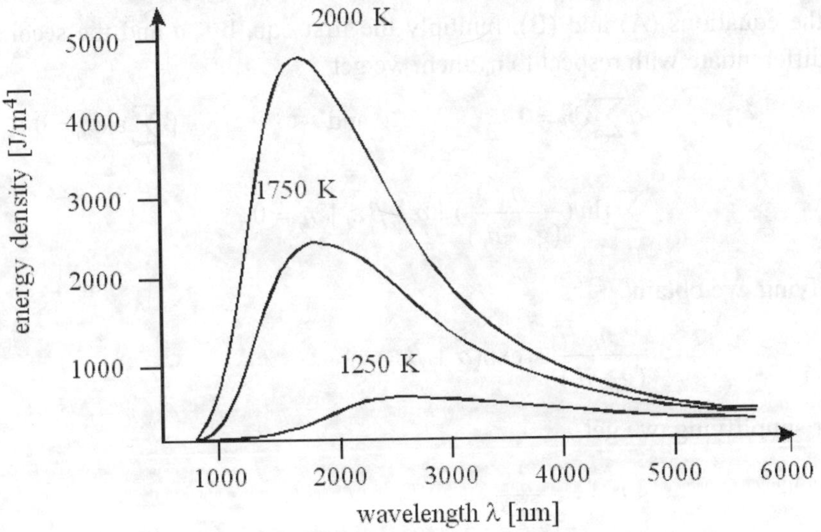

Fig. 5.4. Spectral emittance of a glowing body.

The radiation spectrum that is emitted from a body is a result of the processes like the vibrations of atoms that take place inside the body. The emittance is a result of attainment of thermal equilibrium with the atoms. This equilibrium processes, in which absorption and emission take place continuously, shapes the continuous spectrum of radiation. Only a fraction of radiation comes out through the surface of the body. The process of absorption and emission is a common feature of all bodies. A good absorber is a good emitter, and a poor absorber is a poor emitter.

A body with a perfectly absorbing surface is called a black body. It is a hypothetic body that does not reflect light and therefore, appears black, if the temperature is low enough so as not to be self luminous. All black bodies heated to any temperature emit thermal radiation.

The ideal characteristics of a black body are achieved in a cavity radiation. A cavity is a hallow cube with a tiny hole in one side of the cube as shown in Fig. 5.5 The walls of the cavity are made of suitable materials (tungsten, tantalum, molybdenum etc). When the cavity is heated, the hole acts like a black body and the radiation is emitted through the hole. In a black body, the spectral radiancy R_λ is a universal function of wavelength λ, and temperature T, and of nothing else.

The energy density emmited (represented by R_λ) depends only on the temperature as $R_\lambda = \sigma T^4$, σ is a constant called Stefan-Boltzmann constant. The value of R_λ is 5.67×10^{-8} watts/m^2 K^4. The above relation is called Stefan-Boltzmann law.

Fig. 5.5 A Black body.

5.6.2 Wien's Displacement law

Wien showed that the product of the wavelength of maximum emission (λ_{max}) and the absolute temperature (T) is constant and is equal to equal to 0.29 cm K. This observation of Wiens is called Wien's displacement law. Wien also showed on empirical basis that spectral radiation density varies with wavelength and temperature as given by the relation

$$R_\lambda \, d\lambda = \frac{c_1 \lambda^{-5}}{e^{\frac{c_2}{\lambda T}}} \, d\lambda \quad \text{.....(5.38a)}$$

where λ is the wavelength, T, the absolute temperature, c_1 and c_2 are two adjustable constants. Wiens formula was obeyed upto long wavelength range. The Wien's displacement is given as

$$\lambda_m T = \text{Constant} \quad \text{.....(5.38b)}$$

5.6.3 Rayleigh-Jeans Radiation Law

Rayleigh and Jeans considered the black body radiation in a cavity as consisting of a large number of electromagnetic standing waves. The black body cavity is filled with radiation field. Each stationary wave mode corresponds to two degrees of freedom. Each mode of vibration has an average thermal energy, kT. (k is the Boltmann's constant).

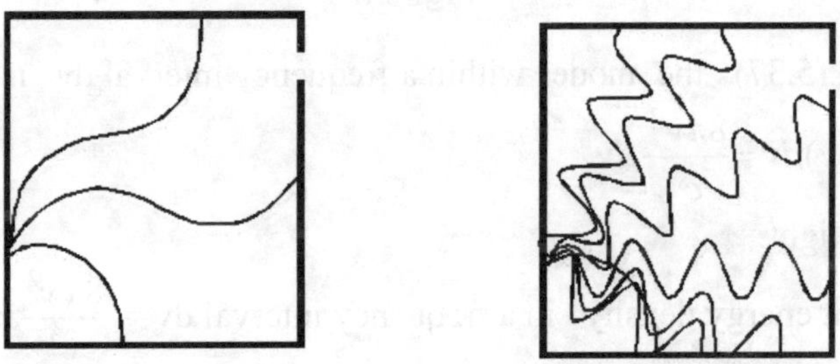

Fig. 5.6 The black body Cavity and standing waves.

Consider the cavity to be a box of edge, L. If a node has to occur at the walls, then, the cube edge should be integral multiple of half wave lengths. T he stationary waves follow the rule

$$n^2 = n_x^2 + n_y^2 + n_z^2 = (\frac{2L}{\lambda})^2 = \frac{2L}{c^2}\nu^2 \qquad .....(5.39)$$

where $n_{x,y,z} = 0,1,2,.........$

To count the number of standing waves, g (ν)d ν within the cavity whose wave frequencies lie between ν and ν + dν, we have to count the permissible sets of n_x , n_y ,n_z values for the frequencies. Think of a n-space in which whose coordinates are n_x, n_y, n_z. Then, the permissible values are g(n)dn in one quadrant of the sphere (Fig. 5.7) and given by the expression

$$g(n)dn = 2(\frac{1}{8})4\pi n^2\, dn = \pi n^2\, dn \qquad .....(5.40)$$

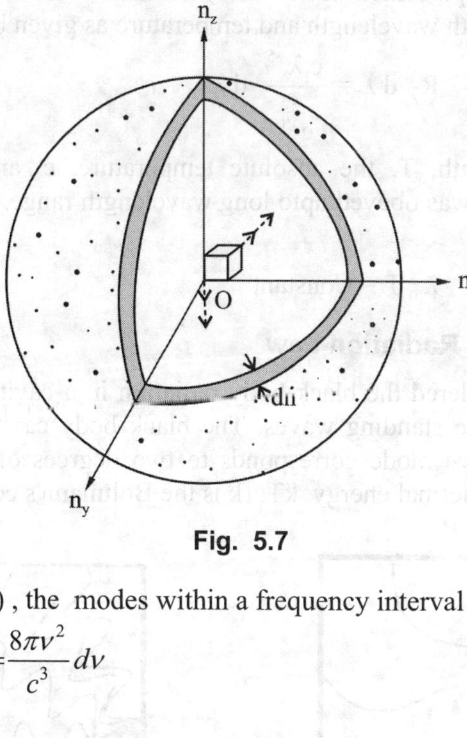

Fig. 5.7

Then, using eqn.(5.37) , the modes within a frequency interval dν is obtained as

$$g(\nu)d\nu = \frac{8\pi\nu^2}{c^3}d\nu \qquad .....(5.41)$$

c is the velocity of light.

Then, the radiant energy density in a frequency interval dν = $\frac{8\pi \nu^2}{c^3}$kT dν .....(5.42)

According to energy of equivalence, an energy (1/2)kT is associated.

In terms of wavelength, this expression is written as

$$R_\lambda\, d\lambda = 8\,\pi\, k\, T\, \frac{d\lambda}{\lambda^4} \qquad .....(5.43)$$

The above expression between R_λ and λ is depicted in the form of a plot as shown in Fig. 5.8.

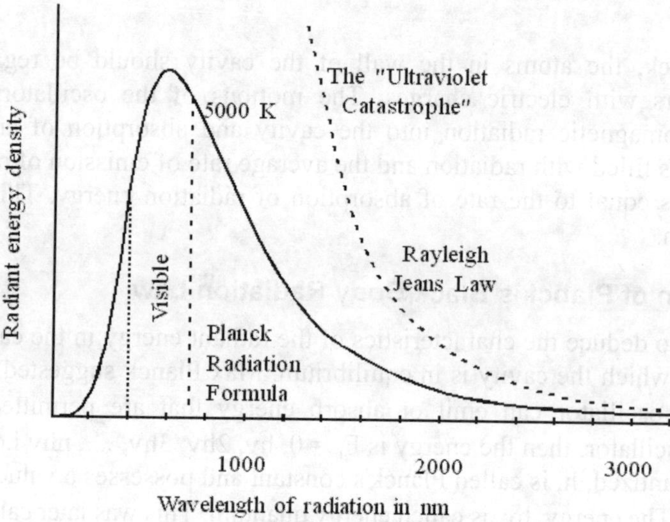

Fig. 5.8 General form of black body radiation spectrum.

5.6.4 Failure of Rayleigh-Jean's Law

The equation fits only at long wavelengths, at short wavelengths or high frequencies; it leads to infinite values of radiant energy. This deviation is called **ultraviolet catastrophe**. The Rayleigh's law was derived on the basis of classical physics. It does not possess any arbitrary constants. It fits exactly at long-wavelengths.

The failure of Rayleigh's law at short wavelengths presented a crisis on the very concepts on which it was based. The classical theory was unable to account for the observed radiation spectrum. This led Max Planck to look into the phenomenon afresh.

5.7 Planck's Black Body Radiation Law

5.7.1 Max Planck's Modification

Max Planck (1900) abandoned the wave concept of radiation. He suggested a revolutionary concept, namely, the quantization of energy. Planck first modified Wien's formula by putting a mere, – 1, into the denominator of the Wien's formula to get

$$R_\lambda \, d\lambda = \frac{c_1 \lambda^5}{e^{c_2/\lambda T} - 1} \, d\lambda \qquad \text{.....(5.34)}$$

and he adjusted c_1 and c_2 to suit Rayleigh's law at long wavelengths. He found that his formula fitted the experimental observations in every wavelength range.

Rayleigh's derivation never had a flaw in the derivation and Max Planck thought that the flaw must ultimately lie in the concepts involved in the classical physics itself. It should be remembered that Planck's hypothesis of quantization of radiation field gave birth to quantum theory in physics which became the spring board for many concepts and theories that followed in twentieth century in atomic physics, spectroscopy, nuclear physics, solid state physics etc.

According Planck, the atoms in the wall of the cavity should be regarded as small harmonic oscillators with electric charges. The motions of the oscillators result in the emission of electromagnetic radiation into the cavity and absorption of energy from the cavity. The cavity is filled with radiation and the average rate of emission of radiation energy by the oscillators is equal to the rate of absorption of radiation energy. This is the sate of thermal equilibrium.

5.7.2 Derivation of Planck's Black Body Radiation Law

It is now possible to deduce the characteristics of the radiant energy in the cavity from those of oscillators with which the cavity is in equilibrium. Max Planck suggested two postulates. They are: 1. The oscillator can emit or absorb energy that are permitted. If, ν, is the frequency of the oscillator, then the energy is $E_n = 0, h\nu, 2h\nu, 3h\nu, ... nh\nu$ i.e., the energy of the oscillator is quantized, h, is called Planck's constant and possesses a value, and n is called 'quantum number'. The energy, hν, is called energy quantum. This was later called as 'photon'.

The energy values of the oscillator can be written as En = nhν ; n, is an integer. The oscillators do not, thus, radiate energy continuously but only in quanta. This quantum of energy is emitted from the oscillator when it goes from higher energy state to lower energy state. With this quantization condition, Planck calculated the average energy of the oscillators and from that he derived his radiation law.

Applying MB distribution function, the average value of energy per oscillator is given as

$$\overline{\in} = \frac{\sum_{n=0}^{\infty} n h \nu N \exp\left(-\frac{n h \nu}{kT}\right)}{\sum_{n=0}^{\infty} N \exp\left(-\frac{n h \nu}{kT}\right)} \quad .....(5.\,45)$$

Since n is an integer, the above equation can be written as

$$\overline{\in} = \frac{0 + h\nu \exp\left(-\frac{h\nu}{kT}\right) + 2h\nu \exp\left(-\frac{2h\nu}{kT}\right) + 3h\nu \exp\left(-\frac{3h\nu}{kT}\right) + ...}{1 + \exp\left(-\frac{h\nu}{kT}\right) + \exp\left(-\frac{2h\nu}{kT}\right) + \exp\left(-\frac{3h\nu}{kT}\right) + ...}$$

Suppose, $x = \exp\left(-\frac{h\nu}{kT}\right)$

then, $\overline{\in} = h\nu x \frac{1 + 2x + 3x^2 + ...}{1 + x + x^2 + ...}$

Both the numerator and denominator are the convergent series (x < 1).

$$\overline{\in} = h\nu x \frac{\frac{1}{(1-x)^2}}{\frac{1}{(1-x)}} = \frac{h\nu x}{1-x} = \frac{h\nu}{\frac{1}{x}-1}$$

$$\therefore \quad \overline{\in} = \frac{h\nu}{\exp\left(\frac{h\nu}{kT}\right)-1} \qquad \text{.....(5.46)}$$

The number of modes of vibration, in a unit volume of cavity space, is equal to $\frac{8\pi\nu^2}{c^3}\,d\nu$ (from eqn. 5.41).

Thus, the radiant energy density in a frequency range dν is given as

$$R_\nu\, d\nu = \frac{8\,\pi\,\nu^2}{c^3}\;\frac{h\,\nu}{\exp\left(\frac{h\nu}{kT}\right)-1}\,d\nu$$

$$= \frac{8\pi\,\nu^3 h}{c^3}\;\frac{1}{\exp\left(\frac{h\nu}{kT}\right)-1}\,d\nu \qquad \text{.....(5.47)}$$

or in terms of wavelength λ, the radiant energy density in wavelength range dλ is given as

$$R_\lambda\, d\lambda = \frac{8\pi h c}{\lambda^5}\;\frac{1}{\exp(hc/\lambda KT)-1}\,d\lambda \qquad \text{.....(5.48)}$$

The above relation is Planck's law for energy density of black body or cavity radiation. This law also contains Wien's displacement law as well as Stefan-Boltzmann Law. By integrating Plank's law, we obtain the total flux or power per unit area, emitted by the black body at all wavelengths

$$R = \int_0^{\infty} R_\lambda\, d\lambda = \int_0^{\infty} \frac{8\,\pi\,hc}{\lambda}\cdot\frac{1}{\exp(hc/\lambda kT)-1}\,d\lambda = \sigma T^4$$

where $\sigma = \frac{2\pi^5 k^4}{15c^2h^3} = 5.67 \times 10^{-8}\ \text{W/m}^2\,\text{K}^4$

5.7.3 Wien's Displacement Law

The Wiens displacement law is obtained by finding the wavelength for maximum total energy radiated per unit time per unit area at a fixed temperature, i.e.,

$$\frac{d}{d\lambda}\left[\frac{8\pi c h \lambda^{-5}}{\exp(ch/\lambda kT)-1}\right] = 0$$

then, we get $\lambda_m T = \dfrac{ch}{4.97k}$ .....(5.49a)

eqn (5.49a) represents Wien's displacement law as given in eqn. (5.38b).

At longer wavelength, $\dfrac{hc}{\lambda} >> kT$, i.e., $e^{\frac{ch}{\lambda kT}} >> 1$

Then, $R_\lambda d\lambda = 8\pi h c \lambda^{-5} \exp(-ch/\lambda kT)d\lambda$– Wien's law

This is shown in Fig. 5.9 in which the validity of black body radiation is given. Fig. 5.10 represents the validity of the radiation law where radiated energy density is polled against frequency of radiation.

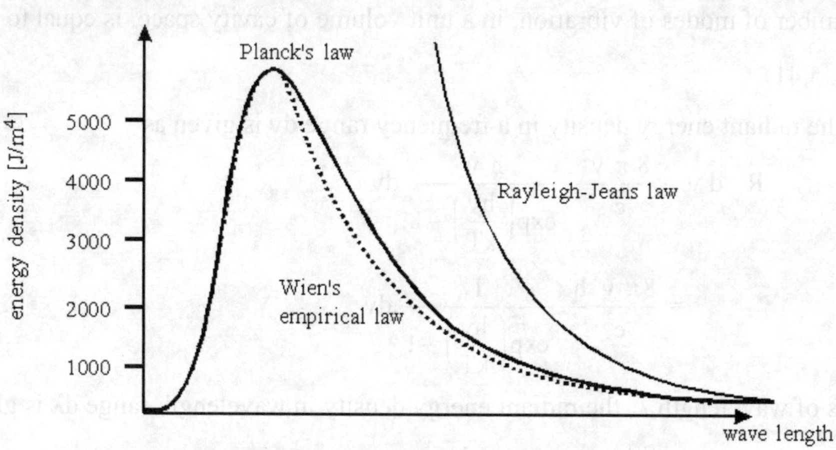

Fig. 5.9 The validity of the three laws of black body radiation.

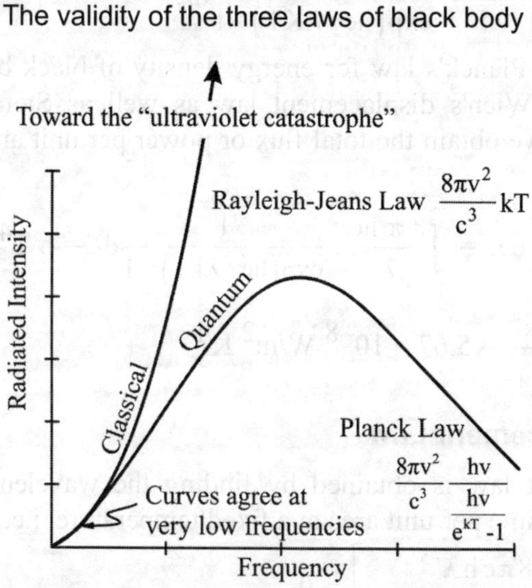

Fig. 5.10 Black body radiation with X-axis as frequency axis..

5.8 Planck's Black Body Radiation Law on the Basis of BE Statistics

5.8.1A Characteristics of Photon gas:

1. *Photons are zero-mass* bosons.
2. Photons are the force carriers of the electromagnetic field.
3. Photons enclosed in a container of volume *V*, maintained at temperature *T*, can readily be absorbed or emitted by the walls of the container. Photons obey a simplified form of Bose-Einstein statistics in which there is an unspecified total number of particles. This type of statistics is called *photon statistics*.
4. In a photon gas, there will also be an equilibrium distribution, but photons do not collide with each other (except under very extreme conditions) so that the equilibrium distribution must be established by other means.

The most common way that an equilibrium distribution is established is by the interaction of the photons with matter. If the photons are absorbed and emitted by the walls of the enclosure containing the photon gas, and the walls are at a particular temperature, then the equilibrium distribution for the photons will be a black body distribution at that temperature.

There are two features that are specially associated with photon gas.

Two aspects have to be considered for photon gas:

(a) A very important difference between a gas of massive particles and *a photon gas with a black body energy distribution is that the number of photons in the system is not conserved.*

A photon may collide with an electron in the wall, exciting it to a higher energy state, removing a photon from the photon gas. This electron may drop back to its lower level in a series of steps, each one of which releases an individual photon back into the photon gas.

There fore, the condition given in (A) earlier is not valid for Photons. *It means* $\sum \partial n_i \neq 0$ *for photons.*

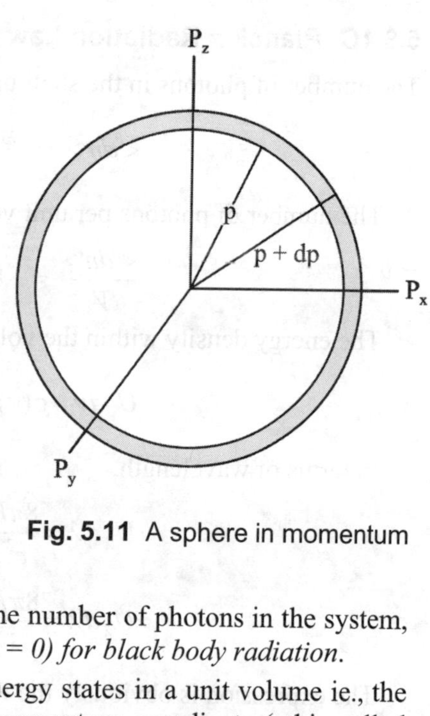

Fig. 5.11 A sphere in momentum

(b) Although the sum of the energies of the emitted photons is the same as the absorbed photon, the number of emitted photons will vary. It can be shown that, as a result of this lack of restriction on the number of photons in the system, *the chemical potential of the photons must be zero (α = 0) for black body radiation.*

What is needed now is to calculate the number of energy states in a unit volume ie., the values of g. Consider a sphere that is hollow in momentum coordinates(this called momentum space) whose total volume is V. the photons can be found anywhere in this space.

In the sphere under view, consider a space enclosed between p and p+dp momenta values. In this volume there will a finite number of energy states. The volume of this shell is given as $4\pi p^2 dp$.

According to Heisenberg's uncertainty principle, the minimum volume in momentum space is given as h^3/V, where 'h' is the Planck's constant.

5.8.1B Number of Energy States

The number of energy states in the momentum values between p and p +dp is given by g(p)dp.

$$\frac{g(p)dp}{h^3/V} = 4\pi p^2 dp \qquad \text{.....(5.50)}$$

There are two modes of propagation for photons. The above expression, therefore, should be multiplied by 2.

We know that, $p = {hv}/{c}$ or $dp = {hdv}/{c}$

Therefore, $$g(v)dv = 8\pi V \frac{v^2}{c^3} dv \qquad \text{.....(5.51)}$$

5.8.1C Planck's Radiation Law

The number of photons in the shell under consideration between p and p+dp is

$$< dn > = \frac{g(v)dv}{e^{\alpha+\beta\varepsilon} - 1} = 8\pi V \frac{v^2}{c^3} \frac{dv}{e^{hv/kT} - 1} \qquad \text{.....(5.52a)}$$

The number of photons per unit volume is given by the expression

$$\frac{< dn >}{V} = 8\pi \frac{v^2}{c^3} \frac{dv}{e^{hv/kT} - 1} \qquad \text{.....(5.52b)}$$

The energy density within the volume between v and v +dv is given as

$$U_v dv = g(v)dv \times \frac{< dn >}{V} = 8\pi \frac{hv^3}{c^3} \frac{dv}{e^{hv/kT} - 1} \qquad \text{.....(5.53)}$$

In terms of wavelength,

$$U_\lambda d\lambda = \frac{8\pi h\, c^3}{c^3\ \lambda^3} \left(-\frac{c}{\lambda^2}\right) \frac{d\lambda}{e^{hc/\lambda cT} - 1}$$

$$U_\lambda d\lambda = \frac{8\pi hc}{\lambda^5} \frac{d\lambda}{e^{hc/\lambda kT} - 1} \qquad \text{.....(5.54)}$$

The expression is obviously the Planck's radiation law of a black body.

The Planck radiation formula is an example of the distribution of energy according to Bose-Einstein statistics and the law is derived on the basis of photon gas.

5.9 Electron Theory

5.9.1A Metals

Metals occupy a rather special position among solids, because of their variety of striking properties like good conduction of electricity and heat, malleability , ductility, and a striking luster on a freshly cut surface.

Many of the elements in the periodic table belong to the category of metals, as they favour this physical state. Even to understand non-metallic behaviour, one must first understand metallic behaviour for explaining why copper conducts so well, one begins to learn why common salt does not. As any other solid, the coherence of metal atoms takes place because the total energy becomes lowered in the solid form. The electron's energy potential energy is lowered in a solid than it would be if it belonged to an isolated atom. The decrease of potential energy possibly results in increase of kinetic energy. The energy levels become closely spaced and there would as many energy states as the total number of electrons present.

During twentieth century, physicists have tried to explain through simple theories the metallic behaviour both qualitatively and quantitatively. During this process, success and failure were tasted time and again.

5.9.2B The Free Electron

According to Drude (1900), the following features are associated with electrons in metals.

1. The valence electrons of the constituent atoms of a metal move about freely through the volume of a metallic specimen. The electrons are considered as particles which move in a container, the motion being random in all directions. They possess momentum. Without the external electric field, there are no internal forces on them. The valence electrons are considered 'free' like the molecules of a gas, with a small difference that these electrons move between the ion cores entirely free in the space bounded by the surface of the specimen. These ion cores and electronic gas (cloud) are held together due to the electrostatic interaction between the positively charged ion cores and the negatively charged electronic cloud.
2. The electrons after collisions with the ion cores, gets scattered in random directions as shown in Fig. 5.12. The ion cores are heavier and immobile. When no electric field acts on a metallic sample, the electrons are confined inside the metal only and the net movement of the electrons is zero.

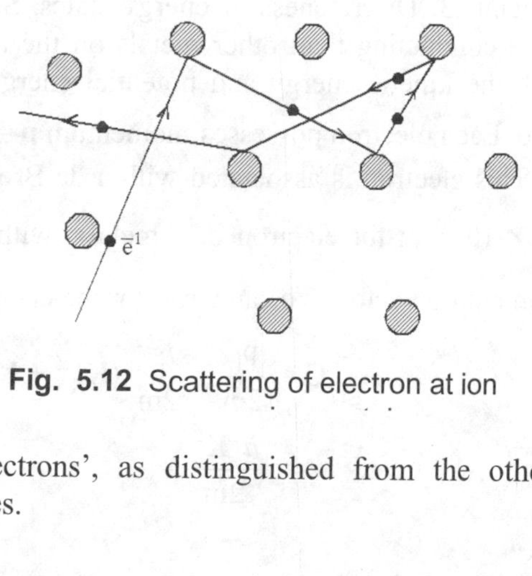

Fig. 5.12 Scattering of electron at ion

3. The valence electrons are responsible for the conduction of electricity by metals, and for this reason these free electrons are termed as 'conduction electrons', as distinguished from the other electrons of the filled shells of the ion cores.

4. The conduction (Valence) electron in the metal finds itself in a uniform field of positive ion cores and that of the other electrons in a given volume. As the repulsive force between the electrons is neglected, it is assumed that the electrons sees uniform potential even inside the metallic solid, i.e. the conduction electrons move freely every where within the specimen without any change in their energy.

   In addition to the above features, the free electrons follow the following characteristics:

5. We may consider that the potential energy of an electron at rest inside the metal should be lower than that of an electron at rest outside the metal. In such a case, we may assume arbitrarily, that the potential outside the metals is very large or infinite. It is this high potential, which restricts the electrons from leaving the surface at lower temperatures. This can be viewed as though, there is a sharp potential barrier of height φ_s (E_s is the height of the barrier), at the surface as shown in Fig. 5.13. The electrons are bound to the metal by the potential barrier. E_1 represents top most energy level of free electron inside the metal. This potential at the surface seen by an electron is generally known as "work function" of metal.

   All the free electrons are assumed to move in potential well as described in Fig. 5.13.

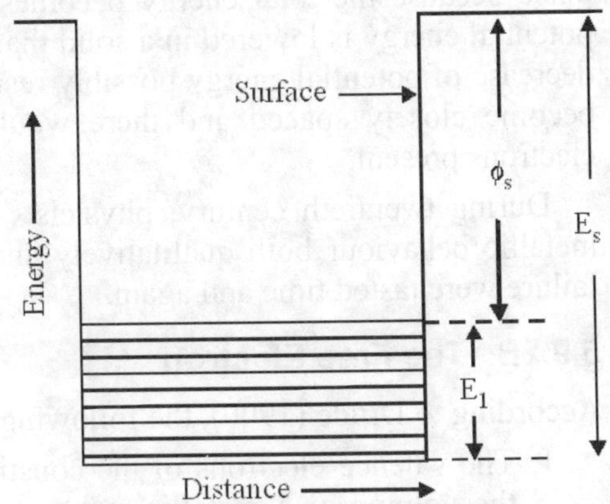

Fig. 5.13 Potential barrier for electrons in a well.

5.9.2 Electrons are Quantum Particles

Sommerfeld treated electron as a quantum particles. The electrons follow the principles of quantum theory, namely 1. Pauli's exclusion principle, 2. Heisenberg's uncertainty principle,3. Discreteness of energy states. Sommerfeld could explain why some metals are more conducting than other metals on the basis of quantum theory. The electrons possess both the kinetic energy and potential energy. The potential energy is arbitrarily taken as zero. Each electron possesses momentum $\mathbf{p} = mv$ and has an energy, $\varepsilon = p^2/2m$.

This electron is associated with a de Broglie wave (the de Broglie wave length is about 15×10^{-10} m for electron in a metal) with a wave vector, $\mathbf{K} = \frac{2\pi}{\lambda} = \frac{\mathbf{p}}{\hbar}$, where **p** is the momentum vector. We can write for the energy of the electron as

$$\epsilon = \frac{|p|^2}{2m} = \frac{\hbar^2}{2m}(\mathbf{K_x}^2 + \mathbf{K_y}^2 + \mathbf{K_z}^2)$$

$$= \frac{\hbar^2 \mathbf{K}^2}{2m} \qquad \text{.....(5.55)}$$

The wave function associated with the electron should satisfy the time independent Schrödinger equation. The eigen values associated for the electrons in potential well are given as

$$\epsilon_n = \frac{n^2\pi^2\hbar^2}{2mL^2}$$

where n = 0, 1, 2, 3…., L is the width of the potential well.

The energy is dependent on wave vector **K**. The electron energy is plotted against the wave vector associated with the electrons and we get a plot as shown in Fig. 5.14.

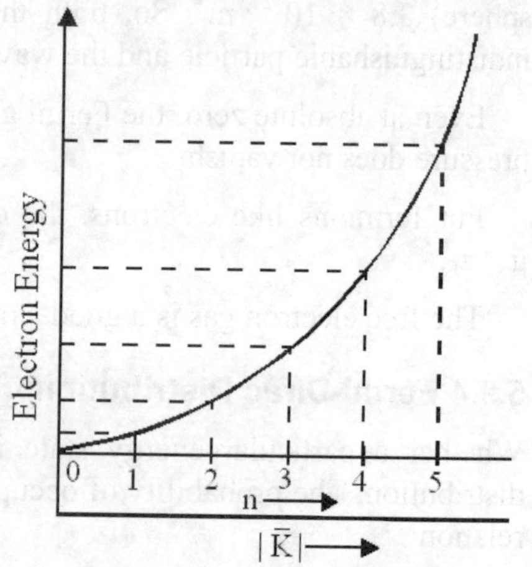

Fig. 5.14 Variation of energy of electrons with quantum number 'n' and wave vector K.

5.9.3 Electrons as Fermions

Electrons in a metal are free electrons. They move freely inside the metal. They are considered just like the particles in an ideal gas. The electrons in a metal form 'electron gas'. The electrons are quantum particles and the de broglie wave length of the electrons is much greater than the average interatomic separation. The quantum effects become very important in solids. The energy states are spaced closely in the case of the electron gas.

All the energy states are equally accessible to the electrons. But, Pauli's exclusion principle has to be followed. That is, no two electrons can have the same spin quantum number. If one electron has 1/2 spin and the second electron will have −1/2. In the ground state of the metallic system, the particles occupy the lowest possible energy levels up to a finite energy level, called 'Fermi energy. In momentum space, the electrons are filled a sphere of radius, p_F, the surface of which is called 'Fermi surface'. The electrons are associated with wave functions that are antisymmetric.

Suppose we associate a wave function $\Psi_s(1)$ with one electron and $\Psi_s(2)$ with another in an energy level. Then, a wave function is anti symmetric if we interchange any pair of particles. This is expressed as

$$P\psi_s(1,2) = -\psi_s(2,1)$$

P is called the exchange operator. The type of particles that are follow anti symmetric rule are known as fermions and obey FD statistics. The electrons come under that class. The wave functions overlap. Why the fermions are considered indistinguishable can be known from the volume of the electron and the free volume that is available. Consider that each atom contributes one free electron to conductivity in a metal. For a normal metal, the density of electrons is of the order of $10^{28}/m^3$ .Then the volume available for the electron is 10^{-28} m^3. The momentum of the electron with 1.6ev energy is 6.8×10^{-25}. Therefore, the uncertainty in its position is 0.97×10^{-9} m. Then, the possible volume associated is (treating electron as a sphere) 3.8×10^{-27} m^3. So, both the volumes are comparable. The electron becomes an indistinguishable particle and the wave functions overlap.

Even at absolute zero, the Fermi gas is contained with externally fixed barrier because the pressure does not vanish.

For fermions like electrons, the chemical potential μ is equal to the Fermi energy i.e., $\mu = \varepsilon_F$.

The free electron gas is a good example of a Fermi gas.

5.9.4 Fermi-Dirac Distribution in the Case of Electron Gas

Whether a particular energy state is occupied or not, can be known from Fermi-Dirac distribution. The probability of occupation of a given quantum energy state E is given by the relation

$$F(\varepsilon) = \frac{1}{\exp[(\varepsilon - \varepsilon_F)/kT] + 1} \quad \text{.....(5.57)}$$

where ε_f is the Fermi energy. F(ε) is called Fermi function. ε is the energy of the energy state whose occupancy is considered.

Fig. 5.15 shows the variation of F(ε) with energy 'ε' at different temperatures. At T = 0K, we observe the probability of occupancy of energy levels below E_f is maximum, that is, equal to unity. That means, all the energy states below Fermi level are occupied at absolute zero. Fermi energy, F_f represents the maximum energy that an electron can have, below which all the energy states are occupied. In the above function eq. (5.28) if $\varepsilon = \varepsilon_F$, we get F(ε) = 1. Thus the Fermi level is the topmost level that can be occupied at 0K.

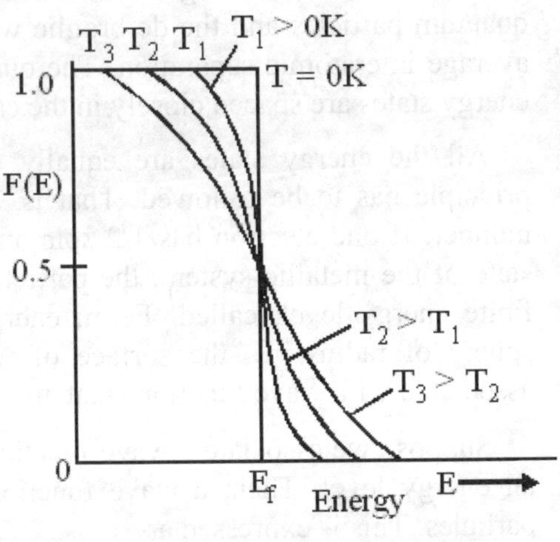

Fig. 5.15 Probability of occupation of different energy states with the increase of temperature.

When kT << ε_F, i.e., the thermal energy is smaller than the value of ε_F, (the thermal energy at room temperature is about 0.025 ev), the value of F(ε) becomes 1/2. We can define the Fermi energy $\varepsilon = \varepsilon_F$ as that energy which has 50% probability of occupation by electrons. At 0K, F(ε) = 1, at , F(ε) sharply falls to zero with increasing upto $\varepsilon = \varepsilon_F$; as shown in the Fig. 5.9.

At temperatures above 0K under the condition, kT << ε_F, some electrons are excited to higher levels. The F(ε) vs ε curve crosses beyond E_F. As temperature is increased, more and more electrons go into empty states beyond ε_F.

5.9.5 Density of States

The density of states (DOS) of a system describes the number of states at each energy level that are available to be occupied. A high density of states at a specific energy level means that there are many states available for occupation. If the density of states is zero, no states can be occupied at that energy level. The number of electrons at each energy level is then obtained by multiplying the number of states with the probability that a state is occupied by an electron. Since the number of energy levels is very large and dependent on the size of the semiconductor, we will calculate the number of states per unit energy and per unit volume.

Density of states in an electron gas: we have to find the number of available states at within interval k and k+dk. Consider a sphere of radius k in k-space. The number of points in this space gives the eigen state as each eigen state is represented by a point in k-space (shown in Fig. 5.16.

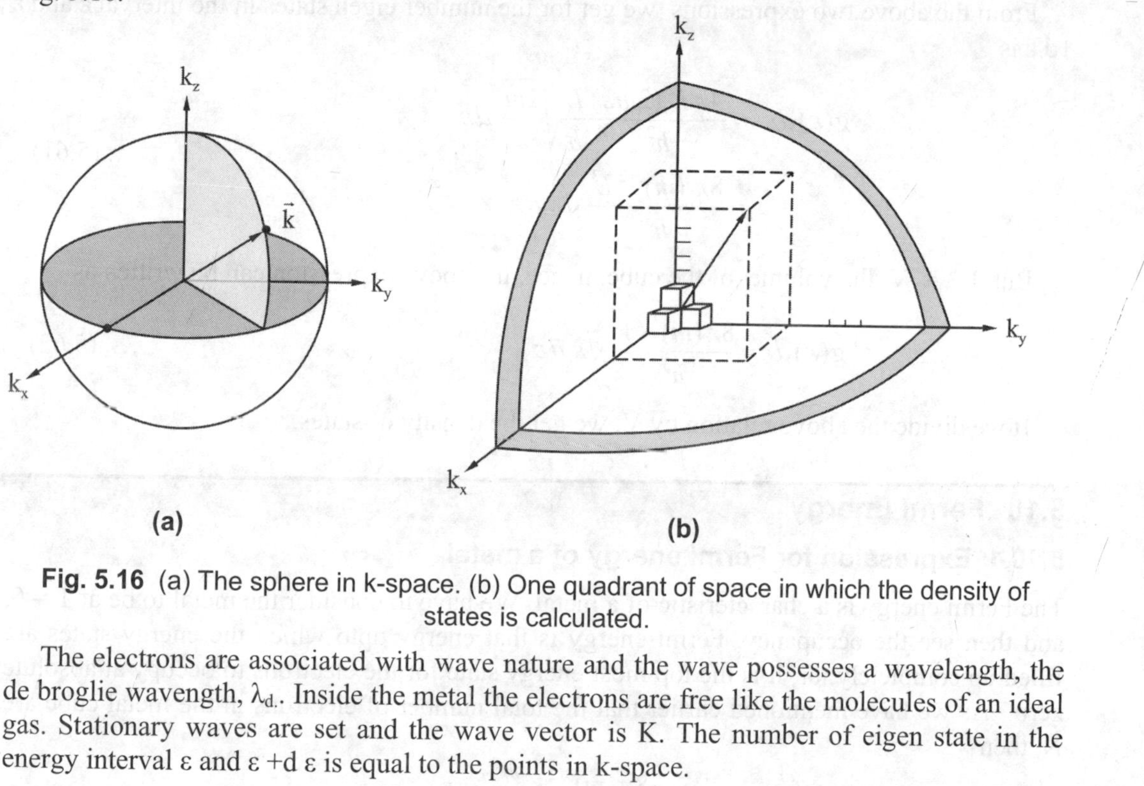

Fig. 5.16 (a) The sphere in k-space. (b) One quadrant of space in which the density of states is calculated.

The electrons are associated with wave nature and the wave possesses a wavelength, the de broglie wavength, $\lambda_{d.}$. Inside the metal the electrons are free like the molecules of an ideal gas. Stationary waves are set and the wave vector is K. The number of eigen state in the energy interval ε and ε +d ε is equal to the points in k-space.

Consider that the is cube with side, L. the waves associated will set up standing waves at the boundaries of the cube. The wave vector is given as in the condition given in eqn.(5.29).K is already defined. the total number of standing waves will be same as the number of points in k-space whose distances from the origin lie between K and K+dK. Consider a spherical shell of radius K and thickness dK in the sphere shown in Fig. 5.10(a). The volume of the spherical shell of radius dK is $4\pi K^2 dK$.We look into only the octant of the shell as shown in Fig. 5.16(b) that includes non negative values of K. For each standing wave, there are two independent directions of polarization for other wise identical waves. This is because, the electron has two spin states, ± ½. Therefore, the total number of independent stationary waves is given as

$$g(K)\,dK = 2(\frac{1}{8})(4\pi k^2)\,dK = \pi K^2\,dK \qquad .....(5.58)$$

K satisfies the following condition. Λ is the wave length of the wave (de broglie wavelength)

$$K = (\frac{2L}{\lambda}) = (\frac{2L}{h}p) = (\frac{2L}{h})(\sqrt{2m\varepsilon}) \qquad .....(5.59)$$

$$\therefore \quad dK = (\frac{L}{h})\sqrt{\frac{2m}{\varepsilon}}\,d\varepsilon \qquad .....(5.60)$$

From the above two expressions, we get for the number eigen states in the interval ε and ε +d ε is

$$\therefore \quad g(\varepsilon)\,d\varepsilon = \frac{4\pi L^2 (2m\varepsilon)}{h^2}\frac{L}{h}\sqrt{\frac{2m}{\varepsilon}}\,d\varepsilon$$

$$= \frac{8\pi\,(m)^{3/2}\,L^3}{h^3}\sqrt{\varepsilon}\,d\varepsilon \qquad .....(5.61)$$

But, L^3 = V, the volume of the cube. hence, the above expression can be written as

$$g(\varepsilon)\,d\varepsilon = \frac{8\pi (m)^{3/2} V}{h^3}\sqrt{\varepsilon}\,d\varepsilon \qquad .....(5.62)$$

If we divide the above relation by V, we get the density of states.

5.10 Fermi Energy

5.10.1 Expression for Fermi energy of a metal

The Fermi energy is a characteristic of a metal. We have to consider the metal to be at T = 0, and then see the occupancy. Fermi energy is that energy upto which the energy states are filled up completely, or, it is the top most energy state for the electrons to occupy at absolute zero. As we have mentioned earlier that the total number of electrons in the metal cube are N, then

$$N = \int_0^{\varepsilon_F} g(\varepsilon)\,d\varepsilon = \frac{8\pi (m)^{3/2} V}{h^3}\int_0^{\varepsilon_F}\sqrt{\varepsilon}\,d\varepsilon$$

$$= \frac{16\sqrt{2}\,\pi V m^{3/2}}{3h^3}\varepsilon_F^{3/2} \qquad .....(5.63)$$

$$\therefore \; \varepsilon_F = \frac{h^2}{2m}(\frac{3N}{8\pi V})^{2/3} \qquad .....(5.64)$$

(N/V) represents the density of the metal.

5.10.2 Electron – Energy distribution:

We can obtain the electron distribution in the eigen states in the energy interval dε ,by multiplying the number of states in that energy interval ,with the distribution function and we obtain the expression given below.

$$n(\varepsilon)d\varepsilon = g(\varepsilon)d\varepsilon \times \frac{1}{e^{(\varepsilon-\varepsilon_f)}+1} = \frac{(3N/2)\varepsilon_F^{-3/2}\varepsilon d\varepsilon}{e^{(\varepsilon-\varepsilon_f)}+1} \quad .....(5.65)$$

5.10.3 Occupation of Energy Levels

According to Sommerfeld Theory: The energy levels in the potential well are shown in Fig. 5.17. They are shown as discrete energy levels. Suppose we drop electrons into the box, the energy levels in the well will be occupied by electrons. The first electron goes to occupy the lowest energy state (E_0) available. The next one would occupy the same energy state. But the third electron has a tendency goes to occupy the second level (E_1). Subsequent electrons occupy the higher states as shown. At 0K, all the electrons occupy energy states below a level E_{FO}. Above 0K, the electrons can occupy energy states above E_{FO}. Some of the lower energy states become empty. It is difficult to say which energy state is empty and which energy state is occupied. Many electrons tend to go into empty states well above 0K.

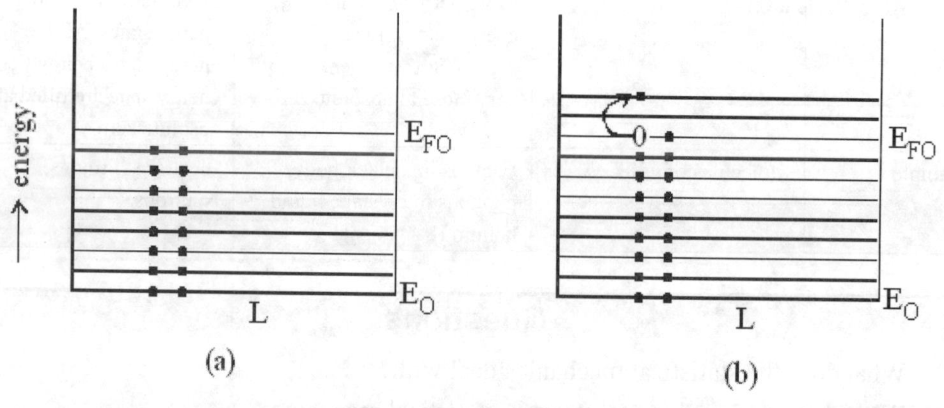

Fig. 5.17 Occupation energy levels (a) at 0K (b) Above 0K

The above distribution of electrons in energy levels is in contradiction with the classical theory. According to classical theory, any number of electrons can occupy the same energy state without any restriction. Application of an electric field greater than the height of the potential box, will act as driving force for some electrons to occupy higher energy states and then move freely. Many electrons can possess the same energy.

5.11 Summary of the Three Statistical Functions – A Comparison

	Maxwell-Boltzmann	Bose-Einstein	Fermi-Dirac
Particles	Identical, distinguishable	Identical, indistinguishable	Identical, indistinguishable
Type	classical	Quantum	Quantum
Name of particles	Classical	Bosons	Fermions
Particle nature, Wave functions	Any spin is possible. on overlapping wave functions.	Spin 0.1.2……	Spin 1/2,3/2,5/2……Overlapping wave functions, anti symmetric
Density	low	dense	Highly dense
Distribution functions	$A e^{-\varepsilon/kT}$	$\frac{1}{e^{\alpha} e^{\varepsilon/kT} - 1}$, α = 0 for photons.,	$\frac{1}{e^{\alpha} e^{\varepsilon/kT} + 1}$.,., $\alpha = -\varepsilon_F/kT$ ε_F = Fermi energy
Particle distribution	No limit to the number of particle in any energy level	No limit to number of particles in any energy state. More particle s in lower energy states. Ideal gas suffers Bose condensation very close to absolute zero.	More than one particle per energy state is not allowed. Fewer particles at low energy states. All the energy states below Fermi energy state are filled at absolute zero.
Example	Molecules of gases	Photons in a black body, phonons in crystals, liquid helium II,	Electrons. Protons, Neutrons.

Questions

1. What does the statistical mechanics deal with?
2. What do you mean by a 'system' in statistical mechanics?
3. Explain 'microstates' and a macrostate?
4. The basic postulates of Maxwell-Boltzmann statistics.
5. Obtain the expression for the probability of arranging n_1, distinguishable particles in $\in_1$ energy levels, n_2 particles in $\in_2$ energy level …..n_i particles in $\in_i$ energy level on the basis of MB statistics.
6. Write the expression for Bose-Einstein distribution function? Under what condition does it approach MB distribution?
7. What are the basic postulates of Bose-Einstein statistics?

8. What are the characteristics of 'Bosons'.
9. What is meant 'degeneracy' of an energy level?
10. How does BE statistics differ from MB statistics?
11. What are the basic postulates of Fermi-Dirac static's?
12. How does the spin equal to half in + give the lead to FD statistics?
13. Write the distribution function of FD statistics? Explain its behaviour?
14. What are the differences between FD postulates and DE postulates?
15. What are fermion's?
16. Mention three features that make the fermions differ from Bosons.
17. What is photon gas? What are its characteristics?
18. There are two particles that are in distinguishable. Suppose that they have to be distributed in 3 possible quantum states. Show in how many ways you can distribute them according to (a) Bose-Einstein statistics (b) Fermi-diriac statistics.
19. Discuss about the wave function of two particles that are interchanged in the lens of (a) Bose-Einstein statistics (b) FB-statistics.
20. Describe Wien's displacement law.
21. What is a black body? Give a figure of radiance spectrum of a black body?
22. Explain Rayleigh-Jean's law of radiation. Derive the expression for radian energy of a blackbody on the basis of Rayleigh-Jeans.
23. Discuss the executed feature of Plank's theory of slack body radiation.
24. What was the failure of Rayleigh-Jean's theory of black-body radiation law.
25. Explain how photon moral can lead to Plank's black body radiation law.
26. Reduce the Plank's law to Rayleigh-Jean's law.
27. Under what condition does Plank's law of radiation lead to Wein's displacement law.
28. Write short note on (a) Fermi energy of metal (b) Wien's displacement law.
29. What are the number of energy states that would be a row table under Bose-Einstein reliable statistics.
30. Obtain the expression for the distribution of particles under Bose-Einstein statistics.
31. Obtain the expression for the distribution of distinguishable particle under statistics.
32. What is an electron gas.
33. What are the characteristics of electrons that make them fermions?
34. Under what condition does the FD distribution law reduce to MB distribution law.
35. Obtain the expression for the number of energy states between the energies $\in$ and $\in$ $+d \in$ under FD distribution.
36. Obtain the expression for Fermi energy of electron. Define Fermi energy of electron gas?

37. What are the two features that are associated with photon gas ?
38. What is free electron gas model of metals. Discuss salient features of Sommerfeld's modified free electron gas model and its success in explaining some physical properties of metals.
39. (i) What is Fermi energy? Discuss the distribution of electrons in different energy levels based on Fermi-Dirac statistics.
    (ii) How does the distribution vary with temperature ?
40. Discuss the variation of electron energy with quantum number and wave vector.
41. Define the term 'density of states.
42. Does Fermi energy of a metal change with temperature?
43. Write down the 'Schrodinger equation' for a free electron in a potential well. Give the energy spectrum of the electrons in the well.
44. What is the relation between Fermi energy of a metal and the velocity of electrons?

Objective Questions

I. Choose the correct choice in the following objective questions.

1. Many properties of metal can be explained on the basis of
   (a) Weiss theory
   (b) BCS theory
   (c) free electron theory
   (d) tightly bound electron theory

2. The wave vector associated with an electron is (**p** is the momentum vector)
   (a) $\mathbf{p}\hbar$ (b)
   (c) $\mathbf{p}/\lambda$ (d) $\mathbf{p}/\hbar$

3. The energy associated with an electron in K - space is ky
   (a) $[k_x^2+k_y^2+k_z^2]^2$
   (b) $\frac{\hbar^2}{2m}(k_x^2+k_y^2+k_z^2)$
   (c) $\frac{\hbar^2}{2m\left(k_x^2+k_y^2+k_z^2\right)}$
   (d) $\hbar^2\left(k_x^2+k_y^2+k_z^2\right)$

4. The Fermi - Dirac distribution function in the following choices is
   (a) $\exp[(E-E_F)kT]-1$
   (b) $\frac{1}{\exp[(E-E_F/kT]-1}$
   (c) $\frac{(E-E_F)}{\hbar k}\exp$
   (d) $\frac{1}{\exp[(E-E_F/kT]+1}$

5. The mean speed of an electron as per the Maxwell – Boltzmann distribution is
   (a) $\frac{1}{2}k_BT$ (b) $\frac{3k_BT}{m}$
   (c) $\sqrt{\frac{3k_BT}{m}}$ (d) $\sqrt{\frac{8k_BT}{m}}$

6. Free electron model can not explain
   (a) the conductivity behavior of metals
   (b) the positive temperature coefficient in metals
   (c) the conduction in semiconductors
   (d) the thermal conductivity behavior of metals

7. The velocity of a free electron in terms of the wave vector **k** is
   (a) $\hbar \mathbf{k}$ (b) $\frac{\hbar}{\mathbf{k}}$
   (c) $\frac{\hbar \mathbf{k}}{m}$ (d) $\frac{\hbar m}{\mathbf{k}}$

8. The Fermi level for potassium is 2.1ev. Then, the velocity of the electron at the Fermi level is (effective mass of the electron is 8.9×10^{-31} kg)
   (a) 8.7×10^6 m/s (b) 0.87×10^6 m/s
   (c) 8.7×10^4 m/s (d) 8.7×10^4 m/s

9. Select the correct statement
   (a) Fermi energy is the highest level that the electron can occupy at 0 K
   (b) Fermi energy is the minimum energy that the electron should possess to stay in the atom
   (c) Fermi energy is the energy possessed by a metal at its melting point
   (d) The Fermi energy is the lattice energy of metals

10. The relation between free electron velocity V and propagation vector K is
   (a) VK = constant
   (b) V/K = constant
   (c) $\frac{VK}{m^2}$ = constant
   (d) $\frac{V}{K^2}$ = constant

11. The energy of a free electron in the lowest energy state when placed in a potential box of width a is
   (a) $\frac{h^2}{ma^2}$ (b) $\frac{h^2}{8ma^2}$
   (c) $\frac{4h^2}{8ma^2}$ (d) $\frac{2h}{ma^2}$

12. The particles of MB distribution are ____________.
   (a) identical, indistinguishable and overlapping wave function.
   (b) identical, distinguishable overlapping wave functions
   (c) identical, distinguishable and non overlapping wave functions.
   (d) identical, but too big and non distinguishable

13. A photons in a black body
   (a) possess lot of chemical potential
   (b) posses lot of energy that is equal to Fermi energy
   (c) are non conservative and possess zero chemical energy
   (d) possess half integral spins

14. The wave functions of particles of MB distribution.
   (a) do not overlap
   (b) do overlap
   (c) particles fo not have wave function
   (d) the above all three are wrong

15. Many properties of metal can be explained on the basis of
   (a) Weiss theory
   (b) BCS theory
   (c) free electron theory
   (d) tightly bound electron theory

16. The wave vector associated with an electron is (**p** is the momentum vector)

(a) $\mathbf{p}\hbar$ (b) $\hbar/\mathbf{p}$

(c) **p**/l (d) $\mathbf{p}/\hbar$

17. The energy associated with an electron in K - space is ky

(a) $[k_x^2+k_y^2+k_z^2]\hbar^2$

(b) $\frac{\hbar^2}{2m}(k_x^2+k_y^2+k_z^2)$

(c) $\frac{\hbar^2}{2m\left(k_x^2+k_y^2+k_z^2\right)}$

(d) $\hbar^2\left(k_x^2+k_y^2+k_z^2\right)$

18. According to the classical model of free electron,

(a) An electron can have any wave vector, there is no restriction on the number of electrons with the same vector

(b) the electrons can have restricted values of wave vector

(c) only two electrons can have the same wave vector

(d) the electrons can have spin and can have only two values of the wave vector

19. An electron is confined to a box of infinite height, and length L. Then, the potential energy V(x) is under the boundary conditions given as

(a) $V(x)=0$ for $0<x<\infty$

(b) $V(x)=\infty$ for $0<x<L$ $V(x)=0$ for $x\leq 0; x\leq L$

(c) $V(x)=\infty$ for $0<x<L$
$V(x)=0$ for $0<x<L$

(d) $V(x)=0$ for $x\geq 0, x\leq L$
$=\infty$ for $x\leq 0, x\geq L$

20. The wave function y associated with a free electron is represented by the relation.

(a) $\frac{d^2\psi_n}{dx^2}+E_n\psi_n=0$

(b) $E_n\frac{\partial^2\psi_n}{\partial x^2}+\psi_n=0$

(c) $\frac{d^2\psi_n}{dx^2}+\psi_n=0$

(d) $\frac{d^2\psi_n}{dx^2}+\left(\frac{2m}{\hbar^2}\right)E_n\psi_n=0$

21. The Fermi - Dirac distribution function in the following choices is

(a) $\exp[(E-E_F)kT]-1$

(b) $\frac{1}{\exp[(E-E_F)/kT]-1}$

(c) $\exp\frac{(E-E_F)}{\hbar k}$

(d) $\frac{1}{\exp[(E-E_F)/kT]+1}$

22. The Fermi-Dirac distribution function is reduced to the Boltzmann distribution function under the condition.

(a) $E_{F0}<E<kT$

(b) $E_{F0}=$ infinity

(c) $(E-E_F)<kT$

(d) when $(E-E_F)>kT$

23. The mean speed of an electron as per the Maxwell – Boltzmann distribution is

(a) $\frac{1}{2}k_BT$ (b) $\frac{3k_BT}{m}$

(c) $\sqrt{\frac{3k_BT}{m}}$ (d) $\sqrt{\frac{8k_BT}{m}}$

24. The current density in a metallic sample in terms of drift velocity V_d, and n, the concentration of electrons per unit volume is

(a) $-\dfrac{ne}{V_d}$ (b) $-\dfrac{V_d}{ne}$

(e) $-ne\,V_d$ (d) $\dfrac{eV_d}{n}$

25. The change of conductivity of a metal specimen is accounted by

(a) the change of concentration of electrons per unit volume with temperature
(b) the change of volume of the specimen
(c) the change of mass of the electron with temperature
(d) the change of mobility of the electron with temperature

26. The velocity of a free electron in terms of the wave vector **k** is

(a) $\hbar\mathbf{k}$ (b) $\dfrac{\hbar}{\mathbf{k}}$

(c) $\dfrac{\hbar\mathbf{k}}{m}$ (d) $\dfrac{\hbar m}{\mathbf{k}}$

27. Choose the correct statement

(a) Even pure metals show always some residual resistivity at absolute zero
(b) Only the elements with odd number of valence electrons can be metals
(c) The resistivity of metals decreases with the impurity concentration
(d) The resistivity of a pure metal is proportional to the square root of absolute temperature

28. The Fermi level for potassium is 2 . 1eV. Then, the velocity of the electron at the Fermi level is (effective mass of the electron is 8.9×10^{-31} kg)

(a) 8.7×10^6 m/s (b) 0.87×10^6 m/s

(c) 8.7×10^4 m/s (d) 8.7×10^4 m/s

29. Select the correct statement

(a) Fermi energy is the highest level that the electron can occupy at 0 K
(b) Fermi energy is the minimum energy that the electron should passess to stay in the atom
(c) Fermi energy is the energy passessed by a metal at its melting point
(d) The Fermi energy is the lattice energy of metals

30. Experiments show that the resistivity of normally pure metal

(a) decreases to a small but finite value below 10 K
(b) increases to a finite value below 10 K
(c) goes to zero at 0 K
(d) depends strongly on the electric field below 10 K

31. The relation between free electron velocity V and propagation vector k is

(a) Vk = constant
(b) V/k = constant
(c) $\dfrac{Vk}{m^2}$ = constant
(d) $\dfrac{V}{k^2}$ = constant

32. When an electrical field applied across a metallic ample is switched off

(a) the drift velocity decreases linearly with time
(b) the drift velocity decreases to zero instantly
(c) the drift velocity decreases with time exponentially
(d) the drift velocity attains a constant value

33. The energy of a free electron in the lowest energy state when placed in a potential box of width a is

(a) $\frac{h^2}{ma^2}$ (b) $\frac{h^2}{8ma^2}$

(c) $\frac{4h^2}{8ma^2}$ (d) $\frac{2h}{ma^2}$

II. Fill in the Blanks

1. Most probabilities of distribution of particles in a system becomes ________ when the system is in thermal and statistical equilibrium.
2. The MB distribution function is expressed as ________.
3. Particles with integral spins are called ________.
4. Particles with integral spin do not follow ________.
5. Particles with odd half-integral spins are called ________.
6. Fermions obey ________ principle.
7. The density of standing waves in a black body cavity is dependent on ________.
8. The density of states in a black body cavity is independent of ________ of the cavity.
9. Phonons are the particles that follow ________ statistics.
10. The peak in the black body spectrum shifts to ________ wavelengths as the temperature is increased.
11. Wien's displacement law is given as ________
12. Electrical conduction in metals is explained by ________ theory.
13. Electron gas exists in

(a) sodium chloride (b) copper

(c) quarter (d) silicon

14. The number of electron states in a Fermi gas between the energy states $\in$ and $\in + d\in$

(a) is proportional to the square of energy $\in$

(b) is proportional to the square of mass and square root of energy $\in$.

(c) is proportional to cube root of mass and square root of energy $\in$

(d) is independent of mass of the electron

15. At high temperatures, the system of fermions approach ________ statistics.
16. Phonos are the particles that follow ________ statistics.
17. The Fermi energy of an electron in an metal depends on the ________ of the electron gas.
18. Rayleigh-Jeans' law for radiation density is expressed in terms of wavelength as ________.
19. Wein's law as deduced from plank's radiation is given as ________.
20. Conduction electrons in a simple metal are known as ________ electrons.
21. The conduction electron in a metallic crystal finds itself in a uniform field of positive ion cores and the rest of the electrons in a given volume. Therefore, the electron is supposed to be in a ________ of certain width and height.
22. The energy of an electron in k-space is given as ________.
23. The electrons in a potential well are associated with discrete ________ states.
24. The Fermi-Dirac distribution function is valid for indistinguishable particles that have ________ spin.
25. The highest energy level that can be occupied by an electron in a metal at 0 K is called ________ level.

26. The value of Fermi energy in a metal depends on the number of ________ per unit volume.

27. The number of energy states available per unit volume, between the energies lying between E and E + d E, is called ________. (dE is infinitesimally small energy interval).

28. The number of electrons per unit volume per unit energy range about an energy E, is given by the product of ________ and ________.

29. The energy levels in a free and isolated atom are ________.

30. The energy levels in an atom can accommodate as many electrons as allowed by ________ principle.

31. The Fermi energy of copper that has electron density $85 \times 10^{27}/\ m^3$ is ________.
(**Hint :** Mass of the electron $= 9.1 \times 10^{-31}$kg and $(3\pi^2)^{2/3} = 9.56$)

32. The energy of a free electron in the case of Sommerfeld free electron model is ________, where K is the wave vector associated.

33. The ________ of the electron is proportional to K, the wave vector.

34. Conduction electrons in a simple metal are the ________ electrons in the metal.

35. The conduction electron in a metallic crystal finds itself in a uniform field of positive ion cores and the rest of the electrons in a given volume. Therefore, the electron is supposed to be in a ________ of certain width and height.

36. The electrons in a potential well are associated with discrete ________ values.

37. The Fermi-Dirac distribution function is valid for indistinguishable particles that have ________ spin.

38. The highest energy level that can be occupied by an electron in a metal at 0 K is called ________ level.

39. At 0 K, all the energy levels less than the Fermi energy are ________ and all those energy levels above fermi level are ________.

40. The value of Fermi energy in a metal depends on the number of ________ per unit volume.

41. For electron energies, $E - E_f >> kT$, (k is the Boltzmann constant), the Fermi – Dirac distribution becomes closer to ________ distribution.

42. The number of energy states available per unit volume, between the energies lying between E and E + d E, is called ________. (dE is infinitesimally small energy interval).

43. Fermi-Dirac distribution function is applicable only for particles that follow ________ principle.

44. The Fermi-Dirac particles possess ________ of spin, follow ________ principle and they are indistinguishable.

45. If g (E) represents the density of states, and f(E) is the Fermi-Dirac energy distribution function between E and E + dE, then the number of particles between energy values E and E + dE, is ________.

III. True / False

1. The molecules of a diatomic gas can be treated as Max well-Boltzmann particles.

   [True/False]

2. The Maxwell-Boltzmann distribution of particles of a gas indicates that the most probable velocity is possessed by maximum number of molecules at any temperature.

   [True/False]

3. $e^{+\beta\varepsilon_i}$ in a Boltzmann factor.

   [True/False]

4. MB particles are indistinguishable particles and their eigen values are restricted.

[True/False]

5. BE statistics is applicable to indistinguishable particles that obey Pauli's exclusion principle.

[True/False]

6. BE statistics are followed by photons and phonons.

[True/False]

7. Bosons tend to cluster in low energy states at low temperatures.

[True/False]

8. BE statistics is applicable to electrons if they could be treated as distinguishable particles.

[True/False]

9. FD statistics is applicable to protons

[True/False]

10. Fermions can not exist in metals.

[True/False]

CHAPTER 6

Principles of Quantum Mechanics

6.1 Introduction

The wave nature of light was established by experiments on interference, diffraction and polarization of light. It was envisaged that light is a part of electromagnetic spectrum. However, the electromagnetic theory could not explain the observation of the spectral distribution of radiant energy from a black body. Max Planck proposed the concept of 'quanta of energy' and discreteness of radiant energy. Planck's concept content of energy quanta was supported by experimental results on photo electric effect in metals and Compton effect. Later, Neil Bohr proposed the model of an atom on the basis of his postulates that the angular momentum of the electron is quantised, and the electron revolves in discrete energy (stationary) states around the nucleus. The transition of electron from higher energy state to lower energy state results in a spectral emission.

Louis de Broglie, in an attempt to understand the meaning of the postulates of Bohr, proposed the hypothesis of matter waves. He found an analogy between the set of discrete energy levels of electrons in an atom and the stationary vibrations of a string. On the basis of the mass-energy relation ($E = mc^2$), Planck's and Bohr's postulates, de Broglie found that a material particle can be associated with wave nature as well as particle nature. The light has dual attributes of wave and particle. In a similar fashion, the particles also exhibit the dual attributes of waves and particles. We must consider them as wavicles. These concepts gave birth to the theory of quantum mechanics.

6.2 Dual Nature of Wave-Particle

6.2.1 de Broglie's Hypothesis

According de Broglie, the Planck's postulate, that the energy of a photon is given as 'hv', and the Einstein's mass-energy relation ($E = mc^2$) can be equivalent.

Then $$h\nu = mc^2 \qquad .....(6.1)$$

i.e., $$\nu = \frac{mc^2}{h} = \frac{mc\lambda\nu}{h} \qquad (\because c = \nu\lambda)$$

then $$\lambda = \frac{h}{mc} = \frac{h}{p} \qquad (\because p = mc) \qquad .....(6.2)$$

where 'p' is the momentum of the quantum particle. λ is the wave length associated.

de Broglie wave or wavicle

The electrons in an atom revolve round the nucleus in stationary orbits. de Broglie assumed that the optical properties of atoms are due to some kind of standing waves enclosed within themselves. Then, he came out with a hypothesis that the motion of electron within the atom is associated with a peculiar kind of waves called ' Pilot waves'. Each electron in the orbit can be thought of as being accompanied by a standing wave, as shown in Fig 6.1b, that runs around the electron orbit. This is compared with a standing wave on a thread fixed at two ends as shown in Fig, 6.1(a) Fig. 6.1 (b) gives a 'snapshot' of a wave of an electron.

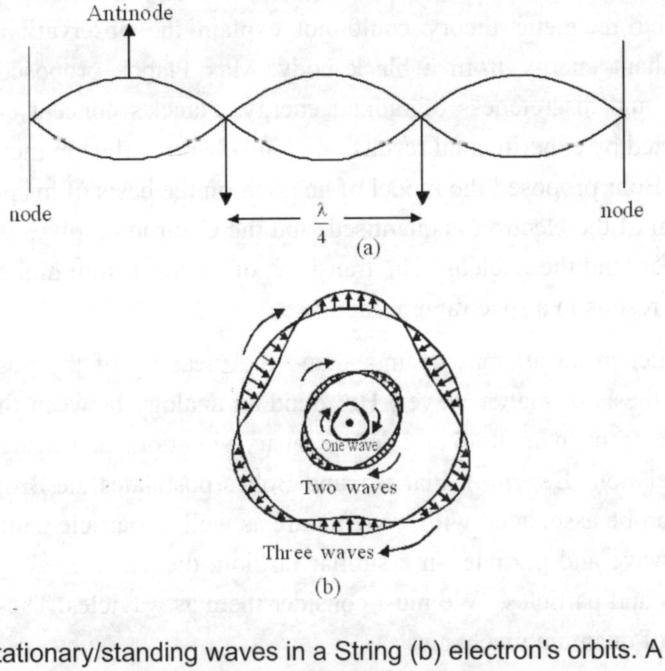

Fig. 6.1 (a) Stationary/standing waves in a String (b) electron's orbits. A 'Snap shot' of a wave at one instant of time.

If this is true, then the only possible orbits would be those whose wavelengths are integral multiples of the wavelength of the corresponding 'de Broglie wave'. Thus, the electron's motion in an orbit was given a new meaning.

The circumference of the nth orbit should be given as

$$2\pi r_n = n\lambda_n \qquad .....(6.3)$$

where r_n is the radius of the Bohr's orbit. The eq. 6.3 is a condition for a standing wave on a string of length $2pr_n$. The entire wave pattern rotates rigidly around the centre.

we know that $$r_n = \frac{n^2h^2}{4\pi^2me^2} \qquad .....(6.4)$$

where n is the quantum number, e, the charge of the election.

The wavelength λ_n is expressed as

$$\lambda_n = \frac{nh^2}{2\pi me^2} = \frac{h}{m} \times \frac{nh}{2\pi e^2}$$

The second term in the above expression is the inverse of the velocity of the electron in Bohr's model. i.e., (where v_n in the velocity of electron in the nth orbit)

$$v_n = \frac{2\pi e^2}{nh} \qquad .....(6.5)$$

$$\therefore \quad \lambda_n = \frac{h}{mv_n} \quad \text{or} \quad r_n p = \frac{nh}{2\pi} \qquad .....(6.6)$$

The relation in eq. 6.6 states the de Broglie's fundamental hypothesis, "the wavelength of the wave associated with a moving particle is equal to Planck's quantum constant divided by the momentum of the particle". The waves associated with a moving particle are called "matter waves".

If the electron behaves like a wave when situated in any atomic orbit, then the same behaviour must be true for any free electron that possesses velocity. If the kinetic energy of the electron is E_k, then, the wavelength associated is given as

$$\lambda = \frac{h}{mv} = \frac{h}{\sqrt{2mE_k}} \qquad .....(6.7)$$

If we calculate the wavelength (l) associated with an electron for a given velocity $v = 10^7$ m/s, and the mass $m = 9.11 \times 10^{-31}$ kg, $h = 6.63 \times 10^{-34}$ JS. We get

$$\lambda = 7.28 \times 10^{-11}\,\text{m} = 0.728\,\text{Å}.$$

Hence, the wavelength comes out to be shorter than the visible light, but is comparable with the wavelength of X-rays (10^{-10} m).Within the atom, electrons always behave very much like waves. But outside, they are treated sometimes as particies and sometimes as waves.

6.2.2 Characteristics of Matter Waves

1. All moving particles can be associated with matter waves. However, large or macroscopic bodies would possess insignificantly small wavelengths. Therefore, it is futile to consider their motion under quantum rules. However, particles like protons, neutrons, pions, muons, and all high energy particles are associated with dual nature. The momentum of the particle determines the wavelength of the de Broglie wave. All the above particles obey the law of quantum mechanics.
2. The de Broglie matter waves are considered as wave packets. It is well known that any source of light emits light in wave packets. A wave packet consists of many waves (like a group). The amplitude of such a wave packet is related to the probability of detecting the particle at any particular position at any given time.
3. The particle moves with a velocity, v, where as a wave associated with the packet moves with a velocity, u, called phase velocity. The relation between the phase velocity, u and the particle velocity, v is

$$u = \frac{c^2}{v}$$

As 'v' can never be greater than c (the velocity of light in vacuum), the phase velocity of matter waves can be greater than the velocity of light in that medium. That is how the matter waves differ from electromagnetic waves. An electromagnetic wave has a fixed velocity c.
4. As the waves are associated with a group of waves or wave packet, the group progresses with a velocity v_g, called the group velocity. This group velocity will be same as the particle velocity.

The group velocity v_g is given as

$$\frac{d\omega}{dk} = v_g$$

where 'dω' represents the spread of angular frequency of the waves and 'dk' the spread of a propagation constant of the waves in a packet.

6.3 Relativistic Mass, and Matter Waves

Making use of the mass - energy relation from relativity ($E = mc^2$), we can write,

$$v = \frac{p}{m} = \frac{pc^2}{mc^2} = \frac{pc^2}{E} \qquad .....(6.8)$$

The wave vector k and the angular frequency w can be written as

$$k = \frac{2\pi}{\lambda} = \frac{2\pi p}{h} \text{ and } \omega\, 2\pi = \frac{2\pi E}{h} \qquad .....(6.9)$$

[from $p = \frac{h}{\lambda}$ and $E = h\nu$]

The group velocity v_g can be written as

$$v_g = \frac{d\omega}{dk} = \frac{dE}{dp} \qquad .....(6.10)$$

According to the relativistic mechanics, the energy of a particle is given as

$$E = m_0c^2 + \text{Kinetic energy}$$

$$E = \frac{m_o c^2}{\sqrt{1 - v^2/c^2}}\ ;\ pc = \frac{m_o vc}{\sqrt{1 - v^2/c^2}}\ ;$$

$$E^2 - p^2c^2 = \frac{m_o^2c^4 - m_o^2v^2c^2}{\left(1 - v^2/c^2\right)} = m_0^2 c^4$$

$$\therefore \qquad E^2 = p^2c^2 + m_0^2 c^4,$$

where m_0 is the rest mass, and m_0c^2 is the rest mass energy of the particle.

From the above relation we get,

$$2EdE = 2pc^2\, dp$$

or

$$\frac{dE}{dp} = \frac{pc^2}{E} \qquad .....(6.11)$$

From equations (6.10) and (6.11), we obtain

$$v_g = \frac{dE}{dp} = \frac{pc^2}{E} \qquad .....(6.12(a))$$

We shall now get the relation between group velocity (v_g) and the wave velocity or phase velocity.

The wave velocity $u_p = \frac{\omega}{k}$ and group velocity $v_g = \frac{d\omega}{dk}$, and

$$k = \frac{2\pi}{\lambda} \text{ and } \omega = 2\pi\, \frac{u_p}{\lambda}$$

we have, then

$$\frac{d\omega}{dk} = \frac{d\omega}{d\lambda} \cdot \frac{d\lambda}{dk} = \frac{\frac{2\pi}{\lambda^2}\left[u_p - \lambda \frac{du_p}{d\lambda}\right]}{\frac{2\pi}{\lambda^2}}$$

$$\therefore \qquad \frac{d\omega}{dk} = u_p - \lambda \frac{du_p}{d\lambda} = u_g \qquad .....(6.12(b))$$

The above equation gives the relationship between the group velocity and the phase velocity. For a dispersive medium, $u_p = f(\lambda)$ and $u_g < u_p$. This is the case of de Broglie waves. For a nondispersive medium, $u_p \neq f(\lambda)$ and $u_g = u_p$. This is the case of electromagnetic waves in vacuum. The particle velocity $u = c^2/u_p$.

Both wave and the particle are the models that are constructed in attempting to describe matter. We do not expect the model to give a complete description. The two models complement each other in that, together they give a description of matter. Thus, we should say that the electrons are "waves and particles", not "waves or particles". The same conclusion can be drawn about electromagnetic radiation. During an experiment, the particular model that is used is determined by the apparatus used.

The verification of wave nature of a particle was done through electron diffraction experiments by G.P. Thomson (1928) son of J. J. Thomson and C. J. Davisson and L. H. Germer (1927) separately. These two are discussed in the following sections.

6.4 Davisson and Germer Experiment (1927)

Davisson and Germer were studying the scattering of electrons by a nickel plate that contains many tiny crystals. The experiment was to send a beam of electrons on to the target (Ni block) and study the intensity of the scattered beam of electrons at different angles, as in the case of any scattering experiments similar to that of Rutherford's alpha-particle scattering experiment. Generally, one would expect the intensity of scattered beam to decrease with increasing angle away from the incident beam continuously. Surprisingly, they found that the intensity of scattered electrons was not decreasing continuously but, went through maxima and minima as in the case of X-Ray diffraction from crystals. The electrons were in fact getting diffracted like waves as predicted by de Broglie. They calculated the wavelength of these electrons and it matched with that obtained from de Broglie formula. Thus, the proof for matter waves came in an accidental way. Let us now look into their experiment and the results obtained in detail.

The schematic diagram of the experimental setup used by them is shown in Fig. 6.2. The electron beam produced by the electron gun 'EG' is focused and accelerated by the anode 'A'. A narrow beam of electrons thus accelerated is directed on to a target, a large Ni crystal. The nickel plate could be rotated about an axis parallel to the axis of the incident beam. The electrons which fall on the target are reflected in different directions. The electron collector C, which is connected to a galvanometer (G), is used to collect the electrons coming in different directions from the target. It moves along a circular scale, so that it is able to receive reflected electrons at all angles between 20° to 90°. The accelerating potential used was in the range of 30 to 600 volts. The entire apparatus was enclosed in a highly evacuated chamber. The galvanometer (G) measured the current generated due to the charge collected in C and its flow through G. This current gives a measure of the intensity of the scattered beam of electrons.

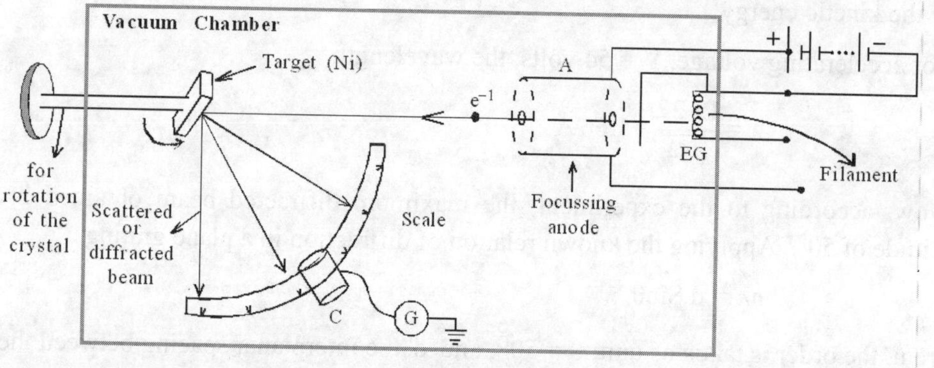

Fig. 6.2 Schematic diagram of Davisson and Germer experimental set up.

With the electron beam incident normally, a diffraction pattern as in the case of X-ray diffraction is produced, where the planes of the crystal act as diffraction elements. For each azimuth (orientation) of the crystal, a beam of electrons is made to fall normally on the crystal. The collector 'C' is moved at all angles and the current recorded by the galvanometer is noted for each position of the crystal.

The current recorded in the form of deflection in galvanometer is plotted against the angle between the incident beam and the beam entering the collector, which is called the colatitude. These observations were repeated for different accelerating voltages and a number of curves are drawn. The graph remained fairly smooth (continuous), till the accelerating voltage is around 44 volts, where small hump like curve is seen. With increasing voltage, the hump became more prominent and completely defined at 54 V with the corresponding colatitude of 50°. Again at higher voltage, the hump gradually started diminishing and for voltages above 68 volts it completely disappeared as shown in the Fig. 6.3.

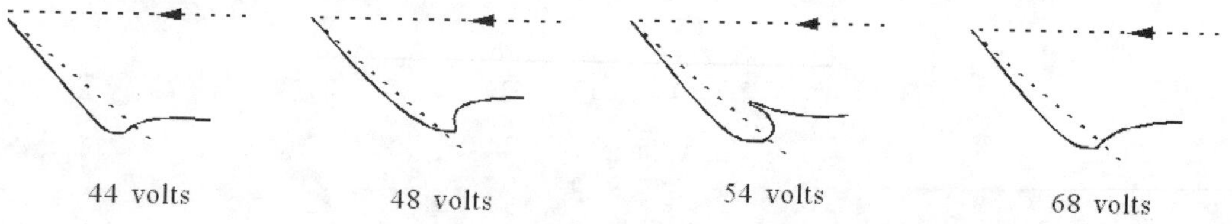

Fig. 6.3

The hump seen prominently at 54 volts at 50^{0} is a convincing evidence for the existence of the electron waves. According to de Broglie's theory, the wavelength for a beam of electrons accelerated with V volts, and velocity is given by

$$\lambda = \frac{h}{m_e v} = \frac{h}{\sqrt{2m_e eV}} \approx \frac{12.27}{\sqrt{V}} \qquad .....(6.13)$$

since the kinetic energy

For accelerating voltage, V = 54 volts, the wavelength

$$\lambda = \frac{12.27}{\sqrt{54}} = 1.66 \overset{o}{A}$$

Now, according to the experiment, the maximum diffracted beam obtained is with a colatitude of 50°. Applying the known relation of diffraction in a plane grating

$$n\lambda = d \text{ Sin}\theta$$

where n, the order is taken as unit; $\theta = 50^o$, and 'd' the interplanar spacing between the nickel atomic planes of (111) is equal to $2.15 \overset{o}{A}$, we get

$$\lambda = 2.15 \text{ Sin } 50 = 1.65 \overset{o}{A}$$

This excellent agreement between the calculated wavelengths, clearly showed that the electrons behave like waves. When the direction of electron beam and the direction of the detector are fixed but the accelerating voltage is varied, it is observed that the current shows maxima at certain accelerating voltages only, corresponding to various orders (n) as shown in Fig. 6.4.

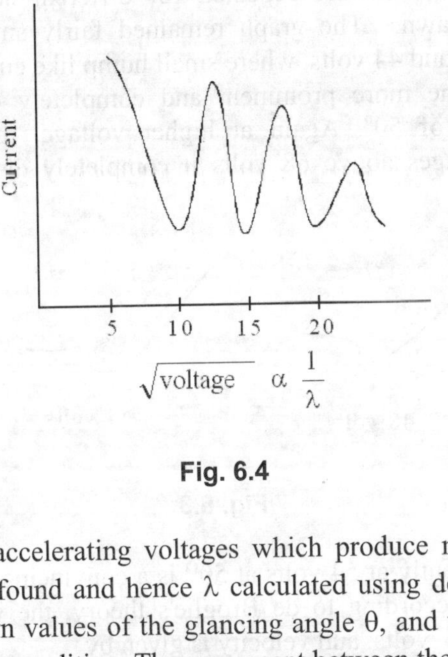

Fig. 6.4

From the curves, the accelerating voltages which produce maximum reflection in the different orders could be found and hence λ calculated using de Broglie relation. On the other hand, with the known values of the glancing angle θ, and the grating spacing 'd', λ is computed using diffraction condition. The agreement between the two results could easily be verified.

6.5 G. P. Thomson Experiment (1928)

G. P. Thomson, son of J. J. Thomson extended the work on electron diffraction to high speed electrons by accelerating them with 10,000 to 50,000 volts and using a very thin metallic film as a target. He used a method similar to Debye Scherrer powder method of X-ray analysis. The experimental set up is shown in Fig. 6.5.

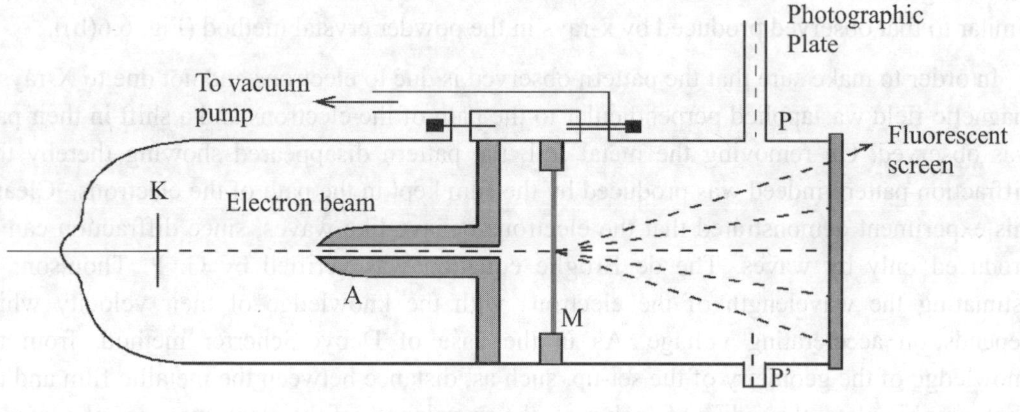

Fig. 6.5 G. P. Thomson's experimental set-up for observing electron diffraction.

As shown in figure, a beam of electrons is produced in a discharge tube with cathode K, by means of an induction coil. In order to obtain a fine pencil of electron beam, electrons were passed through a fine tube 'A' (Anode). This beam was made to fall on a thin metallic (Al or Au) foil (M) of thickness 10^{-6} cm. A photographic plate introduced at the

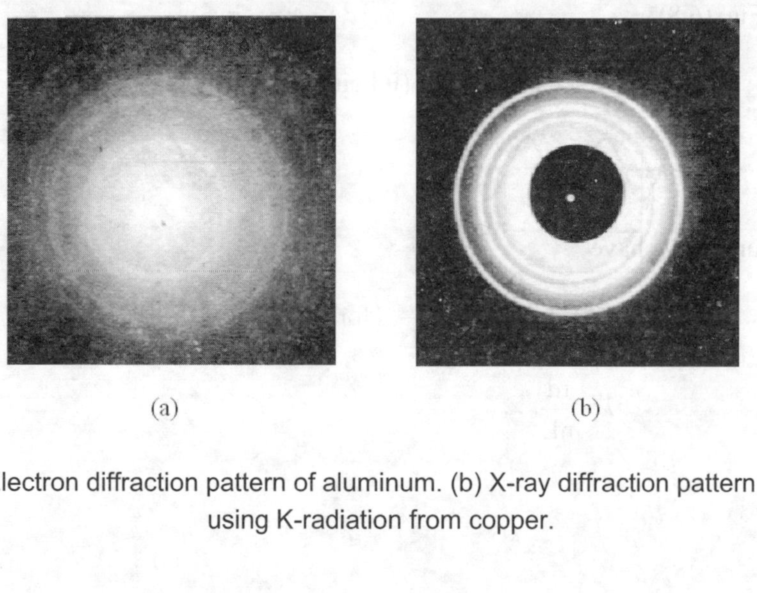

Fig. 6.6 (a) Electron diffraction pattern of aluminum. (b) X-ray diffraction pattern of aluminum, using K-radiation from copper.

points PP' recorded the picture after the electrons pass through the film. A fluorescent screen was used for visual examination, before introducing the photographic plate. The camera part of the discharge tube was maintained at a high vacuum, even when air was allowed to leak through a needle valve into the discharge tube part. Now a beam of electrons of known velocity (for a given accelerating voltage) was made to fall on the photographic plate, after passing through the thin metal foil. When the plate was developed, a symmetrical pattern consisting of concentric rings about a central spot was obtained (Fig. 6.6(a)). This pattern is similar to that observed produced by x-rays in the powder crystal method (Fig. 6.6(b)).

In order to make sure that the pattern observed is due to electrons and not due to X-rays, a magnetic field was applied perpendicular to the path of the electrons and a shift in their path was observed. On removing the metal foil, the pattern disappeared showing thereby that diffraction pattern indeed was produced by the film kept in the path of the electrons. Clearly this experiment demonstrated that the electrons behave like waves, since diffraction can be produced only by waves. The de Broglie equation was verified by G. P. Thomson, by estimating the wavelength of the electrons with the knowledge of their velocity which depends on accelerating voltage. As in the case of Debye-Scherrer method, from the knowledge of the geometry of the set-up, such as, distance between the metallic film and the photographic plate, the radius of each ring, the wavelength of the electrons was calculated as shown :

Consider the Fig. 6.8, OA is the radius (r) of the ring. 'O' is the centre of the ring, Q is the point where the electron strikes a particular plane on the crystal in the metallic film. QO (L) is the distance between the metallic film and the photographic plate. We know that

$$n\lambda = 2d \sin\theta. \qquad \text{(Bragg's law)}$$

From the Fig. (6.8),

$$\frac{r}{L} = \tan\ 2\theta \gg 2\theta \ \left(\theta \text{ being small}\right)$$

$$\therefore \qquad \theta = \frac{r}{2L}$$

For small angle, we have

$$n\lambda = 2d.\theta. = 2d.\frac{r}{2L} \ \left(\text{for small } \theta\right)$$

$$\therefore \qquad \lambda = \frac{rd}{nL}$$

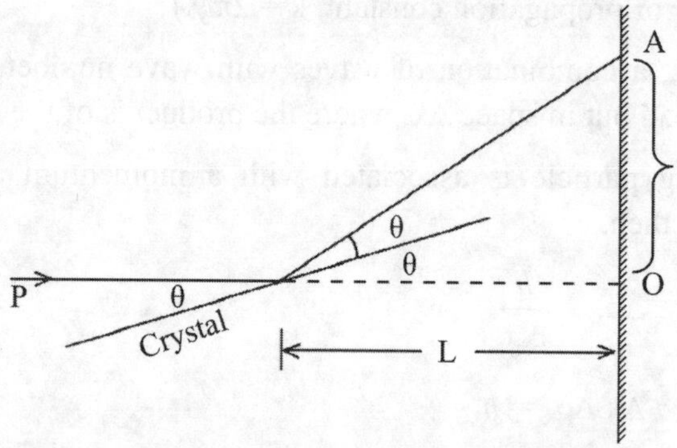

Fig. 6.8 Diffraction of electrons

From the above expression, the wavelength of electrons was calculated which matched well with those calculated from de Broglie equation. This provided the ultimate confirmation for existence of matter waves or wave nature of electrons.

Now a question arises, whether all the particles (objects) in the universe behave as waves and show diffraction. This can be answered by a small calculation of de Broglie wavelength for different types of objects namely (i) a dust particle (ii) a cricket ball (iii) earth etc. In each of these cases, the de Broglie wavelength has approximately a value of 10^{-25} to 10^{-30} m with varying masses and velocities which can't exceed the velocity of light. It is a well known fact that diffraction can be observed only if the grating element, which is 'd', is of the order of wavelength of the wave which is getting diffracted. We can see from $2d \sin\theta = n\lambda$ or $d \sin\theta \approx \lambda$, $\sin\theta$ can't exceed unity, hence $d \approx \lambda$ for $\theta = 90^o$. In the universe, we do not come across grating elements which are of the dimensions 10^{-25} to 10^{-30} m. Hence it is not possible to observe diffraction in the case of particles or objects with larger dimensions than those of electrons.

The electron diffraction is used in electron microscopes to analyse the microstructure of the materials and also to analyse the crystals structure.

6.6 Heiserberg's Uncertainty Principle

In classical mechanics based on Newton laws, the state of a particle in space is uniquely defined by position and momentum coordinates. The particles follow definite trajectories in space. But a quantum particle is different. Each moving particle is associated with a wave packet that extends in space. If a particle is moving in x-direction, its position can be any where within the wave packet. Therefore, we can only define its position to a certain degree of accuracy, Δx, but its momentum can be defined to a greater accuracy. If the wave extends from minus infinity to plus infinity, the position is inaccurate to the level of infinity and the momentum of the particle can be determined precisely.

Suppose, the particle is somewhat confined to within a distance, 'Δx' during our measurement. Then, the momentum of the particle is

$$p = k\hbar$$

(k is the wave vector or propagation constant, $k = 2\pi/\lambda$)

But $\Delta x\ \Delta k \approx 1$, i.e., a combination of waves with wave numbers in the range Δk will yield a wave group spread out in space Δx, where the product is of the order of unity.

Therefore, a moving particle is associated with a momentum and its momentum is confined to a width Δp, then,

$$\Delta p_x = \frac{\hbar}{\Delta x}$$

i.e.,
$$\Delta x\ \Delta p_x = \hbar \qquad \text{.....(6.14)}$$

The above result gives the minimum uncertainty introduced in the measurement. The above expression is normally written as $\Delta x\ \Delta p_x \geq \hbar$.

The above expression is one form of uncertainty relations postulated by Heisenberg in 1927. The above form of inequality means the following :

A particle can not have simultaneously a definite coordinate and a definite momentum (or velocity). If Δx, Δy, Δz and Δp_x, Δp_y, Δp_z are uncertainties in coordinates x, y, z and projections of momentum **p** on to the axes respectively, we have,

$$\Delta x\ \Delta p_x \geq \hbar; \qquad \Delta y\ \Delta p_y \geq \hbar; \qquad \Delta z\ \Delta p_z \geq \hbar. \qquad \text{.....(6.15)}$$

The above inequalities also lead to the following :

The smaller the error in one of the quantities, for instance x, the greater is the error in the other, namely p_x. All dynamical variables like position, momentum, energy, angular momentum etc., form pairs coupled by uncertainty relation. The uncertainty relation of Heisenberg and de Broglie relation are the two basic postulates which are fundamental in quantum mechanics. For energy and time, we have the relation as

$$(\Delta E)\ (\Delta t) \approx \hbar \qquad \text{.....(6.16)}$$

The relation tells us that the uncertainty of energy of the particle is dependent on the time interval used. The particle must be observed for a time Δt to be certain that the particle is at the point of observation.

Example : An electron remains in the excited state for 10^{-8} sec. Then, the energy of the excited state is

$$(\Delta E) \geq \frac{\hbar}{(\Delta t)} \approx 10^{-34} \times \frac{1}{10^{-8}} = 0.66 \times 10^{-8}\ \text{eV}.$$

The above parameter represents the limit of accuracy with which the energy of an excited state can ever be measured. It gives the natural width of the excited state.

We shall consider two examples to illustrate the uncertainty principle.

Microscope :

Consider that we have to measure the position of a moving electron by means of a microscope (γ-ray microscope).

When the electron passes through the field view of the objective, the microscope must be able to resolve between two positions of the electron. The resolving power of microscope is shown in the Fig. 6.9(a).

$$\Delta x = \frac{\lambda}{2\,\text{Sin}\,\theta}$$

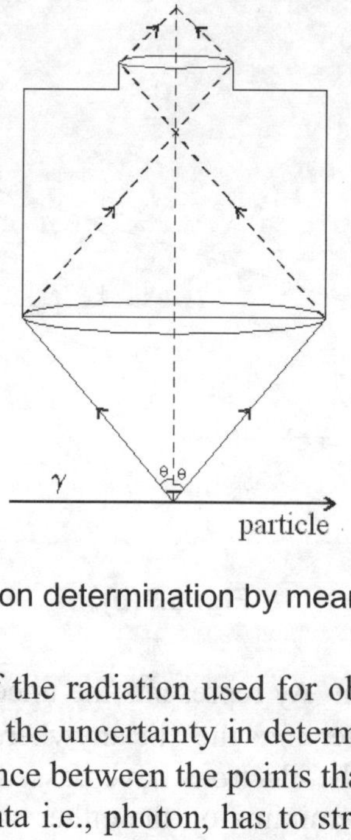

Fig. 6.9(a) Electron position determination by means of a □-ray microscope.

Where 'λ' is the wavelength of the radiation used for observation, 'θ' is the semi-vertical angle of the cone of light and Δx, the uncertainty in determining the position of the particle. 'Δx' represents the minimum distance between the points that can be resolved. If the electron has to be observed, the light quanta i.e., photon, has to strike the electron and get reflected into the objective and eye piece. That means, the photons can be anywhere in the cone of angle 2θ. If p is the momentum of the electron, the photon suffers a recoil with momenta +p Sin θ and –p Sin θ. Therefore, the component of momentum of the electron, perpendicular to the axis of the microscope after the process of recoil is indeterminate to the extent of Δp which is given as

$$\Delta p_y = p\,\text{Sin}\,\theta - (-p\,\text{Sin}\,\theta) = 2\,p\,\text{Sin}\,\theta$$

Hence, $$\Delta p_y\,\Delta x = 2\,p\,\text{Sin}\,\theta \times \frac{\lambda}{2\,\text{Sin}\,\theta} = 2p\lambda = 2p\left(\frac{h}{p}\right) = 2h \geq \hbar$$

∴ $$\Delta x\,\Delta p_y \geq \hbar \qquad \text{or} \qquad \Delta x\,\Delta p_y \approx \hbar$$

Electron Diffraction :

The second illustration is the consideration of diffraction through a single slit. Let a narrow beam of electrons fall on a slit of width Δy as shown in Fig. 6.9(b). The electrons have a velocity i.e., the momentum along the x-direction. There is no component of momentum along y-direction. After the slit is crossed, the electrons will have component of momentum along y-direction. This can be understood from the fact that we get diffraction pattern on the screen S. At C we get maximum intensity, called the principal maximum. At P' we get the first order of the diffraction minimum. The diffraction condition is

$$\Delta y = \frac{\lambda}{\sin\theta}$$

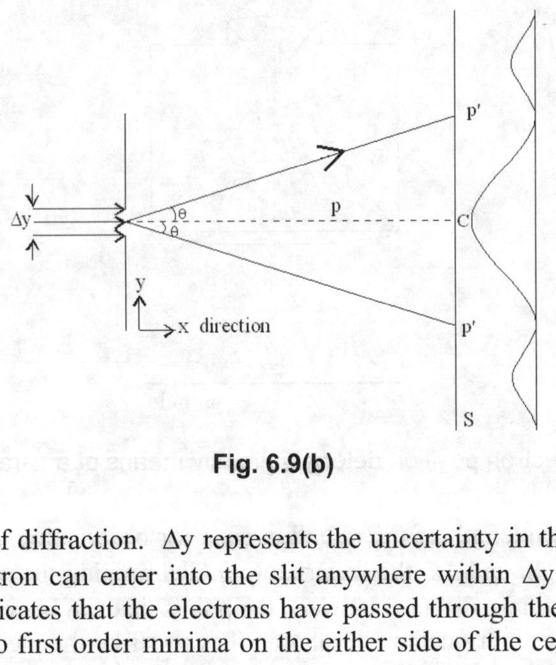

Fig. 6.9(b)

where θ is the angle of diffraction. Δy represents the uncertainty in the position of electron at the slit as the electron can enter into the slit anywhere within Δy. The observation of diffraction pattern indicates that the electrons have passed through the slit. The diffraction pattern consists of two first order minima on the either side of the central maximum (PP'). The momentum component of the electron along the y-axis is now $p \sin\theta$.

Then, $$\Delta y \sin\theta = \lambda = \frac{h}{p} \qquad \left(\text{Since } \lambda = \frac{h}{p}\right)$$

and $$\Delta p_y \sim p\,\frac{\lambda}{\Delta y} = \frac{h}{\Delta y} \qquad \left(\text{Since } \Delta y = \frac{h}{p}\right)$$

$$\Delta x\ \Delta p_y = h$$

Thus, h represents an absolute limit to the simultaneous measurement of coordinate and momentum. The value of h, is very small, therefore, the magnitudes of the experimental variables become minute in the case of atomic-sized particles, the limitations of accuracy become important.

Application of Uncertainty Principle :

We can now explain the absence of electrons in the nucleus of an atom. The radius of a nucleus is of the order of 10^{-14} m. If electrons were present in the nucleus, the uncertainty in its position is of the order of 10^{-14} m. Then

$$\Delta p_x \, \Delta x \approx h$$

$$\therefore \qquad \Delta p_x = \frac{h}{\Delta x} = \frac{1.054 \times 10^{-34}}{10^{-14}}$$

$$\approx 1.1 \times 10^{-20} \text{ Kg} - \text{m/s}$$

For this momentum, the kinetic energy according to relativistic theory is

$$\text{K. E} = \text{p c} = 1.1 \times 10^{-20} \times 3 \times 10^{8} = 3.3 \times 10^{-12} \text{ J}$$

The above result indicates that the kinetic energy must be greater than 20 MeV in the nucleus. But, it was observed that electrons can not have energies more than 4.5 MeV. The energy is far less than 20 MeV. Therefore, the electrons can not exist in a nucleus.

6.7 Schrodinger Wave Equation

Schrodinger, in 1926, using the ideas of matter waves, set up a wave equation to describe a new mechanics called wave Mechanics. Schrodinger used Max Planck's quantum hypothesis of energy quantisation which was proposed for explanation of black body radiation already dealt in fifth chapter.

The Schrodinger equation takes the analogy from partial differential equation of the classical wave equation $dy^2/dx^2 = \frac{1}{v^2}\frac{\partial^2 y}{\partial t^2}$, where 'v' is the velocity of the wave, and y is the displacement of the particle in the medium. The solution of the above given equation is $y = A\,e^{-i\omega(t - x/v)}$ and only the real part of the solution is taken into consideration. Let us see how the Schrodinger equation can be set up in the following section.

6.8 Time-Dependent Schrodinger Equation

Generally, the simplest plane monochromatic wave can be described by,

$$\Psi\,(\mathbf{r}, \mathbf{t}) = Ae^{i(\overline{K}.r - \omega t)} \qquad .....(6.17)$$

where Ψ (**r**, **t**) is the function possessing the characteristics described for a wave function, here 'r' is the position vector and 't' is the time.

$$\mathbf{r} = \mathbf{xi} + \mathbf{yj} + \mathbf{zk}$$

The above equation, if written in one dimension, say for motion along x-axis, is,

$$\Psi\,(x, t) = Ae^{i(\overline{K}.r - \omega t)} \qquad .....(6.18)$$

where $K = \frac{2\pi}{\lambda}, \quad \omega = 2\pi f$

From de Broglie relationship, $\lambda = \frac{h}{p}$, we get, $p = \hbar K$ and for energy $E = \hbar\omega$

where $\hbar = \frac{h}{2\pi}$.

From the relations given above, equation (6.15) can be written as

$$\Psi(x,t) = A\exp\left[\frac{i}{\hbar}(px - Et)\right] \quad .....(6.19)$$

Differentiating the eq. 6.19 with respect to 't', we get

$$i\hbar\frac{\partial\Psi}{\partial t} = E\Psi(x,t) \quad .....(6.20)$$

Similarly differentiating eq. (6.19) with respect to 'x', we get,

$$i\hbar\frac{\partial\Psi}{\partial x} = p\Psi(x,t) \quad .....(6.21)$$

once again differentiating eq. (6.21) we get,

$$-\frac{h^2}{2m}\frac{\partial^2\Psi}{\partial x^2} = \frac{p^2}{2m}\Psi(x,t) \quad .....(6.22)$$

A Non-Relativistic Free Particle Energy :

The energy is expressed as

$$E = \frac{p^2}{2m} \quad \text{(only Kinetic Energy)}$$

A free particle is one which moves only with kinetic energy. Therefore, from equations 6.20 and 6.21 we get,

$$i\hbar\frac{\partial\Psi}{\partial t} = -\frac{\hbar^2}{2m}\frac{\partial^2\Psi}{\partial x^2} \quad .....(6.23)$$

this equation is called one-dimensional, time-dependent, Schrodinger wave equation for a free particle. Ψ is a function of time and space.

The equations (6.20) and (6.21) suggest that Energy E, and momentum p can be represented by the operators

$$E \longrightarrow i\hbar\frac{\partial}{\partial t} \quad \text{and} \quad p \longrightarrow -i\hbar\frac{\partial}{\partial x} \quad .....(6.24)$$

Hence, equation (6.23) can be written as

$$E\Psi(x,t)=\frac{p^2}{2m}\Psi(x,t) \quad .....(6.25)$$

where E & p are operators. The total energy of the particle is

$$E=\frac{p^2}{2m}+V\text{, where V denotes the potential energy.}$$

V can be a function of both x and t. Therefore, now (6.23) can be given in a more generalised Schrodinger equation as

$$i\hbar\frac{\partial\Psi}{\partial t}=-\frac{\hbar^2}{2m}\frac{\partial^2\Psi}{\partial x^2}+V(x,t)\Psi(x,t) \quad .....(6.26)$$

The above equation governs the behavior of all microscope particles. Extending the arguments to a three-dimensional system, we get,

$$i\hbar\frac{\partial\Psi}{\partial t}=-\frac{\hbar^2}{2m}\left(\frac{\partial^2\Psi}{\partial x^2}+\frac{\partial^2\Psi}{\partial y^2}+\frac{\partial^2\Psi}{\partial z^2}\right)+V(r,t)\,\Psi(\mathbf{r},t) \quad .....(6.27)$$

or

$$i\hbar\frac{\partial\Psi}{\partial t}=\left[\frac{p^2}{2m}+V(\mathbf{r},t)\right]\Psi(\mathbf{r},t)$$

$$i\hbar\frac{\partial\Psi}{\partial t}=-\left[-\frac{\hbar^2}{2m}\nabla^2\Psi+V\Psi\right] \quad .....(6.28)$$

H, the Hamiltonian is given by,

$$H=\frac{p^2}{2m}+V=-\frac{\hbar^2}{2m}\nabla^2+V \quad .....(6.29)$$

Then, finally, we get the three dimensional time dependent Schrodinger equation as,

$$i\hbar\frac{\partial\Psi}{\partial t}=H\Psi=\left[-\frac{\hbar^2}{2m}\nabla^2+V\right]\Psi \quad ....(6.30)$$

or in one-dimension.

$$i\hbar\frac{\partial\Psi}{\partial t}=H\Psi=\left[-\frac{\hbar^2}{2m}\frac{\partial^2}{\partial x^2}+V(x)\right]\Psi \quad ....(6.31)$$

In deriving the above equation, we have made use of wave and particle nature of matter. As mentioned earlier, the wave function Ψ is in general a complex function. Schrodinger equations are valid for particles moving witn velocities much less than the velocity of light. Hence eq. (6.31) is known as non-relativistic Schrodinger Equation.

6.9 The Time-independent Schrodinger Equation

The wave function Ψ (x, t) shown in eqn. (6.31) can be written as,

$$\Psi(x, t) = \psi(x)\, \phi(t)$$

then, by the method of separation of variables, we can solve $\Psi(x, t)$ by solving for $\psi(x)$ and ϕ (t) separately. Substituting the above equation in time dependent Schrodinger equation (6.31) we get,

$$-\frac{\hbar^2}{2m}\frac{\partial^2(\psi\phi)}{\partial x^2} + V\psi\phi = i\hbar\frac{\partial(\psi\phi)}{\partial t}$$

where we have dropped the arguments i.e., the terms in the brackets, and we write ψ instead of $\psi(x)$ and ϕ for ϕ (t). Since ψ is a function of x only and independent of 't', and similarly ϕ is independent of x and dependent on 't' only, the partial derivatives can be replaced with ordinary derivatives. Thus, we get,

$$-\frac{\hbar^2}{2m}\phi\frac{d^2\psi}{dx^2} + V\psi\phi = i\hbar\psi\frac{d\phi}{\partial t} \quad .....(6.32)$$

If this equation is divided by $\Psi = \psi\phi$, we get,

$$-\frac{\hbar^2}{2m}\frac{1}{\psi}\frac{d^2\psi}{dx^2} + V = i\hbar\frac{1}{\phi}\frac{d\phi}{\partial t} \quad .....(6.33)$$

We assume that V, the potential, is independent of time and is dependent on x only. Then the left hand side of the above equation is independent of time, and right hand side is independent of 'x'. Since x and 't' are independent variables, the only way a function of 'x' can always equal a function of 't' is to have both functions equal to a constant. The constant, that each side of equation (6.33) must be equal, is called the separation constant E, and we write two equations for eq. (6.33) as,

$$-\frac{\hbar^2}{2m}\frac{1}{\Psi}\frac{d^2\psi}{dx^2} + V = E \quad .....(6.34)$$

and

$$i\hbar\,\frac{1}{\phi}\frac{d\phi}{dt} = E \quad .....(6.35)$$

That is, we have separated the partial differential equation into two ordinary differential equations which can be written as,

$$-\frac{\hbar^2}{2m}\frac{d^2\psi}{dx^2} + V\psi = E\psi \quad .....(6.36)$$

called the time-independent Schrodinger equation, and

$$i\hbar\frac{d\phi}{dt} = E\phi \quad .....(6.37)$$

Since 'V' is assumed to be independent of 't' (time), hence, eq. (6.37) can be solved by writing it as,

$$\frac{d\phi}{dt} = -\frac{iE}{\hbar}\phi \qquad .....(6.38)$$

and the solution is

$$\phi = e^{-iEt/\hbar} \qquad .....(6.39)$$

by integrating on both sides with 'ϕ' and 't'. Eq. (6.39) gives the time dependence of the product solution, ψϕ, and we have

$$\Psi = \psi(x)\, e^{-iEt/\hbar} \qquad .....(6.40)$$

Because this is similar to a sinusoidal equation, ψ is called the amplitude and the differential equation (6.40), is called amplitude equation.

Since ϕ must be well behaved, as per definition of wave function, the separation constant E is restricted to a real number. If E were complex, the exponential would have a real part and would approach infinity as t approached infinity. In order to know what 'E' represents, we shall calculate its expectation value from

$$<E> = \int_{-\infty}^{\infty} \psi^* \left(i\hbar\frac{\partial}{\partial t}\right)\psi\, dx$$

$$= \int_{-\infty}^{\infty} \psi^* e^{(iEt/\hbar)}\left(i\hbar\frac{\partial}{\partial t}\right)\psi e^{-(iEt/\hbar)} dx = E\int_{-\infty}^{\infty} \psi^*\psi dx \qquad .....(6.41)$$

Since the wave function ψ is normalized, that is $\int_{-\infty}^{\infty} \psi^* \psi dx = 1$, $<E> = E$ and similarly $<E^2> = E^2$. Normalisation gives a single value for the probability, and hence it is an observable. The E ultimately gives the total energy (Kinetic + Potential). The above Schrodinger equation (6.36) describes systems which are said to represent stationary states, there by indicating that the system is stationary in time, as potential V is independent of time. In these states, the position probability density $|\psi|^2$ at every point 'x' in space remains independent of time, so also the expectation values of the all dynamical variables. The energy of a particle described by a wave function has a definite value and in general shows discrete values. Let us see this in the case of a particle present in a box in a later section.

6.10 Physical Significance of Wave Function

1. Ψ(x, t) is the wave function. V(x) is the potential function assumed to be independent of time, and m is the mass of the particle. The wave function Ψ(x, t) is used to

describe the behaviour of the system and, mathematically, $\Psi(x, t)$ can be a complex quantity. H and E are called Hamiltonian and energy operators respectively.

2. Max Born postulated in 1926 that function $|\psi(x, t)|^2$ dx is the probability of finding the particle between x and x + dx at a given time, or that $|\psi(x, t)|^2$ is a probability density function. This is the function that has a physical meaning. The probability density $P = |\psi|^2 = \psi\psi^*$, where ψ^* is the complex conjugate of Ψ.

3. Any wave function can be represented as

$$\Psi(x, t) = \psi(x) . \phi(t)$$

Where $\psi(x)$ is independent of time and $\phi(t)$ is independent of x, the probability density can be expressed as

$$(|\psi(x, t)|)^2 = (|\psi(x)|)^2 \qquad ....(6.43)$$

and is, therefore, independent of the time. The wave function can be factored by separation of variables, as shown above in section (6.9).

4. The time independent Schrodinger wave equation can be written as

$$H\,\psi(x) = E\,\psi(x)$$

or

$$\frac{d^2\psi(x)}{dx^2} + \frac{2m}{\hbar^2}(E - V(x))\psi(x) = 0 \qquad ....(6.44)$$

Since the function $|\psi(x)|^2$ represents the probability density function, then for a single particle, we must have that

$$\int_{-\infty}^{\infty} (|\psi(x)|)^2\, dx = 1 \qquad ....(6.45)$$

The above relation is called normalization condition. The wave function that satisfies the relation (6.44) is called a normalized wave function.

The probability of finding the particle some where in the space specified is certain. This equation allows us to normalize the wave function. The wave function Ψ should be finite, single valued, continuous and well behaved.

6.11 Postulates of Wave Mechanics or Quantum Mechanics

Wave Mechanics, the mechanics of quanta (discrete packets of energy), can be developed from a set of postulates and this is known as axiomatic formulation of quantum mechanics. Following are the fundamental postulates of quantum mechanics:

1. Dynamical states of a particle or a quantum mechanical system can be described by a wave function which contains all the information about the particle. The wave

function should be single valued, finite and continuous. It should satisfy orthonormality condition i.e.

$$\int \Psi_1^* \Psi_j d\tau = \delta_{ij} \qquad .....(6.46)$$

where, Ψ^* = complex conjugate of Ψ

$\delta_{ij} = 0$ for $i \neq j$;

$= 1$ for $i = j$

$d\tau = dx\, dy\, dz$ = volume element.

2. Wave functions Ψ_i (Ψ_1, Ψ_2, Ψ_3,........) represent the physical states of a system and satisfy linear super position theorem.

i.e., $$\Psi = \sum_i C_i \Psi_i = C_1 \Psi_1 + C_2 \Psi_2 + ... \qquad .....(6.47)$$

Ψ_i = discrete states

C_i = constants

3. Schrodingers wave equation describes the behaviour of wave function in time and space (i.e., position).

$$H\Psi(x, t) = ih \frac{\partial \Psi}{\partial t} \qquad .....(6.48)$$

Time dependent equation is

$$-\frac{\hbar^2}{2m}\frac{\partial^2}{\partial x^2}\Psi(x,t) + V(x)\Psi(x,t) = i\hbar\frac{\partial \Psi}{\partial t}(x,t) \qquad .....(6.49)$$

$$\frac{d^2\psi(x)}{dx^2} + \frac{2m}{\hbar^2}(E - V(x))\psi(x) = 0 \qquad .....(6.40)$$

4. In classical physics, certain definite functions are associated with observable quantities like the position, momentum, energy etc. In wave mechanics or quantum mechanics, certain operators are associated with the observable quantities like momentum, angular momentum, kinetic energy, etc. The operators are given in the Table given below. Every observable (physical parameter) in quantum mechanics is represented by operators.

Variable	Operator
Momentum p_x	$-i$
Energy	$\hbar\partial / \partial t$

5. The only possible value, which a measurement of a variable p can yield, is given by the equation

$$P\Psi = \lambda\Psi \qquad .....(6.51)$$

where $P \rightarrow$ = operator of variable p

$\psi \rightarrow$ = wave function

$\lambda \rightarrow$ = eigen value

Therefore, a linear eigen value equation is linked with each operator. ψ is called the eigen function of the operator $i\left(\frac{h}{2\pi}\right)\frac{\partial}{\partial t}$ and λ is called the eigen value. The word, eigen value means " proper or characteristic value". The eigen values are not restricted to energy alone. They refer to all dynamical variables.

6. When a given system is in a particular state, the expected mean of the sequence of measurements of the observable represented by operator A is given by expectation value of operator A i.e.,

$$\langle A \rangle = \frac{\int \Psi^* A \Psi d\tau}{\int \Psi^* \Psi d\tau} \qquad .....(6.52)$$

where, Ψ^* = Complex conjugate of Ψ

$d\tau$ = volume element.

6.12 Particle in One Dimensional Potential Box

We shall now take up an example of particles constrained to remain localized in some part of space. Let us assume that particles are trapped in a potential well with very steep sides as shown in the Fig. 6.7. A particle of mass m moving in the x-direction width 'a' with potential barriers at x = 0 and x = a. The barriers are artificially made steep and high and are infinite in extent. Therefore, no particle can escape by surmounting the barriers.

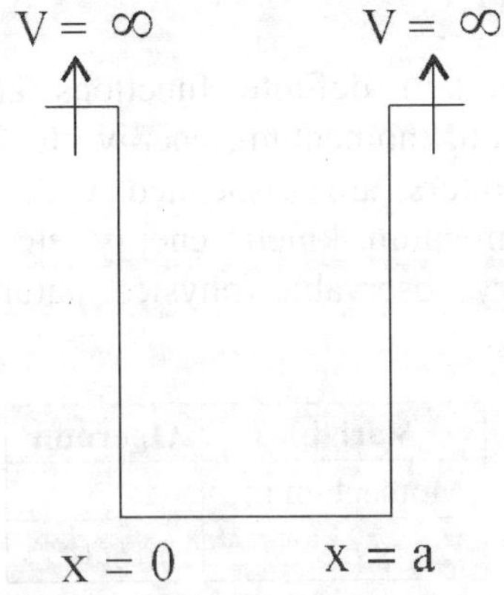

Fig. 6.7 Potential box

The potential well can mathematically be described as

$$\begin{aligned} V(x) &= \text{infinity} && \text{for } x < 0, \\ &= 0 && \text{for } 0 < x < a \\ &= \text{infinity} && \text{for } x > a \end{aligned}$$

The time independent Schrodinger wave equation is

$$\frac{d^2}{dx^2}\psi(x) + \left(\frac{2\pi}{\hbar}\right)^2 2m\,[E - V(x)]\,\psi(x) = 0 \qquad .....(6.53)$$

Where E is the total energy of the particle. If E is finite, the wave function must be zero in the regions outside the box i.e.,

$$\psi(x) = 0 \qquad \text{for } x = 0 \text{ and } x = a$$

$$\psi(x) = \text{finite} \qquad \text{for } 0 < x < a$$

Inside the box, we take the potential energy of the particles as zero. Then

$$\frac{d^2}{dx^2}\psi(x) + \frac{2mE}{\hbar^2}\psi(x) = 0 \qquad \text{for } 0 < x < a$$

To find general solution, the above equation can be written as

$$\frac{d^2\psi}{dx^2} + k^2\psi = 0,$$

where $$k^2 = \frac{2mE}{\hbar^2}$$

This equation has a solution of the form

$$\psi(x) = A\cos(kx) + B\sin(kx) \qquad .....(6.54)$$

Applying the boundary conditions, we get,

$$\psi(0) = A\cos(k \times 0) + B\sin(k \times 0) = 0$$

$\therefore$ The relation of eq. (6.51) is valid only if we take A = 0. Hence, the solution becomes

$$\psi(x) = B\sin kx$$

At $x = a,\ \psi(a) = 0$

$\therefore$ $B\sin ka = 0$

i.e., $\sin ka = 0$

i.e., $ka = n\pi$

$$k = \frac{n\pi}{a}$$

The solution can now be written as

$$\psi(x) = B \sin\left(\frac{n\pi}{a}x\right) \qquad .....(6.55)$$

Since $$k = \frac{2\pi}{h}\sqrt{2mE}$$

we get energy $$E_n = \frac{h^2k^2}{8\pi^2 m} = \frac{n^2h^2}{8ma^2} \qquad .....(6.56)$$

We can get the energy spectrum of the particles in the potential well from the above equation. The ground state energy is obtained for n = 1. Its value is $h^2/8m\,a^2$. The second level (n = 2) has an energy $E_2 = 4\,h^2/8ma^2$.

Similarly $E_3 = 9\,h^2/8ma^2$, $E_4 = 16\,h^2\,/\,8ma^2$ and so on. The energy spectrum is given in the Fig. 6.8.

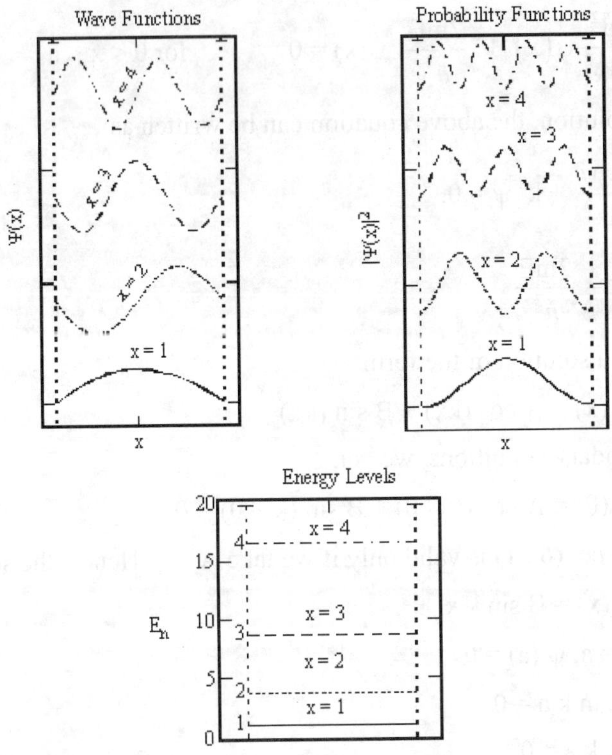

Fig. 6.8 (a) The wave function (b) the probability density, (c) energy levels of a particle in a potential well.

The total energy of the particle is quantized. That is, the energy of the particle can only have certain discrete values. The discrete values of energy are called eigen values. Note that the ground state energy is not zero. The quantization of the particle energy is contrary to the concepts in classical mechanics where particles can have continuous energy values.

The wave function is given by

$$\psi = B \sin\left(\frac{n\pi}{a} x\right)$$

The constant B is to be determined. It can be obtained from the fact that the probability that the particle is located somewhere in the well must be unity. The probability that a particle is located in length dx is $\psi\psi^*$ dx. Then

$$\int_0^a \psi\,\psi^* dx = 1$$

The above condition is called the normalization condition. Substituting for $\psi(x)$, and since $\psi(x)$ is real ($\psi = \psi^*$), we get,

$$\int_0^a |\psi(x)|^2 dx = 1$$

i.e.,

$$\int_0^a B^2 \sin^2 (n\pi x / a) dx = 1$$

$\therefore$

$$B = \sqrt{2/a}$$

Hence, the normalized solution of ψ becomes

$$\psi = \sqrt{2/a}\ \sin\left(\frac{n\pi x}{a}\right) \qquad .....(6.47)$$

Therefore, the wave function of a bound particle in a potential well is one of a set of discrete values, each one corresponding to a different value of the integer 'n'. This is true for all bound particles. Each allowed value of the wave function is called an eigen function. The diagram of Fig. 6.10 shows that a particle in the ground state is most probably located at the centre of the well. For higher energy state like E_2, the particle can be located at some position removed from the centre of the well. There are positions at which there is a zero probability of finding the particles.

Exercise 1 :

Determine the wave length associated with an electron having kinetic energy equal to 1KeV.

$$\text{Energy} = 1 \times 10^3 \text{ ev} = 10^3\ 1.6 \times 10^{-19}$$

$$p = \sqrt{2\times10^3\times1.6\times10^{-19}\times9.1\times10^{-31}} \qquad (\because p = \sqrt{2mE})$$

$$\lambda = \frac{6.6\times10^{-34}}{\sqrt{2\times10^3\times16\times10^{-20}\times9\times10^{-31}}} = 0.389\ \mathring{A} \qquad (\because \lambda = \frac{h}{p})$$

Exercise 2 :

Determine the wavelength associated with a particle like an electron having energy much greater than its rest mass energy.

The rest mass energy of an electron

$$mC^2 = 9.1\times10^{-31}\times(3\times10^8)^2$$
$$= 8.2\times10^{-14}\ J$$
$$= 0.512\times10^6\ eV \qquad (\because 1J = 1.6\times10^{-19}\ eV)$$

Let us take the example of an electron that has an energy 1 MeV.

The total energy = $(0.512\times10^6 + 1\times10^6)$ eV = 1.512 MeV

$$\therefore \quad m = \frac{1.512\times10^6\times16\times10^{-20}}{(3\times10^8)^2} = 2.69\times10^{-30}\ Kg$$

We have for the relatively particle

$$1 - \frac{v^2}{c^2} - \frac{m_o^2}{m^2}$$

i.e.,

$$v = \sqrt{(1-m_0^2/m^2)c^2}$$

$\therefore$ In the present case

$$V = \sqrt{\left[1-\left(\frac{9.1\times10^{-31}}{2.69\times10^{-30}}\right)^2\right](3\times10^8)^2} = 2.82\times10^8\ m/s$$

$$\lambda = \frac{6.6\times10^{-34}}{2.69\times10^{-30}\times2.82\times10^8} = 0.00087\ \mathring{A} \qquad (\because \lambda = \frac{h}{p} = \frac{h}{mv})$$

Exercise 3 :

An electron beam associated with a de Broglie wavelength of 0.2 $\mathring{A}$ falls on a thin film of nickel. The inter planar distance for (1 1 1) planes in nickel is 2.04 $\mathring{A}$. What is the angle of diffraction for the first order reflection from (1 1 1) planes of a nickel film?

$$2d\sin\theta = \lambda$$

$\therefore \quad 2 \times 2.04 \sin\theta = 0.2$

i.e., $\quad \sin\theta = \dfrac{0.2}{2 \times 2.04} = 0.049$

$\therefore \quad \theta = 2.8^0$

Exercise 4 :

A free electron is confined to a potential well of infinite height and width 1. Compute the energy difference between the first two energy states.

$$E_1 = \frac{h^2}{8ma^2} = \frac{\left(6.6\times10^{-34}\right)^2}{8\times9\times10^{-31}\times10^{-20}\times16\times10^{-20}} = 37.81 \text{ eV}.$$

$$E_2 = 4 \times E_1 = 151.24 \text{ eV}$$

$$\therefore \quad E_2 - E_1 = 113.43 \text{ eV}.$$

Questions and Problems

1. Explain the dual nature of light, both as a particle and wave.
2. What are matter waves ? Who proposed this concept ? How did he come to this conclusion ? What is its importance ?
3. Calculate the de Broglie wavelength of a dust particle of mass 0.001mg and moving with velocity 10^4 m/s. Can we observe diffraction due to dust particle wave ? Give reasons for your conclusion.
4. Calculate the de Broglie wavelength of a cricket ball of mass 500 gm and moving with velocity 150 km/hr. What should be the size of the grating element to observe diffraction with a cricket ball ?
5. Describe an experiment to establish the wave nature of electron?
6. How did Davisson and Germer detect the presence of matter waves ?
7. What are the important conclusions of G. P. Thomson experiment ? How did he arrive at them ?
8. If an electron is accelerated with a potential difference of 10,000 volts, what would be its de Broglie wavelength ? Take standard data for the quantities required.
9. What is the difference between any function y = x(t) and a wave function Ψ(x, t). What are the characteristics of a wave function ?
10. Calculate the energies that can be possessed by a particle of mass 9×10^{-31} kg which is placed in an infinite potential box of width 10^{-8} cm.
11. Write the time-dependent Schrodinger wave equation and obtain the time-independent Schrodinger wave equation from that.

12. Obtain the energy spectrum of a free particle moving in a potential box of height $V = \infty$.

13. A body has a mass of 0.66 gm and moves with a velocity 2×10^5 cm/s. What is the de Broglie wavelength associated with it, can you get diffraction with a beam of such bodies from a nickel foil?

14. A particle of mass 1 gram is placed in a box of side 10^{-5} m. A free electron ($m = 9.1 \times 10^{-31}$ kg) is placed in a box of side 1 $\overset{o}{A}$. What are the differences of energy in the first two levels in both the cases? Which case is possible?

15. A moving electron has a kinetic energy 64 eV. Compare it with the rest mass energy of the electron. What is the de Brogile wavelength of the electron waves (mass of the electron 9×10^{-31} kg)?

16. An α - particle is accelerated by a potential difference of 25 KV. What is the de Broglie wavelength if mass of the α-particle is $4 \times 1.67 \times 10^{-21}$ kg?

    [**Hint :** use the relation : $\lambda = h / \sqrt{4eVm_\alpha}$]

17. Show that if the uncertainty in the location of a particle is equal to the de Broglie wavelength, then the magnitude of uncertainty in its velocity is equal to the magnitude of the velocity of the particle.

18. What are the conclusions one can draw from the solutions of Schrodinger wave equation for a particle in a potential box of width a ?

19. What do you understand by the terms 'eigen value' and 'eigen function' ?

20. Show that a particle of mass m in a potential well of infinite height and width 'a' possesses discrete energies.

21. Neutrons are in thermal equilibrium at 3000 ^{o}K. Calculate the energy in eV and the de Broglie wavelength.

22. Write a short note on 1. de Broglie wavelength 2. Heisenberg's uncertainty relation 3. electron diffraction

23. A quantum particle in a one-dimensional potential box of width '*l*'. Find the eigen value of the energy of the particle?

24. Show that a particle like electron is confined to a potential value of width 1°A will have an energy in a second energic state as 151 eV.

25. Give at least one illustration of Heisenberg's uncertainty principle.

26. An electron is confined in a one-dimensional potential well length 0.1 nm. What is the kinetic energy of the electron in the ground state.

27. Prove that the wave function of a particle in a potential well is one of a set of discrete values.

28. What is meant by normalization condition?

Objective Questions

I. Choose the correct choice

1. The wave nature of matter is not apparent in heavy masses because
   (a) the 'λ' associated is very very small
   (b) the 'λ' associated is in the long wavelength region
   (c) they do not have symmetry
   (d) the momentum associated is too small

2. A particle possesses a kinetic energy E and mass m, then its deBroglie wavelength is

   (a) $\frac{h}{\sqrt{2mE}}$ (b) $h\sqrt{2mE}$

   (c) $\sqrt{\frac{2mE}{h}}$ (d) $\frac{1}{h\sqrt{2mE}}$

3. An electron is accelerated through a potential difference of V volts. The deBroglie wavelength of the electron is

   (a) $\frac{12.27}{V}$ (b) $\frac{12.27}{\sqrt{V}}$

   (c) $12.27\sqrt{V}$ (d) $12.27V$

4. Read the statements (A and B) and choose the answer. (A) The deBroglie Wave length of a moving particle is inversely proportional to its momentum (B) Only a charged particle in motion is associated with matter waves.
   (a) A and B are correct
   (b) A and B are wrong
   (c) A is wrong and B is correct
   (d) A is correct and B is wrong

5. An α- particle has a mass M and a proton has a mass m_p. When they are accelerated by the same potential

   (a) $\lambda_p^2 m_p = \lambda_d^2 2M$ (b) $\lambda_p = 4\lambda_d$

   (c) $\lambda_p M = \lambda_d m_p$ (d) $\lambda_p m_p = \lambda_d M$

6. If the momentum of a particle is doubled, then its deBroglie wave length
   (a) halves
   (b) doubles
   (c) quadruples
   (d) remains the same

7. In Davission-Germer experiment, the nickel crystal acts as
   (a) an absorber
   (b) polarizer
   (c) three dimensional grating
   (d) a metallic reflector

8. The masses of neutron and electron are m_n and m_e respectively. If they should have the same deBroglie wave length, then their velocities should be in the ratio

   (a) 1:1 (b) $\frac{m_e}{m_n}$

   (c) $\frac{m_n}{m_e}$ (d) $\frac{m_e^2}{m_n}$

9. Increasing the potential difference between anode and filament in Davisson - Germer experiment
   (a) causes an increase in the wavelength of the electron-waves
   (b) causes a decrease in the wave velocity
   (c) causes a decrease in the wavelength of the electron-waves
   (d) causes a decrease in the momentum of the electron

10. Increase in the velocity of electrons in Davission-Germer experiment,
    (a) results in the steady increase of the intensity of diffraction line
    (b) results in the increase of intensity of the diffraction line and then a decrease in intensity at a certain velocity
    (c) results in a decrease of the intensity of diffraction

11. An electron has a kinetic energy of 32 eV. Then, its deBroglie wavelength is (mass of the electron $\approx 9 \times 10^{-31}$ kg)

   (a) 1.1 Å (b) 2.2 Å

   (c) 3.3 Å (d) 4.4 Å

12. An electron (e) and a proton (p) are accelerated such that they possess the same deBroglie wave length ($m_p = 1834\, m_e$), Then

   (a) $V_p = 1834\, V_e$ (b) $(KE)_p = 1834\,(KE)_e$

   (c) $V_e = 1834\, V_p$ (d) $(KE)_e = 1834\,(KE)_p$

13. An α - particle has a mass $4m_p$ and a proton m_p. If they possess the same kinetic energy, then, The ratio of the deBroglie wave lengths is

   (a) 1 : 4 (b) 4 : 1

   (c) 2 : 1 (d) 1 : 2

14. The deBroglie wave length of an electron with a kinetic energy of 1 MeV is (in angstroms)

   (a) 50 (b) 0.5

   (c) 0.008 (d) 0.005

15. If V_g is the group velocity, V is the phase velocity and then V_g is expressed as (if λ is the wave length of particle waves)

   (a) $V_g = V - \lambda \dfrac{dV}{d\lambda}$ (b) $V_g = V + \lambda \dfrac{dV}{d\lambda}$

   (c) $V_g = V \dfrac{d\lambda}{dV} V$ (d) $V_g = V - \dfrac{d\lambda}{dV} V$

16. Choose the correct statement

   (a) The group velocity of the particle waves depends on the way the frequency varies with the propagation constant k.

   (b) The group velocity and the particle velocity in particle waves are different

   (c) Davisson and Germer experiment proves the uncertainty principle

   (d) Davisson and Germer experiment proves that diffraction effects are not produced by matter waves

17. The speed of an electron is 0.01 times that of the speed of light. Then, its deBroglic wavelength approximately is

   (a) 4. 84 Å (b) 3.0 Å

   (c) 24.2 Å (d) 2.42 Å

18. A wave equation is given as

   (a) $\dfrac{\partial^2 y}{\partial x^2} = V^2 \dfrac{\partial^2 y}{\partial t^2}$

   (b) $\dfrac{\partial^2 y}{\partial x^2} = \dfrac{1}{V_2} \dfrac{\partial^2 y}{\partial t^2}$

   (c) $\dfrac{\partial^2 y}{\partial x^2} = \dfrac{1}{V^2} \dfrac{\partial^2 y}{\partial x^2}$

   (d) $\dfrac{\partial^2 y}{\partial x^2} = V \dfrac{\partial y}{\partial t}$

19. $P\psi = -i\hbar \dfrac{\partial \psi}{\partial x}$ can be expressed as

   (a) $\dfrac{\partial \psi}{\partial t} = -i E \hbar \psi$

   (b) $-\dfrac{\hbar^2}{2m} \dfrac{\partial^2 \psi}{\partial x^2} = i\hbar \dfrac{\partial \psi}{\partial x}$

   (c) $\dfrac{\partial \psi}{\partial t} = -\dfrac{i}{\hbar} E \psi$

   (d) $\dfrac{-\hbar^2}{2m} \dfrac{\partial \psi}{\partial t} = \dfrac{\partial^2 \psi}{\partial x^2}$

20. The differential operator $i\hbar \left(\dfrac{\partial}{\partial t}\right)$ operating on ψ is equivalent to

   (a) The momentum (b) The velocity

   (c) mass (d) The energy

21. The momentum P_x is represented by the operator

(a) $-i\hbar\dfrac{\partial\psi}{\partial t}$ (b) $\hbar\dfrac{\partial\psi}{\partial t}$

(c) $-i\hbar\left(\dfrac{\partial}{\partial x}\right)$ (d) $\hbar\left(\dfrac{\partial}{\partial x}\right)$

22. If ψ is the wave function associated with a particle, the probability density of finding the particle within x and x + dx is

(a) ψdx (b) $\left(\dfrac{\partial\psi}{\partial t}\right)dx$

(c) $\left(\dfrac{\partial\psi}{dx}\right)$ (d) ψψ* dx

23. $\dfrac{\hbar^2}{2m}\nabla^2 + V = H$, gives the total energy of a particle, Then

(a) $\dfrac{-\hbar}{2m}\dfrac{\partial^2\psi}{\partial t} = V_c\,\hbar\dfrac{\partial\psi}{\partial t}$

(b) $\dfrac{-\hbar^2}{2m}\dfrac{\partial^2\psi}{\partial x^2} + V = i\hbar\dfrac{\partial\psi}{\partial t}$

(c) $\dfrac{\hbar^2}{2m}\dfrac{\partial^2\psi}{\partial x^2} = V - i\hbar\dfrac{\partial\psi}{\partial t}$

(d) $\dfrac{\hbar^2}{2m}\dfrac{\partial^2\psi}{\partial t^2} + V = i\hbar\dfrac{\partial\psi}{\partial x}$

24. The Schrodinger's amplitude equation for one dimensional matter waves is given as [ψ is a function of (x, t)] is

(a) $\dfrac{-\hbar^2}{2m}\dfrac{\partial^2\psi}{\partial x^2} = i\hbar\dfrac{\partial\psi}{\partial x}$

(b) $\dfrac{-\hbar^2}{2m}\dfrac{\partial^2\psi}{\partial x^2} + V = i\hbar\dfrac{\partial\psi}{\partial t}$

(c) $\dfrac{-\hbar^2}{2m}\dfrac{\partial^2\psi}{\partial x^2} + V\psi = i\hbar\dfrac{\partial\psi}{\partial t}$

(d) $\dfrac{-\hbar^2}{2m}\dfrac{\partial^2\psi}{\partial t^2} + V = i\hbar\dfrac{\partial\psi}{\partial x}$

25. The characteristics of wave function ψ are

(a) Real function, finite and discontinuous

(b) Complex, single valued, finite and continuous function

(c) Complex, infinite and discontinuous function

(d) Complex, single valued and infinite

26. The operator form of kinetic energy of a particle of mass m is

(a) ∇^2 (b) $\dfrac{\nabla^2}{2m}$

(c) $-\dfrac{\hbar^2}{2m}\nabla^2$ (d) $-\dfrac{\hbar}{m}\nabla^2$

27. The wave function for the motion of a particle in a potential well of width a is given as $\psi_n = B\,\mathrm{Sin}\left(\dfrac{n\pi}{\alpha}x\right)$, then B is

(a) $\dfrac{1}{a}$ (b) $\sqrt{\dfrac{1}{a}}$

(c) $\dfrac{2}{a}$ (d) $\sqrt{\dfrac{a}{2}}$

28. When the wave function is normalized, then

(a) $\psi\psi^* = 1$ (b) $\psi\psi^*\,dx = 1$

(c) $\int_{-1}^{+1}\psi\psi^*\,dx = 1$ (d) $\int_{-\alpha}^{+\alpha}\psi\psi^*\,dx = 1$

29. For a particle of mass 'm' confined to a cubical box of side L, the allowed values of energy E are given as

(a) $\dfrac{h^2}{2mL^2}n^2$, where n = 1, 2, 3 .....

(b) $\dfrac{h^2}{8mL^2}n^2$, where n = 1, 2, 3 .....

(c) $\dfrac{h^2L^2}{2mn^2}$, where n = 1, 2, 3 .....

(d) $\dfrac{h^2n^2}{mL}$, where n = 1, 2, 3 .....

30. The Hamiltonian of a free particle is

(a) $H = -\frac{\hbar^2}{2m}\nabla^2$

(b) $H = V + \frac{h^2}{2m}\nabla^2$

(c) $H = -\frac{i\hbar}{2m}\nabla^2$

(d) $-\frac{h^2}{8\pi^2 m}\nabla^2 + V(x)$

31. When an electron is confined to a rectangular potential well of width 1 Å, Then, the minimum energy is ($h = 6.6 \times 10^{-34}$ J sec)

(a) leV (b) 2400 eV
(c) 1200 eV (d) 37.8 eV

32. When an electron is confined to a rectangular potential well of width 1 Å, the minimum linear momentum of the electron is

(a) 3.3×10^{-34} kg – m /sec
(b) 2.2×10^{-24} kg m / sec
(c) 3.3×10^{-24} kg m / sec
(d) 1.1×10^{-24} kg m / sec

33. An electron is confined to a potential box of infinite height and width 10 Å. The probability of finding the particle in a small interval Δx at the centre of the box for the energy state immediately above the ground state is

(a) zero (b) 0.5
(c) 0.9 (d) 1

34. The general solution of the wave equation for a free particle is

(a) $\psi_n(x) = e^{ikx}$

(b) $\psi_n(x) = A_1 e^{ikx} + A_2 e^{-ikx}$

(c) $\psi_n(x) = k^2 E_n$

(d) $\psi_n(x) = \frac{2m}{\hbar^2} E_n \psi_n(x)$

35. The wave function $\psi_n(x)$ associated with a free particle present in a potential box of width 'L' is

(a) $A \sin \frac{n\pi x}{L}$ (b) $A \sin(Lx)$

(c) $A \sin\left(\frac{2Lx}{\hbar}\right)$ (d) $A \sin \frac{\hbar x}{L}$

36. The Hamiltonian of a free electron is (m is the mass of the electron)

(a) $\frac{\hbar^2}{2m}\nabla^2$ (b) $-\frac{\hbar^2}{2m}\nabla^2$

(c) $-\frac{\hbar^2}{2m}\nabla^2 + V$ (d) $-\frac{\hbar^2}{2m}\nabla^2 - V$

37. The wave function of a free particle in its lowest energy state in a potential well of width a is

(a) $\sqrt{\frac{a}{2}} \text{ Sin } \frac{\pi x}{4}$ (b) $\sqrt{\frac{2}{a}} \text{ Sin } 2\pi x$

(c) $\sqrt{\frac{2}{a}} \text{ Sin } \frac{\pi x}{a}$ (d) $\sqrt{\frac{4}{a}} \text{ Cos } \frac{4\pi x}{a}$

II. Fill in the blanks :

1. The wavelength of the waves associated with particles is given by the relation __________.
2. The wave length of an electron accelerated from rest through a potential difference of V volts is __________.
3. The wave packets associated with a particle moving with a velocity V, The group velocity of the waves is equal to V_g. Then V_g = __________.
4. The diffraction of electrons was demonstrated for the first time by __________.
5. The energy of neutrons is 1eV. Then the wavelength associated is __________, (mass of neutron is 1.67×10^{-27} kg)
6. The electron diffraction is based on the __________ properties associated with electrons.
7. Compared to macroscopic bodies, the microparticles display __________ de Broglie wavelengths.

8. The quantum mechanical particles obey the laws of propagation of __________ waves.

9. The probability of finding a particle in an element of volume dV, W dV = __________ where Ψ and Ψ* are the wave function and its complex conjugate respectively.

10. The schrodinger amplitude equation for a free particle is __________ .

11. The Schrodinger amplitude equation describes the __________ states of particles.

12. The eigen values of the particles are also called __________ states.

13. The values of energy E_n for which the Schrodinger amplitude equation is solved are called __________ values.

14. The differential operator, $-i\hbar\dfrac{\partial}{\partial x}$, when operating on the wave function Ψ, is equivalent to the __________ of the particle.

15. The total energy E of a particle is equivalent to a differential operator __________ which operators on the wave function.

16. The acceptable wave function Ψ and its derivatives should be __________, __________ and __________ and Ψ should also be normalizable.

17. The Schrodinger wave equation for a free particle in a potential well of infinite height is given as __________ .

18. If the width of the potential well is of atomic dimensions, the energy levels are separated by __________ distances.

19. The probability of finding the particle between two points x = 0 and x = L, where L is the width of the potential well, is __________ .

20. A particle electron in a potential well is represented by a wave function ψ_n, where 'n' is called the quantum number. The corresponding energy E_n values will exist for __________ values of n.

21. The energy levels in a free and isolated atom are __________ .

CHAPTER 7

Band Theory of Solids

7.1 Introduction

We have learnt earlier that metals have free electrons and the electrons in metals are considered to be in a potential well. The conductivity of metals depends on the concentration of electrons and the mobility. Different metals have different conductivities. For example, divalent (Cd, Zn etc) and trivalent (Al, In etc) metals should have more conductivities than the conductivities of mono valent (Cu, Ag, etc) metals. However, the experimental observation was contrary to the above. There are a number of insulators like Al_2O_3, MgO, ZnO etc. The resistivity of a good insulator may be as high as $O^{20}\ \Omega - m$ while the resistivity of a good metal may be as low as $10^{-10}\ \Omega - m$. Besides, we have elements like silicon and germanium which are semiconductors. The conductivity mechanism in these semiconductors could not be explained on the basis of just free electron theory. Therefore, the theory had to undergo modifications and such modified theory is now known as the band theory of solids. The constant potential well in free electron gas model is replaced by a periodic potential in the crystal. This consideration of the electron being present in a periodic potential and its quantum nature gives remarkable aspects of the conduction and magnetism of solids. We shall briefly study the introductory aspects of band theory of solids and see how the insulators, semiconductors and metals are distinguished.

7.2 Electron in a Periodic Potential

In order to understand the concept of energy band formation, let us first recall the formation of energy levels in an isolated atom. When atoms are at large distances of separation as in the case of gas molecules, the mutual interaction between the electron orbits of the neighbouring atoms can be neglected. The atoms have independent energy levels permitted by quantum rules on the basis of Sommerfeld atomic model. Each energy level accommodates as many electrons as allowed by Pauli's exclusion principle. The energy levels are degenerate energy levels. However, when the atoms are brought closer to each other, the electrons respond to the influence of the other nuclei, and the electron orbits of the atoms over lap when the atoms are brought still closer as in the case of a solid. The Pauli's exclusion principle will then apply as a whole to the system of atoms that are brought together. Each energy level associated with the individual atom now repeats as many times as the number of atoms in the solid. What was an isolated level in the single atom broadens out into a large number of closely spaced levels of the atoms (Fig. 7.1). Such a set of energy levels that has a finite width is called an energy band. The electrons which originally have the same energies now get their energies modified so that the exclusion principle is not violated.

7.2.1 Formation of Energy Bands

The Fig. 7.1(a) depicts the energy band formation in a solid or crystal. At large interatomic distances (represented by r), different energy levels are shown as discrete energy levels. But, as the distance between the atoms is reduced, it can be observed that each set of energy level combines together to form a band of closely spaced levels under the influence of neighboring atoms. We can see from the Fig. 7.1(a), that the spreading of levels starts at different values of r for different energy levels. The spreading of levels starts at larger distances for outer shells as the electrons in the outer shells interact first with neighbours compared to the inner shell electrons. At a particular distance 'r_o', the inter atomic distance in a solid, each energy band is separated by an finite energy gap E_g.

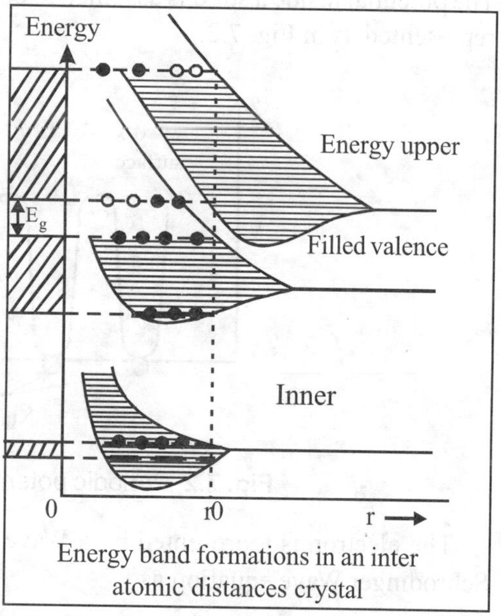

Energy band formations in an inter atomic distances crystal

Fig. 7.1 (a) Formation of energy bands in a solid

Let us consider silicon atom. The isolated silicon atom possesses discrete energy levels. Ten of the fourteen silicon atom electrons occupy deep-lying energy levels close to the nucleus.

The four remaining valence electrons are weakly bound relatively. Fig. 7.1 (b) shows the band splitting of silicon. We have to consider the n = 3 level for the valence electrons, since the first energy states have electrons that are bound tightly to the nucleus. The 3S state corresponds to n = 3 and $l = 0$, and contains two quantum states per atom. This state will contain two electrons at T = 0K. The 3P state corresponds to n = 3 and $l = 1$, and contains six quantum states per atom. This state will contain the remaining two electrons in the individual silicon atom.

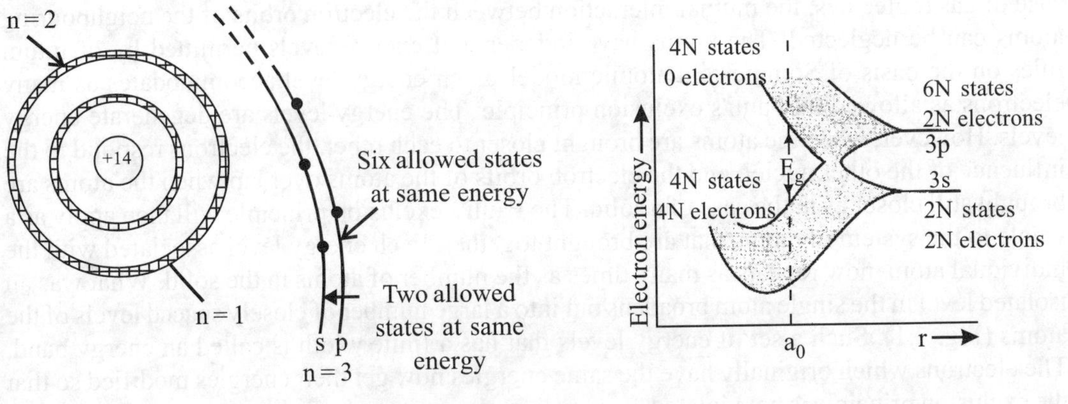

Fig. 7.1 (b) Band splitting in silicon.

As the distance between atoms decreases, the interaction of 3S and 3P states interact and overlap. At equilibrium distance, the bands split such that there are four quantum states per atom in the lower band and four quantum states per atom in the upper band. The lower band is called 'the valence band' and at 0K, it is full. The upper band is called 'the conduction band' and is completely empty at 0K. Both are separated by an energy gap, E_g.

7.2.2 Periodic potential and Bloch Theorem

The potential inside a solid is assumed to be periodic with the periodicity of the lattice, roughly represented as in Fig. 7.2.

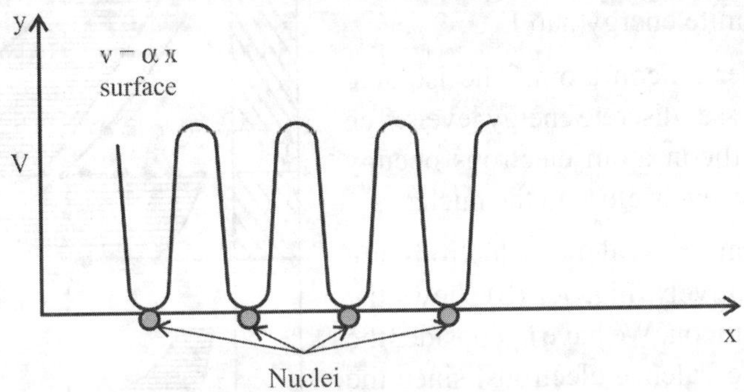

Fig. 7.2 Periodic potential (1-Dimensional) in a crystal.

The electron is represented by a Wave function Y with a corresponding one dimensional Schrodinger Wave equation as

$$\frac{d^2y}{dx^2}+\frac{8\pi^2m}{h^2}\left[E-V(x)\right]\psi=0 \qquad .....(7.1)$$

V(x) is the periodic potential and satisfies the equation

$$Vt(x) = V(x + a) \quad .....(7.2)$$

a represents the periodicity

Suppose the wave equation has the solution of the form

$$\psi(x) = 4(x)e^{\pm ikx} \quad .....(7.3a)$$

The solution given in equation 7.3a is known as Bloch function. In three dimensions, it is written as

$$\psi_k(\bar{r}) = U_k(r)e^{\pm(ik\bar{r})} \quad .....(7.3b)$$

But, we shall restrict ourselves to one dimensional case only. From the above form of the solution, we can see that the free electron wave is modulated by the periodic function $U_k(x)$.

Now let us assume that they are N atoms in the lattice row and they are even: Let the length of the chain be L, then

$$U_k(x) = U_k(x + Na) \quad .....(7.4a)$$

And above

$$U_k(x) = U_k(x + a) \quad .....(7.4b)$$

Then substituting the proposed solution in equation (2), we have

$U_k(x)$ has the same periodicity as the lattice. This is known as Bloch Theorem.

The Bloch function

$\psi(x) = \exp(ikx)\, U_k(x)$ has the properly

$$\psi(x + Na) = U_k(x + Na) \exp ik(x + Na)$$

Writing in simple form

$$\psi(x + Na) = U_k(x + Na)e^{\pm ikx}.e^{\pm ikNa}$$

$$= U_k(x)e^{\pm ikx}e^{\pm ikNa} \quad .....(7.5)$$

$$= \psi(x)e^{\pm ika} \quad .....(7.6)$$

$$= \phi(a)\,\psi(x) \quad .....(7.7)$$

where

$$\phi(a) = e^{\pm ika} \quad .....(7.8)$$

It is evident from above, that the electron wave is represented by the Schrodinger equation and the solution should satisfy the condition given in eqn. (7.7) and has the property given in eqn. (7.8). This is the essence of Bloch Theorem. Let us consider only one type of the solution i.e

$$\psi(x + Na) = \psi(x)e^{\pm ikx}$$

$$= \phi(a)e^{+ikx}$$

Taking the complex conjugate of the above, we get

$$*\psi(a+Na)=\phi(x)e^{-ikx}$$

$$=\phi(x)\psi*_k(x)$$

Then,

$$\psi(x+Na)\psi*(x+Na)$$

$$=\phi(a)\phi*(a)\psi_k(x)\psi*_k(x)$$

$$=\psi_k(x)\psi*_k(x) \qquad .....(7.9)$$

The deduction from the above equation is that the probability of finding the electron is same everywhere in the entire row of the atoms. Obviously, the electron is not localized around any particular atom. The electron belongs to all the atoms (i.e shared by all atoms) in the row of atoms.

If the above condition has to be satisfied, we should have

$$e^{ikNa}=1$$

This can be true only if

$$kNa=n2\pi$$

$$k=\frac{n2\pi}{Na}=\frac{2\pi n}{Na}=\frac{2\pi n}{L} \qquad .....(7.10)$$

where L is the length of the row of atoms. The integer 'n' can take values 0, 1, 2, 3, ... If we consider that N = 2n then $k=\frac{\pi}{a}$; This represents the boundary and is called the first Brillouin Zone. The allowed values of k within the first zone is equal to N.

With this back ground, let us take up kronig – Penney model, a mathematical model to describe the occurrence of energy bonds and is a quantitative description of the consequence of periodic potential.

A crystal is a combination of atoms arranged in a regular fashion by bringing atoms to closer spacings.

This results in the formation of energy bonds separated by a forbidden Kronig – Penney model shows the forbidden gap regions intervening the regions of allowed energies. A simple one dimensional model is considered.

7.3 The Kronig - Penney Model

A one dimensional single crystal lattice is represented by one dimensional Kronig-Penny model of periodic potential function. It is assumed here that the potential energy of an electron, when it moves in one-dimensional perfect crystal lattice, is represented in the form of rectangular wells as shown in Fig. 7.3. The period of the potential is (a + b); in regions where $0 < x < a$, the potential energy is assumed to be zero and in regions such as $-b < x < 0$, the potential energy is V_o.

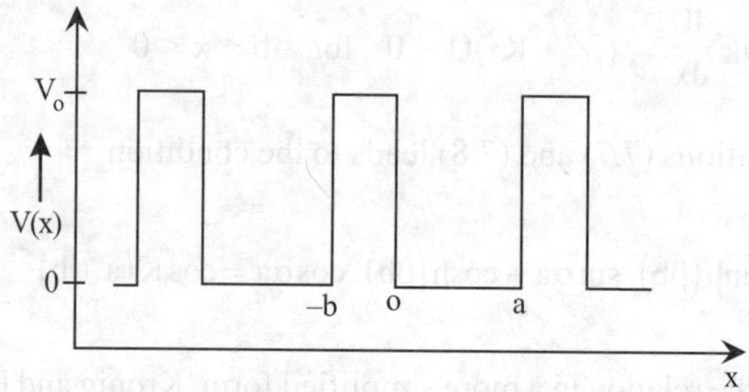

Fig. 7.3 Kronig-Penney one-dimensional Periodic potential well.

Though this is a hypothetical model, far from truth, it illustrates many important features of the quantum behaviour of electrons in a periodic lattice. Through this model, the Schrodinger equation can be solved explicitly in terms of elementary functions. The Schrodinger equation for the two regions can be written as

$$\frac{d^2\psi}{dx^2} + \left(2m/\hbar^2\right)E\psi = 0 \qquad \text{for } 0 < x < a \qquad .....(7.11)$$

and

$$\frac{d^2\psi}{dx^2} + \left(2m/\hbar^2\right)(E - V_o)\psi = 0 \qquad \text{for } -b < x < 0 \qquad .....(7.12)$$

Let us assume that the energy E of the electrons under consideration is less than V_o.

Let us put $\alpha^2 = \dfrac{2mE}{\hbar^2}$ and $\beta^2 = 2m(V_o - E)/\hbar^2$ .....(7.13)

in the above equations, where $\propto$ and β are two real quantities. On substitution of (7.13) in (7.11) and (7.12), we have

$$\frac{d^2\psi}{dx^2} + \alpha^2\psi = 0 \quad \text{for} \quad 0 < x < a \qquad .....(7.14)$$

$$\frac{d^2\psi}{dx^2} - \beta^2\psi = 0 \quad \text{for} \quad -b < x < 0 \qquad .....(7.15)$$

The Bloch theorem states that the solution of wave equation for a periodic potential will be of the form of a plane wave modulated with the periodicity of the lattice of the form.

$$\psi = U_K(x)\, e^{iKx} \qquad .....(7.16)$$

where $U_k(x)$ is the periodic function in x with period (a + b). K is constant of motion.

Substituting these in (7.14) and (7.15), we get

$$\frac{d^2U}{dx^2} + 2iK\frac{dU}{dx} + (\alpha^2 - K^2)U = 0 \quad \text{for } 0 < x < a \qquad .....(7.15)$$

$$\frac{d^2U}{dx^2} + 2ik\frac{dU}{dx} - (\beta^2 + K^2)U = 0 \quad \text{for } -b < x < 0 \qquad \text{.....(7.16)}$$

Solution of the equations (7.7) and (7.8) leads to the condition

$$\frac{\beta^2 - \alpha^2}{2\alpha\beta}\sinh(\beta b).\ \sin\alpha a + \cosh(\beta b).\ \cos\alpha a = \cos K(a+b) \qquad \text{.....(7.19)}$$

In order to express the relation in a more simplified form, Kronig and Penney considered the case, for which the potential barrier becomes periodic delta function, i.e., V_o tends to infinity and b approaches zero, but the product V_ob or $\beta^2 b$ (since as $V_o \to \infty$, $\beta^2 \to \infty$) remains finite. With these limits imposed, eqn. (7.19) becomes

$$\frac{mV_ob}{\hbar^2\alpha}\ \sin\alpha a + \cos\alpha a = \cos Ka \qquad \text{.....(7.20)}$$

Let us define a quantity P such that

$$P = \frac{mV_oba}{\hbar^2} = \lim_{\substack{b\to 0\\ \beta\to\infty}} \ \lim \frac{\beta^2 ab}{2} \qquad \text{.....(7.21)}$$

which is clearly a measure for the 'area' V_ob of the potential barrier. In other words, increasing P means binding a given electron more strongly to a particular potential well. It also tells us that when $P \to 0$, the electron behaves like free electron.

Then, finally eqn. (7.20) can be written as

$$P\ .\ \frac{\sin\alpha\ a}{\alpha\ a} + \cos\alpha a = \cos Ka \qquad \text{.....(7.22)}$$

The eq. (7.8) gives relation between the parameter K and the potential barrier V_{ob} if we assume $\alpha a = Ka$,

then, from (7.3) $\quad \alpha^2 = \dfrac{2mE}{\hbar^2} = K^2$

and hence $\quad E = \dfrac{K^2\hbar^2}{2m}$

which was the result obtained in the case of Sommerfeld free electron model.

If a graph is plotted representing the left hand side of the eq. (7.12) on the y-axis as a function of αa, for a value $P = 3\pi/2$, we get a curve as represented in Fig. 7.4.

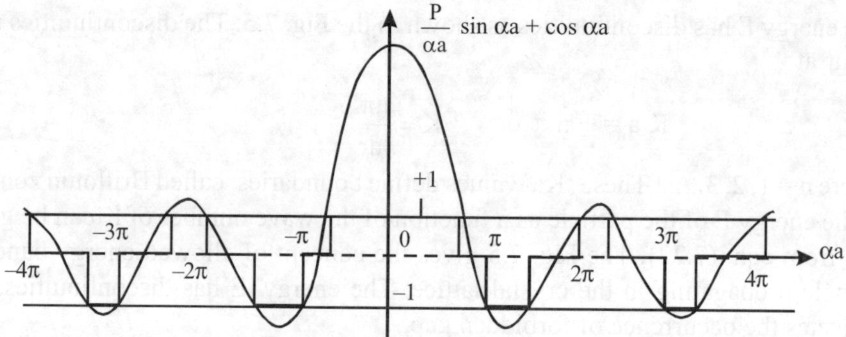

Fig. 7.4 Plot of the function $P\frac{\sin \alpha a}{\alpha a} + \cos aa$ for $p = \frac{3\pi}{2}$. The allowed values of the energy E are given by those ranges of $a = \left[\frac{2mE}{\hbar^2}\right]^{\frac{1}{2}}$ for which the function lies between +1 and –1.

Since, $\alpha^2 = 2mE/\hbar^2$ the abscissa is a measure of energy. Further from the eqn. (7.12), one can see that due to the restriction imposed by the right hand side of the equation, cos Ka, which can take values only between +1 and –1.

7.4 Origin of Energy Band Formation in Solids

1. The parameters α is related to the total energy E of the electron. A plot of the energy E of the particle as a function of the wave number K can be generated from Fig. 7.5. Fig. 7.6 shows the plot giving the concept of allowed energy bands for the electrons propagating in the crystal lattice.

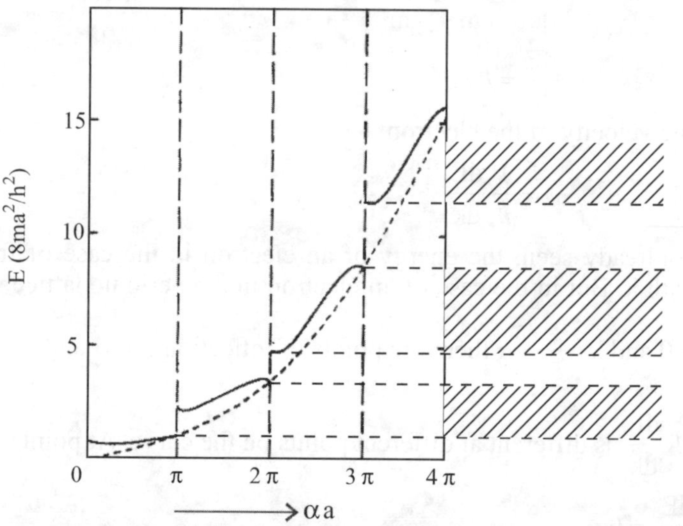

Fig. 7.5 Plot of energy vs wave number for the Kronig-Penney potential. The energy bands indicated.

The energy E has discontinuities as shown in the Fig. 7.5. The discontinuities in energy occur at

$$K a = \pm n \pi \text{ or } \quad K = \frac{n\pi}{a} \qquad .....(7.23)$$

where n = 1, 2, 3....... These 'Ka' values define boundaries, called Brillouin zones. A plot of the energy E of the particle as a function of the wave number of K can be generated, and from eqn. (7.22). The Fig. 7.5 gives the concept of allowed energy bands for the particle propagating in the crystal lattice. The energy, E has discontinuities, and this indicates the occurrence of forbidden gap.

2. The width of the allowed energy bands increases when 'αa' increases. Then, the forbidden regions become narrower.
3. The width of the allowed band, increases with the increasing value of P. P measures the 'binding energy' of the electrons to the lattice.

   When P tends to be infinite, the band reduces to discrete energy levels. The electron becomes a free particle in a potential box of atomic dimensions.

7.5 Concept of Effective Mass of an Electron and Hole

7.5.1 Energy and Velocity of an Electron

In the case of a free electron model, we have learnt that the energy and wave vector k are related as

$$E = \frac{\hbar^2 k^2}{2m}$$

or $$\frac{dE}{dk} = \frac{\hbar^2 k}{m} = \frac{\hbar p}{m} \qquad (p = \hbar k) \qquad .....(7.24)$$

or $$\frac{dE}{dk} = \hbar v$$

where v is the velocity of the electron.

$$\therefore \qquad v = \frac{1}{\hbar}\frac{dE}{dk} \qquad .....(7.25)$$

As we have already seen, the energy of an electron in the case of free electron varies parabolically with k. But in the case of an electron in the periodic lattice varies as shown in Fig. 7.6. At k = 0 and $k = \pm \frac{\pi}{a}$, there are points of inflection (at M and N).

The value of $\frac{dE}{dk}$ is different at different points on the curve. At points M, O, N, the curve is flat, hence $\frac{dE}{dk} = 0$.

So, at points M, O, N, the velocity of electron is zero. At points A and B, $\frac{dE}{dk}$ is maximum and the velocity is maximum. The slopes at A and B are negative and positive respectively and let $k = k_0$ at A and B. The velocity decreases on either side of $k = k_0$. This is completely different from that of a free electron. The variation of velocity with k-value is shown in Fig. 7.7.

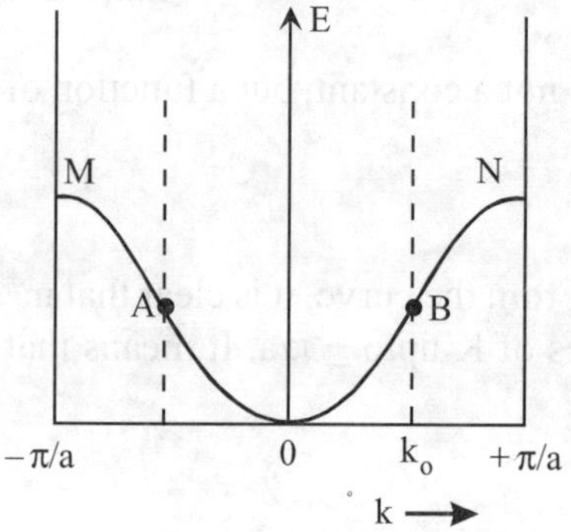

Fig. 7.6 E vs K curve.

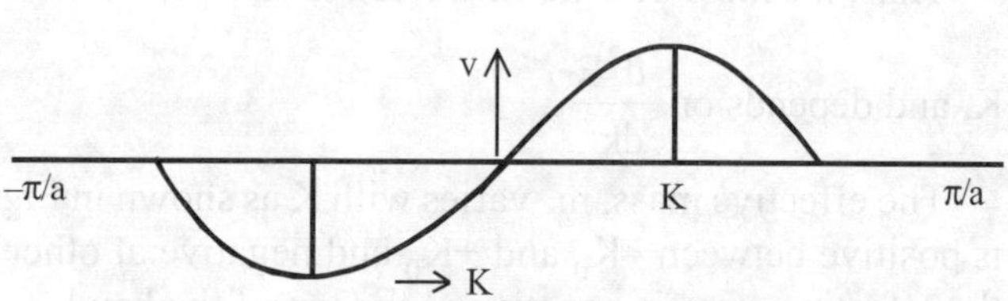

Fig. 7.7 v vs K curve.

7.5.2 Effective Mass of the Electron

Suppose, we apply an electric field (ξ) on the electron which is moving in one dimensional periodic potential. Suppose, it is initially at the state k. Then, the energy gained, dE in time dt is given as

$$dE = e\xi\, v\, dt \qquad .....(7.26a)$$

$$= e\xi\left(\frac{1}{\hbar}\frac{dE}{dk}\right)dt \qquad \text{(from eqn.7.25)}$$

Writing $dE = \left(\frac{dE}{dk}\right)dk$, we get

$$\frac{dE}{dk} = \frac{e\xi}{\hbar} \qquad .....(7.26b)$$

Applying Newton's second law,

$$a^* = \frac{dv}{dt} = \frac{d}{dt}\left(\frac{1}{\hbar}\frac{dE}{dk}\right) = \left(\frac{1}{\hbar}\right)\frac{d^2E}{dk^2}.\frac{dK}{dt}$$

i.e.

$$a^* = \left(\frac{e\xi}{\hbar^2}\right)\frac{d^2E}{dk^2} \qquad .....(7.27)$$

This is different from what is true for a force electron $\left(a = \frac{e\xi}{m}\right)$

Comparing the two equations, we notice that the electron in a periodic lattice moves as if it has an effective mass, m* given by

$$m^* = \hbar^2 \Big/ \frac{d^2E}{dK^2} \qquad .....(7.28)$$

Thus, the mass of a moving electron in a periodic lattice is not a constant, but a function of K, and depends on $\frac{d^2E}{dk^2}$.

The effective mass, m^* varies with K as shown in Fig 7.8. From the curve, it is clear that m^* is positive between $-K_0$ and $+K_0$ and negative at other values of K upto $\pm\,\pi/a$. It means that the effective mass is negative at the top of the band.

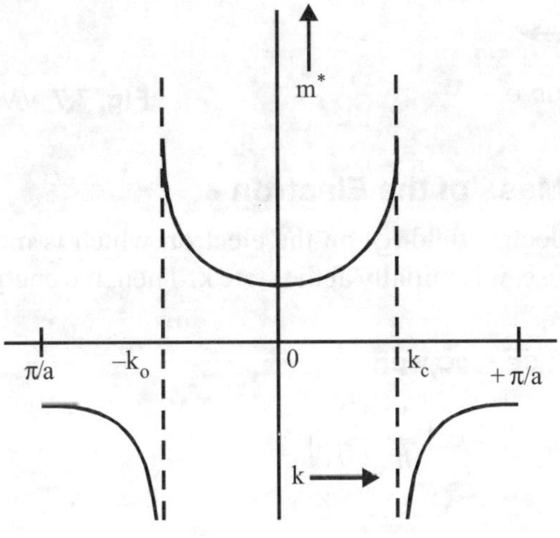

Fig. 7.8 Variation of m* with K.

The electron gets decelerated beyond K_0. An electron near the top of an allowed energy band behaves as if it has a negative mass. This motion is compared to the movement of small ice cubes dropped in a water container.

The force on the electron is given as

$$F = \hbar\frac{dk}{dt} \qquad .....(7.29)$$

The force produces a change in the velocity in a direction other than the direction of F.

7.7 Classification of Metals, Conductors Semiconductors and Insulators

The conductivity of a crystal can be obtained from the knowledge of the effective number of free electrons in a band. Consider a band and assume that it is filled upto k_o, as shown in Fig. 7.10. The following points have to be noted.

(i) The effective number of (free) electrons in a completely filled band vanishes because dE/dK is zero at the top of the band.

(ii) The effective number of (free) electrons reaches a maximum for a band field upto the inflection point of the E (K) curve, because there dE/dK reaches a maximum.

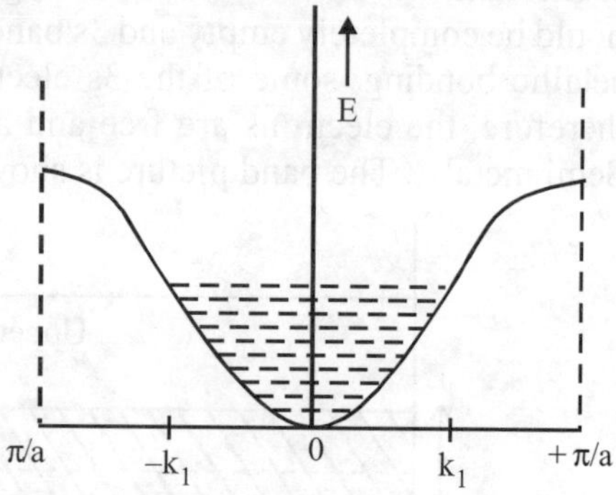

Fig. 7.10 Energy Band filled upto k_1

From the point of view of electrical nature, materials are classified as (1) metals (2) insulators and (3) semiconductors. The differences among them can be understood on the basis of their band structure only. We will only mention here how the band structure is different for the three category of materials without going into details

The number of energy states in a band depends on the crystal as calculated on the basis of number of primitive cells in a statistical mechanics. The number of electrons available depends upon the electron configuration of the atoms. It is the valence electrons that ultimately fill the top most band. The atoms should have odd number of electrons to partially fill the band.

Metals: The first type of metals are the elements that have odd number of electrons in the atoms. The monatomic elements like sodium, rubidium copper, silver, gold and tri-valent elements like aluminum are the examples. Consider the example like sodium. It has an electron configuration ($1s^2\ 2s^2\ 2p^6$) $3s^1$. The inner bands are formed by bracketed electron states and $3s^1$ electrons group together to form half filled band. The electrons with old number of valence electrons give partially filled top most band. There fore, they are good metals. The band structure is shown in Fig 7.11(a) & (b)

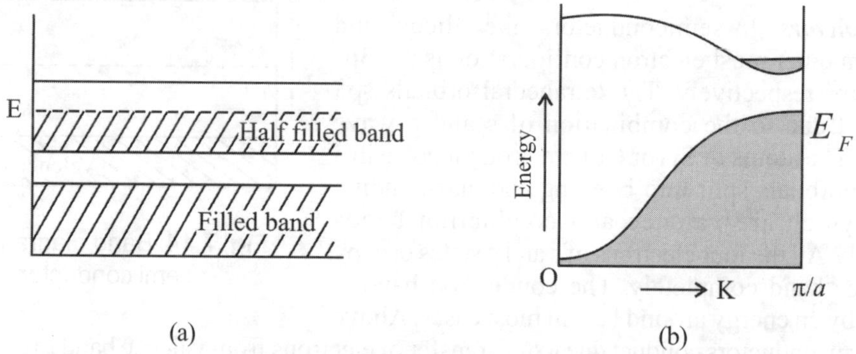

(a) (b)

Fig. 7.11 Band diagram of metals

Semi metals: There are some elements like alkali earth metals (like barium, magnesium, calcium etc.) in which there are even number of valence electrons. As a rule, they should have a filled topmost band and the element should be insulator. But this is not so. Let us take the example of magnesium, that has an electron configuration $1s^2 2s^2 2p^6 3s^2$. It means that 3p electron band should be completely empty and 3s band should be completely filled. However, for the sake of metallic bonding, some of the 3s electrons go into 3p bands and these two bands overlap. Therefore, the electrons are free and are available for conduction. These metals are called 'Semi metals'. The band picture is shown in the Fig. 7.12 (a) and (b).

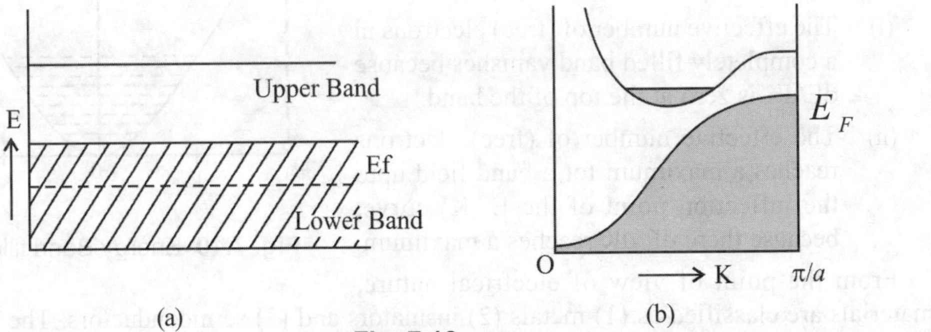

Fig. 7.12

Insulators: Insulators resist the flow of charges. They do not possess any free charges. They are highly resistive. Diamond is a good example for insulating substance. Other examples are SiO_2, TiO_2, ZrO_2, MgO, Al_2O_3, GaN etc.

Diamond is made up of carbon atoms with electron configuration $1s^2 2s^2 2p^2$. In this sp3 hybrid splits into two bands, each of the bands can accommodate four electrons. The electrons belonging to 2s and 2p shells fill the lower bands. The upper band is left totally empty. The edge of (bottom) of the upper band called conduction band is well above the top of the filled (called valence band) band. In diamond, the band gap is 5 eV. The representative band diagram is shown in Fig. 7.13.

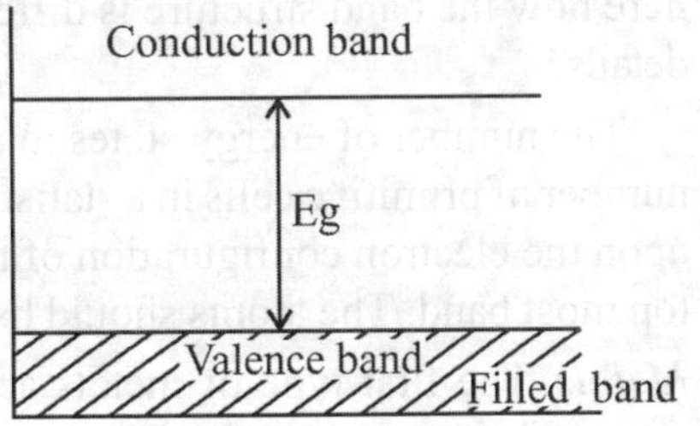

Fig. 7.13 Band diagram of an insulator

Semiconductors: In semiconductors like silicon, and germanium outer most electron configuration is $3s^2 3p^2$ and $4s^2$, $4p^2$ respectively. The tetrahedral orbitals sp3 are formed due to the combination of s and p wave functions. The atoms of Si (or Ge) are brought colse in a solid. The orbitals split into bonding and anti-bonding orbitals which are valence and conduction bands respectively. All the four electrons of s and p states occupy the valence band completely. The conduction band is separated by an energy around l eV in most cases. Above 0K, the semiconductors conduct due to the transfer of electrons from valence band to conduction band. Any electron in conduction band is a free electron. The band diagram is shown in Fig 7.14

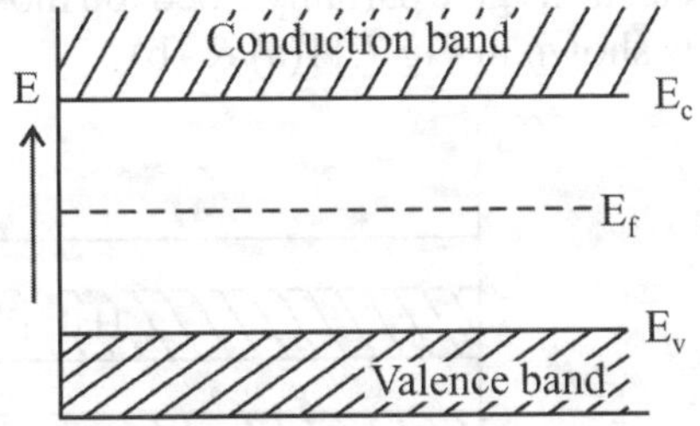

Fig. 7.14 Band diagram of a semi conductor

Example :

Calculate the conductivity of Copper at 300K. The Collision time for electron scattering is 3×10^{-14}s at this temperature.

Solution :

The number of free electrons per m^3 of Copper is given by;

$$n = \frac{Nd}{A}$$

$$n = \frac{6.023 \times 10^{23} \times 8960}{0.06354} = 8.5 \times 10^{28} m^{-3}$$

$$\sigma = \frac{ne^2\tau}{m} = \frac{8.5 \times 10^{28} \times (1.602)^2 \times 10^{-38} \times 3 \times 10^{-14}}{9.1 \times 10^{-31}}$$

$$= 7.19 \times 10^7 (\text{Ohm m})^{-1}$$

N = Avogadro number = 6.023×10^{23} mol^{-1}

A = At. wt = 63.54 amu

d = density = 8960 kg/m^3

e = charge of electron = 1.602×10^{-19}C

m = rest of mass of electron = 9.1×10^{-31}kg.

Problems

1. Ignoring the random velocities of electrons, calculate the Hall coefficient of copper which has a FCC structure. If the conductivity of Cu is 10^8(Ohm m)$^{-1}$, estimate the electron mobility.

   (Hint : FCC structure has 4 effective lattice points)
2. Assuming one free electron per atom in Copper whose conductivity at room temperature is 10^8(Ohm m)$^{-1}$, calculate the relaxation time and mean free path for electrons at the Fermi level.
3. The Fermi level for Potassium is 2.1eV. Calculate the velocity of the electron at the Fermi level.
4. Assuming the free electron theory, calculate the Fermi energy of bcc lithium and fcc aluminium. The dimensions of the unit cells are 3.50A and 4.04A respectively.

Questions

1. Show that the Kronig-Penney model leads to energy band structure in solids.
2. Discuss the consequences of periodic potential on the variation of velocity and mass of an electron in a solid under the influence of an electric field.
3. On the basis of band theory, explain how materials are divided into metals (conductors), semiconductors and insulators. How does the density states curve change in the above materials?
4. What are the conclusions that are arrived at from Kronig-Penny model?

5. What is meant by 'Effective Mass' ?
6. Show how does velocity of the electron change with K-Vector's magnitude?
7. Give the energy vs K on the basis of Kronig-Penny model.
8. Discuss how the velocity of an electron change with |K| value.
9. Mention the points at which the velocity of the electron change within the first Brillouin zone
10. Compare the velocity of a free electron and the velocity of an electron in a periodic potential
11. What is an effective mass of an electron? Discuss the variation of effective mass of an electron with in a zone
12. What is meant by negative mass of an electron. Can really there be a negative mass for any particle?
13. Analyse the relation $P\frac{\sin\alpha a}{\alpha a} + \cos\alpha a = \cos Ka$ from the point of view of the origin of energy band.
14. How does the width of an energy band change with d a $\left(\alpha = \frac{\sqrt{3mE}}{\chi};\right)$ and $P\left(\frac{mV_0 b}{\chi d}\right)$.
15. On what factors does the band width depend?
16. Explain the concept of 'hole'
17. Discuss the effective mass of hole with reference to that of electron.
18. Discuss the band structure of metal.
19. Explain the band structure of semi metal
20. Explain the band structure of an insulator like diamond.
21. Explain the band structure of semiconductor in comparison to that of an insulator.
22. How does the band model of materials help us to distinguish metals, insulators and semiconductors?
23. Distinguish between metal, insulator and semiconductors
24. Discuss the motion of an electron in a periodic potential field and explain the formation of energy bands.
25. Discus the motion of an electron in a periodic Lattice.
26. Discuss the kronig-Penney model for the motion of an electron in a periodic potential.
27. Explain Bloch theorem.
28. Explain the term 'effective mass' of an electron. Explain the conditions under which the effective mass of an electron become positive and negative.
29. What in first Brillouin zone.
30. Distinguish between 'electron' and 'hole'.
31. Explain why a metal is a good conductor and insulator is not, on the basis of band diagram.

Objective Questions

I. Choose the correct choice

1. Free electron model can not explain
   (a) the conductivity behavior of metals
   (b) the positive temperature coefficient in metals
   (c) the conduction in semiconductors
   (d) the thermal conductivity behavior of metals
2. The main out come of Kronig - Penney model for electron in a periodic lattice is
   (a) the energies of electrons lie in the for bidden gap
   (b) the energy spectrum consists of an infinite number of allowed bands separated by gap,
   (c) the electron occupies energy levels separated by gaps
   (d) the width of the allowed energy band decreases with the increasing binding energy of electrons
3. According to band theory the velocity of the electron in a periodic lattice is
   (a) zero at the edge of the band
   (b) infinite at the edge of the band
   (c) dependent on the shape of the crystal
   (d) finite only at the edge of the bands
4. Read the statements and answer
   (A) An energy band is nothing but a band of closely spaced energy levels of atoms in the solid
   (B) The effective number of free electrons in a completely filled band is zero
   (a) (A) and (B) is wrong
   (b) (A) is correct and (B) is wrong
   (c) (A) is wrong and B is correct
   (d) (A) and (B) are correct
5. According to the band theory of solids,
   (a) an insulator possesses over lapping bands
   (b) an intrinsic semiconductor possesses overlapping bands
   (c) an intrinsic semiconductor contains few electrons in conduction band at any temperature above 0 K,
   (d) above 0 K, insulator becomes a semiconductor by a movement of energy bands
6. The effective mass of an electron is considered as (a + b = periodicity)
   (a) positive between $k = \pm \frac{\pi}{a}$
   (b) positive between $k = \pm \frac{\pi}{2a} = k_o$
   (c) negative between $k = \pm \frac{\pi}{2a} = k_o$
   (d) negative between $k = \pm \frac{\pi}{a}$
7. The expression for effective mass of an electrons is
   (a) $\hbar^2 (d^2E / dK^2)$ (b) $h^2 (d^2E / dK^2)$
   (c) $\frac{\hbar^2}{d^2E / dK^2}$ (d) $\hbar^2 k^2$
8. Choose the correct statement
   (a) If the binding energy of electron is increased the width of the allowed band also increases
   (b) The width of allowed band decreased if the binding energy of electron inceases
   (c) The region between $\frac{2\pi}{a}$ to $\frac{-2\pi}{a}$ is called first Brilloin zone.
   (d) Find Band theory concludes that the energy of an electron in a solid is $\frac{\hbar^2 K^2}{2}$

9. The acceleration of electron in a periodic Lattice is
   (a) dependent directly on dE / dK
   (b) dependent directly on d^2E / dK^2 and the electic field
   (c) is no dependent of the electric field
   (d) is dependent on temperature alone
10. Choose the correct statement
   (a) The mass of a bound electron in a periodic lattice is constant
   (b) Insulators possess over lapping bands.
   (c) Semi metals and semi conductors are one and the same
   (d) Semi metals have a band gaps of 1 ev.

II. Fill in the blanks :

1. The nature and quantity of conductivity differences between insulators and conductors can be explained by __________ theory
2. The energy levels in an atom can accommodate as many electrons as allowed by __________ principle.
3. When atoms are brought closer to each other as in a solid, the isolated energy levels group together to form an energy __________ consisting of closely spaces sets of __________.
4. The solid consists of energy bands, each energy band is separated from the next energy band by a finite __________ .
5. If m = mass of the electron, n = the number of electrons per unit volume, $\tau(E_F)$ is the collision time, then, the current density J_x in the metal as a result of the electric field E_x is __________.
6. If E_F is the Fermi energy, m* is the effective mass of the electron, then the velocity of the electrons is given as __________.
7. According to ohm's law, the current density J, the conductivity σ and the electric field E are related as __________.
8. In metals, there is an over lapping of __________ and __________ bands.
9. The band theory of solids deals with electron motion in a __________ field of a crystal.
10. Near the bottom of band, the effective mass of the electron is __________ in sign.
11. The potential energy of an electron that moves in one dimensional perfect crystal lattice is represented in the form of __________ wells.
12. The energy of a free electron in the case of Sommerfeld free electron model is __________, where k is the wave vector associated.
13. The drift velocity V_D of the electron is much __________ than the velocity V_F of the electron at the Fermi surface.
14. The __________ of the electron is proportional to k, the wave vector.
15. The electrons in a periodic lattice move with __________ mass.
16. The binding energy of the electrons in a periodic lattice is infinitely large, i.e. the electrons behave like particles in a constant __________ of atomic dimensions.
17. The width of the allowed band __________ with increasing binding energy of the electrons.
18. Any solid for which, a given band is completely filled and the next allowed band is completely empty is called __________.
19. According to Kronig - Penney model, the boundaries of allowed energy range correspond to k values given by __________.
   (**Hint :** (a+b) is the periodicity, a >> b)
20. The value fo E_F is determined by the condition that the total number of particles is __________ at a given temperature T.
21. The effective mass of the electron in a band is expressed as __________.
22. When the electron is near the top of the band, then its effective mass is __________.
23. A hole near the top of an allowed band has __________ effective mass.
24. Semi metals possess __________ bands.
25. The number of energy states in a band is equal to the __________ of states.
26. The upper most band is filled with __________ electrons.

27. If an element has odd number of electrons, the upper most band will be __________ filled.
28. The effective mass of hole is __________ that of electron in magnitude
29. Bolck theorem states that $u_K(x) =$ __________ where 'a' represents the periodicity.
30. An electron in a band moves as if it has a mass given by the expression __________.
31. The expression $\hbar \frac{dK}{dt}$ give the ________ on the electron in a lattice.

III. True or False

1. When an electron moves in periodic potential of lattice of a solid, their, then amplitude of the wave associated it has the form $u_k(x) = u_k(x + a)$. [a represents the periodicity]

   [True/False]
2. An Electron in a potential of V (x) is represented by

$$\frac{d^2\psi}{dx^2} + \frac{8\pi^2 m}{h^2}[E - V(x)]\psi = 0$$

   [True/False]
3. In the equation

$$P\frac{\sin\alpha a}{\alpha a} + \cos\alpha a = \cos Ka$$ P represents the scattering power of the potential barriers.

   [True/False]
4. A E-K diagram for an electron in a periodic potential shows that E–K plot is a continuous parabola.

   [True/False]
5. The region between $\frac{-\pi}{a}$ to $\frac{+\pi}{a}$ is called first Brillouin zone.

   [True/False]
6. When the scattering power of the potential barriers in a solid tends to become zero, the solid behaves like a metal.

   [True/False]
7. The velocity of an electron in a periodic lattice is given as $\frac{h^2 K}{m}$.

   [True/False]
8. The maximum velocity that an electron can attain is at the inflection on E-K curve.

   [True/False]
9. The effective mass of electron in determined by factor dE/dK.

   [True/False]
10. In the upper half of the band, the electron behaves like one having a negative mass.

    [True/False]
11. These in no effective mass for holes.

    [True/False]
12. Metals show discontinuous E-K diagram.

    [True/False]
13. The electron energies in a soild are such that electrons tend to stay in the forbidden gap.

    [True/False]
14. At temperature above 0K, the probability of occupancy of Fermi level is 50%.

    [True/False]
15. The parabolic relation between E and K is valid in solids other than metals.

    [True/False]
16. The band of energy levels higher to the valence band will be vacant at absolute zero.

    [True/False]
17. Electrons can not be excited across a forbidden band with thermal energy.

    [True/False]
18. Metals like beryllium are conductors because of band over lapping and formation of a partially filled band at the top.

    [True/False]
19. The effective mass of a 'hole' is not different from the actual mass of an electron in a solid.

    [True/False]

CHAPTER 8

Semiconductor Physics

8.1 Introduction

We have studied in the earlier section that semiconductors are associated with relatively narrow forbidden band gaps. for example; germanium, silicon, and gallium arsenide are associated with band gaps of 0.67, 1.01 and 1.43 eV respectively. Indium arsenide has a band gap of 0.17 eV. The resistivities of semiconductors are low; the resistivity of pure silicon is about 10^3 ohm- m, while that of germanium is 10^{-1} ohm-m. Semiconductors are classified as elemental semiconductors and compound semiconductors. Silicon, germanium, and grey Se are elemental semiconductors. The compound semiconductors are made of two or three elements. The difference in electro-negativity leads to a combination of covalent and ionic bonding in compound semiconductors. Some of the simple compound semiconductors are: 1. group III –V compound semiconductors like GaAs, GaP, GaN, InSb, 2. group II-VI compound semiconductors ZnS, CdSe, PbTe, and mixed semiconductors like Si-Ge. There are oxide semiconductors like Cu_2O, ZnO, and also semiconductors like PbS, SiC. A list of semiconductors is given in the Table8.1 along with their band gaps and the type.

On the basis of band structure, the semiconductors are classified as direct (example-GaAs, InSb, InP etc), and indirect band gap semiconductors (example- Si, Ge).

Ternary semiconductors are formed by the addition of a small quantity of a third element to the compound semiconductors, for example $Al_x\,Ga_{1-x}\,As$. The subscript x refers to the alloy content of the material, what proportion of the element is added and what proportion is replaced by the alloy material. The addition of alloys to semiconductors can be extended to include quaternary materials such as $Ga_x\,In_{(1-x)}\,As_y\,P_{(1-y)}$ or GaInNAs.

Table 8.1 Some semiconductors with their band gaps.

Crystal	Type of gap	E_g (eV) At 300K	Crystal	Type of gap	E_g (eV) At 300K
Si	I	1.11	GaAs	D	1.43
Ge	I	0.67	CdTe	D	1.44
GaP	I	2.25	InAs	D	0.36
Grey Tin	D	-	ZnO	-	3.2
InP	D	1.27	ZnS	-	3.6

I- refers to Indirect, D- refers to direct band gaps.

The semiconductors are divided into two categories. They are 1.Intrinsic, and 2.Extrinsic semiconductors. The intrinsic semiconductors are the pure semiconductors. The extrinsic semiconductors are impurity doped (doped deliberately) semiconductors. The conductivity of semiconductors is a strong function of temperature. Pure semiconductors are rarely used in devices, but their characteristics should be known thoroughly for any device fabrication. They are the starting materials for making the extrinsic semiconductors.

Intrinsic Semiconductors: Highly pure semiconductors function as intrinsic semiconductors. Their impurity level is so low that the contribution of the impurity to overall behavior is negligibly low. Even compound semiconductors, like basic gallium arsenide are intrinsic semiconductors. The formation of a crystal lattice of crystals like Ge is shown in Fig.8.1a.

The intrinsic semiconductors are insulators at 0K. The valence band is completely filled, conduction band is completely empty (See the Fig. 8.1.). The Fermi level in these semiconductors lies midway between the bottom (called band edge) of the conduction band and top of the valence band.

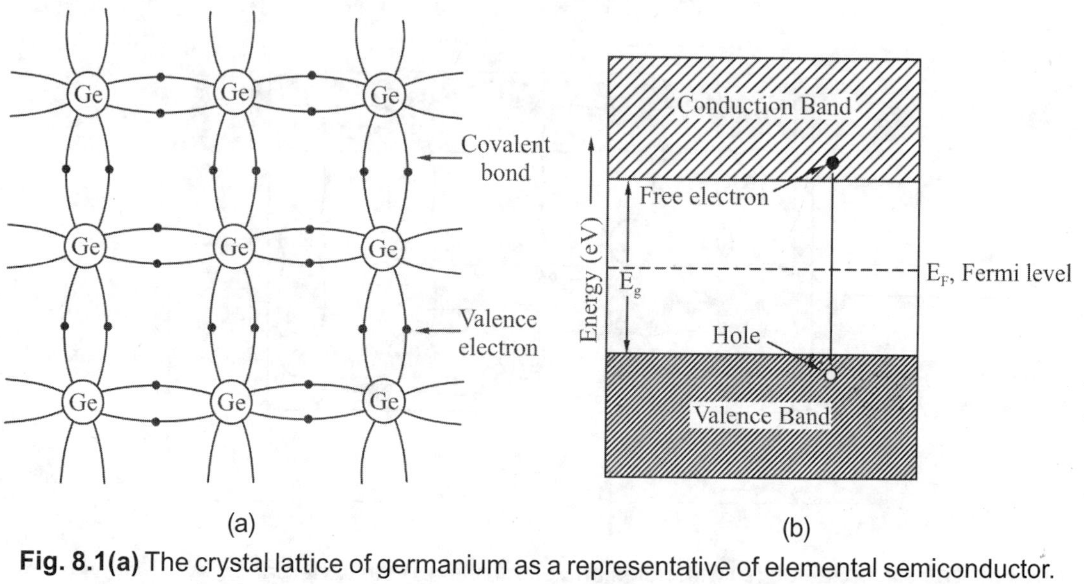

Fig. 8.1(a) The crystal lattice of germanium as a representative of elemental semiconductor. **(b)** the band picture of an intrinsic semiconductor. Electron-hole pair in the semiconductor.

At any temperature more than 0K, some electrons are transferred to empty levels of conduction band from valence band by the aid of thermal energy. The formation of electron-hole pairs is a characteristic of semiconductors (Fig. 8.1). The electrons tend to fall back to valence band and this may result in energy emission. At any temperature, some electrons would be present in the conduction band and equal number of holes would be present in the valence band. It means that there would be electron- hole pairs. When a voltage is applied, a current in the external circuit exists as shown in Fig. 8.2(b). The conductivity of the semi conductor is due to movement of holes in valence band and an electron in conduction band.This is shown in silicon crystal in Fig. 8.2 (a) and (b). We will study details in later sections.

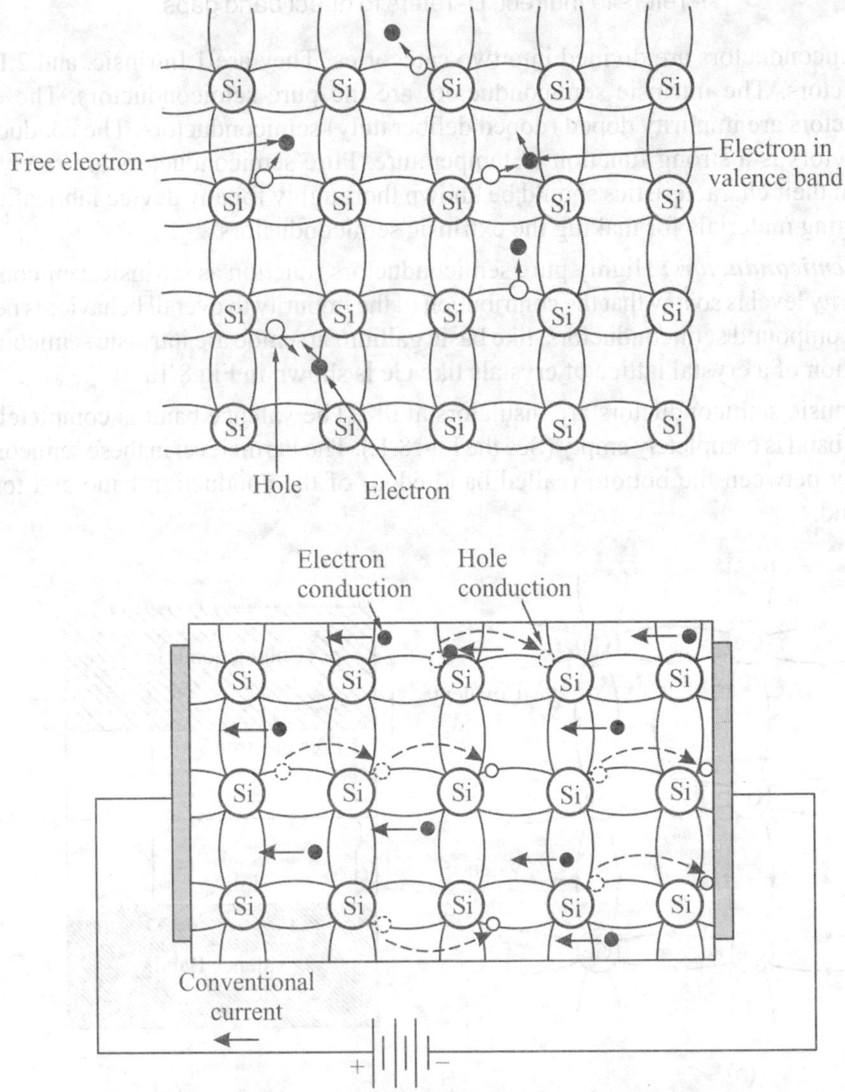

Fig. 8.2. The silicon lattice (a) with holes and electrons, (b) the schematic view of movement of electrons and holes. Electrons move opposite the field.

Extrinsic Semiconductors: Extrinsic semiconductors are of two types. They are 1. n-type and p-type semiconductors.

The n-type semiconductors are obtained by doping the intrinsic semiconductor with pentavalent atoms like arsenic, phosphorus and antimony. Band satisfaction needs only 4 electrons to be shared and the extra electron stays attached to the lattice. It is shown as orbiting. It requires very little energy to take an electron into the conduction band. This energy level is represented by E_d. And if electrons go into the conduction band, the donor atoms are ionized indicated by Nd^+. The atoms which contribute electrons are called donors (donor impurities). The donor levels are placed just below the conduction band as shown in the Fig.8.3. When electron is transferred to the conduction band, it leaves an ionized donor N_d^+.

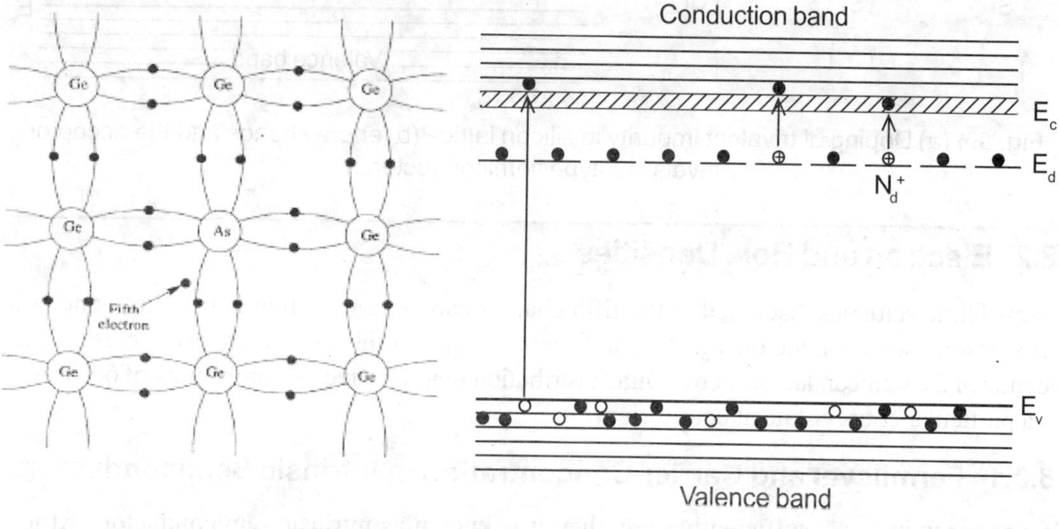

Fig. 8.3 (a) Doping of pentavalent impurity in germanium lattice (b) energy bands and the donor levels in n-type semiconductors.

The p-type semiconductors are obtained when trivalent atoms like Boron, Aluminum, Gallium, or Indium are doped. The Fig. 8.4(a) shows the effect of doping of trivalent atoms. These elements have three electrons in their outermost orbit and they occupy the substitutional positions. Then, only three of the four bonds needed are satisfied and the missing electron is a hole. The impurities accept electrons and therefore are called acceptors. The acceptors effectively introduce holes and their energy levels are just above the valence band as shown in Fig. 8.4 (b).The energy of the levels represent the binding energy of acceptor ion,(N_a^-).The Fermi level for n-type semiconductors lie below the conduction band edge and the fermi level for p-type semiconductors lies just above the top of valence band.

In III-V semiconductors, atoms from group II act as acceptors when occupying the site of a group III atom, while atoms in group VI act as donors when they replace atoms from group V. The dopant atoms from group IV have the property that they can act as acceptors or donor depending on whether they occupy the site of group III or group V atoms respectively.

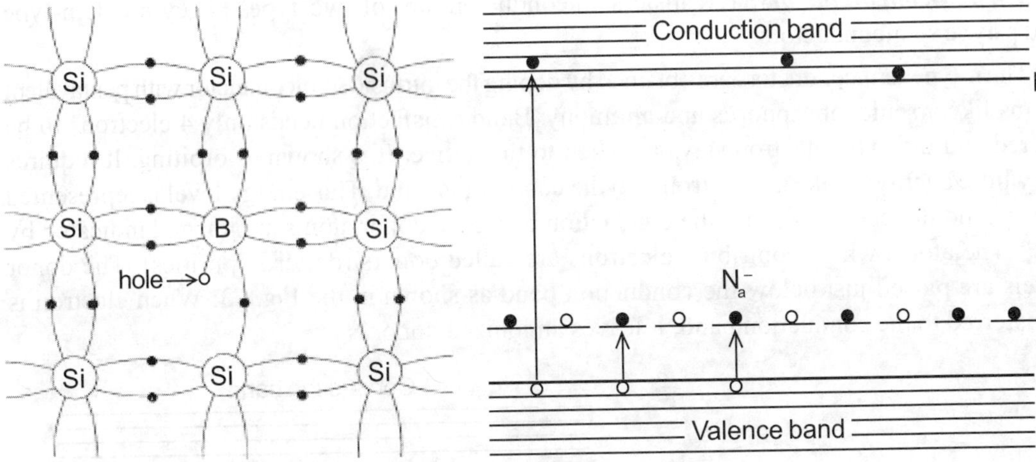

Fig. 8.4 (a) Doping of trivalent impurity in silicon lattice (b) energy bands and the acceptor levels in p-type semiconductors.

8.2 Electron and Hole Densities

Fermi-Dirac statistics discussed in the fifth chapter can be used to find out electron and hole densities in a semi conductor. Electron/hole density is the number of electrons or holes per unit volume of the semiconductor. Fermi Dirac distribution function gives the probability of occupancy of a particular energy state at equilibrium.

8.2.1 Fermilevel and Carrier Concentration in Intrinsic Semiconductors

If a semiconductor is sufficiently pure, then it is known as intrinsic semiconductor. At low temperatures all the valence electrons are bound to the atoms through covalent bonds. But at higher temperatures when thermal energy (kT) becomes comparable with the bond energy, some of the covalent bonds will be broken and electron hole pairs will be generated. The electrons can occupy conduction band and holes will be created in the valence band. Each broken covalent bond generates one electron and a hole. Thus the electron hole density in an intrinsic semiconductor will be equal and will be depending on the number of broken covalent bonds. The reason for rapture of covalent bond can be anything besides thermal excitation. Thus electron density $n = p =$ hole density $= n_i =$ Intrinsic density.

Under equilibrium conditions, the number of electrons per unit volume (n) present in the conduction band will be equal to the number of holes per unit volume (p) in the valence band.

The number of electrons dn_e per unit volume of a solid in the energy interval dE between the energies ε and $\varepsilon + d\varepsilon$ is given as

$$dn_e = \rho_e d\varepsilon \quad .....(8.3)$$

ρ_e is the density of electrons in the energy interval $d\varepsilon$, and is given by

$$\rho_e = N(\varepsilon)\ P_e(\varepsilon) \quad .....(8.4)$$

$N(\varepsilon)$ is the number of states available per electron volt per unit volume centered around ε and $P_e(\varepsilon)$ is Fermi Dirac distribution function for electrons.

The density of states $N(\varepsilon)$ is proportional to the square root of energy and is expressed by

$$N(\varepsilon) = \eta\varepsilon^{1/2}$$

where $\eta = \frac{4\pi}{h^3}\left(2m_e^*\right)^{3/2}$ is the proportionality constant

Here, m_e^* is the effective mass of the electron.

$$\therefore \quad N(\varepsilon) = \frac{4\pi}{h^3}\left(2m_e^*\right)^{3/2} \varepsilon^{1/2} \qquad .....(8.5)$$

$N(\varepsilon)$ is the number of electrons per unit volume and unit energy.

Therefore,

$$dn_e = \frac{4\pi}{h^3}\left(2m_e^*\right)^{3/2} \frac{\varepsilon^{1/2} d\varepsilon}{1+e^{(\varepsilon-\varepsilon_F)/kT}} \qquad .....(8.6)$$

At T = 0 K, $P(\varepsilon) = 1$, hence n_e is given by

$$n_e = \int_0^{\varepsilon_F} dn_e = \frac{2}{3}\frac{4\pi}{h^3}\left(2m_e^*\right)^{3/2} \varepsilon_F^{3/2} \qquad .....(8.7)$$

$$\varepsilon_F^{3/2} = \frac{3}{2}\frac{h^3}{4\pi}\left(2m_e^*\right)^{3/2} n_e$$

The above equation indicates that Fermi energy in a solid is dependent only on free electron concentration n_e at 0 K.

$$\varepsilon_F = \left(\frac{3}{2}\frac{h^3}{.8\pi}\right)^{2/3} \left(2m_e^*\right)^{-1} n_e^{2/3} = 3.64 \times 10^{-19} n_e^{2/3} \qquad .....(8.8)$$

8.2.2 Carrier Concentration

To find out the concentration of electrons in conduction band we have to integrate eqn. (8.6) from ε_c to ∞.

$$\therefore \quad n_n = \int_{E_c}^{\infty} \frac{4\pi}{h^3}\left(2m_e^*\right)^{3/2} \frac{\left(E-E_c\right)^{1/2} dE}{\left[1+\exp\left(E_F-E\right)/kT\right]} \qquad .....(8.9)$$

E_c = energy of conduction band edge ;

E_F = Fermi energy.

If $E_c - E_F >> KT$, one in the denominator can be neglected in comparison with exponential term and eqn. (8.9) can be rearranged as

Let $$\gamma = \frac{4\pi}{h^3}\left(2m_e^*\right)^{\frac{3}{2}} \qquad .....(8.10)$$

$$n = \gamma \exp\left[\left(E_c - E_F\right)/kT\right]\int_0^{\infty}\left[\left(E - E_c\right)^{1/2}\exp\left(E - E_c\right)/kT\right]dE$$

$$= \gamma \exp\left[\left(E_c - E_F\right)/kT\right]\left(kT\right)^{3/2}\int_0^{\infty} x^{1/2}e^{-x}dx$$

(Here $x = (E - E_c)/kT$)

$$n = \gamma\left(kT\right)^{3/2}\exp\left[-\left(E_c - E_F\right)/kT\right]\frac{\sqrt{\pi}}{2}$$

$$n = N_c \exp\left[-\left(E_c - E_F\right)/kT\right] \qquad .....(8.11)$$

$$N_c = \gamma\left(kT\right)^{3/2}\frac{\sqrt{\pi}}{2} = 2\left(\frac{2\pi m_e^* kT}{h^2}\right)^{3/2} \qquad .....(8.12)$$

N_n is the effective density of the states in the conduction band. Equation (8.12) gives the concentration of electrons in the conduction band as a function of temperature. The proportionality constant N_n is called the effective density of states.

Similarly the hole concentration in the valence band is given by

$$p = \int_0^{\infty}\gamma\left(E_v - E_c\right)^{\frac{1}{2}}\exp\left[-\left(E_F - E\right)/kT\right]dE$$

$$p = N_v \exp\left[-\left(E_F - E_v\right)/kT\right] \qquad .....(8.13)$$

where $$N_V = 2\left[\frac{2\pi m_p^* kT}{h^2}\right]^{\frac{3}{2}} \qquad .....(8.14)$$

Here N_v is the density of states in the valance band. m_p^* is the effective mass of hole.

8.2.3 Fermi Level Intrinsic Semiconductors

For an intrinsic semiconductor, $n = p$, hence

$$N_n \exp\left[-\left(E_c - E_F\right)/kT\right] = N_v \exp\left[-\left(E_F - E_v\right)/kT\right]$$

$$\frac{N_n}{N_v} = \frac{\exp\left[-\left(E_F - E_v\right)\right]/kT}{\exp\left[-\left(E_c - E_F\right)\right]/kT}$$

$$= \exp(-E_F + E_v + E_c - E_F)/kT$$

$$= \exp(E_c + E_v - 2E_F)/kT$$

$$E_F = \frac{E_c + E_v}{2} + \frac{kT}{2}\ln\left(\frac{N_v}{N_n}\right) \quad .....(8.15)$$

$$\frac{N_v}{N_n} = \frac{m_p^*}{m_e^*} \quad \text{(from eqns. 8.12 and 8.14)}$$

If the effective masses of electrons and holes are equal, then $N_v = N_n$, the effective density of states in conduction band and valance band is equal, in such a case one gets from eqn. (8.15)

$$E_F = \frac{E_c + E_v}{2} \quad .....(8.16)$$

i.e., Fermi level lies exactly in the middle of energy gap (Fig.8.5). Otherwise Fermi level varies with temperature as per the above eqn. (8.15). Normally, when the density of states and the two effective masses are unequally, the Fermi level is no longer symmetrically placed. The fermi level adjusts itself to maintain the charge neutrality condition that demands, n = p.

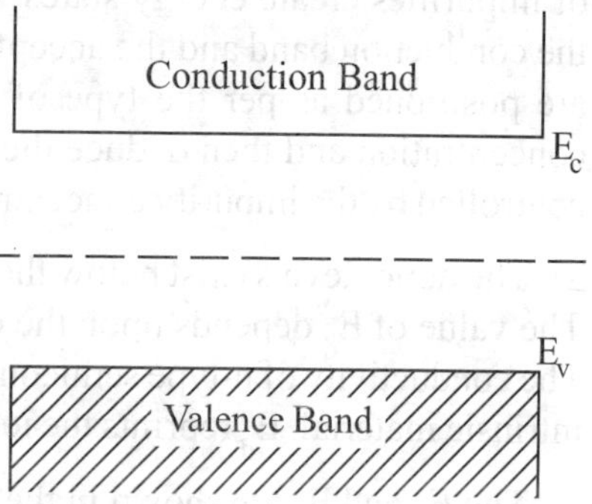

Fig. 8.5 Fermi Energy in Intrinsic Semiconductor.

8.2.4 Intrinsic Concentration

The product of equilibrium concentration of electrons in conduction band and holes in valance band at a given temperature is constant and is equal to the square of intrinsic concentration. This is called law of mass action. In an intrinsic semiconductor electron concentration is equal to hole concentration, but in semiconductors doped with impurities (extrinsic semiconductors) this is not the case. However, the product of electron and hole concentration in both types of semiconductors is always constant and it is always equal to the square of the intrinsic concentration. Product of electron and hole concentrations np is :

$$np = n_i^2 = N_c N_v [\exp - (E_c - E_F)/kT] [\exp - (E_F - E_v)/kT]$$

$$= N_c N_v \exp[(-E_c + E_v)/kT]$$

$$= 4\left(\frac{2\pi kT}{h^2}\right)^3 \left(m_e^* m_p^*\right)^3 \exp\left(-E_g / kT\right) \quad .....(8.17)$$

where $E_g = E_c - E_v$ is the band gap of the semiconductor.

Hence, taking square root of the equation (8.17), we get

$$n_i = 2\left(\frac{2\pi kT}{h^2}\right)^{3/2}\left(m_e^* m_p^*\right)^{3/2} \exp\left(\frac{-E_g}{2kT}\right)$$

$$= A\,T^{3/2} \exp\left(-E_g/2kT\right) \qquad .....(8.18)$$

Here $$A = 2\left(\frac{2\pi k m_e^* m_p^*}{h^2}\right)^{3/2} \qquad .....(8.19)$$

The above equation implies that intrinsic concentration of electrons in a semiconductor is independent of Fermi level. It is also evident that it depends strongly on temperature for a particular semiconductor.

8.3 Carrier Concentration in Extrinsic Semiconductors

The number of charge carriers, their type and electrical properties such as conductivity, capacitance etc., are extremely sensitive to the doping concentration. As already stated, doping of impurities create energy states within the forbidden band. The donor levels are just below the conduction band and the acceptor levels are just above the valence band. The Fermi levels are positioned as per the type of impurity. We shall first obtain the expression for carrier concentration and then deduce the expression for Fermi energy. The electrical properties are controlled by the impurity concentration.

The donor levels exist below the conduction band, E_d within about a few milli electron volts. The value of E_d depends upon the effective mass of the electron and static dielectric constant. The conductivity of n-type semiconductor is several orders bigger than that of the corresponding intrinsic material. E_d reprints the ionization energy of donor level.

The E_d and E_a are shown in the Fig. 8.6 (a) and (b). E_d and E_a denote the ionization energy of the donor and acceptor atoms. We shall first give the meaning of the symbols we are going

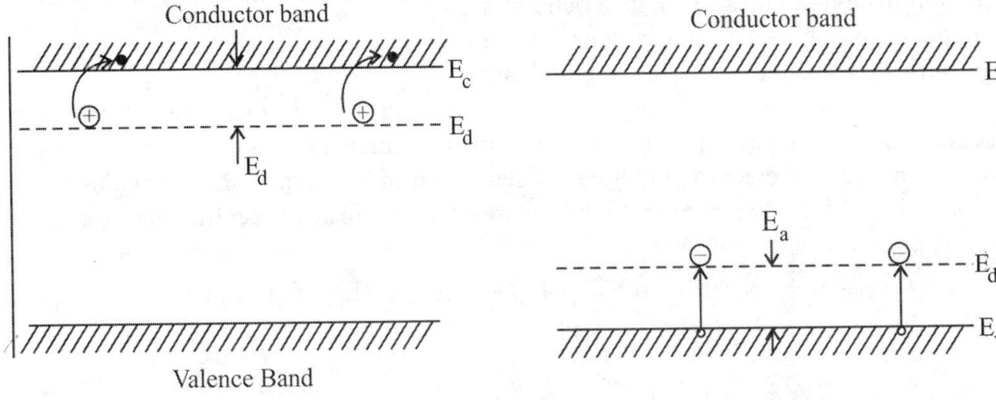

Fig. 8.6 (a) The position of donor level E_d relative to the conduction band edge E_c.
(b) Acceptor levels E_a relative to valence band edge E_v.

to use. N_d and N_a represent the density of all available donors and acceptors respectively. N_d^+ and N_a^- represent the density of ionized donors and acceptors respectively. n and p represent density of electrons in the conduction band and holes in the valence band. We consider that the concentration of impurity is small (such semi conductors are called degenerate semiconductors). We consider first that both donor atoms as well as acceptor atoms are present in the crystal and the impurity atoms are uniformly distributed in the crystal. Then, the condition for charge neutrality is given as

$$n + N_a^- = P + N_d^+ \qquad .....(8.20)$$

where $$N_d = N_d^0 + N_d^+ \quad \text{and} \quad N_a = N_a^0 + N_a^- \qquad .....(8.21)$$

when electrons go the conduction band, they leave ionized donor levels N_d^+. N_d^0 levels are the remaining levels. Similarly, when electrons from conduction band go to acceptor levels, they make acceptor atoms ionized equal in number to N_a^-. The remaining levels are N_a.

we have $$n = N_d - N_d^0$$

$$= N_d\left(1 - \frac{1}{1+\exp\left[(E_d - E_F)/kT\right]}\right) \qquad .....(8.22)$$

$$= \left(\frac{N_d}{1+\exp\left[-(E_d - E_F)/kT\right]}\right) \qquad .....(8.23)$$

But, we have for the concentration of electrons as

$$n = N_c \exp\left[-(E_C - E_F)/kT\right] \qquad .....(8.24)$$

then, $$N_c \exp\left[-(E_C + E_F)/kT\right] = \frac{N_d}{1+\exp\left[-(E_d - E_F)/kT\right]}$$

$$\therefore \quad N_d = N_c \exp\left[-(E_c + E_d)/kT\right]\exp(2E_F/kT) + N_c \exp(-E_c/kT)\exp(E_F/kT) \qquad .....(8.25)$$

or $$\frac{N_d}{N_c} = \exp\left[-(E_c + E_d)/kT\right]\exp(2E_F/kT)$$

solving for $e^{EF/kT}$, then

$$\exp(E_F/kT) = \frac{-1 + \sqrt{1 + 4\frac{N_d}{N_c}\exp(E_c - E_d)/kT}}{2\exp(-E_d/kT)} \qquad .....(8.26)$$

***Case* I :** suppose $(N_d / N_c) \exp [E_c - E_d]/kT >> 1$, then, the Fermi level has to coincide with the bottom of the conduction band. At these temperatures (low), the n–semiconductor behaves like an insulator.

$$\exp(E_F / kT) = \frac{2(N_d/N_c)^{1/2} \exp((E_c - E_d)/kT)}{2\exp(-E_d / kT)}$$

$$= (N_d/N_c)^{1/2} \exp((E_c + E_d)/kT) \qquad .....(8.27)$$

Taking logarithms on both sides and adjusting the terms, we get

$$E_F = \frac{(E_c + E_d)}{2} + \frac{1}{2} kT \log_e (N_d/N_c) \qquad .....(8.28)$$

$$\text{when } T = 0K \; ; \; E_F = \frac{E_c + E_d}{2} \qquad .....(8.29)$$

The above expressions give the Fermi level position in the forbidden gap. As the temperature is increased from 0k or impurity concentration is changed, the Fermi level starts shifting thereby changing the conductivity. We can now write for electron concentration as we substitute for E_F in eqn. (8.26)

$$n = (N_c N_d)^{1/2} \exp[(E_d - E_c)/2kT] \qquad .....(8.30)$$

***Case* II:** suppose, $(N_d/N_c)\exp[(E_c - E_d)/kT] << 1$, the concentration of the impurity is low such that $N_d << N_c$ or the temperature is high. Then $(E_c - E_d) << kT$. ie.

the term in the square root in eqn (8.26) is given as

$$4(N_d / N_c)\exp(E_c - E_d / kT) << 1$$

$$\therefore \quad \exp(E_F / kT) = (N_d / N_c) \exp(E_c / kT)$$

Taking logarithms wc get

$$E_F = E_c + kT \log_e \left(\frac{N_d}{N_c}\right)$$

The Fermi energy is slightly below the conducting band and it will come down as the temperature is decreased. Substituting the expression for E_F in the above expression, we get

$$n = N_d \qquad .....(8.31)$$

At temperatures indicated in the condition under case II, the density of electrons reaches the maximum with thermally created electrons from valence band give negligible contribution. Similar calculations performed for p-type semi conductors yield the concentrations in the valence band as

$$p = N_v \exp[(E_v - E_F) / kT] \qquad ......(8.32)$$

with

$$N_v = 2\left(\frac{m_p^* kT}{2\pi h^2}\right)^{3/2} \qquad ......(8.33)$$

$$\frac{P}{N_v} = \exp\left(\frac{-E_v}{kT}\right)\exp(E_F / kT)$$

$$\therefore \qquad \exp(E_F / kT) = \frac{P}{N_v}\exp(E_v / kT)$$

taking logarithm on both sides, we have

$$E_F = kT \ln\frac{p}{N_v} + E_v \qquad .....(8.34)$$

we have $p \simeq N_a$ at temperatures where the thermally created holes become negligible compared to the holes created by acceptor atoms. All the acceptor levels are occupied by electrons. The position of E_F in both n-type as well as p-type semiconductors is between band edge and the energy level (E_d or E_V) corresponding to either donor or acceptor atoms.

At very high temperature, the thermally created charge carriers, increase significantly to very high level. The extrinsic semi conductor behave like an intrinsic semiconductor.

It should be remembered that the law of mass action is equally applicable in extrinsic semi conductors. We know that we have

$$p_n n_n = n_i^2 \qquad ......(8.35)$$

the above equation is valid for n-type semiconductor in which electrons (n_n) are majority charge carriers, and holes (p_n) are minority charge carriers. But $n_n = N_d$ (the concentration of electrons in n-type semi conductor)

$$\therefore \qquad p_n \simeq \frac{n_i^2}{N_d} \qquad .....(8.36)$$

For a p-type semiconductor

$$p_p n_p = n_i^2 \qquad .....(8.37)$$

p_p and n_p represent the concentration of majority carriers and minority carriers respectively.

$$\therefore \qquad n_p = \frac{n_i^2}{N_a} \qquad .....(8.38)$$

at moderate temperatures. It should be remembered that the 'hole' movement is from atom to atom in the valence band as shown in Fig. 8.7. Effectively, the electron moves in the opposite direction (this is also shown in Fig. 8.2 (b)).

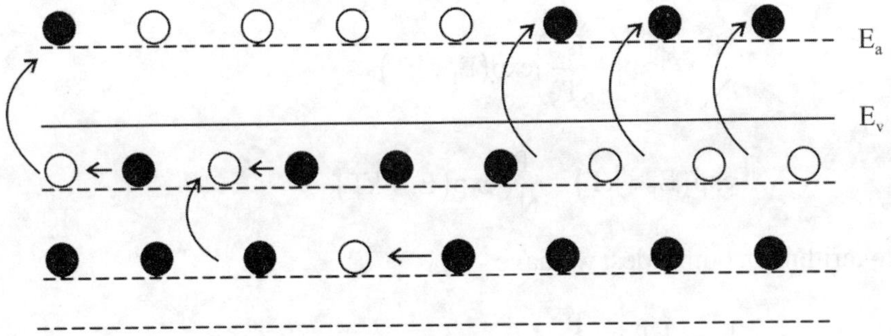

Fig. 8.7 Movement of hole takes place from atom to atom.

At higher temperatures, both n-type as well as p-type semiconductors are characterized by intrinsic behaviour as the impurity based charge carriers attain the saturation level ($n_n = N_d$ and $p_p = N_a$). The intrinsic carries will be dominating. Thus, all extrinsic semiconductors become intrinsic in nature and Fermi level drops to intrinsic fermi level value. The Fig. 8.8 shows the variations of electron density and energy as functions of temperature in the case of n-type semiconductor.

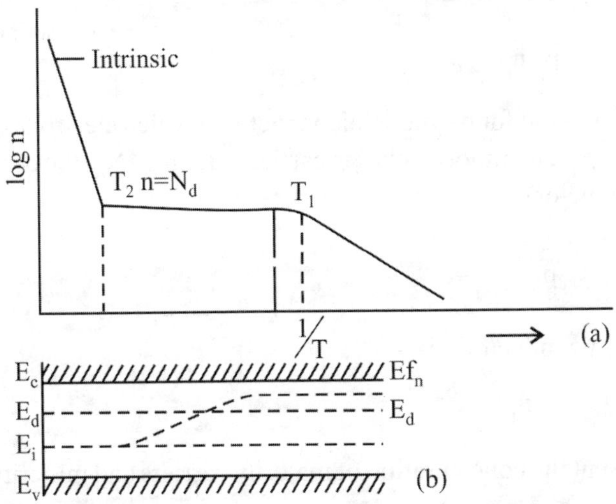

Fig. 8.8 Variation of electron concentration in the conduction band with temperature. The shifting of Fermi level with inverse of temperature is shown in Fig (b).

Table 8.2 gives the ionization energies of a few donors and acceptors in silicon.

Table 8.2

Impurity Donors	E_d (eV) (below conduction band edge)	Impurity acceptors	E_a (eV) (above valence band edge)
Sb	0.045	B	0.039
P	0.067	Al	0.045
As	0.072	Ga	0.054
		In	0.016

8.4 Currents in Semiconductors

8.4.1 Drift Current

An electric field applied to a semiconductor will produce a force on electrons and holes so that they will experience a net acceleration and net movement, provided that there are available energy states in the conduction and valence bands. This net movement of charge due to an electric field is called the drift. The net drift of charge gives rise to drift current.

To illustrate the concept of drift, let us consider a schematic model of the random thermal velocity and position of a hole in a semiconductor.

The Fig. 8.9(a) shows such a semiconductor with zero applied electric field. Let the motion of an electron start from a point A (Fig. 8.9(a)). It drifts to point B and undergoes a collision. The moving electron can collide with another electron, impurity atom and atoms of the host lattice. As a result of the collision, the direction of the motion of the electron gets altered and it moves towards C. The electron keeps on colliding and after a number of collisions, its average displacement over a time interval is zero. In the Fig. 8.9(a) the electron comes to a point (E) after suffering four collisions. These collisions or scattering events alter the velocity characteristics of the particle. However, on the average, the displacement of the particle over a time interval remains zero.

If a small electric field (E-field) is now applied to this material, as indicated by Fig. 8.9(b), there will be a net drift of the electron opposite to the direction of E-field, although the net drift velocity will be only a small perturbation on the random thermal velocity. There will be a net linear resultant displacement of the electron. A similar analysis is applied to hole, but the hole motion will be in the opposite direction to that of electron.

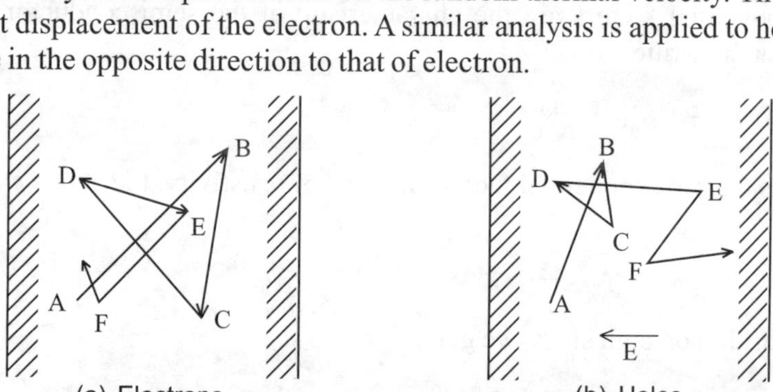

Fig. 8.9 Motion of electrons in absence (a) and the presence of electric field, E_a.

Both electrons and holes contribute to the drift current, so the total drift current density is the sum of the individual electron and hole drift current densities, and is given by

$$J_{drf} = e(\mu_n n + \mu_p p)E = \sigma E_a \qquad .....(8.39)$$

where μ_n and μ_p are the mobilities of the electrons and holes respectively; n and p are electron and hole concentrations respectively and e is the magnitude of the electronic charge.

Here $$\sigma = e(\mu_n n + \mu_p p) \qquad .....(8.40)$$

where $$\mu_n = \frac{-v_d}{E_a} = \frac{e\tau_e}{m_e} \qquad .....(8.41)$$

and $$\mu_p = \frac{-v_d(\xi)}{E_a} = \frac{e\tau_p}{m_p} \qquad .....(8.42)$$

where τ_e and τ_p are average time of electron and holes between collisions. The drift velocity, v_d is proportional to the electrical field. The proportionality constant is called the mobility of the charge carrier.

8.4.2 The conductivity of and intrinsic semiconductor

The conductivity of a semi conductor is expressed as

$$\sigma = |e|(n\mu_n + p\mu_p) \qquad .....(8.43)$$

Where n and p are electron and hole densities respectively, μ_n and μ_p are the nobilities of electron and hole respectively.

For an intrinsic semi conductor n = p, hence

$$\sigma = n_i(\mu_n + \mu_P)$$

$$= 2e\left(\frac{2\pi kT}{h^2}\right)^{3/2}\left(m_e^{\alpha}\ m_p^{\alpha}\right)^{3/2}(\mu_n + \mu_p)\exp\left(\frac{-Eg}{2kT}\right)$$

The preexponential factors are the characteristic of the semi conductor. Therefore, the conductivity can be written as

$$\sigma = \sigma_0 e^{-Eg/2kT} \qquad .....(8.44)$$

Where σ_0 is the pre exponential factor. In terms of resistivity, the above expression can be written as

$$\rho = \rho_0\ e^{Eg/2kT} \qquad .....(8.45\ (a))$$

Taking logarithm on both sides, we get

$$\ln\rho = \rho\ \ln\rho_0 + \frac{Eg}{2kT} \qquad .....(8.45\ (b))$$

8.4.3 Diffusion Currents

Diffusion of charge carriers takes place from higher concentration region to lower concentration region. As the particles are charged particles, a current called diffusion current will arise.

Diffusion is due to nonuniform distribution (concentration gradient) of charge carriers. As per the Fick's first law of diffusion, diffusion flux F is proportional to and directed opposite, to the concentration gradient of charges (Fig. 8.10).

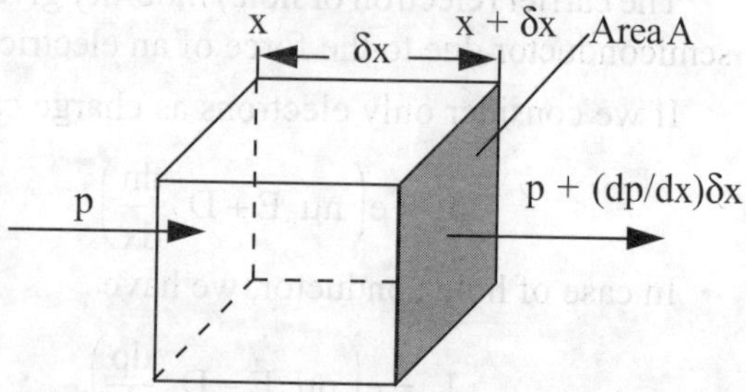

Fig. 8.10 Diffusion of Charges.

The migration of charge carriers from one region of higher concentration to another of lower concentration, is called diffusion. As the carriers are charged, a current called diffusion current will arise.

The diffusion current density is proportional to the concentration gradient, $\frac{dn(x)}{dx}$. The diffusion current component due to electron motion is given as (along the x-direction)

$$J_e(diff) = eD_e\left(\frac{dn}{dx}\right) \qquad .....(8.46)$$

where D_e is the diffusion coefficient of electron.

The hole diffusion takes place in the opposite direction, that is

$$J_p(diff) = -eD_p\left(\frac{dp}{dx}\right) \qquad .....(8.47)$$

The diffusion process is shown in Fig. 8.10. The concentration of holes entering into the volume is shown in P. The concentration leaving the volume from the opposite face is $p+\left(\frac{dp}{dx}\right)\delta x$, where $\left(\frac{dp}{dx}\right)$ gives the concentration gradient of holes. For an intrinsic semi conductor

$$J_{diff} = J_n\,(diif) + J_p\,(diif) \qquad .....(8.48)$$

The process of diffusion also leads to the recombination process in which an electron meets a hole and recombines. This leads to the elimination of electron - hole pairs.

The total current density in a semi conductor is the sum of the drift and diffusion current densities

$$J = J_{drf} + J_{dff} \qquad .....(8.49)$$

The total current density in a semiconductor is the sum of the drift and diffusion current densities :

$$J = J_{drf} + J_{dif}$$

The carrier (electron or hole) mobility gives an indication of how well the carrier moves in a semiconductor due to the force of an electric field.

If we consider only electrons as charge carriers, then,

$$J_e = e\left(n\mu_e E + D_e \frac{dn}{dx}\right) \quad .....(8.50)$$

In case of hole conductor, we have

$$J_p = e\left(p\mu_p E - D_p \frac{dp}{dx}\right) \quad .....(8.51)$$

8.5 Equation of Continuity

When a semiconductor is under the action of an external agency like radiation, then, excess carriers in addition to the thermally created carriers (n_o/p_o) will be created. The equation of continuity governs how the injected carriers (because of external force) settle with time when they are injected into the semiconductor.

The equation contains the terms indicting the rates of the processes involved. They are the rate of flow of charge carriers equated against the electron-hole pair generation rate, electron-hole recombination rate, and the gradient of current density along the semi conductor.

Consider a semi conductor of cross sectional area A and consider and infinitesimal volume of width dx as shown in the Fig. 8.11. Suppose a potential V is applied across it and then an elctric field acts at every point.

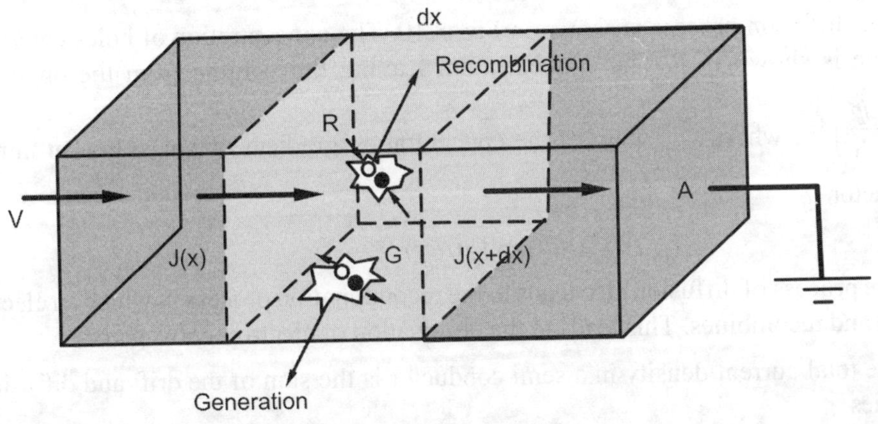

Fig. 8.11

Consider first the electron flow. The rate of change of electron flux in the volume is given as

$$\frac{\partial}{\partial t}(n)Adx = \left[\frac{J_n(x)}{-q} - \frac{J_n(x+dx)}{-q}\right]Adx + G_n\, Adx - R_n\, Adx \qquad .....(8.52)$$

where J_n represents the current density, G_n is the electron-hole pair generation rate, R_n is the electron-hole recombination rate.

i.e.

$$\frac{\partial}{\partial t}(n) = \frac{1}{8}\left[\frac{+d\,J_n}{dx} + G_n - R_n\right] \qquad .....(8.53)$$

where $J_n\,(x + dx) = J_n\,(a)\ \frac{+\partial\, J_n}{\partial x}\delta x$; e can replace the 'q'

Similarly, for holes, the equation is given by

$$\frac{\partial}{\partial t}p = -\frac{1}{q}\frac{\partial\, J_p}{\partial x} + \left(G_p - R_p\right) \qquad .....(8.54)$$

The current density J_n is a combination of drift current and diffusion currant as a already explained in the earlier section. J_n is given as

$$J_n = e\mu_n\, n\, E + eD_n \frac{dn}{dx} \qquad .....(8.55)$$

and

$$J_p = e\mu_p\, p\, E - eD_p \frac{dp}{dx} \qquad .....(8.56)$$

Substituting the above equations in eqn. (8.53) and (8.54) we get

$$\frac{\partial n_e}{\partial t} = n\mu_n \frac{\partial E}{\partial x} + \mu_n E \frac{\partial n}{\partial x} + D_n \frac{\partial^2 n}{\partial x^2} + G_n - R_n \qquad .....(8.57)$$

and

$$\frac{\partial p}{\partial t} = -p\mu \frac{\partial E}{\partial x} - \mu_p E \frac{\partial p}{\partial n} + D_p \frac{\partial^2 p}{\partial x^2} + G_p - R_p \qquad .....(8.58)$$

The recombination rate (R) is dependent on the number of charge carriers over the thermally created number and the time the electron spends before recombination. This time is called the relaxatron time, τ_e for electron and τ_p for hole.

Diffusion length

We have discussed the aspects of movement of excess carriers. The excess carriers are created at x = 0. The charge carriers, holes diffuse along the field and electrons opposite to the filed. The excess carriers concentration falls off exponentially with distance from the junction as

$$\delta\, p(x) = p\,(o)\, e^{-x/L_p}$$

where p(o) is the hole concentration in n-type semiconductor at x = 0. L_n is called diffusion length for holes. The diffusion length is the average distance covered by an excess carrier during a life time τ. In terms of diffusion coefficient and life time, the diffusion length is given as $L_p = \sqrt{D_p \tau_p}$ for n-type semiconductor.

Similarly, for electrons, the diffusion length is given as $L_n = \sqrt{D_n \tau_n}$ for a p-type semiconductor.

8.6 Einstein's Relation

The carrier diffusion coefficient gives an indication of how well the carrier moves in a semiconductor as a result of a density gradient. The mobility and diffusion coefficient are not independent parameters, and are related via the Einstein's relation :

$$\frac{D_n}{\mu_n} = \frac{D_p}{\mu_p} = \frac{kT}{e} \quad .....(8.63)$$

Note that this relation between the mobility and diffusion coefficient contains temperature. It is important to keep in mind that, as the mobilities are strong functions of temperature (due to lattice scattering and ionized impurity scattering processes), the diffusion coefficients are also strong functions of temperature. The constant kT/e is called thermal voltage. These relations are true only for non-degenerate semiconductors.

Electron in the valence band may be excited into the conduction band when, for example, high-energy photons are incident on a semiconductor. When this happens, not only is an electron created in the conduction band, but a hole is created in the valence band; thus an electron-hole pair is generated. The additional electrons and holes created are called excess electrons and excess holes.

8.7 Direct and Indirect Band-Gap Semiconductors

In energy (E) versus (k) wave vector diagram, if the minimum of the conduction band is directly above the maximum of valance band, then, the material is called direct band gap material (Fig. 8.12). The minimum in the conduction band is important because electrons excited from the valance band fall to this point much more rapidly than they fall from a point high in conduction band to the band minimum. Also holes move very quickly to the maximum of the valence band.

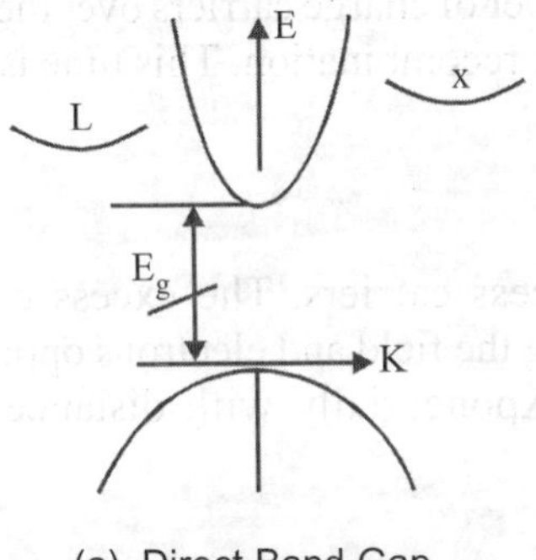

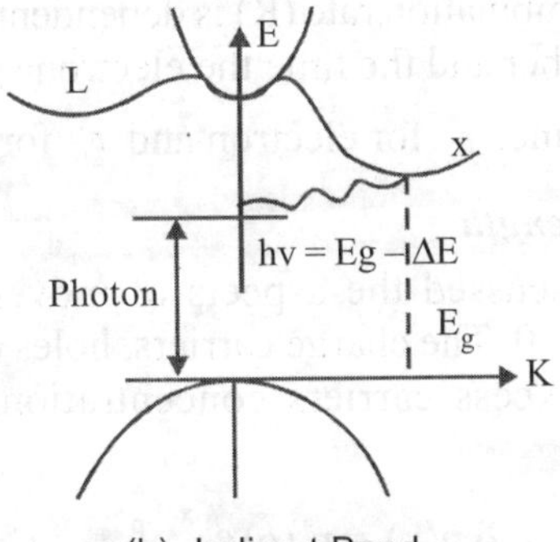

Fig. 8.12 Direct and Indirect band gap semiconductors.

The maximum of valence band and minimum of conduction band occur at the same K value in K-space. A photon is absorbed directly resulting in the transfer of electron from the top of the valance band to the bottom of conduction band.

In indirect band gap materials (like Si, Ge, GaP), the conduction band minimum is shifted from K = 0, and when transition of electron or hole occurs, the wave vector must be conserved and hence only those photons with certain discrete values of K are absorbed. This requires simultaneous collision between electrons, holes and the lattice. This results in the creation of a phonon and heat is generated. A three-body collision is required for transitions in indirect band gap materials and two-body collision is required for transitions in direct band gap material.

Band to band transitions have a much higher probability of occurrence in a direct band gap material because only an electron and hole need to collide. Therefore, the elemental semiconductor materials like Si or Ge are not used for optoelectronic device fabrication as they are indirect band gap materials. They have the following drawbacks:

1. Fundamental band gap in elemental semiconductors is indirect.
2. They have low optical absorption coefficients with absorption edges in infrared region.
3. As a consequence of above properties they have low conversion efficiencies.

8.8 Hall Effect

In a semiconductor, both electrons and holes are the charge carriers. Knowledge of charge carrier concentration in semiconductor is necessary for calculating the current flow in it. The current density resulting from the net drift of the average charge carrier in response to the external voltage is given as

$$J_s = -qn\langle v_d \rangle \tag{8.64}$$

where $[n\langle v_d \rangle]$ is the number of charge carriers crossing a unit area per unit time. 'n' represents the concentration of electrons. When a current flows along a semiconducting strip with its plane perpendicular to an external magnetic field, a potential difference is developed across the strip as a result of interaction of charge carriers with the magnetic field. This is called "Hall effect".

Consider a strip of a semiconductor as shown in Fig. 8.13. In the absence of the magnetic field, the current I just flows as a result of the applied voltage.

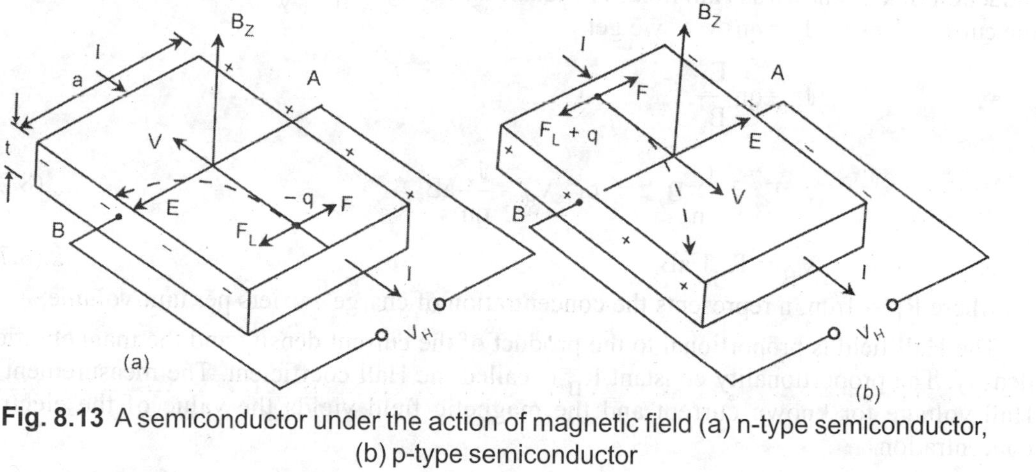

Fig. 8.13 A semiconductor under the action of magnetic field (a) n-type semiconductor, (b) p-type semiconductor

If a magnetic field is applied now whose induction B is perpendicular to the direction of the current, potential difference V_H is developed between the ends A and B, It is found experimentally that V_H is proportional to the magnetic field induction B and current I, and is inversely proportional to the plate thickness t.

$$V_H = R_H \frac{BI}{t} \quad .....(8.65a)$$

$$V_H = R_H J\, B\, a \quad .....(8.66b)$$

J is the surface current density, a is the width of the specimen. R_H is called the Hall constant or Hall coefficient and is a constant of the material.

First consider an n type semiconductor. The current flows in it in the x-direction and therefore, the electrons drift with a velocity - v_x the opposite direction. Each electron experiences a force called Lorentz force due to the magnetic field. It is given as

$$F_L = q\,\overline{v} \times \overline{B} \quad .....(8.67)$$

where q is the charge of the electron. The direction of the force is determined by the right hand rule. As the magnetic field is in the z-direction, the Lorentz force can be written as

$$F_L = -q\, v_x\, B_z \quad .....(8.68)$$

this force tends to deflect the electron in the y-direction, uncompensated positive charges pile up on the opposite surface. An electric field E is established between the faces A and B. Therefore the electron is under the action of two forces, the net force in the y-direction is given as (the sum of electric and magnetic forces)

$$F_y = q(E_y - v_x B_z) \quad .....(8.69)$$

but a steady flow of electrons has to be maintained along the x-direction, the electric field generated (E_y) counter balances the effect of the magnetic field so that the net force on the electron becomes zero as they drift along the x-direction. The pile up of charges on the lateral faces ceases. In such a case

$$qE_y = qv_x B_z \quad .....(8.70)$$

The establishment of this electric field E perpendicular to the current flow and the magnetic induction field is known as Hall field. The resulting voltage $V_H = E_y$ a is called Hall voltage. As the current density, $J_x = qn\langle v_x \rangle$, we get

$$J_x = qn \frac{E_y}{B_z}$$

$$\therefore\ E_y = \frac{J_x}{qn} B_z \qquad \text{or} \quad V_H = \frac{J_x}{qn} aB_z \quad .....(8.71)$$

or $$V_H = R_H J_x a B_z \quad .....(8.72)$$

where $R_H = 1/qn$, n represents the concentration of charge carriers per unit volume.

The Hall field is proportional to the product of the current density and the magnetic field density. The proportionality constant R_H is called the Hall coefficient. The measurement of Hall voltage for known current and the magnetic field yields the value of the electron concentration.

$$n = \frac{I_x}{tq} \cdot \frac{B_z}{V_H} \qquad .....(8.73)$$

All the quantities on the right hand side of the above equation can be measured to a very good accuracy and Hall effect observation gives a very accurate value of carrier concentration. Thus, Hall effect is an experimental means for determining the properties of a semiconductor.

For n type semiconductor, the sign of q is negative. But for p-type semiconductor, (Fig.8.12b) the $R_H = 1/qp$ where p is the equilibrium concentration of holes. The charge carriers drift in the direction of current, the Lorentz force points from A to B. The point B is at the positive potential with respect to the point A. Thus, the sign of Hall Voltage can determine the nature of the charge carriers in a semiconductor. It is customary to take the Hall coefficient R_H positive when the current is due to holes, and negative when the current is due to electrons. The mobility of the charge carriers is given as

$$\mu_n = \frac{\sigma}{qn} \text{ and } \mu_p = \frac{\sigma}{qp}$$

where σ is the conductivity of the semiconductor

$$\therefore \quad \mu_n = \frac{\sigma}{qn} = \frac{1/\rho}{q\left(\frac{1}{qR_H}\right)} = \frac{R_H}{\rho} \qquad .....(8.74)$$

where is ρ the resistivity. The mobility is a function of temperature. Measurements of Hall voltage and resistivity over a range of temperatures yields plots of majority carrier concentration and mobility as a function of temperature. Such information is important for the analysis of a semiconductor material.

Example 1

What is the minimum energy released in electron-hole recombination in Si at 300^0K. Find the wave length of emitted radiation $E_g = 1.12$ eV.

$$\frac{hc}{\lambda} = E_g = \text{minimum energy released.}$$

Substituting $E_g = 1.12$eV, $h = 6.6 \times 10^{-34}$ J-sec, c is the velocity of light $= 3 \times 10^8$m/sec, we get $\lambda = 1.1 \times 10^{-6}$m. The minimum energy released is 1.12 eV.

Example 2

Calculate the probability that an energy state 3KT above Fermi level E_F, is occupied by an electron at 300^0K.

$$f(E) = \frac{1}{1 + \exp\frac{(E - E_F)}{kT}} \quad (E - E_F = 3kT)$$

$$= \frac{1}{1 + \exp(3)} = 0.0474$$

Example 3

Calculate the volume density of quantum states in the energy interval 0 to 1 eV.

$$dE = 1 - 0 = 1\ \text{eV}\ ;\ m_e = 9.11\ \times 10^{-31}\ \text{kgs.}$$

$$\int_0^1 dn_E = \frac{4\pi}{h^3}(2m_e)^{\frac{3}{2}}\int_0^1 \frac{E^{\frac{1}{2}}dE}{1+e^{-(E-E_F)/KT}} \simeq \frac{4\pi}{h^3}(2m_e)^{3/2}\int_0^1 E^{1/2}dE$$

$$= 4.5 \times 10^{27}\ \text{m}^{-3}$$

Example 4

For an intrinsic semiconductor of Ge, energy gap is 0.7eV, determine the position of Fermi level and electron concentration at 300^0K if effective mass of electron is six times that of hole.

Fermi level in intrinsic semiconductor is given by

$$E_F = \frac{E_c + E_v}{2} + \frac{KT}{2}\ln\left(\frac{N_v}{N_c}\right)$$

Here $\quad E_g = E_C + E_V = 0.7\text{eV}$

Temperature equivalent energy, kT = $(1.38 \times 10^{-23}) / (1.6\times 10^{-19}) \times 300 = 0.026$eV

Given $\quad 6N_v = N_c$

Substituting the parameters in the expression for E_F, we get $E_F = 0.385$eV.

Carrier concentration is given by

$$n = 2\left(\frac{2\pi kT}{h^2}\right)^{3/2}\left(m_e^* m_p^*\right)^{3/2}\exp\left(\frac{-E_g}{2kT}\right)$$

$$= AT^{3/2}\exp\left(-E_g/2kT\right)$$

where A = $4.83 \times 10^{21}\ K^{-1}$

Hence n = $3.5 \times 10^{19}/m^3$.

Example 5

Plot the Fermi-Dirac probability function for $-5kT \le (E - E_F) \le 5kT$

The Fermi-Dirac distribution is given by

$$f_F(E) = \frac{1}{1+\exp\left(\frac{E-E_F}{kT}\right)}$$

Where E_F is the Fermi energy, k is the Boltzmann's constant, and T is the temperature in Kelvin's.

$$\beta = \frac{E-E_F}{kT}$$

So we need to plot the function

$$f_F(\beta) = \frac{1}{1+\exp(\beta)}$$

plotting the graph in the interval $\beta = (-5, 5)$

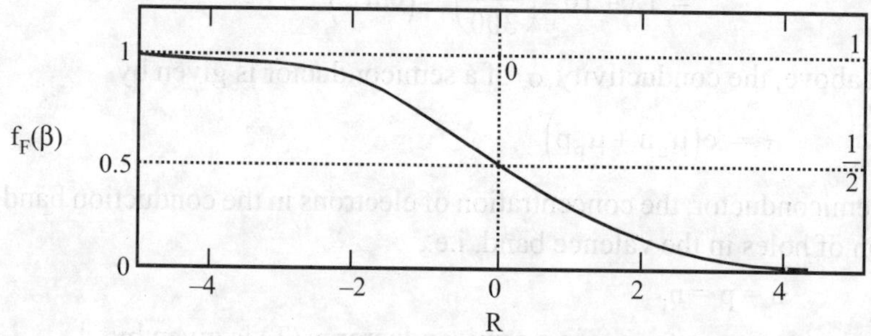

Fermi-Dirac probability function

Note that the probability of a state being occupied at $E = E_F$ ($\beta = 0$) is one half.

Indeed, $f_F(E = E_F) = \frac{1}{1+\exp(0)} = \frac{1}{1+1} = \frac{1}{2}$

Example 6

Obtain an expression for electric field as a function of distance in a semiconductor. Also find an expression for total current in the semiconductor.

The electron current has both diffusion and drift current components, so the electron current density for this material is given by

$$J_n = e\mu_n n(x)E + eD_n\left(\frac{d}{dx}n(x)\right)$$

solving this equation for the electric field E yields:

$$E(x) = \frac{J_n - eD_n\left(\frac{d}{dx}n(x)\right)}{e\mu_n n(x)} \text{cm}^{-1}$$

This electric field distribution can be readily plotted against x :

The total drift current density is equal to the sum of the individual electron and hole drift current densities, and is given by

$$J_{drf} = e\,(\mu_n n + \mu_p P)\,E$$

where μ_n and μ_p are the mobilities of the electrons and holes, respectively, n and p are electron and hole concentrations, respectively, and e is the magnitude of the electronic change. Then,

$$\sigma = e\,(\mu_n n + \mu_p P)$$

where σ is the conductivity of the semiconductor materials. Since the mobilities are functions of temperature and impurity concentrations, conductivity can be a rather complicated function of temperature and impurity concentration also.

The effective density of states function $N_c(T)$ in silicon can be written in the form:

$$N_c(T) = 2.8\ 10^{19}\left(\frac{T}{300}\right)^{3/2}\left(cm^{-3}\right)N_v(T)$$

$$= 1.04\ 10^{19}\left(\frac{T}{300}\right)^{3/2}\left(cm^{-3}\right)$$

As discussed above, the conductivity σ of a semiconductor is given by

$$\sigma = e\left(\mu_n n + \mu_p p\right)$$

for an intrinsic semiconductor, the concentration of electrons in the conduction band is equal to the concentration of holes in the valence band, i.e.,

$$n = p = n_i$$

The intrinsic carrier concentration in a semiconductor $n_i(T)$ is given by

$$n_i(T) = \sqrt{N_c(T)\ N_v(T)\exp\left(\frac{E_g}{k.T}\right)}$$

where N_c and N_v are the effective density of states functions in the conduction and valence bands respectively, E_g is the band gap energy, and k is Boltzmann's constant. Substituting this equation for $n_1(T)$ into the expression for conductivity, we get

$$\sigma(T) = en_i(T)\left[\mu_n(T) + \mu_p(T)\right]$$

The conductivity in intrinsic silicon is rather low. However, conductivity can be increased by many orders of magnitude by introducing impurities (dopants) into the lattice of the material. A typical value of conductivity at T= 300K of a silicon crystal with impurity (donor or acceptor) concentration of 10^{18} cm^{-3} is $(1\Omega - cm)^{-1}$, six orders of magnitude higher than the intrinsic value.

Example 7

Find out the resistivity of a semiconductor for which $n_i = 2.4 \times 10^{19}$ per m^3, $\mu_n = 0.39\ m^2/V\text{-}s$ and $\mu_p = 0.19\ m^2/V\text{-}s$.

Solution :

Since $\sigma = e\left(\mu_n n + \mu_p p\right)$

Resistivity is $\rho = \frac{1}{\sigma} = \frac{1}{e\left(\mu_n + \mu_p\right)n_i}$

$$= \frac{1}{1.6\times10^{-19}(0.39+0.19)\times2.4\times10^{19}} = 0.449\ \text{ohm} - \text{m}.$$

Example 8

Calculate the drift velocity of holes and electrons at 300K if the applied electric field is 100V/m in silicon and germanium. Also find diffusion coefficient at 300K in these materials.

Solution :

Drift Velocity = $\mu E = 0.0038 \times 100 = 0.38$ m/sec for electrons

Diffusion coefficient D = $(KT/e)\mu = 0.93$ m/sec for electrons.

Similarly we can find out drift velocity and diffusion coefficient for holes.

Example 9

Calculate the density of donor atoms which have to be added to produce n-type material of resistivity 0.19 ohm-m, given that mobility of electrons is 0.325 m^2/v-s.

Solution :

We know that $\sigma = e \cdot (\mu \cdot n)$

$$n = \frac{\sigma}{e\mu}$$

$$= 1/(1.6 \times 10^{-19} \times 0.19 \times 0.325) = 1.01 \times 10^{18} \text{ per m}^3.$$

Example 10

Conductivity of a semiconductor at 300K is 2 mho per m. Find its conductivity at 600K, if its Fermi energy does not change with temperature.

Solution :

We know that
$$\frac{\sigma_{T1}}{\sigma_{T2}} = \frac{e\mu AT_1^{3/2} \exp\left(\dfrac{E_F - E_c}{kT_1}\right)}{e\mu AT_2^{3/2} \exp\left(\dfrac{E_F - E_c}{kT_2}\right)}$$

substituting the values given and simplifying, conductivity at 600K can be evaluated.

Example 11

Find the diffusion coefficient of electrons in silicon at 300K if μ_e is 0.19m^2/V-sec.

Solution :

From Einstein's equation,

$$\frac{D}{\mu} = \frac{k.T}{e}$$

$$D = (1.32 \times 10^{-23} \times 300 \times 0.19) / 1.69 \times 10^{-19}$$

$$= 0.00445 \text{m}^2/\text{sec}$$

Example 12

A sample of silicon is doped with 10^{23} phosphorus atoms/m^3. The mobility is 0.07 m^2/V-s. What is its resistivity? Suppose the sample is 100μm thick and a current of 1 mA passes through it. What is the Hall voltage at a field of 1.2 KG ?

Solution :

$$\sigma = nq\mu_n = 10^{17} \times 1.6 \times 19^{-19} \times 0.07 = (0.112\Omega - cm)^{-1}$$

$$\therefore \quad \rho = \frac{1}{0.112} = 8.9 \times 10^{-4}\Omega - cm. B_z = 1.2 \times 10^{-1}\,Wb/m^2$$

$$R_H = -(qn)^{-1} = -(1.6 \times 10^{-19} \times 10^{23})^{-1} = 66.6 \times 10^{-6}\,m^3/C$$

$$\therefore \quad V_H = \frac{I_n B_z}{t} R_H = \frac{1x10^{-3} \times 1.2x10^{-1} \times 66.6 \times 10^{-6}}{100 \times 10^{-6}} = 79.9\mu V$$

Questions

1. With the help of energy level diagrams explain insulators, semiconductors and conductors.
2. With necessary theory, obtain a relation between energy gap and temperature in intrinsic semiconductors.
3. Derive the expression for Fermi level in an intrinsic semiconductor.
4. Sketch the energy band diagrams of (a) an intrinsic (b) N-type (c) P-type semiconductors. Indicate Fermi, Donar and acceptor levels wherever present.
5. What is the significance of volt equivalent of temperature. Obtain an expression for Fermi energy in terms of this parameter.
6. Obtain an expression for the hole concentration in a non-degenerate semiconductor.
7. State the expression for the density of electrons and holes in an intrinsic semiconductor.
8. Explain affect of temperature and impurity ions on the Fermi level in a semiconductor.
9. Derive an expression for the number of electrons per unit volume in conduction band of an intrinsic semiconductor.
10. The band gap of a semiconductors is 2eV. Find out wave length of radiation emitted when electrons of this semiconductor recombine directly.
11. Electrons can be excited across forbidden energy gap in a semiconductor by photons of sufficient energy. What is the maximum energy of a photon to create electron-hole pairs in a) germanium, b) silicon, c) diamond.
12. The relative density of a semiconductor is 18.8, its atomic weight is 184 and there are two free electrons per atom. Find out the density of free electrons and Fermi energy.
13. In a given n-type semiconductor material, the donor concentration is 1 atom per every 2 x 10^8 semiconductor atoms. Assuming that the effective mass of electron is equal to its true mass, find the value of temperature at which Fermi level coincides with the edge of conduction band.
14. Using expression for concentration of holes in a semiconductor i.e.,

$$p = \int \left[(E_v - E)^{1/2} \exp(E - E_f)/kT \right] dE$$

Show that dp/dE is maximum at $E_v - E = kT/2$.

15. What is equation of continuity for semiconductors. What physical law leads to the equation of continuity.
16. What are drift and diffusion currents ? Obtain expression for each of these.
17. What is carrier scattering ? Discuss the reasons for carrier scattering.
18. Define carrier life time and diffusion length.
19. What is Hall effect? Explain.
20. The resistivity of an intrinsic semiconductor at 270K is 3000 ohm–m. Calculate intrinsic carrier concentration if $\mu_n = 0.17 m^2/V\text{-sec}$ and $\mu_p = 0.0035 m^2/V\text{-sec}$.

    (Answer : *1×10^{16} per m^3*)
21. The conductivity of a semiconductor varies as exp(– 4350/kT), find its band gap.

    (Answer : *0.67 ev*)
22. A sample of germanium has an intrinsic resistivity of 0.6 ohm-m. Calculate the value of current density in an applied field of 1V/m, if there are 10^{18} donors and 4×10^{17} acceptors in each m^3 of the specimen and if mobilities of electron and hole are $\mu_n = 0.0018 m^2/V$ –sec and $\mu_p = 0.0038 m^2/V$ –sec.

    (Answer : *3.65×10^6 A/m^2*).
23. Find the conductivity of intrinsic silicon at 300 k. It is given that n_i at 300 k in silicon is $1.5 \times 10^{16}/m^3$ and the mobilites of electrons and holes in silicon are 0.13 m^3/v-s and 0.05 m^2/v-s respectively.

    (Answer : 4.3×10^{-4} Ω–m)
24. The R_H of a specimen is 3.66×10^{-4} m^3/c. Its resistivity is 8.93×10^{-3} Ω-m. Find μ and n.

    (Answer : 4.1×10^{-2} m^2/ V–S).
25. State the expression for the density of electrons and holes in an intrinsic semi conductor. Derive the expression for Fermi level.
26. Explain Hall effect. Show that for an n-type semiconductor the Hall coefficient is $R_H = -\frac{1}{ne}$.
27. Describe drift and diffusion currents in a semiconductor. Derive their expressions and deduce Einstein relation.
28. Explain the effect of temperature on the Fermi level in semiconductors.
29. Write a note on diffusion length.
30. Show that the Fermi level is nearer to the conduction band in n-type semi conductor.
31. Explain the effect of temperature on resistivity of a semiconductor.
32. Derive the relation between the intrinsic carrier concentration and absolute temperature.
33. Distinguish between intrinsic and extrinsic semiconductors with suitable examples.
34. Explain the origin of Hall effect in semi conductor.
35. Explain how Hall effect is useful in distinguishing 'hole conductor' and 'election conductor. (or n p - type and n - type semiconductor).

Objective Questions

I. Choose the correct choice in the following objective questions.

1. A pure or intrinsic semiconductor at absolute zero is
   (a) perfect conductor
   (b) super ionic conductor
   (c) perfect insulator
   (d) super metal

2. A pure elemental semiconductor
   (a) possesses infinite resistance at room temperature
   (b) shows a positive co-efficient of resistivity
   (c) is metallic at high temperature
   (d) shows a negative temperature co-efficient

3. A small amount of Indium is incorporated in Germanium, then it is
   (a) n-type semiconductor
   (b) p-type semiconductor
   (c) an insulator
   (d) a semiconductor with positive temperature co-efficient of resistance

4. One of the following materials has equal concentration of holes and electrons at room temperature
   (a) pure silicon
   (b) silicon dropped with antimony
   (c) a semimetal
   (d) silicon dropped with arsenic

5. The electric current in a pure semiconductor when an a voltage is applied, is due to
   (a) electrons
   (b) holes and electrons
   (c) electrons in the valence band
   (d) holes only

6. Fermi energy represents
   (a) the maximum energy possessed by electrons at 0^0 K
   (b) the minimum energy that electrons can have at 0^0 K
   (c) the top of the valence band
   (d) the energy of the electron in conduction band

7. In an intrinsic semiconductor
   (a) σ
   (b) $\sigma = p\,\mu\,e$
   (c) $\sigma = e\,(\mu_n n + \mu_p p)$
   (d) $n\mu_n e > p\,\mu_p e$

8. In an intrinsic semiconductor, the electrical conductivity 'σ' is proportional to
   (a) $T^{1/2}$
   (b) $T^{3/2}$
   (c) $T^{-3/2}$
   (d) $T^{-1/2}$

9. The movement of a hole is brought about
   (a) by atoms being broken from it
   (b) by the electron jump into it from conduction band
   (c) by a valence electron jump into it from the neighbouring atoms
   (d) by the thermal agitation of holes into the for hidden gap.

10. Choose the correct statement
    (a) an electron in the conduction band is bound strongly to the parent atom
    (b) donor type impurities create excess hole in Si
    (c) an electron in the conduction possess more energy than that in valence band
    (d) the fermi energy level contains all electrons in it

11. If we want to inject excess charge carriers that can become mobile
    (a) we have to drill holes in semiconductor and put acceptors in it
    (b) we have to apply reverse bias
    (c) we have to produce cavities on the surface of the semiconductor
    (d) we have to irradiate the semiconductor with photons of energy equal to the energy gap

12. One of the following is a direct-bond gap semiconductor
   (a) Silicon
   (b) Germanium
   (c) gallium arsenide
   (d) Si - Ge mixed semiconductor

13. In a indirect band- gap semiconductor
   (a) Radiative transitions take place readily
   (b) Radiative transitions can not take place efficiently
   (c) Valence band maximum and conduction minimum occur at the same K - value
   (d) Radiative transitions are forbidden

14. The concentration of free electrons
   (a) is greater than that of holes is p-type semiconductor
   (b) is equal to that of holes in intrinsic semiconductors
   (c) is less than that of holes in n-type semiconductor
   (d) depends inversely on temperature of the semiconductor

15. Read the statements and answer
   (A) The conductivity of silicon is less than that of germanium at room temperature.
   (B) The diffusion current in a semiconductor is due to an applied voltage
   (a) (A) and (B) are correct
   (b) (A) is wrong, B is correct
   (d) (A) is correct, (B) is wrong
   (d) (A) and (B) are wrong

16. An extrinsic semiconductor
   (a) consists of minority carriers and majority carriers
   (b) consists of immobile holes and mobile electrons
   (c) consists of immobile electrons and mobile holes
   (d) consists of free electrons only

17. An extrinsic semiconductor
   (a) does not conduct till a critical temperature is reached
   (b) behaves like n – type semiconductor if it has acceptors
   (c) behaves like an intrinsic semiconductor beyond a c critical temperature
   (d) behaves like an insulator at room temperature

18. If N_d and N_a are the donor and acceptor impurities densities respectively, n and p are the electron hole densities respectively in an extrinsic semiconductor, then
   (a) $N_d - N_a = p + n$
   (b) $n^2 - (N_a - N_a)p = 0$
   (c) $(N_d - N_a) = p^2 + n^2$
   (d) $(N_d - N_a) + p$

19. When a silicon crystal is doped with trivalent in purity atoms, then, the energy levels of the in purity atoms are
   (a) just above the top of valence band
   (b) just below the bottom of the conductor band
   (c) at the fermi level
   (d) exactly at thc middle of the band gap

20. In an N – type semiconductor
   (a) the donor levels are at the middle of the band gap
   (b) the election density is equal to the hole density
   (c) concentration of conductor electrons called majority carriers is equal to the concentration of donors
   (d) concentration of conduction electrons is equal to the hole concentration

21. Choose the correct statement
   (a) conduction electrons in n – type semiconductor move with in valence band
   (b) At high temperatures, the carrier concentration in n – type semiconductor is proportional to the square root of donor concentration

(c) At low temperatures, n – type semiconductor become intrinsic semiconductor

(d) The fermi level in n – type semiconductor is just above valence band

22. In a semiconductor diode, arrow represents
(a) p – type material
(b) intrinsic material
(c) n – type material
(d) cathode

23. The conductivity of an intrinsic semiconductor is given by the expression
(a) $\sigma_i = q\,n\,p$
(b) $\sigma_i = q\,\mu\,n\,p$
(c) $\sigma_i = q\,n_i\,(\mu_n + \mu_p)$
(d) $\sigma_i = q\,(\mu_n + \mu_p)\,n\,p$

24. One of the following expressions gives the fermi level in n – type semiconductor.

(a) $E_F = E_c\,kT\,ln\,\frac{N_c}{N_d}$

(b) $E_F = \frac{E_c}{kT}\,ln\,\frac{N_c}{N_d}$

(c) $E_F = \frac{kT}{E_C}\,ln\,\frac{N_d}{N_c}$

(d) $E_F = E_c - kT\,ln\,\frac{N_c}{N_d}$

25. The fermi level in a p-type material is given by
(a) $E_F = kT\,ln\,N_v$
(b) $E_F = kT\,ln\,N_A\,N_V$
(c) $E_F = E_V + kT\,ln\,\frac{N_V}{N_A}$
(d) $E_F = kT\,ln\,\frac{N_V}{N_A}$

26. For germanium at room temperature, the intrinsic concentration is2.5 × 10^{13} cm^3. The donor impurity concentration is 5 × $10^{15}$$cm^3$. Then the hole concentration in the semi conductor is given (per cm^3) as
(a) 2.5 × 10^{11}
(b) 12.5 × 10^{10}
(c) 12.5 × 10^{26}
(d) zero

27. Read the statements
(A) The product of the number of holes and the number of electrons is constant in an intrinsic semi conductor
(B) The product of the number of holes and the number of electrons is dependent on the donor and acceptor in purity doping
(a) A and B are correct
(b) A is wrong , B is correct
(c) A and B are in correct
(d) A is correct and B is in correct

28. In an n type semiconductor
(a) the fermi level is between donor levels and the bottom of the conduction band
(b) the fermi level lies deep in the conduction band
(c) the fermi level lies above the top of the valence band
(d) the fermi level lies exactly as the middle of the energy gap

29. The band gaps, at 300 K, of Si, Ge, and Sn and carbon are 1.12, 0.67, 0.1 and 5.5eV respectively. The crystal that will have the highest electrical conductivity at 300 K is
(a) Ge
(b) Sn
(c) Si
(d) C

30. The ratio of diffusion coefficient to mobility of carriers is proportional to
(a) $\frac{1}{T^2}$
(b) $\frac{1}{\sqrt{T}}$
(c) $\sqrt{T}$
(d) T

31. If the Hall coefficient is negative, then the semiconductor is

(a) intrinsic

(b) n-type

(c) compound semiconductor

(d) p-type

II. Fill up the blanks

1. In semiconductors, the valence band is ________ and the conduction band is ________ at low temperatures.

2. A low electron effective mass means ________ electron mobility.

3. The conductivity of p - type silicon is 4 ohm^{-1} m^{-1} , The mobility of holes is 0.05 m^2 / V- S. Then the density of impurities is (atoms/m^3) ________ .

4. A pn junction diode, when forward biased, behaves likes a ________ .

5. Fermi energy is the ________ energy that can be possessed by an electron at 0 K.

6. When forward biased, the forward current increases ________ with forward bias.

7. The concentration of charge carriers in an intrinsic semiconductor is proportional to ________ . (Let T = the absolute Temperature)

8. In a p - type semi conductor, the mobility of holes is about 0.2 m^2 / V A Sec, and the concentration of acceptor impurity is 5×10^{22} atoms m^3 , then conductivity σ is ________ $\Omega^{-1}\,m^{-1}$.

9. The mobility of electron is ________ than that of a hole.

10. In an intrinsic semiconductor at a temperature T, the product of election and hole densities is given by the relation ________, where E_g, is the forbidden gap, and A, is a constant.

11. The position of the Fermi-level in an inpurity doped semiconductor depends on ________ and the concentration of ________ .

12. If N_a is the density of acceptor atoms in a p - type semiconductor, the electrical conductivity σ of the semiconductor is equal to ________, where μ_p is the mobility of holes.

13. The diffusion hole current density (J_p) in a semi conductor is proportional to the ________.

14. When an external electric field E is applied to a semiconductors, the total current density J_p due to holes is given by relation ________ , where D_p is the diffusion coefficient, $\frac{dp}{dx}$ is the concentration gradient and p is the density of holes, μ_p is the mobility.

15. When an electric field E is applied to an n- type semiconductor the drift current is given as ________ , where m_n is the mobility of electron.

16. If n_i is the intrinsic charge density, μ_n, μ_p are the mobilities of electron and hole, the intrinsic conductivity of a semiconductor is given as ________.

17. If N_v represents the density of states, N_a represents the density of acceptor atoms, then the Fermi level in a p-type semiconductor at temperature T is given by the relation ________ .

III. True or False

1. Gallium Arsenide is an indirect band gap semiconductor.

[True/False]

2. Intrinsic semiconductors have a completely filled valence band at 300 K.

[True/False]

3. Intrinsic semiconductors have some electron-hole pairs at any temperature above 0 K.

[True/False]

4. Boron doped germanium is a n-type semiconductor.

[True/False]

5. Arsenic doped silicon is an n-type semiconductor.

[True/False]

6. The Fermi level in n-type semi conductors lies between donor levels and conduction band edge.

[True/False]

7. The acceptor energy level represents the binding energy of acceptor atom to the lattice

[True/False]

8. The intrinsic concentration of electrons in a semiconductor is dependent on the Fermi energy.

[True/False]

9. The donor levels exist below the conduction band by a few milli electron volts.

[True/False]

10. All extrinsic semiconductors become intrinsic at high temperatures.

[True/False]

11. The drift current is independent of temperature in a semiconductor.

[True/False]

12. The mobility of a carrier gives an indication of how well the carrier moves in a semiconductor.

[True/False]

13. When the holes are majority carriers in a semiconductor, the Hall coefficient is negative.

[True/False]

14. Silicon in an indirect band gap semiconductor.

[True/False]

15. Silicon is extremely useful as an opto electronic material.

[True/False]

Chapter 9

Physics of Semiconductor Devices

9.1 Introduction

Pure semi-conductors have very few device application possibilities. Extrinsic or impure semiconductors have relatively many applications possible. As already discussed in the previous sections, extrinsic semiconductors can be p – type or n-type semiconductors. Most of the useful and interesting semiconductor device applications use an intelligent combination of these impure semiconductor materials. Semiconductor junction devices have a p – type of semiconductor placed beside an n-type of semiconductor material. Examples of semiconductor junction devices are diodes, transistors, Silicon controlled rectifier etc. A pn junction may be formed from a single semiconductor material by doping a part of it with n-type (pentavalent or donor) impurity and other part with p-type (trivalent or acceptor) impurity. Fig. 9.1 gives the classification of semiconductor devices.

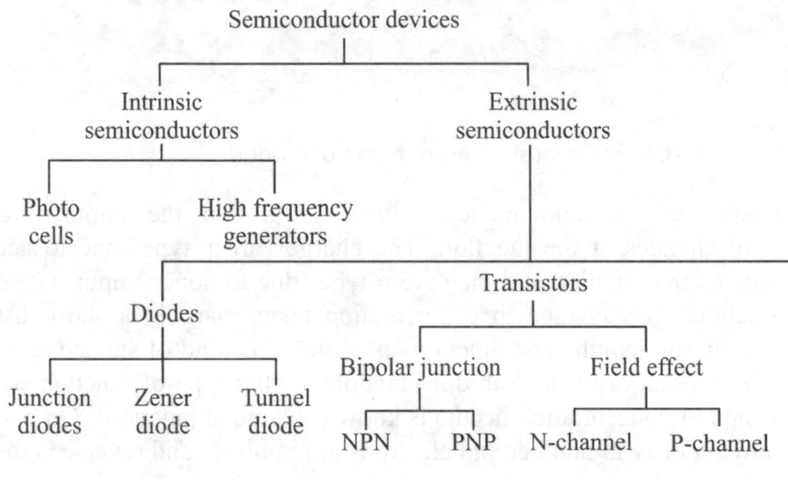

Fig. 9.1 Semiconductor Devices

Diodes are fabricated from semi-conducting material. The first letter in their identification is A for germanium diode or B for silicon diode. The second and third letters specify the type and usage of the diode. For example BY125 diode is a silicon diode and Y indicates that the diode is regulating diode. They can be encased in glass, metal or a plastic housing. They have two leads: cathode (k) and an anode (A). The most important property of all diodes is their resistance is very low in one direction and very large in the opposite direction. When a diode is measured with a multimeter and it reads a low value of ohms, this is not really the resistance of the diode. It represents the voltage drop across the junction of the diode. This means a multimeter can only be used to detect if the junction is not damaged. If the reading is low in one direction and very high in the other direction, the diode is operational.

When a diode is placed in a circuit and the voltage on the anode is higher than the cathode, it acts like a low value resistor and current will flow. If it is connected in the opposite direction it acts like a large value resistor and current does not flow. In the first case the diode is said to be "forward biased" and in the second case it is "reverse biased."

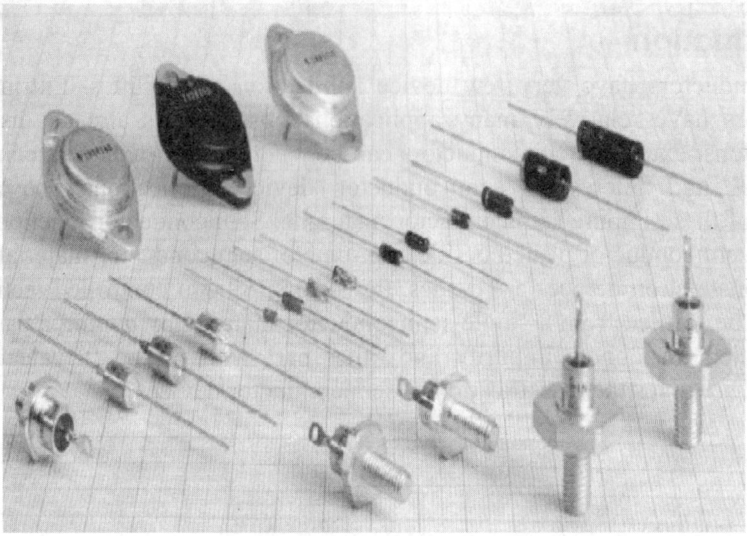

Fig. 9.2 Several different types of diodes

The essential feature of all semiconductor junctions is that the impurity charge concentration suddenly changes at the junction. The charges are p-type (due to acceptor impurities) on one side of the junction and they are n-type (due to donor impurities) on the other side of the junction. This change in concentration takes place in a short distance, typically of about one micron length. The junctions in which this kind of sudden change of carrier concentration is present are called abrupt junctions. Other type of junction where a relatively gradual change of concentration occurs is known as graded junctions. On p-side of the junctions, holes are majority in number and electrons are minority and reverse is the case in n–side of junction diode.

9.2 Fabrication of pn Junction Diodes

Diodes can be fabricated by inducing a transition of the material from p-type to n-type to occur with in a semiconductor material. This can be done in different ways:

1. Grown junction
2. Alloy junctions
3. Diffused junctions
4. Epitaxial junctions

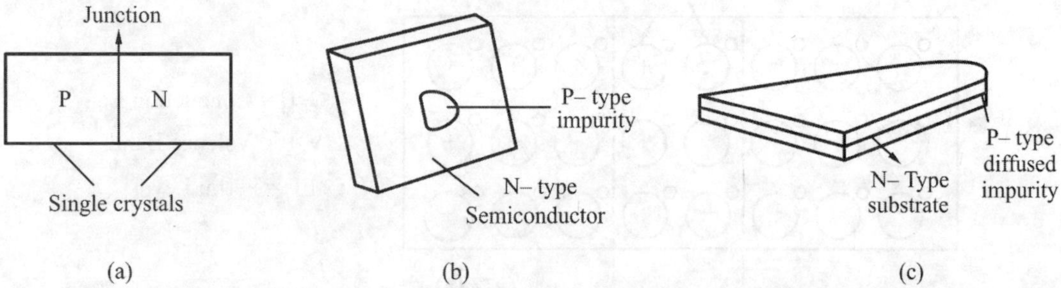

Fig. 9.3 Fabrication of p-n junction diodes (a) grown junction (b) alloy junction (c) diffused junction

Formation of P-N Junction

When n and p-type semiconductor materials are brought into contact with each other, a contact potential V_o is created. The potential barrier is created because the Fermi energies of the separate n- and p-type materials are different and when they are brought into contact, they must reach equilibrium. Before the semiconductor materials are brought into contact, n material has large electron concentration and p material has large hole concentration. Upon joining the materials, charge diffusion takes place between the two materials because of the concentration gradient of the charges. The holes diffuse from p side to n side and electrons diffuse from n-side to p-side. Diffusion current cannot build up indefinitely, but will reach equilibrium. The uncompensated ions are left behind by the charge diffusion from n to p-side or vice-versa. This will result in development of positive space charge near the n-side of the junction and a negative space charge near the p-side of the junction. Thus an electric field is generated in a direction opposite to the diffusion current. The electric field builds up till the equilibrium is reached.

Fig. 9.4 illustrates the depletion region, potential distribution and energy band bending in an abrupt junction to equilibrate the Fermi energy. In an abrupt junction, doping of p and n-sides is uniform. The Fermi energy equilibrates by a transfer of negative charge from n-type to p-type material.

Formation of depletion layer

The electric field appears in a region W about the junction, and there is an equilibrium potential difference V_o across W. The region W is called transition region and contact potential across W is built in a potential barrier and it is necessary to maintain the equilibrium at the junction. Contact potential cannot be measured with a meter as no current results from it. Within the transition region, since free charges are depleted or removed, an electric field is created. This region is also known as depletion region.

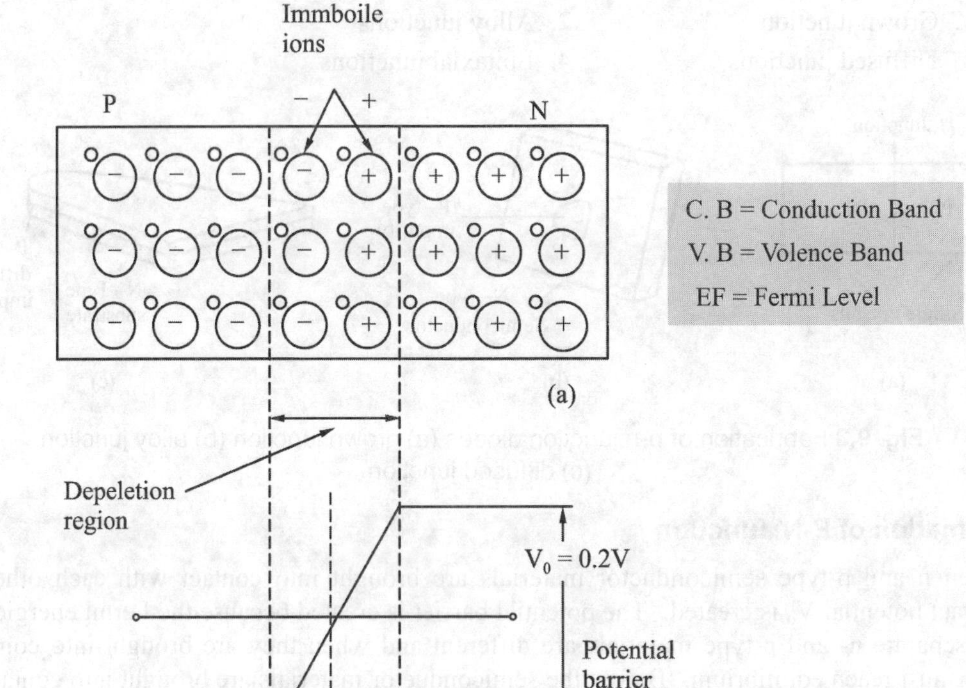

Fig. 9.4 (a) n and p materials are brought into contact, (b) A potential barrier appears across the junction, (c) The Fermi energy equilibrates across the junction.

Equilibrium in a pn Junction Diode

We assume an abrupt junction, which means that the doping density is constant at N_a on the p-type side and constant at N_d on the n-type side, and that the transition from N_a to N_d is *abrupt*, i.e., the depletion layer width is infinitesimally small.

Balance of drift and diffusion

Current generated by an electric field is drift current. Current generated by a diffusion gradient is diffusion current.

- In a pn junction diode, holes diffuse from the p-type side of the diode to the n-type side, and electrons diffuse from the n-type side to the p-type side.
- When holes and electrons diffuse across the junction, they leave behind ionized donors (on the n-type side) and acceptors (on the p-type side). The resulting space charge sets up an electric field that opposes diffusion.
- At equilibrium, the net movement of electrons and holes across the junction is zero, because movement by diffusion balances movement by drift. The net current (for electrons and for holes separately) across the diode is therefore zero.
- The ***depletion region*** is that region over which there is a space charge (ionized donors and acceptors). In the Fig. 9.4(b) given x refers to the width of depletion layer from the center of the depletion region on p-side and n-side :($-\mathbf{x_{po}} < \mathbf{x} < \mathbf{x_{no}}$.). The space charge region that is built at the junction is also known as depletion region, because this region is depleted of mobile charges. It is also called the barrier layer, as there is a potential barrier. This potential barrier prevents further diffusion of charges across the junction
- The regions outside the depletion region are the ***neutral regions***, or the ***bulk semiconductor***.

Important properties of pn junction diodes under equilibrium are as follows (the derivations of the following equations is outside the scope of this book):

Built-in potential

$$V_o = \frac{kT}{q}\ln\frac{N_d N_a}{n_i^2}.$$

The ***depletion region width***

$$w = x_{no} + x_{po} = \sqrt{\frac{2\varepsilon}{q}\left(\frac{1}{N_d}+\frac{1}{N_a}\right)V_o}.$$

Relative size of ***n- and p-side junction depths*** is

$$\frac{N_a}{N_d} = \frac{x_{no}}{x_{po}}.$$

Electric field *E(x)* and ***potential*** V(x) in the depletion region on the n-type side:

$$E(x) = \frac{qN_d}{\varepsilon}\left(x - x_{no}\right)$$

$$V(x)=V_{no}-\frac{qN_d}{2\varepsilon}\left(x-x_{no}\right)^2 \quad V_{no}=\frac{1}{q}(E_F-E_i)_{bulk}=\frac{kT}{q}ln\frac{N_d}{n_i}$$

and on the p-type side:

$$E(x)=-\frac{qN_a}{\varepsilon}\left(x+x_{po}\right)$$

$$V(x)=V_{po}+\frac{qN_a}{2\varepsilon}\left(x-x_{po}\right)^2 \quad V_{po}=\frac{1}{q}(E_F-E_i)_{bulk}=\frac{kT}{q}ln\frac{N_a}{n_i}$$

In these equations, the symbol E is being used for both electric field and E_F and E_i is energy.

Calculation of Diode Current-Voltage Relationship

We assume that the electric field in the neutral regions of the diodes is zero, so the only current in those regions is diffusion current (we can't have a drift current without electric field). We calculate the excess minority carrier density at the edge of the depletion region on each side of the diode.

The excess minority carriers are coming from the other side of the diode. (It is convenient to think in terms of a forward-biased diode, but the idea works for reverse bias, too, since excess carrier densities can be negative.) The result is the solution to the diffusion equation. (see chapter 8 for details) For the hole density coming over the depletion region to the n-type side, we have

$$\delta p(x_n)=\Delta p e^{-x_n/L_p}$$

where $L_p=\sqrt{D_p\tau_p}$ is the diffusion length. This is the excess minority carrier density in the neutral region of the n-type side of the diode. It is a decreasing function of distance, because as the carriers diffuse away from the junction, they recombine (and are thus lost) after an average time τ_p and after traveling an average distance L_p.

The minority carrier density at $x_n = 0$ is

$$\Delta p = p_n\left(e^{qV_a/kT}-1\right)=\frac{n_i^2}{N_d}\left(e^{qV_a/kT}-1\right)$$

where p_n is the minority carrier density on the n- side. We then calculate the hole diffusion current density by differentiating the above equation with respect to time. Then, we get for the hole current density as

$$J_p(x_n)=-qD_p\frac{d\delta p(x_n)}{dx}=q\frac{D_p}{L_p}\delta p(x_n)=qn_i^2\frac{D_p}{L_pN_d}\left(e^{qV_a/kT}-1\right)e^{-x_n/L_p}.$$

Similarly we can also get an expression for electron current density on the p- side of the junction is given as

$$J_p(x_n) = -qD_p \frac{d\delta n(x_p)}{dx} = q\frac{Dn}{L_n}\delta n(x_n) = qn_i^2 \frac{D_n}{L_n N_a}\left(e^{qV_a/kT} - 1\right)e^{-x_p/L_n}$$

If there is no generation of current in the depletion region, then the total current across the diode is given by

$$J = J_p(x_n = 0) - J_n(x_p = 0) = qn_i^2\left(\frac{D_p}{L_p N_d} + \frac{D_n}{L_n N_a}\right)\left(e^{qV_a/kT} - 1\right).$$

This is the **ideal diode equation**. It is also called rectifier equation.

Majority and Minority Carrier Currents

The total diode current J was obtained by adding the minority carrier currents on either side of the diode. These are functions of distance – they decrease with increasing distance away from the junction. Since the total diode current is constant, the *difference* between the *total* current and the *minority* carrier current is the *majority* carrier current. So, for example, on the n-type side of the diode, the minority carrier current is $J_p(x_n)$ (given above), and the majority carrier current is $J_n(x_n) = J - J_p(x_n)$. There are equivalent equations and arguments, not given here, for the other side of the diode.

9.3 Volt-ampere Characteristics of pn junction Diode

An ideal junction diode is the one, which has no reactive components (capacitance or inductance associated with it). The working of the pn junction diode can be understood with the help of energy band picture. It is necessary that in the two semiconductor materials which are in contact (p-type and n-type), the Fermi level E_F should be at the same energy level on either side of the junction under equilibrium condition.

When p-type material is brought into a contact with the n-type material for the formation of pn-junction diode, immediately charge transfer takes place in such a way that the Fermi level gets adjusted. The process of charge transfer continues till the Fermi levels in both the materials get aligned to the same energy level. After this equilibrium condition is reached, no more charge transfer takes place.

In n-type material, the Fermi level is near conduction band and in p-type semiconductor material, it is near the valance band. Energy levels on p-side of the junction will be at a higher level compared to the energy levels on n-side of the junction. When equilibrium is established the potential barrier that is formed at the junction prevents further transfer of charges (electron or holes) across the potential barrier or hill. Hence, under the equilibrium condition, no current would flow across the diode.

In order for the current to flow through the diode, voltage has to be applied across the diode. Applying voltage to the diode is called biasing the diode. Diode can be biased in two ways:

1. Forward bias
2. Reverse bias.

Forward biased diode:

In forward bias, the positive terminal (p-side) of the diode is connected to the positive of the voltage source and negative terminal (n-side) of the diode is connected to the negative terminal of the voltage source. The connection is shown in the circuit given in Fig. 9.5(a). For the current to flow through the diode, a voltage should be applied in such a way that the potential barrier present in the diode at the junction should be reduced. This is accomplished in the forward bias. The potential barrier height decreases by applied voltage. If, barrier height is V_o, and the applied forward voltage is V, the effective barrier height after forward bias would be $(V_o - V)$. Hence ,the current can flow through the junction because the drift and diffusion currents are no longer balanced. Thus, the reduction in the barrier height by applied bias voltage causes holes from p-side of the junction to be injected into n-side of the junction and electron move form n-side to p-side of the junction. Thus, there is a current flowing through the external circuit.

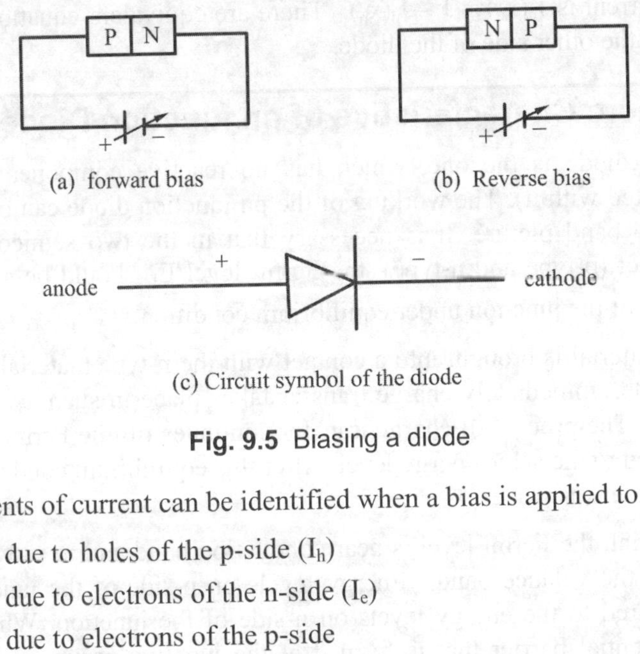

Fig. 9.5 Biasing a diode

Four components of current can be identified when a bias is applied to a diode. These are:

1. Current due to holes of the p-side (I_h)
2. Current due to electrons of the n-side (I_e)
3. Current due to electrons of the p-side
4. Current due to holes of the n-side

The first two currents are the currents due to the motion of majority carrier (I_h, I_e) and later two are the minority carriers. Minority carrier current is a result of thermally activated process. On the other hand majority carrier current due to holes and electrons is due to the diffusion of the charges across the junction due to the reduced barrier height because of the forward bias. Since hole and electron have opposite charge, they move in opposite directions, by the application of voltage. Thus the total current flowing through the diode I is equal to the sum of electron and hole currents and is given by:

$$I = I_0[\exp(\frac{eV}{kT}) - 1]$$

where I_O in the above equation is called saturation current density and it is equal to the sum of the minority charge currents across the junction, e is the electron charge, V is the applied voltage, k Boltzmann constant and T temperature in degree Kelvin. This is the diode equation. It is also called the rectifier equation. It indicates that current exists in forward bias only.

When forward bias is applied to a diode, the exponential term in the above equation rapidly increases, and becomes much larger than unity, and then, the above equation can be written as:

$$I = I_0[\exp(\frac{eV}{kT})]$$

Here kT is the temperature equivalent of voltage and is equal to 26 mV at $300^{O}C$. The above equation gives voltage - current characteristics of a forward biased junction diode. It indicates that the current across the junction increases exponentially with the applied voltage. Hence, if a voltage greater than a fraction of volt is applied ,then a large current flows across the junction, and the junction width and barrier height will be reduced. Very large currents result in ohmic heating .

Bias simply means ***applied dc voltage*** (for our purposes) and we define it as being applied to the p-type side of the diode. So, for example, a positive bias is a voltage that is positive at the p-type side with relative to the n-type side as shown in Fig.9.6a.. Applied bias V appears across the depletion region only, and there is no voltage drop across the neutral regions. In all our equations, V_o becomes $V_o - V$. Summarising, we have

Under forward bias

- the depletion region is reduced in size, and the energy barrier to electron flow is reduced, since $V_o > V_o - V$ and V is positive as shown in Fig.9.6b.
- the diffusion currents of electrons and holes increase exponentially because the lowered barrier height increases the probability that electrons and holes can diffuse past the barrier as shown in Fig. 9.6c. However, the drift current stays about the same. Therefore the total diode current increases exponentially.

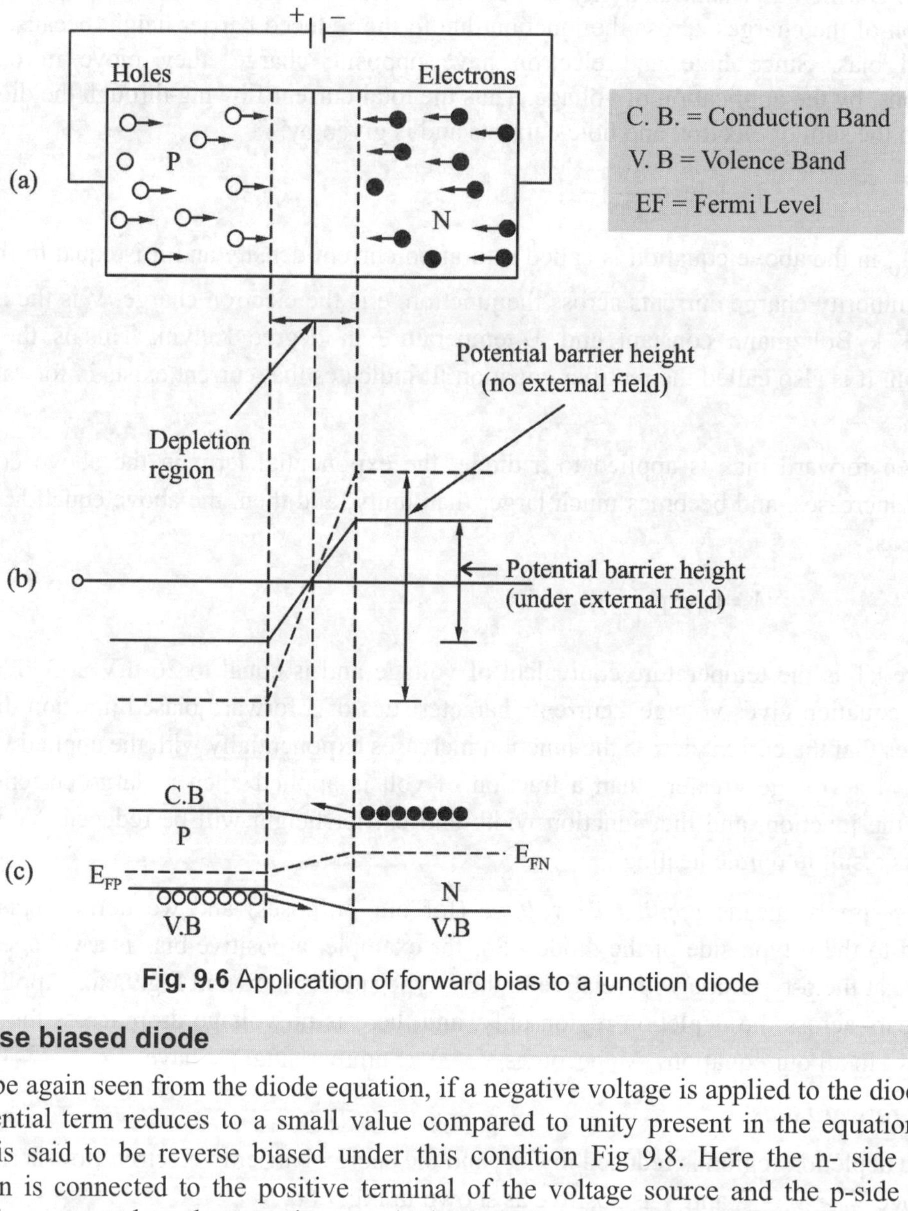

Fig. 9.6 Application of forward bias to a junction diode

Reverse biased diode

It can be again seen from the diode equation, if a negative voltage is applied to the diode, the exponential term reduces to a small value compared to unity present in the equation. The diode is said to be reverse biased under this condition Fig 9.8. Here the n- side of the junction is connected to the positive terminal of the voltage source and the p-side of the diode is connected to the negative terminal of the battery. Under limiting condition, the current I gets saturated and is equal to $-I_O$. This current is called the reverse saturation current.

The reverse saturation current is decided by the diode parameters only such as the doping concentrations of the diode and is independent of the applied voltage. This current depends on the minority charge carriers crossing the junction and does not depend on the barrier height of the junction. Diode is a bipolar device since both electrons and holes are responsible for currents in the diode. Summarising, we have

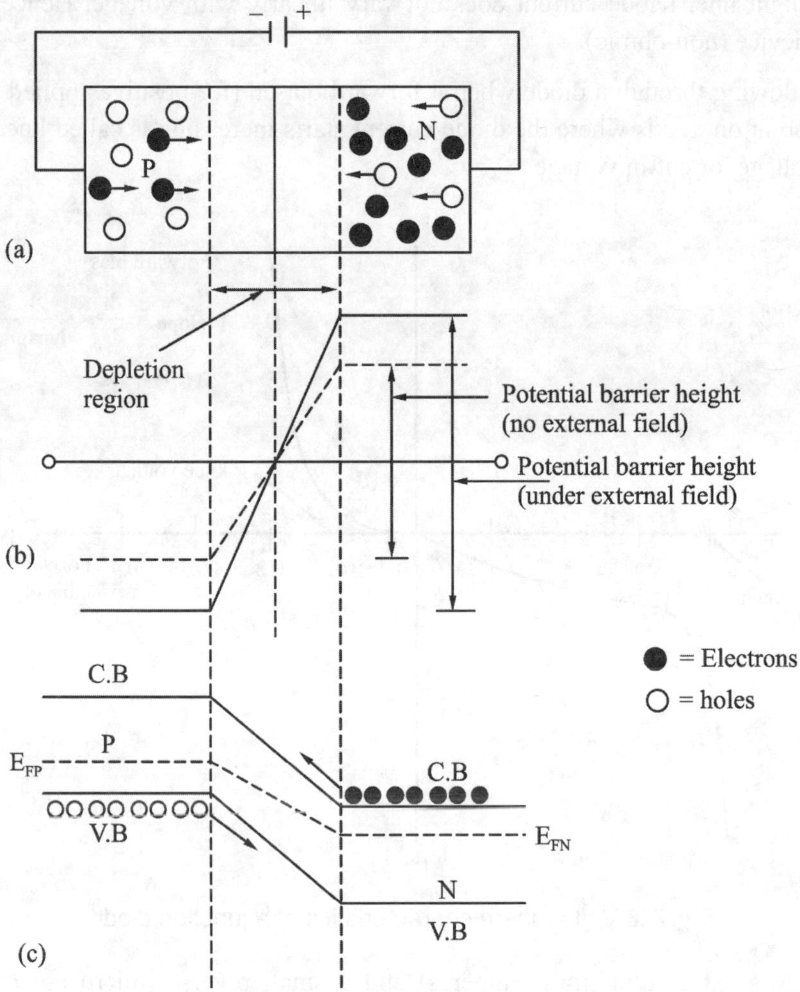

Fig. 9.7 Application of Reverse bias to a junction diode.

Under reverse bias

- the depletion region increases in size (9.7b), and the energy barrier to electron and hole flow is increased; V is negative;
- the diffusion currents of electrons and holes reduce dramatically due to the increased barrier height, but the drift current stays about the same (9.7c). Therefore, the total diode current is negative, and approximately constant with voltage

Voltage-Current characteristic of a diode:

The Fig. 9.8 gives volt - ampere characteristic curve of the junction diode. One can see that it is not a straight line. Diode current does not vary linearly with voltage. Hence, diode is a non-linear device (non-ohmic).

Current flowing through a diode when it forward biased (for positive applied voltages) is large. The point on x-axis where the diode current starts increasing is called knee voltage or threshold voltage or cut-in voltage.

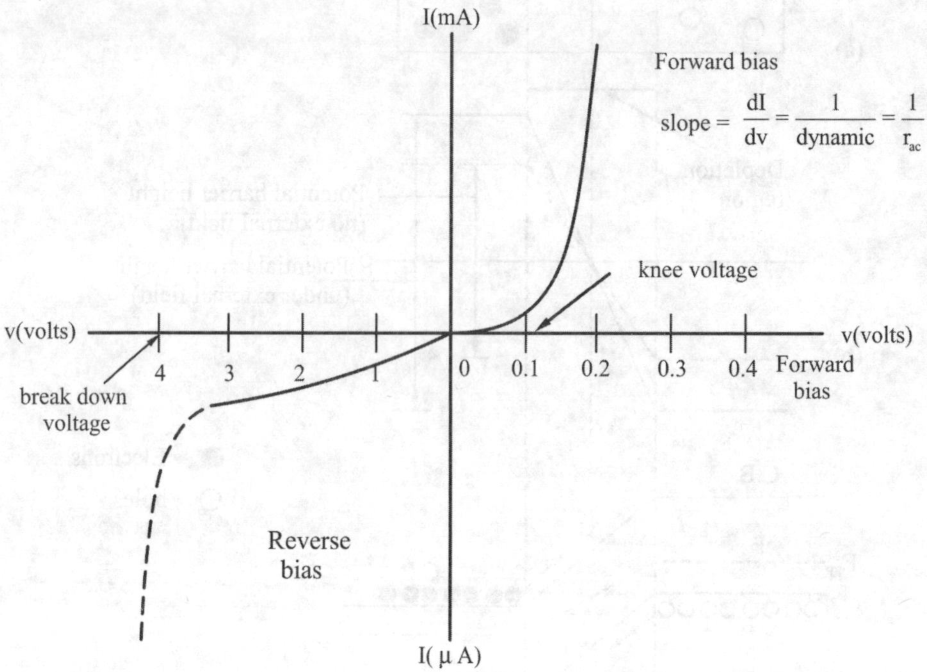

Fig. 9.8 Volt ampere characteristics of a junction diode

A large forward current (milli amperes) and a small reverse (micro amperes) current allowed by the junction diode is the characteristic of the diode, which makes it useful in rectifier applications. This property of the diode is called unilateral conduction of the diode. Hence a diode will have small resistance when forward biased and high resistance when it is reverse biased. An ideal diode must have zero resistance when forward biased and infinite resistance when reverse biased.

Static resistance or dc resistance of the diode is give by

$$r_{dc} = \frac{V}{I} = \frac{V}{I_0\left[\exp(eV/kT) - 1\right]}$$

Dynamic resistance or ac resistance of the diode is give by

$$r_{ac} = \frac{dV}{dI} = \frac{kT}{I_0[\exp(eV/kT)]} = \frac{kT}{(I + I_0)}$$

Break down voltage in the reverse bias:

When a large reverse voltage is applied to the diode, the electric field gradient at the junction becomes large and the covalent bonds in the material get disrupted and a number of charges will be set free. This results in increase in the reverse current of the diode. This is called breakdown of the junction diode. This excessive reverse current of the diode damages it. Hence, a diode should not be operated in the break down region of reverse bias voltage.

The diode resistance for a reverse biased diode extremely large. At high temperatures, the diode shows a linear variation of resistance with temperature. Hence diode becomes ohmic at higher temperatures.

There are two breakdown mechanisms: *Zener* and *avalanche*.

Zener breakdown occurs at relatively low voltage (5 – 10 V), and in diodes for which the doping densities on both sides of the diode are large (more than 10^{17} cm^{-3}).

Avalanche breakdown occurs at relatively high voltage (~100 V or more), and in diodes for which the doping densities are moderate or low. There will be avalanche multiplication of free charges.

A large band gap material has a larger breakdown voltage at any doping density than a small band gap material. This is because, for a given doping density, a large band gap material has a smaller probability of generating carriers that can contribute to avalanche.

Approximate diode models:

As discussed above, a diode is a non-linear and unilateral device. There are two models of a diode. One is small signal model and the other one is large signal model. When a diode is being operated at a point close to its knee voltage, small signal model can be used. If it is being operated at a much higher current compared to the current corresponding to the knee voltage large signal model should be used. Large signal model of a diode is described as follows:

1. Since a diode has small resistance when forward biased and large resistance when reverse biased, a diode acts as a closed switch when forward biased and open switch when reverse biased.
2. There is a knee voltage or threshold voltage for the diode. This voltage is the minimum voltage that must be applied to the diode to enable conduction of the current. Hence a diode can be approximated to a switch which is in series with a battery whose voltage is equal to the knee voltage.

3. Diode is made out of p and n type semiconductor materials. The bulk of the material will have certain resistance. This bulk resistance of the diode acts as if it is in series with the switch, battery that are associated with diode.

In view of the above model, the equivalent circuit of the diode is shown in the Fig. 9.9. In the small signal model of the diode, in addition to the bulk resistance of the diode, the dynamic resistance of the junction is also included.

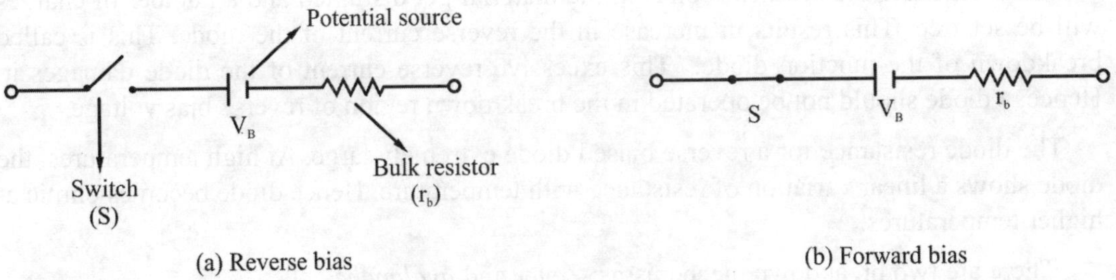

Fig. 9.9 Equivalent circuit of diode

Photo diodes

These devices, which include some of the best light detectors, are based upon *p-n junctions*: a single crystal doped partially with acceptors, partially with donors, with a sharp interface separating the two. Some of the acceptors near the interface attract electrons from the donors on the other side, and for a range about the interface no neutral impurities are left. This place where the *holes* and electrons neutralized each other is called the *space charge,* or *depletion, region*. The width of the depletion region depends upon the densities of the dopants on either side of the interface.

PN junction has a built-in polarization, and consequently a built -in electric field. It either takes work or produces work to move a charge across an electric field region, so the effect of the *p-n* junction on the energies of electrons in the valence and conduction band is as illustrated in Figure 9.10 and 9.11: the bands bend so that one side has higher energies than the other. This effect makes it a useful circuit element whether or not it's detecting light - it's a *diode*. The photodiode structure works as a detector whether or not there's an additional voltage put on it. It's usually used with a reverse bias or zero bias (in the latter condition it's often called a *photovoltaic* detector).

If a photon with energy $h\nu \geq E_g$ is absorbed in the depletion region, an electron - hole pair is produced, and these are swept in opposite directions by the built-in field. These charges can be collected at the ends, and the result is that exactly one electric charge made it all the way through the detector ($G \equiv 1$), or

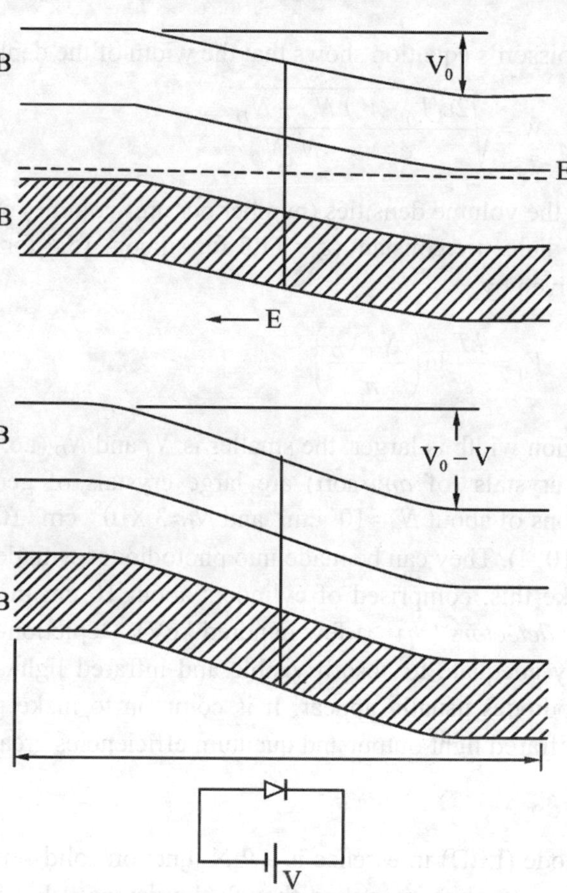

Fig. 9.10 Electric field and valence and conduction band energies in a *p-n* junction

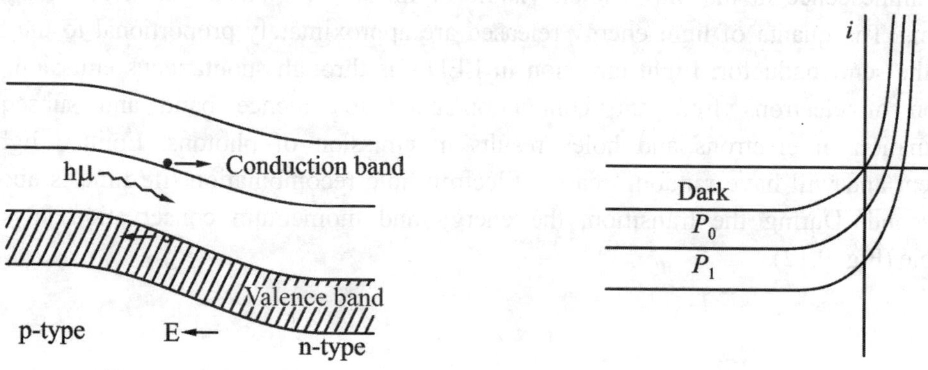

Fig. 9.11 Detection of light in a semiconductor *p-n* junction (left), and current-voltage characteristics (right).

The solution of Poisson's equation shows that the width of the depletion region is

$$w = \sqrt{\frac{2\varepsilon(V_0 - V)}{e} \frac{N_A + N_D}{N_A N_D}}$$

where N_A and N_D are the volume densities (number per unit volume) of acceptors and donors, respectively, ε is the dielectric constant ($\varepsilon = 11.3$ for silicon, 15.4 for germanium), and V_0 is the built-in voltage given by

$$V_0 = \frac{kT}{e} \ln\left(\frac{N_A N_D}{n_i^2}\right)$$

The depletion-region width is larger, the smaller is N_A and N_D (i.e. in a *very* pure crystal). The world's purest crystals (of *any* sort) are large crystals of germanium with residual impurity concentrations of about $N_A \approx 10^9\ cm^{-3}$ and $N_D \approx 3 \times 10^8\ cm^{-3}$ (that is, germanium pure to one part in about 10^{13}!). They can be made into photodiodes with depletion regions as long as 25 cm. Diodes like this, comprised of cylinders about 10 cm in diameter, are the most sensitive *gamma-ray detectors* ($h\nu \gtrsim$ a few hundred keV). Depletion-region widths of a few microns are perfectly adequate to absorb visible and infrared light, though, and that's the form in which photodiodes usually appear. It is common to make photodiode arrays with visible and/or near-infrared light output and quantum efficiencies greater than 0.8.

Light emitting diode

A Light-Emitting Diode (LED) in essence is a P-N junction solid-state semiconductor diode that emits light when a current is applied though the device. LED is a semiconductor p-n junction diode (Fig 9.12) that under forward bias conditions can emit radiation by electroluminescence in the ultra violet, visible or infrared regions of the electromagnetic spectrum. The quanta of light energy released are approximately proportional to the band gap of the semiconductor. Light emission in LED's is through spontaneous emission. The transition of electrons from the conduction band to valence band and subsequent recombination of electrons and holes results in emission of photons. Emitted light is incoherent and will have random phases. Electron hole recombination life time is about 1 nano second. During the transition, the energy and momentum conservation laws are applicable (Fig. 9.12).

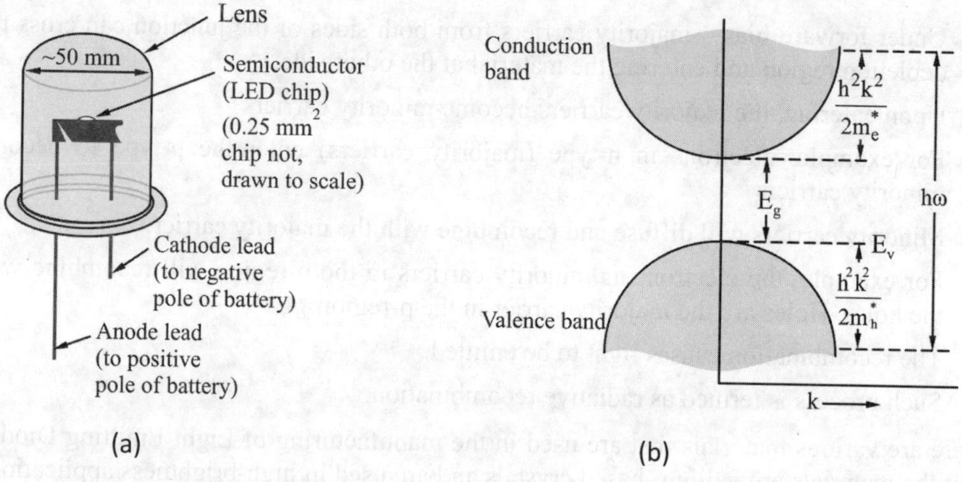

Fig. 9.12 (a) Light emitting diode (b) spontaneous emission in a LED

Important features of light emission in light emission by LED are illustrated in Fig. 9.13 and can be summerised as follows:

- Luminescence is a term used to describe the emission of radiation from a solid when the solid is supplied with some form of energy.
- Electroluminescence occurs when excitation results from the application of an electric field

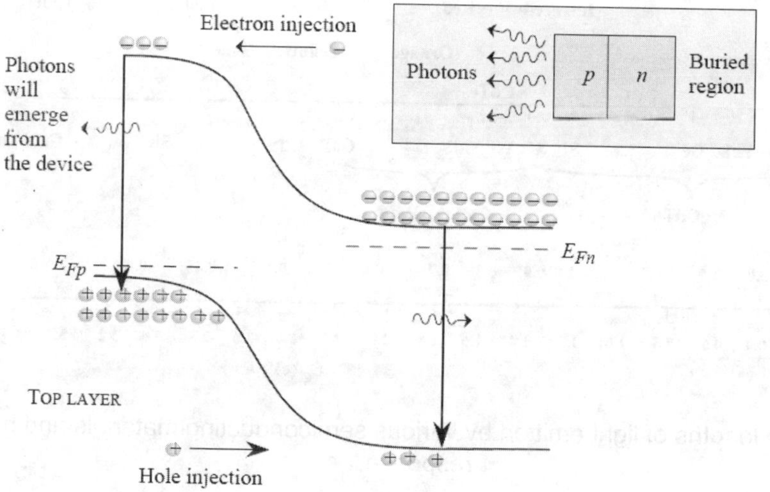

Fig. 9.13 (a) Energy band bending and charge injection in a pn junction diode

- In a p-n junction diode injection electroluminescence occurs resulting in light emission when the junction is forward biased

- Under forward bias – majority carriers from both sides of the junction can cross the depletion region and entering the material at the other side.
- Upon entering, the majority carriers become minority carriers
- For example, electrons in n-type (majority carriers) enter the p-type to become minority carriers
- Minority carriers will diffuse and recombine with the majority carrier.
- For example, the electrons as minority carriers in the p-region will recombine with the holes. Holes are the majority carrier in the p-region.
- The recombination causes light to be emitted
- Such process is termed as radiative recombination.

There are various materials that are used in the manufacturing of Light Emitting Diodes. Most of the materials are gallium-based crystals and are used in high-brightness applications. Among these include AlGaAs (Aluminum-Gallium-Arsenide), a semiconductor that typically generates the red spectrum, often used in signs, displays and electronic equipment.

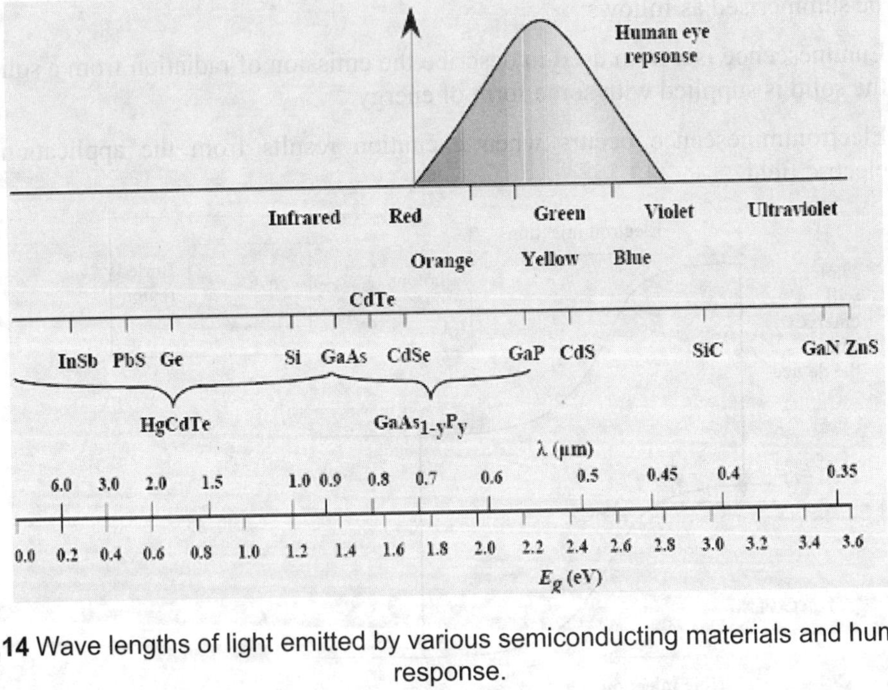

Fig. 9.14 Wave lengths of light emitted by various semiconducting materials and human eye response.

InGaAlP(Indium-Gallium-Aluminum-Phosphide) produces the yellow-green wavelength to red are often used in signs, auto interior as well as exterior, traffic signals and cellphones. InGaN (Indium-Gallium-Nitride) typically generates Blue, Green and white spectrums and are used most often in full color signs, cell-phones, auto interior, traffic signals. InGaN LEDs too has been made the light source of choice for many diagnostic and photo-therapy applications from the Ultra-violet to the near Infrared. Light-emitting diodes (LED) emit light in proportion to the forward current through the diode. Some of the semiconductor materials and the wave lengths of their emission are summerised in figure 9.14.

Liquid crystal display: Principles and Applications

Liquid Crystal Displays (LCD's) are used in many applications. We describe characteristics of different types of LCD's, including Twisted Nematic (TN) and Super Twisted Nematic (STN). We define operating parameters, illumination modes, legibility (brightness, contrast and viewing angle).

Liquid crystals are materials which combine the properties of both liquids and crystals. Rather than having a melting point, they have a temperature range, known as a mesophase, within which the molecules are almost as mobile as they would be in a liquid. Unlike a liquid, the molecules are in an ordered state similar to a solid crystal. It was realised that thin layers of a particular type of liquid crystal can be switched from transparent to opaque, or vice-versa, by the application of a voltage. This is the fundamental property of LCD's. LCD's are flat and compact. They use virtually zero power and are extremely reliable. They are customisable and easily driven by readily available electronic circuits. LCD's are a passive device and hence do not emit light, and consume no power. They can also be back-lit for pleasant night viewing. An LCD consists of two glass plates which are sealed together with a gap of 6-10 μm as shown in Fig. 9.15. The inner surfaces of the glass plates are coated with transparent electrodes which define the characters, symbols, or other patterns to be displayed. The electrode material is usually indium-tin oxide.

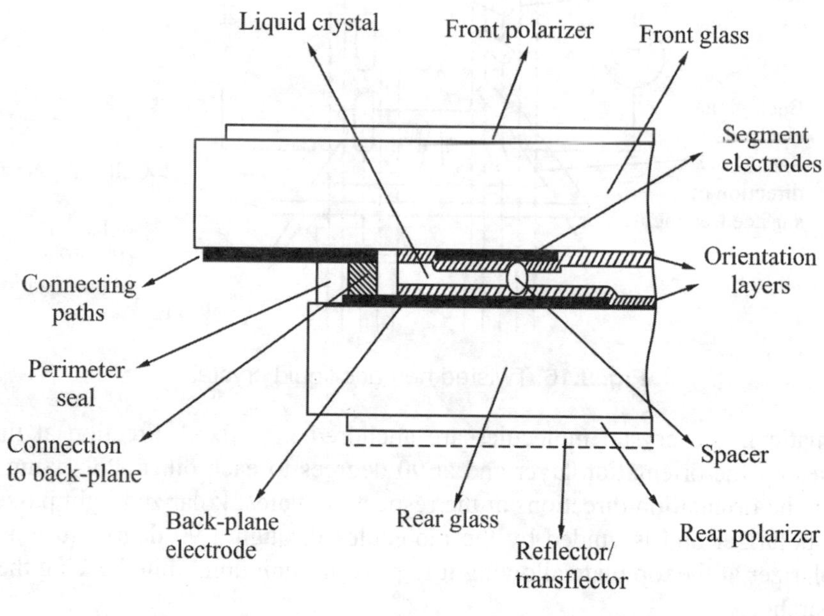

Fig. 9.15 Structure of a liquid crystal

Between the electrodes and the liquid crystal there are polymeric layers which are treated in a way that induces the adjacent liquid crystal molecules to maintain a defined orientation angle. For this reason, the polymeric layers are also known as orientation, or alignment layers.

The most common type of liquid crystal used in display technology is nematic. Nematic liquid crystal has rod like molecules which align themselves spontaneously parallel to each other, and gives the material anisotropic optical and electrical properties. This means it has different properties in different directions. The operating principle of TN LCD's is illustrated in Fig. 9.16.

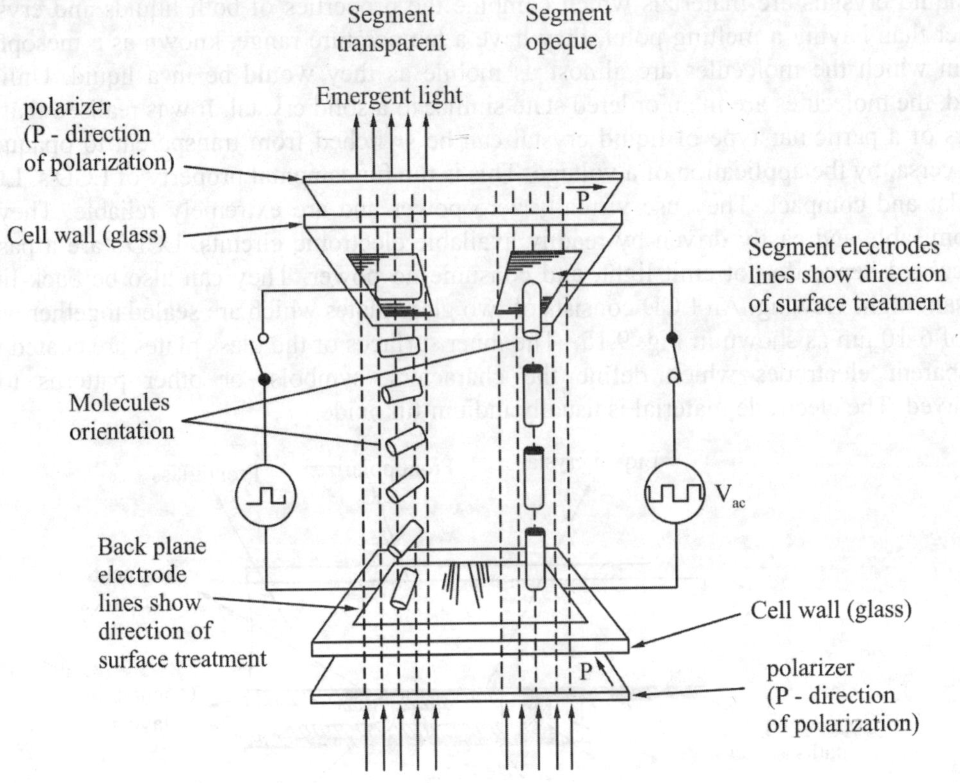

Fig. 9.16 Twisted nematic liquid crystal

The nematic liquid crystal molecules are anchored in a fixed direction at the top and bottom plates by the orientation layer and at 90 degrees to each other. Polarizing filters are aligned with the orientation directions at the respective plates. Polarized light passes through the bottom polarizer and is guided by the molecules through a 90 degree turn and aligned with the polarizer at the top plate allowing it to pass through uninhibited giving the display a bright appearance.

Conversely if a voltage is applied across the electrodes, they will lose heir ordered form and hence their 90 degree twist from upper to lower plates. The light passing through the lower plate will not be rotated by the molecules and will be absorbed by the top polarizer giving the display a dark appearance. Under these conditions the display will appear black when on and bright when off. This is known as a positive display. If one polarizer is rotated through 90 degrees the effect will be reversed and the display will appear black under no voltage field conditions and bright when a voltage is applied. This is known as a negative image. If the electrodes were to cover the top and bottom plates the LCD would act as a light shutter. The usual arrangement is for the electrodes to be patterned in such a way that specific segments can be switched to form characters, graphics or symbols such as a seven segment display. It should be noted that visible segments are only formed where the segment electrodes and the common back-plane electrode overlap. The remaining (unwanted) parts of the electrodes are required for connections to the outside.TN displays require a large change in the drive voltage to switch from the off state to the on state (or vice versa). These displays have a very vague and gradual slope in their contrast-voltage curve. This makes them impractical for use in multiplexing mode when the rate is high; say above 1:4. The solution is to use an STN display. It has been discovered the transition from low to high contrast - i.e. from OFF to ON states - can be made much sharper by increasing the twist angle of the crystal from 90 degrees to a value between 180-270 degrees. A small change in voltage can then result in a a large change in contrast. This can be used to produce a bright display with a wide viewing angle, at high multiplex ratios up to 1:100. STN display manufacturing is similar to TN, but has much tighter display tolerances and requires special alignment layers, with a subsequent increase in cost. LCD's can be operated in one of three modes:

Lighting is from behind the LCD. These displays must always be back-lit.

The LCD is backed by a diffuse metallic reflector, such as brushed aluminium foil, that reflects ambient light through the display and hence no back-lighting is possible. acked by a partly transmissive display reflector that reflects light in addition to transmitting diffused back lighting for night use Back lighting an LCD not only allows it to be viewed in the dark, but it adds a tremendous amount of aesthetic appeal. There are three methods used for back-lighting as illustrated in Fig. 9.17.

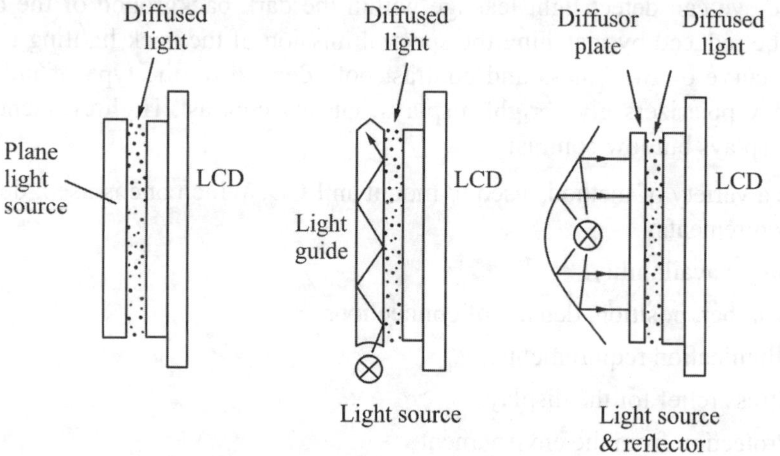

Fig. 9.17 Back lighting of liquid crystals

A constant plane of light is generated evenly behind the display .A true plane light source is the electroluminescence (EL) panel which emits diffused light. The drawbacks of this type of lighting include low luminance level and high operating voltage. A point source of light such as a fluorescent light or LED is positioned on one side of the display and a diffuser is used to distribute the light evenly under the LCD.

A light source is placed behind the LCD. The light is distributed by a reflector and diffuser producing very effective illumination. The light can be a halogen lamp or a series of LED's. The legibility of an LCD depends on a variety of factors:

(a) Pattern layout.
(b) Type of display.
(c) Driving condition.
(d) Illumination.
(e) Viewing angle.
(f) Operating temperature.

There are three optical specifications by which legibility is expressed:

Brightness of an LCD is the ratio of the luminance of the reflected or transmitted light to the luminance of the incident light. The brightness of a TN LCD cannot be higher than 50% since even an ideal polarizer transmits only half the incident light. Reflective displays will, therefore, tend to appear rather grey. A brighter picture is obtained through the use of back-illuminated displays. Contrast ratio (CR) of an LCD is the ratio of the brightness of light areas (B_L) to the brightness of dark areas (B_D)

For a TN LCD typical maximum contrast ratios can range between 5 and 50. In a reflective display, the maximum contrast ratio that can be detected by the eye is normally 10 and the limit of legibility about 2. As an example the contrast ratio of this page is about 6. For back-illuminated displays, a higher contrast ratio is required, especially for a negative image, as the eye can detect light leakage within the dark background of the display. This leakage can be reduced by matching the spectral mission of the back lighting to the spectral transmission curve of Brightness and contrast both depend on the type of polarizers used. Low efficiency polarizers give bright displays but low contrast. High efficiency polarizers give bright displays but low contrast.

There are a variety of methods used to mount an LCD. Which one is used, depends on the following requirements:

(a) Space availability.
(b) Number, position, density of connections.
(c) Illumination requirements.
(d) Stress relief for the display.
(e) Protection from the environment.

Reflective and transflective displays should be mounted as close as possible to the front, to maximise their exposure to ambient light. Mounting should also be in a position to allow the optimum viewing angle. To prevent the display from environmental and mechanical stresses, a transparent protective cover plate must be used, constructed from glass or other transparent material. If it's frosted, it will help minimize shadows and reflections.

Questions

1. Explain various fabrication methods of junction diode.
2. Describe the variation of electric field, space charge, and potential as a function of distance from the junction of a diode.
3. What is the significance of barrier potential in a diode?
4. Explain the unilateral conduction of a diode.
5. With necessary circuit diagrams, explain the working of a diode in forward and reverse bias.
6. Explain the working of a diode with the help of diode equation and hence describe its volt-ampere characteristics.
7. Sketch the energy band diagram of a diode in forward bias and reverse bias conditions. Explain the effect of applied bias voltage on the energy band diagram of the diode.
8. An ideal diode has reverse saturation current of 30 microamperes at 125^0C. Find the dynamic and static resistance of the diode when a bias of 0.2 volts is applied in forward and reverse bias.
9. Explain volt ampere characteristics of a diode
10. Describe various breakdown mechanisms of diode.
11. Explain the diode parameters a) knee voltage b) reverse saturation current c) diode resistance
12. Explain the working of a photo diode.
13. What is the difference between the working of a photo diode and light emitting diode?
14. Explain the physical process of emission of light in LED.
15. Explain the process of electroluminescence in LED light emission.
16. Describe the physical principle of working of liquid crystal displays.
17. What are the characteristics of LCDs?
18. Describe different mounting methods and method of back lighting of liquid crystals.
19. Obtain the diode equation.

Objective Questions

I. Choose the correct choice among the four alternatives.

1. Choose the correct statement
   (a) The most important characteristic of a p – n junction is its ability to conduct current in one direction only
   (b) The depletion layer contains only electrons and holes
   (c) The barrier potential in a silicon based p – n junction is 7.0 V.
   (d) Under for ward bias condition, the depletion layer is stretched into n – region

2. Under a reverse bias condition in a p n junction
   (a) minority carriers are arrested in depletion layer
   (b) a reverse saturation current is produced
   (c) The majority carriers diffuse across the junction
   (d) potential drop across the diode remains constant

3. The characteristic of a p – n diode is
   (a) it can act as an amplifier
   (b) it can act like an oscillator
   (c) it acts like a rectifier
   (d) the cut - in voltage is the function of the for wared current

4. The depletion layer in a p – n junction diode is
   (a) depleted of electrons
   (b) depleted of holes
   (c) a neutral zone
   (d) a charged region of positive and negative ions

5. The correct statement in the following is
   (a) In an unbiased p – n junction, the junction current at equilibrium is due to drift current only
   (b) In an unbiased p – n junction, the junction current at equilibrium is zero because equal and opposite charges cross the junction
   (c) In an unbiased p – n junction, the minority carriers are injected into other region across the junction
   (d) The depletion layer consists of electrons

6. In an ideal p – n junction diode
   (a) a reverse bias does not give any current
   (b) a reverse bias gives rise to a very high current
   (c) a forward bias is applied by connecting p - side to the negative of the battery
   (d) a reverse bias is applied by connecting p side to the positive of the battery

7. In forward bias of a p – n junction diode,
   (a) the current increases with voltage linearly
   (b) the electron from n - part are injected into p - part of the device
   (c) the holes from n - part are injected into p - part
   (d) the holes from n part are injected into p - part across the junction

8. A Zener diode
   (a) Shows extremely large reverse saturation current
   (b) does not show any forward current
   (c) possesses a sharp break down in the reverse bias
   (d) possesses a negative resistance

9. The light emitting diode works
   (a) under no bias condition
   (b) under forward bias condition
   (c) under reverse bias condition
   (d) under avalanche condition

10. (A) Any light emitting diode should preferably be indirect band gap semiconductor
    (B) Silicon based p n junction diodes are ideal for light emission
    (a) A and B are incorrect
    (b) A is incorrect, B is correct
    (c) A and B are incorrect
    (d) A is correct, B is incorrect

11. The light emission in L E D is due to
   (a) the recombination of electron - hole pairs near the junction
   (b) the electrons from the n - region jumping over the barrier into p - region
   (c) non radiative recombination in the p - region
   (d) the sparks created at the junction
12. A Varactor diode
   (a) is a reverse biased p n junction which acts like a charged capacitor
   (b) is made up of p and n semi conductors possessing very low majority carriers
   (c) has its capacitance as a linear function of the voltage across it
   (d) has its capacitance as directly proportional to the square root of the voltage across is

II. Fill up the blanks

1. A p n junction diode, when forward biased behaves likes a __________ .
5. Fermi energy is the __________ energy that can be possessed by an electron at 0 K.
2. When forward biased, the forward current increases __________ with forward bias.
7. The concentration of charge carriers in an intrinsic semiconductor is proportional to __________ . (Let T = the absolute Temperature)
8. In a p - type semi conductor, the mobility of holes is about 0.2 m^2 / V A Sec, and the concentration of acceptor in purity is 5×10^{22} atoms m^3 , then conductivity σ is __________ $\Omega^{-1} m^{-1}$.
9. The mobility of electron is __________ than that of a hole.
10. In an intrinsic semiconductor at a temperature T, the product election and hole densities is given by the relation __________, where E_g, is the forbidden gap, and A, is a constant.
11. The position of the Fermi-level in an impurity doped semiconductor depends on __________ and the concentration of __________ .
12. If N_a is the density of acceptor atoms in a p - type semiconductor, the electrical conductivity σ of the semiconductor is equal to __________, where μ_p is the mobility of holes.
13. The diffusion hole current density (J_p) in a semi conductor is proportional to the __________.
14. When an external electric field E is applied to a semiconductors, the total current density J_p due to holes is given by relation __________ , where D_p is the diffusion coefficient, $\frac{dp}{dx}$ is the concentration gradient and p is the density of holes, μ_p is the mobility.
15. When an electric field E is applied to an n- type semiconductor the drift current is given as __________ , where m_n is the mobility of electron.
16. If n_i is the intrinsic charge density, μ_n, μ_p are the mobilities of electron and hole, the intrinsic conductivity of a semiconductor is given as __________.
17. If N_v represents the density of states, N_a represents the density of acceptor atoms, then the Fermi level in a p-type semiconductor at temperature T is given by the relation __________ .
3. When a light beam is incident on a photodiode, the current that increases is __________.
4. The minimum energy of a photon that is required for excitation in a semiconductor photo device is equal to __________ of the semiconductor.
5. In LED, the bias that is applied is __________ bias.
6. In Ga As, the band gap energy E_g = 1.9 eV, then emitted wave length is __________.
7. The accumulation of electrons and holes on the two sides of the junction in a solar cell give rise to __________ voltage.

CHAPTER 10

Dielectic Properties

10.1 Introduction

Dielectrics are basically electrical insulating materials. This is the definition one finds for most of the engineering practices. However, the more general definition for a dielectric is given as a substance whose basic electrical property is the ability to be polarized and in which an electrostatic field can exist. The electrical insulator is an engineering material possessing the properties of a dielectric and preventing the leakage of electrical charges in electrical devices. Such substances as bakelite, PVC used in electrical wiring and tubes, many other polymeric materials, come under this category.

But in the modern technology, the term 'dielectric' has a broader meaning and they are considered as materials in which electrostatic fields can persist for a long time. The electrical insulators are passive materials while many dielectrics in use are active and play a vital role in many electronic applications. Such materials as capacitive materials in capacitors, ferroelectric, antiferroelectric, piezoelectric, and territes magnetic materials come under the umbrella of dielectrics. No electronic circuit can be free from dielectric devices. Over the years, the importance of dielectrics has over grown in electronic industry on par with semiconductors from the days of simple dielectrics like gutta-percha, natural rubber, paper, or vegetable oils. The development of a dielectric is linked now with the progresses that are taking place in radio electronics, nuclear engineering, miniaturization of electronic circuits, computer hardware etc.

Dielectrics are electrical insulators. They possess high resistivity values in the range $10^6\ \Omega - m - 10^{16}\ \Omega - m$. Under high voltage bias, they allow very little current (in the range

10^{-6}A – 10^{-14}A). They withstand very high voltages. The conduction phenomenon is mostly associated with ionic motion through defects or hopping of charges. They have no free charges. The electrical properties of a dielectric are associated with its inherent property of possessing electric dipoles. If such a material is placed between the plates of a parallel plate capacitor (Fig. 10.1), its capacitance increases.

Definitions

An electric dipole is a combination of a positive and negative charges seperated by a distance. If two charges +q and –q are separated by a distance **r** then dipolemoment is defined as

$$\mathbf{p} = q\,\mathbf{r}$$

If a solid has n dipoles in it, then the dipole moment is defined as

$$p = \sum_{i=1}^{n} q_i\, r_i$$

Polarisation **p** is defined as dipole moment per unit volume

$$\mathbf{P} = \frac{\mathbf{p}}{V} = \frac{\text{dipole moment}}{\text{volume}}$$

Dielectric constant $\in$ of a dielectric is defined as the ratio of capacitance of a parallel plate capacitor with and without dielectric

$$\in_r = \frac{\text{Capacitance with dielectric}}{\text{Capacitance without dielectric}}$$

$$= \frac{C}{C_0}$$

Polarisation per unit applied electric field is defined as polarisability.

$$\alpha = \frac{\text{Polarisation}}{\text{Electric field}} = \frac{\mathbf{P}}{\mathbf{E}}$$

or $$\mathbf{P} = \text{polarisation} = \propto \mathbf{E}$$

Elcctric susceptability χ is related to dielectric constant as

$$\in_r = 1 + \chi$$

χ is characteristic of the dielectric material.

10.1.1 Dielectric Medium Between the Plates of a Capacitor

Consider that the capacitor (Fig. 10.1) comprises rectangular, parallel plates of area of A $(\text{meter})^2$ each a distance d apart.

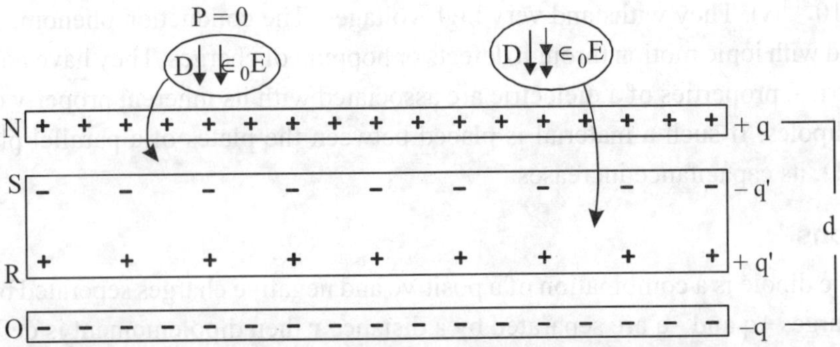

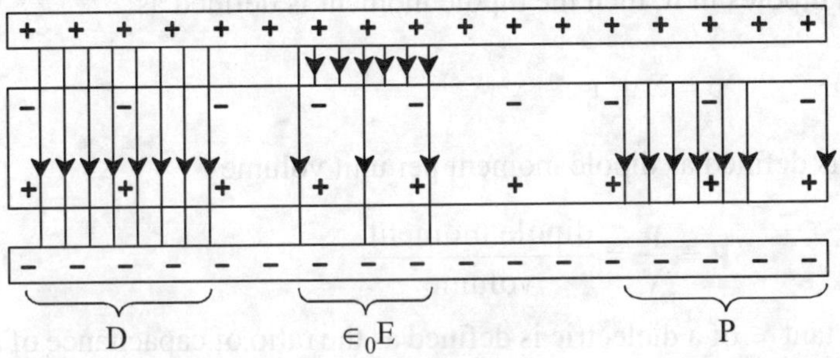

Fig. 10.1

Let the charge across each plate of the capacitor be Q in coulombs due to the potential difference between the plates. Then the charge is proportional to the voltage between the plates of the capacitor i.e. the electric field strength between the plates (V/d, volts/m).

$$Q = CV = \varepsilon_r C_o V$$

Where $C_o = \varepsilon_r A/d$, and $\varepsilon_0 = 8.854 \times 10^{-12}$F/m, the capacitance when free space is present between the plates, ε_r is the relative permittivity of the medium. The relative permittivity is called the dielectric constant. The electric field E at every point in the medium falls by a factor $1/\varepsilon_0\varepsilon_r$ of the field value with free space between the plates. The new capacitance in the presence of the dielectric medium between the plates is $\varepsilon_r C_o$. The factor ε_r is called specific inductive capacitance and is independent of the shape and size of the capacitor. The purpose of the dielectric medium is to maintain two large plates at a very small separation, increase the potential difference such that the capacitor can withstand without break down and then to increase the capacitance. ε_r is always a dimensionless quantity. The parallel plate capacitor as shown in the Fig. 10.1. exhibits the surface charge density on the plates. Let σ be the surface charge density on the plates, the voltage across the plate is given as,

$$V = \int_0^N E \bullet dx \quad \text{.....(10.1)}$$

$$= \int_0^R E\, dx + \int_R^S E\, dx + \int_S^N E\, dx$$

$$= \frac{\sigma}{\varepsilon_0} x_o + \frac{\sigma}{\varepsilon_0 \varepsilon_r} t + \frac{\sigma}{\varepsilon_0}\left[d - (t + x_o)\right]$$

$$= \frac{\sigma}{\varepsilon_o}\left[d - t\left(1 - \frac{1}{\varepsilon_r}\right)\right]$$

$$C = \frac{\sigma A}{V} = \frac{A\varepsilon_o \varepsilon_r}{[\varepsilon_r d - t(\varepsilon_r - 1)]} \qquad .....(10.2)$$

The magnitude of electric field intensity E is same at all points inside the dielectric which is placed between the two electrodes provided the dielectric under consideration is homogeneous. In free space, the electric field is defined as the force per unit charge. The electric field E arising from a point charge q is given as $E = \frac{q}{4\pi\varepsilon_o r^2}\mathbf{r}$ Volt/metre; where r is the distance between the test charge and q. **r** is the unit vector in the direction of r. The electric field is the negative gradient of the potential distribution in the region i.e., E = –dV/dr. It is a measurable quantity in free space. But, the measurement of electric field inside a dielectric is impractical. Therefore, a theory or model can be formulated for the behaviour which relates the externally measurable dielectric constant to the internal structure of a dielectric. The externally applied electric field is related to the ability of the dielectrics to become electrically polarized under the action of the field.

10.1.2 Atomic View of a Dielectric Polarization

A dielectric is a substance in which the charges or electrons are not free. They are tightly bound to the nucleus of the atom. An ideal dielectric does not possess free charges. The molecules or atoms in a dielectric contain positive and negative charge species. Under no electric field condition, the dielectric is neutral and the positive charge - negative charge species called dipoles are randomly oriented and they compensate each other. When an electric field is applied, the molecules or atoms are acted upon by a force. The positive and negative charges are pushed in directions opposite to each other as shown in Fig. 10.2. These displacements are extremely small, very small fractions of the molecular diameter. The charge configuration is altered and there exists a displacement of positive charge in the dielectric relative to the negative charge. This results in the creation of electric dipoles. This process is called polarization. The polarization is a molecular phenomena.

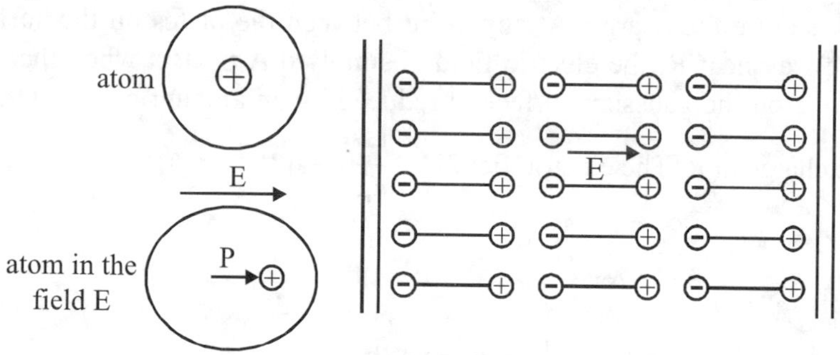

Fig. 10.2

Definition of Polarization

A dielectric has a large number of atoms or dipoles in its volume. In the absence of an external electric field, the volume of the dielectric does not have a net dipole moment. This is because the algebraic sum of the charges for all the molecules in the given volume is equal to zero (The dielectric is assumed to consist of very large number of molecules and the dimensions of the dielectric are large compared to the dimensions of the molecule). When an electric field is applied, the action of the field now brings the charges of the molecules into certain order and molecular dipoles are produced. There is now a net dipole moment associated with the dielectric. Each molecule contributes to the dipole moment.

The molecules become polarized in the presence of the electric field. If there are N molecules and each molecule acts like a dipole with a dipole moment **p**. The algebraic sum of the dipole moments is equal to $\Sigma \mathbf{p}_i$, i referring a molecule 'i'.

$$\sum_{i=1}^{N\Delta V} \mathbf{p}_i = N\,PV = \mathbf{P}\Delta V$$

where, **P** represents the average dipole moment of each molecule, V is the volume element.

$$\mathbf{P} = \sum \frac{\mathbf{p}_i}{\Delta V} \mathrm{C/m^2}$$

P is a vector quantity and is called polarization density. It is a macroscopic description of molecular phenomena, the tendency of molecules gives rise to a dipole moment. The proportional **P** is proportional to electric field strength to **E**.

$$\therefore \qquad \mathbf{P} = \alpha \mathbf{E} \qquad .....(10.3)$$

This is true for isotropic linear dielectrics. For isotropic dielectrics the directions of the vectors **P** and **E** coincide at each point and the proportionality constant in the eq. (10.3) is a scalar quantity and is called polarizability. For anisotropic dielectrics, **P** and **E** differ in directions and the proportionality constant a is tensorial in nature.

10.2 Electric Flux Density, Electrical Field and Polarization in a Dielectric Medium

Consider once again the parallel plate capacitor as shown in Fig.10.1. As a result of the electric field, the plates acquire the charge. At any point between the plates on the surface (called gaussian surface) say near R, the electric field is simply q/A ε_o. But when the dielectric is present, the charge on the gaussian surface is reduced by an amount q – q', where q' is the induced surface charge at R. The electric field is $\frac{1}{A\varepsilon_r}(q-q')$.

Then
$$\frac{q}{A} = \frac{q}{A\varepsilon_r} + \frac{q'}{A}$$

$$= \varepsilon_o\left[\frac{q}{A\varepsilon_0\varepsilon_r}\right] + \frac{q'}{A} = \varepsilon_0 E + P$$

The quantity q/A is equal to D called electric flux density

$$\mathbf{D} = \varepsilon_0 \mathbf{E} + \mathbf{P} \quad \text{........(10.4)}$$

when there is no dielectric,

$$\mathbf{D} = \varepsilon_0 \mathbf{E}$$

when there is a dielectric,

$$\mathbf{D} = \frac{q}{A} = \varepsilon\mathbf{E} \qquad \text{where } \varepsilon = \varepsilon_0 \varepsilon_r$$

$\therefore$

$$\varepsilon\mathbf{E} = \varepsilon_0 \mathbf{E} + \mathbf{P}$$

i.e.,

$$\varepsilon_0(\varepsilon_r - 1)\mathbf{E} = \mathbf{P} \quad \text{........(10.5)}$$

i.e.,

$$(\varepsilon_r - 1) = \frac{\mathbf{P}}{\varepsilon_0 \mathbf{E}} = \chi \quad \text{........(10.6)}$$

where χ is the electric susceptibility of the dielectric medium. $\chi\varepsilon_0$ is called absolute dielectric susceptibility and has units F/m.

The quantity **D** is introduced and is conveniently used while referring to different media. '**D**' is called electric displacement (electric induction). Suppose there is a charge placed at the centre of a sphere of radius r. The charge q will emanate q lines of force and these lines cross the surface of the sphere. The number of lines of force received by unit area is called electric displacement. **D** depends only on the distribution and magnitudes of charges which produce an electric field, but is independent of the nature of the medium. In the Fig. 10.1, the field lines of the three vectors **D**, **E**, **P** are shown. It indicates the variations of the field in the dielectric medium.

'**D**' has the units of coulomb/m^2 and **D** and **E** have the same direction in isotropic homogeneous medium.

Gauss Law

The permittivity of any material is greater than unity and for vacuum $\varepsilon_r = 1$ and **P** is equal to zero. The dielectric susceptibility is equal to zero in such a case. The most important relation between **D** and the charge in a volume is given by Gauss theorem. Gauss law states that the total normal electric induction from the charge q through a closed-surface is equal to the charge enclosed. The flux density **D** at any point on the surface is q/π r^2. Suppose the electric flux is like a flow across the surface d**S** as shown in Fig. 10.3. Then total flux passing through d**S** is

$$d\psi = \mathbf{D} \bullet d\mathbf{S} = D\cos\alpha\, dS = q\cos\alpha \frac{dS}{\;}.$$

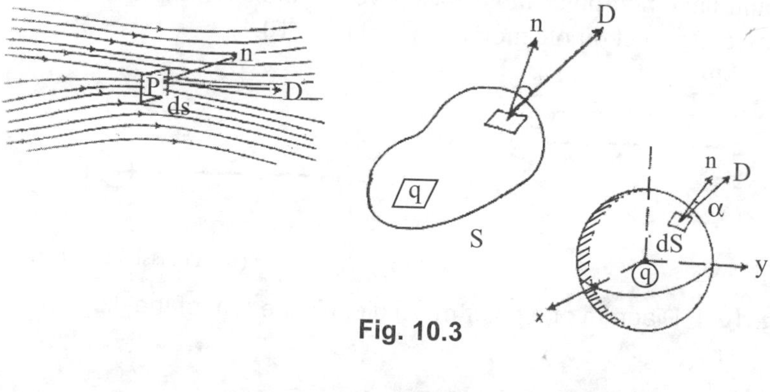

Fig. 10.3

But dS cos α/r^2 is the solid angle dΩ subtended at q by the elemental area dS. The electric flux is given as $(q/4\pi)\, d\Omega$. The sum of the flux over the whole surface will be the integral of the solid angle over the whole surface which is 4π steradians. The total flux is therefore $(q/4\pi)\, 4\pi = q$. Thus for a closed surface

$$\oint \mathbf{D}_n \bullet d\mathbf{S} = q \qquad \text{.......(10.7)}$$

This is called Gauss law. The surface integral of the electric flux lines through a surface is equal to the charge that the surface encloses. The electric flux is a property of any vector field. The electric flux is measured by the number of lines crossing or cutting through a surface. In terms of electric field, the Gauss law is written as $\epsilon_o \oint \mathbf{E} \bullet d\mathbf{S} = q$. If a surface encloses equal and opposite charges, the flux is zero. A charge outside a surface does not make contribution to the value of q.

D is a parameter which is important for a dielectric, which is made up of several layers. **D** is same in all the layers but **E** is different. The permittivity of a dielectric determines its ability to form an efficient capacitance. A capacitor stores energy for all engineering purposes, normally permittivity is considered as dielectric constant especially in the case of solids. Gauss law is used to evaluate electric field if the charge distribution is symmetrical and the integral can be evaluated. Coulomb's law can be deduced from Gauss law. The student should take this as his assignment and do. Gauss law is one of the fundamental equations of electromagnetism

10.3 Polar and Non-Polar Dielectrics

The dielectrics can be divided into two classes, (1) Polar (dipolar) dielectrics, (2) Non-Polar dielectrics.

It is already mentioned that the polarization is a molecular phenomena. A molecule consists of atoms or groups of atoms or ions. Each molecule may possess a net dipole moment. Each one possesses a charge. The spatial arrangement of charges in the molecule may be different in different molecules. All the positive charges and negative charges in the molecule can be replaced by one positive and one negative charge. They are mutually equal in magnitude and located at the centres of gravity of separate positive and corresponding negative charges. The centres of gravity may not coincide in some molecules. Then one would have polar molecules. If they coincide, one would have non-polar molecules. Even in the absence of an external field, a polar molecule possesses a net dipole moment p = ql (Fig. 10.4).

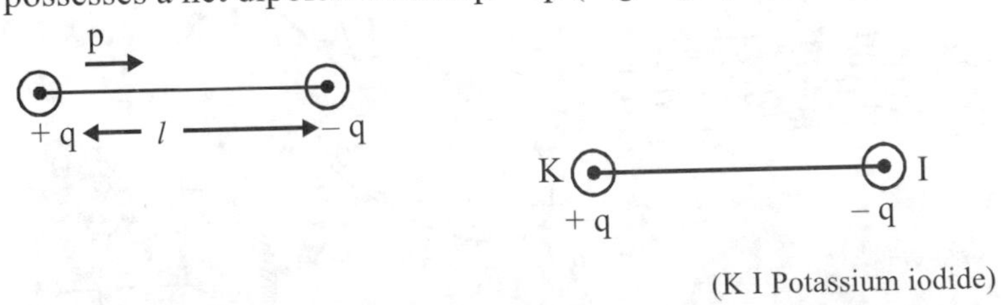

Fig. 10.4 Diagram of a polar molecule, l is the arm of the dipole.

The dipole moment would be the order of 10^{-30}C-m. All hydrocarbons are non polar or weakly polar i.e. (**P** = 0). The other examples for non polar type are monatomic molecules (He, Ne, Ar, Kr). The molecules consisting of two identical atoms linked by a homo polar bond like H_2, N_2, $C1_2$, etc are also non polar. CO_2 is non polar. (Fig. 10.5)

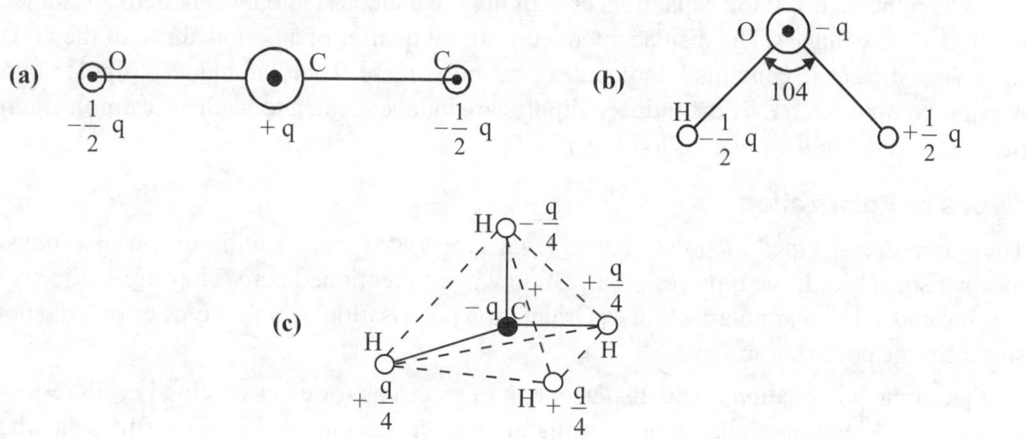

Fig. 10.5 (a) CO_2 (b) H_2O (c) CH_4.

Methane CH_4, Benzene C_6H_6 are non polar substances. But H_2O, CH_3Cl, CH_2CI_2, $CHCI_3$ etc. are asymmetric and polar. They have a net dipole moment. For example water molecule has a net dipole moment 6.1×10^{-30}C-m.

Hydrocarbon compositions like polyethylene, palatine, ceresui, poly iso bulytene, Polypropylene, polystyrene, rubber, electrical insulating, petroleum oils etc. are non polar. But polyvinylchloride is a strong polar substance. Polyvinyl alcohol, cellulose, phenol formaldehyde, etc., are polar substances. Polytetraflouro ethylene (Teflon) is non-polar, but polytriflouro ethylene is a polar substance.

10.4 Electronic, Ionic and Orientation Polarization

Polarizability

The polarization in a linear dielectric medium is expressed as **P** = α**E**. **P** represents the induced dipole moment. The value of **P** is measured in Coul.m. and **E** in V/m. Therefore the units of α are F.m^2. In CGS system, α is expressed in cm^3. The above relation is strictly valid only if **E** is small. In case **E** is large or very large, **P** does not increase at the same rate as earlier and the rate of increase of **P** will be smaller and this leads to saturation effects. At a particular value of electric field the polarization reaches a saturation value. Many dielectrics saturate at the fields of the order of 10^7Volts/m.

The polarizability α of a dielectric represents the nature of individual atoms or molecules. It is an important microscopic electrical characteristic of a dielectric. Polarization takes place in all the molecules of a dielectric. It is its bulk property . The displacement of charges in the presence of the electric field is very small and elastic in nature. If the field is switched off, the displaced charges may tend to return to their original states. The displacement of charges in the presence of an electric field can be in atoms, ions or dipolar molecules of substances. There arises a displacement current (capacitive current) in a dielectric when the direct voltage applied is taken off or when the magnitude of the voltage is changed with time. This current can exist only when the plates of the capacitors containing the dielectric in between them are subjected to alternating voltages. The displacement current is a quarter of a period ahead of the voltage. Suppose a dielectric contains n molecules per unit volume. Then the induced dipole moment per unit volume is n $\alpha \mathbf{E}$. If the induced dipoles are independent, a total electric dipole moment per unit volume or the polarization P is n $\alpha \mathbf{E}$.

Types of Polarization

They are several types of polarization. Each type is associated with an intrinsic physical mechanism. Three basic types of polarization will be mentioned here. They are 1. Electronic, 2. Ionic and 3. Dipolar polarisation or Orientation polarisation. Another type of polarization is space charge polarization.

Electronic polarisation is the displacement of electrons (or electron cloud) with respect to the nucleus. Electronic polarization occurs in all atoms or ions and is always present when a dielectric is under the action of an electric field. This type of polarization occurs during a very short interval of time ($\approx 10^{-15}$ sec) which is the time period of oscillations of uv rays. Homo polar neutral dielectrics possess this type of polarization only.

Ionic polarization is the mutual displacement of ions forming ionic molecule. This type of polarization occurs during a short time ($\approx 10^{-13} - 10^{-12}$ sec) and this time is longer than that for electronic polarization. It occurs in some substances where ionic bonding between the constituent ions exists.

Dipolar polarization is associated with polar molecules. The tendency of the dipolar molecules to rotate into the field direction gives rise to this type of polarization. It is temperature dependent i.e., it is controlled by the thermal motion of molecules. It occurs in gases, liquids and amorphous solids.

Both electronic and ionic polarizations are caused by deformation of an atom or molecule, which is a displacement of charges with respect to each other in the direction of electric field. The polarizations are set very fast. The temperature has no significant effect on their magnitude. Once the electric field is removed, the polarizing species return the electrical energy to the source. Therefore, it is not associated with any losses.

10.5 Internal Field in a Dielectric Medium and Clausius - Mosotti Equation

It has been said that if a molecule attains an induced electric moment **P** and this induced electric moment is equal to the product of the polarizability of the molecule and the intensity of electric field acting on the molecule. A dielectric consists of a large number of molecules. Each molecule is associated with a dipole moment **p**. Each molecule is surrounded on all sides by other polarized molecules of the dielectric. All the other molecules now impose an internal electric field on the molecule under consideration in addition to the external electric field. Therefore, at a given point in the dielectric, there exists an effective field which is equal to the sum of the external field and the internal field due to all other molecules around it. The total intensity of electric field (**E'**) will be the sum of the external field intensity **E** and the components of the internal field caused by the action of other polarized molecules on the molecule under consideration.

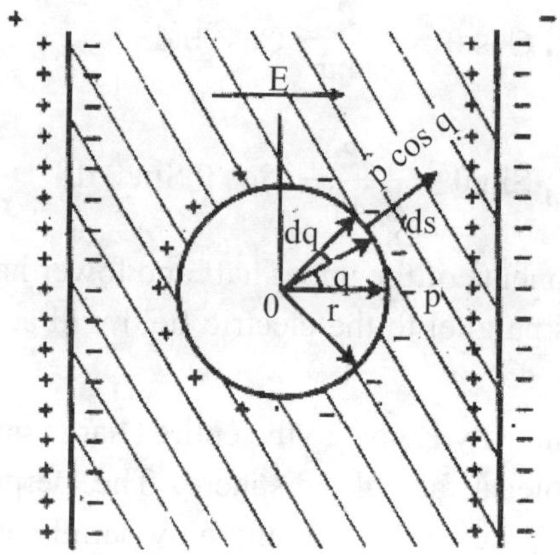

Fig. 10.6 A dielectric is placed between two electrodes.

Consider a dielectric placed between two electrodes as shown in Fig. 10.6. Let a sphere in a dielectric be described containing a molecule considered at its centre. Let r be the radius of the sphere and let it be sufficiently large compared to the dimensions of the molecules, but sufficiently small compared to the dimensions of the dielectric placed between the electrodes. On the molecule at the centre, there are three fields acting.

1. The external electric field **E**,
2. The internal field $\mathbf{E_1}$, due to the molecules that are present outside the sphere.
3. The field $\mathbf{E_2}$ due to the molecules inside the sphere. Then the total internal field intensity **E'** is

$$\mathbf{E'} = \mathbf{E} + \mathbf{E_1} + \mathbf{E_2} \quad .....(10.8)$$

Let it be imagined that the dielectric be removed from the sphere. As the pattern of the electric field is not disturbed, this is effectively equivalent to placing a charge on the surface of the sphere. Let the charge density on the surface of the sphere be σ . Then,

$$\sigma = P \cos \theta \quad .....(10.9)$$

Where θ is the angle between radius vector r and the direction of **E**. Let an elemental area **dS** is considered on the surface of the sphere. Then charge dq due to this surface **dS** is

$$dq = \sigma\, ds = \mathbf{P} \operatorname{Cos} \theta\, d\mathbf{S} \quad .....(10.10)$$

As a result of this charge, an electric field of intensity $\mathbf{dE_1}$ is created at the centre of the sphere. Where

$$dE_1 = \frac{dq}{4\pi\varepsilon_0 r^2} = \frac{P}{4\pi\varepsilon_0 r^2} \text{Cos } \theta dS \quad \text{......(10.11)}$$

This can be resolved into two components (1) $\mathbf{dE_1}$ Cos θ parallel to the direction of **E** and (2) $\mathbf{dE_1}$ Sin θ the component perpendicular to **E**. The two components are

$$dE_1 \text{ Cos } \theta = \frac{P}{\varepsilon_0 4\pi r^2} \text{Cos}^2 \theta \, dS \quad \text{......(10.12a)}$$

and

$$dE_1 \text{ Sin } \theta = \frac{P}{\varepsilon_0 4\pi r^2} \text{Cos } \theta \text{ Sin } \theta \, dS \quad \text{.....(10.12b)}$$

The perpendicular components of the upper half and lower half of the sphere compensate each other. The components parallel to the electric field will actually contribute to the total electric field.

The total electric field intensity **E'** due to the entire charge on the sphere is determined by integrating dE_1 over the whole surface of the sphere. The perpendicular components in the upper and lower halves cancel out each other and only parallel components contribute to the total field.

∴ Integrating (10.12 a), one gets

$$E_1 = \int dE_1 \text{Cos}\theta \, dS = \int_0^{\pi} \frac{-P}{\varepsilon_0 4\pi\pi^2} \text{Cos}^2\theta \, dS \quad \text{.....(10.13)}$$

But $dS = 2\pi r^2 \text{ Sin}\theta \, d\theta$. Therefore equation (10.13) becomes

$$E_1 = \frac{P}{2\varepsilon_0} \int_0^{\pi} \text{Cos}^2\theta \text{ Sin}\theta \, d\theta = \frac{P}{3\varepsilon_0} \quad \text{.....(10.14)}$$

The direction of $\mathbf{E_1}$ is same as that of **E**.

From the considerations of symmetry, $\mathbf{E_2}$ is taken as zero. Hence the total internal field is given as

$$\mathbf{E'} = \mathbf{E} + \frac{\mathbf{P}}{3\varepsilon_0} \quad \text{.....(10.15)}$$

but $\mathbf{P} = \varepsilon_0(\varepsilon_r - 1)\mathbf{E}$

Inserting the above relation in (10.14), we have

$$\mathbf{E'} = \mathbf{E} + \frac{\mathbf{P}}{3\varepsilon_0} = \mathbf{E}\left(1 + \frac{\varepsilon_r - 1}{3}\right) = \mathbf{E}\frac{(\varepsilon_r + 2)}{3} \quad \text{.....(10.16)}$$

Clausius-Mosotti Relation

From the above discussion, it is clear that the internal field **E'** depends on the nature of the dielectric material as the expression for **E** contains ε_r. In the case of free space, **E'** coincides with **E**. Consider a unit volume of a dielectric, then the value of polarization is given as

$$\mathbf{P} = \mathrm{Np} = \mathrm{N}\,\alpha\,\mathbf{E'} \qquad .....(10.17)$$

where N represents the number of molecules per unit volume of dielectric.

$$\varepsilon_0(\varepsilon_r - 1)\mathbf{E} = \mathrm{N}\alpha\frac{(\varepsilon_r + 2)}{3}\mathbf{E} \qquad .....(10.18)$$

or

$$\frac{(\varepsilon_r - 1)}{(\varepsilon_r + 2)} = \frac{\mathrm{N}\alpha}{3\varepsilon_0} \qquad .....(10.19)$$

The above equation is called 'Clausius-Mosotti equation'. It relates the macroscopic quantity, the dielectric constant and the polarizability, (the microscopic parameter) and N.

Multiplying the equation (10.19) by M/ρ, where M = molecular weight and ρ = density of the dielectric material, one gets

$$\frac{(\varepsilon_r - 1)}{(\varepsilon_r + 2)}\frac{\mathrm{M}}{\rho} = \frac{\mathrm{N}\alpha}{3\varepsilon_0}\frac{\mathrm{M}}{\rho} = \frac{\alpha \mathrm{N}_A}{3\varepsilon_0} \qquad .....(10.20)$$

where N_A is the Avogodros number, 6.023 x 10^{26}k mole^{-1}, is expressed in F.m^2, ε_0 in F/m (8.854 × 10^{-12} F/m).

The quantity $\frac{(\varepsilon_r - 1)}{(\varepsilon_r + 2)}\frac{\mathrm{M}}{\rho}$ is called Molar polarisation of a dielectric. Its dimension is equal to that of the ratio of volume to mass.

The above expression is effectively valid only for all neutral molecules i.e. neutral liquids and especially for gases in which molecules are far apart. The expression is not valid for strongly polarized substances. The above expressions do not indicate how the permittivity depends on the external parameters. In reality the permittivity depends on the frequency of the voltage applied on the dielectric, temperature, pressure, humidity etc.

10.6 Electronic Polarisability

The process of electronic polarization will now be considered. Consider an atom with atomic number Z, then the charge on its nucleus is + Ze. When this atom is subjected to electric field, the nucleus and the electron cloud are pulled apart by an amount x as shown in Fig. 10.7. Then the force along the direction of the field is

$$\mathbf{F}_1 = \mathrm{Ze}\,.\,\mathbf{E} \qquad .....(10.21)$$

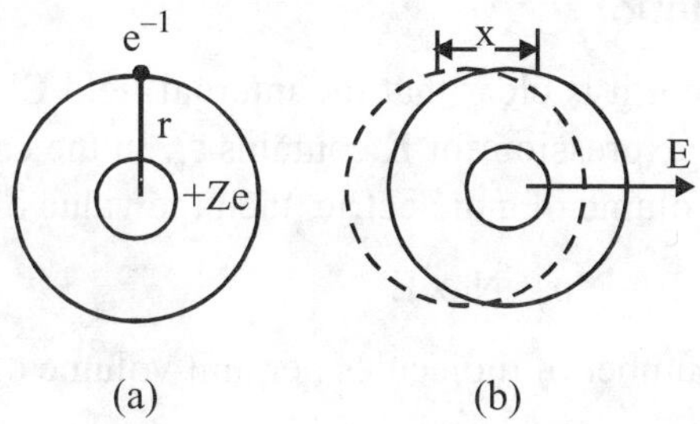

Fig. 10.7 (a) An atom without field, (b) with the field.

The nucleus is surrounded by electron cloud of charge -Ze distributed over a sphere of radius r. the charge density is given as

$$\rho = \frac{-Ze}{\left(\frac{4}{3}\right)\pi r^3} = -\left(\frac{3}{4}\right)\left(\frac{Ze}{\pi r^3}\right) \qquad .....(10.22)$$

Assuming uniform charge distribution of electron cloud, the total charge in the electron cloud is

$$\left(\frac{4}{3}\right)\pi x^3 \rho = \left(\frac{4}{3}\right)\pi x^3 \times \left(\frac{-3Ze}{4\pi r^3}\right) = -\frac{Zex^3}{r^3} \qquad .....(10.23)$$

The force (F_2) due to the attraction between the nucleus and the electron cloud now is, i.e., the Coulomb force,

$$F_2 = \frac{Ze}{4\pi\varepsilon_0 x^2} \times \left[-\frac{Zex^3}{r^3}\right] = -\frac{Z^2e^2x}{4\pi\varepsilon_0 r^3} \qquad .....(10.24)$$

In equilibrium condition, the forces F_1 and F_2 are equal.

$$+ZeE = \frac{-Z^2e^2x}{4\pi\varepsilon_0 r^3}$$

i.e.,

$$x = \frac{4\pi\varepsilon_0 r^3}{Ze}E \qquad .....(10.25)$$

Electronic dipole moment induced P_e is

$$\mathbf{P_e} = Zex = 4\pi\varepsilon_0 r^3 \mathbf{E} \qquad .....(10.26)$$

Therefore the electronic polarizability α_{el} is given as

$$\alpha_{el} = 4\pi\varepsilon_0 r^3 \qquad .....(10.27)$$

Consider the atom of hydrogen. It has a radius, r = 0.053 nm. Hence, it has an electronic polarizability $\alpha = 4\pi\ (8.85 \text{ x } 10^{-12}) \times (0.53 \times 10^{-10})^3 = 1.67.10^{-41} \text{Fm}^2$. The electronic polarizability for halogen atoms are, 1. Fluorine $\alpha = 4.45 \times 10^{-41}$ Fm^2, 2. Chlorine $\alpha = 2.67 \times 10^{-41}$ Fm^2, 3. Bromine $\alpha = 40.05 \times 10^{-41}$ Fm^2, 4. Iodine $\alpha = 64.5 \times 10^{-41} \text{Fm}^2$.

The polarization $\mathbf{P} = n\alpha_{el}\mathbf{E}$, where n is the number of atoms per unit volume. The homopolar dielectrics (i.e. without ionic structure) possess only one type of polarisation i.e. the electronic polarisation.

Lorentz-Lorentz Relation

An optical wave is an electromagnetic wave. When it passes through a dielectric medium, it undergoes a phase change as well as velocity change. In an electromagnetic wave, a changing electric field produces a changing magnetic field which in turn generates an electric field and so on with a resulting propagation of energy in the forward direction. Maxwell developed Maxwell equations as generalizations of circuit relations from Gauss law, non-existence of monopoles, Ampere's law and Faraday's law of electric induction. Maxwell's equations are point relations i.e., they are valid at every point in the medium. Based on his equations, Maxwell developed the wave equations for electromagnetic waves in a medium. He showed that the velocity of propagation of such a wave for unbounded medium is given by $V = \frac{1}{\sqrt{\mu\varepsilon}}$, where μ is the permeability of the medium, and 'ε' the permittivity of the medium. Normally for dielectrics which are nonmagnetic 'μ' can be taken as μ_0, the permeability for free space. The velocity of waves in the free space $\mathbf{V_0} = \frac{\mathbf{1}}{\sqrt{\mu_0\varepsilon_0}}$. As the value of ε for a medium is greater than ε_0, the velocity of an electromagnetic wave (optical wave) in a medium is less than that in free space. The origin of the refractive index of the medium is seen from the point of permittivity of the medium. The refractive index is defined as the ratio of velocity of electromagnetic waves in free space to their velocity in the medium. Thus the refractive index for a medium is greater than that of the free space. The refractive index for a medium is equal to the square root of relative permittivity of the medium. Therefore, Clausius-Mosotti equation can be expressed for an optical medium as

$$\frac{\eta^2 - 1}{\eta^2 + 2} = \frac{N\alpha_{el}}{3\varepsilon_0} \qquad .....(10.28)$$

where η is refractive index of the medium. The above relation is called Lorentz-Lorentz equation. The determination of α_{el} is possible from the determination of the refractive index. The dielectric constant at optical frequencies arises almost totally from the electronic polarizability. Lorentz-Lorentz equation can also be written as:

$$\left(\frac{\eta^2 - 1}{\eta^2 + 2}\right)\frac{M}{\rho} = \frac{N_A\alpha_{el}}{3\varepsilon_0} \qquad .....(10.29)$$

The term on the left side is called molar refraction. Using the Clausius-Mosotti equation throughout the visible and ultra violet range of electromagnetic spectra where in the proper vibrational frequencies of electrons lie, the polarizabilities can be calculated. The Table 10.1. gives the values of the square of the refractive indices and the relative permittivity for a number of compounds. The elemental solid dielectrics exhibit the electronic polarizability only.

Ionic Polarization

In the case of substances with ionic bonding like Na*Cl*, K*Cl*, $CaCO_3$, Rutile etc. there appears lot of discrepancy between η and ε_r values. Therefore, it is obvious that a different polarization mechanism is possible in these compounds apart from electronic polarization and that polarization is ionic polarization. For these compounds, normally $\varepsilon_r > \eta^2$. Table 10.1. gives the static and optical dielectric constants for a few compounds.

Table 10.1 Static and Optical dielectric constants for a few compounds A comparison

Sl.No	Substance	$\eta^2 = \varepsilon exp$	E_r
1.	Hydrogen	1.00028	1.00027
2.	Oxygen	1.00054	1.00055
3.	Ethylene	1.00130	1.00138
4.	Polystyrene	2.4	2.6
5.	Diamond	5.76	5.7
6.	NaCl	2.37	6.0
7.	KCl	2.22	4.8
8.	Calcite	2.78	6.1
9.	MgO	2.95	9.8

The processes of electronic and ionic polarization have common features. Both are caused by deformation of charge distribution. The deformation is a displacement of charges with respect to each other in the direction of the electric field. The state of polarization is set in very fast in both cases. More or less, both are of the same magnitude. The polarization is practically unaffected by variations in temperature of the dielectric. Normally the process is reversible and does not suffer from dielectric losses. Both are present in ionic dielectrics.

10.7 Orientation Polarization

The molecular solids like polymers and other polar dielectrics exhibit a tendency towards dipole or orientation polarization. A molecular dipole as shown in Fig.10.8 tends to orient (rotate) into the direction of the external field.

Debye, in his theory proposed that the orientation of the dipole is identical with a sphere rotating in a viscous medium. Each molecule in the polar dielectric is a dipole having a permanent dipole moment. The polar molecular dipoles are randomly distributed in space in the absence of an electric field (Fig.10.9 (a)). The net dipole moment of the dielectric is zero (here the contribution from the internal field and from other dipoles is neglected). But when the dielectric is kept under electric field, (Fig. 10.9 (b)) the field produces a torque in individual dipoles and there is a tendency for the field to align them with the field and a net dipole moment per unit volume is originated in the dielectric. If the field is strong enough, the dipoles may completely be aligned along the field direction and there will be a saturation polarization in the dielectric.

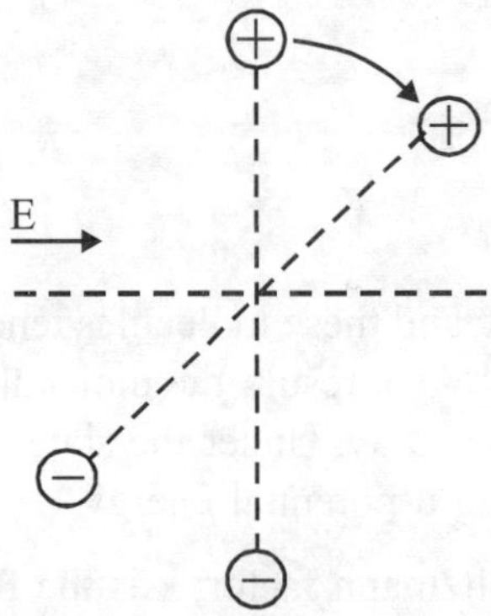

Fig. 10.8 Orientation of dipole molecule

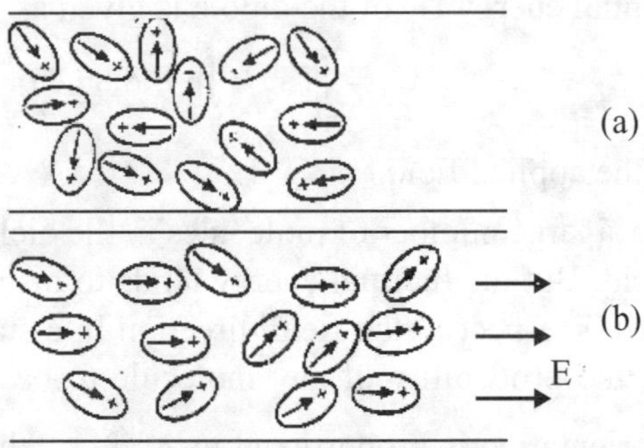

Fig. 10.9 Dipole orientation (a) In zero field (b) In the presence of strong field

The rotation of polar molecules under the action of an electric field is treated as the system of randomly oriented dipoles going into an orderly arrangement of dipoles in the field. In pure form, it can appear only in gases, liquids and amorphous viscous bodies like polymers and glassy dielectrics. In many crystalline solids below their melting points, the dipoles are said to be frozen. Therefore, orientation polarization is ruled out.

Langevin - Debye Equation :

It is possible to determine the dielectric susceptibility of polar substances supposing that the molecular dipoles can rotate freely. Let a unit volume of dielectric be considered and this volume contain n molecules. Each molecule is associated with a permanent dipole moment **p**, represented by a vector (Fig.10.10). When a static field **E** is applied to the dielectric, there is a tendency to orient the dipoles in the field direction. The thermal energy works against this and tries to disorganise the orderly state. Consider a sphere of radius r in the dielectric and a small surface dS on the sphere.

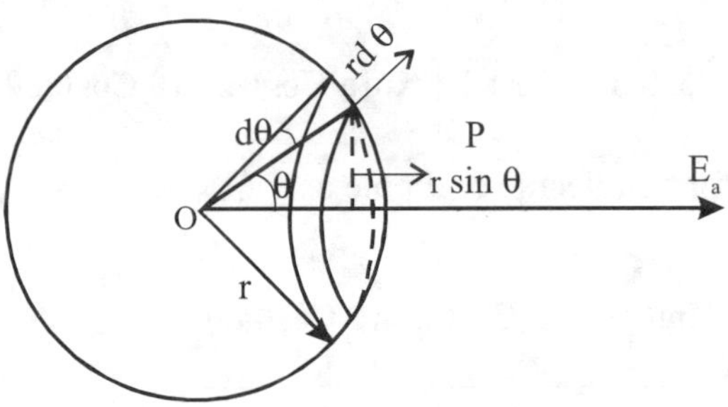

Fig. 10.10

Torque on the dipole **T = p x E**.

The potential energy U_p of the dipole is given as

$$U_p = \int_{\theta} T d\theta = \int_{\theta} p\, E\, Sin\theta\, d\theta = pE_a Cos\theta \quad .....(10.30)$$

where E_a is the applied field.

There are a large number of molecules in the dielectric and these molecules tend to orient along the field. But the thermal energy tends to disturb it. As a result, the molecules will be distributed with respect to the field direction in a statistical way. Under the state of thermal equilibrium, the probability of any molecule to be having a potential energy U_p due to its position or orientation is proportional to $e^{-Up/kT}$, the Boltzmann factor; k is the Boltzmann constant. The number of dipoles per unit volume per unit solid angle making an angle θ with the direction of the electric field may be given as

$$dn(\theta) = K \exp\left(\frac{\mathbf{pE_a} Cos\theta}{kT}\right) d\Omega \quad .....(10.31)$$

where K is a constant of proportionality and k is the Boltzmann constant. The conical solid angle dΩ included between θ and θ + dθ is equal to

$$\frac{2\pi r\, Sin\theta\, r\, d\theta}{r^2} = 2\pi\, Sin\,\theta\, d\theta \quad .....(10.32)$$

The number of dipoles per unit volume lying in this conical solid angle is given by the relation

$$n(\theta)\, d\theta = dn = Ke^{pE_a Cos\theta/kT}\, 2\pi\, Sin\,\theta\, d\theta \quad .....(10.33)$$

The total number of molecules per unit volume over all the values of θ i.e. from 0 to π is

$$\int dn = n = \int_{\theta}^{\pi} K \exp\left(\frac{pE_a Cos\,\theta}{kT}\right) 2\pi\, Sin\,\theta\, d\theta \quad .....(10.34)$$

Let $$a = \frac{pE_a}{kT} \quad .....(10.35)$$

Then the total polarization (P) acquired is

$$P = \int_0^n p\, Cos\theta\, dn = 2\pi KP \int_0^{\pi} \exp(a\, Cos\,\theta) Sin\,\theta\, Cos\theta\, d\theta \quad .....(10.36)$$

This can be written for a molecule or dipole as

$$P = \frac{2\pi Kpn \int_0^{\pi} \exp(a\, Cos\theta) Sin\,\theta\, Cos\theta\, d\theta}{2\pi K \int_0^{\pi} \exp(a\, Cos\theta) Sin\,\theta\, d\theta} \quad .....(10.37)$$

Let x = Cosθ; i.e. dx = Sinθ dθ, then x changes from –1 to +1, then

$$P = \frac{np\int_{-1}^{+1} x\exp(ax)dx}{\int_{-1}^{+1}\exp(ax)dx} \quad \text{......(10.38)}$$

i.e.,

$$P = np\left[\frac{e^a + e^{-a}}{e^a - e^{-a}} - \frac{1}{a}\right] = np\left[\text{Coth}a - \frac{1}{a}\right] \quad \text{......(10.39)}$$

But 'np' is the maximum possible dipole moment that the substance can have. It corresponds to a saturation value (P_s) and all the molecular dipoles are aligned parallel to the applied electric field. This represents a state of complete saturation. The above expression (10.39) is the Langevin-Debye expression for the mean effective polarization of a polar dielectric in a field E_a at temperature T.

Let np = P_0, Then

$$\frac{P}{np} = \text{Coth}\,a - \frac{1}{a} = \frac{P}{P_0} \quad \text{.....(10.40)}$$

Let Coth a – 1/a = L(a) where L(a) is called Langevin function. L(a) can be expressed as

$$L(a) = \frac{a}{3} - \frac{a^3}{4^5} + \frac{2a^3}{4^5}$$

If $a \to \infty, L(a) \to 1;\ a \to 0,\ L(a) \to 0$

For small values of a, the L(a) vs a plot has a slope of 1/3. (See Fig. 10.11). Here L(a) curve is practically a straight line. This is true for field strengths of 10^8 V/m. The dielectric strengths of many common dielectrics are smaller than this value.

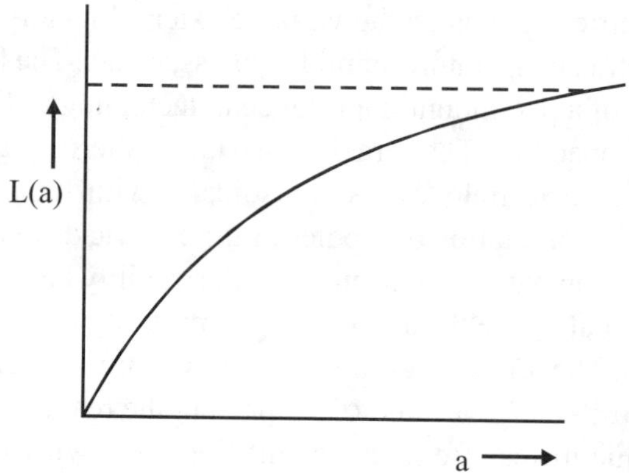

Fig. 10.11 Langevin function Vs a

When $$L(a) = \frac{1}{3}, \quad \mathbf{P} = np\frac{a}{3} = \frac{np^2}{3kT}\mathbf{E_a} \qquad .....(10.41)$$

The polarization is proportional to the field. The polarization depends inversely on the temperature. The susceptibility

$$\chi = \frac{P}{\rho E} = \frac{np^2}{3\rho kT} \qquad .....(10.42)$$

where ρ is the density of the dielectric, n represents the number of atoms per unit volume.

Then $n = \frac{N_A \rho}{A}$ Where A is the atomic weight and N_A is the Avogadro's number.

$$\chi = \frac{N_A P^2}{3AkT} = \frac{C}{T} \qquad .....(10.43)$$

If P is of the order of 10^{-30} C-m, then P/kT is of the order of 10^{-9}m/volt.

Then α_a , the orientation polarizability is

$$\alpha_a = \frac{P_a}{3} = \frac{p^2E}{3kT\varepsilon_0} \qquad .....(10.44)$$

The substance shall have a total polarizability,

$$\alpha_{Total} = \alpha_e + \alpha_i + \frac{P^2}{3kT} = \alpha_d + \frac{P^2}{3kT} \qquad .....(10.45)$$

where α_d is called deformation polarizability and is equal to $\alpha_\iota . + \alpha_e$. Equation (10.44) is known Langevin-Debye equation. It relates orientation polarizability to the temperature. The dielectric constant of polar substances depends on temperature.

10.8 Behaviour of Dielectric in Alternating Fields

The pcrmittivity of a dielectric depends on the variable external factors like frequency of the voltage applied to the dielectric, temperature, humidity, pressure etc., The frequency dependence of the dielectric constant is of a paramount importance in technology. The dielectric are used as materials for insulatio or capacitors. They are operated under alternating fields. An alternating field is one in which the electric field varies sinusoidally with time. The electric field is represented as $E = E_0 e^{iwt}$. The orentation of dipoles or the electric displacement in a dielectric tries to follow the field variation with time. Then, the polarizability becomes a function of time or the frequency of the oscillating field. The relative permittivity as well as the polarzability become complex quantities. The relative permittivty is now represented as $\varepsilon^*_\alpha = \varepsilon'_\alpha - i\,\varepsilon''_*$ and the polariziability as $\alpha^* = \alpha'_* - i\,\alpha''_* \cdot \varepsilon'_a$ and α'_* represent the real quantities and ε''_α and α'' represnt the imiganary quantities. Both the quantities vary with the frequency of the alternating field.

10.8.1 Dielectric Loss and Dielectric Relaxation

Dielectric Loss :

Consider that a dielectric sample is placed between two paralel plates that work as electrodes. An oscillating field E is applied across the plates. The field is described as $E = E_0 e^{j\omega t}$. All dielectrics possess finite conductivity or resistivity values. A real dielectric sample now shows the displacement field D with a phase difference with respect to the E-field.

Then, the displacement field D is

$$D = D_0 e^{j(\omega t - \delta)}$$

i.e.,
$$D = \varepsilon_0 E + P = \varepsilon_0 E + \varepsilon_0 (\varepsilon_r - 1) E$$

The relative permittivity ε_r consists of two components: (1) a real component ε'_r and (2) an imaginary component ε''_r and ε^*_r is a complex quality represented by

$$\varepsilon^*_r = \varepsilon'_r - j\varepsilon''_r$$

then,

$$D = \varepsilon_0 E + \varepsilon_0 [\varepsilon'_r - j\,\varepsilon''_r - 1]E$$
$$= \varepsilon_0 \varepsilon'_r E - j\varepsilon_0 \varepsilon''_r E$$

i.e.,
$$D_0 e^{j\omega t} e^{-j\delta} = \varepsilon_0(\varepsilon'_r - j\varepsilon''_r) E_0 e^{j\omega t}$$

$\therefore$
$$\frac{D_0}{E_0} e^{-j\delta} = \varepsilon'_r - j\varepsilon''_r$$

$\therefore$
$$\varepsilon'_r = \frac{D_0}{E_0} \cos\delta$$

and
$$\varepsilon''_r = \frac{D_0}{E_0} \sin\delta$$

and
$$\tan\delta = \frac{\varepsilon''_r}{\varepsilon'_r}$$

The factor 'tan δ' is called the loss tangent or dissipation of the dielectric. It indicates the amount of energy that is dissipated in the dielectric when an electric field is applied across it. The dielectric power loss is given as

$\omega\varepsilon_0\varepsilon'_r V^2 \tan\delta$ watts / cycle, when the AC Voltage V is applied. As a result of this loss, dielectrics absorb energy and get heated up. The dielectric loss is a funciton of temperature. It increases with temperature in solids. The loss is also a function of frequency.

In polar dielectrics, there will be a resultant polarization to start with when voltage is applied across two electrodes between which a dielectric is placed, the dipoles tend to orient along the field direction. Suppose the field is switched on, there will be a built up of charge or polarization in the dielectric with time. The polarizaticn attains an equilibrium value after a long time. Similarly when the field is switched off suddenly, the polarization will not instantaneously become zero. This type of delay is due to the fact that the dipoles take certain time for their rotation. Therefore, the polarization in the dielectric is a function of time. In fact, a dipolar substance possesses two terms of polarization as soon as the electric field is applied, the first term is the initial value of polarization P_0, and a second term, the time dependent polarization $P_1(t)$. The dependence of polarization on time gives rise to relaxation. This sort of relaxation is responsible for dielectric loss variation with frequency.

The change in polarization with time is decribed as

$$P(+) = P_\alpha (1 - e^{-t/n})$$

Pa represents the polarization after a lapse of very longtime. The variation of with frequency of the alternating field is described by the expression

$$P = \left[\chi_0 \varepsilon_0 + \frac{\chi_1 \varepsilon_0}{1 + j\omega\tau} \right] \mathbf{E} = \chi a \, E$$

χ_o = represents the dielectric susceptibility n state field (i.e., the frequency = 0), χ_1 represents the time dependent susceptibility. χ_a is the total dielectric susceptibility. In such a case, we have

$$\chi_a = \chi_0 \varepsilon_0 + \frac{\chi_1 \varepsilon_0}{1 + j\omega\tau}$$

Then $$\varepsilon^*_\alpha - 1 = \varepsilon_0 \chi_0 + \frac{\chi_1 \varepsilon_0}{1 + \omega^2 \tau^2} - j \frac{\chi_1 \varepsilon_0 \omega\tau}{1 + \omega^2 \tau}$$

The real part of the permittivity is

$$\varepsilon^*_a = \varepsilon_0 \chi_0 + \frac{\chi_1 \varepsilon_0}{1 + \omega^2 \tau^2}$$

and the imaginary part of susceptibility is

$$\varepsilon''_\alpha = \frac{\chi_1 \varepsilon_0 \omega\tau}{1 + \omega^2 \tau^2}$$

The variation of the relative permittivity is similar to what we have shown in Fig. 10.12.

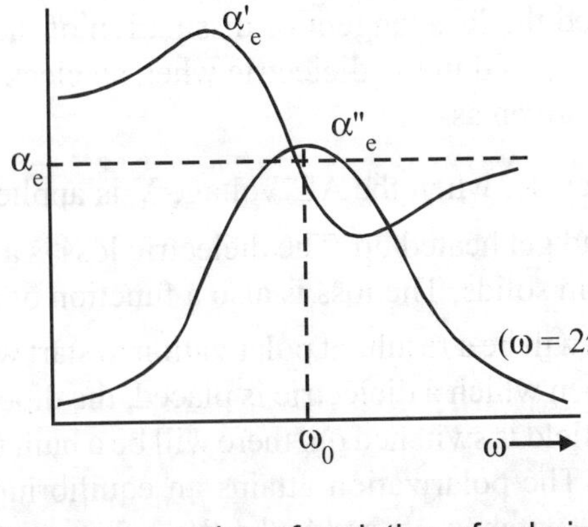

Fig. 10.12 Schematic representation of variation of polarizability with frequency.

10.8.2 Frequency Dependence of Electronic Polarizability

An alternating field is one in which the electric field varies sinusoidally with time. An electromagnetic wave can be made to be incident on a dielectric. As a result of the time varying electric field, the component of the transverse electromagnetic wave, the electron in the atom is forced to carry out motion. The electron is displaced by an amount x relative to the nucleus of the atom and the system is then left to itself. The restoring force which tends to drive the electron to the nucleus is given by

$$F = \frac{e^2 x}{4\pi\varepsilon_0 R^3} = -K'x \quad \text{........(10.46)}$$

where x is the displacement and K' is the restoring force constant. In the absence of the damping force, the equation of motion of the electron is written as

$$m\frac{d^2x}{dt^2} = -K'x + e\mathbf{E}_0 \exp(j\omega t) \quad \text{.....(10.47)}$$

where the alternating electric field is expressed as $\mathbf{E} = \mathbf{E}_0 \exp(j\omega t)$. This is the equation of motion of a forced harmonic oscillator. Apart from the external field, the electron is also under the action of a local field. Therefore the above equation can be written as

$$m\frac{d^2x}{dt^2} + K'x = e\mathbf{E}_{local} \quad \text{.....(10.48)}$$

or $$\frac{d^2x}{dt^2} + \frac{K'}{m}x = \frac{e}{m}\left(\mathbf{E} + \frac{\mathbf{P}}{3\varepsilon_0}\right)$$

or $$\frac{d^2x}{dt^2} + \omega_0^2 x = \frac{e}{m}\left(\mathbf{E} + \frac{\mathbf{P}}{3\varepsilon_0}\right) \quad \text{.....(10.49)}$$

or $$\frac{d^2x}{dt^2} + \left(\omega_0^2 - \frac{Ne^2}{3\varepsilon_0 m}\right)x = \frac{e}{m}\mathbf{E} \quad \text{.....(10.50)}$$

where $\mathbf{P} = Nex$. N represents the total number of electrons. The solution of the equation is of the form

$$x = x_0 \text{ Sin } (\omega_0 t + \delta) \quad \text{.....(10.51)}$$

Then $$x_0 = \frac{eE_{ap}}{m\left(\omega_0^2 - \omega^2 - \frac{Ne^2}{3\varepsilon_0 m}\right)} \quad \text{.....(10.52)}$$

x_0 is the maximum displacement. Then the dipole moment, $p = ex_0$.

The electronic polarizability is given as

$$\alpha_e = \left(\frac{ex}{E_{loc}}\right) = \frac{Ne^2/m}{\left(\omega_0^2 - Ne^2/3\varepsilon_0 - \omega^2\right)} \quad .....(10.53)$$

ω_0 is the natural or resonance angular frequency. For an atom R $\approx 10^{-10}$m, m = 0.9 × 10^{-31}Kg, N = 10^{29}m^{-3}, $\omega_0 \approx 10^{16}$ radians sec^{-1}. α_e is of the order of 10^{-40}Fm^2. The frequencies at which resonance takes place lie in the vacuum ultraviolet range of electromagnetic spectrum.

The frequency dependence of the real and imaginary parts of α'_e are shown schematically in Fig. 10.12. The polarizability represents the dielectric constant. The y-axis can as well be taken as dielectric constant.

The above curves represent also the variation of permittivity arising from electron polarizability. The imaginary part vanishes at $\omega = 0$. However the real part (α'_e) has a value $\frac{Ne^2}{m\omega_0^2}$ and this is the value of ω_e under the static field (i.e. dc) condition.

The real part is positive for all values of $\omega < \omega_0$ and negative for all values of $\omega > \omega_0$. The real part is zero at $\omega = \omega_0$. The damping factor is normally very small and the real part of polarizability does not change with frequency from zero frequency upto a certain frequency close to ω_0.

Suppose $\omega_0 - \omega_0 = \Delta\omega$ and $\Delta\omega << \omega_0$

$$\therefore \quad \omega^2{}_0 - \omega^2 = (\omega_0 + \omega)(\omega_0 - \omega) \approx 2\,\omega\Delta\omega$$

$\therefore$ The real of polarizability is

$$\alpha'_e = \frac{Ne^2}{m} \times \frac{2\omega_0\Delta\omega}{4\omega^2(\Delta\omega)^2 + \gamma^2\omega_0^2} = \frac{2\Delta\omega/\omega_0}{4(\Delta\omega)^2 + \gamma^2} \quad .....(10.54)$$

Then α'_e is maximum when $4(\Delta\omega)^2 = -\gamma^2$

Therefore the damping factor is a measure of the separation between maximum and minimum of the polarizability α'_e.

As $\omega \to \infty, \alpha''_e \to 0$ and around $\omega = \omega_0$ it is

$$\alpha'_e \approx \frac{Ne^2\gamma/m\omega_0}{4(\Delta\omega)^2 + \gamma^2} \quad ....(10.55)$$

The α''_e exhibits maximum at $\omega = \omega_0$. This imaginary part of polarizability gives rise to absorption of energy by the dielectric from the field.

The electronic dielectric constant gives the square of the refractive index. The incident electric field is the radiation wave. The refractive index also will have a real component and an imaginary component.

At $\omega \sim \omega_0$, the real part of polarizability or permittivity becomes negative. At these frequencies, the radiation or the field will be reflected. The materials appear opaque for these fields.

10.9 Frequency Dependence of Dielectric Constant Over a Wide Frequency Range

We have seen how individual polarization mechanisms give rise to frequency dependence of polarizability. It reflects directly the frequency dependence of dielectric constant of a solid.

The effects of frequency are especially important because dielectrics are often used in alternating fields. The polarization depends on the ability of the fundamental particles to move relative to each other. At high field frequencies, the more massive molecular particles can not move rapidly enough to follow the direction of the electric field. Therefore, at high frequencies, dipoles can not rotate rapidly enough to contribute to polarization. They remain, on the average, at an angle 90^0 to the field direction. The electrons are capable of relaxing rapidly, the electron cloud distortions follow the field direction more easily at high frequencies. The Fig. 10.13. schematically shows how the polarizability might vary with frequency. At zero frequency, the polarization is static. The polarization is due to space-charge built up on the surfaces of the dielectric or at the interface of the electrode and the dielectric. At radio wave frequencies, the effective polarization starts to decrease. At infrared frequencies the ionic polarization drops off. This then leaves just the electronic contribution at the optical frequencies.

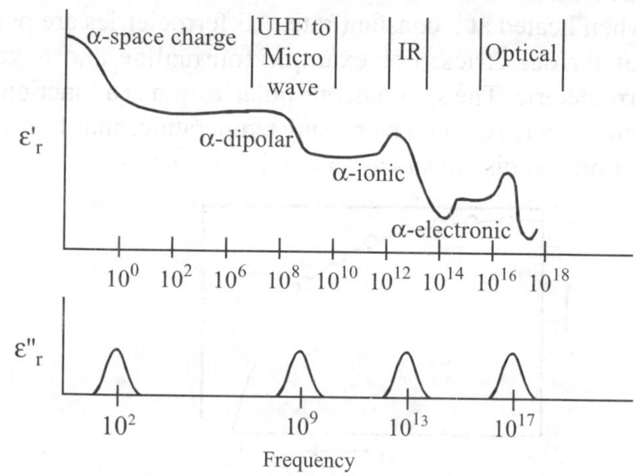

Fig. 10.13 The frequency dependence of polarizability.

Temperature Dependence of Dielectric constant

The electronic polarizability and ionic polarizability do not change with temperature. However dipolar polarizability depends on temperature. Normally, the dielectric constant in linear dielectrics decreases with temperature. The more interesting part comes from the loss variation with temperature. The dielectric loss increases with temperature. This is because, the dielectrics become conducting as temperature is increased. The conductivity of a dielectric material increases with temperature. This produces an absorption of energy i.e., the dissipation. Therefore the loss will increase with temperature. The dielectric relaxation time will decrease with temperature. In non linear dielectrics, the loss may exhibit a maximum at a certain temperature; A Ferroelectric is one such example. The defects in the crystals also affect the dielectric constant and dielectric loss variation with temperature. The defect complexes (like ion-vacancy complexes, or vacancy-impurity associations) play a vital role in crystals in relaxation phenomena, dielectric loss and dielectric variation with temperature.

10.10 Ferroelectricity and Piezoelectricity

Ferroelectricity, pyroelectricity and piezoelectricity phenomeon can be exhibited by a certain class of crystals. These crystals do not possess a centre of symmetry. The positive ions and the negative ions in these crystals are displaced with respect to each other. This displacement give rise to a permanent dipole moment. We shall study the ferro electrics and piezoelectrics in the following.

10.10.1 Ferroelectricity

Ferroelectric crystals are those crystals which exhibit spontaneous polarization i.e. electric polarization without any external electric field. Each unit cell of a ferroelectric crystal carries a reversible electric dipole moment, spontaneously oriented parallel to the dipoles of neighbouring cells. Ferroelectricity is defined as the reversibility, in a polar crystal, of the direction of the electric dipole by means of an applied electric field. Since it requires polarity in a crystal, it can occur only in pyroelectric crystal classes. The pyroelectric crystals exhibit a dielectric polarization when heated at a constant rate. All ferroeletrics are pyroelectrics. But, all pyroelectrics are not ferroelectrics. For example tourmaline and hexagonal CdS are pyroelectrics, but not ferroelectric. The spontaneous polarization is a function of temperature. The spontaneous polarization decreases with increasing temperature, and at a certain temperature the spontaneous polarization just disappears as shown in Fig. 10.14.

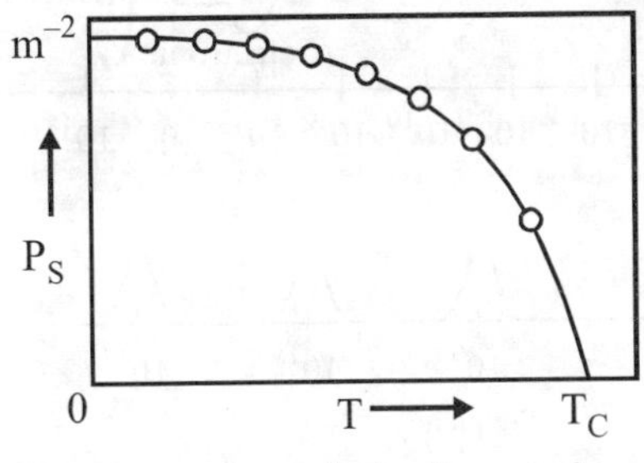

Fig. 10.14 Spontaneous polarization varies with temperature.

The ferroelectric crystals possess high dielectric constants. They also show dielectric anomalies i.e., the dielectric constant reaches a maximum value at the transition temperature called the Curie point (T_c). The crystal is ferroelectric below the Curie point and above the transition temperature the crystal becomes paraelectric in nature. In paraelectric phase, the dipole moments in the crystal are randomly oriented. Above the transition temperature, T_c the dielectric constant (ε_r) changes as per the Curie-Weiss law, expressed as

$$\varepsilon_r = \varepsilon_0 + \frac{C}{T - T_c} \quad \text{.....(10.56)}$$

where C is called the Curie constant. T is the temperature and T_c the Curie point. C is of the order of $10^3 - 10^5$ K.

It has been pointed out earlier, that Mosotti expression for the internal field in a dielectric could be applied only to non-dipolar substances. But it is interesting to apply Mosotti expression for a dipolar substance. The polarizability and polarization are related as

$$P = N\alpha E_{loc} \quad .....(10.57)$$

where $E_{loc} = E + P/3\varepsilon_0$ and the terms involved have the usual meanings.

$$\therefore \quad P = \frac{N\alpha E_{loc}}{1 - \dfrac{N\alpha}{3\varepsilon_0}} \quad .....(10.58)$$

From the above expression, it is seen that p tends to infinity when $(N\alpha/3\varepsilon_0)$ tends to one. As the substance is dipolar, the polarizability is expressed as

$$\alpha = \alpha_e + \alpha_d \quad .....(10.59)$$

Suppose the displacement between positive charge and negative charge is x, then the dipole moment is

$$P = qx = \alpha E_{loc} = \alpha F/q \quad .....(10.60)$$

where F is the restoring force trying to bring the two charges of the dipole together. The work done in creating N such dipoles is given as

$$W_1 = N\int F dx = \left(\frac{Nq^2}{\alpha}\right)\int x.dx$$

$$= \frac{Np^2x^2}{2\alpha} = \frac{Np^2}{2\alpha} = \frac{p^2}{2N\alpha} \quad .....(10.61)$$

the energy of electric displacement is

$$W_2 = \int E_{loc} dp = \int (E + p/\varepsilon_0) dp = \frac{p^2}{6\varepsilon_0} + \int E\, dp \quad .....(10.62)$$

Then the net energy associated with a polarized dielectric is

$$W_2 - W_1 = \left(\frac{p^2}{2N\alpha}\right)\left[\left(\frac{N\alpha}{3\varepsilon_0}\right) - 1\right] + \int E\, dp \quad .....(10.63)$$

Hence, the net energy is positive even for zero external field. This energy is stored in the crystal. It is energetically favourable to create highly polarizable dipoles in an orderly fashion. At sufficiently high temperatures, this orderly arrangement is destroyed. Thus, there exists a critical temperature above which spontaneous polarization disappears in the crystal, but still the crystal may possess a high dielectric constant. For example (Ba, Sr) TiO_3 ceramic used as capacitor material possesses a dielectric constant of about 6000 above the transition temperature.

The general requirement for the establishment of a ferroelectric state lies in the fact that an ordered array of induced ionic dipoles must be able to have a smaller total energy than the crystal with zero dipole moment. This requires the presence of a very large number of extremely polarizable ions in the crystal. Above the critical temperature, called Curie point T_c, the amplitude of thermal vibrations of atoms is strong enough to prevent the ordered dipole array to exist. A solid which is ferroelectric below T_c, is paraelectric above T_c with susceptibility varying as $(T - T_c)^{-1}$.

10.10.2 Important Characterstics of ferroeletric Materials

We will now see some of the important characteristics of ferroelectrics :

1. As already stated, Ferroelectrics exhibit spontaneous polarization. Every ferroelectric substance has a characteristic Curie point (T_c) which can be influenced by impurities. The transition from ferroelectric to paraelectric state may be accompanied by change in crystal structure (example barium titanate) or change in space group of the crystal structure.
2. They possess very high dielectric constants. The dielectric constant is large even in paraelectric phase.
3. The spontaneous polarization becomes zero at the transition temperature.
4. All ferroelectrics exhibit polarization reversal. Each unit cell of a ferroelectric crystal carries a reversible dipole moment, spontaneously oriented parallel to the dipoles of the neighbouring cells. The commonly accepted criteria for ferroelectricity is a hysteresis loop on D-E studies (Fig.10.15). The loops can be square in some cases. Above the transition temperature, the hysteresis loop disappears and the crystal behaves like an ordinary dielectric.

   The P-E curves suggest that there is a net remnant polarization P_r. The spontaneous polarization P_s is obtained by extrapolating the polarization at high fields back to zero field along a tangent. P_{sat} itself is a function of the peak field. P_r and P_s are nearly equal in single crystals.

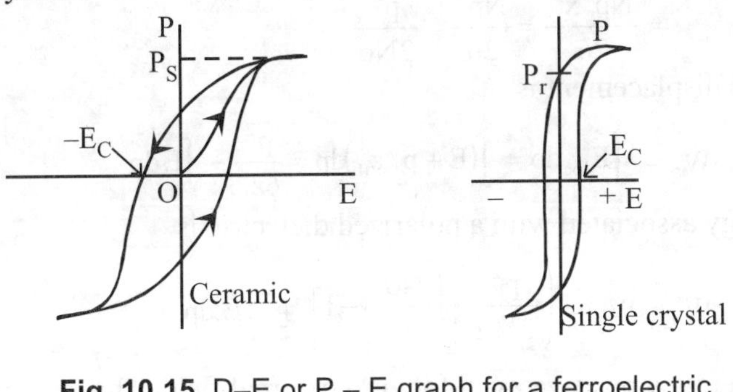

Fig. 10.15 D–E or P – E graph for a ferroelectric.

5. All ferroelectrics are pyroelectric and piezoelectric but the reverse may not be true. That is, all pyroelectrics are not ferroelectrics, and all piezoelectrics are not ferroelectrics. Quartz is peizoelectric but it is not ferroelectric. Ferroelectrics show reversible piezoelectric and pyroelectric effects. The observation of hysteresis loop in D-E studies indicates the presence of domains, called ferroelectric domains. Each domain is spontaneously polarized.
6. Ferroelectric crystal contains an array of domains with different directions of polarization. Normally, neighbouring domains in which the direction of P_s is different will surround a domain. When an external field is applied, the domain wall may move to permit an enlargement of domains in which the direction of P_s nearly becomes parallel to the applied field. At the saturation, all domains have the polarization vectors in the direction of least energy i.e., in the direction of the applied field.

7. The ferroelectric to paraelectric transition is marked by a dielectric peak in the dielectric constant vs. temperature plots (Fig. 10.16) for a ferroelectric crystal.

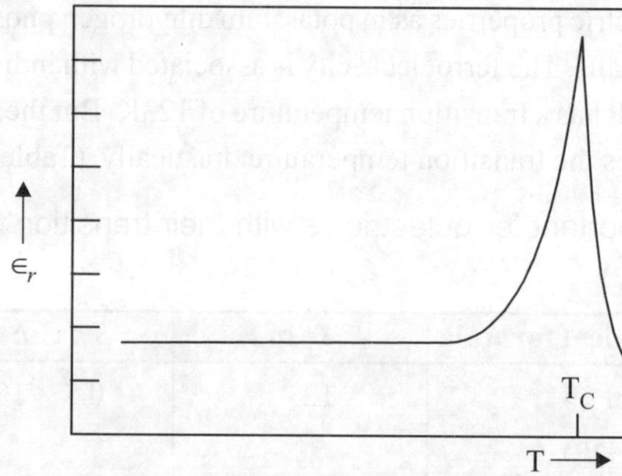

Fig. 10.16 Dielectric constant vs temperature plot for a ferroelectric, Ba TiO_3.

8. Ferroelectric crystals exhibit birefringence, i.e. double refraction when a plane polarized light is passed through them.
9. Both the specific heat and thermal expansion co-efficient exhibit a discontinuity at the transition temperature.

Ferroelectric Materials

There are a very large number of crystals which exhibit ferroelectricity. The ferroelectricity was first discovered in Rochelle Salt (Na K $C_4H_4O_6$, $4H_2O$). It exhibits spontaneous polarization over a range of temperature (–18^{0}C to 20^{0}C). The variation of spontaneous polarization is shown in Fig.10.17. The Fig. 10.17 (a) displays the variation of spontaneous polarization in Rochelle salt. The cutting points on temperature axis give the range of temperature in which the Rochelle salt is ferroelectric. The deuterated samples show an increased range of temperature (curve b). The deuterated samples also show the increased spontaneous polarization.

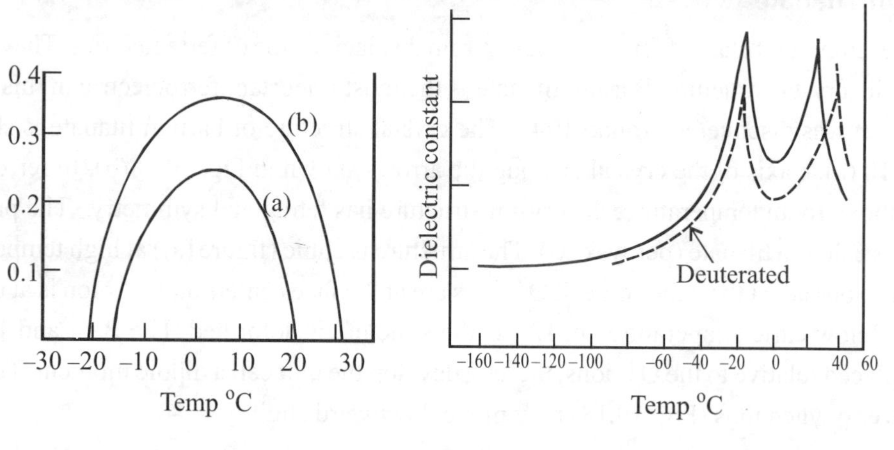

Fig. 10.17 The variation of spontaneous polarization and dielectric constant in Rochelle salt.

Some ferroelectric crystals are classified into order-disorder type and displacive type. The order-disorder class of crystals includes crystals with hydrogen bonds in which the motion of protons is related to the ferroelectric properties as in potassium dihydrogen phosphate (KDP or $KH_2 PO_4$) and its isomorphous salts. The ferroelectricity is associated with individual ordering of ions in the crystal lattice. KDP has a transition temperature of 123K. But the replacement of hydrogen by deuterium increases the transition temperature drastically. (Table 10.2).

Table 10.2 List of a few important ferroelectricals with their transition temperatures

Substance with chemical formula	T_c in K	Ps Cm^{-2}
Pottassium dihydrogen Phosphate KH_2PO_4(KDP)	123	0.5×10^{-2}
Deuterated KDP KD_2PO_4	213.22	4.5×10^{-2}
Triglycine sulphate (TGS)	322	3×10^{-2}
Triglycine solenate	295	3.2×10^{-2}
Barium Titanate PbTiO3	393	2.6×10^{-2}
Lead Titanate PbTiO3	763	3×10^{-1}
Potassium niobate $KNbO_3$	712	
Lithium niobate LINbO3	1,470	
Lithium Tantalate $LiTaO_3$	890	2.3×10^{-1}

Barium Titanate

The most important class of ferroelectrics is from displacive type of ferroelectrics. They possess perovskite crystal structure. Barium titanate is the most important ferroelectric in this class of crystals. It was discovered around 1943. The crystal structure of barium titanate is shown in Fig.10.18 (a). c axis of the crystal is along the arrows shown in Fig. 10.18(b). In ferroelectric phase above room temperature, the crystal structure has tetragonal symmetry. The prototype crystal is calcium titanate (perovskite). The structure is cubic (figure (a)) at high temperatures. The Ba^{2+} ions are at the cube comers, O^{2-} ions are at the face centers and Ti^{4+} ion is at the body center. Below curie temperature i.e. 120°C the structure is deformed. The Ba^{2+} and Ti^{4+} ions are displaced relative to the O^{2-} ions, thereby develop the unit cell a dipole moment. The upper and lower oxygen ions (Fig. 10.18) may move downward slightly.

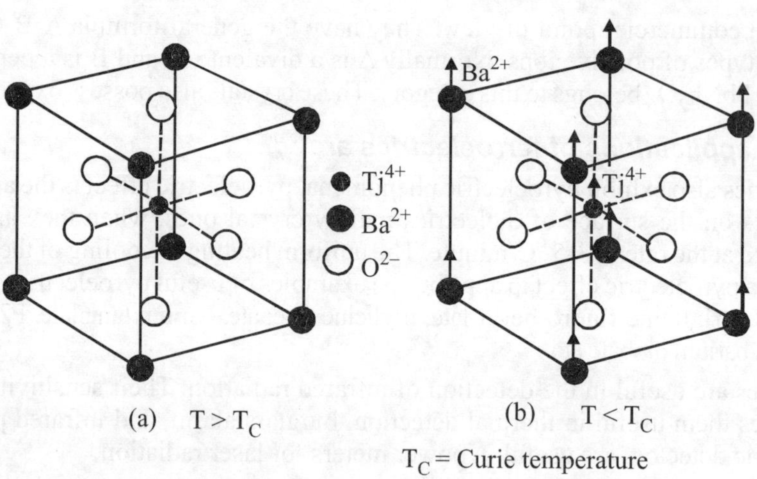

Fig. 10.18 The Barium titanate structure.

The volume of the cell is 64×10^{-30} m^3, so that dipole moment of the cell is 2×10^{-29}C.m. The crystal structure below transition temperature is tetragonal, it is actually a tetragonal distortion in which the unit cell elongates by about 10% along one pseudo-cubic direction (C-axis) and contracts by about 0.5% along each of the axes perpendicular to this direction. The sub lattice containing Ba^{2+} and Ti^{4+} ions shifts upwards along the C-axis with respect to the Oxygen sub lattice. The relative displacement of the two sub lattices is 0.01 nm. This is responsible for the spontaneous polarization. The Ti^{4+} ion is present at the center of an octahedral formed by oxygen ions. Barium titanate also exhibits two more transitions one at 0°C and another at –80 °C. They are shown in the Fig. 10.19.

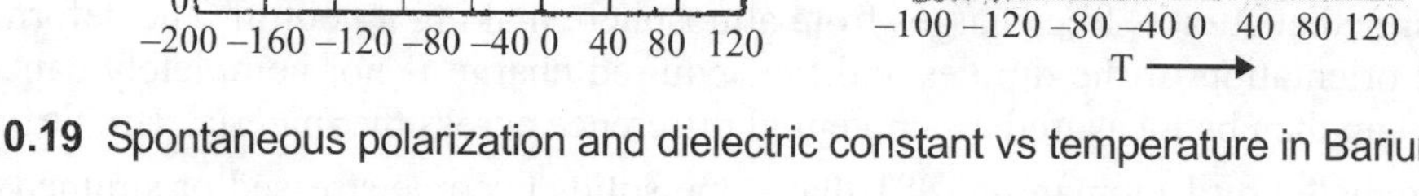

Fig. 10.19 Spontaneous polarization and dielectric constant vs temperature in Barium Titanate.

There are a number of other ferroelectrics similar to $BaTiO_3$ with perovskite structure and these are mentioned with their transition temperatures in Table 10.2. They have a general formula ABO_3 (A = divalent, B = Tetravalent). The most important ferroelectrics, other than barium titanate, are lead titanate, lithium niobate, potassium niobate, etc. But, there are hundreds of ferroelectrics today which have been synthesized, but, it is not possible to give a list of all those ferroelectrics here. The ferroelectrics with potassium tungsten bronze structure are also important from commercial point of view. They have the general formula $A_xB_2O_6$, where A and B are two types of positive ions. Normally A is a divalent ion and B is a pentavalent ion (Nb^{5+} or Ta^{5+}). $PbNb_2O_6$ belongs to this category. These crystals also possess oxygen octahedra.

Some of the applications of ferroelectrics are :

The ferroelectrics also exhibit pyroelectric phenomena. pyroelectric effect is the appearance of electric charges on the surface of dielectrics mainly crystal ones, when they are uniformly heated or cooled at the rate of 3-5 ^{0}C/minute. The uniform heating or cooling of the pyroelectric is essential for a pyroelectric effect to appear. The examples of useful pyroelectrics are triglicine sulphate (TGS), triglicine fluoro beryl late, triglicine selenate, barium tantalate, PZT in ceramic form, strontium barium niobate etc.

Pyroelectrics are useful in the detection of infrared radiation. Their sensitivity to infrared radiation makes them useful in thermal detection, burglars alarm, and infrared photography. The pyroelectric detectors are useful in power meters for laser radiation.

The ferroelectrics which exhibit a square loop in polarization vs electric field characteristic are useful in memory technology. The Ferroelectric RAMS (FRAM) are developed. Barium-strontium titanate, strontium bismuth titanate are useful here. The high retention and capacity, endurance, access time and stability are the key factors in utilising square loop ferroelectrics in RAMS.

There are ferroelectrics which exhibit diffuse phase transition i.e., their phase transition temperature is not sharp. The transition temperature shifts to higher temperature with the frequency. This behaviour is called relaxor behaviour. Such ferroelectrics in thin film form are useful in multi layer capacitors. Their size is very small. They are useful in microelectronic circuits.

10.10.3 Piezoelectricity

In some ionic crystals, the center of positive charge in the unit cell of the crystal does not coincide with the center of the negative charge. There is a net polarization associated with the unit cell of a crystal. Pyroelectricity, Piezoelectricity and Ferroelectricity are the properties of these solids. The first experimental demonstration of piezoelectricity was done by Jacques and Pierre Curie in 1880 in tourmaline, quartz and Rochelle salt. There is one important property that a material must have in order to be piezoelectric or ferroelectric. The unit cell of the crystal exhibiting the piezoelectric or ferroelectric property should have a structure lacking the center of inversion. Each unit cell of piezoelectrics develops an electric polarization when subjected to elastic strain or deformation. In the absence of stress, the free piezoelectric crystal has surface, charge that attracts free charges from atmosphere making it neutral. The deformation disrupts the orientation of the dipoles and the acquired charge is not completely cancelled. This charge can then be measured as a potential difference across the material sample.

It was predicted by Lippman in 1881 that some solids become stressed or strained when subject to an external electric field in certain directions.Curie verified this converse or reciprocal effect. The strain produced as a result of the electric field is due to the electromechanical coupling in the crystal.

Consider a row of ions in an ionic crystal as shown in Fig. 10.20. It shows a one dimensional rigid ion spring model of the crystal lattice. The springs represent equivalently the cohesive force resulting from the electrostatic Coulomb energy and the quantum mechanical repulsive energy. In a noncentro symmetric lattice, the springs joining the ions are different i.e. harder or softer. When an electric field is applied, the cations are drawn in the direction of the electric field and the anions in the opposite direction leading to the relative change in the interionic distance. Depending on the direction of the electric field, the soft spring expands or contracts more than the contraction or expansion of the hard spring causing a strain 's' (i.e. change in length of the chain). This strain is proportional to the electric field applied.

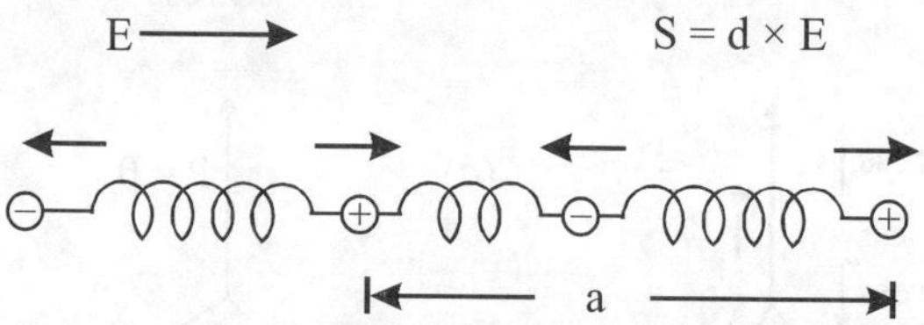

Fig. 10.20 Converse piezoelectric effect.

In ordinary solids a stress T causes a proportional strain 's' related by an elastic modulus; T = Y s. Piezoelectricity is the additional creation of the electric charge by an applied stress. This is the direct piezoelectric effect. In terms of the dielectric displacement D, (charge Q per unit area) and stress T, we may write D = Q/A = d T, where d is expressed in Coulombs/Newton. 'd' is called piezoelectric constant. The value of 'd', should be high for applications. In the converse effect, an electric field E produces a proportional strain S, (expansion or contraction) where S = d E (d is expressed in meters/volt). Jaffe defines the piezoelectric constants as partial derivatives evaluated at constant stress T or a constant electric field E given as

$$d = \left(\frac{\delta S}{\delta E}\right)_T = \left(\frac{\delta S}{\delta T}\right)_E$$

In crystals the stress and strain are tensors. Therefore the piezoelectric coefficients are tensors.

Piezoelectricity and ferroelectricity are exhibited by a few classes of crystals. Out of thc 20 classes of crystals which can show piezoelectricity, ten crystal classes have a unique polar axis in the unstrained condition. These crystals can develop an electric charge when uniformly heated. There will be a change in the magnitude of the dipole moment of the crystal with the change in the temperature. Such crystals are called pyroelectrics. When heated, they give rise to a pyrocurrent under short circuit condition. All pyroelectrics are piezoelectric but all piezoelectrics are not pyroelectric. For example, quartz is a nonpolar piezoelectric crystal. But it is not ferroelectric or pyroelectric. Such crystals do not contain a net crystal dipole moment. But a crystal like tourmaline has a net internal dipole moment (polarization). Its polarization is as high as $80 \times 10^{-10} C/m^2$ and gives rise to pyroelectricity .

As already stated, piezoelectricity is a property which is possible only for ionic solids with crystalline structures lacking a center of inversion. Consider a molecule in a hypothetical ionic solid which in equilibrium gives three electric dipoles of equal magnitude at 120^0 intervals, as shown in Fig. 10.21. This molecule has a point group 3m. The net dipole moment of the molecule vanishes by symmetry. But a net dipole moment will appear if the molecule is stretched or compressed along a direction parallel to, or perpendicular to, one of the three vertices. Similarly, an electric field applied parallel to one of the three vertices will enforce a distortion of each molecule. This results in a contraction or elongation.

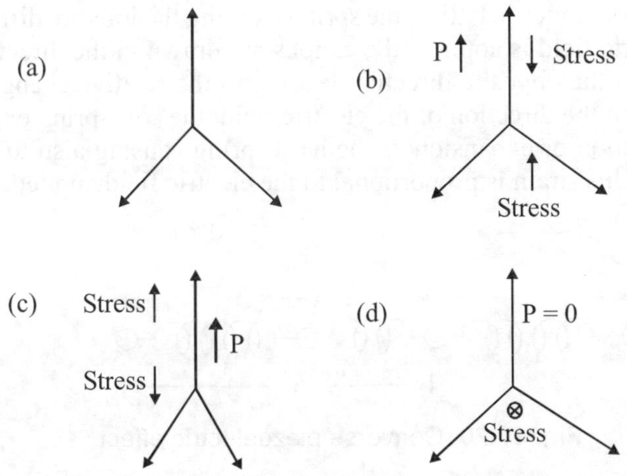

Fig. 10.21 A piezoelectric molecule under strain (a) molecule in equilibrium (b) Net polarization developed in vertical tension or horizontal compression (c) net polarization developed by vertical compression or horizontal tension of the crystal parallel to the field direction, accompanied by a length change of opposite sign in the perpendicular direction. (d) But, an electric field applied perpendicular to the direction of a vertex will not change the crystal dimensions, because the field is perpendicular to a reflection plane of symmetry.

Most of the commercial gadgets based on ferroelectricity and piezoelectricity are in polycrystalline form. Quartz, Rochelle salt and KDP are the substances which are used in single crystal form. Some of the piezoelectrics are mentioned in Table 10.3.

Table 10.3 Piezoelectric coefficient of a few compounds

Sunstance	$d_{33} 10^{-12} CN^{-1}$
Ba TiO_3 Single Crystal Ceramic	 85-120 191
PbTiO3 Single Cryastal Ceramic	 117 60
$SrBi_2Ta_2O_9$	0.23
LINbO3	6.0
PLZT $(Pb1_{-x}\ La_{2x-3})\ (Zr_yTi1_{-y})O_3$ PLZT means La-modified Pb(Zr, Ti)O_3 Ceramics	38-550

Energy can be given to a piezoelectric body either mechanically by stressing it or electrically by charging it. All the energy given to it is not converted into the effect. Therefore the piezoelectrics are characterized by a strength of piezoelectric effect. This strength is measured by electromechanical coupling factor K. When an electric field is applied to the piezoelectric crystal or ceramic, it measures the fraction of the electrical energy converted to mechanical energy (or vice versa when a crystal is stressed). The actual relationship is given as

$$K^2 = \frac{\text{electrical energy converted to mechanical energy}}{\text{Input electrical energy}}$$

or

$$K^2 = \frac{\text{mechanical energy converted to electric energy}}{\text{Input mechanical energy}}$$

10.10.4 Piezoelectric Materials

All the commercial piezoelectrics are basically ferroelectric except a few like quartz (SiO_2). Quartz has a piezoelectric coefficient 2.2 pC/N, along the piezoelectric axis, which is low, but it has a good mechanical and thermal strength. It is highly useful in the generation of ultrasonic waves. When an alternating electric field is applied to a piezoelectric element or plate, it is compressed and extended thereby vibrating. The quartz oscillators are useful in radio transmitter, quartz clocks and quartz wrist watches.

Piezoelectric effect is known for approximately 1500 materials. Choosing the suitable material for required application is a difficult problem of materials engineering. Piezoelectric ceramics are widely used. But ceramics are fragile, difficult to fabricate in different shapes and possess a high acoustic impedance. Therefore piezoelectric ceramic polymers or piezoelectric-polymer composites are used. They can be shaped easily. A piezoelectric ceramic is characterized by an electromechanical coupling factor (K_{33}) and piezoelectric coefficient d_{33}.

Barium titanate was the first ceramic developed and is widely used. It has high coupling factors (49%), piezoelectric coefficient (19.7×10^{-12}C/m). It is highly stable chemically. Lead titanate ($PbTiO_3$) has polar axis along 6 directions and it shows a spontaneous polarization of nearly 80μC/cm^2. Pure lead titanate cracks easily under the action of the field. Stabilization can be obtained by the addition of impurities like La^{3+}. When properly poled, the ceramic gives a d_{33} of 40×10^{-12}C/N. Lead zirconate-lead titanate solid solutions are widely used as they show strong piezoelectric effects, especially those with around 0.48 lead zirconate. The ceramic when poled properly has d_{33} more than 230pC/m and K_{33} about 0.67–0.75. They have low loss and high coercive field. The solid solution is also modified by the addition of La^{3+}, Si^{3+} or Fe^{3+}. They are widely used commercial ceramics.

Piezoelectricity is also found in polymers, for example polyvinylidine fluoride (PVDP) polymer possesses ferroelectricity but small d constant. The advantages are lightweight, high elasticity and lead free. They are compatible with human body and water. They are useful as sensors and actuators. They have low quality factor but a wide frequency band resonance. In commercial applications, polymer- piezoelectric composites are used.

10.10.5 Applications of Piezoelectrics

All commercial piezoelectric materials used today are ferroelectrics. Then the applications can also be treated as the applications for ferroelectrics. Piezoelectric bodies provide a coupling between electrical and mechanical forces and hence serve as transducers that convert or detect electrical or mechanical signals. Hence they are highly useful in converting very small mechanical energies into electrical fields. Hence, they find utility in producing very small mechanical displacements and small amounts of electric charge per cycle. Several applications are available for piezoelectrics and we will discuss a few of them.

High frequency electric oscillations can be transformed into mechanical oscillations in the production of ultrasonic waves. Converse piezoelectric effect is used here. Quartz is a hexagonal crystal. The symmetry considerations will give three fold polar axes in this crystal. When the crystal is placed in an oscillating electric field, the direction of the field being along one of the polar axes, the crystal will oscillate. The oscillations are due to changes in thickness (contraction and expansion). Because of the alternating strains, oscillations are produced in the medium to which the crystal makes contact. The frequency $f = \frac{n}{2l}\sqrt{Y/\rho}$ where n = 1,2,3.......... The value of Y/ρ depends on the polar axis chosen. The frequency depends on the thickness of the plate. Frequencies as high as 50 MHz can be achieved by proper selection of thickness. Today we have piezo electrics which can give much higher frequencies.

1. **Phonograph pick-ups:**

   Prior to 1950, Rochelle salt was used but the ceramics are much superior in their mechanical and chemical stability . A typical ceramic phonographs pick up elements is shown in Fig. 10.22.

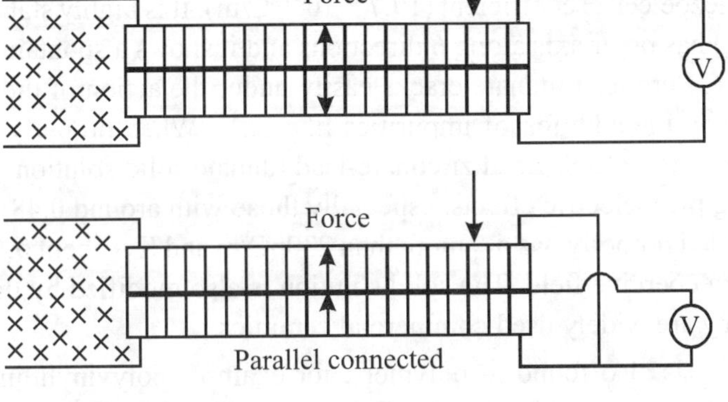

Fig. 10.22 Phonograph pick up. The arrow on the element indicates the polarization direction.

The ceramics plates are cemented on brass vane and are electrically poled in opposite directions. When the pin attached at the end falls in a groove, there is compression in one plate and extension in another plate of piezoelectric ceramic. Therefore, an electric field is generated. Both the ceramic plates give the electric field in the same direction. PZT modified by impurities is largely used here nowadays. Another important application is their use as vibrators. When an alternating voltage is applied across a piezoelectric element, it vibrates and at a particular frequency of the field, it resonates. The resonant frequency of the element depends on the thickness. The mechanical vibrations can be transferred into solids or liquids as ultrasonic waves. This frequency range lies from 1000 Hz to 10 MHz. Hard type piezoelectric ceramics are used. The speakers and buzzers utilize the piezoelectric vibrations. One such vibrator is given in Fig. 10.23.

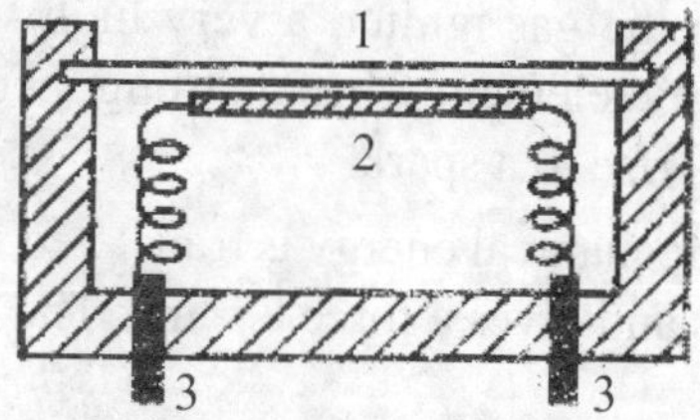

1. Eastic membrane 2. Piezoelectric element 3. Electrodes

Fig. 10.23 Piezoelectric vibrator

Piezoelectric ceramics are useful in generation and reception of sound waves in water. These ceramics in this sense are called smart materials because they are actuators as well as detectors. They are useful as ultrasonic cleaning transducers, (frequency range 20-100 KHz), under water detecters of sounds. Under water applications comprise of the passive listening (hydrophone) as well as active method of producing sound pulses and timing their echo from obstacles. This is useful in fish finding and in sonar. These devices operate at high power levels for maximum range. They are also useful for submarine detection. Ultrasonic sources and detectors are extremely useful in factories. Ultrasonic detectors are useful in medical diagnosis. Piezoelectrics are broadly utilised to induce ultrasonic oscillations. The ultra sound is used for fault detection i.e. for finding internal cracks and other hidden defects in solid bodies like bars, rods, plates etc. The simplest application of piezoelectrics one finds at domestic level is the gas igniter which is shown schematically in Fig. 10.24.

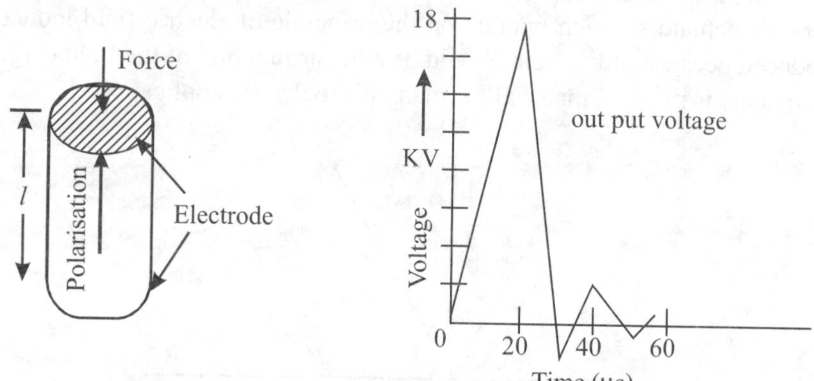

Fig. 10.24 Piezoelectric igniter. The arrow on piezoelectric element indicates the polarization direction.

In the gas igniter, a very high voltage is generated spontaneously across a piezoelectric ceramic element when a strong mechanical stress pulse is given to the element. This voltage gives rise to a spark.

Mechanical energy is translated into electric signal. The mechanical energy is generated by mass and gives a force F = ma. The Fig. 10.25 gives a schematic diagram of an accelerometer.

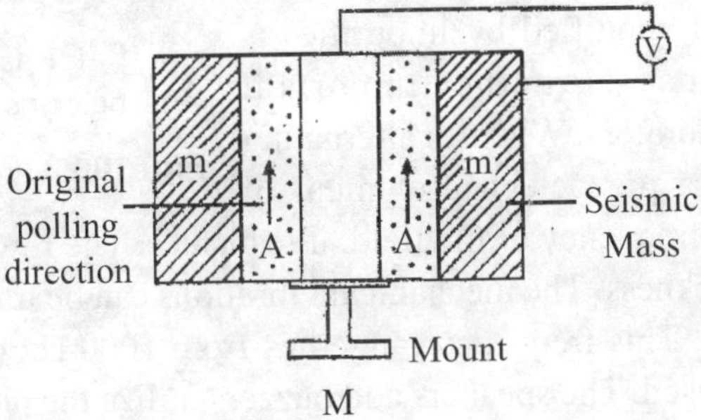

Fig. 10.25 A schematic diagram of Accelerometer.

The piezoelectric element in tubular shape A is kept around a mount M. Around the element, a dense metal with a mass m is kept. The whole assembly is encapsulated in a protective case. As a response to acceleration, the mass imparts a force to the element which is in turn generates a voltage proportional to the stress. The voltage is calibrated against acceleration.

The beautiful aspect of piezoelectrics is that they can be used as sources as well as detectors. The vibrations produce alternating electric field. The alternating electric fields produce vibrations in them. Here both piezoelectric effect and converse piezoelectric effect are used. Such materials which have this property of actuators (Producers), as well as detectors are called smart materials.

Today we require very fine displacement elements in fields of optics, precision machinery, small motors, and micro electronics. Positioning of various parts of machinery in micromechanics is needed. Piezoelectrics are developed into micrometer displacement gadgets, called ceramic actuators which operate on the principle of electric field induced strain. The strain produced because of the field should be a linear function of the field (Fig. 10.26). Lead zirconate titanate is primary material for many piezoelectric applications.

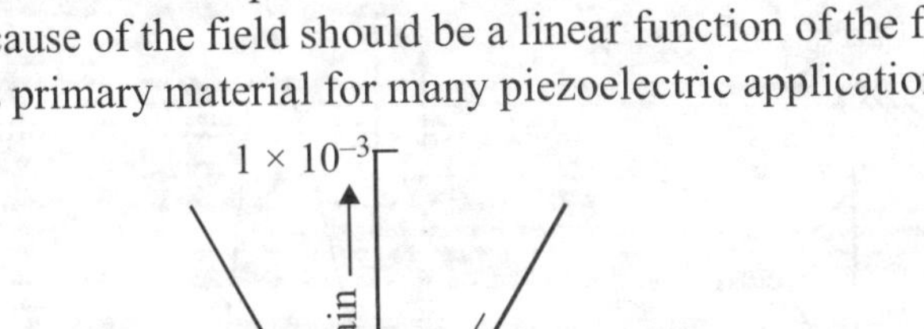

Fig. 10.26 Field induced strain.

Dot matrix printer heads have piezoelectric displacement transducers. The advantages are high printing speeds, low energy consumption and noiseless printing. Ink jet printer heads are also based on displacement piezoelectric elements. The piezoelectrics are extremely useful today for the development of microrobots, mini-monitor positioners, speed arresters etc.

10.11 Pyroelectricity

The development of electric dipoles in certain crystalline materials as a result of temperature changes is called pyroelectric effect.

Pyroelectricity is the electrical potential created in certain materials when they are heated uniformly. Although many pyroelectric materials have been engineered these days, the effect was first discovered in minerals such as quartz and tourmaline and other ionic crystals. All pyroelectric materials are also piezoelectric, the two properties being closely related. Pyroelectric charge in minerals develops on the opposite faces of asymmetric crystals. The property of pyroelectricity is the measured change in net polarization (P_s) proportional to a change in temperature. The pyroelectric current, I, measured in the pyroelectric sample is given as

$$I_p = p_i A \frac{dT}{dt} \quad .....(10.64)$$

Where p_i is the pyroelectric coefficient, and A is the area of cross section of the sample, dT/dt is the rate of heating. The normal heating or cooling would be around 3-5^{0}C/min. It should be highly uniform with time.

The pyroelectric coefficient is the change in spontaneous polarisation per unit change of temperature, i.e.

$$p_i = \frac{\partial P_{si}}{\partial T} \quad .....(10.65)$$

where p_i is the pyroelectric coefficient. It has the units C m^{-2} T^{-1}, P_{sp} is spontaneous polarisation.

Pyroelectrics are the bridge between ferroelectrics and piezoelectrics .They possess a spontaneous polarisation which is not necessarily switchable by an electric field. If their polarisation is electrically switchable then they are ferroelectric, a property that may be exploited e.g. for data storage. All ferroelectrics are pyroelectrics, but all pyroelectrics are not ferroelectrics. The best example is tourmaline, it is pyro crystal but not ferroelectric. Ferroelectrics are a subset of Pyroelectrics.

Pyroelectrics fill an entirely separate niche. The pyroelectric effect has been known of for a very long time. The Greek philosopher Theophrastus first noted (in approximately 400 B.C.) that Tourmaline would attract straw and bits of wood when heated. This was due to the fact that the changing the temperature produces surface charges which attract other charged materials.

The polarisation of a pyroelectric sample that ferroelectric in nature shows a a change of polarisation with temperature. At Curie temperature, the pyroelectric coefficient or the pyro current would become zero. Commercial piezoelectrics or pyroelectrics are normally polycrystalline samples in thin film form or ceramic discs, or thin plates. When they are used, they are electrically poled ie., they are kept in between two electrodes and very high voltage is applied across the sample. Electrical poling is necessary to orient ferroelectric domains into one direction. The electrodes are coated on either side of the sample and connections are made to a very sensitive nano or picco ammeter.The current is measured under short circuit condition as shown in the Fig.10.27 below.In the case of single crystal,the electrodes are placed along the principle crystallographic direction

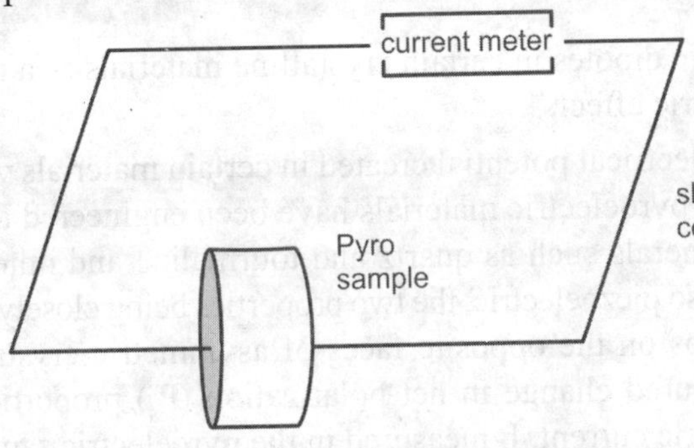

Fig. 10.27 Pyroelectric current measurement.

Examples:

Triglicinesulphate($(NH_2CH_2COOH)_3H_2SO_4$). It has pyro coefficient of -5.5×10^{-4} $Cm^{-2}K^{-1}$ measured at 30^0 C. It is used in pyroelectric 'Vidicon", a device or camera useful for thermal imaging. This camera is highly helpful for disaster teams to locate the trapped people under rubble.

PVDF (poly vinyleden di fluoride has low pyrolelectric coefficient of -0.27×10^{-4} $Cm^{-2}K^{-1}$. But its use lies in the fact that very large thin films can be made and as a result very large charge is generated .The films are useful in measurement of laser power.PVDF is used in burgular alarms that are based on temperature changes. Pyroelectrics are highly sensitive to temperature changes. They can easily detect a temperature change of a micro degree.

Pyrolectrics are excellent detectors of infra red radiation and they make excellent devices for infrared photography and night photography in the dark. They make perfect devices for testing the level of IR radiation that passes through a gas sample. Pyroelectrics are now used to monitor levels of pollution through IR detection.

Exercise 1

The diameter of an argon atom is 3.84×10^{-10}m. What is polarizability at STP. It's dielectric constant is 1.00044.

$$\text{The number of molecules per unit volume} = \frac{6\times10^{26}}{22.4} = 2.6 \times 10^{25}$$

$$\alpha = \frac{\varepsilon_0(\varepsilon-1)}{N} = \frac{8.85\times10^{-12}\times0.00044}{2.6\times10^{25}}$$

$$= 1.43 \times 10^{-40} \text{ farad m}^2.$$

Exercise 2

The polarizability of Oxygen atom is 1.9×10^{-47} Fm^2. On being polarized it has a dipole moment 4.8×10^{-23} C–m. What is the displacement of the centre of negative charge cloud from the nucleus.

$$\alpha = 1.9 \times 10^{-47} F - m^2$$

$$P = 4.8 \times 10^{-23} \text{ C-m}$$

$$\alpha = \frac{P}{E}; E = \frac{P}{\alpha} = \frac{4.8\times10^{-23}}{1.9\times10^{-47}} = 2.526\times10^{+24} \text{ Volt/m}$$

Lorenz Force = 8 eE

This is equal to the Coulomb force between nuclear and electric charges

$$\therefore \quad 8 \text{ eE} = \frac{8e\times 8e}{4\pi\varepsilon_0 x^2} \qquad (\because e = 1.6 \times 10^{-19} \text{ Col.})$$

$$x^2 = \frac{8e}{4\pi\varepsilon_0 E} = \frac{8\times1.6\times10^{-19}}{4\times3.14\times8.85\times10^{-12}\times2.53\times10^{24}}$$

$$= \frac{12.8\times10^{-19}}{281.22\times10^{+12}}$$

$$= 0.0455 \times 10^{-31}$$

$$= 45.5 \times 10^{-34}$$

$$= \sqrt{45.5\times10^{-34}}$$

$$\therefore \; x = 6.74 \times 10^{-17} \text{ m.}$$

Exercise 3

Carbontetrachloride contains 74 electrons in its molecule. Its relative permittivity is 2.26, when its density is 1.68×10^3 Kg/m^3. If the field acting on the liquid is 5×10^6 V/m, what is its electronic polarizability and average electron displacement.

$$N = \frac{6.02\times10^{26}\times1.68\times10^3}{156} = \frac{10.11}{156}\times10^{29} = 6.5 \times 10^{27} \text{ molecules/m}^3$$

$$p = \frac{p}{N} = \frac{5\times10^6\times8.85\times10^{-12}\times1.26}{6.5\times10^{27}} = 8.57\times10^{-33}\text{C} - \text{m}$$

$$x = \frac{p}{ne} = \frac{8.57\times10^{-33}}{74\times1.6\times10^{-19}} = \frac{8.57}{118.4}\times10^{-14} = 0.724\times10^{-15}\text{m}$$

Exercise 4

A monatomic gas contains 3×10^{25} atoms/m^3 at a certain temperature at one atmosphere pressure. The radius of the argon atom is 0.19 nm. What is the relative permittivity of the gas at the given pressure and temperature. What is the polarizability of the atom. What is the intensity of polarization.

We know that $\varepsilon_0 (\varepsilon_r - 1) = N \alpha_c$

$$N = 3 \times 10^{25} \text{ atoms/m}^3 \; ; \; \varepsilon_0 = 8.85 \times 10^{-12} \text{ F/m}$$

The dipole moment per unit field $P = 4 \pi\varepsilon_0 R^3$

$$4 \times 3.14 \times 8.85 \times 10^{-12} \times (0.19 \times 10^{-9})^3$$

The polarization $p = 3 \times 10^{25} \times 4 \times 3.14 \times 8.85 \times (0.19)^3 \times 10^{-39}$

$$\varepsilon_r - 1 = 3 \times 4 \times 3.14 \times (0.19)^3 \times 10^{-2}$$

$$\therefore \quad \varepsilon_r = 1.0026$$

$$\text{Polarizability} \quad \alpha_e = \frac{0.0026 \times 8.85 \times 10^{-12}}{3 \times 10^{25}}$$

$$\alpha_e = \frac{\varepsilon_0(\varepsilon_r - 1)}{N} = \frac{26 \times 8.85 \times 10^{-16}}{3} \times 10^{-25} = 7.66 \times 10^{-40} \text{ farad m}^2.$$

Exercise 5

A Water molecule has a dipole moment 6.2×10^{-30} C.m. What is the polarization intensity of a water drop of 0.1 cm radius polarized in the same direction.

$$\text{The Volume of the water drop} = \frac{4}{3} \times \pi \times 10^{-12} \text{ m}^3$$

$$\text{The number of molecules in the drop } N = \frac{6 \times 10^{26} \times 4\pi \times 10^{-12}}{18} \times 10^3$$

$$= \frac{4 \times 3.14}{3} \times 10^{-17}$$

$$P = Na = \frac{4 \times 3.14 \times 10^{+17}}{3} \times 6.2 \times 10^{-30} = 26.38 \times 10^{-13} \text{ C/m}^2.$$

Questions

1. Define : 1. Electric monopole, 2. Dipole, 3. Quadrupole, 4. Dipolemoment, 5. Polarizability.
2. Establish a relation between the electric field strength **E**, polarization **P**, and electric displacement **D**. Write down the units of **E**, **P** and **D**.

   Find the magnitude of **D** in a dielectric material for which $\chi = 2$, and $\mathbf{P} = 3.08 \times 10^{-7}$ C/m^2.

3. What are polar molecules and non polar molecules ? Give examples.
   Derive the expression for the polarizability of hydrogen atom assuming its negative charge as distributed in a spherical volume of radius r around the nucleus.
4. Derive an expression for the total electrical field on a molecule inside an isotropic dielectric medium, Hence, show that

$$\frac{\varepsilon_r - 1}{\varepsilon_r + 2} = \frac{N\alpha}{3\varepsilon_0}$$

5. Derive and discuss Clausius-Mosotti expression. Discuss various sources of polarization in molecules.
6. What are dipolar substances ? Explain the temperature variation of their susceptibility and polarizability. What is saturation polarization ?
7. What is deformation polarizability ? Does it vary with temperature ?
8. Discuss the frequency variation of electronic polarizability.
9. Explain how electric polarizability is correlated to refractive index of a material.
10. What is ionic polarizability ? Is it sensitive to temperature ?
11. A solid contains 5 × 1028 identical atoms per m^3. Each atom has polarizability of 2 × 10^{-40} Farad/m^2. Assume that the internal field is given by Lorentz relation. Calculate the ratio of the internal field to the applied field.
12. Obtain the expression for dipolar polarizability of a dipolar system of molecules. Discuss the temperature variation of dipolar polarizability.
13. What is "Complex permitivity" ? What is "loss factor" ?
14. Discuss the frequency dependence of electronic polarizability.
15. Explain the phenomenon of "dielectric relaxation". Hoe does this relaxation depend on the frequency of the applied field across a dielectric ?
16. Discuss the frequency dependence of complex permitivity ?
17. Discuss how the total polarizability of a dielectric substance show the marked difference when studied as a function of the frequency of the applied field.
18. The polarizability of oxygen atom is 1.85 × 10^{-47} F.m^2. The atom is being polarized by an electric field of 2.5 × 10^{24} Volt/m. Calculate the distance of the centre of negative charge cloud to the nucleus. The concentration of the gas is 2.7 × 10^{25}/m^3. Calculate the induced dipolemoment per unit volume.
19. The Ammonia molecule has a polarizability 2.38 × 10^{-39} F.m^2 and 1.87 × 10^{-39} F.m^2 at temperatures 280K and 450K respectively. Calculate the value of deformation polarizability and dipolar polarizability and dipole moment of the molecule.
20. Two identical atoms of polarizability 2.5 × 10^{-40} F.m^2 are separated by a distance 3 × 10^{-10}m. A homogeneous field of 10^5 V/m is applied along the line joining the centres of the atoms. Calculate the magnitude of internal field.
21. A dielectric which is ionic in nature has a dielectric constant of 5.2 and the refractive index 2.56. what is the ratio between electronic and ionic polarizabilities of the dielectric?

22. In a free space $\mathbf{D} = 2\mathbf{i} - 4\mathbf{j} + 1.5\mathbf{K}$ C/m^2. What is **P** in a dielectric which has relative permittivity $\varepsilon_r = 5.2$. The dielectric is placed in the free space.
23. A parallel-plate capacitor, for which $C = \varepsilon A/d$, has a static voltage V across the plates. What is the stored energy in the electric field ?
24. Find the magnitudes of **D**, **P** and ε_r for a dielectric material in which $\mathbf{E} = 0.20$ MV/m and $\chi_e = 4.75$.
25. What class of crystals exhibit piezoelectricity, ferroelectricity and pyroelectricity.
26. Explain the origin of piezoelectric effect and inverse piezoelectric effect.
27. What is polarization catastropy ? Define a ferroelectric crystal.
28. What are the various important properties of ferroelectrics.
29. Mention the ferroelectric characteristics of barium titanate.
30. Give examples of crystals which are ferroelectric.
31. Give examples of piezoelectric materials.
32. Discuss the applications of piezoelectrics and ferroelectrics.
33. Discuss the general dependence os dielectric constent or polarizability on the frequency of the applied electric field.
34. What is Pyroelectricity? Give some examples.
35. What is pyroelectric effect?
36. Which type of crystals show pyroelectricity?

Objective Questions

I. Choose the correct choice

1. Dielectrics are
   (a) electric insulators
   (b) electronic conductors
   (c) hole conductors
   (d) materials that work under low voltages
2. When a dielectric is placed between the plates of a capacitor
   (a) its capacitance decreases
   (b) the electric field at every point falls by a factor, $\frac{1}{\epsilon_o \epsilon_r}$, ϵ_r is the relative permittivity
   (c) the electric field at every point increases by a factor $\frac{1}{\epsilon_o \epsilon_r}$
   (d) the electric field becomes negative
3. A dielectric
   (a) does not contain molecules
   (b) when placed between the plates of a capacitor, increases the electric field at every point in the medium
   (c) contains free charges or electrons
   (d) does not contain free charges or electrons
4. When an electric field is applied to a capacitor containing dielectric medium
   (a) the positive and negative charges are pushed to one side
   (b) the positive and negative charges are pushed in oppossite directions
   (c) all the dipoles will disappear
   (d) the electric dipoles become radomly oriented

5. The process of polarisation in a dielectric is
   (a) the displacement of positive and negative charges in the oppossite directions in the presence of an electric field
   (b) the vibration of the atoms of the dielectric with the electric field
   (c) the attainment of zero dipole moment
   (d) association of molecules with one another

6. Read the statements and answer
   (A) The polarization density in a dielectric is proportional to the applied electric field
   (B) The electric displacement D has the units 'Coulomb'
   (a) **(A)** and **(B)** are wrong
   (b) **(A)** is wrong and **(B)** is correct
   (c) Both **(A)** and **(B)** are correct
   (d) **(A)** is correct, **(B)** is wrong

7. Choose the correct relation for a dielectric
   (a) $\mathbf{D} = \epsilon_o \mathbf{E}$ (b) $\mathbf{D} = \epsilon_o \mathbf{E} + \mathbf{P}$
   (c) $\mathbf{D} = \epsilon_o \mathbf{P}$ (d) $\mathbf{D} = \mathbf{P} \times \mathbf{E}$

8. If ϵ_r is the relative permittivity of an isotropic dieletric medium, then the electric susceptibility is
   (a) $\epsilon_r \epsilon_o$ (b) ϵ_r
   (c) $\epsilon_r - 1$ (d) ϵ_r / ϵ_o

9. In an isotropic dielectric medium
   (a) **D** and **E** are perpendicular to each other
   (b) **D** and **E** are inclined at 45^0 to each other
   (c) **D** and **E** have the same direction
   (d) **D** and **P** are perpendicular to each other

10. The relation between D_n and v, the charge in a dielectric medium of volume V is given by
   (a) $\oint \mathbf{D_n} \cdot \mathbf{dS} = q$
   (b) $\oint \mathbf{D_n} \cdot \mathbf{dS} = V/q$
   (c) $\oint \mathbf{D_n} \cdot \mathbf{dS} = q$
   (d) $\mathbf{D_n} = q$

11. In the following, the molecules which are non polar are
   (a) $CHCl_3$
   (b) Polyethylene
   (c) Petroleum oils
   (d) CO_2

12. The polarization that is affected by temperature is
   (a) ionic polarization
   (b) dipolar polarization
   (c) electronic polarization
   (d) deformation polarization

13. The Clausius- Mosotti equation among the following is
   (a) $\epsilon_r - 1 = \dfrac{Nd}{3\epsilon_o}$ (b) $\dfrac{\epsilon_r - 1}{\epsilon_r + 2} = \dfrac{Nd}{3\epsilon_o}$
   (c) $\epsilon_r + 2 = \dfrac{Nd}{3\epsilon_o}$ (d) $\epsilon_o (\epsilon_r - 1)\, \mathbf{p} = \mathbf{E}$

14. The internal field in a dielectric that has a polarization **P** under the action of an external field **E** is
   (a) $3\epsilon_o \mathbf{P}$ (b) $3\epsilon_o \mathbf{E/P}$
   (c) $\mathbf{P}/3\epsilon_o$ (d) **P/E**

15. When a dielectric medium of relative permittivity ϵ_r is under the action of an external electric field **E**, the total effective field intersity **E'** is
   (a) $\mathbf{E}\left(\dfrac{\epsilon_r - 1}{3}\right)$ (b) $\mathbf{E}\left(\dfrac{\epsilon_r - 2}{3}\right)$
   (c) $\mathbf{E}\dfrac{(\epsilon_r - 1)}{(\epsilon_r + 2)}$ (d) $\dfrac{3(\epsilon_r - 2)}{(\epsilon_r + 1)}\mathbf{E}$

16. For a dielectric medium of permittivity ϵ_r is under the action of an external electric field **E**, the polarization **P** is given as
   (a) $\mathbf{P} = \epsilon_o (\epsilon_r + 2)\, \mathbf{E}$
   (b) $\mathbf{P} = \epsilon_r \mathbf{E}$
   (c) $\mathbf{P} = \epsilon_o \mathbf{E}$
   (d) $\mathbf{P} = \epsilon_o (\epsilon_r - 1)\, \mathbf{E}$

17. Read the statements and answer
   (A) Strongly polarized substances do not follow Claussius- Mosotti equation
   (B) The homopolar substances possess ionic polarization only
   (a) (A) is correct, (B) is wrong
   (b) Both (A) and (B) are wrong
   (c) Both (A) and (B) are correct
   (d) (A) is wrong, (B) is correct

18. The atom of hydrogen has a radius r as 0.053 nm. Then its electronic polarizability is
   (a) 1.67×10^{-41} Fm^2
   (b) 4.45×10^{-41} Fm^2
   (c) 3.34×10^{-41} Fm^2
   (d) 2.27×10^{-41} Fm^2

19. If μ_0 and $\in_o$ represent the permeability and the permittivity of free space, then the velocity of the electro magnetic waves in the free space is
   (a) $\sqrt{\mu_o \in_o}$ (b) $\dfrac{\mu_o}{\in_o}$
   (c) $\sqrt{\dfrac{\mu_o}{\in_o}}$ (d) $\dfrac{1}{\sqrt{\mu_o \in_o}}$

20. If 'η' represents the refractive index of a medium, α_{el}, the electronic polarizability, and N, the total number of atoms in the medium, then
   (a) $\eta^2 + 1 = N\alpha_{el}$
   (b) $\eta^2 = \dfrac{N\alpha_{el}}{3\in_0}$
   (c) $\dfrac{\eta^2 - 1}{\eta^2 + 2} = \dfrac{N\alpha_{el}}{3\in_0}$
   (d) $\eta^2 + 2 = \dfrac{N\alpha_{el}}{3\in_0}\left(\eta^2 - 1\right)$

21. For an optical medium, the electronic polarizability is such that
   (a) $\eta^2 = \sqrt{\in_r}$ (b) $\eta^2 = \in_r$
   (c) $\eta^2 = \sqrt{\mu_o \in_o}$ (d) $\eta^2 - 1 = (\in_r)$

22. The common feature between iomc polarization and electronic polarization is
   (a) both are strongly dependent on temperature
   (b) both are responsible for infrared absorption of a dielectric
   (c) both are caused by electronic displacements
   (d) both the polarizations are unaffected by variations in temperature of the dielectric

23. The observed relative permitivity of a material is 9.27. The value of (refractive index)^2 is 1.92. Then , the polarization in it is
   (a) entirely electronic
   (b) party electronic and party ionic
   (c) entirely dipolar in nature
   (d) the material can not transmit visiblerays

24. Molecular dipoles in polar dielectrics
   (a) exhibit a tendency towards orientation polarization
   (b) do not exhibit electronic polarizability
   (c) do not exhibit orientation polarization
   (d) do not exhibit dielectric nature

25. The Langevin's – Debye equation is valid for
   (a) non polar solids
   (b) only for solids exhibiting ionic polarizability
   (c) only for solids exhibiting deformation polarizability
   (d) is valid for materials exhibiting orientation polarizability

26. Carbon tetrachloride has the electronic polarizability of 8.57×10^{-33} C–m.Then, the average electron displacement is (the number of electrons = 74; charge of electron = 1.6×10^{-19} C)
   (a) 7.24×10^{-8} m
   (b) 0.724×10^{-15} m
   (c) 724×10^{-15} m
   (d) 72.4×10^{-13} m

27. A water drop contains 3×10^{17} molecules. The dipolemoment of the water molecule is 6.2×10^{-30} C.m. Then, the polarization intensity of the water drop is (Hint : polarization intensity = number of water molecules × dipole moment)
   (a) 26.4×10^{-13} C / m^2
   (b) 2.06×10^{-47} C. m
   (c) 18.6×10^{-13} C / m^2
   (d) 0.5×10^{-47} (C– m)$^{-1}$

28. The relative permittivity of a dielectric is 5.75. It is placed in an electric field of 2×10^5 V/m. Then, the polarization is ($\in_o = 8.85 \times 10^{-12}$ F/m)
   (a) 101.8 C – m^{-2}
   (b) 11.5×10^{-5} C – m^{-2}
   (c) 10.5×10^{-6} C – m^{-2}
   (d) 16.8×10^{-18} C – m^{-2}

29. The susceptibility of a dielectric material is 2. The polarization is 3.08×10^{-7} C / m^2. Then magnitude of **D** vector is
   (a) 1.54×10^{-7} C / m^2
   (b) 6.16×10^{-7} C / m^2
   (c) 6.5×10^{-7} C / m^2
   (d) 4.59×10^{-7} C / m^2

30. Choose the correct statement in the following
   (a) The permittivity of a dielectric depends on the frequency of the ac voltage, temperature, humidity etc.,
   (b) The permitivity of all soild dielectrics is independent of temperature
   (c) The square of the relative permitivity is equal to the refractive index of a dielectric
   (d) The semiconductors possess infinite dielectric constants

31. Select the correct statement
   (a) The ratio of real part of the permitivity to its imaginary part is called the dielectric loss factor
   (b) Polar dielectrics do not show any dielectric loss
   (c) At optical frequencies, ionic polarizability becomes dominant
   (d) The ratio of the imaginary part and the real part of the permitivity is equal to the loss factor

32. The loss factor in a dielectric depends on
   (a) its conductivity only on
   (b) the real part of the dielectric constant
   (c) the thickness of the dielectric
   (d) the initial polarization of the dielectric

33. When a piezo electric crystal is stressed, it
   (a) gives rise to ferroelectricity
   (b) gives rise to orientation polarizability
   (c) gives rise to an electric charge in the crystal
   (d) gives rise to strain that has no relation to stress

34. Verify the correctness of the statements
   (A) All piezo electrics are ferro electrics
   (B) All piezo electrics are pyro electrics
   (a) (A) is correct, B is wrong
   (b) (A) and (B) are correct
   (c) (A) is wrong, (B) is correct
   (d) (A) and (B) are wrong

35. All pyro electrics are
   (a) piezo electric
   (b) not piezo electric
   (c) ferro electric
   (d) anti ferro electric

36. All piezo electrics are
   (a) characterized by the dielectric curie temperature
   (b) characterized by electro mechanical coupling factors
   (c) ferro electrics
   (d) pyro electrics

37. Choose the correct statement in the following
   (a) The electrical poling of a piezo electric sample results in the orientation of polarization vectors into the direction of the electric field
   (b) An unpoled poly crystalline piezo electric sample exhibits maximum electro mechanical coupling
   (c) It is not necessary for a ferro electric crystal to show polarisation reversal
   (d) The hysteresis loop is not exhibited by ferro electrics

38. One of the following is a ferro electric crystal
   (a) quartz
   (b) tourmaline
   (c) lead zirconate titanate
   (d) polyvinyl acetate

39. One of the following is pyro electric
   (a) quartz
   (b) polyvinyl chloride
   (c) cubic phase of barium titanate
   (d) solid solutions of lead titanate and lead zirconate

40. Choose the correct statement
   (a) all piezo electrics are pryo electrics
   (b) the para electrics exhibit extremely large spontaneous polarization
   (c) the spontaneous polarization of a ferro electric crystal becomes infinity at the curie temperature
   (d) all ferro electrics are pyro electrics

41. The barium titanate below curie tempreature has
   (a) tetragonal structure
   (b) cubic structure
   (c) hexagonal structure
   (d) mono climic structure

42. The spontaneous polarisation in barium titanate arises as a result of
   (a) the charges on barium Ba^{2+} ions
   (b) the excess charge of Ti^{4+} ions
   (c) the displacement of the Ba^{2+} and Ti^{4+} relative to the 0^{2-} ions in the unit cell

43. Piezo electric ceramics are useful as
   (a) excellent light intensity detectors
   (b) excellent capacitor materials
   (c) generators of infrared radiation
   (d) generators and receptors of ultrasonic waves

44. One of the following material is useful in electrical insulation
   (a) TiO_2 (b) Freon
   (c) Acrylic plastic (d) Iron oxide

45. A very useful refractory material in the following
   (a) Polystyrene (b) Al_2O_3
   (c) Sn O_2 (d) Kapron

46. A certain ferro electric sample exhibits a square hysteresis loop with very low coercivity, such a sample is useful as
   (a) capacitor material
   (b) infrared detector
   (c) ultrasonic wave generator
   (d) memory element

47. Ferno electrics behave like normal dielectrics
   (a) under high electric fields
   (b) above curie temperature
   (c) below curie temperature
   (d) under low electric fields

II. Fill in the blanks

1. In an isotropic dielectric medium, the directions of polarization vector P and the electric field intensity E are in the ________ at every point in the medium.

2. The polarization density P in a dielectric medium is proportional directly to the ________.

3. The polarizability of a dielectric is the ratio of the ________ and the ________

4. The electric field displacement vector D in free space is related to the electric field E as ________.

5. The electric susceptibility χ of a dielectric medium is related to the relative permitivity as ________.

6. The polarization density P is related to the electric field intensity E through the relative permitivity ϵ_r. The relation is ________

7. The units of D, the electric displacement, are ________.

8. Gauss law in electrostatics is ________.
9. The surface integral of the electric flux density through a surface enclosing a volume is equal to the ________ that the surface encloses
10. All hydrocarbons are ________ or weakly ________.
11. The distance between two charges + q and –q is *l*. Then, the net dipole moment is ________.
12. The units of polarizability α are ________.
13. When a dielectric containing 'n' molecules per unit volume is placed in an electric field E, the induced dipolemoment per unit volume is equal to ________, where 'α' is the polarizability.
14. Dipolar polarization is associated with ________ molecules.
15. The polarizations that are caused by the defor mation of an atom or molecule are ________ and ________.
16. For a dielectric with relative permittivity and having 'n' molecules perunit volume, the polarization α is given by the relation α = ________.
17. $\frac{(\epsilon_r - 1)}{(\epsilon_r + 2)} \frac{M}{\rho\alpha}$ = ________ (M = molecular weight, the density, ρ, the polarizability in the dielectric medium)
18. The electronic polarizability α_e of an atom of radius r is equal to ________ (in mks units)
19. The homopolar dielectrics possess only one type of polarization, that is ________ polarization
20. The dimensions of the quantity $\frac{1}{\sqrt{\mu_o \epsilon_o}}$ are ________.
21. The refractive index of a dielectric is equal to ________ of the relative permitivity at optical frequencies.
22. The Lorentz – Lorentz equation for a dielectric is ________.
23. The iomic polarization in an ionic solid is unaffected by the variations in ________.
24. The molecular dipoles in substances like polar dielectrics or amorphous solids are ________ distrubuted in space in the absence of the electric field.
25. When a dielectric is placed in an electric field, the dipoles align along the field as a result of a ________ acting on the dipoles.
26. The orientation polarization in a dielectric is strongly influenced by ________, and the electric field intensity.
27. The Langevin-Debye equation relates the orientation polarizability to the ________.
28. The total polarizability of a substance like a polar polymer is equal to ________.
29. A synthetic oil used as capacitor is ________.
30. A good dielectric should have ________ conductivity and ________ dielectric loss.
31. Materials like TiO_2, SnO_2 ZnO_2 are useful as ________ materials.
32. A gas that has high dielectric strength and is used in refrigerators is ________.
33. The mineral insulating oil that has a high dielectric strength is ________.
34. In gases, the dielectric break down is followed by ________ of gases and ________.
35. The insulating material that is used in spark plug is ________.
36. The piezo electric materials are useful in the production of ________ waves.
37. The imaginary part of the dielectric constant represents the ________ in the dielectric.
38. The polarizability that is not affected by temperature changes is ________ polarizability.
39. Above Curie temperature, the polarization vs electric field hysteresis loop becomes ________ passing through the origin.

40. The break down electric field intensity is known as __________ of the dielectric.
41. The spontaneous polarization of a ferro electric crystal becomes __________ at curie temperature
42. Pyro electrics are useful as __________ detectors
43. All ferro electrics are __________ electric and __________ electric.
44. In the para electric state of a ferro electric, the electric susceptibility is __________ proportional to absolute temperature.
45. All ferroelectrics are __________.
46. Triglicinesulphate is an example of both ferroelectricity and __________.
47. The development of __________ in certain crystalline materials as a result of temperature changes is called pyroelectric effect.
48. The units of pyrorelectric coefficient are __________.
49. Pyroelectric materials can be used as detectors for
    (a) nuclear energy,
    (b) IR radiation,
    (c) X-rays,
    (d) electrons.

II. True or False

1. The unit of polarizability is Farad. [True/False]
2. The internal field in a dielectric that has a polarisation P under the action of an electric field E is $P/3\ \varepsilon_0$. [True/False]
3. Electronic polarizability is strongly dependent on temperature. [True/False]
4. At optical frequencies, conic polarisation becomes dominant. [True/False]
5. All pyroelectrics are piezoelectrics. [True/False]
6. Barium titanate has cubic structure above curie temperature. [True/False]
7. The dimensions of the quantity $\frac{1}{\sqrt{\mu_0 \varepsilon_0}}$ are same as that of velocity. [True/False]
8. For a dielectric, $P = \varepsilon_r/\varepsilon_0$. [True/False]
9. A ferroelectric is a normal dielectric above curie temperature. [True/False]
10. Quartz is both ferroelectric and piezoelectric. [True/False]
11. In an isotropic dielectric medium, the electric displacement D and E are perpendicular to each other. [True/False]
12. The ratio of real part of permittivity to the imaginary part of permittivity of a dielectric is called dielectric loss. [True/False]
13. The polarization density of a dielectric is proportional to the applied electric field. [True/False]
14. The distance between two charges (each of charge q) is x. Then, the dipole moment is q/x. [True/False]
15. For a dielectric, Langevin-Debye equation leads to x = C/T. [True/False]
16. Piezoelectrics are useful in position monitors. [True/False]
17. A substance has a dielectric constant 1.0018. The number of molecules per unit volume are 8.85×10^{25}. Then, the polarizability is 1.8×10^{-40} F.m^2. [True/False]
18. Tourmaline is an example for ferro electricity. [True/False]
19. The main contribution to polarizability of a dielectric at optical frequency is due to electronic polarizability. [True/False]
20. CO_2 molecule is polar molecule [True/False]

CHAPTER 11

Magnetic properties

11.1 Introduction

One of the most important and fascinating branches of Physics is the Physics of magnetism and electromagnetic phenomena. The electric fields and magnetic fields can be separate and the two are very dissimilar in many respects. But when they are time dependent, there is a unified blend of the electric and magnetic fields in the form of electromagnetic waves. The energy of an electromagnetic wave concentrates alternately in the electric and magnetic fields. However, in this chapter, we are concerned with the origin of magnetic dipole moment in materials, the material classification and their utilization. The fundamental entity responsible for electric field or magnetic field is charge. The static electric field is associated with the charge. But we do not have magnetic charge in nature. A moving charge gives rise to a magnetic field. The magnetic field has its microscopic sources. The microscopic sources are the electrons, protons and the neutrons. All these particles have a magnetic moment. But, all materials are not magnetic in the sense that all materials do not show the same type of magnetic moment. Different materials possess different types of magnetism. That is inferred from the fact that different materials behave differently when placed in a magnetic field. In these materials, there is a way of interactions of the charged particles which will either give or not give magnetic properties to the material body. Even if there is an interaction among the magnetic dipoles (each moving charge in an atom is associated with a magnetic dipole moment), the interactions differ from material to material giving rise to different types of magnetic phenomena. The magnetic properties of materials are widely used in many applications starting from permanent magnets used as a toy by a child or a permanent magnets in loud speakers to computer memory systems. The magnetic materials find wide applications in all types of motors, dynamos, telephone receivers, ultrasonic generators, transformer cores, antennas in telecommunication, microwaves, etc.

The magnetic materials dispersed on a substrate or made into thin film coatings on a tape are used for audio, video, and digital tape recording. Data storage in floppies is nothing but recording on magnetic tapes. The utility of cylindrical magnetic domains as bubble memories is very important to achieve high writting density in memory systems.

Magnetiostriction associated with magnetic materials is used in ultrasonic applications or in strain gauges. The subatomic particles possessing magnetic moments and the subatomic particles present in the tissue of living beings respond to the external magnetic fields. The magnetic resonance methods are useful in medical diagnosis. Therefore, the study of the basics of magnetism, its origin in different materials and the utility of magnetic materials is essential for electronics engineer. We will discuss the above aspects in the present chapter.

It is known that a moving charge constitutes an electric current and gives rise to a magnetic field. Consider a long thin conductor like a wire carrying a current I. The conductor is surrounded by free space.

When a compass needle is kept near the wire (Fig. 11.1) the needle gets deflected indicating that the conductor produces a magnetic field around it. When the compass needle is moved around the wire, the locus of all points at which the needle shows the same deflection will be a circle. We get different concentric circles, around the wire, which represent the positions of the same deflection. The relation between the direction of the magnetic field and that of current is given by the right hand rule with the thumb pointing in the direction of the current, the fingers circling the conductor or wire.

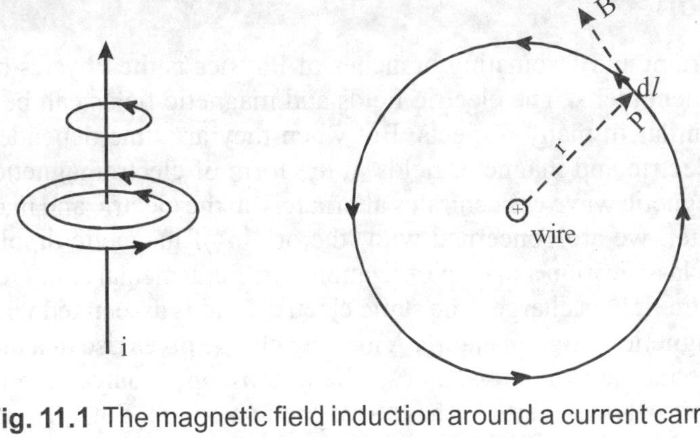

Fig. 11.1 The magnetic field induction around a current carrying wire.

Consider a wire in which a current I exists, then

the magnetic field density B at any point on the circle is

$$\mathbf{B} = \frac{\mu_o}{2\pi r} I \qquad \text{........(11.1)}$$

$$\oint \bar{B} \bullet dl = \mu o I \text{ or } \oint \bar{H} \bullet dl = I$$

The above rectum is called ampere's law.

The magnetic flux over a closed surface is zero i.e.,

$$\int_s \bar{B} \bullet ds = 0$$

i.e., B is solenoidal. There is no magnetic mono pole.

where B is called the induction field. μ_0 is called the permeability of free space, its value is $4\pi \times 10^{-7}$ Henry/m.

Consider a current carrying loop as shown in Fig. 11.2. Then, the magnetic field density B is given as

$$B = \frac{\mu_o I_r}{2\left(r^2+z^2\right)^{3/2}} \qquad ....(11.2)$$

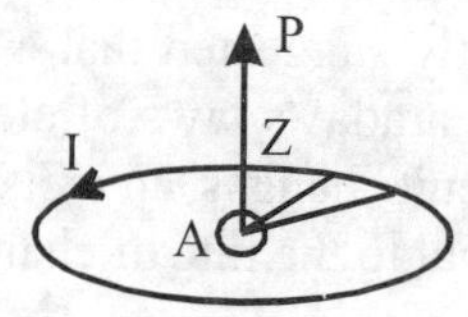

Fig. 11.2

z is the distance between P and the centre of the loop. At the centre of the loop, the magnetic flux density due to the current in the loop is

$$B = \frac{\mu_o}{2r} I \qquad ....(11.3)$$

The magnetic flux, ψ_m is given as $\psi_m = BA$, where A is the area and the magnetic field lines are perpendicular to A. The SI unit of the magnetic flux is weber. If B is not uniform over an area, then we consider an infinitesimally small element of area dS, then, the flux across this area is **B**. d**S**.

Then $$\psi_m = \iint_S B \bullet dS \qquad ....(11.4)$$

dS is a vector. The magnetic flux over a closed surface is

$$\oint_s B \bullet \mathbf{dS} = 0 \qquad ....(11.5)$$

i.e., **B** is solenoidal. This is the representation of Gauss law in magnetostatics. It indicates that there is no magnetic monopole.

A current carrying loop acts like a magnetic dipole. We have seen earlier that the flux density at a distance r from a straight conductor is given by the equation (11.1). B is integrated around the path of radius r enclosing the wire once. This can be expressed as

$$\oint \mathbf{B} \bullet dl = \frac{\mu_0}{2\pi r} \oint \mathbf{d}l = \frac{\mu_0 \mathbf{I}}{2\pi r}.2\pi r = \mu_0 \mathbf{I} \qquad ....(11.6)$$

If the integration is taken over a singly closed path $\mathbf{H} = \mathbf{B}/\mu$. H is called the magnetic field intensity and

$$\oint H \bullet dl = \mathbf{I} \qquad ....(11.7)$$

The above relation is called Ampere's law. It states that the line integral of **H** around a single closed path is equal to the current enclosed. The units of **H** are $\frac{\text{Webers/m}^2}{\text{Weber / amp.m}} = \text{amp.m}^{-1}$.

The quantity H is also called magnetomotive force (mmf). The sign of the integral depends on the direction in which the current is enclosed. The sign is positive if the path followed by the line integral is parallel to **B** and negative, if the path followed is antiparallel. Ampere's law holds for a closed path of any shape.

According to Oersted, a magnetic field is associated with an electric current. In a similar way, Faraday and Henry independently suggested that a magnetic field changing with time should give rise to an electric field. Faraday's laws of electromagnetism are stated as: (1) A magnetic flux linked with a closed circuit changes, an induced emf is set up in the circuit whose magnitude at any instant is proportional to the rate of change of magnetic flux linked with the circuit. If ϕ is the induced emf, then

$$\xi \ \alpha \ \frac{d\phi}{dt} \qquad \text{.......(11.8)}$$

(2) The direction of induced emf, ξ is such that it opposes the change in flux that produces it. This law is known as Lentz's law.

$$\therefore \qquad \xi = -\frac{d\phi}{dt} \qquad \text{.......(11.9)}$$

11.2 A Magnetic Material in an External Magnetic Field: The Three Vectors B, H and M

A loop of area A with a current I is associated with a magnetic moment of IA. The fields at large distances from this loop are identical with those of a bar magnet of dipolemoment m*l*, where m is the pole strength and *l* is the pole separation, provided the magnetic moment of the bar is equal to that of the loop. Thus, ml = IA. A bar magnet has atoms inside it. These atoms are associated with magnetic moments that are aligned in the same direction in the bar. The magnetic bar consists of tiny magnets or a miniature current loops. The bar as a whole should consist of a magnetic moment m*l*. The effect of atomic magnets can be described by a quantity called magnetization **M**, defined as the magnetic dipole moment per unit volume analogous to the polarization **P** in a dielectric. Thus $M = m_i / V = ml / V$, where $m_i = ml$, V is the volume of the magnet.

(i) ***Magnetization:*** Magnetization has the dimension of both magnetic dipole moment per unit volume and of magnetic pole strength per unit area. It is expressed in amperes per meter. We can define M at any point in a magnetic bar or rod that has a continuous distribution of infinitesimal magnetic dipoles i.e. a continuous magnetization where as the dipoles are discrete in size. The following expression gives the definition.

$$M = \underset{\Delta V \to 0}{\text{Lt}} \ \frac{m_i}{\Delta V} \qquad \text{......(11.10)}$$

or $$m_i = \int_V M dV \text{ Am}^2$$

where the integration is carried over the volume of the bar. A magnetized bar or rod is equivalent to uniformly wound coil like solenoid in which a current, I exists.

(ii) ***Magnetic induction:*** Consider (as shown in Fig. 11.3) a coil in the form of a toroid of area A and radius R, that has N turns and initially has air as core, then the magnetic flux density B_o in the toroid is

$$\mathbf{B}_o = \mu_o \frac{NI}{2\pi R} = \mu_o \mathbf{H}$$

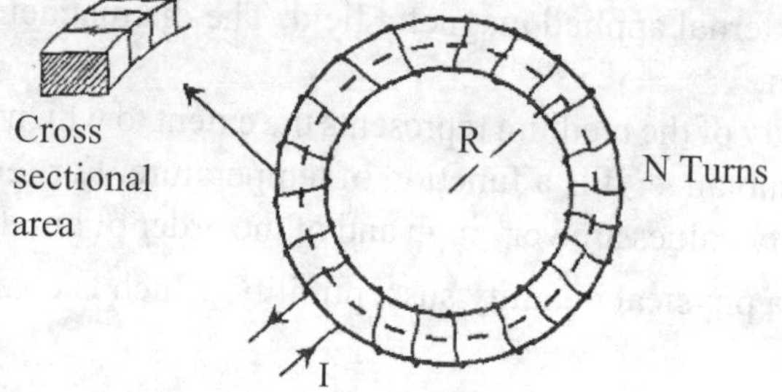

Fig. 11.3 Toroid with N turns.

B_0 arises due to the applied magnetic field intensity, H.

Suppose the same winding is placed on a ring of a magnetic material and the core has the same radius R, then the magnetic flux density from the windings is

$$\mathbf{B}_m = \mu_o \frac{NI}{2\pi R} = \mu_0 \mathbf{M} \quad \text{......(11.11)}$$

Then the total **B** is given as

$$\mathbf{B} = \mathbf{B_0} + \mathbf{B_m} = \mu_0 (\mathbf{H} + \mathbf{M}) \quad \text{......(11.12)}$$

where **B** is the magnetic flux density (T), **H**, the magnetic field intensity, (Am^{-1}), and **M**, the magnetization (Am^{-1}). The above expression is developed for a toroid. But it is valid in general for the materials kept as cores in any coil. The expression (11.12) can be written as

$$\mathbf{B} = \mu_o \mathbf{H}\left(1 + \frac{\mathbf{M}}{\mathbf{H}}\right)$$

Then $$\mu_{eff} = \mu_0\left(1 + \frac{\mathbf{M}}{\mathbf{H}}\right) \quad \text{......(11.13)}$$

μ_{eff} is called the permeability of the magnetic substance. We define relative permeability as

$$\mu_r = \frac{\mu_{eff}}{\mu_0}$$

then, the relative permeability is given as

$$\mu_r = \left(1 + \frac{M}{H}\right) \qquad \text{......(11.14)}$$

(iii) ***Permeability:*** The relative permeability for dramagnetic substance is independent of temperature. The diamagnetic substance has tendency to repel the magnetic lines of force due to external applied magnetic field. The dia magnetism will be discussed in another section.

The permeability of the medium represents the extent to which field flux lines penetrate the magnetic subtance. It is a function of temperature. For ferromagnetic substances, the permeability values are very high and of the order of hundered to a few thousands.

(iv) We introduce a physical quantity susceptibility, which is characteristic of a magnetic substance.

We define a parameter, χ, such that

$$\chi = \frac{M}{H} = \mu_r - 1 \qquad \text{......(11.15)}$$

χ is called the magnetic susceptibility and is a dimensionless quantity . In isotropic media, **M** and **H** are in the same direction, so that this quotient is a scalar and μ_r is a scalar. Both μ and χ are the physical characteristics of a magnetic material. Either of them can be measured in a laboratory

By the magnitude and sign of magnetic susceptibility, all solids can be divided into the three groups, diamagnetic, paramagnetic and ferromagnetic. If the susceptibility is negative, the material is diamagnetic. If the susceptibility is positive, it is paramagnetic. The value of susceptibility is low for these materials.If the susceptibility is positive and very high, the material is ferromagnetic. Paramagnetic substances possess very low value of permeability.

The relative permeability for diamagnetic substance is independent of temperature. At diamagnetic substance has tendency to repel the magnetic lines of force due to external dippled magnetic field. The dia magnetism willbe discussed in another section.

11.3 Origin of Magnetic Moment of Atoms

(a) Bohr Magneton

We have learnt that a current carrying conductor gives rise to a magnetic field around it. Only moving charges induce magnetic field in the sorroundings. A bar magnet gives rise to induction field around it. However, we do not see any visible current circuit in a bar magnet. Then, where are those currents in a magnet ? To answer this question, we have to study the magnetic nature of atoms and the reasons for this magnetic nature. We shall study this in the following.

An atom of any element consists of electrons revolving round its nucleus. Each electron orbitting (Fig. 11.4) around the nucleus produces a loop current, I = eν, where ν is the frequency of rotation of the electron in the orbit. The electron possesses an angular momentum L_l. The magnetic moment μ_{lm} of the loop of current is given as

$$\mu_{lm} = (\text{area of the loop}) \times (\text{current})$$

$$\text{current} = \frac{\text{Charge} \times \text{Angular Frequency}}{2\pi}$$

$$\therefore \quad \mu_{lm} = \frac{e\omega r^2}{2}$$

$$\therefore \quad \mu_{lm} = \pi r^2 \times \frac{e\nu}{2\pi r} = \frac{e\nu r}{2} \qquad \text{.....(11.16)}$$

The symbol μ_{lm} should not be confused with the symbol μ for the permeability of a medium.

The magnetic moment of the electron, μ_{lm} due to its motion around the nucleus is called the orbital magnetic moment. The direction of μ_{lm} is perpendicular to the plane of the orbit (in accordance with right screw rule).

The angular momentum of the electron L_l = mvr, where m and v are the mass and velocity of the electron. But, according to the quantum theory, the angular mementum is an integral multiple of $\frac{h}{2\pi}$

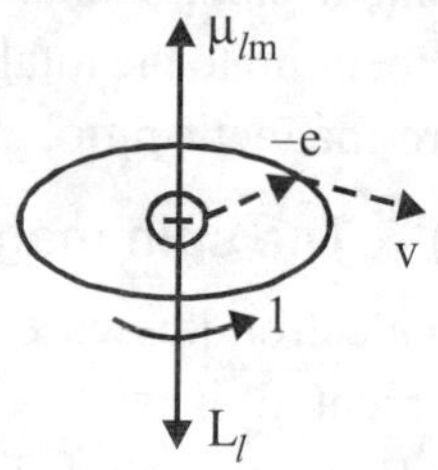

Fig. 11.4 Electron orbital.

$$L_l = \frac{h}{2\pi}\sqrt{l(l+1)} \qquad \text{....(11.17)}$$

where l is the orbital quantum number. It can assume any value of the integral series of n, the principal quantum number. l can take values, 0, 1, 2, 3.... The other electrons with l = 1, 2, 3,... (i.e. p, d, f, g..) have nonzero angular momentum and hence possess magnetic moment given by

$$\therefore \quad \mu_{lm} = -\left(\frac{e}{2m}\right)L_l = -\left(\frac{eh}{4\pi m}\right)\sqrt{l(l+1)}$$

$$= -\mu_B\sqrt{l(l+1)} \qquad \text{.....(11.18)}$$

where

$$\mu_B = \frac{eh}{4\pi m} \qquad \text{......(11.19)}$$

μ_B is a unit measure of atomic magnetic moment and is called the Bohr magneton. One 'Bohr magneton' is equal to 9.27408×10^{-24} Amp m^2. The negative sign in eqn. 11.12 indicates that the magnetic moment μm is anti parallel to the angular momentum L_I.

(b) Quantum numbers, the magnetic quantum number:

The orbit angular momentum can orient itself about the chosen direction, such as the direction of a magnetic field, H only in such a manner that its projection on H direction is an integral multiple of $h/2\pi$ i.e.,

$$L_{lH} = m_l h/2\pi \qquad .....(11.20)$$

Here, 'm_l' is called the magnetic quantum number. It can take all integral values from $-l$ to $+l$.

$m_1 = -l, -(l-1), ...0, 1, 2, ...1$, in total, $(2l+1)$ values as shown for p electron in Fig 11.4. From equation (11.14), we get the orbital magnetic moment as

we get $$\mu_{lH} = -m_l \mu_B \qquad .....(11.21)$$

An electron in p level has three possible components of angular momentum along the direction of the field, $h/2\pi$, 0, $-h/2\pi$. The magnetic moment components are μ_B, 0, $-\mu_B$. The ratio of magnetic dipole momentum to the total angular momentum of the electrons is called the gyro-magnetic ratio, g.

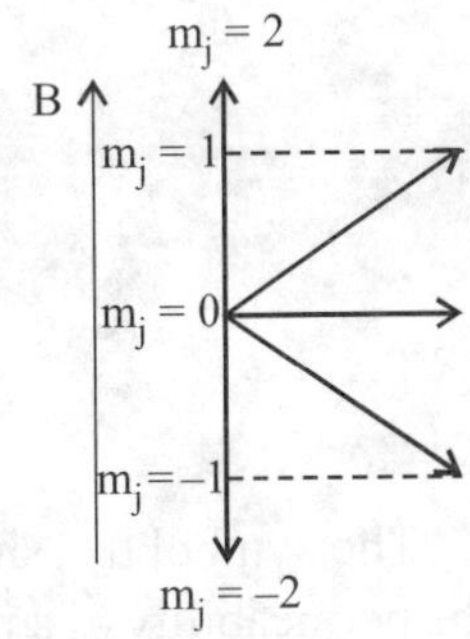

Fig. 11.5

(c) The spin magnetic moment:

An electron possesses an intrinsic angular momentum of spin S to which the spin magnetic moment, μ_s is related. The projection of this moment on the direction of the H field is equal to the Bohr magneton. The magnetic moment due to spin in the direction of the field is

$$\mu_{SH} = \frac{-e}{m}\frac{h}{2\pi}.m_s \qquad ......(11.22)$$

where m_s is called spin magnetic quantum number associated with spin. We have $m_s = \pm 1/2$. The spin gyro-magnetic ratio is determined by the electron's interaction with other electrons. For a free electron, the ratio is equal to 2. Remember that $\mu_s = \mu_{B/2}$. (This is an approximate relation)

The net spin magnetic moment of atoms containing paired electrons or filled electron shells is zero. Hund's rule states that the electron spins always add together in such a way as to give maximum spin angular momentum. Suppose an atom has 10 electrons in its shell, the first five electrons give a moment $5eh/4\pi m$. The next five electrons with opposite spin will nullify this moment.

Summarizing, the magnetic moments, via substance arise due to the following aspects:

1. The orbital magnetic moment of the electrons
2. The spin magnetic moment of the electrons
3. The spin magnetic moment of the nucleus.

When an atom has odd number of electrons, one of them will become unpaired, so the atom on the whole will possess a permanent magnetic moment.

When an atom has an even number of electrons, two cases are possible (1) all electrons are paired and the resultant spin moment is zero. (2) two or more electrons can be paired parallel and the atom then possesses a permanent magnetic moment.

(d) Classification of materials on the basis of magnetic moments:

Solids whose atoms do not possess a permanent magnetic moment or a net magnetic moment are diamagnetic. Orbital magnetic moment is responsible for diamagnetic effects.

Solids with atoms in which there is a net magnetic moment can be paramagnetic, ferromagnetic or antiferromagnetic.

If the interaction between magnetic moments of atoms does not exist or is negligible, the solid is said to be paramagnetic (Fig. 11.6(a)).

If all the neighbouring magnetic movements tend to align themselves parallel to each other, the body is ferromagnetic (Fig 11.6(b)).

If the neighbouring magnetic moments tend to align themselves antiparallel, the body is antiferromagnetic. (Fig. 11.6(c)). If the anti parallel magnetic moments are different in magnitude in a solid, then it is said to be ferrimagnetic (Fig.11.6(d)).

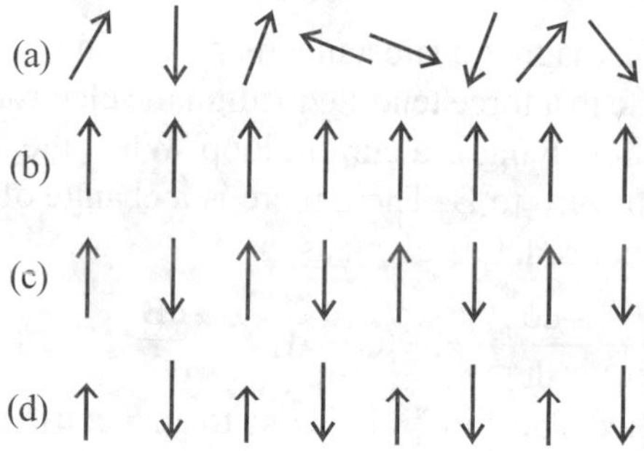

Fig. 11.6 Schematic representation of magnetic moments in materials.

11.4 Diamagnetism

(a) What is a diamagnetic material?

A diamagnetism substance possesses a net negative magnetism. Even though the substance consist of atoms that have no net magnetic moment, there is a tendency of electric charges partially to shield the interior of a body from an applied magnetic field. The solid responds to an external magnetic field. The negative susceptibility arises from Lentz's law. There is a tendency to shield the interior of the substance from the externally applied magnetic field. Therefore, an induced current gives rise to a magnetic field that opposes the applied field. A diamagnetic sample moves away from the magnetic field.

(b) Expression for change in magnetic moment: Lansevins classical picture of Diamagnetism.

A fast spinning top revolves about a vertical axis along with the rotation about its own axis. If the axis does not coincide with the vertical, it precesses (Fig. 11.7(a)). A similar motion occurs for an electron in an external magnetic field. An electron revolving in its orbit is equivalent to a closed loop of current possessing a magnetic moment. The magnetic field strives to turn the electron orbit normal to **H**, causing a precessional motion of the orbit (Fig. 11.7(b)) around the field direction. The precessional motion of a 'top' stems from the action of gravitational force and precesses about the direction of that force.

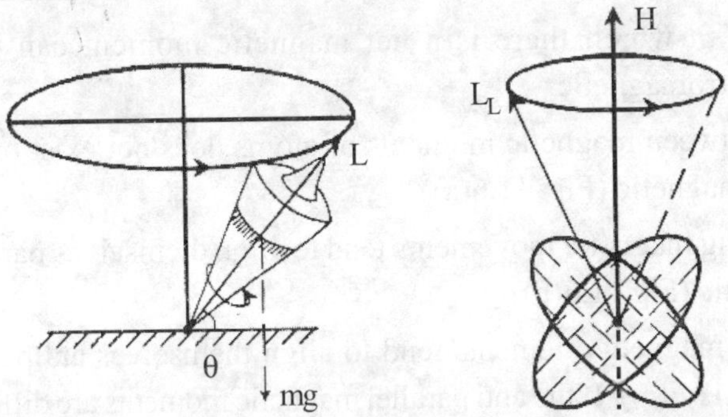

Fig. 11.7 (a) Spinning top, (b) An electron and orbital magnetic moment.

Suppose the atom in a diamagnetic substance is placed in a magnetic field of strength, **H**. This magnetic field gives rise to a force tending to align the electron orbit normal to **H** as shown in Fig. 11.7.b. The electron revolving is a current loop. When the magnetic field is applied, the induction field changes from zero to B. Then, there is a change of flux through the area of the loop. Hence, an emf ξ is produced. ξ is given as

$$\xi = \frac{-d\phi_m}{dt} = -\mu_0 \frac{d}{dt}\mathbf{H} = \frac{-d\mathbf{B}}{dt} \quad .....(11.23)$$

where ϕ_m is the magnetic flux. The emf ξ gives rise to an electric field, **E** acting longitudinally to the direction of the electron motion. Then, the electric field, **E** is given as (assuming the electron's orbit to be circular)

$$\mathbf{E} = \frac{\xi}{2\pi r} \quad .....(11.24)$$

The electric field gives rise to a force on the electron

Force on the electron, **F** = ma = e**E**

i.e.,

$$ma = \frac{e\,\xi}{2\pi r} = -\frac{e\left(\frac{d\phi_m}{dt}\right)}{2\pi r}$$

$$\therefore \qquad a = -\frac{e}{2\pi\, r\, m}\left(\frac{d\phi_m}{dt}\right) \qquad .....(11.25)$$

The above expression gives the acceleration associated with the motion of the electron. Due to the applied field, obviously, the electron suffers a change in its velocity. Its magnitude of the change in its velocity depends on the change in magnetic flux in a unit time. Therefore, the change in velocity is given as,

$$\Delta v = \left(\frac{-e}{2\pi\, r\, m}\right)\frac{d\,\phi_m}{dt} \times dt$$

$$= \frac{-e}{2\pi\, r\, m}\,\pi r^2\, B = \left(\frac{-er}{2\, m}\right)\mathbf{B} \qquad .....(11.26)$$

In time dt, the magnetic induction changes from 0 to B. The corresponding change in the angular frequency of the electron is

$$\Delta\omega = \frac{\Delta v}{r} = \left(\frac{-e}{2\, m}\right) B \qquad .....(11.27)$$

This can be expressed as

$$\Delta\omega = \omega \pm \omega_0 = \frac{-eB}{2\, m} \qquad .....(11.28(a))$$

$$\omega = \pm\,\omega_0 - \frac{eB}{2\, m} \qquad .....(11.28(b))$$

$$= \pm\,\omega_0 - \omega_L$$

where $\omega_L = eB/2m$ .....(11.28(c))

The frequency, ω_L is called the Larmor frequency. It characterises the change in the velocity of the electron. This is the frequency of precession. This change gives rise to a change in the orbital magnetic moment.

The electrons whose orbital moments are parallel to the field are slowed down. The electrons whose orbital moments are anti parallel are speeded up. This is called Larmor theorem.

The angular momentum of the electron = $m\,\omega^2\, r_0$

Associated current with the electron $= \dfrac{e\omega}{2\pi}$

$\therefore$ the magnetic moment of the orbiting electron $= \dfrac{e\omega}{2\pi} \times \pi\, r_0^{\,2}$

$$= \frac{er_0^2\,\omega}{2} \qquad .....(11.29)$$

As the field is applied, there is a change in angular frequency, hence a change in magnetic moment. This is given as

$$\Delta\mu_m = \frac{e\, r_0^2}{2}\left(-\frac{eB}{2m}\right) = \frac{-e^2 r_0^2 B}{4m} \qquad .....(11.30)$$

The negative sign indicates that the change opposes the field.

An atom normally possess many electrons. We have to take the mean square distance of the electrons from the field axis instead of simple r_0^2. Then,

The orbit of an electron would be in a plane i.e., $r_0^2 = r_x^2 + r_y^2$

$$\therefore \qquad \Delta\mu_m = \frac{-e^2 B}{4m} < \bar{r}_0^2 > \qquad .....(11.31)$$

where $< \bar{r}_0^2 >$ is the average distance from the field axis.

Suppose there are 'n' atoms per unit volume in the diamagnetic substance.

Then, the change in magnetic moment.

$$\Delta\mu_m = \frac{-e^2 B}{4m} < \sum_i r_i^2 >$$

The summation $\sum_i$ is overall atoms. It represents the averaging of the sum of the squares of the distances from the electrons to the Larmor precession axis.

Consider the random orientation of atoms in space, then, the mean square distance of the electrons from the nucleus, $r^2 = < r_x^2 > + < r_y^2 > + < r_z^2 >$ for the three coordinate axis. For symmetrical atoms, we have

$$r_x = r_y = r_z = \frac{r_0}{3}$$

then $$(r^2) = \frac{r_0^2}{3} + \frac{r_0^2}{3} = \frac{2r_0^2}{3} \qquad .....(11.32)$$

There are n atoms per unit volume and each atom has an atomic number z. Then, the number of electrons per unit volume is equal to n z.

$\therefore$ The magnetic momentum

$$\Delta\mu_m = \frac{-nz\, e^2}{4\, m} B\left(\frac{2}{3}\right) r_0^2$$

$$= \frac{-n\, z\, e^2\, \mu_0 H\, r_0^2}{6\, m} \qquad .....(11.33)$$

This equation is called Langevin's classical relation for the magnetic moment of a diamagnetic substance per unit volume.

(c) The susceptibility of a diamagnetic material:

The susceptibility is given as

$$\chi = \frac{\Delta\mu_m}{H} = \frac{-nz\,e^2\,\mu_0\,r_0^2}{6\,m} \quad .....(11.34)$$

The number of atoms in a substance is given by the expression

$$n = \frac{N_A\,\rho}{A}$$

When N_A is the Avagadro's number, ρ is the density of the substance (Kg/m^3), A is the atomic weight in kg. We get the susceptibility as the mean susceptibility χ as

$$\chi = -\mu_0 \frac{N_A \rho z\, e^2}{A} r_0^2 \quad .....(11.35)$$

In the case of diamagnetic substances, only the outer most electrons give the largest contributions to the magnetic moment.

The diamagnetic susceptibility is independent of temperature and field strength. Monoatomic rare gases, He, Ne, A, Kr, x_e have closed shell electronic structure and are diamagnetic. Similarly, diaatomic gases like H_2, N_2 have filled electronic shells.

Ionic crystals like NaCl, KCl, MgO, covalent crystals like Si, Ge, all organic compounds are diamagnetic in nature.

Most of the simple metals like copper, silver, gold, bismuth, the ions like Li^+, Na^+, K^+, Ca^{2+}, Mg^{2+} etc are all diamagnetic.Super conductors are perfect diamagnetics. The Table 11.1 gives the susceptibilities of few diamagnetic substances.

Table 11.1 gives the susceptibilities of some diamagnetic substances*

Diamagnets	χ	Diamagnets	χ
Gold	-28×10^{-6} Cgs	Water	-13.09×10^{-6} Cgs
Copper	-5.46×10^{-6} Cgs	Helium	-1.88×10^{-6} Cgs
Germanium	-76.84×10^{-6} Cgs	Al_2O_3	-37×10^{-6} Cgs
Silicon	-3.9×10^{-6} Cgs	Diamond	-5.9×10^{-6} Cgs
Selenium	-25×10^{-6} Cgs	Na Cl	-30.3×10^{-6} Cgs

*CRC Handbook of chemistry and physics 59th edition (1978-79), CRC Press.

11.5 Paramegnetism

11.5.1 Magnetic moment

The permanent magnetic moment in atoms arise due to: (1) orbital moment of the electrons, (2) the spin moment of the electrons, and (3) the spin moment of the nucleus.

The magnetic moment of an atom is the vector sum of magnetic meoments of all individual electrons. The orbital angular momentum L is expressed as $\frac{h}{\pi}\sqrt{L(L+1)}$. The maximum value of L is $\sum_{i}^{z} l_i$ and the minimum value is zero. Therefore, L takes values 0,1,2,3... $\sum_{i}^{z} l_i$.

The resultant spin angular momentum, S follows the quantum principle and is expressed as $S = \frac{h}{\pi}\sqrt{S(S+1)}$. The value of S depends on the number of electrons in the atom. If their number is even, S = 0; If their number is odd, $\sum_{i}^{z} S_i = \frac{Z}{2}$.

The spin and the orbital angular momentum are coupled. The total angular momentum is represented by J = |L + S| and is expressed as $\frac{h}{\pi}\sqrt{J(J+1)}$. J can have all values from |L + S| to |L – S| ie (2S + 1) possible values. This is called the multiplicity of energy levels.

$m_{Jh}/2\pi$ is the component of J along the field direction.

The resulting magnetic moment along the direction of an externally applied magnetic field is expressed as

$$\mu_{mj} = g\,\mu_B\sqrt{J(J+1)} = g\,m_J\,\mu_B \quad \text{........(11.36)}$$

m_J takes values from –J to + J. There are (2J + 1) values i.e. orientations for m_j. g is called Lande's splitting factor given in the form

$$g = 1 + \frac{J(J+1)+S(S+1)-L(L+1)}{2J(J+1)} \quad \text{........(11.37)}$$

The total magnetic moment of an atom specificed by the vector J is, therefore, given as

$$\mu_{mH} = -g\frac{e}{2m}\, m_j\left(\frac{h}{2\pi}\right) \quad \text{.....(11.38)}$$

where m_J = J, (J – 1), ...–J

All atoms possessing odd number of electrons exhibit paramagnetism. The examples are many. A few of them are free sodium ions, gaseous nitric oxide, organic free radicals etc., Free atoms or ions with partially filled inner electronic shells, transition elements ions, rare earth ions, actinide elements, etc., are also examples for paramagnetism. There are a number of substances with even number of electrons, including molecular oxygen, that exhibit paramagnetism. The susceptibility of a paramagnetic substance increases with a decrease of temperature. In a non uniform field, they experience a force towards the stronger regions of the field. The examples for paramagnetic substances are Alluminium, Tungsten, Oxygen etc.

11.5.2 Langevin's Classical Theory of Paramagnetism

This theory is based on the fact that the atoms of paramagnetic substance, have a permanent dipole moment. The substance contains a very large number of dipoles per unit volume. They are randomly oriented with respect to each other and the algebraic sum of magnetic moments over the entire volume is zero.

We can treat the assembly of the magnetic dipoles as a paramagnetic gas. When a magnetic field is applied, the magnetic dipoles tend to orient themselves along the field direction. But, the thermal energy gives an agitation to the dipoles and restrict the orientation along the field. At any temperature, the paramagnetic gas has dipoles oriented along all directions with respect to the field, but the majority being along the field direction.

If the μ_m is the magnetic moment of each dipole, the energy of each dipole in the magnetic field is

$$\mu_m = \mu_m \cdot B_o \qquad .....(11.39)$$

This quantity is minimum when the mafnetic dipole is oriented along the field direction. Let the direction of B coincide with the z-axis is as shown in figure given below.

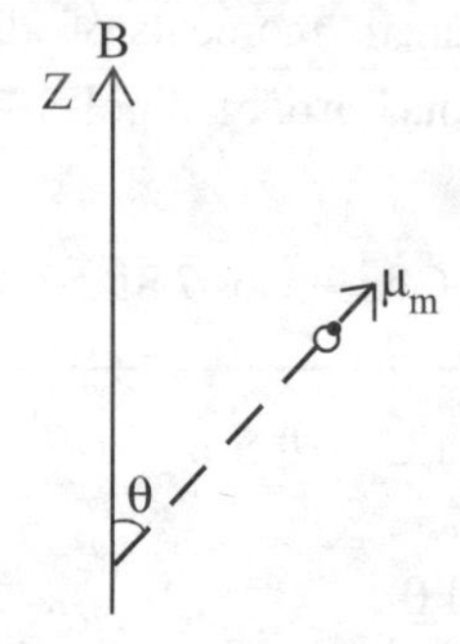

Let 'n' be the number of dipoles oriented along an angle θ w.r.t the field B_o.

The number of dipoles that are oriented along the direction 'θ' at any temperature, T, is proportional to Boltzmann factor $e^{\frac{-\mu_m}{kT}}$.

The number of dipoles, dn (θ) whose magnetic moments lie within a solid angle dΩ is expressed as

$$d\,n\,(\theta) = C \exp\,(-\,U_m/kT)\,d\Omega \qquad .....(11.40)$$

$$= C \exp\,(\mu_m B_o/kT)\,d\Omega \qquad .....(11.41)$$

The solid angle $d\Omega = 2\pi \sin\theta\, d\theta$.

$\therefore\ dn(\theta) = 2\pi\, C \exp(\boldsymbol{\mu_m}\cdot \mathbf{B_o}/kT) \sin\theta\, d\theta$

The total number of dipoles per unit volume that have orientation between 0 to π is given as

$$n = 2\pi \int_0^{\pi} C \exp\left(\frac{\mu_m B_0 \cos\theta}{kT}\right) \sin\theta\, d\theta \qquad .....(11.42)$$

Let $$a = \frac{\mu_m B_o}{kT}$$

then, eq. 11.42 becomes

$$n = 2\pi\, C \int_0^{\pi} \exp(a \cos\theta) \sin\theta\, d\theta.$$

Each dipole has a magnetic moment $\mu_m \cos\theta$ along the field direction. The total magnetic moment of the paramagnetic gas in the presence of the magnetic field induction B_o is given as

$$\left(\overline{\mu_m}\right) = 2\pi\, C \int_0^{\pi} \mu_m \cos\theta\, dn$$

$$= 2\pi\, C\, \mu_m \int_0^{\pi} \exp(a \cos\theta) \cos\theta \sin\theta\, d\theta \qquad .....(11.43)$$

The average magmetisation M of the paramagnetic gas is expressed as

$$\overline{M} = n\, \frac{\text{Sum of the magnetic moments of all the dipoles}}{\text{Total no. of dipoles}}$$

i.e., $$\overline{M} = n\, \frac{2\pi\, C \mu_m \int_0^{\pi} \exp(a \cos\theta) \cos\theta \sin\theta\, d\theta}{2\pi\, C \int_0^{\pi} \exp(a \cos\theta) \sin\theta\, d\theta} \qquad .....(11.44)$$

Let $x = \cos\theta$; i.e. $dx = -\sin\theta\, d\theta$

then, the expression, in (11.44) becomes

$$\therefore \qquad \overline{M} = \frac{n\, \mu_m \int_{-1}^{+1} x \exp(a\, x)\, dx}{\int_{-1}^{+1} \exp(a\, x)\, dx}$$

$$\therefore \quad \overline{M} = n\,\mu_m \left[\frac{e^x + e^{-x}}{e^x - e^{-x}} - \frac{1}{a}\right]$$

$$= n\,\mu_m \left[\coth a - \frac{1}{a}\right].$$

$$= n\,\mu_m\, L(a) \qquad .....(11.45)$$

L (a) is the Langevin's function.

Case 1 :

When $\mu_m B_0 >> kT$

i.e., at low temperatures or large magnetic fields

$$L(a) \rightarrow 1$$

$$\therefore \quad \overline{M} = M_s$$

where M_s is the saturation magnetisation.

Case 2 :

When $\mu_m B_o << kT$

i.e. high temperatures or low fields, 'a' becomes bery small. Then, coth a can be expanded as

$$\text{Coth } a = \frac{1}{a} + \frac{a}{3} - \frac{a^2}{4^5} \; .....$$

Normally a >>1, then

$$\text{Coth } a \approx \frac{a}{3}$$

Then

$$\overline{M} = \frac{n\,\mu_m a}{3} = \frac{n\,\mu_m^2 B_o}{3\,kT} \qquad .....(11.46)$$

The susceptibility, χ of a paramagnetic substance is

$$\chi = \frac{\overline{M}}{H} = \frac{n\,\mu_m^2\,\mu_o}{3\,kT} = \frac{C}{T} \qquad .....(11.47)$$

where $\quad 'C' = \dfrac{n\,\mu_m^2\,\mu_o}{3\,k}.$

The constant 'C' is called the Curie constant. The relation between the magnetic susceptibility and the temperature is called the 'Curie law'.

A paramagnetic substance shows saturation magnetisation as shown in the Fig. 11.8 (a). When $1/\chi$ is plotted against temperature T, a straight line passing through origin is obtained as shown in Fig. 11.8 (b).

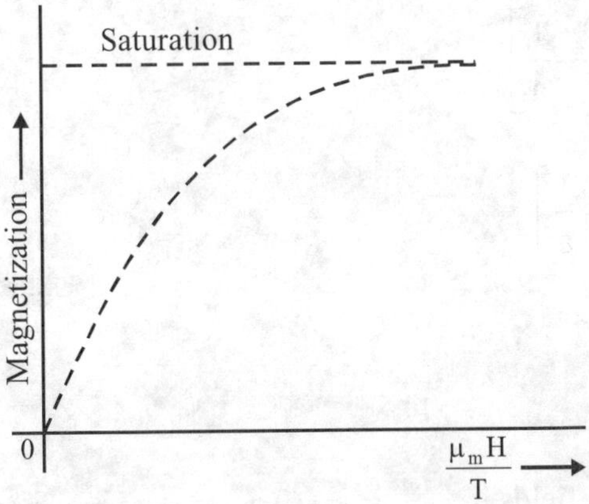

Fig. 11.8 (a) The magnetic moment of a paramagnetic ion as a function of $\frac{\mu_m}{T}H$

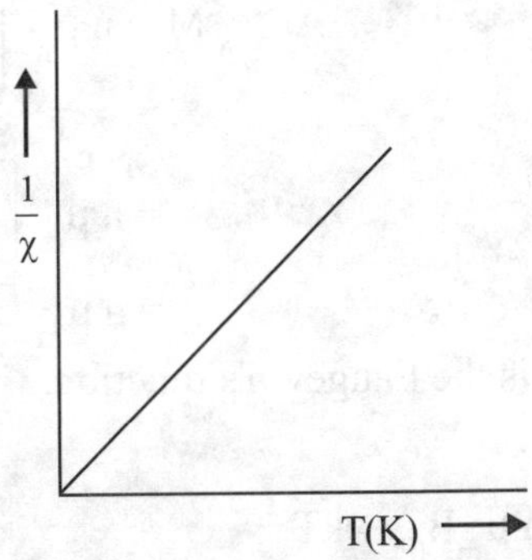

Fig. 11.8 (b) The reciprocal of the susceptibility as a function of Temperature for a paramagnetic substance (Curie law).

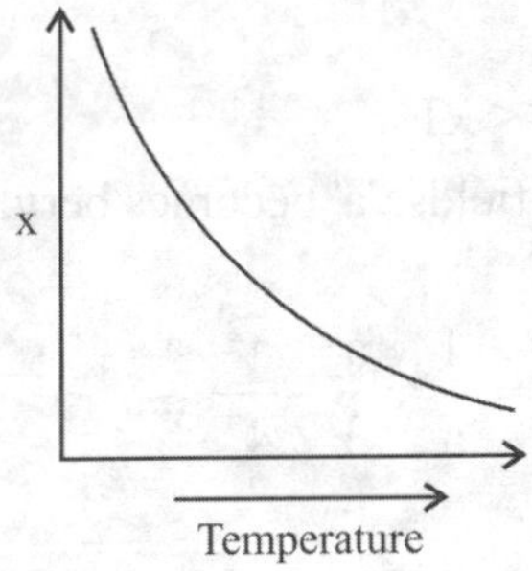

Fig. 11.8 (c)

11.5.3 Quantum approach

Magnetism is due to electrons. An electron is a quantum particle. It is appropriate for us to know about the quantum approach which is so essential these days. We have seen from the eq. (11.38) how the magnetic moment would have discrete orientations with respect to the applied magnetic field.

Suppose we consider that there are n atoms per unit volume in a paramagnetic material. Let these atoms be distributed in two states of spin with $m_l = \pm 1/2$, as shown in the diagram below. Let the population in them be n_2 and n_1 respectively. The populations should be proportional to Boltzmann factor. $e^{-\Delta E/kT}$ where ΔE represents the energy ($\boldsymbol{\mu}_m.\mathbf{B}$)

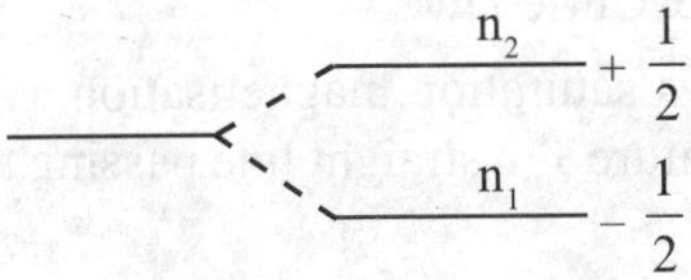

$$\therefore \qquad \frac{n_1}{n} = \frac{e^{m_j \mu_B gB/kT}}{e^{m_j \mu_B gB/kT} + e^{-m_j \mu_B gB/kT}}$$

$$\frac{n_2}{n} = \frac{e^{-m_j \mu_B gB/kT}}{e^{m_j \mu_B gB/kT} + e^{-m_j \mu_B gB/kT}}$$

$\therefore$ The resultant magnetization M is

$$M = e\,(n_1 - n_2)\,\mu_B\, m_j\, g$$

$$= n\,\mu_B\, m_j\, g \left[\frac{e^{m_j \mu_B gB/kT} - e^{-m_j \mu_B gB/kT}}{e^{m_j \mu_B gB/kT} + e^{-m_j \mu_B gB/kT}}\right]$$

$$= n\,\mu_B\, m_j\, g \left[\frac{e^x - e^{-x}}{e^x + e^{-x}}\right] \qquad .....(11.48a)$$

where $$x = \frac{m_j\,\mu_B\, g\, B}{kT} \qquad .....(11.48b)$$

$$\therefore \qquad M = n\,\mu_B\, m_j\, g \tanh x.$$

For x << 1 i.e., $\frac{m_j\, g\, \mu_B\, B}{kT}$ << 1; m_j tanh x function is represented by J B_J (x), the Brillouin function. Then,

$$M = n\, g\, \mu_B\, J\, B_J\,(x) \qquad .....(11.48c)$$

when $$B_J\,(x) = \frac{2J+1}{2J} \coth \frac{(2J+1)\,x}{2J} - \frac{1}{2J} \coth\left(\frac{x}{2J}\right)$$

For a << 1, $\coth a = \frac{1}{a} + \frac{a^2}{3}$

$$\therefore \qquad B_J\,(x) = \frac{(2J+1)}{2J} \times \left(\frac{2J}{(2J+1)\,x}\right) + \frac{(2J+1)^2}{2J^2} \cdot \frac{x}{3} - \frac{1}{2J} - \frac{x}{12J^2}$$

$$= \frac{(J+1)}{3J}\, x$$

$$M = \frac{ng\,\mu_B J\,(J+1)}{3J}\, x$$

$$= \frac{ng\,\mu_B J\,(J+1)}{3J}\left(\frac{m_J\,\mu_B\, gB}{kT}\right)$$

$$= \frac{n\mu_B{}^2 g^2 (J+1) B}{3kT} \quad .....(11.49a)$$

then

$$\chi = \frac{M}{H} = \frac{\mu_0\, n\, \mu_B^2\, g^2 (J+1)}{3kT} = \frac{C}{T} \quad .....(11.49b)$$

Where C is the Curie constant, The factor $p = g(J+1)^{\frac{1}{2}}$ is called effective number of Bohrs magnetons per atom.

The saturation magnetisation is the magnetic moment of a free atom multiplied by the number of atoms i.e., $n\,\mu_B\, g\, J^2$. Eq. (11.49b) is called Curie law.

Paramagntic substances serve as working substances in pramagnetic amplifier and oscillators, and in adiabatic demagnetization. The experimentally observed values of the susceptibility are given in Table 11.2, for a few paramagnetic substances.

Table 11.2

Sustance	χ	Substance	χ
$CrCl_3$	$+ 6890 \times 10^{-6}$ Cgs	$Fe\,SO_4$	$10{,}200 \times 10^{-6}$ Cgs
Cr_2O_3	$+ 1960 \times 10^{-6}$ Cgs	$Fe\,Cl_2$	$14{,}750 \times 10^{-6}$ Cgs
$MnSO_4$	$+ 13660 \times 10^{-6}$ Cgs	$Ni\,SO_4$	$+ 4005 \times 10^{-6}$ Cgs
Fe_2O_3	$+ 3589 \times 10^{-6}$ Cgs	$C_4\,SO_4$	$+ 1330 \times 10^{-6}$ Cgs
CaO	$- 15 \times 10^{-6}$ Cgs	Palladium	567.4×10^{-6} Cgs

11.6 Ferromagnetism

11.6.1 The Characterstics Associated with Ferromagnetism

Ferromagnetism is a phenomenon which occurs in some materials that exhibit spontaneous magnetization. This nature arises in them when the magnetic moments of adjacent atoms are arranged in a regular order i.e. all parallel to each other. The word, 'spontaneous' suggests that the magnetization in the material exists even in the absence of an external magnetic field. The individual localized magnetic moments combine even in the absence of an external field to produce a large value of the bulk magnetization. This is called spontaneous magnetisation. Two important properties are associated with a ferromagnetic substance. They are :

1. A ferromagnetic substance possesses spontaneous magnetization only below a certain temperature T_c, called Curie temperature. Above the Curie temperature, the substance suddenly goes into paramagnetic state. The magnetization in the substance is maximum at 0 K and in principle, this magnetization is present over a long range of temperature. Then, the magnetization decreases with the increase in temperature. Around the transition temperature, it fall off very rapidly reaching zero value at the transition temperature, T_c.

If the substance is cooled from a temperature above the Curie temperature, the spontaneous magnetization reappears at T_c. This property indicates that there is no structural or chemical change in the substance at the transition temperature. The Fig. 11.9 depicts the variation of spontaneous magnetization with temperature in a ferromagnetic substance.

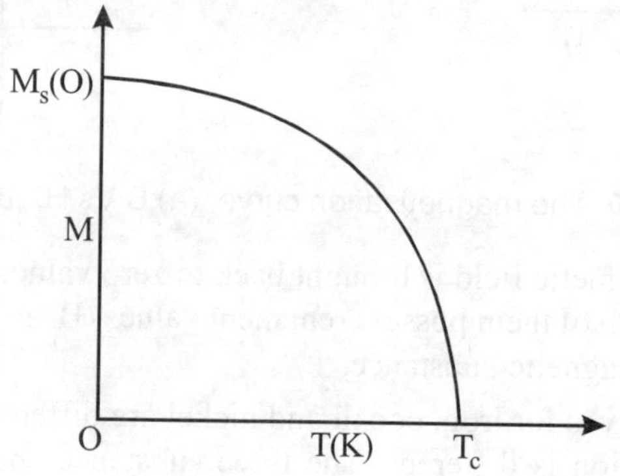

Fig. 11.9 Temperature dependentce of spontaneous magnetization.

At high temperatures, all ferromagnetic substances are paramagnetic. It should be remembered that all paramagnetics do not become ferromagnetic when cooled to a lower temperature. The spontaneous magnetization is a special property of a few substances like iron, cobalt and nickel.

2. Ferromagnetic substances, though spontaneously magnetized, do not show macroscopically observable magnetization. For example, take a piece of iron bar . Iron is basically ferromagnetic in nature, but the iron bar does not show any magnetization. But when it is brought into a magnetic field of a permanent magnet, it shows appreciable magnetization.

   Suppose such an iron piece is taken in the form of an annular ring and a copper wire of a few turns is wound on it. Let another wire be wound on it and be connected to a galvanometer. When a current is passed in the first wire (called primary), a deflection is observed whenever the current is changed in the first coil. The current in the first wire produces a magnetic field of some intensity and then the material possesses a magnetic induction field B. The magnetic induction can be measured and it is a function of the magnetic field intensity H. When H is varied from zero value, B increases nonlinearly and at a certain value of H, it tends to reach a maximum value (Fig.11.10).

   The magnetization at that value of the magnetic field reaches a saturation value. The nature is like any paramagnetc substance, but the magnetization value is enormously large. The saturation is obtained at low values of the field intensities. When the current in the primary is now decreased, there by decreasing the magnetic field intensity, the curve does not retrace its earlier path (Fig.11.10a and b). The curve is called magnetization curve.

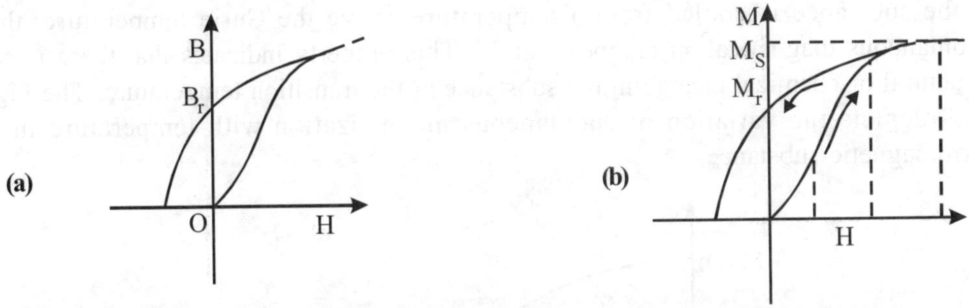

Fig. 11.10 The magnetisation curve. (a) B Vs H, (b) M Vs H.

When the external magnetic field is brought back to zero value, the B as well as M do not reduce to zero values. Both of them possess remanent values (B_r and M_r). This is an important characteristic of a ferromagnetic substance.

The magnetisation curves for iron, cobalt and nickel are different (Fig.11.11). The way in which they reach saturation is different. The three substances possess different saturation values. Ferromagnetic substances are characterised by large positive values of susceptibility. The relative permeability values are very high. The expression $\mathbf{B} = \mu\ \mathbf{H}$ is not valid. The permeablity depends on the field strength in ferromagnetics.

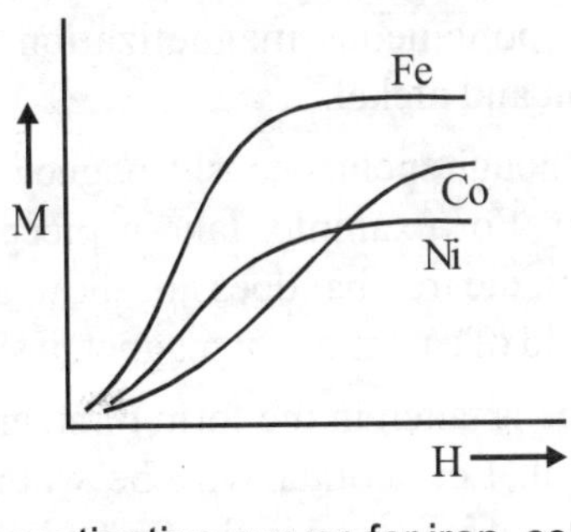

Fig. 11.11 Magnetization curves for iron, cobalt and nickel.

11.6.2 Molecular Field-Weiss Domain Model

The above characteristics prompted Weiss to propose a hypothesis and a model for a ferromagnetic substance. According to Weiss, a ferromagnetic substance consists of an internal molecular field which is responsible for spontaneous magnetization.

He also suggested that a ferromagnetic crystal consists of regions, on microscopic level, in which there is a parallel arrangement of spin moments there by giving a net magnetic moment equal to the spontaneous moment. However these regions themselves are oriented with respect to each other randomly, so that the total magnetization of the ferromagnetic substance like iron piece is negligibly small or zero. These regions of spontaneous magnetization are known as domains. An unmagnetized sample contains domains. Each domain is spontaneously magnetized. A poly crystalline sample like an iron rod contains domains whose magnetic vectors will be randomly oriented to give zero magnetic moment (Fig. 11.12 (a).

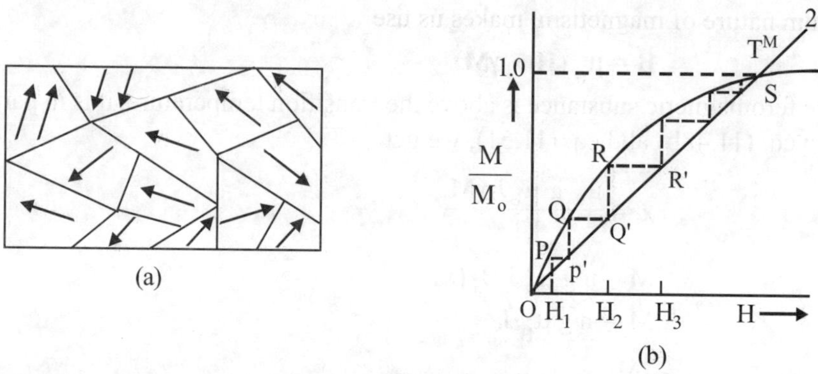

Fig. 11.12 (a) and (b)

Ferro magnetism is due to Spin, Internal field

It is assumed that the magnetic moment of atoms that comprise a ferromagnetic body are of spin (not of orbital) origin. When a ferromagnetic substance is placed in a magnetic field, the spin moments are ordered not only by an external magnetic field **H**, but also by the cumulative action of all magnetic moments of the atoms or spins. This is the basis for the hypothesis of molecular field.

According Curie-Weiss model, a ferromagnetic substance consists of magnetic dipoles and these dipoles interact co-operatively to give an internal molecular field (H_m).

The intensity of the molecular field is directly proportional to the magnetization.

$$\therefore \qquad H_m = \gamma M \qquad .....(11.50)$$

where γ is called the molecular field constant.'γM' is a measure of the tendency of the environment to align a given dipole parallel to the magnetisation already existing. If H_m is ploted aganist M, the plot is a straight line (curve 2 of the Fig. 11.12(b)) curve 1 is eq. (11.48c).

Suppose we take a magnetic sample. If a small magnetic field is momentarily applied, then the magnetization corresponding to P is produced. As a result of the internal field, the point should be p instead of p'. This corresponds to a higher field H_2. The field of this magnitude should produce a magnetization of Q, which should shift to Q'. Thus, the sample will go through O, P, Q, R..... and reaches the point S which is a stable point for the ferromagnetic substance. Thus the internal molecular field sends the sample into the saturation.

11.6.3 Temperature Dependence of Spontaneous Magnetization

A ferromagnetic substance is looked upon as a paramagnetic sample with a very large internal molecular field. The value of M at any temperature may be obtained from the point of intersection of the two plots of equations (11.49) and (11.50). According to Weiss, the total internal field in a ferromagnetic substance placed in an external magnetic field of intensity, H, is

$$\mathbf{H_i = H + \gamma M} \qquad .....(11.51)$$

The tendency of the magnetic field is to align all the dipoles in its own direction.

In quantum nature of magnetism, makes us use

$$\therefore \quad \mathbf{B} = \mu_0 (\mathbf{H} + \gamma\mathbf{M}) \qquad .....(11.52)$$

When the feromagnetic substance is above the transition temperature, it is in paramagnetic state. Taking eq. (11.48b) abd eq. (11.51), we get

$$x = \frac{\mu_0 \, g \, \mu_B J \, \gamma \mathbf{M}}{kT} \qquad .....(11.53)$$

$$M = n \, g \, \mu_B \, J \, B_J (x) \qquad .....(11.54)$$

$$M_s = n \, g \, \mu_B \, J$$

$$\frac{M}{M_s} = B_J (x) \qquad .....(11.55)$$

But, from eq., 11.53, we get

$$\frac{M}{M_s} = \frac{x \, kT}{\gamma \, ng^2 \, \mu_B^2 \, J^2} \qquad .....(11.56)$$

M shouid satisfy both eq. (11.55) as well as (11.56). The values of M/M_s are plotted as function of x as calculated from (11.55) and (11.56) shown in Fig. (11.13).

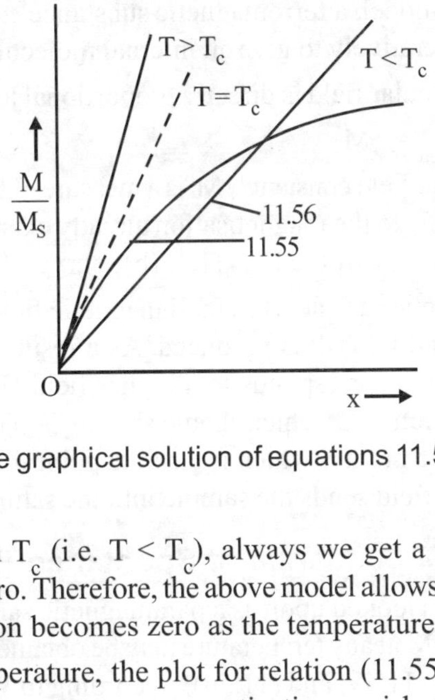

Fig. 11.13 The graphical solution of equations 11.55 and 11.56.

For temperatures less than T_c (i.e. $T < T_c$), always we get a non vanishing value for M although the external field is zero. Therefore, the above model allows spontaneous magnetization. The spontaneous magnetization becomes zero as the temperature rises.

At $T = T_c$, the critical temperature, the plot for relation (11.55) becomes tangential to the curve of relation (11.56). The spontaneous magnetization vanishes.

The maximum magnetization occurs at T = 0K. All magnetic moments line up at that temperature. As temperature is increased, this is disturbed and the magnetization starts changing. Initially the magnetization decreases by small amounts and then drops rapidly when (T/T_c)

reaches more than 0.5 and the magnetization (i.e., spontaneous magnetization) becomes zero at $T = T_c$. This behavior is shown in Fig. 11.14 for iron, cobalt and Nickel at $J = 1/2$. $J = 1/2$ value suggests that only spin is involved in the magnetization process.

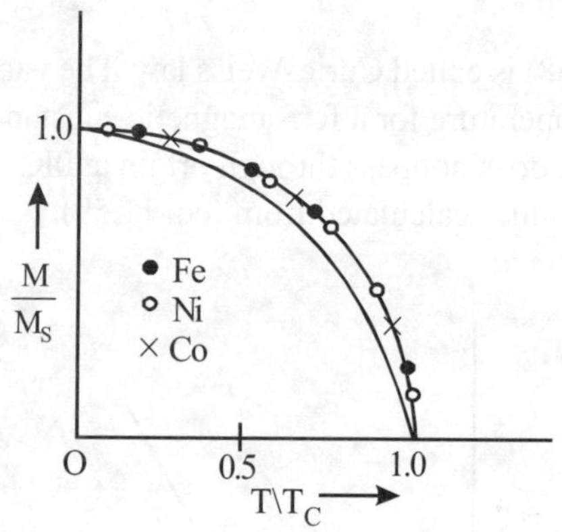

Fig. 11.14 (M / M_s) as a function of (T / T_c) for Fe, Ni and Co. All the experimental points lie on the curve for $J = 1/2$.

11.6.4 Curie-Weiss Law

Let us now consider the magnetization above T_c. As the material will be in the paramagnetic region, it is possible to notice the magnetization only in the presence of the external magnetic field of induction B. Let us assume that the susceptibility associated with it be χ_p. Then

$$\chi_p = \frac{M/H}{1+\gamma M/H} = \frac{\chi_m}{1+\gamma \chi_m} \qquad .....(11.57)$$

where χ_m is the susceptibility of the magnetic material.

$$\therefore \quad (1 + \gamma \chi_m) \chi_p = \chi_m$$

$$\chi_m (1 - \chi_p \gamma) = \chi_p$$

$$\therefore \quad \chi_m = \frac{\chi_p}{1-\gamma \chi_p}$$

$$= \frac{C/T}{1-\gamma C/T}$$

$$= \frac{C}{T-\gamma C} \qquad .....(11.58)$$

$\gamma C = \theta_c$ is called the Curie temperature of the ferromagnetic substance.

At $T = T_c$, the χ_m will have a singularity.

Below $T = T_c$, the ferromagnetic substance is spontaneously magnetised.

Ferromagnetism is basically due to spins of the electrons. The orbital contribution is not present.

The eqn. (11.58) is written as

$$\chi_m = \frac{C}{T-\theta} \qquad \text{.....(11.59)}$$

where $\theta = \gamma C$

The relation (11.59 or 11.58) is called Curie-Weiss law. The variation of susceptebility in the paramagnetic region with temperature for a ferromagnetic substance is given in Fig. 11.15. The variation is linear, but the line does not pass through origin at 0K. There are deviations between experimental values and the values calculated from (eq. 11.59).

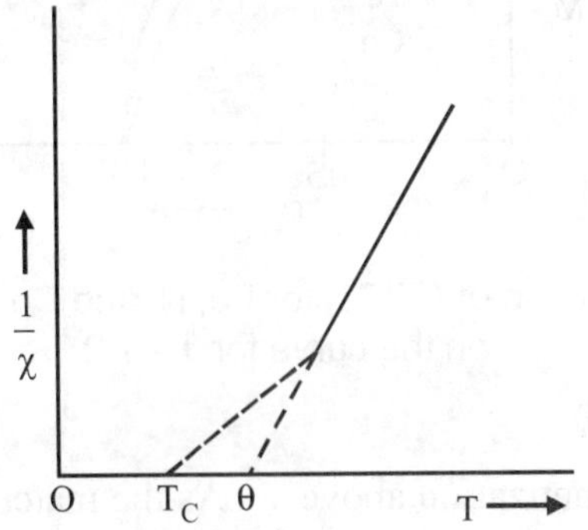

Fig. 11.15 Behavior of the ferromagnetics above the Curie temperature.

The Table 11.3 gives the characteristics of the three basic ferromagnetic substances, namely, iron, cobalt and nickel.

Table 11.3

Material	T_CK	θ K	g	Magnetic moment at T = 0K per atom per unit volume	M_g Amp m^{-1}
Iron (Fe)	1043	1093	1.33	$2.22\mu_B$	1.76×10^6
Cobalt (Co)	1388	1428	1.21	$1.72\mu_B$	1.45×10^6
Nickel	627.2	650	1.35	$0.54\mu_B$	0.5×10^6

11.6.5 Exchange Energy

They are two out comes of the above molecular field theory. Firstly, the molecular field H_m is proportional to the existing magnetization of the sample of the magnetic material. This implies that the magnetization phenomena involved is a co-operative one. Secondly, the magnetization is very strong. But it could not explain the physical basis for the origin of the molecular field. The magnetic properties of ferromagnetism are due to the magnetism arising out of spin interactions of electrons. The orbital motion does not contribute to ferromagnetism. The electrons which are the sources of ferromagnetism are the 3d electrons in the transition elements Fe, Co and Ni. Only those elements containing incomplete 3d shells and rare earths with 4f shell give rise to ferromagnetism. The spin moments for iron group elements are given in Table 11.4

Table 11.4

Element	Sc	Ti	V	Cr	Mn	Fe	Co	Ni
Result spin	1	2	3	5	5	4	3	2
Spin arrangement	↓	↓↓	↓↓↓	↓↓↓↓↓	↓↓↓↓↓	↓↓↓↓	↓↓↓	↓↓

On the basis of Hund's rule, we find 4, 3, 2 unpaired 3d electrons in Fe, Co and Ni respectively. We find that the magnetic moments of the atoms of Fe, Co and Ni respectively should be $4\mu_B$, $3\mu_B$ and $2\mu_B$. But experimentally it was found that these atoms have magnetic moments 2.22 i.e. non-integral.

Heisenberg explained the large value of the molecular field on the basis of exchange interaction. This interaction is purely of quantum nature and arises as a consequence of Pauli's exclusion principle. There exists an exchange interaction between the electrons within the incomplete bands. This interaction depends on the relative orientation of spins and not on the spin magnetic moments. If they are two atoms i, j possessing electrons with the resultant Spins S_i, S_j, then the energy of the interaction is defined as

$$U_{ex} = -2J_c(\mathbf{S_i} \cdot \mathbf{S_j}) = -2J_e S_i S_j \cos\phi$$

where J_e is called the exchange integral. ϕ is the angle between the spins. J_e can be positive or negative, depending on the ratio between the dimensions of the d-shells taking part in the formation of exchange coupling and also on the distance between the atoms in the crystal. For iron group atoms, the exchange integral is positive. For others it is negative. This is shown in Fig. 11.16. Mn and Cr show a negative integral.

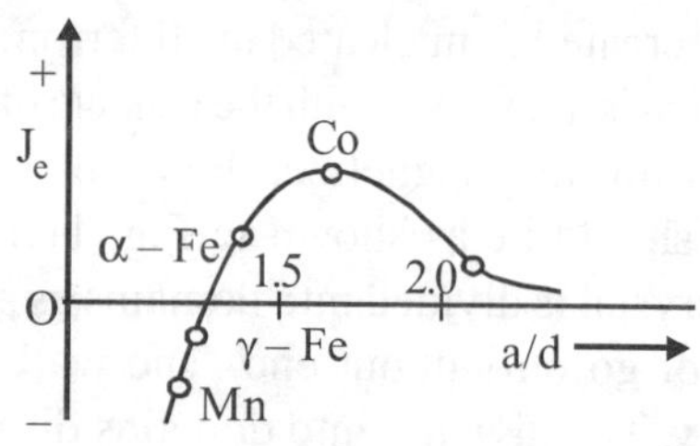

Fig. 11.16 Exchange integral V_s ratio between lattice parameter (a) and diameter (d) of the incomplete inner shell.

The sign of exchange integral indicates which orientation of the spins participating in the formation of exchange coupling is advantageous, namely, parallel or antiparallel. When S_i, S_j are parallel, the energy is negative i.e., S_i ↓↓ S_j as in the case of Fe, Co, and Ni. This gives rise to spontaneous magnetization in the crystals as shown in Fig. 11.17. All the magnetic moments are aligned parallel to each other.

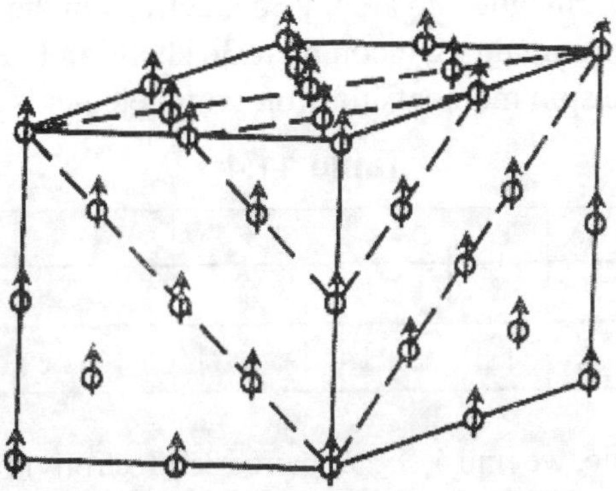

Fig. 11.17 Parallel alignment of magnetic moments in a ferro magnet.

11.6.6 Domains and Magnetization

Origin of Domains :

It has been stated earlier that a ferromagnetic crystal contains magnetic domains. The domains are the spontaneously magnetized regions of the crystal separated by walls called domain walls. In an ordinary iron crystal, the magnetization vectors of the domains are arranged in such a way that their vector sum is zero, so that there are closed magnetic flux paths within the crystal and no net observable magnetization is present in sample. The main domain walls are separated by 180^0 domain walls as shown in Fig. 11.19, where as closure domains (at the surface) have 90^0 walls. The presence of domains and the behaviour of domain walls when subjected to applied field is a fundamentally important feature in understanding the fundamental properties of ferromagnetic materials.

Domain model:

Domains are spontaneously created or nucleated in all ferromagnetic crystals in order to reduce the magnetostatic energy which is associated with the leakage of magnetic flux into the surrounding space. When we take a uniformly magnetized block of a ferromagnet with its associated magnetism, the field lines should be as shown in Fig. 11.19(a). The leakage of flux can be reduced appreciably if the crystal is divided into domains as shown in Fig. 11.19(b) (c) (d). The division of domains can not go on without end. The wall between two domains causes an increase in exchange energy. The division into domains discontinues as soon as a decrease in the magnetic energy due to the domain division becomes an offset by an increase in the exchange energy at the boundary between the domains.

11.6.7 Hysteresis Loop

The most important aspect of ferromagnetic substances is the hysteresis behavior. Consider a ferromagnetic ring placed in a coil or torroidal coil as shown in Fig. 11.18(a). The Secondary coil (S) is connected to a fluxmeter (for the measurement of B). The radius of the ring is kept large compared to its thickness.

Let

The total number of turns in the coil = n

The current in the primary coil = i

The mean radius of the torroid = R

The magnetic field intensity H is given as $H = \frac{ni}{2\pi R}$ amp.turn.m^{-1}.

This magnetic field intensity, H will induce a flux density B in the material of the ring and $B = \mu H$ where μ is the permeability of the material. The flux meter in the secondary coil measures the magnetic flux and the flux density is given as $B = \phi_m / A$, where A is the cross sectional area of the ring sample. Initially, the current in the primary coil is increased from zero in steps so that the magnetic field intensity H increases in steps. B is measured at every step. This is done till the flux density reaches a saturation value, the part OS as shown in Fig.11.18(b) of the magnetization curve is traced.

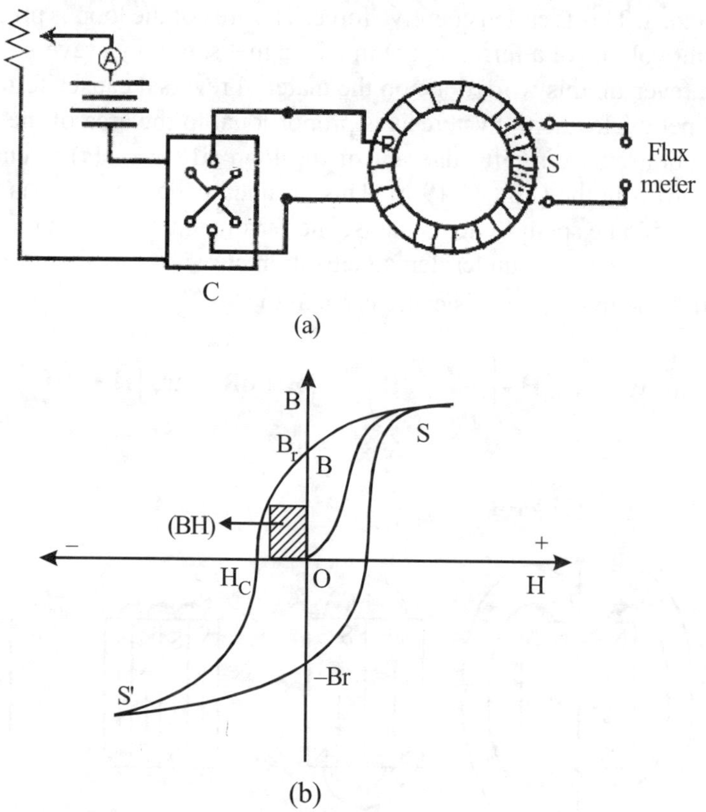

Fig. 11.18 (a) A torroidal coil on a ferromagnetic core. The magnetic flux density is measured with flux meter. (b) Hysteresis loop.

Then, the current in the primary coil is decreased in steps to zero. At H = 0, the flux meter still shows a magnetic flux. Therefore B does not become zero, but reaches a value B_r (part of SB of the curve). B_r is called remanent flux density.

The direction of the current is now changed with the help of a commutator C. Then the current is increased in the reverse direction. At a particular value of H, namely H_c, the flux meter shows a zero flux indicating that the flux density has become zero. H_c is called coercive field. As the current is further increased, the flux meter shows a flux and this continues till the point S' is reached. At this point, the current is decreased till it reaches the zero value. Then, direction of the current is reversed and the current i.e. H increased further till the point S is reached once again. We describe now a loop O S B S' S. This is called the hysteresis loop.

The observation of the hysteresis loop is an indication of the fact that the magnetisation in a ferromagnetic substance is reversible. A ferromagnetic substance is characterized by the saturation magnetisation (M_s), remanent magnetization (M_r), coercivity (H_c). The remanent magnetization is also called retentivity.

If a ferro magnetic substance is magnetized once, one has to apply a demagnetizing field H_c to demagnatize it. This is called coercive force. The area of the loop is proportional to the work done on a unit volume of a ferromagnet in its magnetisation in a reverse sense. In the process of magnetic reversal, this work done on the material reversal causes ferromagnets to heat up. Energy lost per cycle due to hysteresis is proportional to the area of the hysteresis loop. The dismagnetization curve is left quadrant of the loop (B_r to – Hc) is characterised by (BH) product as shown in the (Fig. 11.19 b). This product, called $(BH)_{max}$ is the indicative of the energy that should be spent to demagnetise the ferromagnetic specimen. It is calculated from the largest B – H rectangle under demagnetization curve.

The work done in a hysteresis cycle is equal to

$$W = \int \mu_o \mathbf{H} \bullet \left(\frac{\mathbf{dB}}{\mu_o} - \mathbf{dH} \right) = \int \mathbf{H} \bullet \mathbf{dB} - \mu_o \int \mathbf{H} \bullet \mathbf{dH}$$

$$= \int \mathbf{H} \bullet \mathbf{dB} \quad .....(11.60)$$

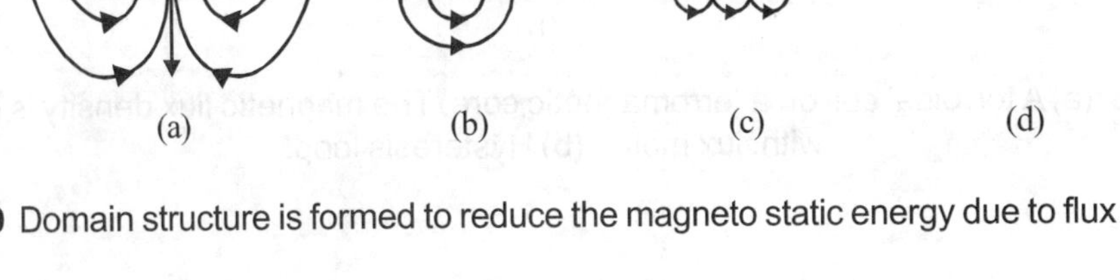

Fig. 11.19 Domain structure is formed to reduce the magneto static energy due to flux leakage.

The domain has a size. The domain size in iron is about 0.1 μm which is less than the size of a grain in polycrystalline iron sample. The transition of spin directions from one domain to another is not, or can not be, abrupt. It is spread over a certain width. Therefore, (shown in Fig. 11.20), domain walls have thicknesses and energy associated with them. This is called wall energy and the boundary between one domain to another is known as Blochwall (δ).

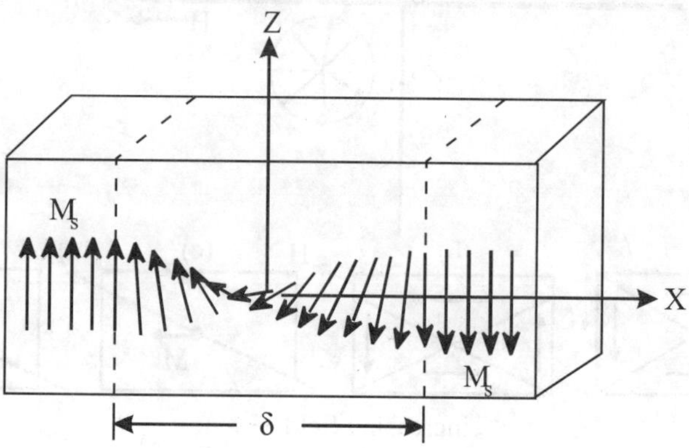

Fig. 11.20 Structure of a domain (180^0) wall, δ is the thickness of the wall.

Domains and Magnetisation :

A quantitative analysis of the magnetization curve can be given in terms of domain movement. Consider a sample with simple domain structure as shown in Fig.11.21(a). There are three stages of magnetization when the sample is placed in a magnetic field. The domain whose magnetization vector is closest to the field direction will grow at the expense of others which are less favorably oriented (Fig. 11.21(b)). Therefore, the process of magnetization is viewed as the one of domain growth and displacement of domain walls which are ultimately pushed out of the material at the magnetization saturation (Fig. 11.21).

Initially when the applied field is weak, the domain wall motion takes place, but it is reversible. That is, when the field is removed, the domain boundaries return to the initial state and the material is demagnatized. At the intermediate field strengths, where the flux density varies more rapidly, the boundary displacements become irreversible. Then, some domains grow at the expense of others (Fig. 11.21(b),(c)). The magnetization remains in the sample even after the removal of the external field. In strong fields, the domains will also be rotated to line up (Fig.11.21(d)) in the external field direction to produce saturation magnetisation. The entire crystal may contain a single domain now. When the external field in reversed, the reorientation does not follow the same path. The process of demagnetization lags behind the field. This results in the hysteresis behavior. There is always an energy loss in this process.

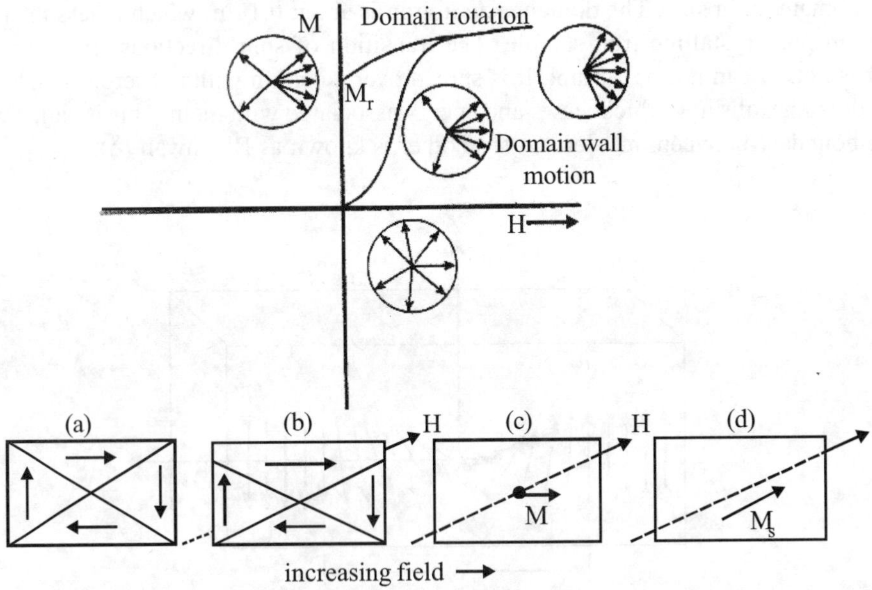

Fig. 11.21 Domain growth in a feromagnetic sample.

11.6.8 Magneto Crystalline Anisotropy

The magnetic substances are crystalline in nature. The observed magnetization in a ferromagnetic crystal depends on the magnitude of the field and also on the crystallographic direction along which the magnetic field is applied. The dependence of magnetic properties on the crystallographic direction is known as magneto crystalline anisotropy. In a crystal cell, the number of atoms along different directions and on different planes will be different. Iron has a cubic structure. The important directions are <111>, <110>, and <100>. Iron goes to saturation easily if the field is in the direction <100>. The crystals in such directions show saturation at a low field value. This is shown in Fig. 11.22 for the crystals of iron, and nickel.

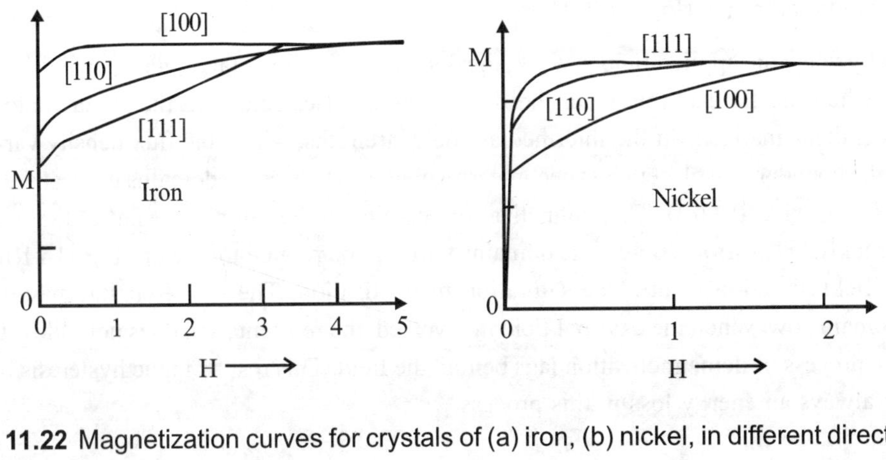

Fig. 11.22 Magnetization curves for crystals of (a) iron, (b) nickel, in different directions.

The magnetization along the easy direction produces a hysteresis loop with very low coercivity and the remanent magnetization, more or less the same as the saturation magnetization. The other directions for magnetization are called hard axes. The crystal magnetized in hard direction possesses the highest magnetic energy.

11.6.9 Magnetostriction

The magnetic substances like ferromagnetic or ferrimagnetic materials undergo a change in dimensions when their state of magnetization is changed by an applied magnetic field. The magnetostriction arises due to the magnetization change due to rotation of the magnetization vector. It refers to a fractional change in a length of a specimen in the presence of the external magnetic field.

The magnetostriction λ, is defined as $\lambda = \Delta l / l$, where l is the original length of the specimen. Δl is the change in length. The effect is strongly dependent on the field strength. λ ranges from 10^{-6} to 10^{-4} for many substances. Some alloys of ferromagnetic metals, ferrites have high values of λ.

The magnetostrictive materials can be used as magneto mechanical transducers for vibrators, ultrasonic machines, sonar, mechanical vibration filters, sound generators etc. They require high saturation magnetostriction (i.e. $\lambda_s > 20 \times 10^{-6}$), high resistivity to reduce eddy currents, high magneto mechanical coupling factors, temperature insensitivity.

11.6.10 Applications of Ferromagnetic Materials

Ferromagnetic materials are useful in very large number of applications. There are many intrinsic properties for these materials. We have discussed most of them. Let us just list out those properties. They are (1) Saturation magnetization M_s. (2) Anisotropy coefficient K_1. K_1 gives the energy required to rotate the magnetization vector 90^0 away from the easy direction. (3) Anisotropy field, it is the field required to rotate the magnetization vector 90^0 away from easy direction. (4) The thermal energy destroys the ferromagnetic alignment in a ferromagnet. The exchange energy parameter is a measure of the strength of coupling between aligned magnetic moment vectors. (5) Domain wall energy (6) Domain wall thickness (7) Permeability. The flux density B is entirely due to the magnetization of the material. The permeability calculated from B-H curve at the point $H \to 0$, is called the initial permeability. It is taken as the standard value for the material. (8) Curie temperature, finally (9) maximum energy product $(BH)_{max}$. This product is given by the maximum area that can be inscribed in the demagnetization quadrant of the B-H loop. The available magnetic energy stored by the ferromagnet is half the $(BH)_{max}$ product.

Ferromagnetic materials are highly useful in a large number of applications. We shall mention some of them in the following.

Permanent magnetic materials (Hard Magnetic Materials) :

The permanent magnets are made from hard magnetic materials. The characteristics of these hard magnetic substances: They have **1.** high saturation magnetization (M_s), **2.** high remanence (M_r), **3.** very high uniaxial magneto crystalline anisotropy energy K_1, **4.** high coercivity. **5.** high maximum energy product $(BH)_{max}$. **6.** High transition temperature (T_c), **7.** good thermal stability, **8.** high mechanical strength, **9.** high machinibility.

Alnico which is an alloy of aluminium, nickel, cobalt and iron. (Fe-Co-Ni-Al) is the best material for permanent magnet applications. The magnetic properties depend on the composition. These days rare earth magnets have become important because large fields can be produced in a small space. Sm Co_5 is a combination of cobalt with samarium. Other impurities are added to improve the properties. Neodymium- iron-boron (Nd - FE -B) magnetic materials are also useful. The Table 11.5 gives the properties of some permanent magnetic materials.

Table 11.5

	B_r (Tesla)	Hc (KAm^{-1})	(BH)max (KJm^{-3})
Alnico	1.30	52	44
Sm Co_5	0.92	690	200
Nd-Fe-B	1.35	780	320
Sm $(Co\text{-}Fe\text{-}Cu\text{-}Zn)_{7.4}$	1.12	560	240

The pernanent magnets are used to produce attractive or repulsive forces without contact, act through vacuum or non ferromagnetic medium. They do not require any power. An ideal permanent magnet is one which exhibits rectangular hysteresis loop with $M_r = M_s$.

Alnico (Ni 18%, Al 10%, Co 12%, Cu 6%, Fe 54%) is the best permanent magnetic material.

High Permeability materials (Soft Magnetic Materials) :

Very high permeability ferromagnetic materials are needed for transformer cores, and inductor cores. The permeability values needed are from a few hunderds to a few thousands.

The characteristics of these high permeability materials are **1.** soft, **2.** high saturation magnetization, **3.** high remanence, **4.** very low coercivmity i.e. domain walls are easily moved or displaced and hysteresis losses are low, **5.** very low magneto Crystalline anisotropy energy values, **6.** high Curie temperatures, **7.** good temperature stability, **8.** good mechanical strength and non-corrosive nature, **9.** high thermal conductivity.

The most useful materials are permendur (alloy of 48% Fe, 50% Co, 2% V, 4% Si-Fe). The alloy is basically an alloy of silicon, iron and cobalt. Alloy with 96% Fe- 4% Si, Iron-nickel alloys of compositions 68% Ni-Fe, 78% Ni-Fe are useful. They are called permalloys. Table 11.6 gives characteristics of some high permeability alloys.

Table 11.6 Magnetic properties of soft high permeability materials.

Materials	Composition (wt%)	μ_i	μ_m	HC(Am^{-1})	B_s(T)
Iron (soft)	99% Fe	200	6000	70	2.16
Si-Fe	96% Fe - 4%Si	500	7000	40	1.95
	97% Fe - 3%Si	9000	40000	12	2.01
Mild Steel	Fe - 01% C - 0.1% Si - 0.4%Mn	800	1100	200	-
78 Permalloy	78% Ni - 22% Fe	4000	100000	4	1.05
Supermalloy	79%Ni - 16%Fe - 5Mo	10000	100000	0.15	0.79
Permendur	50Fe - 50Co	500	6000	160	2.46
7 Perminvas	70Ni - 23Fe - 7Co	850	4000	50	1.25
Aluminium-Iron	90.5Fe-3.5A*l*	500	19000	24	1.90

The high permeability materials are useful as relay switches, magnetic shunts, electromagnets, and indicating meters. The materials with high permeability, very low or zero remanence, good wear resistance, good corrosion resistance, are useful for the magnetic recording heads. All the ferromagnetic materials suffer from eddy current losses. Therefore, they are not useful for high frequency applications. The core materials in the transformers require high permeability and low coercivity.

11.7 Antiferromagnetism

It has been stated earlier that, if the spin moments of neighbouring atoms are oriented in an antiparallel manner, then the antiferromagnetism arises. This suggests that the exchange is of negative sign. The magnetic susceptibility of a magnetic material is given by Curie-Weiss law at high temperatures. If θ is negative for any material, the material is antiferromagnetic in nature. Landau in 1933 proposed the presence of antiferromagnetism in some substances. Some of the substances which exhibit antiferromagnetism are given in Table 11.7.

Table 11.7 Some antiferromagnetic materials or crystals.

Material	T_NK	θK
NiO	525	2000
CoO	291	2000
FeO	198	330
Fe Cl_2	24	570
MnO	116	48
MnF2	67	610
Cr(bcc)	3.8	82
MnS	116	610

Some magnetic materials stay in antiferromagnetic state below a certain temperature, called Neel temperature T_N, named in honor of the French Physicist, Neel who was awarded noble prize for his work on antiferromagnetism.

The antiferromagnetic ordering in solids is pictured as composed of two ferromagnetic sub lattices embedded in each other, interacting via the negative exchange interaction. The Fig. 11.23 depicts the simplest antiferromagnetic structure. What is shown in Fig. 11.23 is a bcc crystal structure with A atoms occupying the corner of the elementary cube and B atoms occupying the centres of the cube. The structure does not exist exactly, but it is a representative model of an antiferromagnetic.

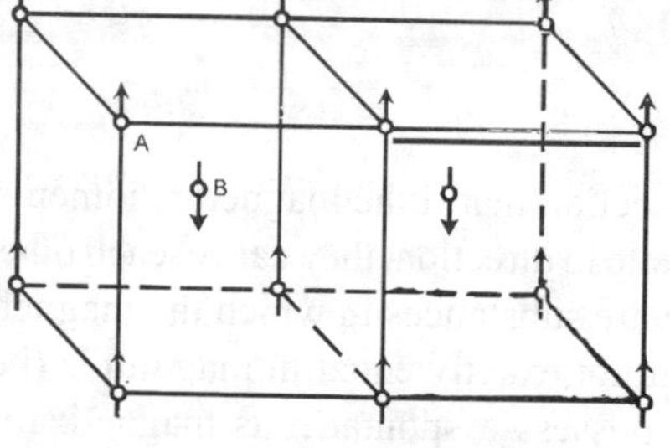

Fig. 11.23 Antiferromagnetic ordering of magnetic moments in a body centred cubic lattice.

An antiferromagnetic unit cell consists of two sublattices A and B that have anti parallel magnetic moments. When there is no external magnetic field on them, the magnetization of the unit cell is zero because the magnetic moments exactly cancel when summed over the crystal cell. When the sample is placed in an external magnetic field, small magnetization takes place. Starting from T = 0K hypothetically, the susceptibility increases as the temperature is increased. When the temperature is equal to T_N, maximum susceptibility is obtained. Above T = 0K, the antiparallel arrangement of spins is gradually upset and the magnetization increases. Once the magnetization reaches a peak at T_N (Neel point) the spin ordering is lost completely and the antiferromagnetic material becomes paramagnetic. The antiferromagnetism susceptibility can be obtained theoretically from molecular field theory. But we shall not do it here as it is outside the scope of the book. An antiferromagnetic substance exhibits a magnetic susceptibility such that the plot of $1/\chi$ vs Temperature has an extrapolated intercept $-\theta$ on T-axis (shown in Fig. 11.24). The Fig. 11.24 shows $1/\chi$ vs T plots for ferromagnetic, antiferromagnetic and paramagnetic materials for comparison.

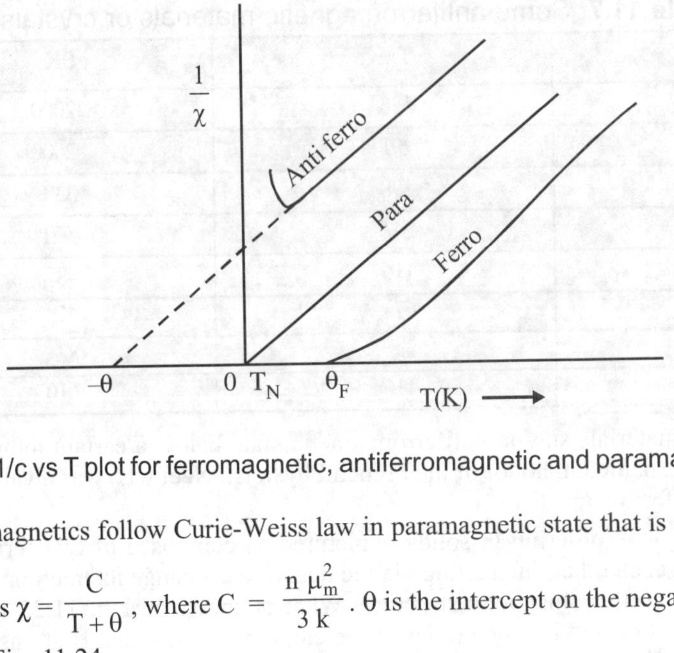

Fig. 11.24 1/c vs T plot for ferromagnetic, antiferromagnetic and paramagnetic materials.

Anti ferromagnetics follow Curie-Weiss law in paramagnetic state that is above T_N. The Curie Law is given as $\chi = \dfrac{C}{T+\theta}$, where $C = \dfrac{n\,\mu_m^2}{3\,k}$. θ is the intercept on the negative side of T axis as shown in the Fig. 11.24.

11.8 Ferrimagnetism

11.8.1 Ferrites

We have seen in the earlier section that if the magnetic moments of sublattices in a crystal cell are equal in magnitude but opposite in direction, they cancel each other giving rise to antiferromagnetism in the crystal cell. But there are substances in which the magnetic moments of the two sublattices are opposite in direction but not exactly equal in magnitude (because of the two types of ions in the lattices). Such crystals possess a spontaneous magnetization and most of the properties of ferromagnets. This uncompensated antiferromagnetism is known as ferrimagnetism. The best example that is known to exhibit ferrimagnetism is the magnetite or lode stone (Fe_3O_4). There are several compounds, all of them mixed oxides of metals and transition metals exhibit ferrimagnetism. The chemical formula of ferrimagnets is MeO, Fe_2O_3, where, Me is a divalent ion like Ba, Sr, Zn, Cd, Mg, Ni, CO, or Cu or Zn, Mg or Cd. A large family of ferrimagnets can be synthesised and used. The ferrimagnets are known as ferrites. The name of the ferrite is usually ascribed to the name of the divalent ion in the lattice.

11.8.2 Structure of Ferrites (for advanced level only)

X-ray diffraction reveals that ferrites have a structure that may be described as a face centered cubic structure of oxygen ions closely packed together with much smaller divalent or trivalent ions placed in the interstitial sites. The basic structure is spinel structure as shown in Fig. 11.25(a), the structure of the mineral spinel ($MgAl_2O_4$).

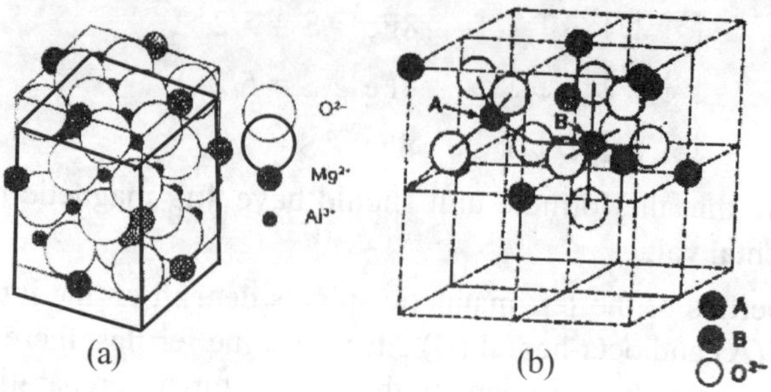

Fig. 11.25 (a) Spinal structure (b) Inverse spinel structure.

The unit cell contains relatively large oxygen atoms forming a f.c.c. lattice in which they are 32 octahedral sites (called B sites) surrounded by six oxygen ions, and 64 tetrahedral sites (A-sites) surrounded by 4 oxygen ions. The unit cell contains 32 oxygen ions. Out of the large number of A and B sites,only 24 interstitial sites are actually occupied by twenty four cations.

Out of these 24 sites, eight are tetrahedral sites, occupied by divalent cations and sixteen are octahedral sites (θ sites) occupied by trivalent Fe^{3+} ions. The ferrites having spinel structure of $MgAl_2O_4$ are known as normal spinels. In some cases, the B-sites are occupied by both divant and ferric ions as $8Me^{2+} + 8Fe^{3+}$, and A sites by Fe^{3+} ions. Such structures are known as inverse spinels (Fig. 11.25(b)). The following Table gives the details of the arrangements.

A sites (8)	B sites (16)	Symbol	Examples	
Normal Spinel	$8Me^{2+}$	$16Fe^{3+}$	$Me_2+(Fe^{3+}{}_2)O_4$	Zn Fe_2O_4, Mn Fe_2O_4, Cd Fe_2O_4
Inverse spinel	$8Fe^{3+}$	$8Me^{2+} + 8Fe^{3+}$		Co Fe_2O_4, Fe Fe_2O_4

The basic characteristics of the ferrite depend on the distribution of the Fe^{3+}, and Me^{2+} ions in A and B sites.

The magnetic super exchange interactions between the octahedral B sites and tetrahedral A sites through the intervening oxygen anions can make the two sub-lattices to be antiferromagnetically ordered. Depending on the type and the preference of cations, a large number of magnetic structures are possible. There exist a very large number of ferrites and mixed ferrites synthesised and used for a number of applications. The arrangement of cations in different sites is important. Let us take a specific example, Fe_3O_4, the magnetite. It can be written as FeO. Fe_2O_3. The Fe^{3+} ion has five 3d electrons. The spin number is 5/2 and the magnetic moment should be $5\mu_B$. There are two such ions in the formula unit. Therefore, there should be a magnetic moment of $10\mu_B$. The Fe^{2+} ion has six 3d electrons. The spin number is four and there it should have a magnetic moment of $4\mu_B$. Hence, the total magnetic moment associated with one formula unit should be $14\mu_B$. But experimentally found value is only $4.1\mu_B$. This discrepancy is explained on the basis of antiparallel arrangement of the iron ions. The magnetic structure is described as $[Fe^{3+}]\ [Fe^{2+}Fe^{3+}]\ O_4$. The sturcture is an inverse spinel. The spin arrangements are given as

Tertrahedral sites A	↑↑↑↑↑↑↑↑	$8Fe^{3+}$, S = 5/2
Octahedral sites B	↓↓↓↓↓↓↓↓	$8Fe^{3+}$, S = 5/2
	↓↓↓↓↓↓↓↓	$8Fe^{2+}$, S = 2

The structure shows that the formula unit should have $4\mu_B$ magnetic moment which agrees well with the experimental value.

The magnetic properties of the ferrimagnetic spinels depend on the interactions between the ions in the tetrahedral (A) and octa-hedral (B) sites. In some ferrites, there is a site preference of particular ions, which is strongly dependent on the temperature of preparation, the rate of cooling and heat treatment. The cation occupation is given below for some important ferrites.

Ferrite	Cation distribution	Spins	Sites
$CoFe_2O_4$ (I S)	$[\overrightarrow{Fe^{3+}}]\ [\overleftarrow{Co^{2+}}\ \overleftarrow{Fe^{3+}}]$	↓↓↓↓↓↓↓↓ ↑↑↑↑↑↑↑↑ ↑↑↑↑↑↑↑↑	A $8Fe^{3+}$ each $5\mu_B$ B $8Fe^{3+}$ $8Co^{2+}$ $2\mu_B$
$NiFe_2O_4$ (I S)	$[\overrightarrow{Fe^{3+}}]\ [\overleftarrow{Ni^{2+}}\ \overleftarrow{Fe^{3+}}]O_4$	↓↓↓↓↓↓↓↓ ↑↑↑↑↑↑↑↑ ↑↑↑↑↑↑↑↑	A $8Fe^{3+}$ B $8Fe^{3+}$ $8Me^{2+}$
$MnFe_2O_4$ (N S)	$[\overrightarrow{Mn^{2+}}]\ [\overleftarrow{Fe^{3+}}]\ O_4$	↓↓↓↓↓↓↓↓ ↑↑↑↑↑↑↑↑↑↑↑↑↑↑↑↑	A $8Mn^{2+}$ $5\mu_B$ B 16 Fe^{3+} $5\mu_B$
γ-Fe_2O_3	$[\overrightarrow{Fe^{3+}}]\ [\square_{1/3}\overleftarrow{Fe^{3+}}]$	↓↓↓↓↓↓ ↑↑↑↑↑↑↑↑↑↑↑	A $6Fe^{3+}$ $5\mu_B$ B $10Fe^{3+}$ $5\mu_B$

Let us consider nickel ferrite $NiFe_2O_4$. The magnetic moment of the Ni^{2+} ion is $4\mu_B$. So there should be a magnetic moment of $8 \times 4\mu_B$ per unit cell. The magnetic polarization J_s should be μ_B. $M_s = (32\mu_B)/a^3 = 0.63$ Tesla; where a is the lattice parameter. This value agrees well with the experimentally observed 0.6 T.

11.8.3 Susceptibility of ferrites

The super exchange inter action accounts for both antiferromagnetic or ferrimagnetic ordering in crystals. The Weiss internal field theory can be applied to ferrimagnetic materials. There are two sublattices A and B, therefore two internal Weiss fields, which are proportional to the sublattice magnetizations. Each, sublattice is associated with a magnetization, M_A, M_B.

According to mean field approximation, in the presence of the applied field or in the presence of a magnetic induction B we have

$$\mathbf{M}_A T = C_A(\mathbf{B} - \nu \mathbf{M}_B) \; ; \; \mathbf{M}_B T = C_B(\mathbf{B} - \nu \mathbf{M}_A) \quad .....(11.61)$$

Solving for M_A and M_B and taking $T_C = \nu(C_A C_B)^{1/2}$, we get

$$\chi = \frac{M_A + M_B}{H} = \mu_o \frac{\left[(C_A + C_B)T - 2T_C^2\right]}{\left(T^2 + T_C^2\right)} \quad .....(11.62)$$

This is a complicated equation for χ i.e. the variation of magnetization with temperature is complicated compared to that in ferromagnets. The variation of magnetic polarization J_s is given in Fig. 11.26(a) for few ferrites.

The temperature variation of $1/\chi$ with T is given in Fig. 11.26(b). The plot is not linear exactly. It has a curvature near the transistion.

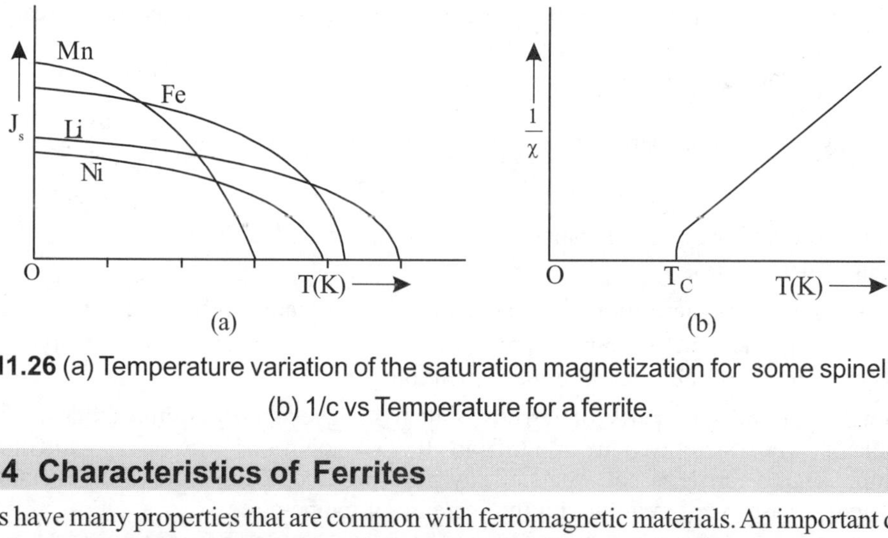

Fig. 11.26 (a) Temperature variation of the saturation magnetization for some spinel ferrites. (b) 1/c vs Temperature for a ferrite.

11.8.4 Characteristics of Ferrites

Ferrites have many properties that are common with ferromagnetic materials. An important difference is that the spontaneous magnetization is the resultant of two oppositely magnetized sublattices, each of which is ferromagnetically ordered. They have become established as one of the most important class of present day materials and are widely used in linear, digital and microwave branches of electrical, electronic and communication engineering. They have a very wide range of properties that are useful in many applications.

1. They exhibit the nonlinear magnetization curve. They exhibit the hysteresis loop. There is a characteristic retentivity and coercivity for each ferrite. The commercial ferrites are in ceramic form i.e. polycrystalline hard samples. The coercivity, retentivity and permeability

depend on the grain size of the polycrystalline samples. Therefore, the preparation of ferrites is a very important aspect in acquiring the required properties. The crystallites are divided into domains separated by domain walls. The number and arrangement of domains in a crystallite is such that the sum of energies, mainly, the wall energy and the demagnetizing field energy, is a minimum. The domain wall movement is responsible for magnetization, as in the case of ferromagnets. Fig. 11.27 shows various types of hysteresis loops observed in Ferrites.

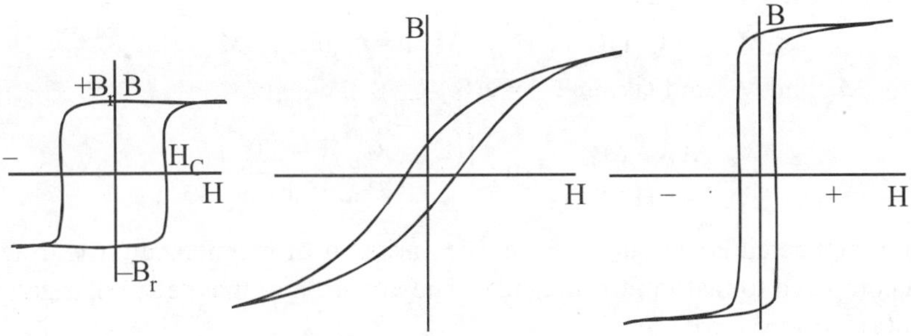

Fig. 11.27 Hysteresis loops of ferrites.

2. Ferrites possess high permeability values normally. The range of permeabilities extends from 15-20 for nickel ferrite to several thousand for some manganese-zinc ferrites. The permeability is defined as $\mu = \mathbf{B}/\mu_0 \mathbf{H}$. As the field strength or flux density increases from zero, the permeability rises until it reaches a maximum value at the knee of the magnetization curve. From there, it falls progressively as the field is increased. At high fields, B-H relation is nonlinear. The initial permeability μ_i under the condition $\mu_i = \underset{H \to 0}{\text{Lt}} \dfrac{1}{\mu_0} \dfrac{dB}{dH}$ is taken as the characteristic of the material. μ_i is a function of temperature. The μ_i increases with increasing temperature and then, at the transition temperature, it suddenly falls to a very low value (paramagnetic state).
3. The saturation magnetization and magneto crystalline anisotropy co-efficients are both strongly dependent on the temperature. The anisotropy energy is proportional to the square or higher power of the saturation magnetization.
4. Another important property of ferrites is their high resistivity. Being oxides, they are insulators. The resistivity values range from $10^2 - 10^{10}$ ohm-m. Ferrites have semiconducting nature. But, a ferrite is not a part of any physical electrical circuit. The resistance of a ferrite decreases very rapidly with the increase of temperature.
5. Ferrites are widely used in many applications such as transformer cores or inductor cores. They are subjected to alternating field bias. The permeability of a ferrite is frequency dependent and is complex. It is given as $\mu = \mu' - j\mu''$, where $j = \sqrt{-1}$. μ' is the real component of the series complex permeability. μ'' is the imaginary component of the series complex permeability. The finite resistivity gives rise to losses. The magnetic field component in the time varying field will induce surface currents in the ferrite. These circulating currents are referred to as eddy currents. They cause unnecessary heat losses. The eddy current loss can be expressed in terms of $\tan \delta = \mu''/\mu'$. Ferrites give rise to very feable eddy currents. In ferromagnetic materials, these eddy current losses are appreciable.

11.8.5 Applications of Ferrites

The practical ferrites are always mixed ferrites. There are three types of mixed ferrites. 1. Mixed normal spinel, $MnFe_2O_4$+ normal spinel $ZnFe_2O_4$. They are called Mn-Zn ferrites. 2. Mixed inverse spinel, $NiFe_2O_4$ and normal spinel ferrite $ZnFe_2O_4$. They are called Ni-Zn ferrites. Several compositions of these ferrites are possible. 3. Mixed inverse spinel ferrites. Inverse spinels form continuous solid solutions. Ni-Co ferrites are examples. The properties can be varied at will and the required material with needed properties can be obtained by proper mixing of the ferrites.

In the Mn-Zn ferrites, the saturation magnetic polarization and coercivities vary from 0.3 to 0.5 T and 4 to 100 Am^{-1} respectively. In the case of Ni-Zn ferrite, the corresponding ranges are 0.1 to 0.4 Tesla and 16-1600 Am^{-1}. The hysteresis loop, the permeability and other parameters are dependent on both intrinsic properties of material, their crystal structure, and composition. But, they are also dependent on the extrinsic parameters like processing parameters, the grain size, density and impurity additions.

Ferrites possess magnetocrystalline anisotropy. They exhibit magnetostriction under the effect of an external magnetic field. Magnetite Fe_3O_4 has a saturation magnetostriction of $+40 \times 10^{-6}$, cobalt ferrite, $Co_{1.1}Fe_{1.9}O_4$, has a saturation magnetostriction as -250×10^{-6}, while simple $CoFe_2O_4$ has magnetostriction of -110×10^{-6}. The soft ferrites possess high permeability and low losses. The characteristics of Mn-Zn and Ni-Zn ferrites are given in Table 11.8.

Table 11.8

Ferrite	μ_i	Saturation magnetization Tesla	Curie temperature T_c	Coercivity Am^{-1}
Mn-Zn	500-20,000	0.3-0.5	100-250 °C	4-100
Ni-Zn	10-2000	0.1-0.36	100-500 °C	16-1600

The mixed ferrites can be made in a variety of grades, each grade having the properties that suit it to particular class of application. The applications were initiated in the field of carrier telephony where the combination of good magnetic properties and high resistivity made these materials very suitable as cores for inductors and transformers. Mn-Zn ferrites are used in television deflection yokes, cores for television line output/eht transformers (permeability should be in between 2000-5000) and standard power supplies. They are useful as induction cores, antennas for medium and longwave broadcasting (permeability should be 500-1000), pulse transformers (permeability should be 3000-20,000). Ferrites are widely used in transductors (variable inductors). They are useful in tuning, automatic frequency control systems, frequency modulation, switching, filters, etc.

Ni-Zn ferrites are also useful in wide band transformers (frequencies 1-300MHz, permeabilities 1000), antennas for medium and long wave broadcasting, power transformers cores (frequencies 0.5-5 MHz; permeabilities 150-500), inductor cores, antennas for short wave broadcasting (frequencies 2-30 MHz, permeability 70-150).

Ferrites like $Mg_{0.85}Mn_{0.45}Fe_{17}O_4$, $Mg_{0.4}Ni_{0.4}Mn_{0.1}Fe_2O_4$, $Cu_{o25}Mn_{0.75}Fe_2O_4$, $Li_{0.47}Ni_{0.006}Fe_{2.47}O_4$, $Co_{0.02}Mn_{0.18}Fe_{0.18}Fe_2O_4$ give rectangular hysteresis loop. They are useful in the construction of computer memory system for rapid storage and retrieval of digital information by switching the direction of magnetization in a very small sized torroidal cores. The squareness of the hysteresis loop is defined by B_r/B_{sat}. It is necessary that the ratio should be close to unity. These ferrites should have very very low coercivity .

Microwave ferrites : Garnets, such as yttrium iron garnet $y_3Fe_5O_{12}$ and few other ferrites like lithium ferrites are useful in microwave applications. Microwave ferrites are those ferrites which are used at frequencies in the range 100 MHz to 500 GHz. The fundamental aspect associated with ferrites is their nonreciprocity. The permeability of the ferrites exhibits anisotropy or the magnetic susceptibility is a tensor. Ferrites possess an internal field. When an electromagnetic wave in the frequency range of microwaves is incident on a ferrite, it is spontaneously magnetized. The ferrite is initially kept in a magnetic field. The aligned magnetic moment vectors of spins precesses about the magnetic field. If the electromagnetic wave that is incident is a plane polarized wave, the plane of polarization of the wave is rotated by an angle θ. When this wave is made to traverse back in the ferrite, the wave suffers another rotation by an angle θ. Therefore, the ferrite is a nonreciprocal device. This property is made use in a variety of devices like isolators, circulators, switches, etc. An isolator is a device in which the incident electromagnetic wave can propagate forward, while there is no reversewave. A circulator is a device that has n ports which are inputs and output terminals. A wave incident at some port like port 1 emerges at port 2. A wave incident at port 2 may not go into port 1, but may emerge from port 3. Thus the circulator provides easy isolation between different channels in a microwave system such as a radar, or transmission linkage. The most important ferrites that are useful are Mg-Mn-Al, Ni-Zn-Al, etc.

Garnets : Garnets are ferri magnetic substances. Garnets have a magnetic structure $C\left[\overleftarrow{Me^{3+}}\right] B\left[\overleftarrow{Fe_2^{3+}}\right] A\left[\overrightarrow{Fe_3^{3+}}\right] O_{12}$ the net magnetic moment per formula unit is $\mu_m = 2 \times 5\mu_m \times 3m_c - 3 \times 5\mu_B$, where m_c is the magnetic moment of Me^{3+} ion (rare earth). The garnets have the general formula $Me_3Fe_2Fe_3O_{12}$. The rare earth ion selected is yttrium. A thin film of ferrimagnetic garnet give rise to bubble domains. The bubble domains are the cylindrical domains of reverse magnetization bounded by a vertical domain wall and surrounded by regions of magnetization and sustained by a bias magnetic field.

The most important property associated with the bubble is the flux gradient across the bubble. If the bubble has to move, there should be a drive field and this drive field should be greater than the coercivity of the bubble wall. The bubble should have very low wall coercivity. The reverse domains (Fig. 11.28) are mutually repulsive and can be moved through the material by very small fields ($< 100 Am^{-1}$).

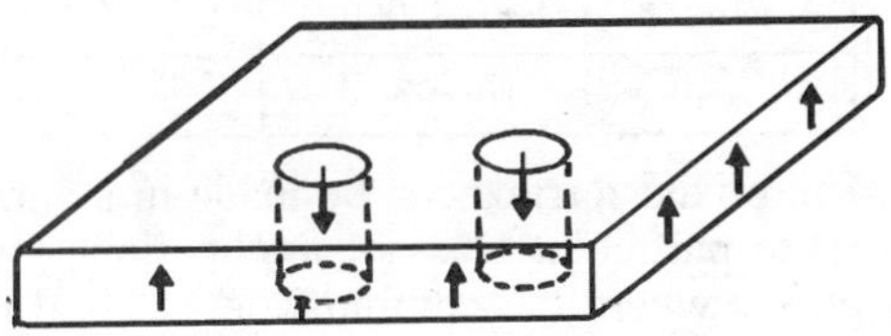

Fig. 11.28 Magnetic Bubbles or Bubble domains in a thin plate of a granite.

The absence of reverse domain may be referred to as zero. The presence of reverse domain may be taken as the stored binary digit 1. Thus a single-crystal platelet of any garnet could be used as the fundamental element of a computer store. Since the reverse domains can be made very small ($5\mu_m$ diameter), one square millimetre of garnet, for example, could store 100000 bits of information.

Magnetic Recording : Magnetic recording of audio, visual and digital information is now one of the most important applications of magnetic materials. The recorded information stores in the form of remanent magnetization of the material such that the magnitude of remanence is proportional to the amplitude of the original electromagnetic input signal. Thus, the material with linear remanent magnetization is needed. As we have to retrieve the recorded information, the remanence should be high to give noise less-play back signal. A material with high coercivity and high magnetization are needed for recording purposes. The amplitude of ac bias is greater than or equal to the coercivity of the tape material. The oxides of γ-Fe_2O_3, Fe_3O_4 and CrO_2 are the most suitable materiats. Most widely used material is cobalt modified γ-Fe_2O_3 and γ-Fe_2O_3. CrO_2 is an unusual oxide compound and it is ferromagnetic rather than ferrimagnetic. Wherever high recording densities are required, CrO_2 tapes are preferred. High quality tapes of audio, video, data tapes and disks are available commercially.

Maxwell's Equations

Introduction

The phenomenon of electromagnetism has its foundation on four laws namely, Gauss law of electric fields, due to a charge distribution, Gauss' law for magnetic fields, Ampere's law for magnetic field induction due to current in a conductor, and finally the Faradays' law of electro magnetic induction. Maxwell treated them as the foundation piller for the developing electro magnetic induction and subsequently the electro magnetic wave model. Maxwell J. clerk looked into the above laws by treating the electric and magnetic fields as physical entities. These physical entites should exist without any necessarily for underlying a medium. It was proved by Faraday's law that charging magnetic field should induce an electromotive force (i.e. an electric field) in a circuit loop. As nature should follow symmetry, it occurred to Maxwell that a changing electric field should induce a magnetic field flux density. He modified Ampere's law so as to suit the circuit loop with a capacitor that contains an insulator between to plates. He introduced the concept of displacement current in the Ampere's original relation. The mutual induction of electric and magnetic field give rise to the phenomenon of electro magnetic oscillations in free space. The electric and magnetic fields create each other perpetually in space, the fields being initated by oscillating charges, dipoles or current elements. Maxwell's equations are the basic equations just like Newton's laws in mechanics and these equations have ultimately are the basic foundations of electro magnetic theory and communication theory.

Electro magnetic fields: The fundamental laws in electro magnetism.

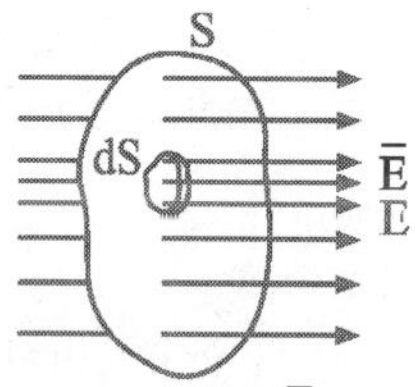

1. **Gauss' Law for the electric field:** The total electric field lines passing through a surface is called the electro flux. We define the flux of the electric field, ϕ_E as

$$\phi_E = \text{Electric field} \times \text{Area} = E \times S \qquad .....(11.63)$$

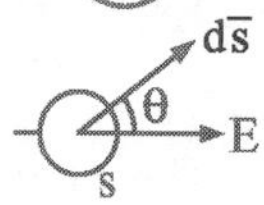

We define a vector dS to be normal to the surface and having a magnitude equal to the area ds. If the angle between **E** and dS is θ, then

$$dS_{11} = dS\cos\theta$$

The flux through surface element dS is

$$d\,\phi_E = E\,dS_{11} = E\,dS\cos\theta = \mathbf{E.dS} \qquad .....(11.64)$$

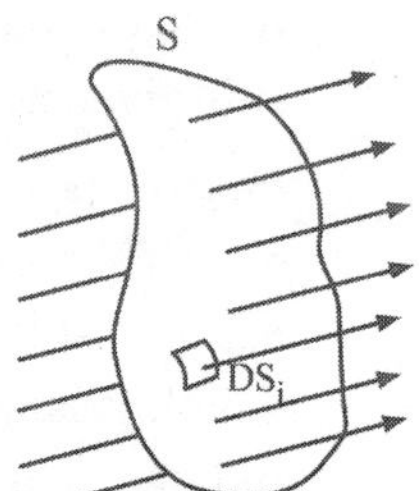

Fig. 11.29

We can consider any surface (Fig. 11.29) ad the electric field lines are as shown in Fig. 11.29. We divide the surface S, into a large number of infitesimelly small elements of areas ds_{11}, then the total flux through the surface is

$$\phi_E = \lim_{\Delta s_i \to 0} \sum \mathbf{E}_i.\mathbf{dS}_i = \int_s \mathbf{E.dS} \qquad .....(11.65)$$

Suppose the surface encloses a volume (like the surface of a baloon), and we have a charge q at the center, then the electric field flux is

$$\phi_E = \mathbf{E.dS} = \frac{q}{4\pi\varepsilon_0}\frac{\hat{\mathbf{r}}.\mathbf{dS}}{r^2} \quad .....(11.66)$$

$\hat{r}.dS$ is the projection of dS on the plane normal to $\hat{r}$.

Then $\mathbf{E} \,.\, \mathbf{dS} = \dfrac{Q}{4\pi\varepsilon_0} d\Omega$

where dw is the solid angle that is made by **dS** at the pint charge at p.

If volume contains large number of charges, q_i, so that $\Sigma q_i = Q$, then the electric field lines may be coming out of the volume or going into the volume through the surface (Fig. 11.30).

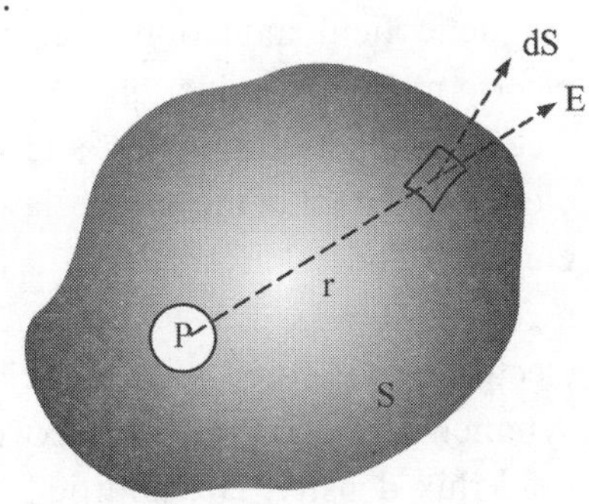

Fig. 11.30

The total flux can be obtained by integrating the expression for E. $d\bar{s}$ over the entire closed surface. Then

$$\phi_E = \int_s \mathbf{E.dS} = \frac{Q}{\varepsilon_o} \quad .....(11.67)$$

So, whenever a charge distribution is present in any region of space, we think of an imaginary closed surface, called a Gaussian surface enclosing the charges. Then Gauss' law states that the total flux through the surface is equal to the net charge, Q enclosed by the surface, given mathematically as

$$\oint \mathbf{E.dS} = \frac{Q}{\varepsilon_o} \quad .....(11.68)$$

The integral is over a closed surface. If the surface does not enclose any charge, then

$$\oint \mathbf{E.dS} = 0$$

The Gauss' law given in eq. 11.68 is the basis for the first Maxwell's equation. The eq. 5 is called the integral form of one of the Maxwell equations that tells about the origin of electric field. We also a define a parameter 'D' called the displacement field. It is given as $\mathbf{D} = \varepsilon_0\mathbf{E}$ and for medium other than free space $\mathbf{D} = \varepsilon\mathbf{E}$

2. **Gauss' law for the magnetic field:** In the case of electric fields, we have isolated electric charges that can generate an electric field. But, in the case of magnetism, there are no isolated magnetic charges. A magnet however you make it infinitesimally small, it is always a dipole. There are no mono poles. A magnetic field is a result of moving charges. If an isolated electric charge is present at some place, then a magnetic field is generated at that charge, only if any other charge moves relativistically with respect to the static charge.

Suppose, we imagine a closed surface encloses an isolated magnetic charge, (imaginary), no field line is set up on the surface. If a dipole is present, then, a closed surface does not have magnetic field lines.

Over any closed surface enclosing a volume, the surface integral of the magnetic field induction B (called magnetic field density) is equal to zero i.e., (**B** is a vector)

$$\oint \mathbf{B}.d\mathbf{S} = 0 \qquad .....(11.69)$$

This is the statement of Gauss' law for magnetic field. It is the integral form of one of the Maxwell equations.

Faraday's Law

Oersted pointed out that the magnetic field is associated with the electric current. In a similar way, Faraday and Henry independently suggested that the changing magnetic field with time should give rise to electric field. Faraday's laws of electromagnetism are: 1. Whenever the magnetic flux linked with a closed circuit changes, an induced emf is setup in the circuit. 2. The magnitude of the emf is proportional to the rate of change of magnetic flux linked with the circuit.

Consider a loop of wire which is closed with a galvanometer as shown in Fig. 11.31. When a magnet is quickly brought towards the coil with its north pole facing the coil, a deflection in the galvanometer is observed. This is due to a momentary current developed in the coil. If the magnet is taken away quickly from the coil, the deflection in the galvanometer will be in the opposite direction.

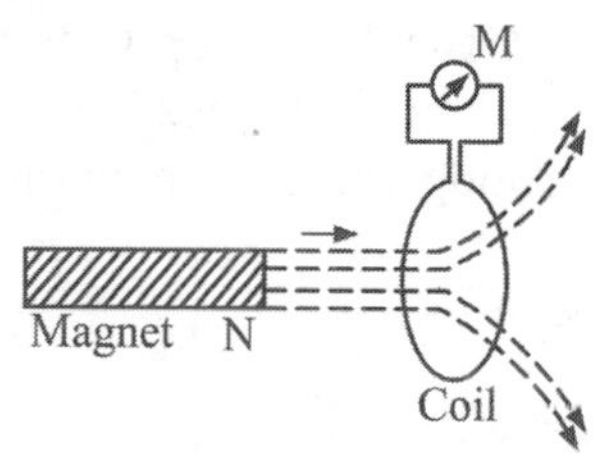

Fig. 11.31

The magnetic flux can be changed through the circuit in a number of ways.

1. By relative motion between a magnet and the circuit Fig. 11.31
2. By changing current in the same circuit
3. By changing current in a neighbouring circuit
4. By rotating the coil in a magnetic field.

Faraday gave the laws for this electromagnetic induction. They are stated as:

1. When a magnetic flux linked with a closed circuit changes, an induced emf is set up in the circuit whose magnitude at any instant is proportional to the rate of change of magnetic flux linked with the circuit. If ξ is the induced emf and ϕ, is the magnetic flux, then

$$\xi \propto \frac{d\phi_B}{dt} \qquad .....(11.70)$$

2. The direction of induced emf is such that it opposes the change in flux that produces it. This law is known as Lentz's law

$$\therefore \qquad \xi = -\frac{d\phi_B}{dt} \qquad .....(11.71)$$

A magnet has a magnetic dipole moment. When the north pole of the magnet is brought towards the coil, the magnetic flux through the coil increases i.e., the flux through the coil is changing with time. It will induce an emf in the coil such that the induced emf will be anticlockwise as seen from the magnet. If a current has to flow in the coil, it will be circulating in such a way as to make the coils north pole face the approaching magnet. That is, there is a force of repulsion between the loop and the magnet. It is left to the student to think and explain the origin of emf because of the other three types of the flux changing mentioned above.

The electric field is irrotational, therefore the electric field can be expressed as the negative gradient of a scalar function ϕ i.e., **E** = – grad ϕ. The electric field can be considered at each point in a medium or space. The electric field is a vector point function or a vector field. Similarly **B** is also a vector field in a medium. An electric field is conservative, hence the work done in moving a test charge around a closed path in the field is zero. This is expressed as $\oint \mathbf{E} \bullet \mathbf{dl} = 0$. But in the case of Faraday effect, the changing magnetic field produces electric field. This electric field is not associated with static charges, but with a changing magnetic field i.e., $\oint \mathbf{E} \bullet \mathbf{dl} \neq 0$.

The electric field produced in Faraday effect is nonconservative. Both conservative and nonconservative fields are identical, but their sources are different. The emf developed in the coil depends on the time rate of change of magnetic induction flux.

Consider a loop C as shown in Fig. 11.32 in time varying magnetic induction fields. Then the induced emf i.e., an induced electric field **E** will set the charge carriers in the loop, there by inducing current.

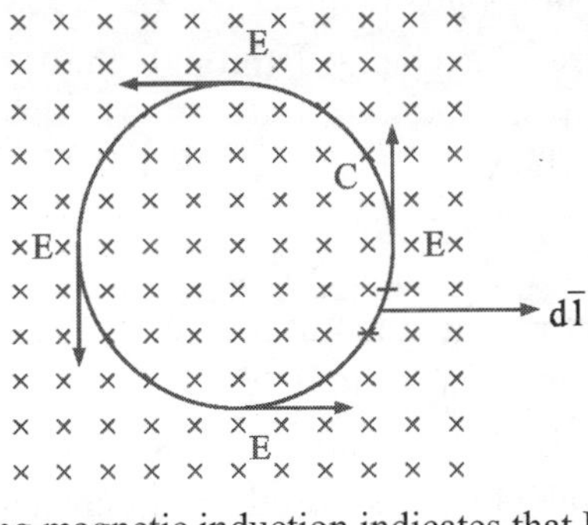

Fig. 11.32 A coil in the changing magnetic induction indicates that **B** is perpendicular to the plane of paper.

The electric field is present at all points in the loop C. This electric field will exert a force **F** = q_o**E** on a test charge q_o tangentially at every point of the loop. The electric field lines form concentric circles. Consider a circle with radius r. Then, the work done in carrying a test charge q_o around the circle is given as

$$W = \oint \mathbf{F} \bullet dl = \oint q_o \mathbf{E} . \bullet dl = q_o E . \bullet 2\pi r$$

$$\therefore \text{ the induced emf } \xi = \bar{\mathbf{E}} . 2\pi \bar{\mathbf{r}} = \oint \mathbf{E} \bullet dl \quad \therefore \quad \xi = \oint \mathbf{E} \bullet dl = -\frac{d\phi_B}{dt} \qquad .....(11.72)$$

Thus Faraday's law establishes a connection between **E** and **B**. The above relation is known as the integral form of Faraday's law. It states that the line integral of E around a closed path is equal to the rate of change of magnetic flux within the boundary of the path. The coil C encloses an area A. If **B** is the induction field at any point P (x, y, z), then the magnetic flux ϕ_B through the area **dS** is $\mathbf{B_n}.\mathbf{dS}$, where $\mathbf{B_n}$ is the normal component of **B**. The total flux in the coil (Fig. 11.33) is

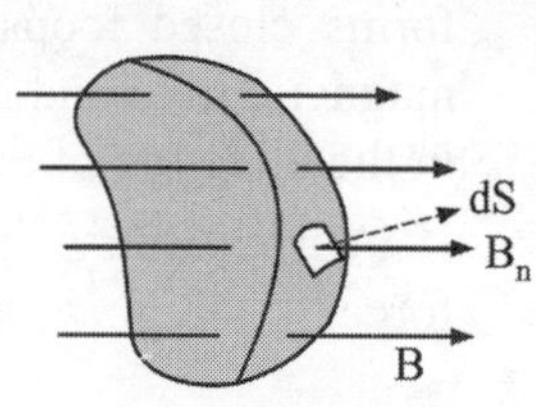

Fig. 11.33

$$\phi_B = \iint_S \mathbf{B}_n \bullet d\mathbf{S} \;\therefore\; \oint \mathbf{E} \bullet d\mathbf{l} = -\frac{d}{dt}\iint_S \mathbf{B}_n \bullet d\mathbf{S} \qquad .....(11.73)$$

The above eq. 11.73, is the integral form of Faraday's law and is the integral form of one of the Maxwells equations that relates varying magnetic field to the electrical field.

3. Magnetic field due to current:

Magnetic field density due to a current: A static charges generate electric field and Gauss law gives a method of calculating electric field for a charge distribution.

A moving charge generates a magnetic field as well as an electric field.

It is known that a moving charge constitutes an electric current. A current element should give rise to a magnetic field. Consider a long thin conductor like a wire carrying a current I. The conductor is surrounded by a medium.

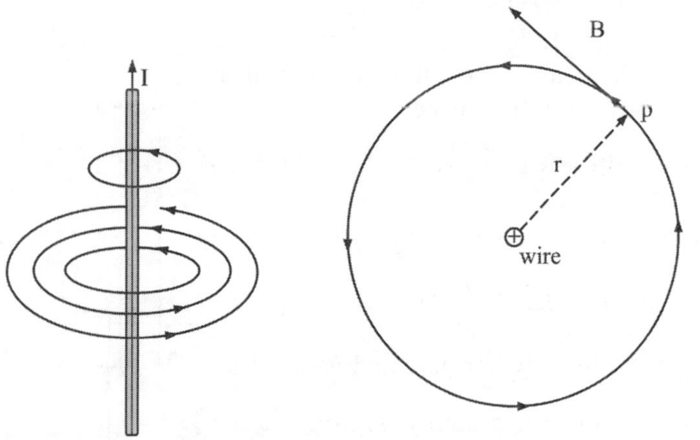

Fig. 11.34 The magnetic field induction around a current carrying wire.

When a compass needle is kept near the wire (Fig. 11.34) the needle gets deflected indicating that the conductor produces a magnetic field around it. When the compass needle is moved around the wire, the locus of all points at which the needle shows same deflection will be a circle. We get different concentric circles, around the wire, which represent the positions of the same deflection. Therefore, it is clear that magnetic field forms closed loops around the conductor. The relation between the direction of the magnetic field and that of current is given by the right hand rule with the thumb pointing in the direction of the current, the fingers circling the conductor or wire.

If a current carrying wire is placed in a uniform magnetic field, then, it is acted upon by a force.

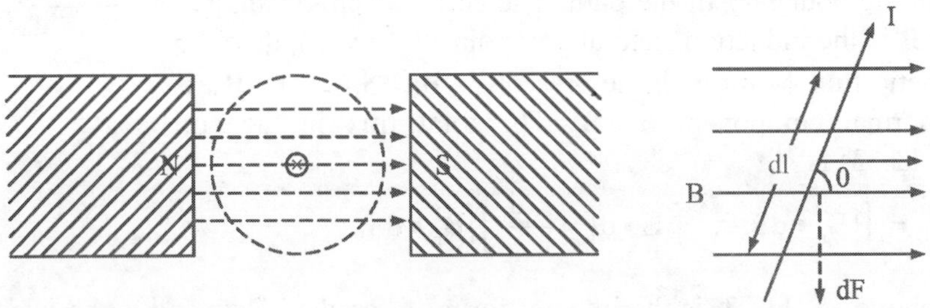

Fig. 11.35(a) & (b) A current carrying conductor is placed in a magnetic field.

Consider a small element of the wire dl. It is in a magnetic field of flux density **B** (Fig. 11.35) and acted upon by a force dF given as

$$\mathbf{dF} = \mathbf{I} \times \mathrm{B}\ \mathbf{d}\boldsymbol{l} \qquad .....(11.74)$$

Both **B** and **I** are vectors, **B** is expressed in Newton amp^{-1} m^{-1}. This unit is called weber m^{-2}. The force on the current element indicates that the magnetic fields are produced by electric currents.

The magnetic flux produced by electric currents at a point near the wire is given by Biot-Savat law.

Once again consider an element **dl** of a wire carrying **I** as indicated in Fig. 11.36. The flux density **dB** contributed by this element at a point P, located at a distance r from the element in free space is

$$\mathbf{dB} = \mathrm{K\,I}\,\frac{d\boldsymbol{l} \times \hat{r}}{r^2} \qquad .....(11.75)$$

The above expression can be written as

$$\mathbf{dB} = \frac{\mu_o}{4\pi}\cdot\frac{\mathrm{I}\ d\boldsymbol{l} \times \mathbf{r}}{r^3} \qquad .....(11.76)$$

When K is a proportionality constant and is equal to $\frac{\mu_o}{4\pi}$. μ_o is referred to the permeability of free space, numerically equal to $4\pi \times 10^{-7} = 1.257 \times 10^{-6}$ Henry m^{-1} (or weber m^{-1} amp^{-1}).

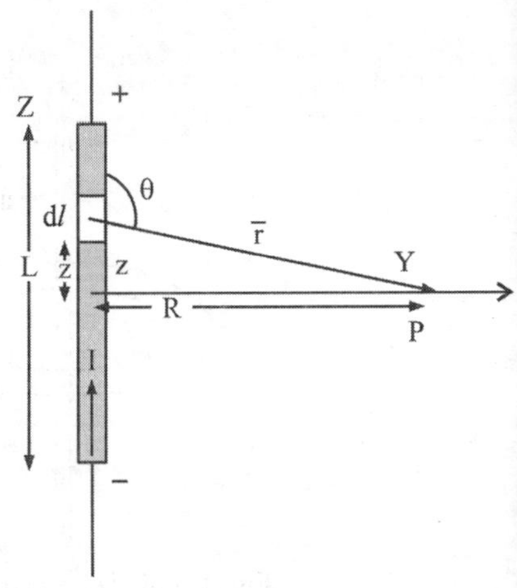

Fig. 11.36

∴ The magnitude of the flux density dB is

$$\mathbf{dB} = \left(\frac{\mu_o}{4\pi}\right)\frac{\mathrm{I}\ |d\boldsymbol{l}|\sin\theta}{r^2} \qquad .....(11.77)$$

The direction of dB is normal to the page, θ is the angle between I and **r**, the radial vector. Suppose now, we consider an infinitely long wire instead of an element dl, then we can consider the wire to be made up of small elements of dl and the magnetic flux density B at the point P due to the wire is (Fig. 11.36).

$d\overline{B}$ is a vector, the direction of the vector is perpendicular to both the current and the radial vector of the point p or the plane consisting of the wire and the radial vector.

If the current is along the Y-direction and the radial vector is along the X-direction, then, the direction of $d\overline{B}$ is along Z-direction. Therefore the express in (4) in written as

$$\mathbf{dB} = \left(\frac{\mu_o}{4\pi}\right)\frac{I\,|dl|\sin\theta}{r^2}\hat{\mathbf{z}} \qquad \text{.....(11.77b)}$$

where $\hat{z}$ is the unit vector along z-axis. θ is the angle between d*l* and the vector δ. Suppose, we take the field due to the entire wire that is straight, the total magnetic field density due to the current is over the entire conductor, therefore, the total magnetic field induction at point P due to the entire wire is obtained by integrating over all current elements $I\,d\overline{l}$:

$$\mathbf{B} = \int d\mathbf{B} = \frac{\mu_o}{4\pi}\int\frac{I\,d\mathbf{l}\times\hat{\mathbf{r}}}{r^2} = \frac{\mu_o}{4\pi}\int\frac{I\,d\mathbf{l}\times\mathbf{r}}{r^3} \qquad \text{.....(11.78\&11.79)}$$

A straight wire Segment

Consider a straight conductor as shown in the Fig. 11.37 placed along the Z-direction. Consider a point P along the y direction.

From the Fig. 11.37 we have

$$r = \sqrt{Z^2 + R^2}$$

$$\sin\theta = \sin(\pi - \theta) = \frac{R}{\sqrt{R^2 + Z^2}}.$$

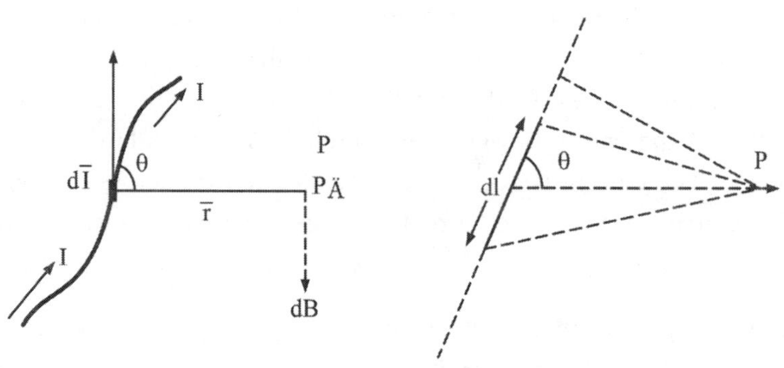

Fig. 11.37

Then, the magnetic field induction at P due for the current element dz is

$$\mathbf{dB} = \frac{\mu_o \mathbf{I}}{4\pi}\frac{\mathbf{dz}\sin\theta}{r^2} = \frac{\mu_o I}{4\pi}\frac{R}{(R^2+Z^2)^{3/2}}\mathbf{dz} \qquad \text{.....(11.80)}$$

The magnitude field induction at P due to the entire wire of very long length L, is obtained by integrating the above expression between the limits $\frac{-L}{2}$ to $\frac{L}{2}$ i.e.,

$$\mathbf{B} = \int d\,\mathbf{B} = \frac{\mu_o IR}{4\pi}\int_{-L/2}^{+L/2}\frac{\mathbf{dz}}{(Z^2+R^2)^{3/2}} = \frac{\mu_o \mathbf{I}}{4\pi R}\frac{L}{(L^2/4+R^2)^{1/2}} \qquad \text{.....(11.81)}$$

If L >> R, or the point p is very close to the wire, then

$$\mathbf{B} = \frac{\mu_o \mathbf{I}}{2\pi R} \qquad \text{.....(11.82)}$$

The flux density at a distance R from a straight conductor is given by the equation (4.82). B is integrated around the path of radius R enclosing the wire once. This can be expressed as

$$\oint \mathbf{B} \bullet dl = \frac{\mu_o \mu_r}{2\pi R}\oint d\mathbf{I} = \frac{\mu_o I}{2\pi R}.2\pi R = \mu_o \mathbf{I}$$

.....(11.83)

where $\mu_o\ \mu_r = \mu$ the permeability of the medium around the wire. If the integration is taken over a singly closed path $\mathbf{H} = \mathbf{B}/\mu$. H is called the magnetic field intensity and

$$\oint \mathbf{H} \bullet dl = I \qquad \text{.....(11.84)}$$

The above relations represent called Ampere's law. The law states that the line integral of H around a single closed path is equal to the current enclosed. The units of H are $\frac{\text{Webers}/m^2}{\text{Weber}/\text{amp.m}} = \text{amp.m}^{-1}$. If a medium is present around the wire, μ_0 is replaced by μ, the permeability of the medium.

The quantity H is also called magnetomotive force (mmf). It is equal to the current enclosed. The sign of the integral depends on the direction in which the current is enclosed. The sign is positive if the path followed by the line integral is parallel to **B** and negative, if the path followed is antiparellel. Ampere's law holds for closed path of any shape. The current is due to the flow of elections across an area per unit time. We define a quantity **'J'** as the current per unit area placed normal to the direction of current. '**J**' is expressed as $I = \iint_S \mathbf{J} \bullet \mathbf{dS}$ where dS is a small element of area of surface. The Ampere's law can be written as

$$\oint \mathbf{B} \bullet dI = \mu_o \iint_S \mathbf{J} \bullet \mathbf{dS} \qquad \text{.....(11.85)}$$

The relations given the most useful forms of Maxwell equations are expressed in differential form. They provide the theoretical understanding of electro magnetism in general. Differential forms of relations always give point to point relations. All the equations were developed to explain the results of laboratory experiments in electro magnetism. Let us obtain the Maxwell equations in differential form.

1. Gauss' law:

$$\oint_S \mathbf{E}.d\mathbf{S} = \frac{Q}{\varepsilon_o}$$

It is always useful to consider a distribution of charges or multiple charges in a region. If the charges are distributed uniformly in a given volume, the charge density 'p' is the charge per unit volume is taken. Then, the expression for electric field at a point ρ from the centre of the volume is

$$\mathbf{E}(\rho) = \frac{1}{4\pi\varepsilon_o} \int_V \frac{\hat{r}}{r^2} \rho \; d\tau$$

l τ is infinitesimally small element of volume, $\hat{r}$ is a unit vector. The charge enclosed in the volume is Q, where

$$Q = \int_V \rho \; d\tau \qquad \therefore \qquad \oint_S \mathbf{E}.d\mathbf{S} = \frac{1}{\varepsilon_o} \int_V \rho \; d\tau$$

The surface integral can be expressed as volume integrated by divergence theorem, as

$$\oint_S \mathbf{E}.d\mathbf{S} = \int_V (\nabla.\mathbf{E}) \; d\tau$$

Where ∇ . **E** is the divergence of the electric field given as

$$\nabla.\mathbf{E} = \left(\frac{\partial E_x}{\partial x}\right)i + \left(\frac{\partial E_y}{\partial y}\right)j + \left(\frac{\partial E_z}{\partial z}\right)k \text{ and } \quad \nabla.\mathbf{E} = \frac{Q}{\varepsilon_0}$$

is the Maxwell equation in differential form. The Gauss' law for magnetic fields is

$$\nabla.\mathbf{B} = 0$$

Let us look into this relation from the point of a steady current. I is a current loop. Then, the magnetic flux density, B is given as

$$\mathbf{B} = \oint \frac{\mu_o}{4\pi} I \frac{d\boldsymbol{l} \times \mathbf{r}}{|r|^3}$$

Then, taking divergence on both sides, we have

$$\nabla \cdot \mathbf{B} = \nabla \cdot \oint \frac{\mu_o I}{4\pi} \frac{d\boldsymbol{l} \times \mathbf{r}}{|r|^3} = \frac{\mu_o I}{4\pi} \oint \nabla \cdot \left(\frac{d\boldsymbol{l} \times \mathbf{r}}{|r|^3}\right)$$

But the divergence of the product of two vectors **A** and **B** is

$$\nabla \cdot (\mathbf{A} \times \mathbf{B}) = \mathbf{B} \cdot \text{ and } \mathbf{A} - \mathbf{A} \cdot \text{ curl } \mathbf{B}$$

$$\therefore \qquad \nabla \cdot \mathbf{B} = \frac{\mu_o}{4\pi} I \oint \left[\frac{\mathbf{r}}{|\mathbf{r}|^3} \cdot (\nabla \times d\mathbf{l}) - \mu \cdot d\mathbf{l} \left\{ \nabla \times \frac{\mathbf{r}}{|\mathbf{r}|^3} \right\} \right]$$

As $d\mathbf{l}$ is not a function of x, y, z coordinates, $\nabla \times d\mathbf{l} = 0$ at the point of observation of **B** Also $\nabla \times \left(\frac{\mathbf{r}}{|\mathbf{r}|^3} \right) = 0$

$$\therefore \qquad \nabla \cdot \mathbf{B} = 0$$

Thus, the divergence of **B** magnetic flux density is equal to zero. The magnetic field lines are solenoidal.

Faraday's Law

Faraday's law establishes a connection between **E** and **B**. It states that the line integral of E around a closed path is equal to the rate of change of magnetic flux within the boundary of the path. The magnetic flux through an area 1 element **dS** is given as $\bar{\mathbf{B}}_n.d\bar{S}$ (Fig. 11.38).

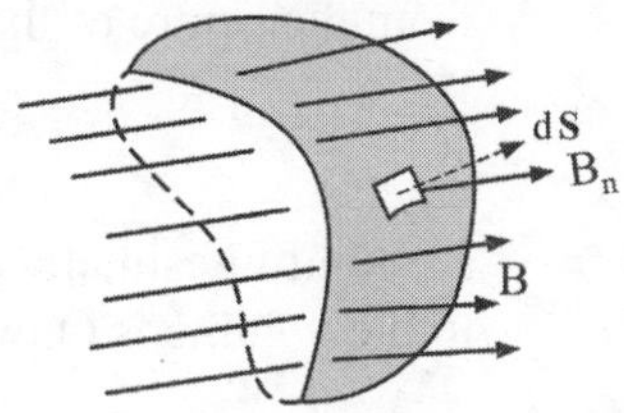

Fig. 11.38

where $\bar{\mathbf{B}}_n$ is the normal component of B.

The total flux

$$\phi_B = \int_S \mathbf{B}_n.d\mathbf{S} \qquad \therefore \oint \mathbf{E}.d\mathbf{l} = -\frac{d}{dt} \int_S \mathbf{B}_n.d\mathbf{S}$$

According to Stokes theorem (Appendix)

$$\oint \mathbf{E}.d\mathbf{l} = \int_S \text{Curl } \mathbf{E} \cdot d\mathbf{S}$$

$$\therefore \qquad \int_S \text{Curl } \mathbf{E} \cdot d\mathbf{S} = \frac{-d}{dt} \int_S \mathbf{B}_n.d\mathbf{S} \qquad \therefore \qquad \text{Curl } \mathbf{E} = -\frac{d\mathbf{B}}{dt}$$

B is a function of x, y, z and t : The above relation is the differential form of Faraday's law. It is one of the Maxwell's Curl equations.

Ampere's Law

Maxwell was trying to establish a firm relation between electric fields and magnetic fields. Faraday's law informs us that in any region, a varying magnetic field can set up electric field. Then, Maxwell imagined that a varying electric field could set up a magnetic field. Simple Ampere's law (Fig. 11.39) discussed earlier is having a missing information and it has to be connected to restore the symmetry between electric and magnetic fields. Let us look into that aspect in which a term called 'displacement current' is introduced apart from conduction current.

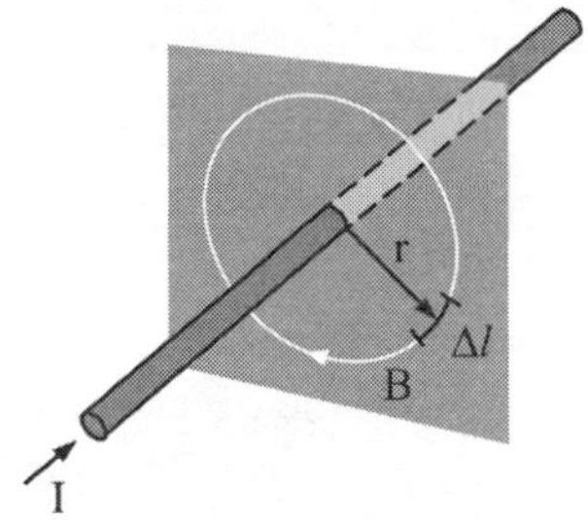

Fig. 11.39

Displacement Current

Consider that a rectangular parallel plate capacitor is kept in an electrical circuit as shown in Fig. 11.40(a).

Ampere's law states that the line integral of the magnetic field around loop is proportional to the total current that passes through the surface bounded by Ampere loop (A). The same loop is stretched upto to the plate P_1 and the Ampere's law gives the same result. Beyond the plate P_1, there is a gap, hence, the Ampere loop appears to be non-existant as there is no possibility of conduction current across the gap. We have to look into this problem in a different way. If Ampere's law still holds, there must be a magnetic field by the changing E-field between the plates. This induced B-field makes it to look like there is a current through the gap, passing through the plates. This is shown in Fig. 11.40 (b)

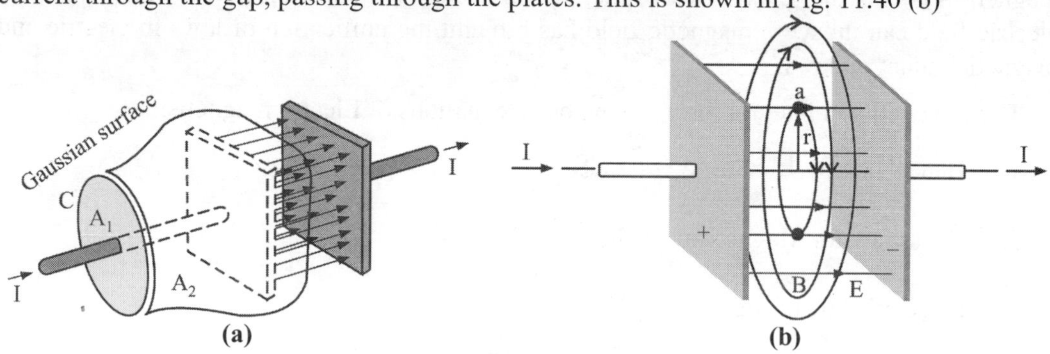

Fig. 11.40

As charge is sent into the capacitor continuously, the electric field in the capacitor changes with time as at a rate $\frac{dE}{dt}$. These electric field lines pass through the medium in the capacitor. The changing electric field gives rise to a changing electric flux at a rate $\frac{d\phi_E}{dt}$. In analogy with Faraday's law, we can write

$$\oint \mathbf{B} \cdot d\mathbf{l} = \mu_o \varepsilon_o \frac{d\phi_E}{dt}$$

Therefore, while a current C (i.e., moving charge) passes through the surface gives rise to a magnetic field, at the same time, changing electric flux through a surface gives a magnetic field also. Then, the complete statement of Ampere's law is given as

$$\oint \mathbf{B} \cdot d\mathbf{l} = \mu_o I + \mu_o \varepsilon_o \frac{d\phi_E}{dt}$$

Comparing the first term and the second terms in the above expression, we find that $\varepsilon_o \frac{d\phi_E}{dt}$ has the dimensions of current and this current is called displacement current, I_d given as

$$I_d = \varepsilon_o \frac{d\phi_E}{dt} \qquad \therefore \oint \mathbf{B} \cdot d\mathbf{l} = \mu_o (I + I_d)$$

For regions between the plates but at radius larger then the plates P_1 and P_2, the B-field would be identical to that at an equal distance from the wire. The displacement current in the gap is equal to the conduction current in the wires. The displacement current, I_d has a magnitude as well as direction. The displacement current density is represented by $\mathbf{J}_d$. Thus Maxwell relation given above is universal.

Thus, Maxwell extended Ampere' law into a more useful or universally acceptable laws.

The form relations given as mentioned below are the Maxwell's equations in integral form. They are the basic equations of electro magnetism.

The concept of displacement current is the outcome of modification of Ampere's law by Maxwell. The Maxwell's relation is not just meant for a capacitor but any electric and magnetic systems that charge with time. A Maxwell's visualization that a time-dependent electric field can induce a magnetic field has brought the unification of laws in electric and magnetic induction laws.

The Maxwell's in integral form are the basic equations of Electro magnetism

1. Gauss' law for Electricity $\oint \mathbf{E}.d\mathbf{S} = \frac{Q}{\epsilon_o}$;
2. Gauss law for Magnetism $\oint \mathbf{B}.d\mathbf{S} = 0$
3. Faraday's law of induction $\oint \mathbf{E} \cdot d\mathbf{l} = \frac{-d\phi_B}{dt}$
4. Ampere's law $\oint \mathbf{B}.\, d\mathbf{l} = \mu_o I + \mu_o \epsilon_o \frac{d\phi_E}{dt} = \mu_o \mathbf{J} + \mu_o \varepsilon_0 \frac{d\mathbf{E}}{dt}$

∴ Differential forms of the Maxwell equations for free space are

1. $\nabla \cdot \mathbf{E} = \frac{\rho}{\varepsilon_o}$; i.e., $\nabla \cdot \mathbf{D} = \rho$ 2. $\nabla \cdot \mathbf{B} = 0$; $\nabla\,.\,H = 0$; $B = \mu_0 H$

3. $\nabla \times \mathbf{E} = \frac{-\partial \mathbf{B}}{\partial t}$ or $-\mu_o \frac{\partial H}{\partial t}$ 4. $\nabla \times \mathbf{B} = \mu_0 J + \mu_0 \varepsilon_0 \frac{\partial \mathbf{E}}{dt}$ or $\nabla \times \mathbf{H} = \left(J_c + \frac{\partial D}{dt} \right)$;

$J_c = 0$ for free space $= \sigma E + \varepsilon_0 \frac{\partial E}{\partial t}$; σ is zero for free space.

Maxwell equations offer a complete description of interactions among charges, currents, electric and magnetic fields. They are the basic tools for finding the electric and magnetic fields at any point in space (like a place away from a communication antenna in communication theory, and responsible for development of quantum electro dynamics).

Maxwell's equations for a medium are:

Given that

$\mathbf{B} = \mu \mathbf{H}$; $\mathbf{J} = \sigma \mathbf{E}$, $\mathbf{D} = \varepsilon \mathbf{E}$; $\nabla\,.\,\mathbf{E} = \frac{\rho}{\varepsilon}$; $\nabla\,.\,\mathbf{H} = 0$; $\nabla\,.\,\mathbf{B} = 0$; $\nabla \times \mathbf{H} = \mathbf{J} + \frac{\partial \mathbf{D}}{\partial t}$

$\nabla \times \mathbf{B} = \mu \sigma \mathbf{E} + \mu \varepsilon \frac{\partial \mathbf{E}}{\partial t}$; $\nabla \times H = \sigma E + \varepsilon \frac{\partial E}{\partial t}$; $\nabla \times \mathbf{E} = \frac{-\partial \mathbf{B}}{\partial t}$

Thus, Maxwell extended Ampere' law into a more useful or universally acceptable law. The Maxwells in integral form are the basic equations of Electro magnetism

1. Gaus's law for Electricity $\oint \overline{E}.d\overline{s} = \frac{Q}{\epsilon_o}$

2. Gauss law for Magnetism $\oint \overline{B}.\ d\overline{s} = 0$

3. Faraday's law of induction $\oint \overline{E}.\ d\overline{l} = \frac{-d\phi_B}{dt}$

4. Ampere's law $\oint \overline{B}.\ d\overline{l} = \mu_o I + \mu_o \in_o \frac{d\phi_E}{dt}$

$$= \mu_o \overline{J} + \mu_o \frac{d\overline{E}}{dt}$$

∴ Summarizing the Maxwell equations are

1. $\overline{\nabla}.\overline{E} = \frac{\rho}{\epsilon_o}$; i.e., $\overline{\nabla}.\ \overline{D} = \rho$

2. $\overline{\nabla}.\ \overline{B} = 0$

3. $\overline{\nabla} \times \overline{E} = \frac{-\partial \overline{B}}{\partial t}$ or $-\mu_o \frac{\partial H}{\partial t}$

4. $\overline{\nabla} \times \overline{B} = B = \mu_o H$

or $\overline{\nabla} \times \overline{H} = \mu_o \left(J + \frac{\partial D}{dt} \right)$

Questions

1. An elctron revolves around a proton in a circular pat of radius 0.35×10^{-10}m. What is the magnetic field at the proton ?
2. What is permeability of substance ?.
3. What is the relation between temperature and the critical magnetic field in a superconductor?
4. Explain how Meissner effect proves the superconductor to be a perfect diamagnetic material? What is the different in critical fields for lead at 7.1 K and 1K?
5. What is Bohr magneton ? A certain element, that has an atomic weight 24.31, has a density 1.74×10^3kg/m^3. In the presence of the magnetic field, all the atomic dipoles align perfectly with the field. What is the magnetic moment ?
6. Explain the origin of magnetic moment in substances.
7. What is diamagnetism ? Give some examples. Explain the theory of diamagnetism.
8. What is the diamagnetism susceptibility of silicon if the electron radius of an atom is of the order of 0,1nm. (The experimental value is -0.4×10^{-5}).
9. Explain the difference between diamagnetism and paramagnetism.
10. Explain the Curie law.
11. Describe how does the magnetic susceptibility vary with temperature for (a) Paramagnetic salt, (b) Diamagnetic material, (c) Ferromagnetic substance.
12. Discuss the salient features of Weiss theory of ferromagnetism. Discuss the expression for Curie constant and Curie temperature.

13. What is spontaneous magnetization ? Explain the origin.
14. A ferromagnetic material is placed in a magnetic field of intensity H. What is the energy stored in the material.
15. Discuss the spin nature of ferromagnetism. Explain why Fe, Co and Ni are ferromagnetic in nature ?
16. What is the hysteresis loop ? What does it represent ? What is the significance ? Describe how it is recorded ?
17. Explain the magnetization curve for a ferromagnetic material on the basis of domains.
18. Nickel is subjected to a magnetic field intensity of 5×10^5 A/m. It gets magnetically saturated. If **B** = 1.27 Tesla, relative density is 9 and atomic weight is 58.71. Compute the magnetic moment per atom.
19. Silicon is subjected to a field of intensity 1000 A/m. The magnetic susceptibility of silicon is -2.9×10^{-5}. What are the magnetization and flux density values ?
20. Discuss Curie-Weiss law for the susceptibility of a ferromagnet.
21. What are magnetic domains ? What role they play in the hysteresis behaviour of a ferromagnet. Give the characteristics of Domains.
22. A polycrystalline iron rod does not show any magnetization. Explain.
23. What are high permeability substances. What are their characteristics?
24. What should be the characteristics of permanent magnetic materials?
25. What is anti ferromagnetism? Discuss its origin in some crystals. Give one example?
26. Give the concept of ferrimagnetism?
27. What are ferrites? Discuss the salient features of normal spinnel and Inverse spinnel structures?
28. Describe the magnetic structures of few ferrites.
29. Discuss the application of Soft ferrites?
30. What are Hard ferrites?
31. What should be the characteristics of recording materials?
32. What are magnetic bubbles?
33. How does the permeability of a ferrite change with temperature. How does its magnetization change with temperature?
34. What is magnetostriction? What are its applications?
35. What will be the induced magnetic moment in the hydrogen atom in a flux density of 1 weber / m^2?
36. A magnetic material is subjected to a magnetic field of strength $200 Am^{-1}$. Its relative permeability is 18. Calculate the magnetization and the flux density.
37. The magnetic flux density in a piece of Ferricoxide is 1.4 Tesla. The magnetic susceptibility at room temperature is 1.6×10^{-3}. Calculate the magnetization, magneto motive force, and the relative permeability.
38. Where are permanent magnets useful?
39. Discuss the general characteristics of ferrites?
40. Discuss the general characteristics of ferromagnetic materials?
41. How does spontaneous magnetization vary with temperature?
42. State Gauss Law for electric fields and explain.
43. What is the meaning of $\oint \mathbf{E} \cdot d\mathbf{S} = 0$
44. What is meaning of $\int \mathbf{B} \cdot d\mathbf{S} = 0$
45. Express the Gauss law for magnetic field.
46. State Faraday's law in integral form and obtain Maxwell's curl question from it.
47. State Ampere's law. Explain integral form of Maxwell's equation relating magnetic field and current in a wire.
48. Obtain Maxwell's equation from Ampere's law.
49. What is displacement current? Discourse the impotence of displacement current introduced in Maxwell's equations.
50. Derive curl equation for magnetic field.
51. State all the four equations of Maxwell in differential form.
52. Discuss the significance of displacement current introduced in Ampere's law.
53. Express all the Maxwell equations for free space.

Objective Questions

I. Choose the correct choice

1. The unit of magnetic induction B is
   (a) Weber (b) Newton-Weber
   (c) Weber/m^2 (d) Newton/amp-m

2. The electron revolves in an orbit of radius r meters inside the atom around the nucleus with a velocity v m / sec. Then , the magnetic moment associated with it is
   (a) e v r (b) $\frac{ev^2}{2}$
   (c) $\frac{ev}{r^2}$ (d) $\frac{evr}{2}$

3. A magnetic dipole of moment **M** is placed in a uniform field of density **B**, Then, its potential energy is
   (a) $-\mathbf{M}.\mathbf{B}$ (b) $\mathbf{M}.\mathbf{B}$
   (c) $\frac{\mathbf{M}.\mathbf{B}}{3}$ (d) $\frac{\mathbf{M}.\mathbf{B}}{\mu_0}$

4. The unit of permeability constant is
   (a) Webers (b) Webers | m^2
   (c) Weber | amp - meter
   (d) Gmp - meter |Web

5. Read the two statements and answer
   (A) Electric fields associated with the stationary charges are conservative, but those associated with changing magnetic fields are non conservative
   (B) The magnetization of a paramagnetic specimen is inversely proportional to absolute temperature
   (a) A and B are wrong
   (b) A is wrong, B is correct
   (c) both A and B are correct
   (d) A is correct, B is wrong

6. If a magnetic field is applied to a diamagnetic substance, then
   (a) a magnetic moment will be induced whose direction is opposite to the field
   (b) the magnetic dipoles tend to point in the same direction as the applied field
   (c) it is attracted towards the higher field – region
   (d) the specimen will develop magnetic domains

7. The change in angular velocity of an electron of an atom when placed in a magnetic field density **B** is
   (a) e **B** (b) $\pm\frac{e}{2m}\mathbf{B}$
   (c) $-\frac{e}{2mB}$ (d) $\frac{B}{2me}$

8. The three vectors **B**, **H** and **M** in a magnetic material are related as
   (a) $\mathbf{B}=\mathbf{H}+\mathbf{M}$ (b) $\mathbf{B}=\frac{\mathbf{M}}{\mathbf{H}}$
   (c) $\mathbf{B}=\frac{\mathbf{M}}{\mathbf{H}}$ (d) $\mathbf{B}=\mu_o(\mathbf{H}+\mathbf{M})$

9. The unit measure of atomic magnetic moment is given as
   (a) 9.27×10^{-20} erg. cm^2
   (b) 9.27×10^{-24} Amp – m^2
   (c) 9.27×10^{-27} Amp m^2
   (d) 9.31×10^{-27} Amp m^2

10. The total number of values the magnetic quantum number 'm$_l$' can take are
    (a) $2l$ (b) $2l+1$
    (c) $l+1$ (d) $l/2$

11. An electron with $l = 0$ always have
    (a) zero angular momentum, and hence zero magnetic moment
    (b) zero angular momentum and eh/2π m magnetic momentum
    (c) angular momentum equal to eh/2π and magnetic momentum $eh/_{2\pi}$
    (d) angular momentum h/2π and zero magnetic momentum
12. Choose the correct statement
    (a) The orbital angular momentum of an electron can have $(l + 1)$ values
    (b) The electrons precess about the nucleus when placed in a magnetic field
    (c) The orbital magnetic moment of the electron for $l = 1, 2, 3$ ..... is given by $-\frac{eh}{4\pi m}\sqrt{l(l+1)}$
    (d) The magnetic quantum number m j can take $(l + 1)$ values where l is the orbital quantum number
13. For an isotropic magnetic medium, M and H are in the same direction and the permeability is a scalar. Then
    (a) the above statement is untrue
    (b) the above statement is true
    (c) the M and H vectors can never be in the same direction
    (d) the permeability can never be a scalar
14. The magnetic susceptibility (χ) for an isotropic magnetic medium
    (a) is a tensor
    (b) is several - hundreds for a para magnetic substance
    (c) have the dimensions Amp. m^{-1}
    (d) is a scalar and is a dimensionless quality
15. The fact that B (the magnetic field density) is solenoidal is represented by
    (a) $\nabla . \mathbf{H} = \mu \ \mathbf{J}$ (b) $\nabla \times \mathbf{E} = \frac{-\partial \mathbf{B}}{\partial t}$
    (c) $\nabla . \mathbf{D} = \rho$ (d) $\nabla . \mathbf{B} = 0$
16. The susceptibility for a diamagnetic substance is
    (a) zero
    (b) negative
    (c) positive and high value
    (d) low but positive
17. When the vectors **L** and **S** (the resultant orbital angular momentum and spin momentum) are combined to give J, then
    (a) The total angular momentum is expressed as $\sqrt{J(J+1)}\ h/2\pi$
    (b) The total angular momentum is given as J (J + 1)h/2π
    (c) J can take values from 0 to ∝ continuously
    (d) J can take (J + 1) values.
18. When a paramagnetic atom is placed in an externally applied magnetic field, the resulting magnetic moment of the atom is
    (a) $g\ \mu_B\ J$ (b) $g\ \mu_B\ (J+1)$
    (c) $g\ \mu_B \sqrt{J+1}$ (d) $g\ \mu_B \sqrt{J(J+1)}$
19. Choose the correct statement
    (a) All atoms possessing odd number of electrons in the atom exhibit diamagnetism
    (b) All atoms possessing odd number of electrons exhibit paramagnetism
    (c) All atoms possessing even number of electrons should necessarily exhibit ferromagnetism
    (d) An anti ferromagnetic unit cell consists of paramagnetic sub lattices
20. According to Curie law
    (a) The magnetic susceptibility of a paramagnetic salt is directly proportional to absolute temperature
    (b) For strong fields and low temperatures, the magnetization attains a minimum value
    (c) The saturation magnetization in a paramagnetic salt is dependent only on the magnetic field and not on temperature
    (d) The magnetic susceptibility is inversely proportional to the absolute temperature

21. Choose the correct statement
    (a) The spontaneous magnetization disappears above Curie temperature for a ferro magnetic substance
    (b) At the curie temperature, the spontaneous magnetization attains a constant value in ferro magnetic substances
    (c) All ferro magnetics are paramagnetic at 0° K
    (d) Palladium and nickel sulphates are the examples for ferro magnetic substances
22. Which one of the following is true in the case of 'domains'
    (a) An unmagnetized iron bar does not contain 'domains'
    (b) A ferro magnetic substance contains domains that are spontaneously magnetized
    (c) A domain moves from one position to another in the presence of an external magnetic field
    (d) In high external magnetic fields, the domains in a ferro magnetic substance just disappear
23. The area of the hysteresis loop of a ferro magnetic or ferri magnetic crystal gives
    (a) the volume of the domains that are present in the crystal
    (b) the number of domains present in the crystal
    (c) the value of spontaneous magnetization of the crystal
    (d) the energy that is consumed in traversing the loop
24. The core of an electromagnet should have
    (a) high permeability and magnetization, low – hysteresis loss and coercivity
    (b) low permeability, hysteresis loss and $(BH)_{max}$ product
    (c) high permeability and coercivity, low $(BH)_{max}$ product, high hysteresis loss
    (d) low permeability and hysteresis loss, high coercivity and $(BH)_{max}$ product
25. The oxygen atom has a net orbital angular momentum $L = \hbar$, then, the orbital magnetic moment is
    (a) 9.27×10^{-34} A.m^2
    (b) 9.27×10^{-24} A.m^2
    (c) 18.54×10^{-24} A.m^2
    (d) 18.54
26. Choose the correct statement
    (a) Anti ferro magnetic substances do not show any magnetization in the presence of an external magnetic field
    (b) The anti ferro magnetics become ferro magnetic above curie temperature
    (c) Ferrites become ferro magnetic above curie temperature
    (d) Anti ferro magnetic ordering in solids is pictured as composed of two ferro magnetic sublattices with anti parallel magnetic moments
27. The extrapolated line of $1/\chi$ vs T plot
    (a) makes an intercept on $1/\chi$ axis on the negative side
    (b) makes a positive intercept at a positive temperature θ_N on T – axis
    (c) makes an intercept on the negative side of the T– axis
    (d) passes through the origin
28. Identity the antiferro magnetic materials in the following
    (a) Fe_2O_3, $Li_2Fe_5O_{18}$, $ZnFe_3O_4$
    (b) $BaFe_{12}O_{19}$, $\gamma-Fe_2O_3$, $SmCo_5$
    (c) $Me_3\ Fe_2\ Fe_3\ O_{12}$
    (d) CoO, MnO, MnS
29. The inverse spinel in the following choices is
    (a) $ZnFe_2O_4$ (b) $MnFe_2O_4$
    (c) $\gamma-Fe_2O_4$ (d) $CoFe_2O_4$

30. The magnetic structure of Fe_3O_4 is

(a) $[Fe^{2+}Fe^{3+}]O_4$

(b) $\left[Fe^{3+}\right]\left[\overset{\rightarrow}{Fe^{2+}}\ \overset{\leftarrow}{Fe^{3+}}\right]O_4$

(c) $\overset{\rightarrow}{Fe^{2}}\left[\overset{\leftarrow}{Fe^{3+}}\right]O_4$

(d) $[\square^{2+}[Fe^{2+}]O_4$

31. Ferrites exhibit

(a) Ferro magnetization

(b) anti ferro magnetism

(c) ferri magnetism

(d) para magnetism

32. The unit cell of the normal spinel structure consists of

(a) 32 tetrahedral sites, 24 octahedral sites

(b) 32 octahedral sites, 64 Tetrahedral sites, 32 oxygen ions

(c) 32 octahedral sites, 64 Tetrahedral sites 64 oxygen ions

(d) 32 octahedral sites, 32 tetrahedral sites, 32 oxygen atoms

33. In 'normal' spinels'

(a) A – sites are occupied by divalent and ferric ions

(b) B – sites are occupied by Fe^{3+} ions

(c) eight A – sites are occupied by eight divalent ions and B – sites by Fe^{3+} ions

(d) eight A – sites are occupied by Fe^{2+} ions, eight – B – sites by Fe^{3+} ions

34. Ferrites possess

(a) high resistivity, low permeability, low magneto crystalline energy

(b) anisotropy energy that is independent of the temperature and field

(c) high permeability values compared to paramagnetism, high resistivity and low eddy currents

(d) high-hysteresis loss, low magneto-crystalline anisotropy, and low permeability values

35. The example for micro wave ferrite is

(a) maganese ferrite

(b) barium ferrite

(c) cobalt ferrite

(d) yttrium iron garnet

36. The material that is useful for magnetic recording is

(a) $\gamma - Fe_2O_3$ (b) cobalt ferrite

(c) nickel – zinc ferrite (d) $y_3Fe_5O_{12}$

37. The magnetic recording material should have

(a) high remanet magnetization, and magnetizations, low coercivity

(b) high remanet magnetization and coercivity

(c) low remanet magnetization and very low coercivity

(d) high $(BH)_{max}$ product, high coercivity and low coercivity

38. The magnetic susceptibility of silicon is -0.4×10^{-5}. Then, the magnetic moment per unit volume in a magnetic field of intensity 10^5 amp m^{-1} is

(a) -0.4 amp m^{-1}

(b) -0.4 amp m^{-3}

(c) 0.124 Wb m^{-1}

(d) -0.124 amp m^{-1}

39. The atomic weight of gadolinium is 157,25. The number of atoms in its cell per unit volume is 29 $\times 10^{27}/m^3$. The value of Bohr magneton is 9.27 $\times 10^{-27}$ amp. m^2. Then, the saturation magnetization is

(a) 5.38×10^3 Amp m^{-1}

(b) 0.538 amp m^{-1}

(c) 538×10^3 Amp m^{-1}

(d) 5376 amp m^{-1}

40. The materials that are useful as induction cores are

(a) Alnico magnets

(b) barium – strontium ferrites

(c) garnets

(d) Ni – Zn ferrites

41. The materials that are used in wind screen wiper motors, door catches, correction magnets, etc are
(a) Alnico (b) Ni – zn ferrite
(c) Steel (d) hard ferrites

II. Fill in the blanks

1. The assigned value of permeability constant is ________ .
2. The quantity $\sqrt{\mu_o \in_o}$ has the dimensions of ________ .
3. A magnetic material possesses magnetization M under the action of a magnetic field of intensity H. Then its effective permeability is ________ .
4. The permeability of a magnetic medium represents the extent to which field flux lines ________ the magnetic medium.
5. The magnetic susceptibility in terms of relative permeability μ_r is given as ________ .
6. In an isotropic magnetic medium the two vectors M and H are in the ________ directions.
7. In a magnetic medium, the H – field originates where the magnetization vector **M** ________.
8. In the case of diamagnetic substances, the susceptibility is ________ .
9. A spinning electron precesses under the action of an external magnetic field about ________ direction.
10. The Larmor frequency of an electron of mass 'm' in the presence of an external magnetic field density B is given by the expression ________.
11. The frequency of motion of an electron in an atom is $\omega_o \approx 10^{16}$ / Sec. If the magnetic field is 1.0 Tesla, then the Larmor frequency of the electron is ________ .
12. The S I unit of magnetic moment is ________.
13. The mass of the electron is m_e. When the electron is in an orbit with a quantum number L.Then the magnetic moment of the electron is ________.
14. The magnetic permeability of diamagnetic substances is independent of ________ .
15. The relative permeability of ________ substance is just greater than one.
16. A ________ magnetic substance works as a working substance in an adiabatic demagnetization.
17. The magnetic susceptibility is ________ proportional to the absolute temperature for a paramagnetic substance.
18. The ratio of magnetic moment to the angular momentum of an electron is called ________.
19. The curie constant of a paramagnetic salt is independent of ________ .
20. A ferromagnetic possesses ________ magnetization only below Curie temperature.
21. The spontaneous magnetization in a ferromagnetic substance approaches zero as the temperature approaches ________ temperature.
22. The magnetization value in every domain of a ferromagnetic substance is equal to ________ magnetization
23. The molecular field in a ferromagnetic lattice is ________ proportional to the magnetization.
24. The total magnetic moment associated with a ferromagnetic crystal having n atoms per unit volume, is given by the relation ________ where g, B (a) are gyro magnetic ratio, and Brilliouin function, and J = L + S.
25. If they are two atoms with S_i, and S_j, then, the energy of interaction is given as ________ .
26. The exchange integral is positive for ________ substance.

27. The exchange integral is negative for __________ .

28. In ferro magnetic substance, the spin moments of all the neighboring atoms are aligned __________ to each other.

29. The area of the hystereses loop of a ferro magnetic substance represents __________ per cycle.

30. The total work done by the magnetizing field per unit volume of the ferro magnetic sample for completing a cycle in M – H hystereses loop is equal to __________.

31. A domain in a ferro magnetic substance is __________ magnetized.

32. In an iron sample, the observable magnetization is zero because the vector sum of magnetizations in all domains is __________ .

33. The boundary between one domain and another domain is called __________

34. The process of magnetization is viewed as the ____________ and displacement of __________ .

35. In strong external magnetic fields, all the domains rotate into the direction of the field to give __________ magnetization.

36. The magnetostriction in ferro magnetic crystal

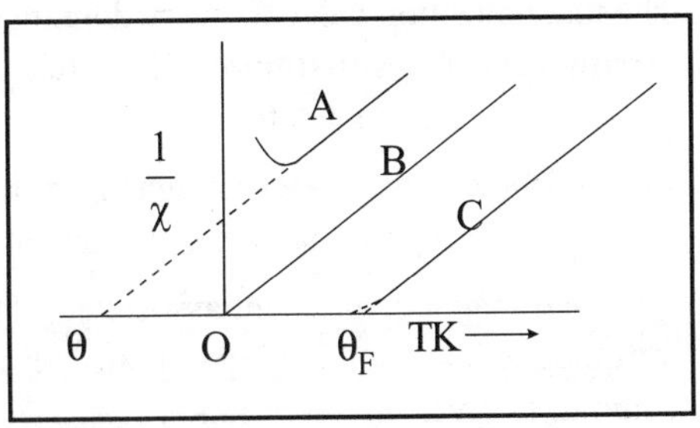

is due to __________ .

37. Magneto mechanical tranducers make use of __________ property of ferro magnetic substances.

38. Permanent magnetic materials should have 1. __________ saturation magnetization, 2. __________ energy product.

39. Realy switches, magnetic shunts made use of __________ permeability ferro magnetics.

40. The ferro magnetic substances are not useful in high frequency applications because of large __________ losses.

41. Anti ferro magnetic substances are transformed into ____________ state above Neel temperature.

42. In Maxwell's Equation $D \times E =$ __________ .

43. $\frac{1}{\chi}$ vs T plots are given for magnetic substances in the following figure, identity the materials.

A________ B ________ C ________ .

44. The uncompensated anti ferro magnetic moments in some crystals give rise to __________ magnetism.

45. In Maxwell's equations for free spaces the differential form of Ampere's law is________

CHAPTER 12

Lasers

12.1 Introduction

In twentieth century, there were a number of outstanding scientific discoveries and inventions. They include, Planck's law, Einsteins explanation of photo electric theory, Bohr's model of atom, the birth of quantum mechanics, uncertainty principle, discovery of high energy particles, positron prediction and its discovery, the inventions of bulb, the discoveries for the improvement of communication systems, semiconductors, transistor, miniarization technologies, optical fibre communication etc. The inventions that had really a major impact on scientific development were the inventions concerning the electric bulb, transistor and laser. They had changed the way of life of a common man. It should be mentioned here that laser is an optical maser. It was in 1954 that a maser (microwave amplification by stimulated emission of radiation) was made. A coherent electromagnetic radiation at frequencies of 10^9 to 10^{11} Hz was generated in a maser. Later in 1958, Schwlow and Towner laid down the theoretical work and the first working laser using synthetic ruby crystal was made in 1960 at Hughes Research laboratory. After few months, Bell Telephone laboratory developed He-Ne laser. C. K. N. Patel of Bell laboratories discovered the infrared carbon dioxide Laser. Since then, laser science has made enormous advances. We will study in the following, a few very fundamental aspects of laser physics.

12.2 Characterstics of Lasers

The word 'Laser' is acronym for "Light Amplification by Stimulated Emission of Radiation". A laser source is nothing but a source of light, but a special source of light that has unique properties. Suppose, we take an ordinary bulb that radiates energy in all directions. It illuminates the entire room and the objects there in. The radiation energy density at any point in the room is extremely small. Suppose, one brings all the beams of light emitted by the bulb into a very narrow pencil of beams, say, of the size of ball pen reefle, then, there will be a tremendous intensity of light concentrated. This intensity would be nearly 10^6 times that of the light from tungsten bulb at any incident point in the room. This is the underlying principle with which a laser source works. The intensity of light is the amount of energy transported by a wave per unit time across a unit area. The laser beam is highly intense as the power of light is concentrated over a width of 1 mm to 3 mm. The laser makes use of processes that increase or amplify signals after those signals are produced by other means. The processes are (i) spontaneous emission of light (ii) Stimulated emission of light (iii) Optical feed back for amplification.

A laser light is 1) coherent 2) Directional and 3) of narrow spectral width. It is a specialised source of light and is used at places where its specific properties are required. We have laser sources that emit light in the range 430-750 THZ. This range spans from far infrared part of the spectrum to the soft x-ray region (λ = 3nm).

Laser wavelengths are mentioned in μm (microns) in the infrared region, (angstroms) Å, in nm (nanometers) in visible, ultraviolet, extreme ultraviolet, and soft x-ray region of the spectrum. For the sake of understanding, we give here the wavelength regions.

1 micron (μm) = 10,000 angstroms = 1,000 nm. A green light has a wavelength of 5000 Å, or 0.5 μm or 500 nm. The infrared radiation spreads from 0.7–1 μm (near infrared), 1 to 10 μm (middle infrared), and 10–1000 μm (far infrared). The following are the steps involved in the lasing action.

1. *Excitation :* The atoms or electrons from lower energy level must be excited or pumped to a higher energy level.
2. *Population Inversion :* The life time of atoms or electrons in the higher energy level should be long enough such that the population inversion can be achieved. When population inversion is attained, more number of electrons will be present in the excited level than in the lower level. For this to occur, a continuous pumping of energy into the system is needed.
3. *Light Amplification :* It is achieved in a resonant cavity where laser action is activated.

12.2.1 Coherence and Monochromaticity

A laser beam is highly monochromatic and coherent. In the following, we shall give the meaning of coherence briefly.

The wave nature of light is responsible for the interference. The interference is nothing but superposition of waves that are generated by a two sources. The two sources should give highly coherent, and mono chromatic light for interfernce. The interference pattern consist of dark and bright fringes. In the formation of fringes, the two wave trains interfere at the point on the screen satisfying the conditions for interference. However, normally the dark fringes spread over are not completely dark and the bright fringe intensity does not correspond to the sum of the squares of the amplitudes of the two wave trains. This is because of the partial coherence of the light. The visibility of the fringes and the contrast depends upon the coherence of the light.

The visibility of fringes is given in terms of the intensity of light at maxima and minima. The visibility is defined as

$$V = \frac{I_{max} - I_{min}}{I_{max} + I_{min}} \quad .....(12.1)$$

Two waves are said to be coherent if they have the same wavelength, amplitude and they have a constant phase relationship both with time (temporal coherence) and in space (spatial coherence).

The visibility is correlated to coherence. If γ_{12} represents the degree of the coherence then,

$$|\gamma_{12}| = \frac{V(I_1 + I_2)}{2\sqrt{I_1 I_2}}$$

where I_1 and I_2 are the intensities of two wave trains that interfere. If $\gamma_{12} = 1$, then we have complete coherence ; if $\gamma_{12} = 0$, then complete incoherence. If $0 < \gamma_{12} < 1$, then partial coherence. When $\gamma_{12} = 0$, no fringes are formed. The parameter γ_{12} is related to the nature of the source.

Why laser beam is coherent?

Laser source is highly coherent because the photon states or the energy states between which the transitions take place are correctly, or perfectly selected. That means, the pumping is suitably selected to induce only those transitions. The very induction of population inversion is responsible for high coherence of the beam. The mirrors are arranged such that the amplification is built up and a direct beam comes out. The selective population of selected photon states is responsible for high degree of coherence. This type of population is possible because of the bosonic nature of photons. In ordinary sources, there is no selectivity of photon states.

Different Lasers emit radiation with different degrees of coherence. There are two parameters associated with coherence. One is, the coherence time τ_o, the second is, the coherence length l_c. The coherence length l_c is related to the monochromaticity of the light as $l_c = c/\Delta\nu$. The coherence time τ_0 is the time interval between two successive wave trains emitted by the source. Then $l_c = c\tau_0$.

The best coherence is found in gas lasers. The coherence time is of the order of 10^{-3} sec. That means the wave train length is about 10^5m (100 km). The ordinary source has coherence length of less than a few centimeters.

The solid state lasers like semiconductor lasers emit radiation with a lower degree of coherence.

12.2.2 Directionality

A good monochromatic source of Sodium lamp available in the laboratories, gives light in all directions. Therefore, there is a divergence of light. The light coming out of a laser source is directional with low divergence. If an ordinary source like an incandescent source has to give a directional light, the filament should be of the length less than a micron (10^{-6}m). It is difficult to produce such filaments. In lasers the divergence will be 6 sec to few minutes of an arc.

The increase in the diameter of laser beam as we move away from the source is a measure of divergence. The gas lasers give highly directive beams with a divergence of about one minute. The solid state lasers give beam with more divergence, the angle of divergence is around 10 minutes.

12.2.3 Speckle nature

Suppose, we focuss a laser beam from a helium-neon laser source or the pen-torch type lasers available in the stationary shops, on to a wall in a dark room. We see a bright red colored beam and the red spot on the wall. Let us keep a good lens and diverge the beam into a bigger spot.

Then, the magnified spot appears to 'speckle', i.e., dark and bright spots appear and disappear continuously. The speckle pattern of the light is caused by the high coherence and low divergence. The dark and bright specks appear because of the interference of coherent beams reflected to our eyes from different points on the focussed spot.

12.3 Principle of Laser Emission

12.3.1 Spontanous and Stimulated Emissions

Let us discuss some fundamental aspects of lasing action. Consider an atom in the ground state designated as E_1. When a radiation is incident on it, the atom absorbs photons. If the energy of the photon is $h\nu_{12}$, the atoms (electrons) in the ground state are excited to higher energy state E_2, called the excited state ($E_2 - E_1 = h\nu_{12}$ (Fig.12.1(a)). This state of the atom is not stable. The tendency of the atom is to come back to the ground state and this can happen by the electron falling back into the ground state level emitting a photon (Fig.12.1 (b)). A substance consists of many atoms and each atom has many electrons distributed in the allowed shells or orbits. At low temperatures, the atoms (electrons) preferably occupy the lowest energy states or the ground state. The tendency of the excited atoms (electrons) is always to fall back into the ground states emitting radiation appropriate to the frequency ν_{12}. The emission of this radiation is a spontaneous process (Fig.12.1(b)). The time that the atoms (electrons) spends in the excited state is extremely short, normally of the order of 10^{-8}sec. The spontaneous emission is a natural process. An atom is described in terms of quantized energy levels labelled E_1, E_2, E_3........ and so forth. The population in any level is the number of atoms per unit volume in that level. Consider energy levels as shown in Fig. 12.1. $E_2 - E_1 = E = h\nu_{12}$. The population ratio of states as per Boltzmann statistics is

$$N_2/N_1 = \frac{e^{-E_2/kT}}{e^{-E_1/kT}} = \exp(-h\nu_{12}/kT) \qquad .....(12.2)$$

At any instant of time, there will be the some atoms (electrons) in the energy state E_1 say, N_1 atoms; some atoms in E_2 state, say N_2, atoms. The population of atoms in any energy state is proportional to Boltzmann factor exp(–E/kT). The ground states are densely populated when compared to the excited state. This is the normal condition for thermodynamic equilibrium.

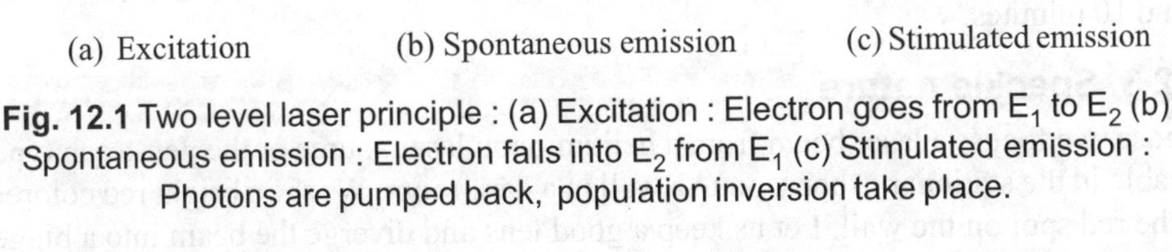

Fig. 12.1 Two level laser principle : (a) Excitation : Electron goes from E_1 to E_2 (b) Spontaneous emission : Electron falls into E_2 from E_1 (c) Stimulated emission : Photons are pumped back, population inversion take place.

The photons emitted spontaneously propagate randomly with different phases. Therefore the light emitted in spontaneous emission is non-coherent. The stimulated emission requires photons to be present already before the process starts (Fig.12.1c). The photon of energy $h\nu_{12}$ that is incident on the system can result in excitation and the generated photon can give rise to a stimulated emission. Let us consider how it is done.

12.3.2 Einstein Coefficients

Einstein in 1917 introduced the concept of stimulated or induced emission of of radiation by atomic systems. He described the interaction of matter and radiation in terms of a process in which an excited atom may be induced, by the presence of radiation to emit a photon, and there by decay to the ground state.

The stimulated emission requires population inversion. The population inversion is possible if the photon that is emitted by spontaneous emission is made to stay within the system to induce, once again, the transitions. This will result in stimulated emission. A stimulated emission is shown in Fig. 12.1(c).

If a photon can stimulate an electron to or atom move from a lower energy state E_1 to a higher energy state E_2 by means of absorption, it can also stimulate an atom to decay from E_2 to E_1. In the case of absorption the atom goes to excited state, photon disappears. When stimulation absorption takes place, the system of atoms should radiate additional photons to conserve energy. This has to occur in order to keep the population of the two energy levels in equilibrium.

The population ratio of the energy states is given by the expression (12.2). Atoms from a higher level fall into the ground state giving rise to spontaneous emission. The probability per unit time with which this process takes place represents the number of transitions.

$$\text{The number of transitions from upper level to lower level} = A_{21} N_2 \qquad \text{.....(12.3a)}$$

These transitions depend on the density of the excited atoms in level 2 (E_2). A_{21} is a constant.

The pumping of energy does two things. (1) it makes atoms to go from lower level (E_1, the ground state), to excited level (E_2).

$$\text{The number of the stimulated upward transitions per second} = N_1 B_{12} U_\nu \qquad \text{......(12.3b)}$$

where U_ν is the density of radiation of frequency ν. B_{12} is a constant. The transition probability is proportional to the radiation energy density (energy per unit area of the incident radiation), and the population in that level from which the transition occurs.

(2) But pumping also activates the transitions from upper level (E_2) to ground level (E_1). The rate at which these induced or stimulated transitions takes place is proportional to the density of radiation and the population in the upper level.

$$\therefore \quad \text{The number of stimulated downward transitions per second} = N_2 B_{21} U_\nu \qquad \text{.....(12.4)}$$

The proportionality constants. A_{21}, B_{12}, B_{21} are called 'Einstein coefficients'.

The net rate of downward transitions must be equal to that of upward transitions, under equilibrium conditions i.e.,

$$N_2A_{21} + N_2B_{21}\,U_\nu = N_1\,B_{12}\,U_\nu \qquad .....(12.5)$$

Solving for U_ν, we obtain

$$U_\nu = \frac{N_2A_{21}}{N_1B_{12} - N_2B_{21}} \qquad .....(12.6)$$

But $$\frac{N_2}{N_1} = \exp\left(\frac{-h\nu_{12}}{kT}\right) \qquad .....(12.7)$$

For simplicity, we replace ν_{12} as ν.

Then, $$U_\nu = \frac{A_{21}}{B_{21}} \frac{1}{\left[\left(\frac{B_{12}}{B_{21}}\right)\exp\left(\frac{h\nu}{kT}\right) - 1\right]} \qquad .....(12.8)$$

According to Einstein, the probability for upward transition and the probability for downward transition will be the same. Then

$$B_{12} = B_{21} \qquad .....(12.9)$$

Then, from eq. (12.8), we get

$$U_\nu = \frac{A_{21}}{B_{21}} \frac{1}{\left[\exp\left(\frac{h\nu}{kT}\right) - 1\right]} \qquad .....(12.10)$$

For atoms in thermal equilibrium with the radiation present, the ratio of stimulated rate to spontaneous emission rate is given by

$$\frac{\text{Stimulated emission}}{\text{Spontaneous emission}} = \frac{B_{21}}{A_{21}} U_\nu$$

$$= \frac{1}{\left[\exp\left(\frac{h\nu}{kT}\right) - 1\right]} \qquad .....(12.11)$$

The above ratio is called 'occupation index'.

According to Max Planck's radiation theory of black body radiation we have

$$\frac{A_{21}}{B_{21}} = \frac{8\pi h\nu^3}{c^3} \qquad ....(12.12)$$

Where c is the velocity of light. The above equation refers actually to a black body radiation.

According eq. (12.12), the rate of induced emission is extremely small in the visible region of the spectrum with ordinary optical sources. ($T \sim 10^3$K). Most of the radiation is emitted through spontaneous transitions. On the other hand, if $h\nu << kT$, the number of stimulated emissions far exceeds the number of spontaneous emissions.

In the above discussion, we have considered that the atoms interact with a photon of single frequency. In practice, atoms can interact with photons of variable energies. Therefore, the occupation index is a function of frequency.

12.3.3 Light Amplification and Laser oscillations

Amplification :

The active medium consists of atoms whose energies are distributed over different levels like E_1, E_2 E_3 and so forth. The population inversion is contrary to normal thermal equilibrium state. The population densities of a two level system are shown in Fig. 12.2 Fig. 12.2(a) gives the normal or Boltzmann distribution and Fig. 12.2(b) gives the inverted distribution. The rate of stimulated downward transitions will exceed that of upward transitions if $N_2 > N_1$.

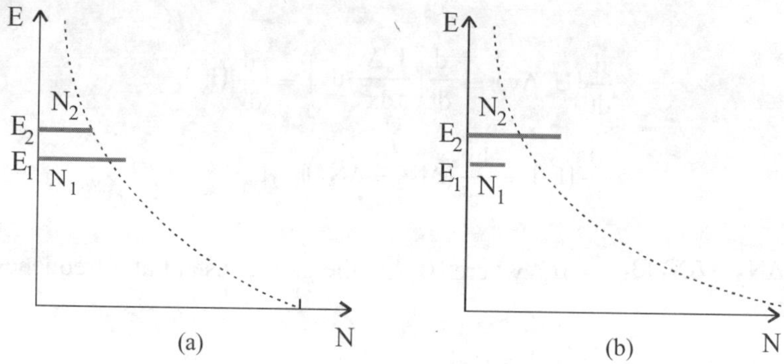

Fig. 12.2 Population Densities in two level system. (a) Normal or Boltzmann distribution (b) Inverted distribution.

In such a case, there would be an exponential increase in the intensity of light beam. This increase of intensity of light is known as amplification. The Fig. 12.2 depicts the population distribution. The population inversion gives an amplification.

Let us see qualitatively, in a very simple way, the magnitude of amplification that takes place in the active medium. Suppose a narrow and parallel beam of light propagates through the medium in which the population inversion is created. The energy density U_ν of the light is related to the intensity (I_ν) of the light beam. If the frequency range of the waves in the beam is equal to $\Delta\nu$, then

$$U_\nu \Delta\nu = \frac{I_\nu \Delta\nu}{c} \qquad \text{........(12.13)}$$

The spectral line has a width $\Delta\nu$ i.e., the frequencies are in the range ν and $\nu + \Delta\nu$. When we give energy to the medium in the process of excitation or pumping, all the atoms do not acquire the same amount of energy. Only certain number of atoms in any energy level acquire sufficient energy to go from one level to another. Suppose, ceratin number ΔN_1 of the N_1 atoms in E_1 are available for absorption. Therefore, the rate of upward transitions (as seen in the earlier section) is

$$B_{12}\, U_\nu\, \Delta N_1 = B_{12}\left(\frac{I_\nu}{c}\right)\Delta N_1 \qquad \text{........(12.14)}$$

Similarly, the rate of downward transition is

$$B_{21}\, U_\nu\, \Delta N_2 = B_{21}\left(\frac{I_\nu}{c}\right)\Delta N_2 \qquad \text{........(12.15)}$$

During the upward transistion, a part of the incident energy hν is absorbed. The downward transition adds the same amount of energy to the beam of light. As more and more transitions take place, the rate of change of spectral energy density in the frequency interval Δν is given as

$$\frac{d}{dt}(U_\nu \Delta\nu) = h\nu(B_{21}\,\Delta N_2 - B_{12}\;AN_1)U_\nu$$

$$= \frac{h\nu}{c}(\Delta N_2 - N_1)B_{12}I_\nu$$

But $$\frac{d}{dt}(U_\nu \Delta\nu) = \frac{d}{dt}\left(\frac{I_\nu \Delta\nu}{dx}dt\right) = \frac{d}{dx}(I_\nu)$$

$$\therefore \quad \frac{d}{dx}(I_\nu) = \frac{h\nu}{c}(\Delta N_2 - \Delta N_1)E_{12}I_\nu$$

Let $\frac{h\nu}{c}(\Delta N_2 - \Delta N_1)B_{12} = \alpha_g$ where 'α_g' is the gain constant at a frequency ν. Then

$$\frac{d}{dx}(I_\nu) = \alpha_g I_\nu$$

$$\therefore \quad I_\nu = I_{o\nu}\, e^{\alpha_g x} \qquad \text{.........(12.16)}$$

we replace ΔN_1 and ΔN_2 by N_1 and N_2 respectively for maximum gain i.e.,

$$(\alpha_g)_{max} = \frac{h\nu}{c}(N_2 - N_1)B_{12}$$

Then from equ. (12.12),

$$(\alpha_g)_{max} = \frac{c^2}{8\pi\nu^2\,\Delta\nu}(N_2 - N_1)A_{12} \qquad \text{........(12.17)}$$

From the above expression (12.17), we can easily understand that the gain constant 'α_g' is positive if $N_2 > N_1$. That means, the population inversion gives rise to the gain or amplification of the beam of light.

12.3.4 Laser Oscillations

The energy due to pumping is confined to the active medium. When sufficient population inversion is achieved, the radiation emitted is built up in intensity in an exponential manner as it propagates in the medium. As the gain increases, it will reach a threshold value when the losses are compensated by gain. This threshold value depends on the life time of the photon, the frequency and the width of the spectral line

The energy losses in the medium are due to the absorption of energy by small amounts at the mirrors, medium, or inhomogeneous portions of the medium etc. When the gain crosses the threshold value, there is a positive feed back, and oscillations will occur as standing waves are set up in the medium. The entire medium is like a cavity resonator that emits energy. The selection of the amplification at a particular frequency takes place. This frequency corresponds to the central point of the spectral line emitted by spontaneous emission. Out of the laser source, we get a coherent radiation. In a laser system, the active medium is placed between reflectors like a pair of mirrors (Fig. 12.3) making the active medium as a sort of cavity resonator. Oscillations are set in and sustained in the cavity. A laser source is a quantum oscillator. The gain depends on the population difference.

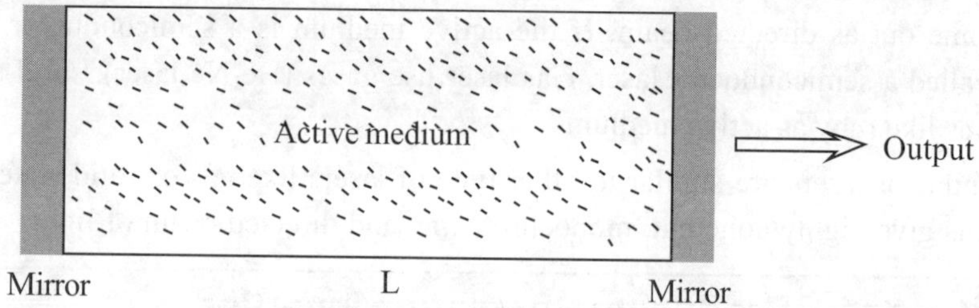

Fig. 12.3 Basic set of mirrors reflectors. In principle, there will be infinite set of reflections between the mirrors before light beam come out.

12.4 Population Inversion

Suppose that the emitted radiation is made to interact with the system of atoms continuously. Then, all the atoms would be present in the radiation field and there are two processes that can occur now. First, absorption of energy resulting in the transition of atoms from lower to higher energy levels and second, the emission of radiation by the transition of atoms from excited state to lower level. The two processes of the absorption of energy and subsequent emission of radiant energy are controlled by the radiant energy density and population difference $[N_2 - N_1]$ in the two levels. Higher the radiation field density, higher is the probability of the occurrence of the above processes. The radiant energy is pumped into the system of atoms.

When the pumped energy is continuously maintained within the system of atoms, we disturb the thermodynamic equilibrium. More and more atoms are pumped into the excited state, and they spend more time in the excited state. Therefore, slowly the population in the excited level will be built up and as nonequilibrium condition proceeds, the population in excited state $[N_2]$ will be greater than the population in ground state $[N_1]$. This condition that N_2 is greater than N_1 is called population inversion. Maintaining larger population in higher energy state is called population inversion. The medium in which this process occurs is known as active medium. The transitions from E_2 to E_1 will prevail over the absorption (transitions E_1 to E_2), when the external signal of energy $h\nu_{12}$ passes through the system continuously. The emission that takes place is the stimulated emission. This process is shown in the Fig. 12.1(c). The difference (δ) in the number of transitions from E_2 to E_1 and E_1 to E_2 is proportional to the probability of the transition P, the number of quanta U (the radiation density) and the difference in populations.

i.e., $$\delta = PU(N_2 - N_1) \quad .....(12.18)$$

The transitions from E_2 to E_1 can amplify the signal if the radiation output is reflected back a number of times into the active system. The laser system acts like an amplifier of electromagnetic radiation. Such an amplifier is called quantum amplifier. In a laser, a system of atoms, electrons and their energy levels is a quantum system. The system can enter into a mode of self-excitation and operates as a quantum oscillator. The oscillations are built up under the actions of random and spontaneously emitted photons. The light can be taken out of the system by following proper reflector configurations so that the light is first confined to the active medium. Then, the light is made to come out as directed beam. If the active medium is a semiconductor, then the laser system is called a semiconductor laser. Gas laser use gases (He-Ne laser), solid-state lasers use special solids like ruby as active mediums.

Semiconductor lasers are similar to other type of lasers like gas or solid state lasers. All the laser systems give highly coherent, monochromatic, and directed beam of light.

12.5 Methods for Producing Population Invertion

There are several ways of producing the population inversions necessary for optical amplification to take place. Some of the methods are :

1. Optical pumping or photon excitation
2. Electron excitation
3. Inelastic atom-atom collisions.

The first process given in Fig 12.1, is followed in solid state laser like Ruby Laser. We have also discussed this in 2 level system. For the second process, we will consider a system consisting of atoms possessing an energy level diagram shown in Fig. 12.4.

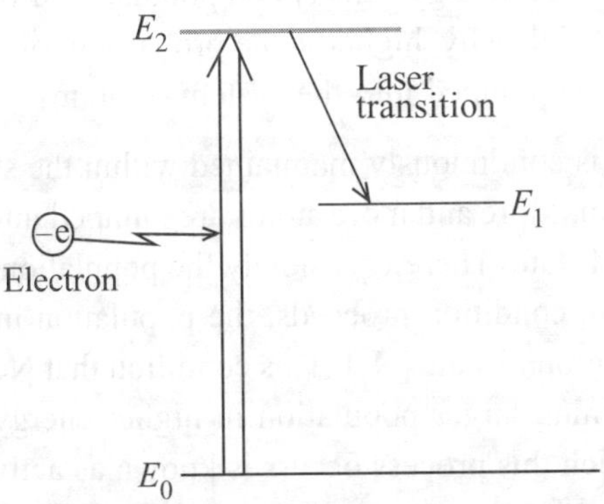

Fig. 12.4 Inelastic atom-atom collisions.

The excitation of an electron can be caused by electric discharge (Fig. 12.4) of a gas. During such an excitation, the laser medium itself carries the discharge in discharge tube, the electrons in the discharge may directly excite the active medium atoms to produce population inversion. This method is used in argon laser which is a gaseous ion laser.

In the third case also, an electrical discharge is used. A suitable mixture of gases is selected such that two different type of atoms, (A and B) possess some excited states A* and B* that coincide or placed extremely close to each other (shown in Fig. 12.5).

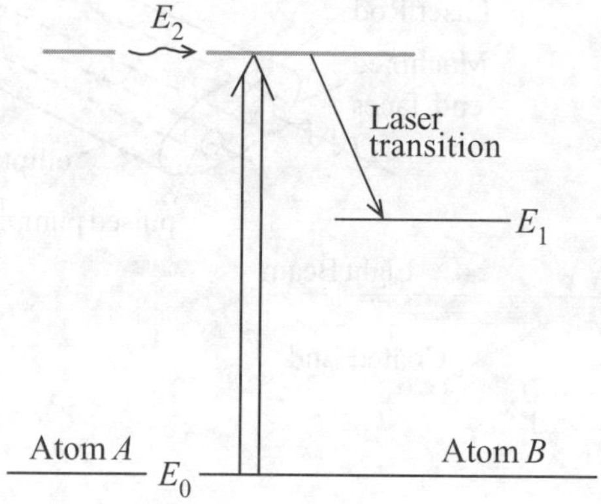

Fig. 12.5 Direct Electron Excitation.

When excited state of atom A, i.e., A*, it occupies a metastable level A*, transfer its excitation to the atom B by collision. Therefore, the population in a level E_B will increase and may become more densely populated than the lower level of B. Then the excited atom B* will go to the excited level or decay by emitting radiation that is stimulated radiation. This process is applied in He-Ne laser.

Types of Lasers :

1. Gas Lasers
2. Solid State Lasers
3. Semiconductor Lasers

We will discuss about gas laser (He-Ne laser) and a semiconductor laser in the following.

12.6 Solid State Laser - Ruby Laser

A Ruby laser system is an example of a three level laser system. Ruby laser is a proto type of solid state lasers. It was first fabricated by Maimann in 1960. The active atoms of the laser medium are embedded in a solid. Both glasses and crystals are used as supporting matrices for solid state lasers. The ruby laser consists of a cylindrical rod as shown in Fig. 12.6. The crystal is aluminium oxide ($Al_2 O_3$). Ruby is doped with chromium (Cr^{+++}) ions (0.05 to 0.5%). The aluminium atoms in the crystal are replaced by chromium ions. The rod is normally 3 to 20 mm in diameter and 30 mm in length. The ends are optically ground and polished to a high degree of parallelism and flatness. The mirrors can be coated at the ends of the rod or can be kept externally to make the rod a resonant cavity.

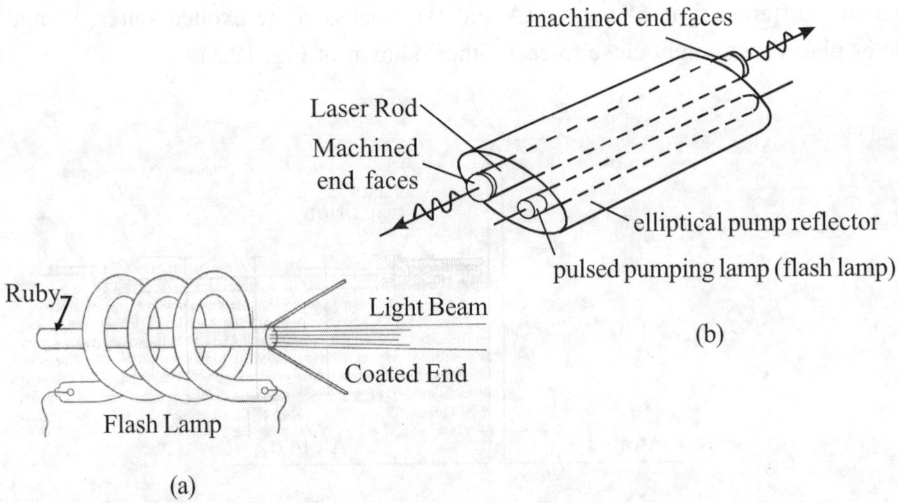

Fig. 12.6 (a) Schematic of a Ruby Laser.
(b) a solid state laser with optical pumping lamp, placed inside a reflector parallel to each other.

The excitation in the laser is caused by optical pumping from an intense source of light like the xenon lamp or high pressure mercury discharge lamps. A helical flash lamp can be used as shown in the Fig 12.6. The laser operates at room temperature and produces high power.

The energy level diagram is given in Fig. 12.7. It should be remembered that the energy level diagram of solid consists of energy bands. The chromium ions absorb the radiation emitted by the flash lamp and get excited to the levels E_2 and E_3 from E_1. The energy level E_1 refers to the ground state.

The states in the energy levels E_2 and E_3 have a very small life time (10^{-9} Sec). There exists a metastable level M . First, the ions decay to the metastable level M by radiationless transition. The metalstable level M has a much longer life time (milli seconds).

Therefore, a population inversion is achieved between the metastable levels and the ground state, i.e., from the population point of view, we have $N_M > N_1$. They are some photons which are spontaneously emitted. They trigger the stimulated emission. As the flash lamp is operated in flashes, the laser gives out pulses of light as a result of the transitions from metastable level to the ground state.

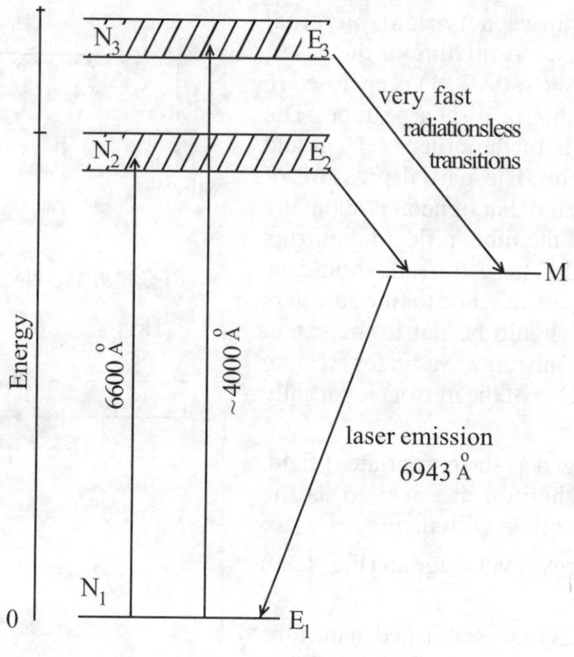

Fig. 12.7 Energy levels of chromium ions in the Ruby crystal.

When the flash lamp operates, the population in the upper level decays very fast and the lasing action stops till the arrival of the next flash from the lamp. The pulses, thus emitted, possess very high power. The threshold power per unit volume required to maintain the population inversion in the three level laser is

$$P_t = \frac{N}{2}\,\frac{h\nu_p}{t_{sp}} = \frac{1.6\times10^{19}\times6.7\times10^{-34}\times6.25\times10^{14}}{2\times3\times10^{-3}} \approx 1100 \text{ W/cm}^3.$$

The average pump frequency used is 6.25×10^{14} HZ. However, the efficiency of the pump source is normally 25% and there are absorption losses in the ruby rod as the light passes through the rod. Even if this loss is about 25%, The power output is about 18 KW/cm^3.

12.7 Gas Lasers

12.7.1 He-Ne laser

Working of He-Ne Laser

The gas laser have the atomic, ionic or molecular gas as active medium with desired set of energy levels. The gases are kept at low densities. The gas is always enclosed in a cylindrical tube arranged between two mirrors.

The first continuous wave (CW) laser that came in 1961 was He-Ne laser. The gas that used was a mixture of helium and neon.

The Fig. 12.8 shows a typical physical arrangement of gas laser. A mixture of the gases, Helium and Neon (in the ratio 7 : 1) is enclosed in a narrow and long quartz discharge tube. The diameter of the tube is of the order of 1 cm and length of about 80 cm. The partial pressure of helium is about 1 torr and that of neon is about 0.1 torr. On either side of the tube, reflecting mirrors M_1, M_2 are arranged. These mirrors should be ideally parallel. They are attached to fine adjusters (a_1, a_2). The mirrors should be flat to the extent of $\lambda/100$ flatness. The mirrors are used for creating an optical resonator. One of the mirrors is partially transparent.

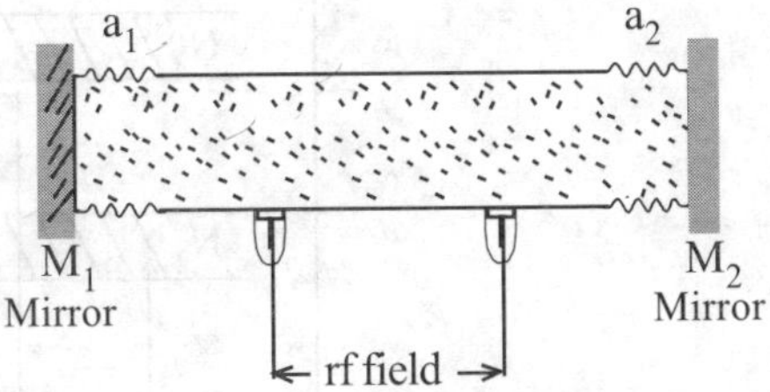

Fig. 12.8 He-Ne laser arrangement.

The gas is discharged at an appropriate rf field. Then, the atoms of helium are excited as the electrons in the tube collide with them.

Let us see the energy level diagram (Fig. 12.9) of helium and neon.

The gas discharge is first established in the tube by applying a high voltage (V~10 kV). The Helium atoms are excited. The excited helium atoms in turn transfer energy to neon atoms.

Helium has energy states 3S states, 2S states and 1S states. Neon has energy levels 2S and 3S levels. When helium atoms are excited by electron impact in the discharge tube, the population of the energy states 3S and 2S states build up. As the energy states of neon are close to the metastable levels of helium, there is a high probability of transfer of energy when the excited helium atoms and electrons collide with neon atoms. Thus, the collisions will result in energy transfer

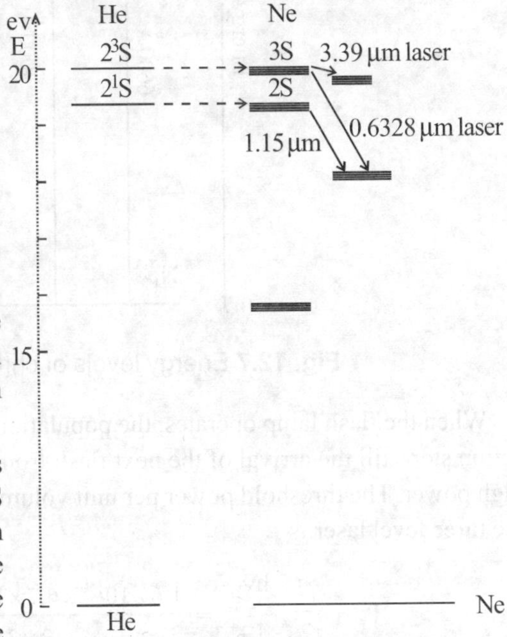

Fig. 12.9 Energy level diagram of He and Ne atoms and laser transitions.

$$\text{He}\ (^3\text{S}) + \text{Ne} \rightarrow \text{He} + \text{Ne}\ (2\text{S})$$

$$\text{He}\ (^1\text{S}) + \text{Ne} \rightarrow \text{He} + \text{Ne}\ (3\text{S})$$

under suitable discharge condition, the population inversion in Ne(2S) and Ne(3S) levels can take place. There is a higher probability of levels 3S being populated more.

The stimulated emission comes as a result of the electron transitions in neon atom which is now in excited state. The transitions and the corresponding wavelengths of the emitted light are given below.

$$3S_2 \rightarrow 2\,P_4 \rightarrow 0.6328\ \mu\text{m}$$

$$2S_2 \rightarrow 2P_4 \rightarrow 1.15\ \mu\text{m}$$

$$3S_2 \rightarrow 3P_4 \rightarrow 3.39\ \mu\text{m}$$

The transitions are shown in Fig. 12.9.

The helium metastable levels in helium-neon laser discharge acquire population densities of the order of 10^{17} per cubic meter. This population density is about three orders of magnitude higher than that of the order of the other excited energy levels of the atoms within the discharge. In the process of collision and the energy transfer to excited states of neon atoms, there is no radiation decay of the helium to ground state. The population densities of excited levels of helium atom is about 10^{17} cubic meter and suppose there are 1.4×10^{22} m^{-3} neon atoms, the excitation rate is of the order of 3.6×10^5 per second. The major spectral line 6328 Å is emitted. All the other spectral lines are eliminated in the process of reflections.

12.7.2 The CO_2 laser:

The carbondioxide laser is a four level laser that gives out infrared radiation. Unlike He-Ne gas laser which is an atomic laser, carbon dioxide laser works with molecular transitions in vibrational and associated rotational levels. CO_2 molecular laser was first demonstrated by C. Patel.

The schematic diagram of conventional carbon dioxide laser and power supply is given in Fig. 12.10.

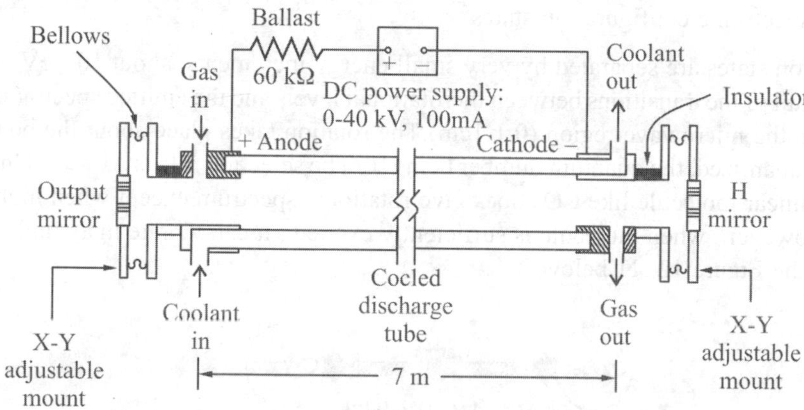

Fig. 12.10 CO_2 Laser unit.

The laser makes use of the discharge of a mixture of gases, namely carbon dioxide, nitrogen and helium roughly in the ratio 15:15:70. A low pressure of the order of 10 – 20 bar is maintained.The helium content is varied a little depending upon the power requirement. The discharge tube is a long narrow tube with the active medium. The electrodes are arranged on either side of the tube as shown in the Fig.12.10 for producing the discharge current. A totally reflecting mirrors made of semiconducting materials like silicon are used for pumping back the radiation into the tube. A partially transmitting windows made of zincselenide,or germanium or NaCl is used to close the tube on either side. The tube behaves now like a resonator. With the proper voltage applied, the laser gives a powerful radiation with nearly 200watts power and a wavelength of 10.6μm comes out. It works as a CW laser. The width of the beam would be 1-7mm with a divergence of 2-6 radians.

Pulsed lasers can be operated by giving pulsed power and the laser is capable of giving pulse widths 0.1 – 1 msec. The peak power would be 5 - 10 times that of CW laser with the same power supply.

As the radiation of the laser is in the infrared region, the discharge produces lot of temperature in the tube. Therefore cooling is necessary. The arrangement is made for it as shown in the fig. 12.10. High power lasers use flowing gas technology to remove hot gas from the discharge tube, and replace it with cool gas What is given in the figure is only a simple model ,the actual lasers are operated in different structures depending on the power requirements. The carbon dioxide laser is a very efficient laser with an efficiency coefficient of 0.4.and the powers to the tune of several kilowatts can be obtained from the laser.

Principle of Laser action:

In an atomic laser, the electronic energy levels are used for pumping ,stimulation and laser out put. In the case CO_2 laser, the molecular energy levels are used.

The molecular energy states arise from the rotation of a molecule as a whole and vibrational energy levels arise from vibration of its atoms relative to the center of mass. Molecules are also subjected to electronic configuration states .

The rotation states are separated by very small energy increments, about 10^{-3} eV. The spectra arises as a result of the transitions between therotational levels and the emitted spectra(or absorbed spectra) lie in the microwave region (0.1-1cm).The rotation takes place about the bond axis and the energy is quantized, the quantum number being 'J'. The selection rule is $\Delta J = \pm 1$. Under normal condition, a linear molecule like CO_2 does give rotational spectrum except when high collisions take place.However , when molecule is sufficiently excited , it can vibrate in a number of modes as shown in the Figure 12. 11 below.

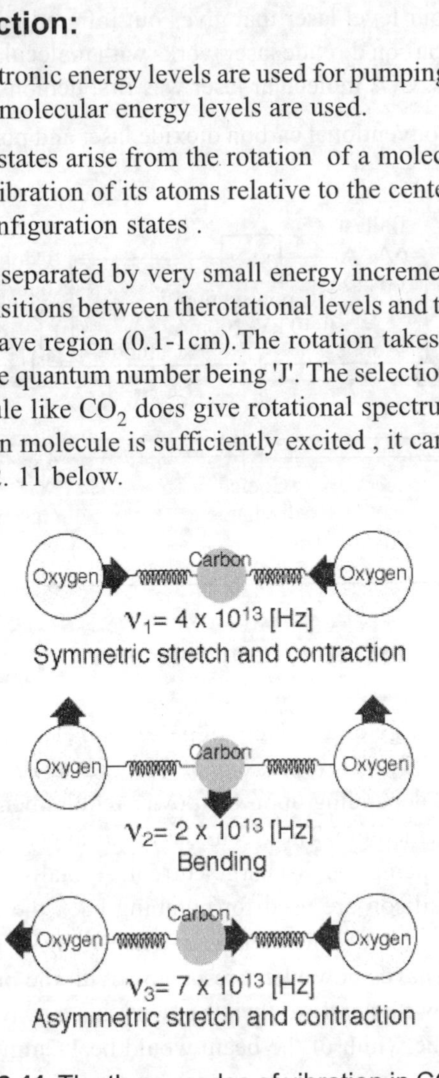

Fig. 12.11 The three modes of vibration in CO_2 molecule.

These vibrtional energies are also quantized and the quantum number associated is 'v'.v can take values 0,1,2,3......The selection rule for transitions is $\Delta v = \pm 1$.The three modes shown vibrate with different frequencies. The three modes are called 1.symmetric stretch and contraction, 2. Symmetric bending, 3. Asymmetric stretching. The ground energy levels are different for these modes.The highest is observed for asymmetric stretching.The lowest energy levels of the vibration are shown in Fig.12.12. nitrogen and carbon dioxide for when excited because of discharge, the nitrogen goes into metastable level and this is on par with (001) level of vibration of CO_2. Collisions take place and the nitrogen transfers energy to the energy level of CO_2 as shown in the Fig. 12.12. As more and more nitrogen goes into metastable level, more and more energy is transferred to (001) state of CO_2. The molecule already excited comes to lower levels.

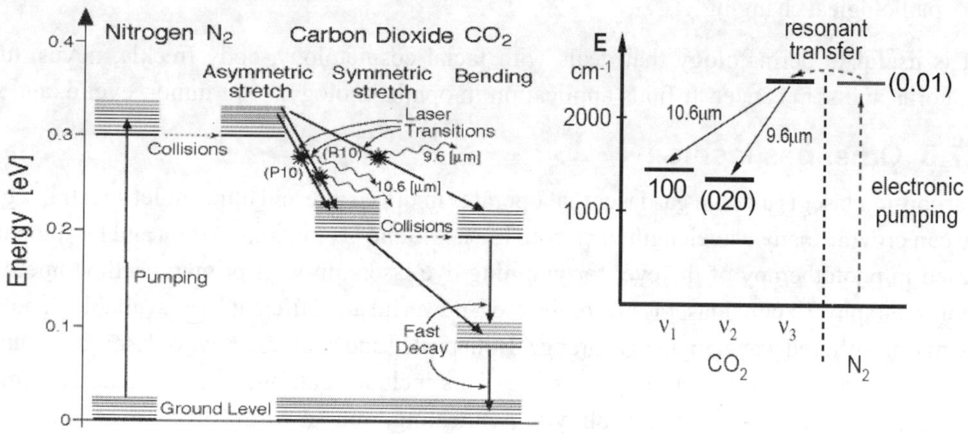

Fig. 12.12 The energy level diagram of the CO_2 laser.

Morc and morc CO_2 molcculcs get excited to the level shown. There are two other levels which support the othertwo modes. , namely, (100) and (020) levels. The molecule tends to come two these levels by emitting 10.600 μm and 9.600 μm radiations that lie in infrared region of spectrum. However, the CO_2 tries to come to levels like (010) level and the dexcited level thereby discouraging the lasing action. Here the helium which in plenty collides with the carbon dioxide molecules and prevents them from coming to lower level. That is the helium atoms cause the depopulation of the lower level. Therefore, the population inversion is achieved efficiently.

Properties and applications of CO_2 Laser:

- The CO_2 laser works in the mid-infrared (IR) range,ie., 9000~11000 A^0 on the basis of vibration–rotation transitions.
- The laser is capable of giving high output power (>10000W)
- The operation of the laser is very simple and the gasses are non-toxic.
- Very high efficiency (up to 40%)
- The advantage of this laser is that it can be operated both in a continuous mode or in pulsed mode.

- The carbon dioxide laser is supposed to have won the war against aging skin. Resurfacing is a treatment to remove acne and chicken pox scars, and changes in the skin due to ageing. Actinic, warts, moles, skin tags, epidermal and dermal, blister and punch grafting, scar treatment, , skin cancer can be treated with this laser.
- These lasers are found to be very useful in surgery. The radiation angle is very small and the energy density (several kilowatts per square centimeter) of carbon dioxide laser is very high. Infra red radiation with wave length, 10600nm, is useful. The beam has good penetration. The laser instrument can excise anything superfluous or useless disorder that the people suffer. It is useful in removing the wrinkles on the face of people and to prevent pathological changing.

It is useful in dermatology that deals with facial cosmetology, body freckle, nevus, tumor, wart, corns and skin cancer. It finds application in ophthalmology: orbit tumor, eyelid, and sty.

12.7.3 Other gas lasers :

The argon-ion laser is a noble gas laser that operates in the visible and ultra-violet spectral regions. They can provide many wavelength lines both in visible and Uv-regions. Argon and Krypton lasers are used in phototherapy of the eye, for pumping dye lasers, in laser printers, cellcytometry and stereolithography. Excimer laser is one of the most powerful and efficient laser available. It operates in the middle infrared wavelength in the range. Both pusled and continuous wave lasers are available. These lasers are used in materials processing. This includes cutting, drilling, material removal, etching, milling, melting, welding, alloying, annealing, and so forth. It has excellent medical applications where it is used for cutting and cauterizing. With these lasers, a very intense heating can be produced at a small area or a point.

12.8 Semiconductor Lasers

A semiconductor is associated with a valence band and a conduction band. Let this band gap energy be E_g and E_{fp} and E_{fn} be the quasi fermi states. In semiconductors like GaAs, GaP, InS etc., that have direct band gap, the electrons have high probability of recombining with holes thereby emitting radiation-photons. This is shown in Fig.12.13. The recombination radiation photons produced may interact with valence electrons and be absorbed. They can also interact with electrons in the conduction band thereby stimulating further photons of the same energy. The injected carrier concentration can become large enough and the stimulated emission can exceed the absorption. The optical gain can be achieved in the active region. In diode lasers, it is necessary to use the external mirrors so that light emitted by the diode is reflected back to provide the positive feed back. The high refractive index of the semiconductor materials ensures that most of the light is reflected into the materials at the semiconductor-air interface. We shall study briefly the working of the laser.

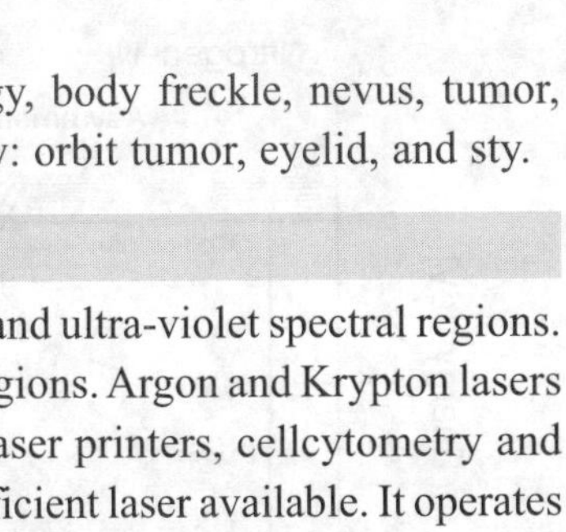

Fig. 12.13 Energy band picture of laser transitions.

12.8.1 Working of a Semiconductor Laser

The principle of working of a Semiconductor laser is similar to LED working. Semiconductor lasers are capable of operating as light source for fiber optical communications. Semiconductor lasers normally operate in 0.8 μm, 1.3 μm 1.55 μm region. Table 12.1 summarizes the emission characteristics of certain semiconductors lasers. Pure semiconductors like GaAs, ZnS, InSb also work as laser sources. But the diodes are prefered for their stability. No laser action has been observed in silicon and germanium.

Table 12.1 Emission characteristics of Laser materials

Material	Wavelength (μm)
Ga As P - GaAs	0.90
$InGa_{0.5}P_{0.5}$ - GaAs	0.89
InGaAsP - GaAsP	0.63
PbSnJ - PbSe	0.91
GaAsP-InGaP	0.68
PbS - PbSe - PbSn	5.48 - 8.95

The essential conditions for lasing action as discussed earlier are :

1. Carrier injection
2. Stimulated emission
   (a) Pumping
   (b) Population inversion
   (c) Presence of photon in the laser
3. Optical feedback - amplification

12.8.2 Laser Diode

Semiconductor lasers work on the same principle as light emitting diode, and the p-n junction provides the active medium. Thus, for obtaining the laser action, we have to provide the condition for population inversion and optical feed back. For a stimulated emission, there must be a region of the device in which there are exited electrons and holes present together. This is achieved by forward biasing a junction formed from very heavily doped p – and n–materials. In such a heavily doped n–material, the Fermi level lies in the valence band. The equilibrium forward bias energy level diagram for a junction formed from such materials is depicted in the figure 12.14. When forward biased, a large concentration of electrons and holes are injected across the junction in opposite directions. Sufficient number of electrons are pushed into conduction band at the junction on n-side. Therefore a population inversion is created in a narrow region or active region. The band diagram of a simple PN junction diode under unbiased and biased conditions is shown in Fig. 12.14.

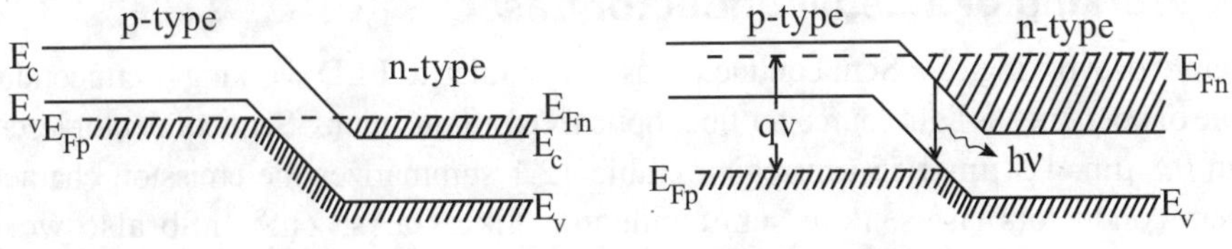

(a) Unbiased laser diode (b) Forward biased laser diode

Fig. 12.14 Band diagram of laser diode.

In most semiconductor lasers, the lasing action is made to occur on the p side of the junction by electron diffusion, as electron-diffusion process has higher efficiency than hole-injection. Over the diffusion length, the injected carrier density is much more than equilibrium charge concentration. This results in population inversion in a narrow region near the junction. In this region, electrons and holes recombine radiatively resulting in the emission of light spontaneously. The photons are made to stay within the region. They give rise to stimulated emission. If the emitted radiation is made sufficiently intense and reflected into the diode junction continuously, a gain in the output beam intensity is achieved. The laser light comes out as a highly directed and narrow intense beam of light. There can be losses due to various reasons but losses are kept at minimum level so that lasing action takes place efficiently.

12.8.3 Example of Laser Diode

The semiconductor diode material is a diode laser that acts as an optical cavity. The active region is parallel to the junction face. The laser beam is taken from the side of the junction. The optical cavity is formed by the face of the crystal itself, which are the cleavage planes in GaAs. The arrangement is shown in Fig. 12.15. The (110) (Fig.12.15) plane of the semiconductor crystal has enough refractive index to confine the light emitted due to recombination of electrons and holes. Multiple reflections take place at the boundary and into the active region. This is a very intelligent manipulation of the optical property of the semiconductor involved. Due to these multiple reflections, the light amplification takes place and gain attained exceeds the losses due to enhanced stimulated emission. The population inversion occurs in active region of thickness d. This thickness will be approximately equal to diffusion length of the injected charges. The injection current at which lasing occurs is called the threshold current (I_{th}) (Fig. 12.16). The power radiated is a function of the forward current I. Below the threshold current, a laser diode operates as a common LED, emitting spontaneous radiation in all directions. Most of the energy is lost in reflections in all directions.

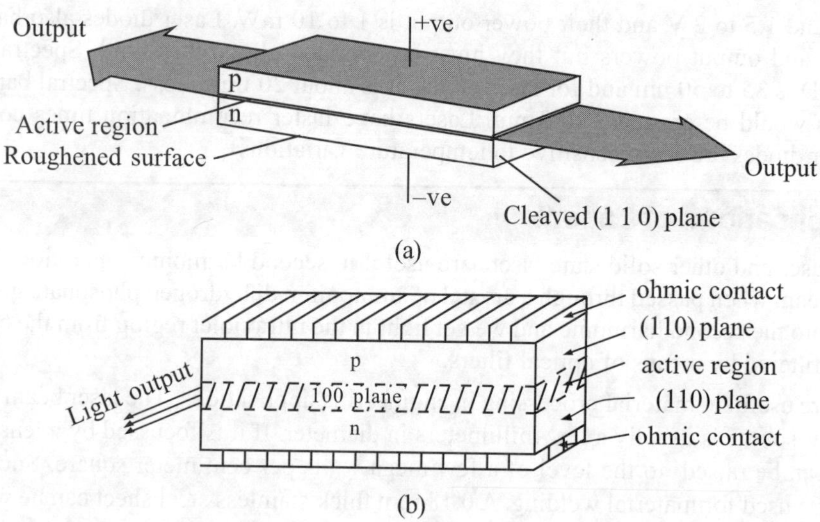

Fig. 12.15 (a) Semiconductor laser diode (b) a detailed figure of (a)

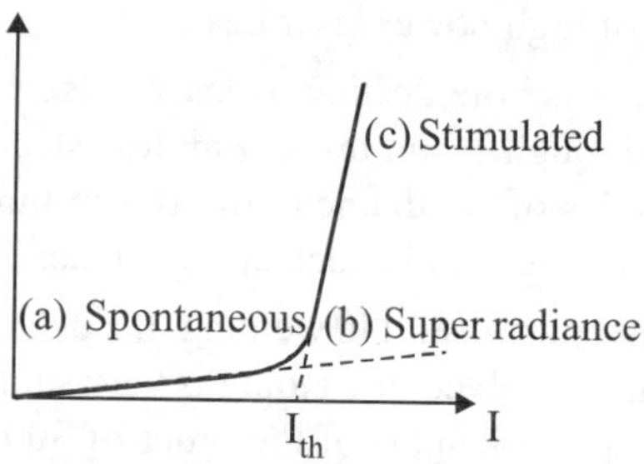

Fig. 12.16 Characteristics of a laser diode.

Above the threshold current (i.e., $I > I_{th}$), the ratio between the probabilities of radiative or nonradiative recombination grows on account of the increased probability of induced optical transitions. This results in a sharp rise in the power output.

The disadvantage of a semiconductor laser stems from the fact that semiconductors are extremely sensitive to temperature changes. With a slight increase in temperature, appreciable increase in forward current takes place and the energy gap of the material is changed, causing a change in the frequency of light emitted. It also results in a change in I_{th}. Due to this, amplification falls off. Therefore heat sinks are provided for the PN junction.

Confinement of light in Homo junction laser discussed above is extremely poor. To improve this, Hetro junction Lasers are used. The efficiency of the laser is increased in hetrojunction lasers.

12.8.4 Comparison of Laser and LED

Laser diodes and LEDs differ in a number of ways, mainly from the point of view of power, spectral width, speed, linearity, temperature sensitivity, reliability etc. LEDs are operated with 50 to 300 mA and 1.5 to 2 V and their power output is 1 to 10 mW. Laser diodes also have similar width input and output powers but they must be operated above threshold. Spectral width of GaA*l*As LED is 35 to 50 nm and for laser diodes it is about 20 times less. Spectral bandwidth of laser diodes would be around 2 to 3 nm. Lasers have faster recombination times compared to LEDs. Laser diodes are more sensitive to temperature variations.

12.9 Applications of Lasers

The Ruby laser and other solid state lasers are useful in second harmonic generation. Ruby laser (red light) beam when passed through a crystal of ammonium di hydrogen phosphate gets partially converted into the second harmonic and we get light in the ultraviolet region from the crystal. The red light is filtered by means of optical filters.

Lasers are useful in material processing in engineering applications. The laser beam that comes out of a laser source is usually a few millimeters in diameter. If it is focussed by a lens to a point, the power can be raised to the level of a few mega watts per centimeter square. Such focussed beams can be used for material welding. A 0.18 mm thick stainless steel sheet can be welded by a Ruby laser pulsed beam of 5 Joules with a pulse length of 5 n sec. This type of precise welding is extremely important in micro electronics in which thin films are used. Thermo-couple wires can easily be welded with the help of high power laser beams.

Another important application is hole drilling. A laser pulse with a width of 10^{-4} Sec and an energy of 0.05 Joule can pass through 1 mm thick strainless steel sheet leaving behind a hole of radius 0.1 mm. One can drill holes of the diameter of 10 μm through very hard substances like diamond. In fact, YAG laser is very useful in such applications.

Ruby laser and solid state lasers are useful in welding of a detached retina. The precise location of the weld can be made, the time involved in welding is very short and the eye does not need any clamping. The pules used have pulse lengths of the order of 300 μsec at an energy level of one joule.

Ruby laser beam with a power output of 100 mega watts with a total energy of 200 Joules are useful in the treatment of skin cancer and tumors.

12.10 Applications of Semiconductor Lasers

The injection lasers discussed earlier, have a disadvantage in the fact that the parameters of the lasers are temperature dependent. As temperature increases, the wavelength of light emitted shifts to longer wavelengths. The temperature also increases the current in the diode, thereby decreasing the amplification. The heat has to be removed always from the laser source to get good efficiency.

The semiconductor lasers are of great use in communication electronics and computer engineering. The driving force in the development of semiconductor lasers is in the field of communication. In the early 1970s, it was shown that optical fibers of good quality could transmit certain wavelengths of light at very low loss and were also relatively inexpensive to fabricate. The laser beam is highly effective for long-range information transmission. In general, the diode lasers have low power. The diode lasers have a special place. They are small in size and are useful in hand-held instruments. They can be integrated with other instruments like in the computer laser printer. Pulsed lasers as well as continuous wave (CW) lasers are available.

The present age belongs to miniaturization. The effective dimensions of the semiconductor devices are coming down from three dimensional to two-dimensional to one-dimensional level to ultimately zero dimensional level.

Quantum dot laser (QD): In these laser systems, the wavelength of the light emitted is not determined by the band gap of the semiconductor material, but by just a dot of a semi conductor that's created between two semiconductor films. The electrons are confined to the dots and occupy sharp energy levels. The quantum dots behave like artificial atoms and the electron is said to experience a quantum confinement. In a quantum dot laser, gallium arsenide dot is created as an ultra thin plane between two different semi conducting thin films, namely the alloy of gallium arsenide and aluminum. As a result of the quantum confinement, the electrons move freely in the dot, forming a two-dimensional electron gas. The size of the device is comparable with de Broglie wavelength. The electron-hole pair recombination takes place in the active region, that is the gallium-arsenide well. These lasers give a coherent radiation that is highly monochromatic with low energy loss and insignificant divergence.

The continuous wave lasers are useful in optical detection and target ranging. They are used in communication systems that involve conveying information at a distance. Laser links are useful in the same way as the microwave links of the coaxial cables. Nowadays, the coaxial cables are being replaced by optical fibers.

The optical fibres, free of impurities can transmit certain wavelengths of light at very low loss. The optical fibres can easily be fabricated at low cost as they are basically silicon dioxide fibres of about 1.3 μm or 1.5 μm in diameter. The fibre optic system is slowly replacing the regular micro wave linking.

The semiconductor lasers are useful in high speed computer networks, avionic systems, satellite networks and high definition television. We can have fibre optic communication like telephone, but also in video, audio and computer signals. High power sources are needed for the above purposes. The fibre optic system will provide the largest market for semiconductor lasers.

Optically coupled lasers are used for the construction of optical logic elements that have very high speed (10^{-10} sec). They are going to be the basic elements for the development of super computers.

The focusing property of laser radiation is also important in low power applications. A laser beam is used for reading the bar code of consumer items in supermarkets. The bar code is a series of parallel bars of varying widths giving information about the consumer product. A semiconductor laser scans these bar codes. The light reflected from this bar code is detected, and, the product and pricing are thereby identified.

Semiconductor laser diodes are also used in as 'needle' or 'probe' preparation and subsequent read out of disks in computer CDROMs (compact disc read only memory). The information to be stored on the disk is imprinted in the disk in digital form by forming small pits (using higher powered laser). A lower power laser that provides a signal for playback of the data can read it. The digitally modulated radiation that is scattered from the disk is collected by an optical sensor and processed through the audio amplifier system. Similar system is used in CD (audio) and VCD (video CD) players.

Applications of He-Ne Lasers :

Helium-Neon lasers emit powers in the range 0.5 mW to 50 mW. The line width of the spectral line (6328 Å) is of the order of a few picometers. In low power tubes the mirrors are fixed to the tube. But in high lower tubes, they are placed away from the tube. He-Ne laser is a common laser that is found in laboratories. The applications include interferometry, laser printing, bar code reading, holography etc. The He-Ne laser provides the stable, single-transverse mode beam necessary for indentifying optical properties of materials such as surface smoothness. By placing the mirrors at different angles, the output can taken to be either 6328 Å or 1.15 μm laser light.

The helium-neon lasers are used for surveying, and the target-aiming devices for guns. These lasers are used for larger distance measurements which is laser modulation telemetry. The laser beam is amplitude modulated and made to fall on a 'target' whose position from a point has to be known. The target reflects light and the reflected light is detected. The phase shift between the reflected and the direct beam is measured to calculate the distance ($\phi = 2\pi/\lambda \times 2n\ D$). The lasers are also useful for surveying and land profiling.

The power of the He-Ne Lasers that are used would be between 2mW to 50 mW. These lasers are highly useful in laboratory experiments in science. They have an operating period of 50,000 hrs. The laser has a coherence length of 300 meters. The lasers are used in holography (three dimensional photography).

The 1523 μm laser is used in optical transmission lines, which have minimum loss in that wavelength region.

Questions

1. Expand 'Laser' and Maser'.
2. Mention some important Laser systems
3. What are the steps involved in lasing action.
4. Discuss in detail the absorption, and emission of spontaneous, stimulated radiation by taking a two level energy level diagram.
5. What is population inversion? Explain.
6. Obtain the expression for occupation index in terms of Einsten's coefficients.
7. Explain the meaning of light amplification. Is population inversion necessary for gain in a laser. If so, why?
8. How is light amplification achieved in a laser system.
9. What is coherence? What is the criteria for estimating the coherence of a light source?
10. Why laser is high coherent source ?
11. What do you mean by 'directionality''? Compare gas lasers and a solid state laser.
12. Mention the three methods of causing population inversion.
13. Explain the method of "inelastic atom-atom collision" for achieving population inversion.
14. Give the physical arrangement of a gas laser. Why are the mirrors used?

15. Give the energy level diagram of helium-neon atoms. Explain the emission of laser light.
16. Give the applications of He-Ne laser.
17. Explain why indirect band semiconductor is not suitable for lasing action.
18. Explain the working of a semiconductor laser.
19. What are the difference between a laser diode and a LED?
20. What is the active region in a semiconductor diode laser?
21. Write a note on the applications of semiconductor lasers.
22. Explain how laser beam is useful in reading CDs.
23. What is meant Laser 'speckle'?
24. Explain the principle of a semiconductor diode laser.
25. What is 'gain'? How it is obtained in a laser ?
26. What is meant by 'Laser Oscillations'?
27. Mention the basic requirements for a 'Laser'.
28. Discuss about the working of a ruby laser.
29. Explain the term 'Visibility' of interference fringes.
30. Distinguish between 'Spontaneous Emission' and 'Stimulated Emission '.
31. Give the diagram of a He-Ne Laser System. Explain the emission of radiation from He-Ne medium.
32. Discuss the working of Ruby Laser. Mention its applications.
33. Give the energy level diagram of CO_2 Laser and explain the working of CO_2 laser.

Objective Questions

I. Choose the correct choice

1. The directionality of a laser beam is measured by
   (a) the visibility of interference fringes
   (b) the size and aperture of the laser source
   (c) the divergence angle of the beam with the distance from the source
   (d) nature of the lasing medium

2. The coherence of light is known by the
   (a) visibility of the interference fringes it produces
   (b) size and nature of the source
   (c) light intensity of the beam
   (d) flickering of the light beam

3. A mercury lamp gives out radiation that is
   (a) spontaneously emitted and incoherent
   (b) spontaneously emitted, 'white' radiation, coherent and directional
   (c) stimulated emission, 'white' radiation, non coherent and directional
   (d) selectively mono chromatic, incoherent and directional

4. The light coming from an ordinary source like a filament bulb is
   (a) coherent, non directional, but powerful
   (b) non monochromatic, non directional, unpolarized and incoherent
   (c) coherent, non monochromatic, non directional but intense
   (d) non coherent, directional, intense and non monochromatic

5. The spontaneous emission is a
   (a) natural and equilibrium process that takes place when atoms are excited
   (b) process that requires population inversion
   (c) negative process
   (d) Non equilibrium process that takes place when atoms are excited.

6. He - Ne Laser is a
   (a) liquid laser (b) organic laser
   (c) solid slate laser (d) gas laser

7. Cutting and Drilling of metals require high energy sources. The Laser that are used are
   (a) He - Ne laser
   (b) semi conductor laser
   (c) solid state lasers
   (d) Co_2 Laser

8. The 'Lasing process' is possible only if there is
   (a) Oscillation of Laser source
   (b) a set of reflecting mirrors
   (c) population inversion
   (d) black body

9. Population inversion can be achieved in atoms if there is
   (a) an extra discharge of the gases in the gas discharge tube
   (b) continuous pumping of energy in to the system by reflectors and amplification.
   (c) Infinite amplification of light in the source by high discharge process
   (d) transfer of energy from Laser substance to the reflecting mirrors.

10. In He -Ne Lasers, the helium and neon are in the ratio
    (a) 1:7 (b) 7:1
    (c) 1:10 (d) 10:1

11. The laser that are used in holography are
    (a) solid state Laser
    (b) Argon pulsed laser
    (c) semiconductor Laser
    (d) He - Ne Lasers

12. The semiconductor lasers are not used in
    (a) bar - code reading
    (b) optical communication
    (c) cutting and drilling of metals
    (d) compute printers

13. The He -Ne laser discharge tube is fitted with $\lambda/100$ flat mirrors for
    (a) good looks and minimum losses
    (b) pumping light in to discharge tube
    (c) getting polarized laser beam
    (d) getting coherent beam.

14. The useful material for semi conductor laser source is
    (a) silicon
    (b) germanium
    (c) organic semi conductors
    (d) gallium arsenide.

15. In semi conductor laser diode the population invasion takes place
    (a) at the junction on n side only
    (b) far away from the junction on the neutral n - side
    (c) far away from the junction on the neutral p - side
    (d) near the junction on the p - side

16. For stimulated emission to take place in an active medium, the condition is
    (a) the production of population inversion in energy levels
    (b) sufficient number of atoms in ground state
    (c) coherent incident radiation on the medium
    (d) large volume of active medium

17. The purpose of the optical resonator in a laser is
    (a) to provide cover to the active medium
    (b) to provide path for stray atoms
    (c) to provide selectivity of photon states
    (d) to send laser radiation in specified direction

18. In He-Ne gas laser
    (a) the laser active centers are neon atoms
    (b) the laser active centers are helium atoms
    (c) the inverted population density is achieved in helium atoms
    (d) the gases are maintained in 1:2 molar ratio
19. Verify the given statements
    (A) The gas lasers normally exhibit greater coherence than solid state lasers
    (B) 'The information capacity' of coherent radiation is tremondously high
    (a) (A) and (B) are wrong
    (b) (A) is wrong or (B) is correct
    (c) (A) is correct, (B) is wrong
    (d) (A), and (B) are correct
20. In He-Ne lasers
    (a) oxgen is used to burn and emit light
    (b) the transfer of energy of excitation takes place with by collission of excited helium atoms with neon atoms
    (c) the helium atoms give the metastable level in between neon energy levels
    (d) the neon atoms under low pressure excite helium atoms
21. Yittrium aluminium garnet is a
    (a) solid state laser source when doped with 3% molybdenum ions
    (b) wonderful superconductor
    (c) very low power laser source
    (d) ferro magnetic material
22. In a ruby laser
    (a) very low peak power pulses of radiation are given out
    (b) a flash lamp is used to excite chromium ions
    (c) a flash lamp is used to excite alumimium atoms
    (d) the chromium ions collide with alumimium ions to give out bursts of energy
23. At thermal equilbrium in a laser source
    (a) the ratio of the number of stimulated emissions to the number of spontaneous emission is equal to Einstein coefficiant for emission
    (b) the ratio of the number of spontaneous to stimulated emissions is given by $\left(e^{\hbar\omega/kT}-1\right)$
    (c) the ratio of the number of the stimulated emission to spontaneous emission is equal to the ratio of their Einstein coefficrents
    (d) the ratio of probabability of stimulated emission to that of spontaneous emission is equal to one
24. The rate at which the stimulated transitions from an upper level E_2 to a lower level E_1 lakes place as a result of an incident radiation of density U_v is proportional to
    (a) The population of E_1 level
    (b) The population on E_2 level and U_v
    (c) inverse of the population in E_2
    (d) inverse of the population in E_2 and U_v
25. Cheek the correctness of the statements
    (A) The gain of a laser source in the active medium depends on the population difference
    (B) The purpose of reflectors in the laser is to deflect the photons from the active medium
    (a) (A) and (B) are wrong
    (b) (A) is wrong and (B) is correct
    (c) (A) and (B) are correct
    (d) (A) is correct and (B) is wrong
26. Choose the correct statement
    (a) The occupation index for two levels E_1 and E_2 is given by $(B_{21}/A_{21})\ U_v$, (U_v is the radiation density)
    (b) The occupation index for two levels E_1 and E_2 is given by U_v
    (c) The rate of induced transitions in a laser is proportional to the energy of the ground level E_1
    (d) cordinary source of light can give coharent beam of light if it is enclosed between mirrors

II. Fill in the Blanks

1. A Laser beam has the characteristics
   1. __________
   2. __________
   3. __________
2. For amplification to take place in Laser source, __________ is a necessary condition.
3. A laser source is a quantum __________.
4. The most important wave length of radiation that is emitted from He - Ne Laser is __________ A^0.
5. Generally,the Lasers that are used in holography are __________ lasers.
6. The pumping process that is used in He -Ne lasers source is __________ .
7. The speckle nature of the laser beam is due to high __________ and __________ divergence of Laser beam.
8. The population inversion in the active medium of Laser source gives rise to __________ of the beam of light.
9. Compared to semi conductor lasers, the gas lasers give __________ coherent beam.
10. In terms of Einstein coefficients B_{21}, A_{21} and the radiation density U_v, the ratio of stimulated emission to spontaneous emission rate is given by the relation __________ in a two level laser system.
11. The lasers that are used for surveying and land profiling are __________ .
12. In He - Ne Laser system, __________ ions give metastable levels.
13. In He - Ne Lasers, the population inversion is achieved through __________ mechanism.
14. The process that can be used for achieving population inversion in Laser systems are 1. __________ 2. __________ 3. __________.
15. The light emission in He - Ne laser takes place as a result of the transition of excited atoms of neon from __________ levels to __________ levels.
16. No laser action is found in __________ semi conductors.
17. Compared to gas lasers, the power output from semiconductor lasers is __________ .
18. The computer printers utilize __________ lasers.
19. The bar code reading is done with the help of __________ laser beams.
20. The bias that is given to a semiconductor junction laser diode is __________.
21. High speed computer network use the __________ lasers.
22. It is convenient to use __________ lasers for optical logic elements in which the lasers are optically coupled.
23. Reading of CD ROMS is done with the help of __________ .
24. Semi conductor lasers can be made from __________ semi conductors.
25. Light emitting diode gives out radiation as a result of __________ emission.
26. When a semi conductor diode is maintained below threshold current, the light emission is __________ emission only.

Chapter 13

Fiber Optics

13.1 Introduction

Optical fibers are the wave-guides through which electromagnetic waves of optical frequencies can be guided through them to travel long distances. Optical fibers allow propagation of light waves in a way, which is very much similar to metallic wave-guides that allow propagation of microwaves through them. Optical fibers work on the principle of total internal reflection of light. Light will propagate for long distances in optical fibers because the absorption coefficient of fibers is exceedingly small. Optical fibers are made with low refractive index materials like glass. Hence optical fibers work as a transmission medium for optical signals. Light is an electromagnetic radiation and its propagation is governed by Maxwell equations. Complete understanding of propagation of light waves through optical fibers needs good understanding of solutions of these equations in the context of optical fibers. Maxwell equations, Fresnel laws and Snell's law can completely explain the propagation of light through optical fiber wave-guides. Since most of these are beyond the scope of the present text, a qualitative treatment of the propagation of light in optical fibers is presented in the next few sections.

13.1.1 Structure of the Optical Fibers

Optical fiber has a transparent core with refractive index n_1 surrounded by a transparent cladding of slightly lower refractive index n_2 (Fig. 13.1). The cladding supports the wave-guide structure and reduces radiation losses into the surroundings. Light energy travels in both the core and cladding, but it does not penetrate into the surroundings. The present day optical fibers have losses of about 4.2 decibel/Km. By refining and purifying glass used for fiber drawing, these losses can be reduced to 1 dB/Km. Optical fibers work as light wave guides. Optical wave-guides can be understood with the help of laws of electromagnetism and ray optics.

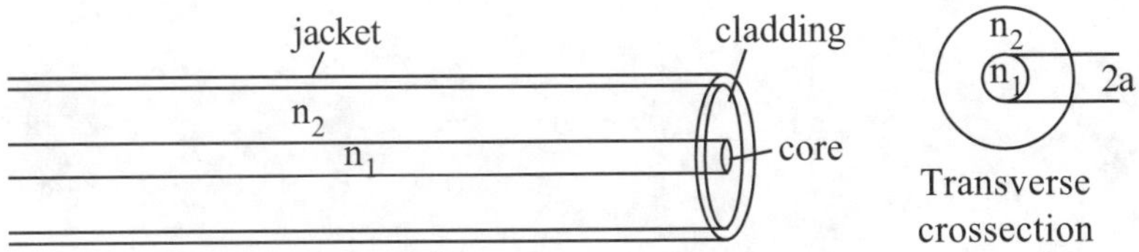

Fig. 13.1 Optical fiber.

Optical fibers are used in optical communication systems. These systems are used in digital telecommunications. The light pulses, which are modulated according to the signal to be transmitted, are guided along an optical fiber. The main advantage of optical fiber communications is its speed. The optical pulse repetition frequencies will be in the range of 1 MBPS (megabytes per second) to 1000 MBPS. Small size, good overall compatibility, relative simplicity is the essential features of fiber optic communication systems. The optical fibers are made out of glass, which is essentially silicon dioxide and is abundantly available and hence is much cheaper than a metal required for metallic wave- guides (copper or silver). The core and clad are placed in a jacket for safety.

13.2 Principle of Optical Fibers

13.2.1 Total Internal Reflection

The propagation of light through a medium is determined by the refractive index of the medium. Refractive index is the ratio of velocity of light in air to the velocity of light in the medium. A light ray travels slowly in denser medium than in rarer medium. When a ray of light is incident on the interface of the two media of different refractive indices, it bends and refraction occurs (Fig. 13.2).

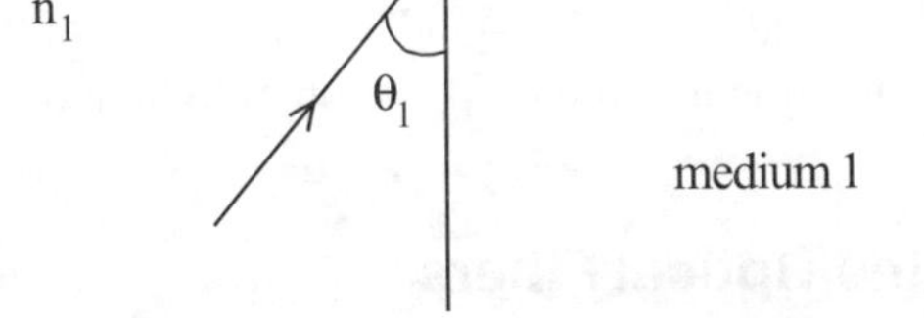

Fig. 13.2 Refraction of light.

Let the refractive index n_1 of the medium 1 be greater than the refractive index n_2 of the medium 2. Let θ_1 and θ_2 be angle of incidence and angle of refraction respectively. Then by Snell's law we have

$$\frac{n_1}{n_2} = \frac{\sin\theta_1}{\sin\theta_2} \qquad \text{........(13.1)}$$

Since n_1 is greater than n_2, we always have θ_2 greater than θ_1 in view of the above equation.

When $\theta_2 = 90^o$, the incident ray emerges parallel to the boundary of the two media. The angle of incidence $\theta_1 = \theta_c$ is called the critical angle. At the critical angle, we have from Snell's law

$$n_1 \sin\theta_1 = n_2 \sin\theta_2$$

i.e.,

$$n_1 \sin\theta_c = n_2 \sin 90$$

$$\sin\theta_c = \frac{n_2}{n_1} \quad \text{........(13.2)}$$

At the angles of incidence $\theta_1 > \theta_c$, the light will be reflected back into the medium 1 and suffers total internal reflection. Hence, total internal reflection occurs at the interface of the two media of differing refractive index and angle of incidence is greater than critical angle. Under these conditions, light can be constrained to move along the optical fiber. Fig. 13.3 illustrates the propagation of light along the fiber via a series of total internal reflections at the interface of core and cladding of optical fiber. Core has slightly higher refractive index than cladding material. The wave is transmitted as guided wave.

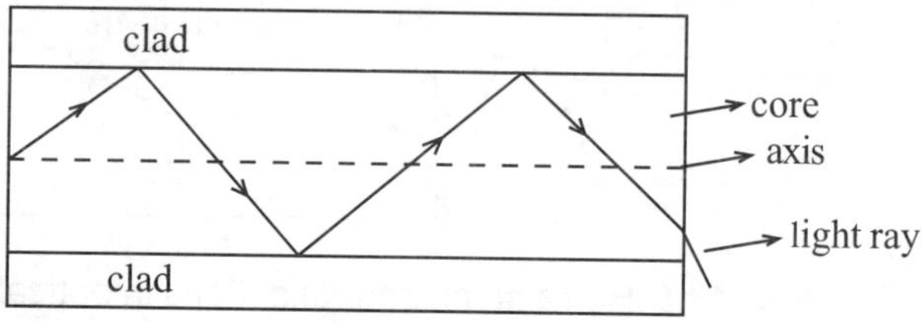

Fig. 13.3 Propagation of light in optical fiber.

13.3 Acceptance Angle

Not all rays falling on the optical fiber will be transmitted through the fiber. Only those rays with sufficiently shallow grazing angle $\theta_1 > \theta_c$ at the core-clad interface are transmitted through the fiber. Also, not all rays entering the fiber will continue to travel along the fiber. This is shown in the Fig. 13.4. Consider a ray A incident at an angle θ_a to the fiber axis. The ray is refracted at the fiber and refracted ray is incident at an angle θ_c on the core-clad interface. Therefore, it suffers total internal reflection. Whereas ray B falls on the optical fiber at an angle greater than θ_a. This ray does not undergo total internal reflection at the core-clad interface and enters the clad material. This ray will eventually get lost in the clad material itself. Thus, for the light rays to propagate through optical fiber by total internal reflection, they must be incident on the fiber core with in acceptance angle defined by conical half angle θ_a. This is the maximum angle to the axis of the optical fiber at which light ray may enter the fiber so that it can be propagated. Hence, this angle is called acceptance angle.

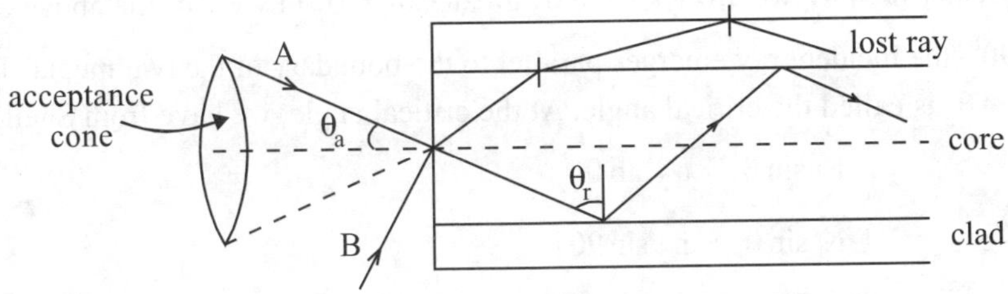

Fig. 13.4 Acceptance angle.

Acceptance angle is the angle at which light should be incident on the fiber for the propagation through it.

13.4 Numerical Aperture

Let a light ray be incident on the fiber at an angle θ_1 to the fiber axis ($\theta_1 < \theta_a$) Fig. 13.5. The ray enters from a medium (air) of the refractive index n_o and fiber core

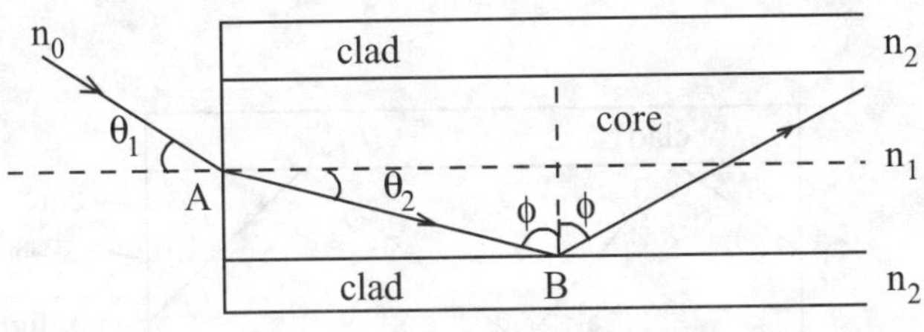

Fig. 13.5 Refractions and reflections in optical fiber.

refractive index be n_1 which is slightly greater than clad refractive index n_2. By Snell's law:

$$n_0 \sin\theta_1 = n_2 \sin\theta_2$$

$$n_1 \sin\theta_2 = n_1 \sin(90+\phi)$$

$$= n_1 \cos\phi = n_1 \left(1 - sin^2 \phi\right)^{1/2}$$

$$= \left(n_1^2 - n_1^2\, sin^2 \phi\right)^{1/2} = \left(n_1^2 - n_2^2\right)^{1/2} = \left(n_1^2 - n_2^2\right)^{1/2} \qquad \text{........(13.3)}$$

= Numerical Aperture.

In the limiting case, $\theta_1 = \theta_a$, $n_o \sin(\theta_a)$ is called numerical aperture (NA). Hence, we have

$$NA = n_o \sin(\theta_a) = (n_1{}^2 - n_2{}^2)^{1/2} \qquad \text{.....(13.4)}$$

In the case of optical fibers used in short communications, the numerical aperture will be in the range 0.4 to 0.5. In the case of long distance communications, the numerical aperture will be in the range 0.1 to 0.3.

The following is the summary of the above discussion :

(a) The optical fiber consists of core, clad (made out of glass) and a safety jacket made out of polymer.

(b) The refractive index of the core glass (n_1) should be greater than the refractive index of clad glass (n_2), i.e., $n_1 > n_2$.

(c) The light is propagated through the fiber as a result of multiple total internal reflections.

(d) Any light wave which is incident at an angle θ greater than the critical angle undergoes total internal reflection.

(e) A light that is incident at an angle less than the critical angle, will be lost in the clad.

(f) The angle at which the light ray is incident on the fiber such that it is transmitted through the fiber is called the acceptance angle. The cone defined by the twice the acceptance angle is called the 'Acceptance Cone'.

(g) The square root of the difference of the squares of refractive indices of the core and of refractive indices of the core and clad [$(n_1^2 - n_2^2)^{½}$] power is called the numerical aperture. It is called the figure of merit of the fiber.

Modes in optical fibers

An optical wave consists of the electric and magnetic field vectors perpendicular to each other and perpendicular to the direction of propagation. This type of wave is known as Transverse Electromagnetic Wave. In the case of bounded medium like an optical fiber, the optical wave is confined to the fiber that acts like a guide. As a result, the electric fields or the magnetic fields are distributed in the guide such that there will be a component of either the electric field or the magnetic field along the direction of propagation of energy. The stable field distribution in the y and z- direction with the periodic dependence in the x-direction is known as 'Mode'. If there is no electric field component along the direction of propagation of energy (i.e., $E_x = 0$), then the mode is called TE mode. If there is component of the magnetic field (H-component) along y and z and $H_x = 0$, then the mode is called TM mode. Both these modes are possible in an optical guide.

The light ray being guided through an optical fiber is a superposition of bound waves of different modes. Each of these guided modes consists of variety of electromagnetic field configurations. These field configurations differ in their sense of polarization, phase, amplitude etc. For a monochromatic light wave of angular frequency $\omega = 2\pi n$, n being frequency representing a mode traveling in positive x direction is given by

$$y = y_0 \exp[i(\omega t - kx)] \qquad .....(13.5)$$

where y is the amplitude of the wave at (x, t). y_o is the maximum amplitude b is the x-component of the propagation vector, $k = 2\pi/\lambda$ .This is the solution of wave equation.

For guided waves, k can have only certain discrete values. These values can be determined by applying boundary conditions at the core-clad interface while solving the differential equation of wave propagation through optical fiber (Maxwell's electromagnetic wave equations). There are three types of modes. They are:

1. Transverse electric (TE) modes ($E_x = 0$)
2. Transverse magnetic (TM) modes ($H_x = 0$)
3. Hybrid modes

Hybrid modes are combination of transverse electric and transverse magnetic modes. Different modes are shown in the Fig. 13.6.

Fig. 13.6 Different modes of electric and magnetic field oscillations in EM waves.

Several low order modes are shown in the Fig. 13.7. The order of the mode is number of zeros crossing of the field. The plots show that the electric and magnetic fields are not completely confined to the center of the fiber. They vary along the diameter of the fiber shown. Lower order modes are concentrated in the center resulting in meridional rays and in higher order modes the fields are normally distributed near edges resulting in skew rays. Increasing the core refractive index, increases the number of propagating modes. Increasing clad refractive index,decreases the number of propagating modes.

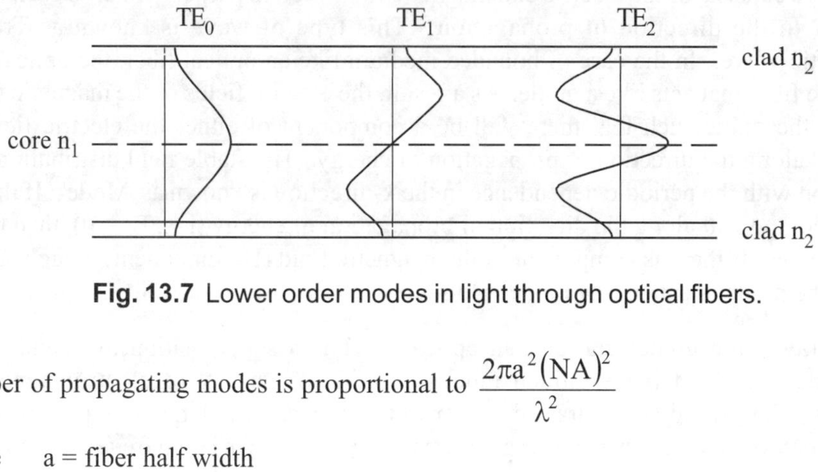

Fig. 13.7 Lower order modes in light through optical fibers.

Number of propagating modes is proportional to $\dfrac{2\pi a^2(NA)^2}{\lambda^2}$

where a = fiber half width

λ = wave length of the light

13.5 Step Index Fiber

Fibers in which core of constant refractive index n_1 is surrounded by cladding of slightly lower refractive index, with a steep refractive index profile at the interface of core-clad is known as step index fiber (Fig. 13.8(a)). The variation of the refractive index of a step index fiber as a function of distance can be mathematically represented as:

$$n(r) = n_1(r < a \text{ inside core}) \quad .....(13.6)$$

$$= n_2(r > a \text{ in clad}) \quad .....(13.7)$$

Core diameter will be of the order of 2 - 10 mm. Step index fibers can be made to be multimode fibers or single mode fibers. In multimode fibers, the core diameter will be large. In single mode fibers, the core has smaller diameter. Step index fibers support the transmission of transverse electromagnetic radiation. Single mode fibers transmit single ray along the axis of the fiber (Fig. 13.8(b)). Single mode fibers have the following advantages:

1. Low inter modal dispersion and lower broadening of light pulses being transmitted.
2. Larger bandwidth can be attained in these. Bandwidth is the difference between the minimum and maximum frequencies of light that can be sent through the fiber.

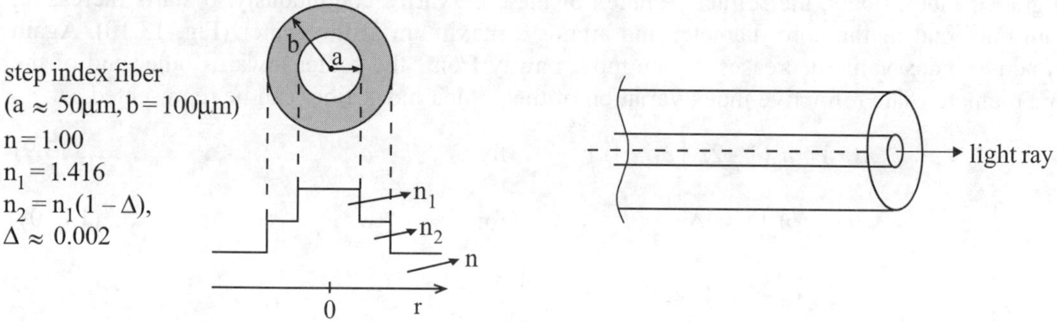

(a) Transverse cross section (b) Longitudinal cross section.

Fig. 13.8 Single mode step index fibre and refractive index profile.

Multimode step index fibers allow finite number of guided modes (Fig. 13.9). Light is an electromagnetic radiation. It has perpendicular electric and magnetic field vectors perpendicular to the direction of propagation. Light can be polarized. The direction of polarization, alignment of electric and magnetic fields will be different in rays of different modes in multi mode fiber. These modes depend on the boundary conditions used in the solution of differential equation for the propagation of light wave in optical fiber. Mode volume of a multimode fiber is the number of modes the fiber can support.

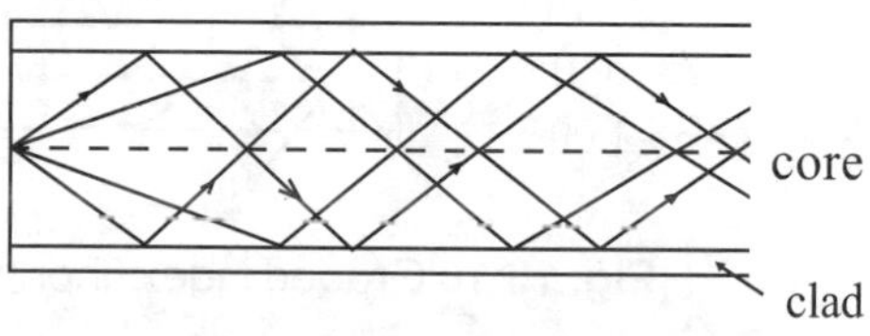

Fig. 13.9 Multimode step index fiber.

Normalized frequency of a multimode fiber is the cut-off frequency below which a particular mode cannot exist. This is related to numerical aperture, radius of the core and wavelength of the light. It is given by:

$$normalised\ frequency = \frac{2\pi}{\lambda} a(NA) \quad .....(13.8)$$

here a = radius of the core

NA = Numerical aperture

λ = Wavelength of the light.

Multi mode fibers have the following advantages:

1. Multimode fibers can use spatially incoherent sources of light like light emitting diodes. Single mode fibers require coherent light sources like laser.
2. These types of fibers have large Numerical aperture and core diameters. Therefore it is easier to couple them with other fibers and sources of light.
3. These fibers have lower tolerance requirements on their properties.

13.6 Graded Index Fibers

In graded index fibers, the refractive index of the core varies continuously. It starts increasing from one end of the core diameter and attains a maximum at the center (Fig. 13.10). Again refractive index value decreases as one moves away from the center towards other end of the core diameter. The refractive index variation of the graded index fiber can be represented as :

$$n(r) = n_1\left(1 - 2\Delta(r/a)^{\alpha}\right)^{r/2} \quad \text{for} \quad r > a \qquad \text{.....(13.9)}$$

$$= n_1(1 - 2\Delta)^{1/2} \quad \text{for} \quad r \geq a \qquad \text{.....(13.10)}$$

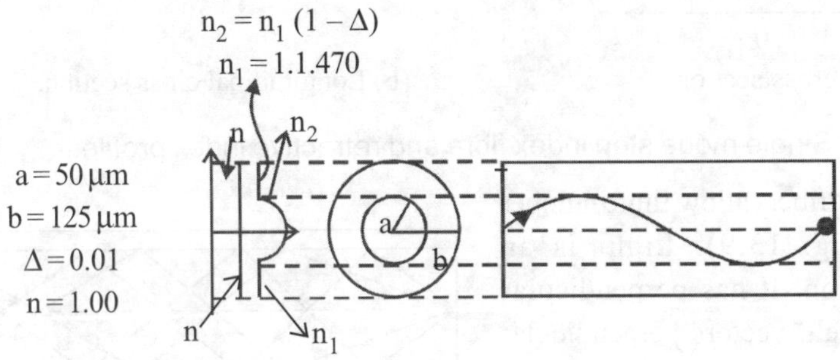

Fig. 13.10 Graded index fiber (multimode) and refactive index profile.

Δ is the fractional difference in the refractive index of the core and cladding = $(n_1^2 - n_2^2)/2n_1^2$. α is the profile parameter. If a is infinity, the fiber will be step index fiber. If $\alpha = 1$ we get a triangular profile of the refractive index variation in the core. Graded index fibers have following characteristics:

1. Refractive index profile is circularly symmetric.
2. Fiber is multimodal with large core diameter.
3. Total internal reflection and refractive index change within core region slowly.
4. Refractive index variations are small over distances comparable with wavelength of light.

Graded index fiber is a multimode fiber that supports propagation of meridional rays. Meridional rays pass through the center of the fiber. (13.11(a)) The gradual decrease in the refractive index from center of the core towards clad results in many refractions (Fig. 13.11(b)) and ray gradually gets bent (curved) until total internal reflection takes place. Multimode graded index fiber exhibits far less inter modal dispersion compared to the step indexed multimode fiber. The light rays which passes through the edge of core and does not pass through the fiber are known as skew rays (Fig. 13.12(a)). Skew rays follow a helical path as shown in the Fig. 13.12(b).

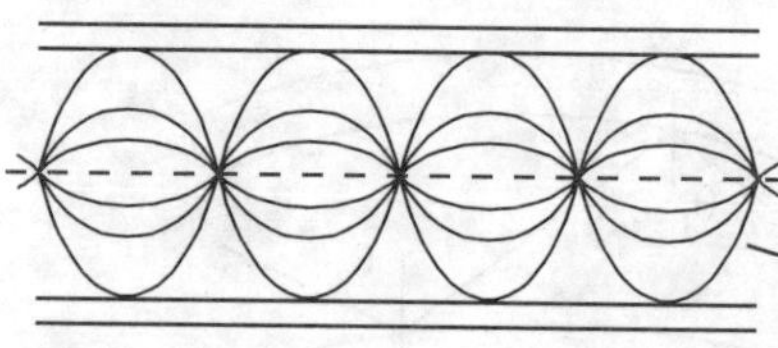

(a) Light Prpopagation in Graded index fiber.

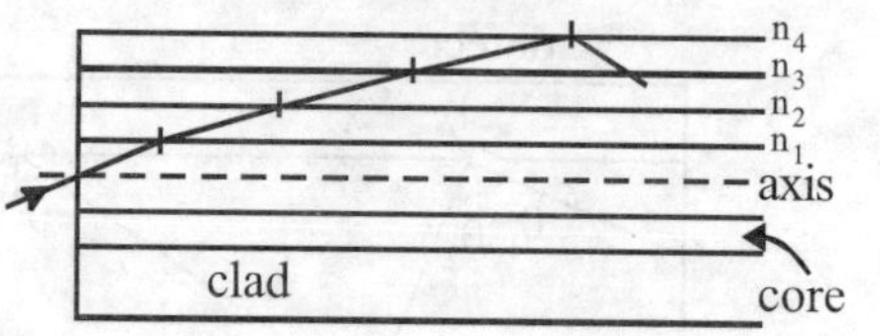

(b) Multiple refractions in graded index fiber with in core material.

Fig. 13.11

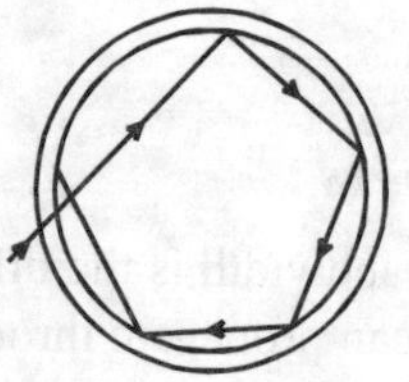

(a) corss-section view

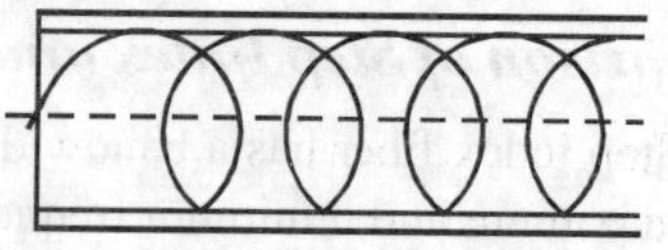

(b) helical path of a ray

Fig. 13.12 Skew rays.

Light is an electromagnetic radiation, which is transverse in nature. Light wave is made of electric and magnetic field oscillations that are perpendicular to each other and also perpendicular to the direction of propagation of the wave. This is the transverse electromagnetic nature of the light. Meridional rays correspond to transverse electric and transverse magnetic modes. Skew rays represent yet another mode of the light waves.

Here is a comparison of different types of fibers:

	Single mode Step index fiber	**Multimode step index fiber**	**Multimode graded index fibers**
Source	Laser	Laser	Laser or light emitting diode
Cost	cheap	costly	cheapest
Coupling	very difficult	difficult	easy
Application	under water cables	Data links	Telephone lines
Losses	least	intermediate	maximum

Mode coupling :

Optical fiber perturbations such as deviation of fiber axis from straight line, change in core diameter, irregularities in core and clad interface, refractive index variations may change the propagation characteristics of the fiber. Individual modes do not normally propagate through out the length of the fiber without large energy transfers to the adjacent modes, even when the fiber is exceptionally good without any bends. This mode conversion or mode mixing is called mode coupling. Mode coupling affects the transmission of light through fibers in several ways. Mode coupling is illustrated in Fig. 13.13.

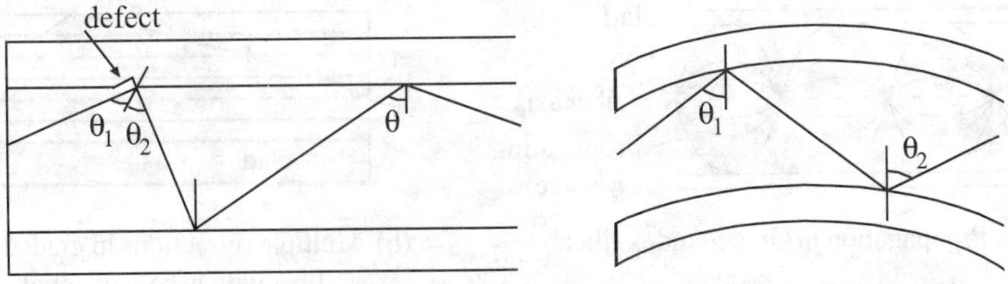

Fig. 13.13 Mode coupling.

Comparison of Step Index and Graded Index Fibers :

1. Step index fiber has a bandwidth of about 50 MHz. Bandwidth is the difference between maximum and minimum frequencies of em-wave that can propagate through fiber. Graded index fibers can have large (theoretically infinite) bandwidth.
2. Graded Index fibers will have lower attenuation than Step Index fibers.
3. Step Index fibers can be single mode fibers where as Graded Index fibers are always multi-mode fibers.
4. The attenuation in Graded Index fibers is of the order of 5 dB/Km and the numerical aperture is 0.16 - 0.2, where as the Step Index fibers of the same physical size show a loss of order of 12 dB/Km and numerical aperture of 0.2 - 0.35.

The following are the properties of Multimode Step index fiber:

Core diameter	50 - 100 mm
Cladding diameter	125 - 500 mm
Buffer jacket	250 - 1000 mm
Numerical aperture	0.16 - 0.5
Attenuation	2.6 - 5 dB/Km.
Band width	6 - 50 MHz.

The following are the properties of Multi mode graded index fiber:

Core diameter	30 - 100 mm
Cladding diameter	100 - 150 mm
Buffer jacket	250 - 1000 mm
Numerical aperture	0.2 - 3
Attenuation	2 - 100 dB/Km.
Band width	300 - 3000 MHz.

Properties of Single mode step index fibers are summarized below:

Core diameter	5 - 10 mm
Cladding diameter	125 mm
Buffer jacket	250 - 1000 mm

Numerical aperture	0.08 - 0.15
Attenuation	2 - 5 dB/Km.
Band width	> 500 MHz.

Preparation of fibers :

The dielectric material (normally glass) used for the fiber drawing should have following properties:

1. The dielectric material normally glass used for core must have a higher refractive index than that of cladding material to ensure guiding the light wave through the fiber.
2. The dielectric material used must have low loss (< 10 dB) in the wavelength region of transmission.
3. Materials and fabrication methods used must be capable of producing graded index fibers.
4. The dielectric used should be malleable. It should have proper elastic properties to enable fiber drawing. The core and clad regions should be compatible.

The glasses used for core, cladding and protective coating materials are synthesized and purified. Coating is used over cladding material to protect the fibers from atmospheric degradation due to chemical, mechanical attacks. It also protects the fibers from humidity. Polymeric materials are used for protective coating. The melts of these materials of core and cladding are taken in concentric conical crucibles hence this method is called double crucible method. The arrangement used for fiber drawing is shown in the Fig. 13.14. The fiber is drawn from the melt. Refractive index grading is achieved through the diffusion of mobile ions across core-clad interface within molten glass. The diameter of the fibers is monitored and controlled. After drawing the fiber protective coating is given to the fiber (Fig. 13.14). Liquid phase methods have a disadvantage of obtaining and maintaining extremely pure glass which limits their ability to produce low loss fibers.

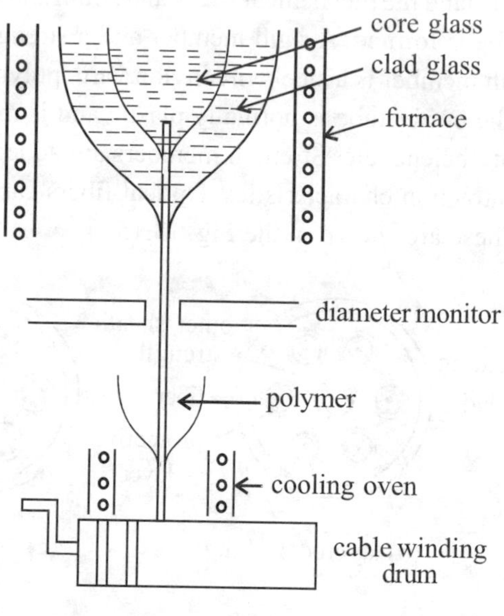

Fig. 13.14 Optical fiber drawing.

Materials used for fiber drawing :

Given below are some of the glass systems used for the optical fiber drawing:

Step index fibers:

Sl.no	Core glass	Cladding glass
1.	Na_2-Bi_2O_3-SiO_2	Na_2-Bi_2O_3-SiO_2
2.	Na_2-Li_2O-CaO-SiO_2	Na_2-Li_2O-CaO-SiO_2
3.	Na_2-CaO-GeO_2	Na_2-CaO-SiO_2
4.	P_2O_5-Ge_2O_3-GeO_2	P_2O_5-Ga_2O_3-SiO_2

Graded index fibers

Sl.no	Core glass	Cladding glass
1.	R_2O-GeO_2 - CaO-SiO_2	Na_2-Bi_2O_3-SiO_2
2.	R_2O-Bi_2O_3-SiO_2	Na_2-Bi_2O_3-SiO_2
3.	Na_2-Bi_2O_3-SiO_2	Na_2-Bi_2O_3-SiO_2

Fiber Packing :

The glass fibers drawn are packed properly to protect them against the fiber damage and breakage during installation and maintenance. Bare glass fibers are brittle and susceptible to damage very easily. It is necessary therefore to cover them properly into a series of protective layers. There is coating and cabling used. Coating is to protect the fibers from chemical and mechanical strains. This enables the cables to withstand the mechanical stress and strain and environmental conditions. Around the fiber a buffer tube is formed. A rigid member and a strength member are fabricated over buffer tube. The strength member is a steel wire, nylon yarn, polymer film, terlene yarn, and Kevlar or fiberglass yarn. The buffer tube is normally plastic and is made of polyvinyl chloride (PVC), polyethylene, poly urethelene, etc. Strength members are added to the cable to minimize the cable elongation and contraction characteristics. Optical fibers can be packed in the layered design or bundled design. These are shown in the Fig. 13.15.

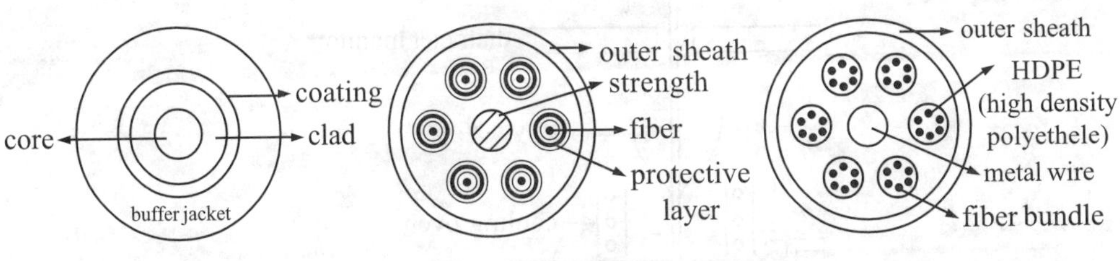

(a) cross-section (b) layered packing (c) bundled design

Fig. 13.15 Packed optical fiber.

13.7 Attenuation in Optical Fibers

A part of the light energy being propagated through optical fiber is lost during its traversal from the source to destination. These losses are due to various reasons. The losses are measured through attenuation constant. Its unit is decibels per kilometer. Attenuation A is defined as

$$A = 10 \log \frac{p_i}{p_o} \quad \text{.....(13.11)}$$

Attenuation constant α is attenuation per unit length. Hence

$$\alpha = A/l = \frac{10}{l} \log \frac{p_i}{p_o} \quad \text{.....(13.12)}$$

where l is the length of the fiber.

The following are the loss mechanisms due to which the intensity of light decreases while propagating through fiber :

1. ***Absorption losses :*** Absorption losses could be intrinsic or extrinsic. Depending the material used intrinsic absorption losses will change. These depend on the absorption characteristics of the material used for fiber drawing. They depend on the absorption of the material in infrared and ultra violet region. Extrinsic absorption losses depend on the purity and perfection of the material
2. ***Scattering losses :*** The glass material used for the optical fibers will have many microscopic inhomogeneities. These are because of the defects and changes in the structure and result in light scattering.
   - **(a) Rayleigh scattering :** Is because of small fluctuation in refractive index and changes in the density. These inhomogeneities are fundamental and intrinsic to the glass used for fiber preparation and cannot be completely avoided.
   - **(b) Mie scattering :** This is due to the irregularities in the core-clad interface and refractive indices, diameter fluctuations, strains, bubbles etc. These defects normally depend on the material, design and method and conditions of manufacture of the optical fibers. These defects can be reduced considerably by proper control of different processes during manufacture of the fibers.
   - **(c) Brillouin Scattering :** Modulation of light through thermal molecular vibrations within the fiber material. The incident photon (light packet) produces a phonon (quantized elastic wave having sound frequency) and a scattered photon. This type of scattering occurs only above certain minimum threshold frequency. Hence it can be completely avoided by selecting proper light frequency below the threshold frequency.

(d) **Raman scattering :** Raman scattering incident photon produces another photon (optical phonon). This can also be avoided as the Brillouin scattering.

(e) **Dispersion losses :** Losses due to dispersion of the transmitted signal causes distortion of the signal. Dispersion losses could cause broadening of the pulses being transmitted resulting in overlap of the adjacent pulses and subsequent distortion. This limits the rate of bit transfer (bits transmitted per second).

(f) **Inter modal dispersion or mode coupling :** This is chromatic dispersion and results in the overlap of signal of different wavelengths. Hence, this will limit the line width of the light source used.

Attenuation coefficient can be expressed as a function of some of the above loss mechanism as follows:

$$\alpha = A\lambda^{-4} + B + C\,(\lambda) \qquad .....13.13$$

A is Rayleigh scattering coefficient

B = Wave length independent losses like dispersions

C = Wave length dependent loss mechanisms like inter modal dispersion

13.8 Applications of Optical Fibers

Optical fibers can be used in a variety of applications. These can be broadly divided in three categories:

(i) *Communication applications* : Light with a large carrier frequency capability has a potential possibility of packing extremely high information content into a single channel. Semiconductor laser and light emitting diodes along with optical fibers can makc this possibility a reality. The main components of fiber optic communication system are:

1. *The optical source :* semiconductor lasers, or LED. The power launched into fiber, rise and fall time are important.
2. *A means of modulating the signal with output of the source :* Intensity modulation is done. Pulse modulation methods for digital or analog transmission are used. Encoding schemes are employed.
3. *Transmission medium :* optical fibers are used. Multi-mode or Single-mode, size, refractive index profile, losses, dispersion, mode coupling, cabling, joints employed are important.
4. *The detector that converts the received light back into electrical wave :* p-n, p-i-n, or avalanche diodes are used.
5. *Amplifiers and signal processor to amplify and process the signals received :* The amplifier design is very important here.

As indicated in the third point above, optical fibers are the transmission media for communication of light signals. Figure 13.16 gives a block diagram of communication using optical fibers.

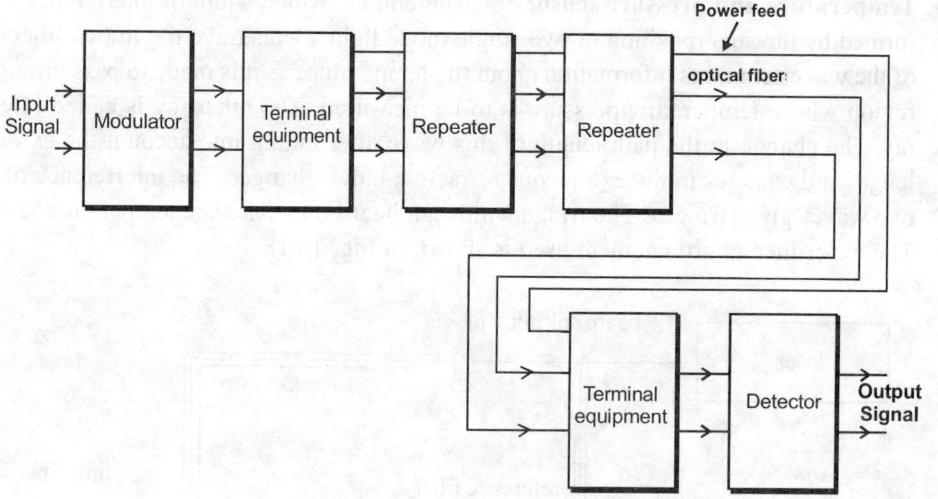

Fig. 13.16 Fibre optic communication system.

In the following communications, the dominant criteria for design in the application, using either digital or analog transmissions, are the required transmission distance and the rate of the information transfer. It is necessary to insert repeaters in the long distance communication. The purpose of the repeater is to convert the optical signal back into the electrical regime. In the analog transmission, it is transmitted and equalized, then it is re-transmitted.

Sensor applications

Many industrial process use optical fiber sensors to monitor displacement, pressure, temperature, flow rate, liquid level, chemical composition etc. These parameters are measured by observing phase, state of polarization, mode coupling and relationship between these properties are used in sensor applications. Single mode fibers are used as sensors by observing phase and polarization modulation, whereas multimode fibers use mode coupling and amplitude modulation. Interference is another important property, which is used in sensing applications.

(a) **Displacement sensors :** Any displacement of a body in the path of the light between two fibers can be detected easily. This application is illustrated in Fig. 13.17.

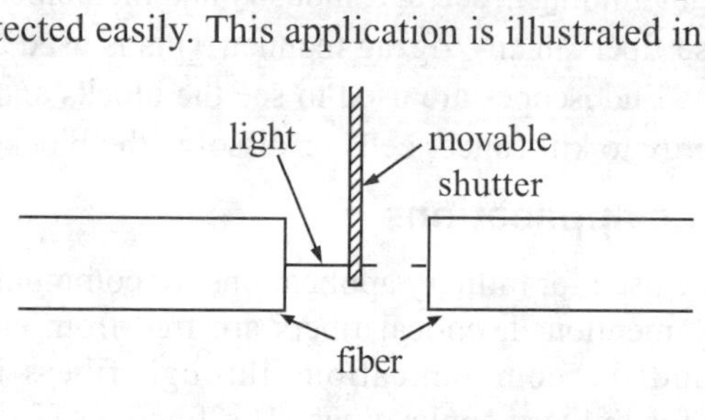

Fig. 13.17 Displacement sensor

(b) Temperature and pressure sensors : Change in the fringe width of interference fringes formed by the superposition of two single mode light waves traveling in two fibers. One of the waves contains information about the temperature as it is made to pass through the region whose temperature/pressure is to be measured. The other ray is called reference ray. The change in the path length of this wave after taking into account the extra fiber length and changes in the extraneous refractive index changes. The interference of these two waves gives fringes. The fringe width can be used to calculate temperature/pressure. The experimental arrangement used is shown in Fig. 13.18.

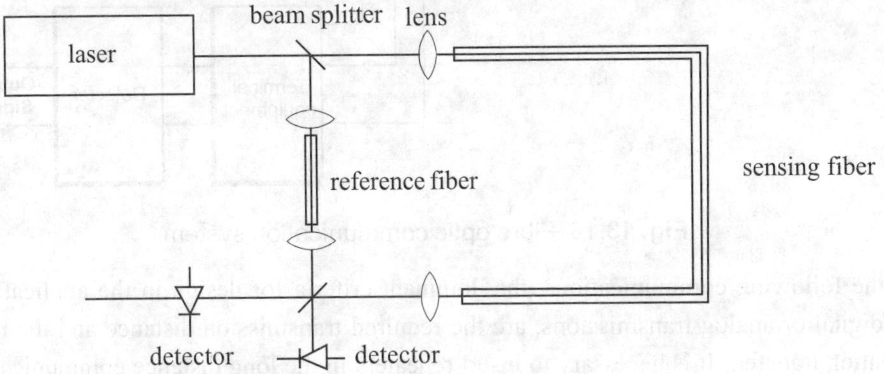

Fig. 13.18 Temperature and pressure sensor.

(c) Chemical sensors : Fiber optic chemical sensors are used. Interference is the technique used in most of these sensing applications. The sensing element is a modified fiber, which can convert concentration of chemical element being sensed into phase changes of the light wave. For example in hydrogen sensor, palladium wire is attached to the sensor. Hydrogen absorption changes the dimension of palladium wire. This results in the introduction of strain in the optical fiber. Strain changes the path length of the light propagating through the fiber.

Light guiding application

Optical fibers are used for light guiding in science, industry and medicine. Since fibers are flexible, light can travel through these fibers in any zigzag manner. This is used in medical diagnostics in endoscopies and angioplasty. Endoscopes are used to see the blocks and tumors in the body and angioplasty uses optical energy to kill cancer cells or dissolve the blocks of an artery.

Military and Aero space applications

Fiber optic technology can be used for military applications for communication, guiding weapons and submarines. As already mentioned, optical fibers are free from electromagnetic and radio frequency interferences and the communications through fibers is secured. Fiber optic communications are suited for military applications. The fibers used must be nuclear radiation resistant.

13.9 Advantages of Fiber Optic Communications

Communication using an optical carrier wave guided through a glass fiber has a number of attractive advantages. Some of them are:

(a) Enormous Potential Bandwidth : The bandwidth of the optical communication channel is very large compared to other communication channels. It spreads form 10^{13} Hz to 10^{16} Hz. Because of larger bandwidth, fiber optic systems have greater information carrying capacity. Bandwidth gives the information carrying capacity of the communication channel. Enhanced bandwidth utilization can be obtained in fiber optic systems by transmitting different signals with different colors (wavelengths) in parallel on the same optical fiber. This is called wavelength division multiplexing. This kind of multiplexing is not possible on other systems of communication (e.g. microwaves), where only one signal can be transmitted at a time.

A fiber has a capacity of 500 channels and its external diameter is not more than 0.5 mm. This is very small compared to 10 mm for a corresponding coaxial cable. They offer low attenuation less than 5 dB/Km of the fiber. The repeaters can be kept at a very long distance like 45 Kms. The coaxial cables require repeater every 1.6 Kms. This saves considerable costs.

(b) Small size and weight : Fiber optic cables have smaller diameters (about 1-1.5 mm). Hence they are compact and little weigh in comparison with copper cables.

(c) Electrical Isolation : Unlike their metallic counterparts (wave guides and co-axial wave guides) optical fibers are electrically insulated. There is no need of any other insulation for electrical isolation of fiber optic cables. This makes optical fibers suitable for use in electrically hazardous environments.

(d) Immune to interference and cross talks : Optical fibers are made out of dielectric materials. Hence, they are free from electrical and electro-magnetic interference (EMI). Electrical noise present does not affect the signals that are being carried in the optical fibers. Hence, there will not be any cross talk in optical fibers unlike their metallic counterparts.

(e) Signal security : Unlike copper cables, a transmitted signal in an optical fiber cannot be obtained without drawing power from the cable. Any such attempt to draw power from the cable can be detected. This feature is attractive for military, banking and general secured data transmission applications.

(f) Low transmission losses : Optical fibers have very low transmission losses.

(g) Flexibility : Metallic wave-guides are not flexible. Optical fibers are small in size and are highly flexible.

(h) Low costs : Optical fibers are made out of glass. Hence, they are cheaper compared to metallic wave-guides and co-axial cables.

Questions

1. Using rectilinear propagation of light and Snell's law explain the transmission of light through optical wave-guides.
2. Define numerical aperture and acceptance angle. Derive a relation between the two.
3. Explain different modes of light waves in optical fibers.
4. What are skew rays and meridional rays?
5. With help of refractive index profile distinguish the step index and graded index fibers.
6. Distinguish between the single mode and multimode step index fibers.
7. Discuss various types of losses in the optical fibers.
8. What are different types of scattering losses in optical fibers?
9. Discuss an arrangement used for fiber drawing and coating using glasses
10. Write short notes on:
    (a) Fiber buffering (b) cable strength and structural members
    (c) Cable sheath
11. Explain various fiber packing methods used.
12. Discuss various types of materials used in fiber optic cable manufacture.
13. Why a fiber is coated with polymer?
14. Explain temperature and pressure sensing applications using optical fibers.
15. What are chemical sensors? How are they different from temperature sensors?
16. Explain the advantages of fiber optic communications
17. Explain the medical applications of fiber optic sensors.
18. Find the numerical aperture and acceptance angle of an optical fiber with core refractive index 1.55 and clad refractive index 1.53.
19. An optical fiber core refractive index is 1.465 and that of clad is 1.45. Find the critical angle at the core-clad interface.
20. A Step Index fiber of core diameter 200 μm and its numerical aperture is 0.2. When a light of 850 nm is propagating through fiber, find the number of propagating modes.
21. Discuss the meaning of a 'Mode'.

Objective Questions

I. Choose the correct choice

1. Optical waveguides guide light waves by
   (a) interference of waves
   (b) causing diffraction
   (c) simple refraction of waves
   (d) by total internal reflection of waves

2. In an optical fibre, the refractive index of the core is n_1, the refractive index of the clad is n_2, then
   (a) $n_1 > n_2$ (b) $n_1 = n_1$
   (d) $n_1 < n_2$ (d) $\frac{(n_1 - n_2)}{n_1} > 1$

3. T M mode of light waves has the following field components, when ray E in the plane of incidence, and x – axis the direction of propagation
   (a) H_x, E_x, E_z (b) H_y, E_x, E_y
   (c) H_x, E_y, E_z (d) H_y, E_y, E_z
4. T E mode involves the following field components for a ray with E in the plane of incidence propagating in x - direction
   (a) E_y, H_y, E_z (b) E_x, H_y, H_z
   (c) E_x, H_y, E_z (d) E_x, H_x, H_z
5. Inter modal dispersion is the
   (a) chromatic dispersion of light rays
   (b) travelling distances by light of different modes in the same time
   (c) pulse broadening
   (d) both (b) and (c)
6. Step index fiber is a
   (a) multimode fibre
   (b) single mode fibre
   (c) both (a) and (b)
   (d) (a) or (b)
7. Angle of acceptance is maximum for a fiber if
   (a) the critical angle is minimum
   (b) the critical angle is maximum
   (c) the critical angle is zero
   (d) the critical angle is negative
8. The total acceptance angle is
   (a) twice the minimum of acceptance angle
   (b) twice the maximum of acceptance angle
   (c) half the maximum of acceptance angle
   (d) half the minimum of acceptance angle
9. Meridinal rays propagate along
   (a) the axis of the fibre
   (b) along the walls of the fiber
   (c) along clad of the fibre
   (d) both (a) and (b)
10. Leaky modes are
   (a) mode which propagate through fibre
   (b) modes which leak out of clad
   (c) modes which penetrate into the clad
   (d) None of the above
11. Single mode step index fibers support propagation of
   (a) meridinal rays
   (b) skew rays
   (c) helical rays
   (d) all the three rays
12. Micro bends and small cracks present in the optical fiber cause
   (a) change in internal reflection angles
   (b) mode coupling
   (c) loss of light intensity
   (d) all the above
13. Loss of light intensity in optical fibres is due to
   (a) Scattering (b) absorption
   (c) refraction (c) a or b
14. Fibre optical communication systems do not have
   (a) EMI (b) RFI
   (c) Cross- talk (d) All the three
15. High information carrying capability of optical fibers is measured with their
   (a) Low losses
   (b) high band width
   (c) Low costs
   (d) high efficiency
16. Preferable source of light used in fiber optic communications is
   (a) LED
   (b) Sodium lamp
   (c) Laser
   (d) Cadmium red light
17. Optical fiber sensing applications mostly use the
   (a) refraction (b) interference
   (c) polarization (d) reflection
18. For wide band width and long distance communication system , the fibers that are needed should be
   (a) high quality, low loss, graded index fiber
   (b) plastic with large cores
   (c) highly dispersive and step indexed
   (d) fluoride fibers working in infrared region

19. Optical fibers are made of
    (a) metallic conductor
    (b) plastics doped with metallic impurities
    (c) dielectric materials
    (d) magnetic oxides
20. Optical fiber applications mostly use the following phenomenon.
    (a) Reflection (b) refraction
    (c) interference (d) polarization
21. In an optical fiber, The refractive index of core is n_1 and that of clad is n_2, then
    (a) $n_1 > n_2$ (b) $n_1 = n_2$
    (c) $n_1 < n_2$ (d) $n_1^2 - n_2^2 < 1$
22. Meridinal rays propagate along
    (a) Walls of the fiber
    (b) axis of the fiber
    (c) both (a) and (b)
    (d) a helical path on the wall

II. Fill up the blanks

1. A glass fiber consists of a ________ glass surrounded by a ________ made of glass of slightly of lower refractive index.
2. If n_1 and n_2 are refractive indices of core and cladding respectively, then, the numerical aperture of the fiber is ________.
3. If 'd' is the diameter of the core, N_A is the numerical aperture of the fiber, then the total number of modes possible for such an optical guide is ________.
4. A single mode fiber will have a ________ core diameter than a multi mode fiber and difference in refractive indices of the core and cladding is very ________.
5. In a single mode fibre, there is ________.
6. The inversion transmission capacity in an optical fiber is ________ proportional to dispersion.
7. In creasing the core refractive index ________, the number of propagating modes in an optieal fiber.
8. In multi mode fibers, the core diameter is ________.
9. Single mode fibers transmit single ray ________ the axis of the fiber.
10. The normalized frequency of a multimode fiber is dependent on ________ and ________ of the core of the fiber.
11. ________ fiber is a multimode fiber.
12. Meridional rays correspond to ________ electirc and ________ magnetic modes.
13. The optical sources that can be used in fiber communications are ________ and ________.

CHAPTER 14

Acoustics of Buildings and Acoustic Quieting

14.1 Introduction

14.1.a Perception of sound

Sound, as we are aware is a form of energy that is transmitted from the source to the detector/ the listener. The sound waves are propagated in the form of longitudinal waves in air. Sound is produced by vibrations caused in a media. The sound waves require a media for the propagation. In the earlier classes, we have learnt that the intensity of sound is proportional to the square of amplitude of the waves that reach the listener. The sound is characterized by pitch or frequency, and intensity or loudness. When sound sources produce waves that are characterized by a number of frequency components (i.e. not just monochromatic), and as they travel they suffer from disturbed intensity, an intensity loss and interference effects. Before learning about he acoustical aspects of rooms, let us just review a few aspects of sound intensity and perception of human ear.

The perception of the sound by a human ear is limited and the frequency of the audible range is supposed to be from 20 Hz to 20,000 Hz. However, the actual hearing of normal person is in the range from 90Hz to 11,200Hz. This range varies from individual to individual, and generally shrinks with age of a person. The perception of sound by a listener depends on the frequency as well as the frequency components, and the amplitude of the wave. The structure of human ear is shown in Fig. 14.1.

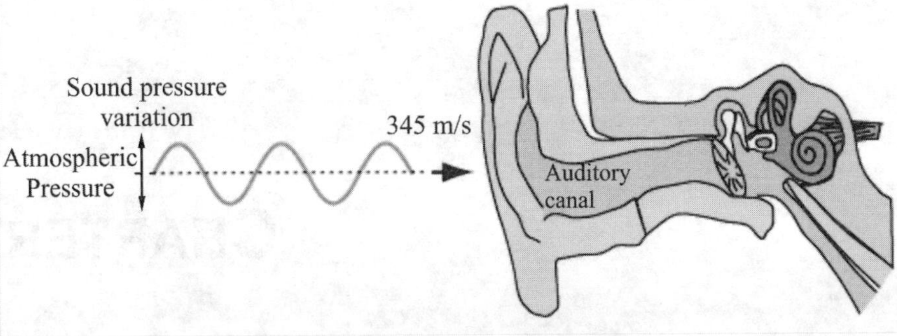

Fig. 14.1 The human ear receiving the sound waves.

The sensitiveness of the ear i.e., the least intensity of sound that is audible, also called the *threshold of audibility* depends on frequency. **Generally, any sound source like musical instruments, human voices, or recorded sound produce waves with lot of frequency components Musical sounds contain overtones. The sensation of sound depends upon the quality of sound (that depends on the overtones) produced by the source**. The amplitudes of these sounds may also defer. The ear should be able to perceive the sound with clarity, other wise, and the sound appears as mixed noise (unnecessary components) to the listener. The human ears can detect not only the changes in the overall sound pressure level but are so sophisticated that they can detect sound, the pressure level of which is well below the background noise level.

The perception of the sound by an ear variant from person to person, but the quality of the sound is perceived by the ear within reasonable range by all ears. The strength of sound is qualified in terms of intensity that is determined by the amplitudes of the waves. Individual perception of the strength of a sound depends on the intensity. The power of sound is the energy density that reaches the ear drum and causes the pressure on it.

In the sounds reaching the ears, the sound with greater power is a dominating one and other sounds with power less than the maximum normally get submerged in the loudest sound. For example, when we stand near an airplane that has just started, the dominating sound from the plane would submerge every other sound that is less than that in that vicinity. The normal human ear is very sensitive and is capable of detecting the difference between 440 Hz and 441 Hz.

14.1.b Sound Intensity and sound power

The calculation of sound power depends on sound pressure, (a sound wave, after all, is a pressure wave) and is proportional to the square of the pressure. A reference pressure is taken as 0.0002 microbar. The ear has an incredible sensitivity. It is capable of detecting pressure variations of less than one billionth of atmospheric pressure. The threshold of hearing corresponds to air vibrations on the order of a tenth of an atomic diameter. This incredible sensitivity is enhanced by an effective amplification of the sound signal by the outer and middle ear structures. The sensitivity is low at low frequencies, being about

10^{-9} watt. At a frequency of 1000Hz, threshold of audibility occurs at an r.m.s pressure variation of approximately 0.0002dynes/cm^2 that corresponds to an intensity (**I_0**) of 10^{-16} watt/cm^2. This is the lowest limit from which the intensity levels are measured.

A sound of intensity that is ten times that of threshold value **I_o** is said to have an IL (Intensity level) of 1 bel or 10 decibels. In terms of intensity level (IL),

$$IL = 10\log\frac{I}{I_0}\ dB \qquad .....(14.1)$$

where the reference intensity level, $I_0 = 10^{-12}$ watts/m^2.

$$\therefore \quad IL \quad = 10\ln\frac{I}{I_0} = \frac{1}{10}$$

$$\frac{I}{I_0} = 1.26 \qquad .....(14.2)$$

Therefore, if we can cause a change in intensity by 2.6%, it enhances the intensity level by 1dB. If I = 100 I_0, then IL = 10 ln 10^2 = 20 dB. If I = 1000 I_0, IL = 10 ln 10^3 = 30 dB. We can also express sound levels in terms of energy density or power. It is given in the following expression. When two sounds differ by 50dB, the louder one is 100000 times more intense. The intensity is proportional to the square of the amplitude of the sound wave. The sound waves are pressure waves and the intensity can be expressed in terms of pressure amplitude as $I = \frac{P_r^2}{\rho_0 c}$, where p refers to rms pressure amplitude.

The intensity and loudness are two different parameters of sound. The intensity refers to the physical picture of sound, while the loudness is a subjective phenomenon, and it directly refers to the sensation produced at the ear. For example, the sounds of a thunder, a bomb, boiler factory come under the deafening sounds. Another property of sound or noise is its loudness. A loud noise usually has a larger pressure variation and a weak one has smaller pressure variation. Pressure and pressure variations are expressed in Pascal, abbreviated as Pa, which is defined as N/m^2(Newton per square meter). The street noises in busy streets of cities, police whisle at close range, are very intense. The sounds on a railway platform or in market streets, modern music programmes at marriage functions, music from high wattage speakers in limited space come under loud and noise- filled sounds.

The sound power is measured in decibel as unit. It is defined as

$$\mathbf{Ps\ (dB) = 10\ Log\ W / W_0} \qquad .....(14.3)$$

where, Ps is the sound power in dB. W is the sound power of source in Watts. Wo, is the reference of sound power, 1E – 12 *watts.*

A low TV sound, or music programme with good quality, quiet home, a corporate office have sound levels that bearable. The unit of loudness is 'phon' and this represents the level of sound of equal loudness at all frequencies. Sound is measured in 'phon' by comparing it actually with a standard tone of thousand cycles per second adjustable intensity.

There is a threshold level zone of hearing The softest sound a normal human ear can detect has a pressure variation of 20 micro Pascals, abbreviated as μPa, which is 20 x 10^{-6} Pa ("20 millionth of a Pascal") and is called the 'Threshold of Hearing'. This is taken as 0dB. The sound start painful beyond 120dB. Actually sounds start becoming painful at 85dB power itself. A continuous hearing of loud sounds degrades the audibility of an ear. This does not depend on the pitch of the sound. The loud and noise full sounds also damage the hearing capacity of an ear.

14.2 Acoustic requirements in Auditoriums or Big Halls

14.2.1 What are auditoriums or halls?

Auditoriums or halls are the places where a large number of listeners sit as participants for listening a musical program, or concert, speeches, lectures etc. The room in which we listen to sounds has an important influence on what we hear. Any audience would like to enjoy the programme that is done on the stage of an auditorium or a big hall in which programmes are held. Even the design of a class room is important for effective sound distribution the room. The design of the room, seating, ceiling, flooring, windows, volume, walls, materials used etc., have a bearing on the quality of sound that one would hear in a hall. The following sections will discuss some of the principles that are important for the design of a hall or an auditorium. However, the design of such spaces is still considered as an inexact science. Let us now briefly see how sound is distributed in a big hall.

14.2.2 Sound Propagation in an Auditorium or a big hall

- As sound waves travel at about 343 meters/second, the sound coming directly from a source within an auditorium will generally reach a listener after a time of anywhere from 0.01 to 0.2 seconds as shown in the Fig. 14.2.
- Shortly after the arrival of the direct sound, a series of semi-distinct reflections (Fig. 14.2.) from various reflecting surfaces (walls and ceiling) will reach the listener. These *early reflections* typically will occur within about 50 milliseconds in a normal hall. Early reflections are not heard as separate from the direct sounds, rather, they tend to reinforce the direct sound. The acoustical property of a hall depends on the multiple reflections that take place in the room or hall.

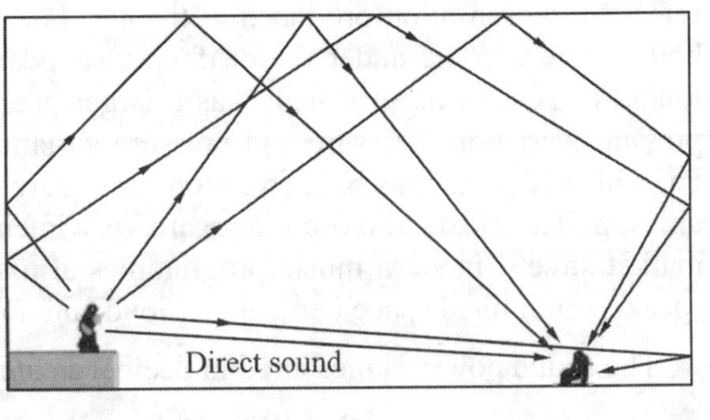

Fig. 14.2 Reflections in a hall to the listener.

- *The reflections which reach the listener after the early reflections are typically of lower amplitude, and very closely spaced in time. These reflections merge into what is called the reverberant sound.*

If the sound source produces a sound that is of single frequency, then, it is likely that stationary waves are set up (echo).This will cause fluctuations in the intensity of sound in the room.

14.3 Reverberation

If the source emits a continuous sound in a room, the reverberant sound builds up until it reaches an equilibrium level. When the sound stops, the sound level decreases at amore or less constant rate until it reaches the level of inaudibility.

After sound has been produced in, or enters into, an enclosed (either partially or fully) space like a hall, it is reflected repeatedly by the boundaries of the enclosure, even after the source ceases to emit sound. This prolongation or persistence of sound after the original source has stopped is called *reverberation*. Reverberant sound is a collection of all the reflected sounds in hall.

Sound waves from the source spread out as spherical waves, strike the walls of the hall. The sound energy is partly reflected, partly absorbed and partly transmitted at the walls. The ratio of sound energy absorbed at a surface to that of incident energy on the surface is called absorption coefficient.

Suppose there is a source of sound of constant power in a hall. This sound travels and falls on the walls, floor, ceiling etc. A large number of reflections take place, the sound becomes diffuse (Fig.14.3a and b) and the energy density in the hall at any point increases exponentially and attains a maximum. The average energy is same throughout the hall or uniformly distributed. We should also understand that a number of sound absorbers like the walls, materials are used for flooring, ceiling, the chairs, audience etc. The total absorption is the summation of all absorptive powers of all those mentioned.

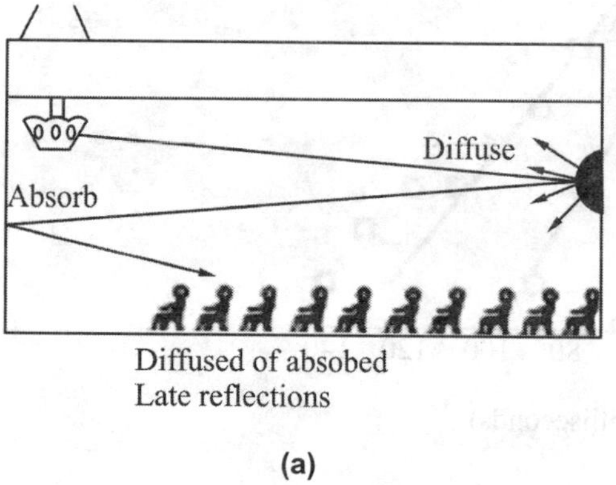

(a)

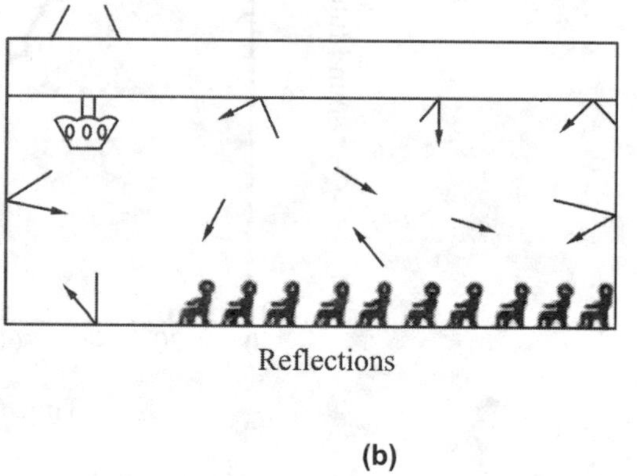

(b)

Fig. 14.3 (a) & (b) Inside a hall, a sound is created and reverberation is set in.

Let the hall contain a number of absorbing materials at the lower portion of the hall with absorption coefficients $a_1, a_2, a_3, a_4, ..,......,$ then, inside the hall, the total absorptive coefficient, $a = \sum_{1}^{n} a_i dS_i$, where dS_i represents a small element of area that absorbs.

Then, $$I = I_m\left(1 - e^{-act/V}\right) \qquad .....(14.2)$$

where I = intensity at time t. The intensity of sound does not increase instantaneously.

After the energy density at a point attains a maximum (at I_m), let the source of sound be switched off. The energy density begins to fall exponentially, that means, the reverberant sound in the hall decays in an approximately exponential fashion as

$$I = I_m e^{-act/4V} \qquad .....(14.3)$$

Where I_m represents the maximum intensity of the source at time t=0, the instant the source is switched off. Whether from a sound impulse or a steady sound which ceases, the reverberant sound in an auditorium decays in an approximately exponential fashion. The Fig.4b shows how the sound decays after it is switched off inside a hall. The decay in an occupied room is faster as compared to the empty room. The growth and fall of sound intensity is shown in Fig.14.7.

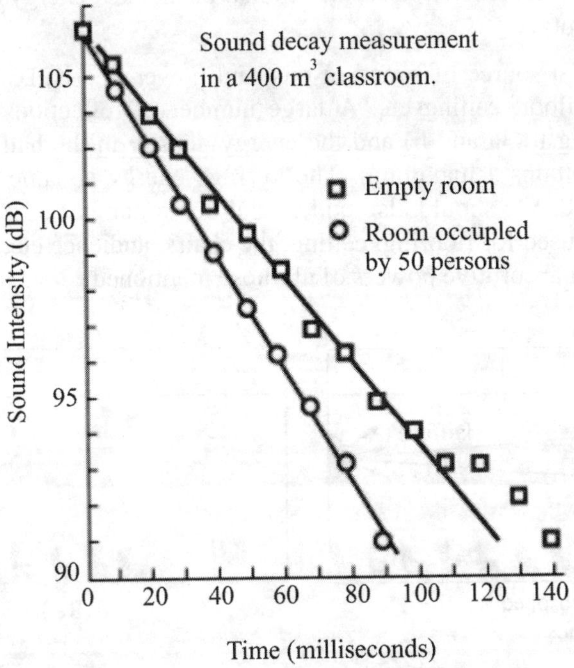

Fig. 14.4 The sound decay in a room. Note that the y-axis represents the logarithmic decrement.. A comparison between an empty room and occupied one is made.

A certain amount of reverberation adds a pleasing characteristic to the acoustical quality. During a continuous sound at a place in auditorium, the reverberant sound level is reached when the rate at which energy is supplied by the source is equal to the rate at which sound is absorbed by the room and its contents (walls, chairs, listeners, flooring etc.)

Reverberation is a desirable property of auditoriums to the extent that it helps to overcome the inverse square law drop off of sound intensity in the enclosure.

14.4 Reverberation Time

The reverberation time of a room characterises how long acoustic energy remains in a room. It is usually defined as the time for the acoustic intensity (or energy density) to decrease by a factor of one million (60 dB).

Since a reasonably loud clap is about 100 dB (SPL) and a whisper is about 40 dB, one can easily estimate the reverberation time for a room by clapping and listening to how long we can still hear some remaining sound from the clap. This assumes that the room is not particularly unusual in its dimensions and that it is reasonably quiet. The measurement of reverberation time is an important aspect of the construction of auditoriums,class rooms, or big halls..

Let us estimate mathematically the reverberation time from fundamentals by considering a simple situation. The principle involved in the mathematical treatment is important for understanding the growth and decay of sound intensity in a room.

14.5 Growth of Sound Intensity in a Room

When a source of sound is put on in an enclosed room, the sound waves spread in all directions. Consider that the enclosure is large such that the sound energy after a time lapse becomes uniformly distributed. The sound waves propagate from the source, strike the walls of the room, and get reflected. As a result of these reflections, the sound intensity at any point in the medium of the room grows with time. We make a few approximations to get a mathematical relation between the sound intensity and the rate of flow of sound energy from the source.

The volume of the medium just near the source is not considered for our understanding of the growth of sound intensity at a point inside the room.

The sound waves spread uniformly in all direction uniformly from the source.

If the sound consist of a few frequency components (especially, of single frequency) only the reflections may set up stationary waves in the room. These standing waves produce large fluctuations in sound energy at any point. However if the frequencies of sound waves consist differ only over a uniform band of frequencies ,then the stationary components will be reduced to a large extent and we can neglect such disturbances from our consideration.

Another approximation we make is that the absorption of sound at the walls (reflectors) is small.

Consider a live room of volume, V. In this room consider a small element of volume, dV. As shown in Fig. 14.5. Let the sound energy density inside the room be uniform and let it be denoted by E .Then, the acoustical energy present in the volume, dV, is EdV.

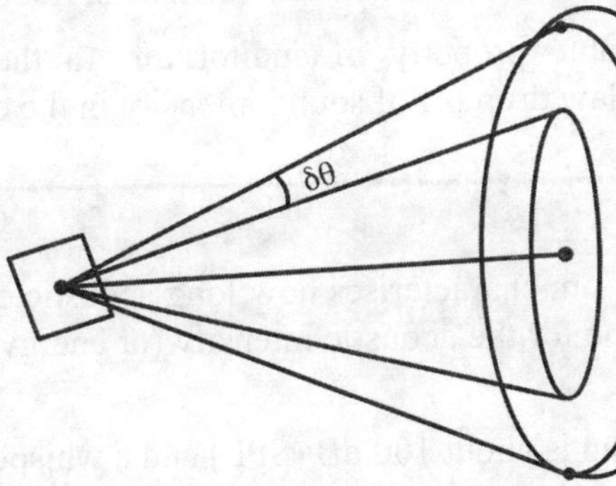

Fig. 14.5 Volume element and surface element considered in the room.

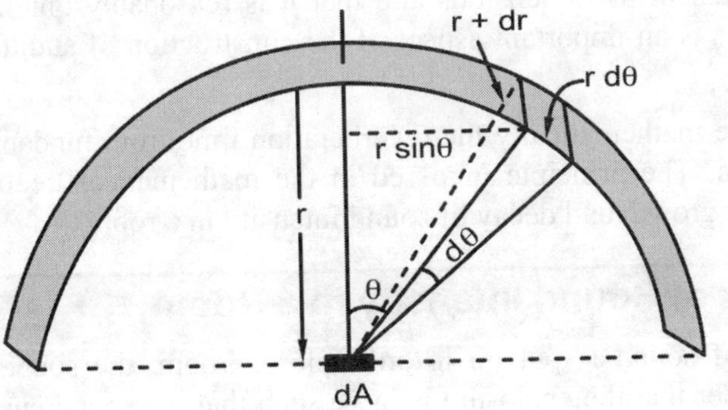

Fig. 14.6 The volume element is enclosed by a sphere of radius r. The distance between the area element dA and the volume element is r.

The surface area of the sphere $= 4\pi r^2$ .....(14.4)

The projected area of the element dA = $dA \cos\theta$ .....(14.5)

The amount of energy dU that strikes the element directly is

The solid angle subtended by the areal element dA at the volume element dV is

$$dU = \frac{da \cos\theta}{4\pi r^2} EdV \quad .....(14.6)$$

$$= \frac{dA \cos\theta}{4\pi r^2} Er^2 \sin\theta \cos\theta \, d\theta dr \quad .....(14.7)$$

The above expression if integrated over the entire shell gives the total energy that is received by the areal element dA

$$\Delta U = \frac{EdA\,dr}{2}\int_0^{\pi/2}\sin\theta\cos\theta d\theta = \frac{E\,dA\,dr}{4} \qquad .....(14.8)$$

Suppose the time taken by the sound energy to arrive at the surface is Δt. The above represents the energy that has arrived at the area dA in the Δt, where

$$\Delta t = \frac{dr}{c} \qquad .....(14.9)$$

The rate of the arrival of sound energy from all directions is

$$\frac{\Delta U}{\Delta t} = \frac{Ec\,dA}{4} \qquad .....(14.10)$$

The intensity of sound that arrives at the dA from all directions is obtained by dividing the above expression by dA

$$I = \frac{Ec}{4} \qquad .....(14.11)$$

A part of the sound energy that falls on the surface is absorbed. All the wall surfaces ,and other surfaces present in side the room absorb sound energy partly. Let the absorption coefficient of all such surface be ' a' such that

$$a = \sum_i a_i\,dA_i$$

Now we can write the equation of state for the sound energy variation with time at any point .

The rate at which the sound energy increases in the medium throughout the medium inside the room = $V\frac{dE}{dt}$.

The rate of absorption of sound energy $= \frac{EcA}{4}$. .....(14.12)

The differential equation governing the growth of sound energy is given as

$$V\frac{dE}{dt} + \frac{Eca}{4} = W \qquad .....(14.13)$$

The solution of the above eqn is

The eqn (14.13) can be written as

$$\frac{dE}{dt} + \frac{Eca}{4V} = \frac{W}{V} \qquad .....(14.14)$$

The boundary condition on the above equation is at time =0, the energy density of sound is zero. The above eqn14.14 can be written as

$$\frac{dE}{dt} = \frac{W}{V} - \frac{acE}{4V}$$

$$\frac{dE}{\frac{W}{V} - \frac{acE}{4V}} = dt \qquad .....(14.15)$$

Integrating the above eqn within the specified limits,

$$\int_0^E \frac{dE}{W - \frac{acE}{4}} = \frac{1}{V}\int_0^E dt \qquad .....(14.16)$$

$$-\frac{4}{ac}\ln[W - \frac{acE}{4}]_0^E = -\frac{4V}{ac}\ln[W - \frac{acE}{4}] - \ln W = \frac{4V}{ac}\ln\frac{W}{W - \frac{acE}{4V}} = \frac{t}{V}$$

$$\therefore \ \ln\frac{W - \frac{acE}{4}}{W} = e^{-\frac{act}{V}}$$

i.e. $$\frac{acE}{4W} = 1 - e^{-\frac{ac}{4}t} \qquad .....(14.17)$$

i.e. $$E = \frac{4W}{ac}(1 - e^{-\frac{act}{V}}) \qquad .....(14.18)$$

At $t = \infty$, $$E = E_{max} = E_m = 4W/ac \qquad .....(14.19)$$

The intensity of sound at any point inside the room is given as

$$I = I_m(1 - e^{-\frac{act}{V}}) \qquad .....(14.20)$$

Where I_m is the maximum intensity of sound that is reached after a long time inside the room.

14.6 Decay of Sound Intensity Inside the Room: Sabine's Formula:

Suppose that when the intensity of sound attains maximum, the source is cut off.

Then the energy at any point starts falling at a particular rate.

When the source is cut off, W = 0.

i.e $$V\frac{dE}{dt} = -\frac{acE}{4} \quad \text{or} \quad \frac{dE}{acE} = -\frac{1}{4V}dt \qquad .....(14.21)$$

Integrating the above eqn.14.21., we get

$$\ln E = -\frac{act}{4V} + \ln C, \qquad .....(14.22)$$

At t = 0, E = E_m, therefore, C = E_m

$$\therefore \quad E = E_m e^{-\frac{act}{4V}} \qquad .....(14.23)$$

The intensity is expressed as

$$\therefore \quad I = I_m e^{-\frac{act}{4V}} \qquad .....(14.24)$$

The eqn(14.24) obtained indicates that the sound intensity falls off exponentially with time from the time the sound is cut off.

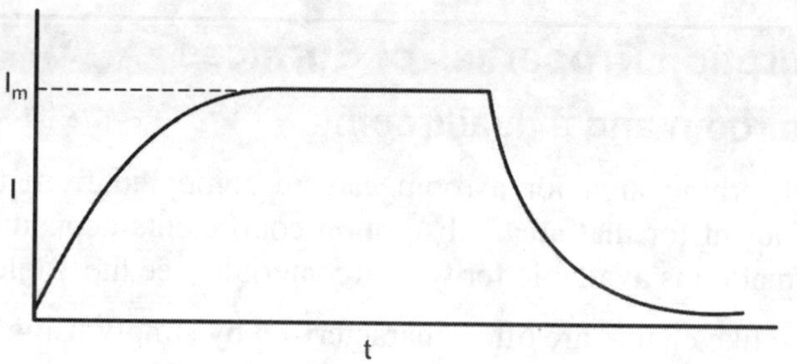

Fig. 14.7 The growth and decay of sound intensity on the basis of discussions given above.

$$\frac{E}{E_m} = e^{-\frac{act}{4V}} \qquad .....(14.25)$$

$$10^6 = e^{\frac{acT}{4V}} \qquad .....(14.26)$$

Taking logs on both sides, we have

$$\ln 10^6 = \frac{acT}{4V} \qquad .....(14.27)$$

Sabine's Formula

The reverberation time is given as

$$T = \frac{4V \ln 10^6}{ac} = \frac{2.303 \times 6 \times 4\,V}{c \qquad a} = \frac{KV}{a} = 0.161\frac{V}{a} \qquad .....(14.28)$$

The velocity is taken as 343 meters / sec. Siben is the unit of absorption. This approximate formula is very close to the result found experimentally by W. C. Sabine,

In 1922, a pioneer in the study of room acoustics, Wallace Sabine came up with the formula which is given below in eqn.(14.28). Reverberation time is typically defined as the time required for the sound level to decrease by 60 dB.

$$\mathbf{RT60 = k\,(V/Sa')} \quad .....(14.29)$$

k is a constant that equals to 0.161 when the units of measurement are expressed in meters, **V** is the volume (m^3) of the room.

- **Sa′** is the total surface absorption of a room expressed in sabins (m^2). It is a sum of all the surface areas(**S**) in the room multiplied by their respective absorption coefficients (**a**). The absorption coefficients express the absorption factor of materials at given frequencies. The absorption coefficient is dependent on the frequency of sound.

  Now, we can define the reverberation time as the time interval with in which, the intensity of sound in a room decays to one millionth of its original intensity when the source of sound is switched off.

14.7 Absorption Properties of Surfaces

14.7.1 A live room and a dead room

The effective absorbing area for a room can be computed from the actual area and the absorption coefficient for that area. Absorption coefficients depend on the frequency of the sound and information is available for typical materials (see the Table 14.1.).

Some objects, like chairs, are often characterized by simply using an equivalent absorbing area based on measurement rather than trying to compute it in detail. The equivalent areas are often reported in units of Sabins. One 'Sabin' is equal to one sq.m. of surface area with coefficient of 1.0 or one sq.m. of open window.

In theory, it is easy to determine the reverberation time of a room. It depends on the volume of the room and the rate at which the sound energy is absorbed by the wall surfaces and the objects in the room. Based on the characteristics and absorption coefficients and reverberations, the halls are classified as a live room and a dead room. They are described as given below.

A live room: A room or hall is a live room if the reverberations cause the loudness of sound and the mean absorption coefficient of materials of the walls of the room is less than 0.4, an auditorium should be *a live room.*

A dead room: If the absorption coefficient of the walls of the hall or room is above 0.4, the room is said to be *a dead room.*

In a bare room, the reverberation time is thus proportional to the ratio of volume to surface area. It is customary to define the reverberation time as the time required for the sound level to decrease by 60 dB (hence the abbreviation **RT60**).

14.7.2 Eyring and Norris formula

Sabine's formula is valid for long rooms. When the absorption coefficient is greater than 0.4, the formula gives erroneous results and the room becomes a dead room. In such halls higher order reflections or late reflections should be taken into consideration. The approach of

Eyring and Norris is given here. Whenever sound falls on a wall, a part of it is absorbed. Let this fraction be a'. Then, a fraction (1-a') is reflected. If successive reflections take place, the intensity is I_0 (1-a'). In subsequent reflections ,the intensity is given as $I_0(1-a')^2$, $I_0(1-a')^3$, $I_0(1-a')^4$ .... $I_0(1-a')^n$. We have considered n reflections taking place when the intensity falls off to 60dB. In such a case, we get

$$n\ln(1-a') = -6$$

$$\therefore \quad n = \frac{2.303\times 6}{\log(1-a')}$$

$$\frac{cT}{\lambda} = n = \frac{2.303\times 6}{\log(1-a')}$$

$$T = \frac{55.5\times 6\,V}{c\,S[-\log(1-a')]} = \frac{0.16\,V}{S[-\log(1-a')]}$$

Where we have substituted the value of velocity c as 345m/s.If the absorption coefficient is small, the above formula reduces to Sabine formula which is valid for reasonably high reverberations.

14.8 Measurement of Absorption in a Room

14.8.1 Characteristics of absorbent materials

In any room, a train of waves do exist when a sound source is put on. The energy of a train of sound waves in the air resides in the regular oscillations of the molecules. The absorption of this energy can occur only by some process by which these ordered oscillations are converted into the random molecular motion. In other words, the absorption of sound is a dissipativc proccss and occurs only when the vibrational motions are damped by the action of the forces of friction or viscosity. The experience shows that only materials which are porous or in-elastically flexible or compressible absorb sound to a considerable degree. For a material to be highly absorbant, the porosity must consist of intercommunicating channels, which penetrate the surface upon which the sound is incident. Cellular products with unbroken cell walls or with an impervious surface do not show any marked absorptive properties. A simple practical test as to whether a material possesses absorbing efficiency because of its porosity is to attempt to force air into it or through it by pressure. If air cannot be forced into it, it will not show high absorbant properties. It is important to calculate or estimate the absorbing characteristics of materials that are used in the construction of auditoriums.

Whether from a sound *impulse* or a steady sound which ceases, the *reverberant* sound in an auditorium decays in an approximately exponential fashion.

14.8.2 Method of measurement of absorption of sound in halls

The study of reverberation time or decay of energy decay is a part of building architecture. The absorption characteristics of materials become very important and essential. The most fundamental method is to measure reverberation time. The reverberation time is a good indicator for the effectiveness of absorption material.

Consider two sources, A and B, that are placed in an enclosure. Let the energy densities emitted by these sources be W_a and W_b respectively.

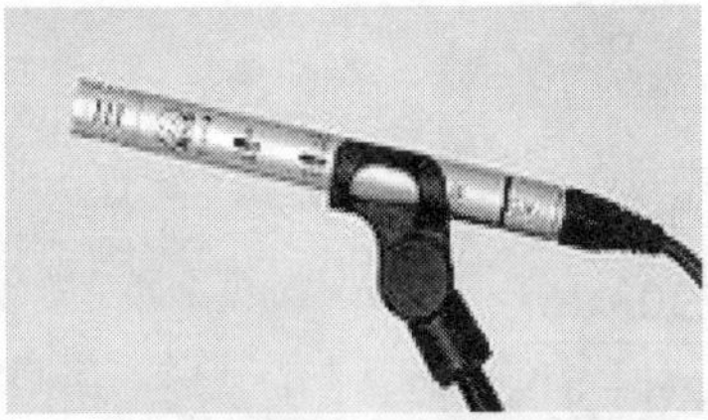

Fig. 14.8(a) A microphone for recording the sound intensity.

Fig. 14.8(b) The room with a micro phone, the source and the material on the wall.

The microphone (14.8a.) is used for recording the sound intensity level..The Fig.14.8b shows the arrangement of a very sensitive microphone opposite the material panel whose absorpyion coefficient has to found. Let T_a and T_b be the reverberation times associated with sources respectively.

First the source, A, is put on and switched off and the time, T_a for which the intensity falls to the threshold audibility is estimated by recording the sound with microphone. After waiting for some time, the second source is switched on and put off and again the reverberation time, T_b is recorded. The process is repeated a number of times and the average values of T_a and T_b are estimated.

Then, $$\frac{4W_a}{ca} e^{-\frac{ca}{4V}T_a} = \frac{4W_b}{ca} e^{-\frac{ca}{4V}T_b} \qquad .....(14.30)$$

$$\frac{W_a}{W_b} = e^{-\frac{ca}{4V}(T_a - T_b)}$$

$$\ln\frac{W_a}{W_b} = \frac{ca}{4V}(T_a - T_b) \quad .....(14.31)$$

$$\therefore \quad a = \frac{4V}{c(T_a - T_b)}\ln\frac{W_a}{W_b} \quad .....(14.32)$$

$$\bar{a} = \frac{\sum_i a_i ds_i}{\sum_i ds_i} = \frac{a}{s} \quad .....(14.33)$$

$$s\bar{a} = \frac{4V}{c(T_a - T_b)}\ln\frac{W_a}{W_b} \quad .....(14.34)$$

By using standard sources of known intensities, the absorption coefficients are accurately measured by the above method.

The following table gives absorption coefficients of a few materials.

Table 14.1 The absorption coefficients of few important materials.

Material	Sound Absorption Coefficient - *a* -
Plaster walls	0.01 - 0.03
Unpainted brickwork	0.02 - 0.05
Painted brickwork	0.01 - 0.02
3 mm plywood panel	0.01 - 0.02
6 mm cork sheet	0.1 - 0.2
6 mm porous rubber sheet	0.1 - 0.2
12 mm fiberboard on battens	0.3 - 0.4
25 mm wood wool cement on battens	0.6 - 0.07
50 mm slag wool or glass silk	0.8 - 0.9
12 mm acoustic belt	0.5 - 0.5
Hardwood	0.3
25 mm sprayed asbestos	0.6 - 0.7
100 mm mineral wool	0.65
Acoustic tiles	0.4 - 0.8
Open window	1.0

14.9 Acoustic Noise

What is acoustic noise?

Acoustic noise is an extraneous sound from low-level, but annoying, to loud and harmful. Acoustic noise is any sound in the acoustic domain both wanted (music, bird song, amplified speech, etc) and unwanted.

The sound of very loud music that has become a common sight nowadays is a noise to many ears that function normally. A jet engine at close quarters, continuous jabbering in an auditorium, continuous sounds of engines, horns in the road traffic, the power transformer near the house, sounds from aircrafts, trains. etc, come under noise that disturbs us. Such a continuous noise can cause *hearing impairment, hypertension, ischemic heart disease, annoyance, sleep disturbance*, and decreased education performance.

They spoil the quietness of the environment. In our homes, or at work places, or at leisure, our lives are constantly affected by an ever increasing number of noise sources. Noise includes low level sounds from appliances like electric air conditioners, televisions, the refrigerator, or high level sounds like factory noise, noise of the crowd in market places, railway platforms, and heavy equipments in industry and so on. Noise is also considered as pollution these days,

Usually the sound of a violin or sitar played by a trained player, referred to as music, is something pleasing. But, the talking of the audience among themselves, the sounds made by fans, while the music is on can be termed as noise. The rate of sound decay is not uniform but fluctuates about an average slope in a room. Too much reverberant sound will result in loss of clarity. Then, the sounds made in that room may work as back ground over the actual signal and noise. In any place noise to actual signal ratio should be kept low. Noise is also mentioned in dB and its level should not exceed 30dB in any room.

The combustion process and exhaust cycle of the engine impart vibrations to the air stream that make themselves known as noise. Noise tends to increase in volume as an engine gets bigger, or as compression increases, or as frequency increases (and volume) or as engine RPM is increased.

There is an optimum time for which one can tolerate noise. The permissible exposure to a continuous noise should not be more than T_N in hours where

$$T_N = \frac{16}{[0.2(L-85)]},$$

where L is the sound level measured in dB.

Table 14.2 gives the sound strength in terms of pressure of some of the sounds from different sources.

Table 14.2 The sound pressures felt on the ear from different sources

Source of Sound/Noise	Approximate Sound Pressure in μPascal
Launching of the space shuttle	2,000,000,000
Full music orchestra on a stage	2,000,000
Diesel Train at high speed at 25 m from it	200,000
Normal Conversation	20,000
Soft Whispering at 2 m in Library	2,000
A nominal sound proof room	200-300
Softest Sound that we can hear	20

14.10 What are the criteria for good acoustics in a hall or auditorium?

Factors affecting the acoustics of a hall and suggested solutions:

I. ***Reverberation:*** If a hall has to be acoustically good, the reverberation time should be neither too short nor too long. The volume of the hall decides this and is approximately proportional to the cube root of the volume rather than just volume. Optimum reverberation time (RT) is a compromise between clarity (requiring short reverberation time), sound intensity (requiring a high reverberant level), and liveliness (requiring a long reverberation time). The optimum reverberation time required for an auditorium is dependent on the use for which it is designed. The seminar or lecture halls require RT of 0.5 sec while musical concerts, RT should be around 1 to 2 sec. For big theaters, RT can be 2-3 sec. Excessive reverberation can ruin the acoustical properties of an otherwise, well-designed room. Reflected sound arriving from the sides plays an important role in the overall reverberation in the room.

Articulation decides the RT. The meaning of articulation is the time gap between the successive syllables produced from sound source. For each syllable, there will be a growth and decay .The second syllable would start before the first syllable decays totally. The absorption in the room is adjusted such that reverberation of the first syllable does not cause a noise for the successive syllables. Rapidity of speech is important factor in reverberation. A speaker who talks slowly in a room requires longer RT.

Remedy:

1. The optimum reverberation is adjusted with the surface areas in walls, ceilings by covering them with suitable absorbing materials.
2. The shape of the hall is very important. The wide walls, high ceilings, rear walls, doors, floor are potential sources of noise. The sidewalls are lined with pillars or rounded pilasters.
3. Ceiling is very important Therefore, false roofing is provided in most of the halls to take care of proper absorption and to *avoid long reverberation time*. There are the good early reflections and the bad late reflections. The early reflections need to be cultivated. The late reflections need to be weeded out. The balcony facing is sculpted to provide early reflections back down to the main floor. The back wall of the balcony and ceiling is sculpted to provide early reflections into the balcony seats. The late reflections are mainly dealt with by combining two features. It begins with having a heavily coffered ceiling. Any sound that is heading upwards eventually hits the coffered ceiling, only to be splintered into a cascade of tiny and off angled reflections. The second factor is the height of the ceiling. A 70ft ceiling with a loudspeaker mounted some 35ft off the floor starts splashing sound back onto the main floor at about 40 ms after the direct signal has passed through the audience. The coffering of the ceiling breaks this reflection up into dozens of low-level reflections with a variety of additional time delays. The high coffered ceiling acts to diffuse and randomize the only possible late reflections in the hall.

Fig. 14.9 A hall with balcony.

4. The hall should be provided with proper absorbing materials for both high frequency as well as low frequency sounds. Composite absorbers should be used.

   The sound absorbing material should be water proof, fire proof, strong, elegant, and economical. Non-perforated panels of wood, pressed wool, or wood fibres, acoustic tiles are good absorbers.

5. In *architectural acoustics*, unwanted sounds should be absorbed rather than reflected inside the room or a hall for an observer. The unwanted sounds are noise from outside, noise made by audience.

6. The stage, the walls, windows etc., are provided with long curtains and they should be in folded forms that the reflection produce diffuse sound quickly and sound is absorbed properly.

   We should provide Classic auditorium design. Elevated central speaker, balconies, coffered ceiling and stage curtains are commonly employed.

   These days, the walls and ceilings in big halls are coated with acoustic plaster which is a polymeric composite.

II. ***The sound intensity in the hall:*** With the absorbers in the hall, sometimes the sound intensity is not sufficient to reach all listeners in the hall especially so in halls with large volume. One has to distinguish between loudness and intensity of sound here. Too much intensity goes into loudness and it is not desirable. Loud sounds produce distorted wave forms.

Remedy:

Distributed sound systems have to be installed in the acoustic ceiling in an effort to help inject life back into the space. These days, stereo phonic sound systems are used and these systems elegantly separate low frequency sounds and high frequency sounds. The special distribution will also be good. A good auditorium will accomplish effective projection of the sound to the rear of the auditorium so that those distant listeners will not experience the extreme loss of sound level caused by the *inverse square law*. That projection is normally achieved by having a sufficiently long *reverberation time*. Another significant contributor will be a high, reflective ceiling to reflect sound to the back of the auditorium. *Good balance of low and high frequencies is made keeping in mind that RT for low freq is longer than for high frequency sounds.*

Spatial impression and early decay time are important. The spatial impression is dependent on contributions to the early reflections from the walls, especially from the sides.

III. ***Echoes:*** The echoes are annoying sounds in a hall because of reflections from the highly reflecting walls. The term echo in a hall or auditorium refers to a reflected sound which dominates over the normal *reverberant sound* of the auditorium. It is not sufficient in this context to just define an echo as a reflected sound, since the entire reverberant sound field is a collection of many reflected sounds. The term echo in auditorium acoustics may be the prominent reflection from a reflective flat back wall, often called a 'slapback' echo. Echo is heard as distinct repeat, 0.1 secs or greater from walls and ceiling with path lengths greater than 15 meters apart.

Remedy: Distracting echoes arise from *focusing surfaces* like domes (egg shaped surfaces), which should therefore be avoided in auditoriums. In fact, it is good practice to include surfaces or structures which break up any large reflective surface. The columns, decorative woodwork, and gives a good mix of reflected sound for more even dispersion of sound. Modern structures tend toward large smooth plaster or concrete surfaces which give problems with echoes.

Flutter echoes, sound focusing, sound shadows, and background noise should be avoided in an auditorium design. Remedy: The walls of the auditorium are not kept smooth, as such walls produce echoes. *Evenly dispersed sound is provided. Absence of echoes can be achieved by corrugated surfaces.* Even high quality amplifiers with proper filters for the carefully mounted speakers bring down the noise due to echoes.

IV. ***Occurrence of dead spots and focusing:*** In the halls where the walls .ceiling are curved or in very big halls, sound concentration takes place at some spots. Naturally, some other spots which are away from such sound foci will be *dead spots*. The setting up of standing waves in a hall because of highly reflecting walls will also give the spots of uneven sound intensity in the room. Sound focusing: The sound intensity may get focused at certain points because of the curved walls and doomed ceiling.

Remedy: Suitable sound diffusers are used in the halls. There should not be any dead spots. Narrow halls are generally preferred to wide ones. There are suggestions that the ceiling should be low. However this not preferred, instead, reflecting clouds are provided. Curved walls and curved ceiling should be avoided.

V. ***Resonance:*** It is an amplification of a sound, e.g. that of an instrument or the human voice, caused by sympathetic vibration in a chamber such as an auditorium. The closed halls function as cavity resonators for sound waves. For certain frequencies, the hall resonates like a hollow cavity. Sound reflections in a hall create standing waves that produce natural resonances that can be heard as a pleasant sensation or an annoying one. Reflective surfaces can be angled and coordinated to provide good coverage of sound for a listener in a concert hall or music recital space. To understand this concept, consider the difference between a modern large office meeting room or lecture hall, and an ordinary *classroom* with all hard surfaces.

Remedy: the acoustic damping is provided. *Green Glue* is the highest performing visco-elastic damping compound available and it is used. on perforated panels of wood, wood fibers, and plastic diaphragms mounted on solid backing separated by air column are used.

VI. Absorption of sound: High absorption of sound in a room causes loss of intensity and distortion. The coefficient of absorption of sound is dependent on the frequency. A balance is made between high and low frequency sounds in the back of big halls or auditoriums. The low frequency sound requires boosting in those corners. Table 14.3 gives the absorption coefficients of some useful materials at different frequencies. Porous materials are effective for higher frequencies.

	125Hz	250Hz	500Hz	1kHz	2kHz	4kHz
Reflective wall materials						
Plasterboard (12mm)panel	0.29	0.1	0.06	0.05	0.04	0.04
Plywood, 5mm panel, 50mm air space, 50mm fiber glass	0.42	0.36	0.19	0.1	0.08	0.05
Concrete blocks	0.1	0.05	0.06	0.07	0.09	0.08
Plywood, 19mm panel, airspace	0.2	0.18	0.15	0.12	0.1	0.1
Absorption wall materials						
Drapers10x350 flat	0.04	0.05	0.11	0.18	0.3	0.35
Fiber glass 150mm	0.18	0.76	0.99	0.99	0.99	0.98
Ceiling materials						
Plaster board,12mm grid	0.15	0.11	0.04	0.04	0.07	0.08
Plaster, gypsumor lime	0.01	0.02	0.02	0.03	0.04	0.05
Sprayed cellulosefiber32mm,solid	0.1	0.3	0.73	0.92	0.98	0.98
Surfaces						
People-adults(1/10 person)	0.25	0.35	0.42	0.46	0.5	0.5
Ventilatingoutlets	0.3	0.4	0.5	0.5	0.5	0.4
Seats						
Wooden bench, occupied	0.5	0.56	0.66	0.76	0.8	0.76
Benches ,cushioned seats and backs, occupied	0.44	0.56	0.65	0.72	0.76	9.67
Cinema seats (fully covered) ,wood	0.5	0.3	0.4	0.76	0.8	0.76
Seats, fabric, upholstered, or occupied	0.6	0.74	0.88	0.96	0.93	0.85
Flooring						
Carpets	0.01	0.02	0.06	0.15	0.25	0.45
Marble, glazed tile	0.01	0.01	0.01	0.01	0.02	0.02
Vinyl tile	0.02	0.03	0.03	0.03	0.03	0.12

Remedy: Discrete sound absorbers are used. Discrete objects like Chairs, free-standing screens or persons are absorbers of sound.

Audience and seating: Seating in a hall is very important. When audience is present, the sound undergoes multiple reflections from the people and seats and it over laps with original and the listener can not distinguish the echo. The audience acts as the acoustic materials to control the reverberation time.

The presence of audience plays a vital role in absorption of sounds. Each person on the average is supposed to have an absorption coefficient of 0.45.Therefore, audience will give good absorption. Many a times, all the seats are not occupied. In such a case, reflection may take place from the seats. Therefore, cushioned and upholstered seats are used. Each upholstered seat without a person is equivalent to a person. Unnecessary reflections from vacant seats can be avoided this way.

The space required by a listener depends upon the purpose. For a musical concert, each person may require about 4.7 m^3, for cine theaters or drama theaters: around 4.2 to 4.9 m^3, for seminar halls : about 3.5-4 m^3.

Direct sound waves, reflected waves, and sound from the speakers some times do not reach all listeners with proper distribution.

Remedy: The Fig.14.10 gives a picture of a modern auditorium. The arrangement and quality of seats provided for the people to sit is very important in high quality halls. Seats are arranged in a circular fashion except the first few rows. Each row has to be elevated by one step. The distance between the rows is also adjusted such that the front row does not become an obstruction. The stage is elevated so that the last row is able to receive the direct sound and also see the stage. Sound shadows may be formed under the balconies and near rear walls. They have to be taken by proper placing of speakers or reflectors of sound. The projection of balconies into the hall should be such that no sound shadows are formed below them. The rear-end walls should be provided with proper absorbers.

Fig. 14.10 The well planned auditorium with people on left side and without people on right side. One can see the balconies, walls, stage, seating arrangement, flooring, doors etc.

The seating, the height and the interior architecture all work in unison to produce the required intelligible acoustic condition for reasonable listening. Important subjective attributes of concert hall acoustics are intimacy, live ness, warmth, loudness of direct sound, reverberant sound level, definition or clarity, diffusion or uniformity, balance and blend, ensemble, and freedom from noise. The architecture of a hall or building has to take care of these aspects in a proper way and provide a pleasant experience.

14.11 Acoustic Quieting and Methods

14.11.1 What is acoustic quieting?

The sounds are produced wherever some vibrating systems exist. Systems like electrical appliances: refrigerators, electric motors, air compressors, transport systems such as vehicles, airplanes, railway engines, transformers, generators, etc. are the sources of sounds that be called as noise.

Acoustic or sound quieting is the process of making machinery quieter by damping *vibrations* to prevent them from reaching the listener.

Machinery vibrates, causing *sound waves* in air, *hydro-acoustic* waves in water, and *mechanical stresses* in solid matter. Quieting is achieved by absorbing the vibrational energy or minimising the sound levels from the source of the vibrations. It may also be redirected away from the listener.

Sound quieting is necessary to reduce the sounds that are produced by objects mentioned above.

When the objective is acoustic quieting, a number of different aspects might be considered. Each appliance or engine is a special case by itself. Each aspect of acoustics can be taken alone or in combination so that the end result is that the reception of noise by an observer is minimised.

Noise suppression is critical in the design of multi-unit houses like flats or modern commercial complexes like malls or business premises that generate significant noise, including music venues like modern restaurants.

Most of the industries or other work places do not take care of noise pollution. This is resulting in major health hazard. Architectural acoustics includes *room acoustics*, the design of recording and broadcasting studios, home theaters, and listening rooms for media playback, class rooms for the colleges, commercial complexes. The main noise paths are *roofs, cleaves, walls, windows, door* and *penetrations*. Sufficient control gives healthy ambience for work.

14.11.2 Methods for acoustic quieting

(a) Limiting the noise at its source itself by keeping sound absorbers and vibration dampers.

(b) *Sympathetic vibrations:* by acoustic decoupling, Acoustic decoupling can be done by building certain parts of a machine to keep the frame, chassis, or external shafts from receiving unwanted vibrations from a moving part.

(c) *Sound transmissions:* by reducing transmission using many methods (depending whether the transmission is through air, liquid, or solid).

(d) *Sound reflections:* by limiting the reflections using many methods, e.g. by using acoustic absorption (deadening) materials, *trapping* the sound, opening a "window" to let sound out, etc.

(e) *Using silencers or mufflers:* The sounds from engines, generators, appliances like air coolers, air conditioners, fans, etc., produce lot of noise that would irritate the ear and brain. So noise reduction is done by keeping silencers or mufflers.

(f) *Mechanical Acoustic Quieting:* staggered studs, double stud walls are used at places where mechanical noise is very high (like heavy machinery rooms, generator rooms). Such studs function by inhibiting the movement of sound from one side of the wall to the other through mechanical paths.

Acoustic damping: Most of the noise in units like electric generators, engines, arises because of vibrations. Therefore, such units are provided with acoustic dampers. Dampers also decrease resonance effects. Green glue is painted on walls to provide damping. Acoustic damping is also useful in *seismic shock* protection of buildings.

(g) *Sound Isolation:* Noise isolation is isolating noise to prevent it from transferring out of one area, using barriers like deadening materials to trap sound and vibrational energy. Example: In home and office construction, many builders place sound-control barriers (such as *fiberglass barrier*) in walls to deaden the transmission of noise through them.

(h) *Noise absorption:* This is useful for noises with no point source and when a listener needs to hear sounds only from a point source and not echo reflections.. For example, submarine hulls would have too much noise, if they are not provided by special coatings. These coatings absorb sounds. Fabric, wood or acoustical tile are used to absorb sounds. Fabric can be wrapped around substrates to create what is referred to as a "pre-fabricated panel" and often provides the good noise absorption if laid onto a wall.

(i) Motors and rotating shafts are commonly fitted with damper mounts at the points where they contact the building or the chassis of a large machine.

14.11.3 Mufflers

Muffler is a device that absorbs noise, especially one used with an internal-combustion engine, a device to deaden sound, especially, one on a car exhaust. When we hear a *car engine* running without a muffler, we hear almost ear breaking noise. The pulsating flow from each cylinder's exhaust process of an automobile engine sets up pressure waves in the exhaust system-the exhaust port and the manifold having average pressure levels higher than the atmospheric. This varies with the engine speed and load. These pressure waves propagate at speed of the sound relative to the moving exhaust gas, which escapes with a high velocity producing an objectionable exhaust boom or noise.

Mufflers are used to suppress the sound from an engine. Inside a muffler, we find a deceptively simple set of tubes with some holes in them. These tubes and chambers are actually are finely tuned as a musical instrument. A muffler is shown in Fig. 14.9. They are designed to reflect the sound waves produced by the engine in such a way that they partially cancel themselves out.

Muffler principles: There are four main performance muffling principles: expansion, noise cancellation, and noise absorption.

Principle of a muffler: A suitably designed exhaust silencer or muffler accomplishes the muffling of this exhaust noise. The basics of silencing are based on the fact that the velocity of sound in the gas at a given temperature is directly proportional to the square root of the product of the pressure and the ratio of the specific heats (at constant pressure to that at constant volume), and inversely to the square root of the density of the gas.

As the temperature varies, the velocity also varies directly as the temperature by another square root law involving the product of the coefficient of thermal expansion of the gas and the temperature. The exhaust noise can be reduced appreciably by providing resonance chambers to offset the noise wave effects.

This is accomplished by the principle of the Helmholtz resonator. In principle, it comprises the exhaust pipe, which goes through the large volume of a chamber. Badic design of a muffler is given in Fig. 14.11.

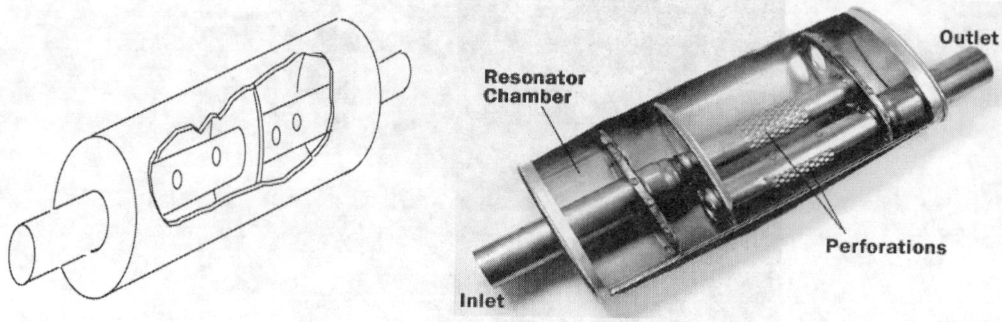

Fig. 14.11 Basic design of a muffler and an actual muffler.

The axial holes in the exhaust pipe enclosed by the chamber allow the gases to vibrate with the large mass of the gases in the chamber (like the vibration of thin sheets) and generate the sound of the same frequency but in opposite phase to that which has to be nullified (called anti-sound).

The muffler volume should be proportional to the engine piston displacement, and inversely proportional to the engine speed and the square root of the number of engine cylinders. The usual length to diameter (l/d) ratio of the resonator is about 4:1 to 8:1.

A small l/d ratio muffler attenuates the sound well for a narrow frequency band, where as the large l/d muffler attenuates the sound to a lesser degree but over a wider frequency band. The effectiveness of the exhaust system in silencing the exhaust depends also on the relative lengths of the exhaust pipe (from the exhaust manifold to the muffler) and the tail pipe. A ratio of 1:2 is better than 4:1, and 1:1 is the poorest ratio. Improvements are done to come over the narrow frequency range of the resonator.

Friction effects and filter off noise effects of other offending frequencies are provided in the resonator. Provision of baffles, resonator mufflers with end baffles, resonator with centre baffle chamber and four-chamber muffler are illustrative examples.

Mufflers are also made using the principle of interference. If the two waves are in phase, they add up to a wave with the same frequency but twice the amplitude. This is called constructive interference. But, if they are exactly out of phase, they add up to zero. This is called destructive interference. At the time when the first wave is at its maximum pressure, the second wave is at its minimum. If both of these waves hit your ear drum at the same time, you would not hear anything because the two waves always add up to zero. The muffler is designed to create waves that cause as much destructive interference as possible. A car is shown with a muffler in Fig. 14.12.

Fig. 14.12 A muffler fitted to a car exhaust.

14.11.4 Sound Proofing

Sound proofing is an isolation process in which a room or a hall is isolated from external noise ie noise from outside. Similarly, the sounds made in the hall are also prevented from reaching the people outside. This sort of sound proofing is needed in seminar halls, music concert halls, radio or cinema dialogue or song recording rooms.

Soundproofing affects sound in two different ways: *noise reduction* and *noise absorption*. Noise reduction simply blocks the passage of sound waves through the use of distance and intervening objects in the sound path. Noise absorption operates by transforming the sound wave. Noise absorption involves suppressing *echoes, reverberation, resonance* and *reflection*.

Soundproofing involves the reduction of the *sound pressure* with respect to a specified *sound* source and receptor. Basically, this can be done by 1. increasing the distance between source and receiver, 2. using *noise barriers* to block or absorb the energy of the sound waves, 3. using damping structures such as *sound baffles*, 4. using active *antinoise* sound generators. 5. Using sound isolators. The first principle of sound isolation is mass. Mass impedes the transmission of sound in a simple way. It is harder for the sound to cause vibration in a very heavy thing than a very light thing. In theory doubling the mass of a panel on the walls and doors without an air cavity will improve things by 6dB.

We face lots of sounds these days due to increased traffic on the roads near the halls, or theaters or auditoriums. Lots of people moving, sounds from neighbours affect the functioning of good rooms that are located near these disturbances. The main focus of residential soundproofing in existing structures is 'the windows'. Long and folded curtains can be used to damp sound. Sounds transmission can be prevented through the use of air chambers known as honeycombs. Single-, double- and triple-honeycomb designs achieve relatively greater degrees of sound damping.

High density sound barrier Mats, acoustic mineral wool to be placed between floor joists and studs in partitions are now a days available for the purpose sound proofing.

It should be remembered that sound proofing requires the arrangement of a room inside a room with acoustic tiles, wooden panel with air gaps, double layered doors, double panel doors for windows, apart from the regular precautions we have discussed earlier for good acoustics.

Noise absorption is achieved by providing dropped ceilings and acoustical tiles in high roof halls as shown in the Fig. 14.13.

Fig. 14.13 A sound proof room with acoustic damping tiles used for noise absorption and soundproofing.

The primary sound proofing limit of curtains is the lack of a seal at the edge of the curtain. Double-pane windows achieve somewhat greater sound damping than single pane windows. The increase in soundproofing with higher energy efficiency windows is primarily due to the better seals on double-pane than single pane windows. Significant noise reduction can be achieved by installing a second interior window. In this case the exterior window remains in place while a slider made of glass is installed within the same wall opening.

The typical sound paths are room partitions, acoustic *ceiling* panels (such as wood *dropped ceiling* panels), *doors*, *windows*, *ducting* and other penetrations.The external noises coming from outside should be eliminated by proper sound proofing. The greater the early decay time (up to two seconds), the greater will be the preference for the concert hall. Above two seconds, the trend it reversed. Interior building surfaces can be constructed of many different materials and finishes.

Fabric covered panels are one way to heighten acoustical absorption. Finish material is used to cover over the acoustical substrate. Mineral fiber board, or *Micore*, is a commonly used acoustical substrate. Wall panels can be constructed to accommodate door frames, baseboard, or any other intrusion. Large panels (generally, greater than 50 square feet) can be created on walls and *ceilings*. Absorption panel - convoluted foam

Fig. 14.14 A room with sound proof panels.

14.11.5 Quieting for specific observers

Quieting is done on the basis of specific cases: example, for observers working near war ships, submarines, war planes, or people working in a factory, near jet engines etc. The submarines run silently. Submarines are used for coast guarding purposes and it should not be detected by a surface ship. Submarines normally operate below depths of about 600 - 1600ft.At these depths, the surface ship using sonar can not detect the submarine. Hull coating, proper propeller design, and quieting the engine parts, pumps, and broad band sources are done. Submarines are operated in sonar channel axis where the speed of ultrasonic waves is lowest. Temperature induced refraction area is a sonic shadow zone, where submarine (observer) hides from acoustic detection.

The same principle of sound *refraction* can be used to prevent certain observers from hearing the noise. For example, an outdoor observer close to the ground will have sound waves refracted toward him when the ground is cooler than the ambient air and away from him when the ground is hotter than the air. It is necessary to place the observer in a zone where the amplitude of sound is least. Take the case of a *jet engine* and make *sound power level* observations along that circle, we find that the sound is loudest directly in line with the jet's exhaust. Observer can work at a place perpendicular to the exhaust as that place would be significantly quieter.

Noise canceling: If both the noise and the signal are received by an electronic or digital medium, noise can be filtered from the signal electronically and retransmitted without the noise. *Helicopter* pilots rely on this technology to speak on the radio to the ground station people.

Exercise 1

A cinema has a volume 7000 sq.m. The total absorption is equal to 180.90 of an open window. What will be the change in reverberation time of an audience fills the hall and therefore increases the absorption to 250 sq.m.

Solution

$$T_1 = 0.16\frac{v}{a} = 0.16 \times \frac{7000}{180.9} = 6.19 \text{ sec}$$

$$T_2 = 0.16 \times \frac{7000}{250.2} = 4.48 \text{ sec}$$

The change is $T_1 - T_2 = 1.71$sec

Exercise 2

A lecture hall of rectangular shape 5.0 m high, 12 m wide and 15m long has walls made of a plaster, wood and glass with an average absorption coefficient a= 0.030. The floor is covered with a carpet (a = 0.20) and ceiling with an acoustic plaster (a = 0.38). What will be the reverberation time of the empty hall? Of the hall with an audience of 30 people?

Solution

Absorption coefficient:

Wood plaster, glass

(Total wall area) × a = 270 × 0.03 = 8.10

Floor carpet with carpet = 180 × 0.2 = 36.00

Ceiling = 180 × 0.38 = 68.40

Persons (30)

Each person absorbs 4.7 units; × 30 people absorb = 4.7 × 30 = 253.5

1. For empty hall

$$T = \frac{0.161 \times 5 \times 12 \times 15}{112.5}$$

$$= \frac{0.161 \times 900}{112.5} = 1.29 \text{ sec}$$

2. For hall with audience

$$T' = \frac{0.161 \times 5 \times 12 \times 15}{253.5}$$

$$= \frac{0.161 \times 900}{253.5} = 0.57 \text{ sec}$$

Questions

1. How is sound intensity estimated? What is the sound intensity corresponding to 80 dB power.
2. What is meant by 'Reverberation'?
3. Explain how the sound intensity level increases when a sound source emits sound in a closed room?
4. How does the sound intensity level fall of when a sound emitting source is switched off? Explain.
5. Discuss the decay of sound intensity at a point in an auditorium.
6. What is an absorption coefficient? What is the absorption of coefficient of a hall?
7. What is reverberation time?
8. Discuss the growth and decay of sound intensity in an auditorium.
9. What are the good characteristics of an auditorium.
10. What is meant by 'optimum reverberation time'?
11. Write down 'Sabine' formula and discuss how volume of a hall and absorption coefficient affect the characteristic of sound in the hall.
12. Define absorption coefficient. Explain how absorption coefficient of a material is measured?
13. What is the function of absorbers in a hall.
14. What is the role of audience and seats regarding sound intensity in a hall?
15. What are the dead spots?
16. Give the expression for the rate of decay of sound in a hall.
17. What is sound proofing?
18. What is meant by acoustic noise? Is acoustic noise good or bad? Mention two methods by which the noise in a hall can be reduced.
19. Write a short notes on

    (a) Optimum reverberation (b) Acoustic noise.
20. What do you mean by 'acousting quieting'?
21. Consider a rectangular room of dimensions 3.5 m × 5 m × 10 m. The interior boundaries of the room have an absorption coefficient a = 0.01. Find the reverberation time. The velocity sound in air is 343 m/sec. What is the intensity level if the sound source 10 microwatt?

Objective Questions

I. Choose the correct answer

1. The reverberation time of a hall,
   (a) Depends only on the volume of the hall
   (b) Depends only on the energy power of the source
   (c) Is short for low frequencies
   (d) Depends on the volume, frequency and total absorption coefficient of the hall

2. Compared to a closed room with audience, the room without audience will have
   (a) Higher absorption
   (b) Less absorption
   (c) Less reverberation time
   (d) More clarity of sound

3. When sound source produces sound in a hall, the intensity of sound at any point grows
   (a) Linearly with time
   (b) At a constant rate as square of the time
   (c) Exponentially with time
   (d) At a rate proportional to $(\text{time})^3$

4. Good reflectors of sound in a hall
   (a) Shorten the reverberation time
   (b) Produces a faster decay of sound energy in the hall
   (c) Suppress high frequency sound
   (d) Produce longer reverberation time

5. The intensity of sound made by a plane on the runway is about 10^5 times the reference intensity. The intensity level in dB is
   (a) 50 dB (b) 5 dB
   (c) 500 dB (d) 10 dB

6. The threshold zone of hearing in principle is approximately
   (a) 10 dB (b) 26 dB
   (c) 120 dB (d) 50 dB

7. One of the following equations represents the growth of Intensity with time when a sound is produced in a hall
   (a) $I = I_m e^{-\frac{act}{V}}$ (b) $I = I_m e^{\frac{act}{v}}$
   (c) $I = I_m \left(1 - e^{-\frac{act}{v}}\right)$ (d) $I = \frac{I_m}{act}$

8. When the sound source in a hall is put off, the intensity of sound in the room
   (a) Decays in logarithmic way with time
   (b) Decays linearly with time
   (c) Decays as the square root of time
   (d) Decays at a rate proportional to $(\text{time})^3$

9. Suppose 60 dB sound is added to 60 dB sound, then the resultant sound intensity level is
   (a) 120 dB (b) 60 dB
   (c) 63 dB (d) 30 dB

10. Acoustic quieting for a car is done
    (a) By keeping the best reflectors of sound
    (b) By keeping mufflers
    (c) By keeping cloth in the exhaust cylinder
    (d) By keeping big tyres

11. Sound proofing for a room can be done by
    (a) Using acousting damping tiles
    (b) Using prefabricated reflectors with high reflection coefficient
    (c) By keeping metallic doors
    (d) By painting green glue on the outside walls of the room
12. The unit of sound intensity is
    (a) dB (b) Watts
    (c) Watts/m^2 (d) Microbars
13. Choose the correct statement
    (a) The audience in an auditorium absorb sound more in high frequency range.
    (b) The audience in an auditorium absorb sound more in low frequency range
    (c) The reverberation time for an empty room is less than that for audience filled room
    (d) Heavy curtains are useful for sound reflections
14. If the reverberation time for a hall is more than 4 or 5 seconds, the acoustic quality of the room is
    (a) Excellent (b) Good
    (c) Satisfactory (d) Very bad

II. Fill in the blanks

1. The unit of absorption coefficient is ________
2. A material has an absorption coefficient, a, of 90.00 square meters of an open window, it means. a = ________
3. The audience in an auditorium absorb more sound in ________ frequency range than in ________ frequency range
4. If a room has the absorption coefficient < 0.2, then room is called ________ room
5. The audible sounds have a frequency range ________ Hz to ________ Hz.
6. The intensity level of a sound is ten times that of threshold value, then the intensity level (II L) in dB is ________ dB
7. Sabine's formula for reverberation time is ________, where V is cubic meters, a is the absorption coefficient of a room and a $= \sum_i S_{a_i}$
8. Erying formula for reverberation time, T is ________
9. The unit of sound noise is ________
10. The dropped ceilings and acoustic tiles in a room reduce ________ in a room
11. The total absorption in a room is expressed in ________
12. A reverberant sound is the collection of all the ________ sound in an auditorium
13. Damping structures are used in rooms for ________
14. In an empty room, the reverberation time is proportional to the ratio of ________ to surface
15. Fabric panels in a hall improve acoustical ________
16. Smooth and high reflecting walls, cause ________ in a hall
17. A speaker who talks slowly in a room requires ________ reverberation time
18. If the absorption coefficient in a room is more than 0.4, the room is called a ________ room
19. The absorption coefficient of an open window is ________
20. Muffler volume should be ________ to the engine piston displacement

III. True or False

1. A sound of intensity that is ten times the threshold volume is said to have an intensity level of 10 decibels.

   [True/False]

2. Reverberant sound in a hall is the sound that is produced due to multiple reflections that occur in a hall after early reflections.

   [True/False]

3. $I = I_m e^{ct/4v}$ represents the intensity of sound that decays with time in an auditorium.

   [True/False]

4. The reverberant sound in a hall decays exponentially with time.

   [True/False]

5. Sensitive microphones are used to measure sound intensity in closed rooms.

   [True/False]

6. Compared to plywood panels, acoustic tiles absorb less sound intensity.

   [True/False]

7. Audience in a room absorb appreciable amount of sound energy made in a closed auditorium.

   [True/False]

8. Electric power generators are provided with silencers to produce acoustic quieting.

   [True/False]

9. Mufflers can not be used in a car for silencing sounds from exhaust.

   [True/False]

10. Wood fibers and acaustic tiles are bad absorbers of sound.

    [True/False]

CHAPTER 15

Nanotechnology

15.1 Introduction to Nano Materials

Nano is an abbreviation of nanometer (nm) which is a unit of length. It is one billionth of a meter, a very small length indeed. Nano is roughly the width of 3 or 4 atoms. Compared to a nano meter, a water molecule is around 0.3 nm across. The diameter of human hair is about sixty thousand to eighty thousand nanometers. If we could put hydrogen atoms together, one nanometer would approximately equal to the length of ten hydrogen atoms in a line. A material is called the nano-material if the size of the material is below 100 nm. You may wonder why 100 nm but not 1000 nm or 10 nm? This is because many physical and chemical properties of a material start changing significantly when the size of grains is below 100 nm.

A human hair is shown in the Fig. 15.1. Red blood cells have a size of about 7000 (7 micro meter) nm width.

The length scale that is of interest in nano science is from 100nm down to the atom sale i.e. 0.2 nm. When one talks of particles of sizes of a few nanometers, then, properties may differ significantly as compared to those properties observed in fine particle or bulk materials. It is this aspect that gives a deep interest in the study of properties of nano materials.

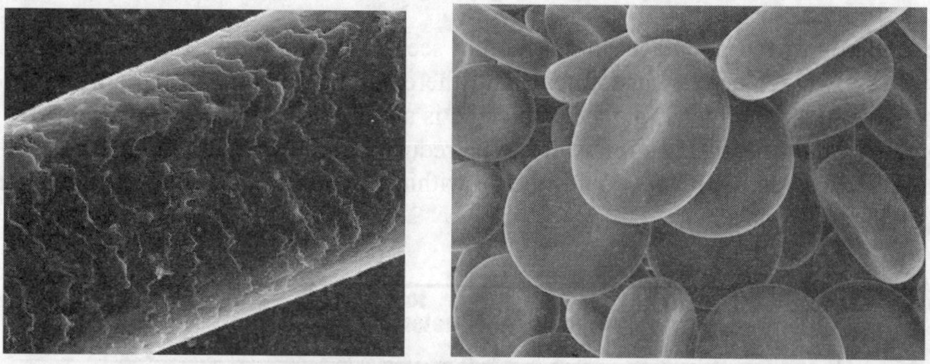

Fig. 15.1 Human hair viewed with electron microscope (left side photo). Right side photo shows the blood cells seen with microscope.

Materials are the basic building blocks of every advanced technology. Every device technology depends on:

1. Type of materials,
2. device structure, design and fabrication, and
3. integration.

In the above, the materials have a vital role to play in device structure, properties and its design. In the modern world, the clients want smaller, swifter and efficient devices. The miniaturization has caught the market and the sizes of various devices have been coming down with time. Earlier, the devices were prepared on the basis of classical materials. The properties of many basic materials have been modified in terms of purity, impurity incorporation, or combination of different materials. Today we have a demand for embedded systems, the systems that can perform many functions. These many functions include detection (sensing), data storage, then, action, and carrying of information. For this purpose we require materials that can perform many functions at a time, or materials that can be incorporated in devices that are a part of embedded systems.

Many functional materials are being prepared and used. The physicists, chemists, or materials scientists have dealt with the understanding of the fabrication, functionalities, structure-property correlations in many material systems like alloys, doped semiconductors, metal wires etc. The particle size or the grain size of material plays a vital role in its properties. The reduction of the size of the particles or grains has resulted in the property enhancement in many cases. The reduction also yields new capabilities to the materials. Naturally, it has occurred to scientists that the reduction of the particle size to nano level could improve the properties of the materials to high level. This has resulted in the birth of 'Nano Science' and 'Nano technology'.

The word nanotechnology comes from the term nanometer, which is just a scale of measurement. Why should a length scale elicit such excitement? First we will attempt to give an indication of how big (or perhaps how small) a nanometer really is and then we will discuss why this length scale (Fig.15.2) is important, and why the potential of its mastery has caught the attention of the populace on a scale not seen since the birth of the space age. A meter is well known scale of measurement, and there are 1,000,000,000 nanometers in a meter. Compared to a nano meter, a water molecule is around 0.3 nm across. A nanometer is one millionth of a millimeter. If people were to be reduced to the size of nanometers, all the people who ever lived on the planet could line up within a parking space. Fig.15.2. gives the size range of nano materials.

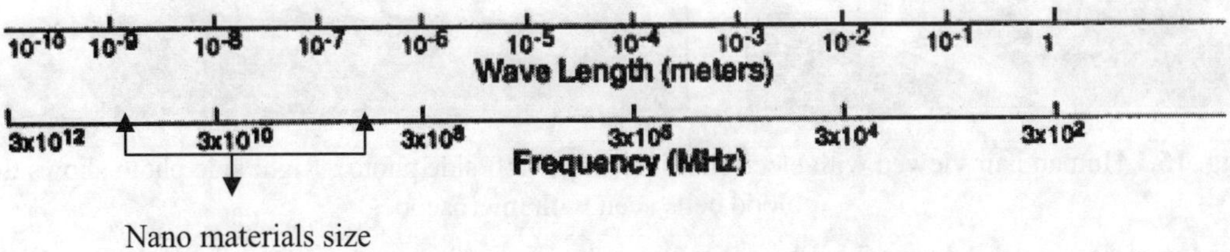

Fig. 15.2 Nano meter scale

In nano technology we are down to atoms or small groups of atoms (molecules) as the building blocks. Molecules and atoms are everywhere on the planet – the rivers and oceans mainly consist of one small molecule: H_2O (water).The size of this molecule is a few angstroms. Sizes below a fraction of an angstrom are impossible to achieve and this requires splitting of individual atoms. It is in this sense that nanotechnology represents a limit, a final frontier in the relentless drive for miniaturization which the semiconductor industry has been doing. But the semiconductor industry is just the most visible manifestation of the importance of miniaturization, virtually all of the physical, biological and engineering sciences are vitally concerned with this subject, although up till now, man made architectures could not really begin to deal with the nanometer length scale.

15.2 Origin of Nanoscience and Nanotechnology

Nano science is the study of matter on an ultra-small scale. One meter is one-millionth of a millimeter and a single human hair is about 80,000 nanometers in width. A nanometer is one billionth of a meter, roughly the width of four atoms.

Nanotechnologies are the design, characterisation, production and application of structures, devices and systems by controlling shape and size at nanometer scale. Nanotechnology tries to build things from molecules, or atoms. For example, think of how your body is made up. Imagine you are made up of different ingredients (which we will call atoms) and the way they are arranged together decide how you look, what color of hair, or skin you have and whether you are tall or short. Nanotechnology works the same way.

Nanotechnology, in its traditional sense, means building things from the bottom up, with atomic precision. This theoretical capability was envisioned as early as 1959 by the renowned physicist Richard Feynman (a noble prize winner at that time). He declared

"I want to build a billion tiny factories, models of each other, which are manufacturing simultaneously. . . The principles of physics, as far as I can see, do not speak against the possibility of maneuvering things atom by atom. It is not an attempt to violate any laws; it is something, in principle, that can be done; but in practice, it has not been done because we are too big."

This visionary physicist Richard Feynman issued a public challenge by offering $1000 to the first person to create an electrical motor "smaller than 1/64th of an inch". Much to Feynman's consternation, the young man who met this challenge, William McLellan, did so by spending many tedious and painstaking hours building the device by hand using tweezers and a microscope.

The field that Feynman hoped to incite has now become the most interesting field of nano technology.

At the nanometer length scale, the laws of physics operate somewhat differently; the classical mechanics that we encounter in everyday life give way to quantum mechanics.

Nano science is a broad term used for the study of materials and processes at the nano scale in a variety of disciplines like biology, chemistry, materials science etc. Nano science is simply the study of phenomena and manipulation of material at atomic, molecular or macro molecular scales, when properties differ significantly from those at a larger scale. Nano science does aim on the study of nanostructures and how such structures can be manipulated physically. Nano science is concerned with understanding of size effects on properties.

One should remember that Nanotechnology is the technological realization of quantum mechanics – the manipulation of materials and processes at the nanoscale level. In 1986, Eric Drexter wrote about 'Engines of creation' and introduced the term nanotechnology. Nanotechnology is being heralded as a new technological revolution, one so profound that it will touch all aspects of human society. The picture below (Fig. 15.3) shows how much influence on society the nanotechnology will have in the time frame.

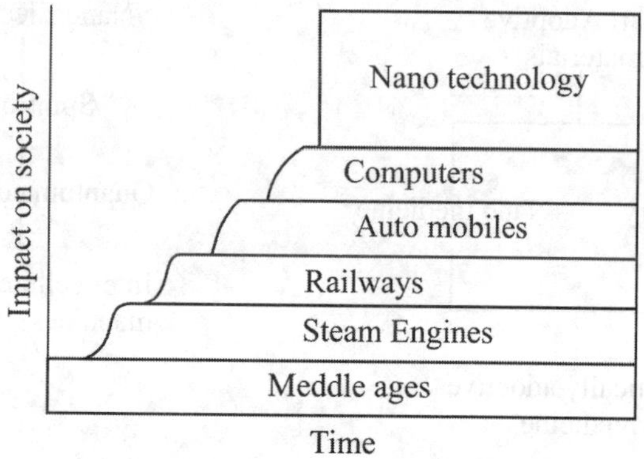

Fig. 15.3 Growth of Nanotechnology

Nano technology is defined as the design, characterization and production and application of structures, devices, systems by controlling shape and size at the nanometer scale. Nano technology aims at providing devices and systems with novel functions as properties. In fact, nanotechnology is an umbrella term that encompasses all fields of science that operate on the nanoscale.

This includes the design and manufacture of ever-smaller computer chips, custom-designed drugs, drug delivery systems (medical), preparation of pure water without contamination (or deflouridation) (health), and materials with vastly increased strength due solely to the arrangement of their molecules (such as carbon nanotubues).

Science and technology of nanomaterials are both vibrant fields, with breathtaking advances on a daily basis. Defining the scope of the subject is difficult. Ultimately the subjects of nanoscience and nanotechnology describe a mode of research and application rather than a unique field of study.

Roughly speaking, the nano technology has been applied to four disciplines: (1). nano electronics: new electronic devices with novel functions and the advantage of high speed and at the same time consuming less energy power. (2). nano material science: new materials with novel properties, which grow perfectly without any impurities and defects, ideal for nano devices. (3). nano biology including mapping the genetic information in DNA and RNA molecules. (4). nano medicine: discover, design, and deliver new drugs on nano level. Fig.15.4 describes different fields of nano technology that are emerging.

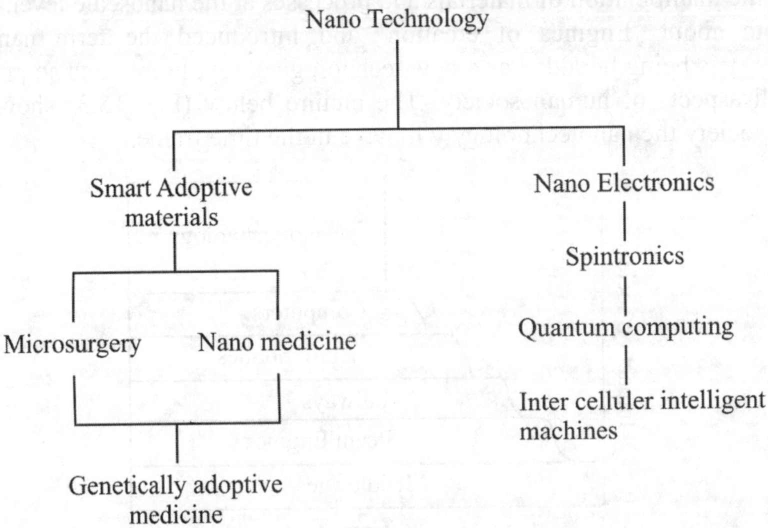

Fig. 15.4 Branches in nanotechnology

15.3 Surface to Volume Ratio of Nanomaterials

Consider a cube, sphere and a cylinder; now let us calculate the surface to volume ratio of these geometries.

Cube

Surface area $S = 6a^2$ where a is the radius.

Volume $V = a^3$

Surface to volume ratio $R = S/V = 6a^2 / a^3 = 6/a$

Sphere

Surface area $S = 4\pi a^2$

Volume of sphere $V = 4/3\ \pi a^3$

Surface to volume ratio $R = S/V = 3/a$

Cylinder

Surface area of cylinder $S = 2\pi al$

Volume of cylinder $V = \pi a^2 l$

Surface to volume ratio $R = S/V = 2/a$

In all cases, the surface to volume ratio follows an inverse power law. But the bottom line is that as the size of the system decreases, the fraction of atoms on the surface will increase. This is both good and bad. In one case illustrated below (Fig. 15.5). It can lead to more efficient catalysts. However in other cases, such as in the optical microscopy of nanostructures it can be a bad thing because of defects at the surface which quench the emission of the sample.

The surface effect will play a major role for a nano-material. Suppose we have a cube with side a, the surface area is $6a^2$. If we divide the cube into 27 ($3 \times 3 \times 3$) smaller cubes each with side a/3, then the total surface area is $18a^2$. In general, we can divide it into $N \times N \times N$ smaller cubes each with side a/N, the total surface area is $N(6a^2)$. We conclude that the smaller the cubes, the larger would be the surface area to volume ratio.

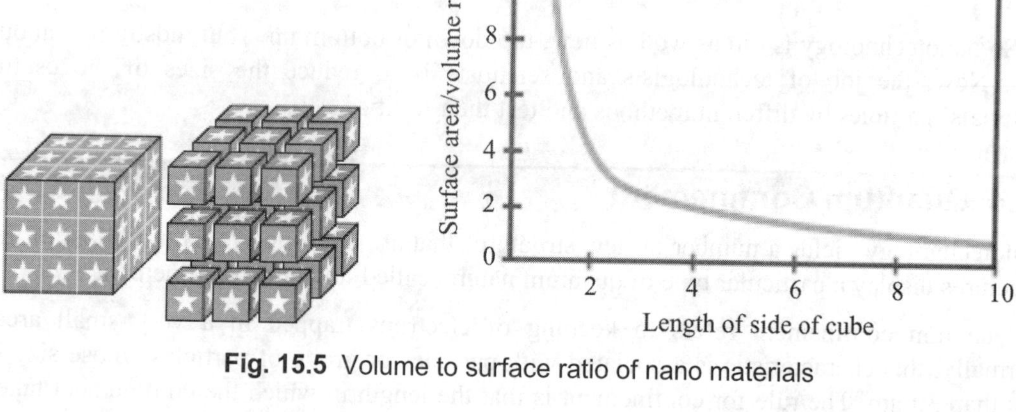

Fig. 15.5 Volume to surface ratio of nano materials

For example, consider a cube of side 1cm

$S = \text{Area} = 6 \times 10^{-4}\ m^2;$ $V = \text{Volume} = 10^{-6}\ m^3$

$S/V = 6 \times 100/m$

A cube of side 1mm

$\text{Area} = 6 \times 10^{-6}\ m^2;$ $\text{Volume} = 10^{-9}\ m^3.$

$S/V = 6 \times 1000/m$

A cube of side 50 nm

$\text{Area} = 6 \times 25 \times 10^{-16}\ m^2;$ $\text{Volume} = (50)^3\ 10^{-27}\ m^3$

$S/V \sim 10^{11}/m$

Therefore, if we have a gram of gold grains whose size is below 100 nm the surface of the grain will become very important. We know that the atoms on the material surface are highly unstable and very active in the chemical reaction. As a result, the increasing of the surface area will greatly enhance:

1. The chemical reaction,
2. the ability of surface absorption, and
3. the catalytic ability of the material.

In addition, the ability of heat conduction, mechanical strength, as well as optical properties of the nano-materials are also very different from that of bulk system. For example, the melting point of metal gold is about 1000°C, whereas for the nano sized gold grains with typical size of 2 nm, the melting point is down to $330\,°C$. The absorption of ultra violet of some nano materials can be enhanced significantly. Furthermore, new effects start to manifest when the size is below 100 nm. Hence a class of new materials emerges at nano scale.

So nanotechnology is old as well as new, top down or bottom up – but mostly it is about size. Now, the job of technologists and scientists is to reduce the sizes of the useful materials' particles by different methods and test their utility.

15.4 Quantum Confinement

Nanotechnology yields a number of new structures that are not found in nature. Most of the structures display a particular type of quantum nature, called 'quantum confinement'.

Quantum confinement refers to keeping of electrons trapped in a very small area. Normally, this characteristic is associated with particles or group of particles whose size is less than 30 nm. The rule for confinement is that the length in which the confinement takes place must be smaller than de Broglie wavelength of electron in the medium, of the length scale should be of the order of Bohr radius of exciton. For example, at 10K, the calculated wavelength in gallium arsenide (a direct band gap semiconductor) is 162 nm for electrons and 62 nm for holes.

We can have quantum confinement in several forms: 2D confinement results in quantum well (or plane). Many lasers are built now a days on this principle.

1-D confinement occurs in nanowires. Examples are InP, and CdSe. In this case the dimension should be less than 10 nm.

0-D confinement occurs in quantum dots only.

Consider a quantum dot. This is normally inserted at the junction in a semiconductor diode. The size of the dot would be around 3-60 nm in diameter. This size is about 30-600 times the size of normal atoms. All the electrons in the dot are confined in three dimensions and this is the characteristic of an atom. That is why quantum dots are called 'artificial atoms'. They exhibit very interesting optical properties. They produce radiation with high efficiency and highly monochromatic and divergence less laser beam in semiconductor lasers.

When the electrons are confined, the particles will have more oscillations and this will result in colour change of the materials. For example, nano gold colloids are dispersed in Ruby glass, the ruby glass exhibits red hue.

15.5 Fabrication of Nano Materials

The production of nano materials or nano crystalline materials requires precise methods. There are various techniques that are capable of creating nano structures.

In general, there two approaches that are classified as 1. Top-down 2. Bottom-up techniques.

1. 'Top-down' techniques involve starting with a block of individual material, etching(removing the surface by dissolution) or milling it down to the desired size.

   The challenge here is to produce smaller and smaller structures. Nano material particles can be made through this method. Let us take the example of turmeric powder. Turmeric pieces are first taken powdered with cylindrical hammers for a long time, sieved and again powdered to a fine powder. This powder is used in our houses. If it is possible to continue the processing of powdering with high energy i.e., pounding the powder in closed vessel with high energy, we get a powder with particles of nano scale.
2. 'Bottom-up' technique involves the assembly of smaller sub unit (atoms or molecules) to make larger structure. The example is honey behave made in nature by honey bees.

The process involves the building of desired structures, atom by atom or molecule by molecule.

Three categories exist in the above process. They are:

1. Chemical synthesis,
2. Self-assemly. Molecules can be transported to surfaces through liquids to form self-assembled monolayers (SAMs) and
3. Positional assembly. Spontaneous organization of molecules into stable, structurally well-defined aggregates (nanometer length scale).

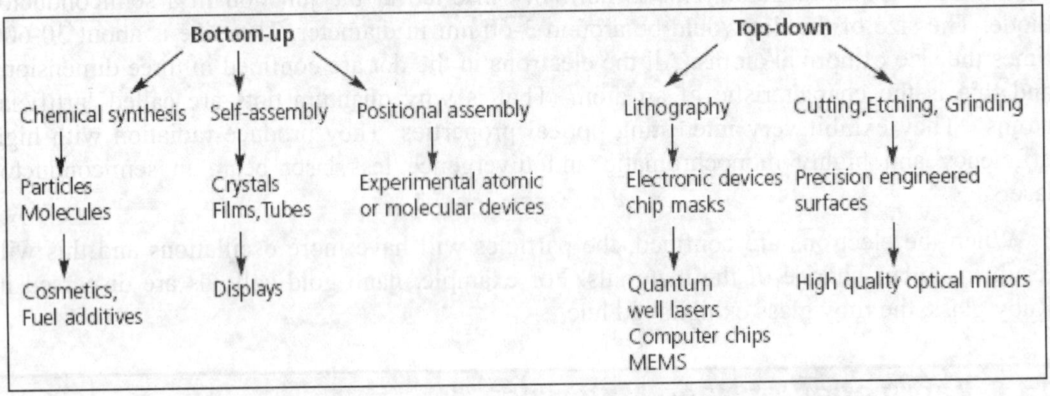

Fig. 15.6 Top down and bottom up methods of nano fabrication

Various methods are used for the preparation of nano materials. Some of the methods used are explained briefly here without going into details. Several methods exist, both physical and chemical, for preparing nano materials. The physical methods can be divided into thermal evaporation and sputtering and the chemical methods can be divided into vapor and liquid phase, which gives rise to different material film deposition techniques.

15.6 Vapour quenching or Vapour condensation methods

A class of methods which constructs the materials atom by atom fashion from the gaseous phase is called vapour quenching methods. These techniques are typically used for producing thin films for electronic and optical applications. The vapour is produced by heating a suitable compound called source and deposited as such without further modification (non reactive deposition) or it is deposited on a substrate where the chemical reaction would intervene to obtain a desired material (reactive deposition). Thermal evaporation, sputtering, r.f. sputtering, other atomic deposition are the various techniques used in this method.

15.6.1 Chemical vapour deposition methods

This method relies on the decomposition of a vapour species (like SiH_4) by thermal energy and r.f. field to heat the substrate upon which the vapour decomposes and deposits itself on the substrate.

MOCVD (metal-organic CVD)

PECVD (plasma-enhanced CVD)

LPCVD (low-pressure CVD)

MOD (metallorganic decomposition, sol-gel)

15.6.2 Electrolytic deposition method

This is achieved from the solutions of the desired material. For example, Ge films can be deposited on Cu cathode by the electrolysis of $GeCl_4$ in glycol.

15.6.3 Physical Vapor Deposition (PVD)

In this method the material/ materials whose nano thin films are to be obtained are evaporated on to a substrate by using different methods. Electron beams, radio frequency waves, resistance heating, laser heating are used to evaporate the materials. There are number of variations in physical vapour deposition methods.

Sputtering (rf-magnetron, dc, ion beam)

Evaporation (e-beam, resistance, molecular beam epitaxy)

Pulsed Laser Deposition (PLD)

As the number of components in a material increase, the degree of complexity for most techniques also increases dramatically. Only PLD, and to some extent sputtering, can be described as techniques with which it is relatively "easy" to produce a target stoichiometry, because both techniques use erosion of a target material to produce the deposition flux. In all other methods, the deposition flux is produced from separate sources, each carrying one of the components of the film. PLD and sputtering have a relatively narrow range to the average energy per atom, but only PLD has an instantaneous value of the deposition rate that can exceed 10,000 Å/s. This fact sets PLD apart from all the other methods. Also the fact that PLD does operate under high background gas pressure is an advantage of the technique over the others that is particularly useful for the insitu production of multicomponent oxides such as high Tc superconducting oxides and complex ferroelectric materials.

PLD is one of the most important applications of the interaction of high-powered laser light with condensed material. It has become increasingly popular within research laboratories as a method of producing thin films of novel materials ever since laser pulses in the nanosecond regime became available in the late 1970s and even more so since its success in the growth of commercial curate superconductor thin film devices in the late 1980s.

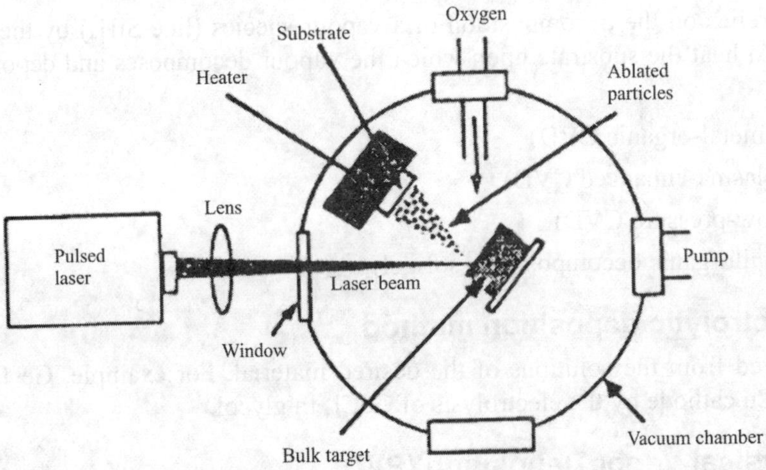

Fig. 15.7 Principles of PLD technique

Pulsed laser deposition is a growth technique in which photonic energy is coupled to the bulk staring material via electronic processes. The principle of PLD is shown in Fig. 15.8. An intense laser pulse passes through an optical window of a vacuum chamber and is focused onto a solid surface (Target), where it is partially absorbed. Above a certain power density, significant material removal occurs in the form of an ejected luminous plume. The threshold power density needed to produce such a plume depends on the target material, its morphology, and the laser pulse wavelength and duration, but might be of the order of 10-500 $MWcm^{-2}$ for ablation using ultraviolet (UV) excimer laser pulses of 10 ns duration. Material from the plume is then allowed to recondense on a substrate, where film growth occurs. The growth process may be supplemented by a passive or reactive gas or an ion source, which may affect the ablation plume species in the gas phase or the surface reaction, in which case one talks about reactive PLD. The physical processes in the PLD are highly complex and interrelated, and depend on the laser pulse parameters and the properties of the target material.

15.6.4 Sol-Gel method

The sol-gel process is a wet-chemical technique (chemical solution deposition) widely used for synthesis of nano materials. Such methods are used primarily for the fabrication of materials (typically a metal oxide) starting from a chemical solution which acts as the precursor for an integrated network (or *gel*) of either discrete particles or network polymers. Typical precursors are metal alkoxides and metal chlorides, which undergo various forms of hydrolysis and polycondensation reactions. The formation of a metal oxide involves connecting the metal centers with oxo (M-O-M) or hydroxo (M-OH-M) bridges, thereby generating metal-oxo or metal-hydroxo polymers in solution. Thus, the 'sol' evolves towards the formation of a gel-like diphasic system containing both a liquid phase and a solid phase whose morphologies range from discrete particles to continuous polymer networks.

Through sol gel process, homogeneous inorganic oxide materials with desirable properties of hardness, optical transparency, chemical durability, tailored porosity, and thermal resistance, can be produced at room temperatures. The specific uses of these sol-gel produced materials are derived from the various material shapes generated in the gel state, i.e., monoliths, films, fibers, and mono sized powders. Many specific applications include optical devices, protective and porous films, window insulators, dielectric and electronic coatings, high temperature superconductors, reinforcement fibers, fillers, and catalysts.

The sol-gel process, as the name implies, involves the evolution of inorganic networks through the formation of a colloidal suspension (sol) and gelation of the sol to form a network in a continuous liquid phase (gel). The precursors for synthesizing these colloids consist of a metal or metalloid element surrounded by various reactive ligands. Metal alkoxides are most popular because they react readily with water. The most widely used metal alkoxides are the alkoxysilanes, such as tetramethoxysilane (TMOS) and tetraethoxysilane (TEOS). However, other alkoxides such as aluminates, titanates, and borates are also commonly used in the sol-gel process, often mixed with TEOS.

Three reactions are generally used to describe the sol-gel process: hydrolysis, alcohol condensation, and water condensation. The characteristics and properties of a particular sol-gel inorganic network are related to a number of factors that affect the rate of hydrolysis and condensation reactions, such as, pH, temperature and time of reaction, reagent concentrations, catalyst nature and concentration, H_2O/Si molar ratio (R), aging temperature and time, and drying. Of the factors listed above, pH, nature and concentration of catalyst, H_2O/Si molar ratio (R), and temperature have been identified as most important. Thus, by controlling these factors, it is possible to vary the structure and properties of the sol-gel-derived inorganic network over wide ranges.

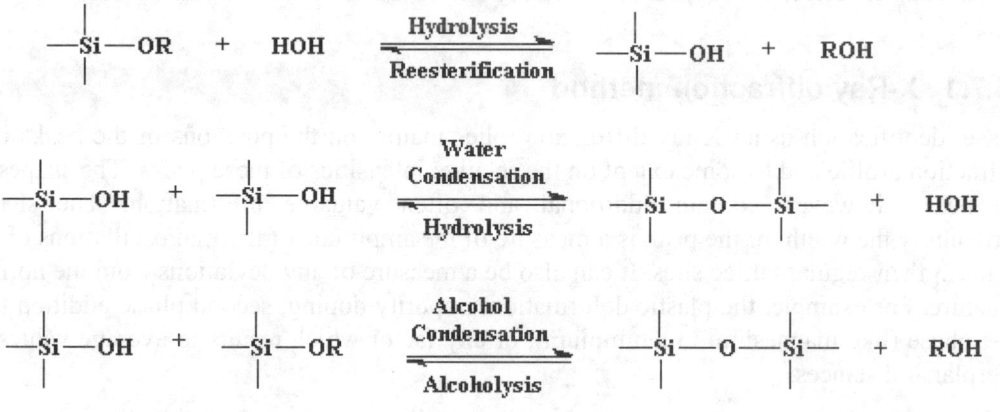

The hydrolysis reaction, through the addition of water, replaces alkoxide groups (OR) with hydroxyl groups (OH). Subsequent condensation reactions involving the silanol groups (Si-OH) produce siloxane bonds (Si-O-Si) plus the by-products water or alcohol. Under most conditions, condensation commences before hydrolysis is complete. However, conditions such as, pH, H_2O/Si molar ratio (R), and catalyst can force completion of hydrolysis before condensation begins. Additionally, because water and alkoxides are immiscible, a mutual solvent such as an alcohol is utilized. With the presence of this homogenizing agent, alcohol, hydrolysis is facilitated due to the miscibility of the alkoxide and water. As the number of siloxane bonds increases, the individual molecules are bridged and jointly aggregate in the sol.

When the sol particles aggregate, or inter-knit into a network, a gel is formed. Upon drying, trapped volatiles (water, alcohol, etc.) are driven off and the network shrinks as further condensation can occur. Sol-gel processes allow the synthesis of ceramics of :

1. high purity, because organometallic alkoxy precursors can be purified by distillation or recrystallisation,
2. high degree of homogeneity, because reagents are mixed at the molecular level,
3. low or high porosity, by using appropriate heat treatment and firing times.

Using sol gel process ferroelectric oxides like barium titante, lead zirconium titanate; high temperature super conductors; TiO_2, ZrO_2 etc have been synthesized. Carbon nano tubes, quantum dots are also made.

15.7 Characterization of Nano Materials

Various techniques are used for characterizing the nano materials. The characterizing technique depends on the type and application of the nano material being characterized. However far all nano materials particle size determination is essential and this can be done using X-ray diffraction and electron microscopy. X-ray diffraction is already discussed in the earlier chapter. Here, we reproduce necessary details of XRD which can be used for nano materials.

15.7.1 X-Ray diffraction method

Phase identification using X-ray diffraction relies mainly on the positions of the peaks in a diffraction profile and to some extent on the relative intensities of these peaks. The shapes of the peaks, however, contain additional and often valuable information. The shape, particularly the width, of the peak is a measure of the amplitude of thermal oscillations of the atoms at their regular lattice sites. It can also be a measure of any deviations from the normal structure. For example, the plastic deformation, impurity doping, second phase addition to a host phase (like magnesium to aluminium), or any factor which results in average values of interplanar distances.

When X-ray powder pattern is taken on plocrystalline powders, the crystallite size causes peak broadening. The well known Scherrer equation explains peak broadening in terms of incident beam divergence which makes it possible to satisfy the Bragg condition for non-adjacent diffraction planes. The crystallite size is easily calculated as a function of peak width (specified as the full width at half maximum peak intensity (FWHM)), peak position and wavelength. The average crystallite size L is given by:

$$\langle L \rangle_{vol} = \frac{K\lambda}{B_{1/2}\cos\theta_B}$$

where θ_B is Bragg angle, $B_{1/2}$ is Full width at half maximum of the diffraction peak, K is a constant and λ is the wavelength of the X-rays used.

15.7.2 Electron Microscopy

Electron Microscopes are scientific instruments that use a beam of highly energetic electrons to examine objects on a very fine scale. This examination can yield the following information:

Topography : The surface features of an object or "how it looks", its texture; direct relation between these features and materials properties (hardness, reflectivity etc.)

Morphology : The shape and size of the particles making up the object; direct relation between these structures and materials properties (ductility, strength, reactivity etc.)

Composition : The clements and compounds that the object is composed of and the relative amounts of them; direct relationship between composition and materials properties (melting point, reactivity, hardness etc.)

Crystallographic Information : How the atoms are arranged in the object; direct relation between these arrangements and materials properties (conductivity, electrical properties, strength etc.)

15.7.3 Transmission electron microscopy (TEM)

In the TEM experiment, a thin or diluted sample is bombarded under high vacuum with a focused beam of electrons. Electrons that are transmitted through the material form contrast patterns that reproduce the image of the sample. This pattern arises from the scattering of electrons. In addition, diffracted electrons give information about the lattice structure of the material. In the case of nanocrystallites, analysis of TEM images is partially responsible for the sizing curves of colloidal quantum dots. The shape of the sample can also be determined from the image. Fig. 5.10 a gives (a) TEM photograph of nano powder.

15.7.4 Secondary electron microscopy (SEM)

The electron microscope cross-section. View and actual SEM is shown in Fig. 15.8 (a & b). In the SEM experiment, an electron beam is focused and raster scanned over the sample. When the incident electrons interact with the sample a number of effects take place, such as the emission of secondary electrons. These effects are highly localized to the region directly under the electron beam and can be used to create an image of the sample. In addition, elemental analysis through energy dispersive or wavelength dispersive techniques can be done using other detectors. SEM photograph of a material is shown in Fig. 15.9 (b).

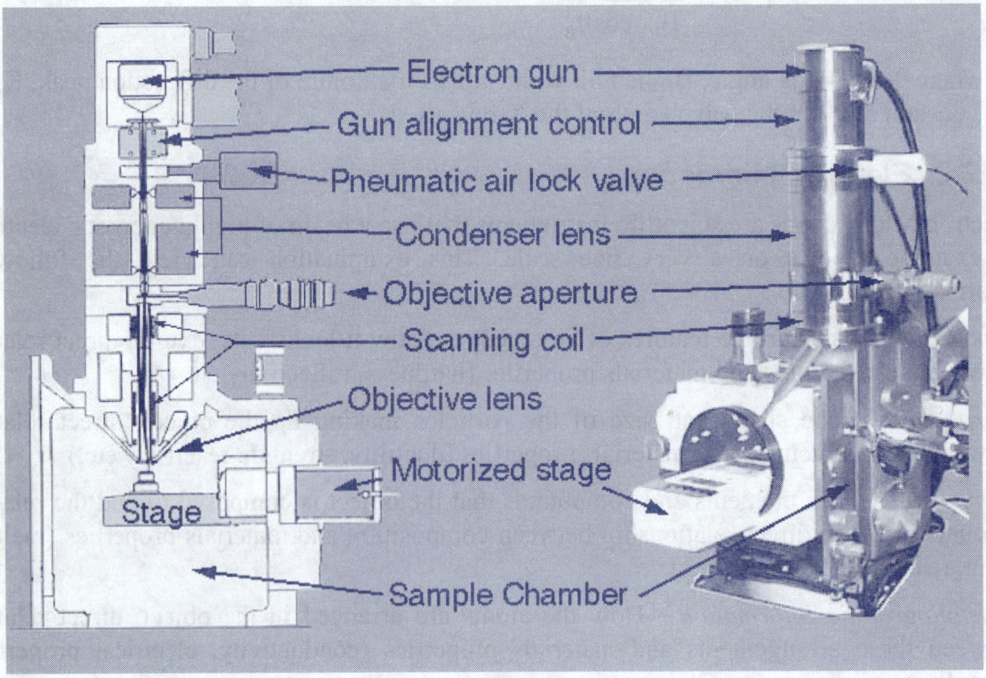

(a) (b)

Fig. 15.8 Electron Microscope

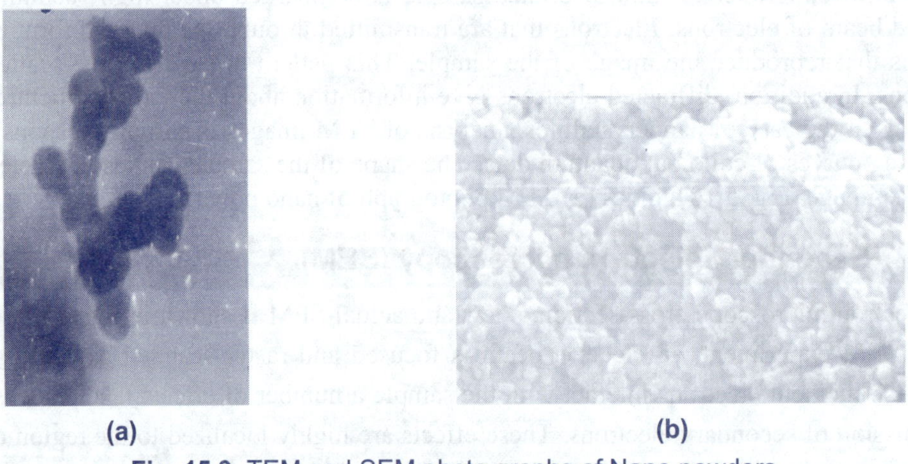

(a) (b)

Fig. 15.9 TEM and SEM photo graphs of Nano powders

15.7.5 Scanning tunneling microscopy

Scanning tunneling microscopy was the original probe microscopy. It was developed by Gerd Binnig and Heinrich Rohrer at IBM Zurich, ultimately leading to the Nobel Prize. The principle of operation is the tunneling of electron from a conductive tip to a conductive substrate or sample through a barrier. The tunneling current is found to be exponentially dependent upon the tip to sample separation allowing for very high sensitivity of sample height. STM has been used to investigate a number of nano related effects such as the discrete atomic like states of colloidal quantum dots and other systems. It has also been used to investigate the coulomb blockade and coulomb staircase phenomena. The STM has also been used to manipulate individual atoms as described in recent work on quantum corrals. One disadvantage of the STM is that it requires conductive samples or relatively conductive samples on a conductive substrate. To circumvent this limitation, the AFM was subsequently developed.

15.8 Carbon Nano Tubes

15.8.1 What are carbon nano tubes?

Carbon nano tube is unarguable poster child of nanoscience. The carbon nanotube is formed from graphene sheets rolled into a perfect cylinder of one atom thick to give a wire of many micrometers in length. It is perfect structure without defects. They were discovered in 1991 by Sumio Iijima. Various methods are suggested for the preparation of carbon tubes. The useful methods are laser ablation, chemical vapour deposition, arc discharge method. The Fig.15.10 shows typical caron nanotubes.

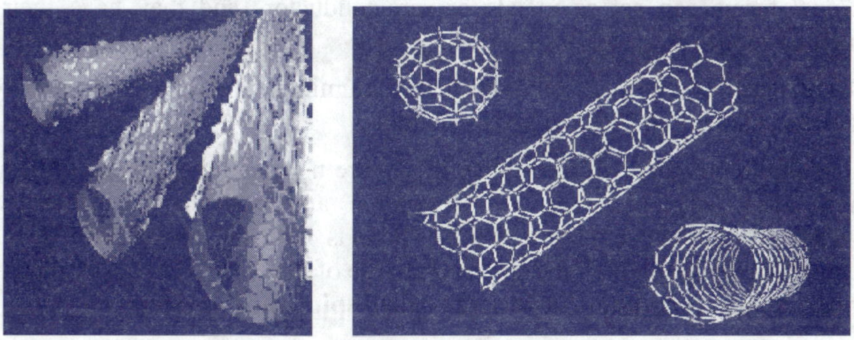

Fig. 15.10 Carbon nanotubes along with Bucky ball –C-60.

The Carbon nanotubes is a man-made nano-material which can not be found in nature. It has many interesting physical and chemical properties and many of them have great potential device application. There are four kinds of structures made of Carbon: (1). diamond structure. (2). graphite (honeycomb) structure: pencil lead. (3). Carbon 60: Buckyball (its discovery won 1996 Nobel Prize in chemistry). (4). Carbon nanotubes (CNT): by rolling up graphite sheet as a cylinder like a straw. Different ways of rolling the graphite sheet give different kinds of CNT: Arm chair nanotubes, zigzag nanotubes, and chiral nano tube. The details of making are not needed at this level, but it is enough to know that depending on different ways of rolling, CNT could be good or poor conductor. It has very good mechanical properties as well.

15.8.2 Properties of carbon nanotubes (CNTs)

They have multitude of important properties that may impart novel applications. Both multi walled as well as single walled CNT structures are made. They offer very good mechanical properties. The strength of sp^2 carbon bonds gives CNTs amazing mechanical properties. They possess remarkable strength. They are found to have high elasticity. The Young's modulus represents the rate of stress with applied strain. The young modulus of nanotubes can be as high as 1000 GPa which is 5 times larger than steel. Their strength is supposed to be 50 to 100 times that of steel. In many CNTs prepared, the tensile stress or breaking stress is of the order of 63 GPa.The elasticity of single walled nanotubes (**SWNTs**) is of the order of 1-1.2 TerraPascal. Imagine a molecule on the atomic scale which is as strong as steel and flexible like a rubber band, such molecular system gives fantastic utility. They will be extremely useful in the preparation of artificial muscles.

SWNTs have thermal conductivity almost as great as twice that of diamond which is the best conductor of heat. The electrical conductivity of SWNTs is of the order of 10^9 Amps per square centimeter, which is about 100 times that is reported for copper. They have two types of structures which give additional electrical characteristics. The structures differ in the alignment of carbon atoms in the cylindrical form. **SWNTs** can be 1. *Achiral* that have atomic uniformity along its axis, 2. *Chiral* that have a twisted alignment from the uniform case. The *Chiral* forms can act as metals or semiconductors and they have their basic nanotube structural motif.

SWNTs have densities approximately half that of aluminum and they are extremely light materials. They are stable upto 2700° C under vacuum.

For the nonvolatile memory, magnetic disk drives are used and these employ the spin of the electron to store large amounts of information and particularly to preserve the information when the computer is shut down. There is now the prospect of developing devices that simultaneously use both the spin and charge of the electron to vastly increase the capability of information processing, in what are called **spintronic devices.**

15.9 Applications of Nanotechnology

The popular perception of Moore's law is that computer chips are compounding in their complexity and the number of transistors are compounding in two dimensions.

The Motorola PowerPC microprocessor contained 7 million transistors, Intel Pentium II microprocessor contains roughly 7.5 million, and Digital Equipment's 64-bit Alpha microprocessor contained almost 10 million. The traditional chip is finally approaching a fundamental physical limit.

The present chip with CMOS silicon will reach its limit. Apart from this; the success of nanotechnology in a few fields is now forcing the semiconductor industry to take notice of the subject. The semiconductor industry will have to look for devices that work on nanoelectronics. The processing unit will have to undergo changes and apply nanomaterials and devices for further growth.

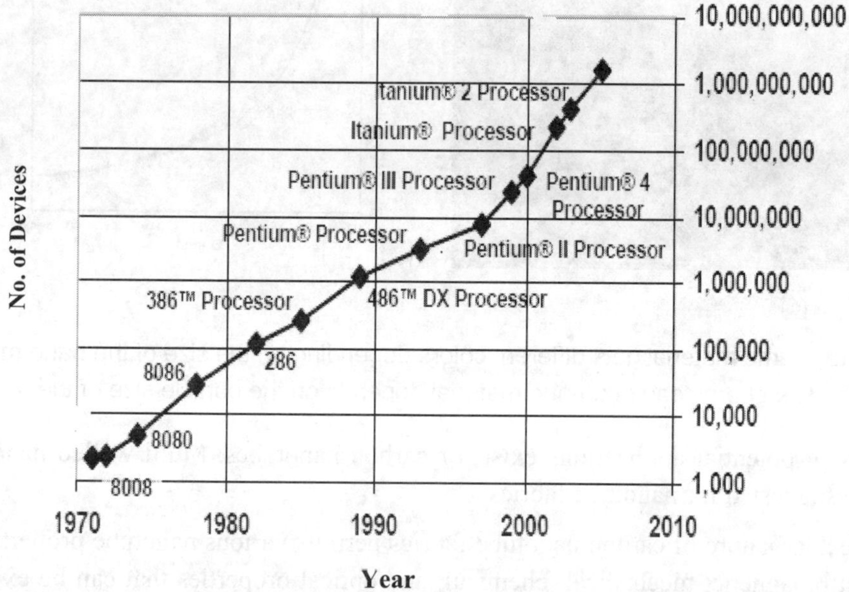

Fig. 15.11 Number of electronic devices (on Y axis) embedded in different Microprocessor chips

Non-silicon computing devices with seamless integrated circuits may emerge very soon. Many technologies descended from conventional ***solid-state silicon methods*** for fabricating *microprocessors* are now capable of creating features smaller than 100 nm, falling under the definition of nanotechnology. *Giant magneto resistance*-based hard drives are already on the market fit this description. Chipmakers in the computing industry are already working at the nanoscale. Many companies are now in the late development stages of processor chipsets of around 60 nm, with Intel being close to market with their 65 nm products. The Intel chipsets feature **greater performance** as a result of a 10-15% improved drive current through the application of nanotechnology. At this nanosize, the chipsets also suffer from less 'leakage', therefore **offering large power savings.**

Compared with bulk materials, nanomaterials have remarkable catalytic, optical, electrical, and mechanical properties. Their potential applications range from probe microscopy tips to interconnects in electronic devices. For example, nanomaterials exhibit different colors that depend on the particle size. The energy gap associated depends on size. This aspect is shown in Fig 15.12.

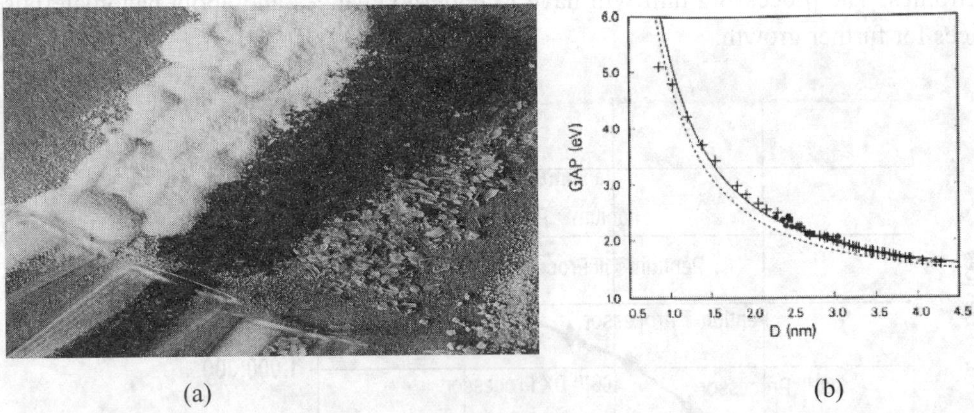

Fig. 15.12 (a) Same material gets different colors depending on the size of the nano material grains (b) The energy gap of a nano material depends on the particle size of the material.

A variety of potential applications exist for carbon nanotubes. Multi walled nano tubes (MWTs) are suggested for nanotube fabrics.

The molecular nature of carbon nanotube fabrics permits various nanotube properties that include electromagnetic, mechanical, chemical, and optical properties that can be exploit to build up integrated electronic devices and physical and chemical sensors. Furthermore, the states in the carbon nanotube are quantized in a particular manner so that the current carrying electrons do not undergo scattering. This allows the nanotube to pass current without resistance (like a superconductor) along its length and with very high current densities. Thus, if the carbon nanotubes could be organized into a seamless integrated circuit, there could be a new generation of computer components comprised of nanoelectronic devices. MWTs are useful in an arena of electronics specifically to create non volatile memory applications in stead of conventional semiconductors. Such systems will be useful in mechanical relays and switches.

It is suggested by many researchers that the carbon nanotubes might be used to replace the conventional materials used in silicon integrated circuits, to fabricate an even smaller generation of nanoelectronic devices that mimic the conventional semiconductor architecture.

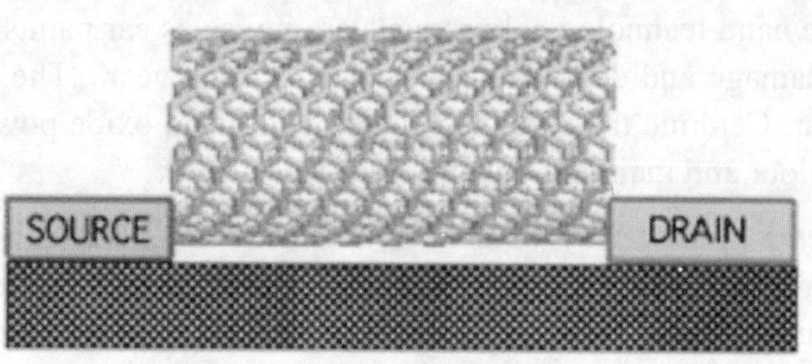

Fig. 15.13 Nano field effect transistor

The best utility of nanomaterials and structures is in the field of Sensors. A carbon tube array is a good sensor. Suppose that a single particle is mounted on one of the carbon tubes. An electric charge is now applied to it. The tube oscillates like a strong flexible spring with a frequency that different from that of the free tube. This frequency change is correlated to the mass of the particle attached. Thus, it is possible to find the mass of the nano particles on molecular level experimentally. SWTs can be used as chemical sensors like hydrogen sensors. They also promise as sensors in DNA detectors.

It is well known that the traditional ceramics is very brittle with poor tenacity and strength. The ceramics made of nano sized material has high hardness and good tenacity. One can even make a spring from the new ceramics material.

15.10 List of Current uses of Nanotechnology

Sunscreens: Sunscreens use fine powders of ZnO, or TiO_2. The nano sized TiO_2 and SiO_2 have strong absorption to the ultra violet. We all know that the ultra violet in the sun light is very harmful to our skin and longer exposures cause irrepairable damage. If these nano sized materials such as TiO_2 are used to make Sun-block lotion or sun-screen make-ups, the damage of the ultra violet radiation to human can be greatly reduced. The micron sized particles of above oxides give whitish colour as we notice on the face of cricket players. However, if nano sized particles of the oxides are used, no such whitish colour appears on the face. Such powders are also clean and more effective.

- Pilkington Company offers protective and glare-reducing coatings for eyeglasses. When UV radiation falls on the coating nano particles become energized and start breaking and loosen organic molecules on the glass. The coating is also hydrophilic and produces water and cleans the glass. Titanium dioxide nanoparticles are used in cosmetics and some food products. Silver nanoparticles are employed in food packaging, clothing, disinfectants and household appliances. Zinc oxide nanoparticles are useful not only as sunscreens and cosmetics, but also as surface coatings, paints and outdoor furniture varnishes; and cerium oxide nanoparticles is used as as a fuel catalyst.

- **Paints:** The nano technological products are in use as car paints that give resistance to scratch damage and degradation of color for long years. The car in the picture is one such car. Ceramic tiles are also coated with nano oxide powders and these tiles do not decolour and maintain dust free surface.

Treatment of cancer: A new way of cancer treatment. In 2002, a hospital in Germany developed a new way of cancer therapy using nano technology. Traditional method is to expose part of human body with cancer tumor in the infrared radiation so that the temperature of the tumor is raised to 40 °C in order to kill the cancer cells. However, large portion of the normal cells around the cancer cells are also damaged. The new method is to inject nano sized particle Fe_2O_3 into the cancer tumor. In the time varying magnetic field, the temperature of the nano sized Fe_2O_3 will rise to 45°C which will kill the cancer cells and leaving the normal cell intact. Recently, it has been found that nanotubes are useful in killing cancer cells. The tubes are attached to an antibody that searches out cancer cells and bonds to the cancer cells. When exposed to IR radiation, the nanotubes cook the cancer cells and kill them. It is also safe way of killing the cells.

Drug delivery: The delivery of drug exactly at the spot that is damaged is an important aspect of treatment or surgery. The drug delivery systems are very important. They have physical and biological usages. Fullerene materials have been explored for use as antioxidants, drug delivery agents and amino acid replacements. Efforts are there to build up efficient devices for drug encapsulation, functional drug carriers. Drug delivery systems like Liposome drug delivery systems are made for use.

Solid silicon micro needles with heights of 150 nm can be inserted painlessly on the skin for transdermal delivery. The needles are shown in the Fig. 15.14. Development of Bio-chip that can be used for controlled release of drug at the site in any part of the body. This is implanted in the body. It can control the delivery and dosage of the drug whenever it is needed. Nano-medical devices implanted in humans will enable significant advances in disease prevention, diagnosis and control.

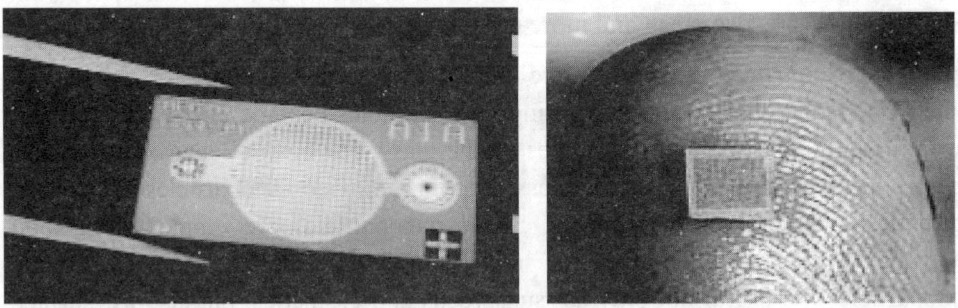

Fig. 15.14

The Fig. 15.14 depicts a bio chip for controlled release of drug or medicine into the body.

One of the most important work that is going on is tissue repair and regeneration scaffolds. This is extremely important for the patients who lost their skin because of severe burns. A sponge-like collagen matrix provides an "open scaffold" structure, which permits dermal cells to migrate into and throughout the sponge and to deposit human collagen and glycosaminoglycans (human skin "biomatrix"). The open pores allow the patient's wound bed fibroblasts to migrate in and rebuild the tissue.

Fig. 15.15

The smart devices for blood sugar level testing and controlling the excess sugar by proper delivery of drug are available. They are based on nanotechnology. Retina implants are also made.

Non-medical Applications

Textiles: Electronic - textiles with incorporated electronic components such as processors, sensors, etc. which can enable wireless communication with other devices. Telecommunications, entertainment, digital assistance, sensing are possible on these clothes. Dirt repellents in clothes are developed. Stain-free clothing and mattresses are possible. Flexible electronic structures are being planned with advent of silicon being made into stretchable wires in recent years. We will be getting self-cleaning textiles by nanotechnology.

Biomemetics: One of the main aims of nano technology is to make devices that are replicas of bio systems. Many of the biological systems work with precision with embedded nanotech systems in their bodies. They include sensors, reflex action activators, protection devices, and wonderful capabilities that are not possible for human beings or the mechatronic devices existing now. This process is called biomemetics meaning imitating biological systems. The future people want Robots smarter than humans (inorganic evolution) mimicking biological creatures, Automated parole, Devices smaller than ants, Talking Clothes, Artificial limbs that work by command from the brain of the person, Artificial muscles as replacement for those who lost the functioning of their muscles, surgery through computer aided hands. Imagine a small living being like an ant. The ant has amazing capabilities of communication, working as a team, excellent detection system, eyes which are natural microscopes. Nano technology aims at making such devices which may possess a few of the capabilities of ants.

Efforts are being to develop new type of surveillance systems, aerospace control systems to meet the demands of today.

Apart from the above mentioned, there are number applications that are being looked into and being slowly implemented.

Water filtration: water filtration is done by filters equipped with nano powders like alumina fiber. The nano alumina (AlOOH) fiber is only 2 nanometers in diameter, smaller than a carbon tube. The nano fibers are highly electropositive. They hold dirt particles very efficiently than general micro glass media. The filtered water would be free of dirt particles and microbes.

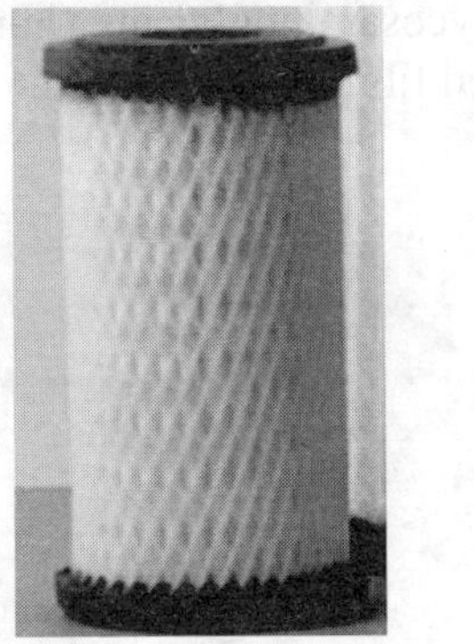

Fig. 15.16 Alumina fiber filter

Nano silver powder is also used for water filtration. Nano silver particles are strategically placed throughout the carbon core to maximise its effectiveness against bacteria and prevents its growth on the filter.

Longer-lasting tennis balls: The goal for tennis ball manufacturers has always been to manufacture tennis balls that will retain their original air pressure and bounce properties for as long as possible. Wilson Sporting Goods introduced the Wilson Double Core tennis ball.

The reason that these ball last twice as long is that the inner core of the balls coated with a nanocomposite coating that inhibits the permeation of air from the inside of the ball. Coatings start with a latex polymer containing butyl microspheres, to which the vermiculite platelets are added. Addition levels up to 50% are used. After application, the coatings dry, with the platelets still aligned, forming a barrier to air escape.

Nanosized phosphors (for electronic displays, catalysts, cosmetics) are developed and used. Nano diode are made and these are found to be very efficient. Display technology for laptop computers, cell phones, and digital cameras are popular.

The object of people working o phosphors is to develop nano powders that will be identical with algue. Nano phosphors which will be like the nano powder coatings on the butterfly wings. The coatings on the wings of the butterfly reflect light giving an interference pattern just like the thin oil films on water on the road giving colours when sun light falls in water.

Miniature nano electromechanical (NEMs) systems: Nano motors and molecular machines are thought about even though it will take time to realize molecular machines. The model one such molecular gear motor, that can be made with bottom up process.

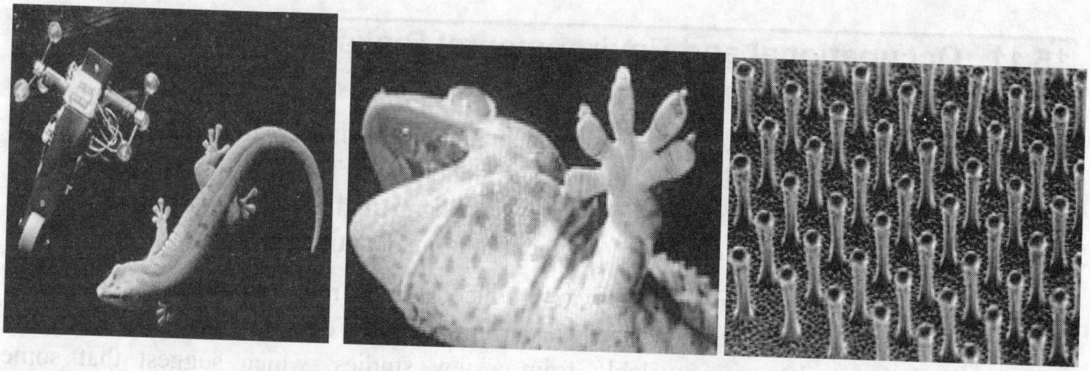

Fig. 15.17 Nano tubes in the feet of gecko.

The Fig. 15.17 gives a very interesting case of nanotubes at work in nature. A gecko lizard shown in the figure possesses millions of nanotubes under its feet. By controlling the air circulation with these tubes, it can climg trees of hundred feet height with ease. The nano tubes are shown in the figure. Now a model-machine is made that can be used to climb the walls of very high multistoried buildings. The machine has a very large number of nanotubes under its legs that can be operated with slight voltages. It is envisaged that the making of spider man capabilities is not far off. A fiction will be coming true with the help of nanotechnology.

Nano powders are being used in a number of applications. Nano zinc oxide and titanium dioxide are used in talcum powders that give resistance to UV light. Biocomposites in toothpaste do not allow easy tooth decay. Fullerenes (football shape carbon molecules) are used in cosmetics. Nano powders of talcum powders can make the face of a person resistant to sun burns. In future, we will be having functional textiles that are based on textile fibres and nano wires of different functional materials. Food protectors are being made out of nano powders. Textiles fitted with flexible electronic gadjets will be our future.

When nano material powder based devices are made and become available to people, we will be seeing a new world that is full of smart and multifunctional devices. We see new homes and offices with talking chairs, comforts at command, kitchen with automatic systems, new type of food packings, new type of air coolers /heaters in the same device, smart windows, and wireless communication for short communications. Materials with exotic properties will be developed. Devices that will be listening to your brain commands are on the way. These gadgets are extremely useful for physically handicapped persons.

The future will be like fantacy and what we read in fiction or comics like spider man or James Bond will be a reality in future. The future technology will be indistinguishable from magic.

15.11 Occupational and Environmental Risks Arising due to Nanotechnology

The impact of nanotechnology on the environment and on the dangers to human health is the most important one that needs to be studied and understood. The commercial synthesis of functional materials with nanometer dimensions has begun, and few data exist on the impact these new materials will have on the environment in large quantities.

In general, nanotechnology is likely to produce a wide range of materials of vastly different structures and natures, and there are currently limited data on potentially damaging effects. Preliminary data are available from a few studies, which suggest that some nanomaterials can cause inflammation of lung tissue and formation of microscopic lesions in rats. However, it is unclear whether these particulates pose actual risks when inhaled. It is not known whether or to what degree different nanometer scale materials may be able to enter the body through skin exposure.

Because the potentially far-reaching consequences of nanotechnology are so diverse, it is essential to continually identify, assess, and respond to new environmental issues as they arise out of emerging scientific and technical research. However, it is important to recognize that many of the new nanotechnologies will not have direct environmental impact

Since the energy required to ignite particles is a function of particle diameter, and nanoparticles are likely to have enhanced electrostatic charging (which may result in their being sufficiently charged as to act as their own source of ignition if dispersed in air), it is important to measure and understand the properties of nanoparticles in the context of their ignition potential.

Sources of risk of human and environmental exposure to nanoparticles are:

- occupational exposure in the workplace (human);
- exposure from deliberate environmental releases, e.g. remediation of contaminated ground waters and land (environment and possibly human);
- exposure from 'unintentional' environmental releases, e.g. from fuel additives and in industrial and domestic waste streams (environment and human);
- exposure from consumer products, such as cosmetics (human); and
- exposure from medical products, including drugs, treatments and devices (human).

Questions

1. (a) Explain Fundamental concept of Nanotechnology.

   (b) What is Quantum Dots?

2. (a) What is Nanotechnology? How is it useful to society?

   (b) Discuss about the properties of Nanomaterials.

3. (a) Discuss in detail about Nanomaterials.
   (b) Mention one of the methods of synthesis of Nanomaterials.
4. (a) What is Bucky ball?
   (b) Mention some of the properties of Carbon Nanotubes?
5. (a) What is a Quantum Dot?
   (b) Briefly discuss about carbon Nanotubes.
6. (a) Discuss about the applications of Nanomaterials.
   (b) What is Bucky ball?
7. (a) Mention the properties of Carbon Nanotubes?
   (b) Explain Fundamental concepts of Nanotechnology.
8. (a) Explain the synthesis of Carbon Nanotubes.
   (b) Explain one of the method of synthesis of Nanomaterials.
9. (a) What is Nanotechnology? How is it useful to society?
   (b) Discuss about Nanomaterials.
10. (a) List a few methods of synthesis of Nanomaterials.
    (b) Mention the applications of Nanomaterials.
11. (a) What are the important properties of Carbon Nanotubes?
    (b) Explain about carbon Nanotubes.

Objective Questions

I. Multiple Choice questions

1. Nanomaterials chemical activities increase due to [ ]
   A) Increment of surface area
   B) Decrement of volume
   C) Both Increment of surface area and Decrement of volume
   D) None

2. Which of the following synthesis methods are used for preparation of Nanotubes [ ]
   A) Discharge
   B) laser Ablation
   C) Chemical vapour deposition
   D) All the above

3. Bond in the carbon nanotube is [ ]
   A) SP B) SP2
   C) SP3 D) None

4. Thermal properties are more in [ ]
   A) Bulk material
   B) Nanomaterial
   C) Same in both bulk and nanomaterials
   D) None

5. In nanometallic material system thermal energy is mainly carried by [ ]
   A) Electrons B) Holes
   C) Phonons D) Ions

6. If a bulk material is converted to a nano size, then the mechanical properties [ ]

A) Increase

B) Decrease

C) Independent if it size

D) First increase and later decrease

7. Who is the first distinguished the concepts of nanotechnology [ ]

A) Richard Feynman

B) Norio Taniguchi

C) Eric Drexler

D) Buckminister Fullerine

8. How many atoms consist in one quantum dot [ ]

A) 1 to 10 B) 10 to 50

C) 10 to 100 D) 10 to 150

9. The Young's modulus of the best anotubes can be as high as [ ]

A) 10 GPa B) 100 GPa

C) 1000 GPa D) 10000 GPa

10. In quantum dots the energy band gap depends on [ ]

A) Surface area B) Volume

C) Size D) All the above

II. Fill in the blanks

1. Nano materials are catalysts, because their enhanced________ activity.
2. Aerogel is ________nano crystalline.
3. Quantum dots are ________ materials.
4. Single walled nanotubes are generally ________ than, the multiwalled
5. Bucky balls consisting ________ structures.
6. The lifetime of a satellites increase by using ________ materials.
7. The strength of the SP2 carbon – carbon bonds gives carbon nanotubes amazing________ properties.

III. True or False

1. One way in which a graphene sheet can be rolled into a nanotube [True/False]
2. Example for direct band gap semiconductor is arsenic [True/False]
3. C-60 is called quantum dots [True/False]
4. The life of a satellites increase by using nanomaterials. [True/False]
5. Nanomaterials area is more volume is less [True/False]
6. One nanometer is equal to 10^{-6} m [True/False]

CHAPTER 16

Physical Optics for Instruments

Introduction to physical optics

Light rays moving in straight lines give the concepts of geometrical Optics and can be used to understand the reflection and refraction phenomena. Light is a transverse electromagnetic wave. Therefore light should display phenomena that are characteristic of waves the interference and diffraction. Geometrical optics cannot account for these wave effects. The study of the interference and diffraction of light is referred to as wave optics or physical optics.

Superposition of waves results in interference. Superposition of two coherent and monochromatic waves is interference. Interference can be constructive or destructive interference. Interference by division wave front and division of amplitude are two methods with which interference can be explained. Thomas Young has contributed for understanding the interference of light waves.

Diffraction can be defined as deviation of a light ray from rectilinear propagation, which is not caused by reflection nor refraction. Bending of light around sharp objects (obstacles) is diffraction. light rays, passing through a small aperture in an opaque screen do not form a sharp shadow on a distant screen. That smooth transition from light to shadow could only be explained by assuming that light has a wave like character. Diffraction theory has been developed by Huygens, Fresnel, Kirchhoff and Summerfield. Depending on the distance between the source and screen diffraction is of two types: Fresnel and Frounhoffer diffraction.

Understanding the interference and diffraction is important for many technologically advanced applications. The main applications of physical optics are in the Fourier-transforming property of lenses or diffraction capability of the lens, and optical information processing (including complex spatial filtering and holography), acousto-optic and electro-optic effects, principles of lasers and photo-detectors, and some topics from nonlinear optics (such as harmonic generation, phase conjugation, and optical bistability). The principal objective of this chapter is to provide a comprehensive approach to the broad field of physical optics and its applications.

Interference

Interference is a wave phenomenon of mechanical waves (sound waves) and electro magnetic waves (light waves, microwaves, radio waves etc). When two waves that originate from a point source travel in the same direction in space, they arrive at a point with a phase difference or without a phase difference. When, the waves superpose with each other and the resultant amplitude at the point is the sum of the individual amplitudes of the waves or difference of the individual amplitudes. It all depends on the factor that is determined by the phases of the individual waves.

Interference of waves that is going to the discussed briefly in the next few sections is based on Huygen's wave theory of light. The light spreads from a point source in all directions as waves. Each point of the wave front of such waves is again a source of secondary wavelets that they spread out further. These wavelets once again have a wave front. The waves spread out in space and the wave fronts can be received on a screen by placing it in their path.

Interference is a very important aspect of wave theory, and it is also highly important in modern technology. The methods of interference are used in many applications like the accurate measurements of radius of curvature of lens, smoothness of surfaces, wave lengths of light emitted by sources etc. Interference effects are in use in the manufacture of optical instruments, optical fibres, integrated chips; in the study of holography, or three dimensional photography, smoothness of surfaces, surface defects of many gadjets etc.

16.1 Principle of Superposition

Suppose we have a stationary tub in which water is filled to the top. Then, suppose a disturbance is caused at two points S_1 and S_2. Then spherical wavefronts travel spreading out as shown Fig 16.1. As the disturbance is caused at S_1 and S_2 continuously; crests (represented by dark circles) and troughs (dotted circles) are created. At some points crest of one wave falls on the crest of another wave (or trough on a trough), the resultant is the sum of the amplitudes due to each wave separately. This is called superposition of waves.

We shall now consider the superposition of two traveling waves in the same direction. They are represented by the expression in terms of the displacement, y, as

$$y_1 = A\sin(kx - \omega t) \qquad .....(16.1)$$

$$y_2 = A\sin(kx - \omega t + \phi) \qquad \text{.....(16.2)}$$

where A is the amplitude, $k = \frac{2\pi}{\lambda}$, the propagation constant;

$$\omega = 2\pi f = \frac{2\pi c}{\lambda}$$

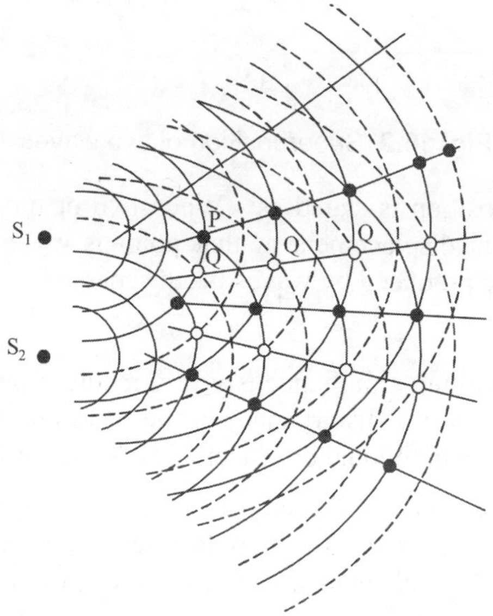

Fig. 16.1(a) Waves spreading out from two sources, S_1 and S_2

The resultant displacement 'y' at any point is given as

$$y = y_1 + y_2 = A\sin(kx - \omega t) + A\sin(kx - \omega t + \phi)$$

$$= 2A\cos(\phi/2)\sin(kx - \omega t + \phi/2) \qquad \text{.....(16.3)}$$

The above expression represents a travelling wave whose amplitude is dependent on the phase difference, between the waves.

When the two waves arrive at a point in phase at the same time, $\phi = 0$. The resultant amplitude is then Fig. 16.2(a). The waves are said to interfere or superpose constructively. The intensity of the wave is maximum. At this point, the crest falls on the crest and trough falls on the trough (as shown in Fig. 16.2(a). This is true for any number of waves that arrive at a point in phase at the same time.

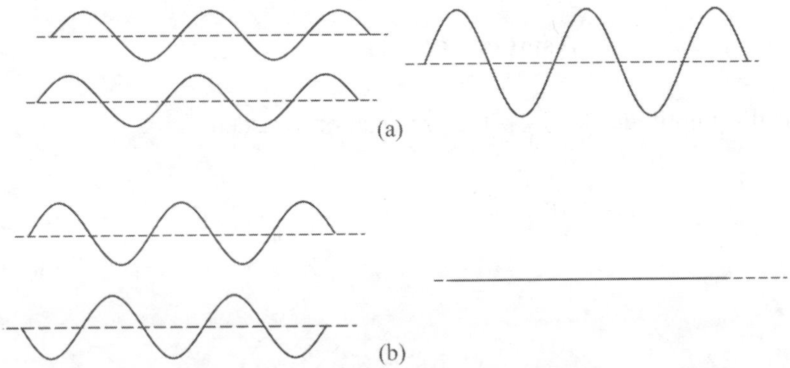

Fig. 16.2 Superposition of two waves

The principle of superposition is stated as: When two or more waves arrive at a point simultaneously, the resultant displacement at that point is given by the algebraic sum of displacements that would be produced by waves taken separately.

It can also be stated as:

A number of waves travelling simultaneously in a medium behave independent of each other and the net displacement (or disturbance) in the medium at any instant of space and time is equal to the sum of the individual displacements due to all waves.

The principle of superposition is valid only for waves of low amplitudes, preferably near wavelengths or same wavelengths. It is not applicable to intense elastic waves like shock waves or light waves of very high intensity. Such waves produce non-linear phenomenon and violate conservation principles.

If the two waves arrive at a point out of phase, they annul each other. The phase difference between the two waves is π. The crest of one wave falls on the trough of the other and vice versa as shown in the Fig. 16.2(b). This is a case of distructive interference.

The interference of light waves is a distinct and fine example of super position principle. It can be clearly understood by studying the interference effect in Young's double-slit arrangement.

16.2 Young's Double-Slit Experiment

A double-slit arrangement is a process of splitting a light beam into two components by two slits and the components of waves interfere to produce an interference pattern on a screen which can be viewed through a micrometer eyepice. The double slit arrangement is shown in Fig. 16.3. The double-slit actually creates two wave fronts that are derived from a single wave front.

S_1 and S_2 are two slits separated by a distance 'd', just comparable with wavelength dimensions. A monochromatic light from a point source 'S' falls on two slits, S_1 and S_2. They emerge from the slits and the interference effects can be seen on the screen MM'. the

Fig 16.3(a) depicts the waves that interfere and Fig. 16.3(b) represents the geometrical picture of the double-slit experiment. Fig. 16.3(b) shows the pattern of interference pattern.

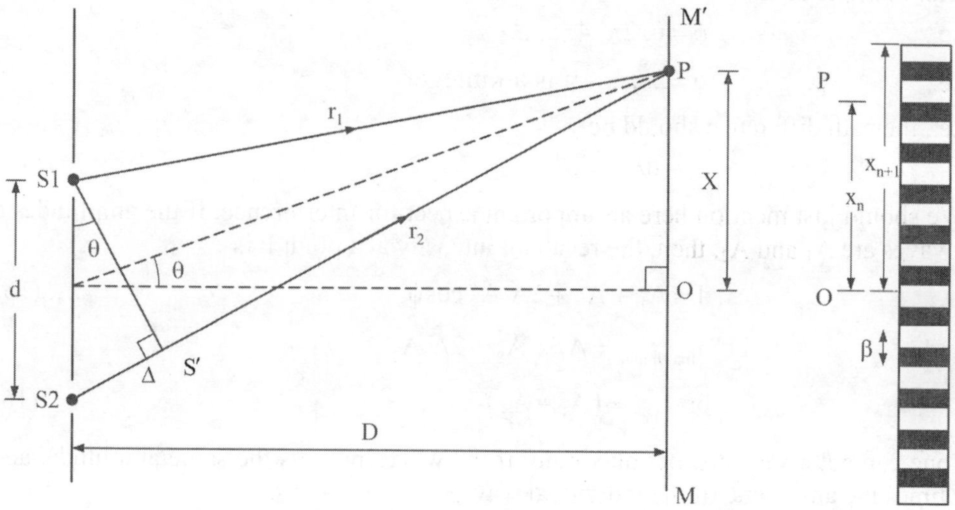

Fig. 16.3

Consider two points, O and P on the screen.

The point O is exactly on the perpendicular drawn from the mid point of S_1 and S_2 and a point P is on the screen at a distance x_n from O.

The resultant at P of the two waves is given as the (eq. 16.4)

$$y = 2A\cos\left(\frac{\phi}{2}\right)\sin\left(kx - \omega t + \frac{\phi}{2}\right) \quad .....(16.4)$$

where ϕ is the phase difference;

Looking at the geometrical figure in Fig. 16.3(b), we can understand that geometrical path difference between the two rays S_1P and S_2P is S_2S'.

We have $S_1S_2 = d$

and $S_2P - S_1P = S_2S' = \Delta$

and the phase difference between the waves at P is given as

$$= \frac{2\pi}{\lambda}(S_2P - S_1P) = \frac{2\pi}{\lambda}\Delta.$$

Condition for Maximum Intensity

For the intensity to be maximum at the a point P, the phase difference should be zero, or the integral multiple of 2π.

i.e., $\phi = 0, 2\pi, 4\pi, ...$

i.e., $\phi = 2n\pi$, n is an integer

i.e., the path difference should be

$$\Delta = n\lambda \quad .....(16.5)$$

We should just mention here an important aspect for interference. If the amplitudes of the two waves are A_1 and A_2. then, the resultant intensity at a point P is

$$I = A_1^2 + A_2^2 + 2A_1A_2 \cos\phi \quad .....(16.6)$$

and $$I_{maximum} = A_1^2 + A_2^2 + 2A_1A_2$$

$$\text{or } I_{max} = (A_1 + A_2)^2 \quad .....(16.7)$$

i.e., one can get a very distinct maximum if the waves possess the same amplitude, and it is four times the amplitude of the individual wave.

Condition for Minimum Intensity

The minimum intensity can occur if $\cos\phi = -1$

i.e. $\phi = (2n+1)\pi$, where n = 0, 1, 2, 3….etc.

and $\phi = \pi, 3\pi, 5\pi, ...etc$.

If the amplitudes of the two waves are A_1 and A_2, then there is a minimum intensity equal to $(A_1 - A_2)^2$. The exact zero intensity is possible only when $A_1 = A_2$. A very good fringe contrast is possible only when the amplitudes of the two waves are exactly equal. This signifies the fact that experiments in utilizing interference of light always derive light beam from the same source, that may act as a point source or extended beam.

The path difference for minimum intensity is given as

$$\Delta = (2n+1)\frac{\lambda}{2} \quad .....(16.8)$$

If the slits are narrow rectangular slits, we observe on the screen a series of bright and dark bands called fringes. The fringes possess equal widths. It is the characteristic of interference fringes. The fringes that can be see are shown in Fig. 16.5. Let us look into the relations between the fringe width (β), and the distance of separation between two slits S_1 and S_1, and the wavelength (λ) of light.

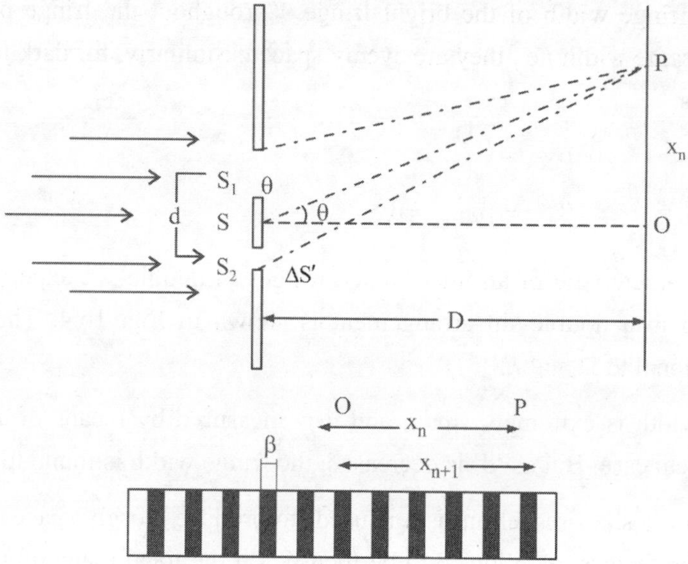

Fig. 16.4 A regular pattern of bright and dark fringes, observed with a monochromatic light.

$$S_2S^1 = \Delta = n\lambda$$

$$\Delta = d\sin\theta$$

We have, $D = n\lambda$ and $\Delta = d\sin\theta$, ...(16.9)

Where x_n locates the position of nth maximum

But $$\frac{x_n}{D} = \tan\theta$$

As θ is very small

$$\sin\theta \approx \tan\theta$$

i.e., $$d\tan\theta = n\lambda$$

$$\frac{x_n d}{D} = n\lambda$$

$$\therefore \quad x_n = \frac{n\lambda D}{d} \qquad .....(16.10)$$

Similarly the position of (n + 1) maximum is

$$x_{n+1} = \frac{(n+1)\lambda D}{d} \qquad .....(16.11)$$

i.e., $$x_{n+1} - x_n = \beta_{bright} = \frac{\lambda D}{d} \qquad .....(16.12)$$

where β is the fringe width of the bright fringe. Throughout the fringe pattern, all bright fringes possess same width i.e., they are evenly spaced. Similarly, for dark fringes, the fringe width is given as

$$\beta_{dark} = \frac{\lambda D}{d}.$$

i.e., $$\beta_{dark} = \beta_{bright} = \beta.$$

The is the characteristic of an interference fringes in double-slit experiment. The fringe pattern obtained in a double-slit arrangement is shown in Fig. 16.4. The fringe width is directly proportional to D and λ.

The fringe width is extremely small, and it is measured by means of a micrometer eye piece in a practical case. But as 'd' is decreased, the fringe width is found to increase.

In the laboratories, a Fresnel biprism is used to produce closely spaced sources of light, S_1 and S_2 and the fringes are allowed to be focused on the focal plane of a micrometer eye-piece. This arrangement is shown in the Fig. 16.5. S is a monochromatic source.

From the eq. 16.12, it is clear that for d <<D, the fringe width can be large even under the condition λ is small. In the laboratories, visible light is used to produce very distinct fringes by a Fresnel biprism. The angle of the prism (obtuse angle) is around 179°.

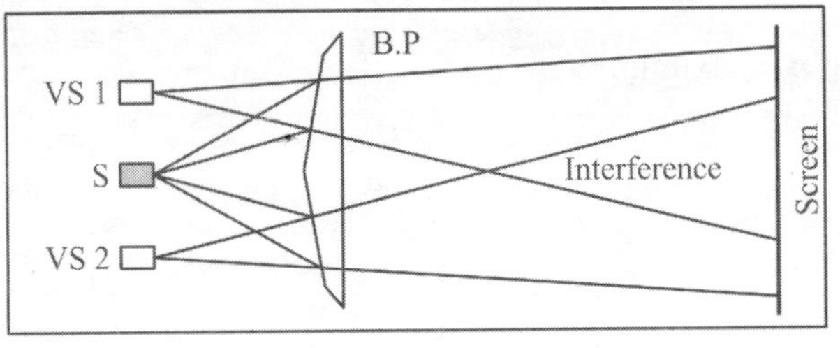

VS are the virtual sources., S-source, a slit., B.P-Biprism

Fig. 16.5 A Fresnel biprism creates two virtual sources to produce interference fringes that are seen in a micrometer eye-piece (shown adjacent to the figure of B.P).

Intensity distribution in double-slit interference

From the geometry of for φ as the triangle S_1 S_2, S' we have

$$\phi = \frac{2\pi}{\lambda}(d \sin\theta) \quad .....(16.13)$$

∴ The intensity at any angle θ is given as

$$I_\theta = 4I_0 \cos^2 \frac{\phi}{2} = 4\, I_o \cos^2 \beta \quad \text{where } \beta = \frac{\phi}{2}$$

$$= 4I_0 \cos^2\left(\frac{\pi d \sin\theta}{\lambda}\right) \qquad \text{.....(16.14)}$$

$$d \sin\theta = n\lambda \text{ for maxima}$$

$$d \sin\theta = (2n+1)\lambda \text{ for minima}$$

The intensity pattern is shown as a function of ϕ, and d sin θ. in Fig. 16.6. the formula in eq. 16.14 describes the intensity distribution and actually energy distribution between dark fringes and bright fringe.

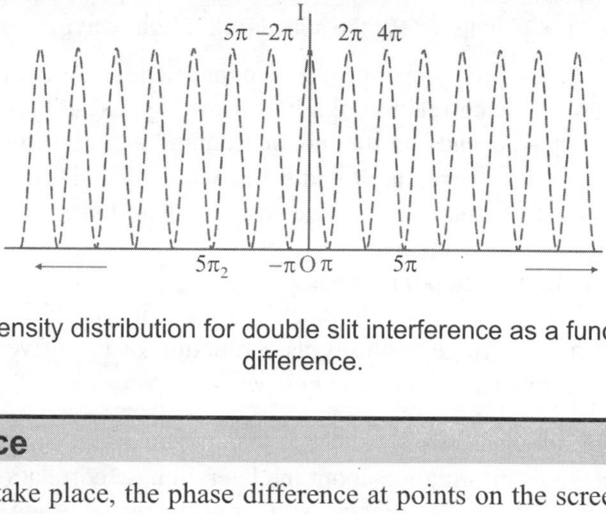

Fig. 16.6 The intensity distribution for double slit interference as a function of the phase difference.

16.3 Coherence

For interference to take place, the phase difference at points on the screen or of observation must not change with time. Then only, we can say, that the waves from the two slits S_1 and S_2 are coherent. It is essential that the two sources created by the double-slit should originate from the source of extremely small dimensions.

Coherence is a measure of the correlation between the phases measured at different (temporal and spatial) points on a wave. For the interference to take place, the phase difference at points on the screen (focal plane of observation) must not change with time. Then only, we can say, that the waves are coherent. Monochromatic nature of light beams is very important criterion for coherence.

Phase correlation between the waves is very important. If the light is derived from two independent sources giving same wavelength, it can not be called coherent light as the interference pattern can not be seen on the screen and the pattern just fluctuates very fast with a period less than 10^{-8} sec. The light originating from any incandescent bulb is incoherent. Coherence describes how similar are two waves coming from a source, regarding amplitude, phase, wavelength, and wave front uniformity. Coherence depends on how the phase relation between the waves remains stable with time and distance of travel. Coherence also indicates the ability of a wave beam to travel as a close packed pencil. Such a condition is achieved more or less to a large extent in the case of Laser sources. Coherence indicates the exact correlation between physical quantities associated with a wave of light. Coherence is specified in two ways .They are: 1. Temporal coherence and 2. Spatial coherence.

A light source, however small it is, has dimensions. What it emits is the bunch of wave trains with a very small gap. This time lag, and gap arises as millions of atoms that are in continuous motion in a random way participate in the emission of light. The temporal coherence is a measure of average phase correlation that is constant over time between two points or over a distance of travel as the wave propagates in space. If a wave is combined with a delayed copy of itself, the duration of the delay over which it produces visible interference is known as the coherence time of the wave, Δt_c. From this, a corresponding coherence length L_c, can be estimated as:

$$L_c = \Delta t_c \times C$$, where C is the speed of the light wave.

Temporal coherence is an indicator of monochromatic nature of a wave. Temporal coherence is a measure of the correlation of a light wave's phase at different points along the direction of propagation – depends on the monochromatic nature of light. Suppose a source gives a light beam that has a wavelength width, $\nabla\lambda$, where λ is the wavelength of the wave ,then ,the propagation constants of the two waves (that have wave lengths differ by $2\pi\ \nabla\lambda/\lambda^2$. This is equal to the inverse of coherence length, L_C. If $\nabla\lambda$ is very small, the temporal coherence is high. Temporal coherence gives the ability to get good interference patterns between the wave and the time-delayed version of itself, as in a Michelson interferometer. Behavior of fringe visibility as a function of the wavelength distribution is the result of the effect known as temporal coherence

Lasers emit light with very high coherence lengths.

Spatial coherence gives information about the lateral phase correlation on the wave fronts at different points of propagation. Waves with uniform wave fronts possess high spatial coherence. Spatial coherence as well as temporal coherence control the visibility of interference fringes in interference observations. An interference pattern possesses dark and bright fringes. The dark fringes may not be exactly dark due to slight incoherence that affects the visibility (distinction of bright and dark fringes).

Spatial coherence is a measure of the correlation of light waves' phase at different points of travel in the lateral direction of propagation of the wave – it is a measure of uniformity of phase at different points on the wave front. In spatial coherence, a lateral coherence length is measured, which is a measure of how far apart two points in a wave front can be such that knowing the phase at one point still lets you confidently predict the phase at the other point. Spatial coherence relates to the ability to focus a beam down to a diffraction-limited spot. In the formation of interference pattern in holography (a three dimensional photography in terms of interference), one has to illuminate a large area. Hence, spatial coherence is very important. Spatial coherence is improved with lenses and apertures. Source size is very important. Normal Helium-neon lasers can produce light with about 20 cm coherence length. The typical red light laser diode (λ= 650 nm) with a frequency stabilizer can have a coherence length of over 1 m. Normal LEDs have a spectra width $\Delta\lambda$ of about 50 nm, and may have a coherence length of 10's to 100's of μm.

16.4 Interference from thin Films

One of the most important and interesting method of using the division of amplitudes is the interference from thin films. A thin film possesses two surfaces separated by extremely small widths and give rise to interference of light. Whenever light rays fall on physical bodies like very thin plates or films, they are partially reflected, and partially refracted into the film. The back surface of the film once again reflects light partially into the film and partially transmits to outside as shown in the Fig. 16.7. The amplitude of the reflected ray depends on the reflection coefficient (r_c) of the surface, and the amplitude of the transmitted ray depends on the transmission coefficient (t_c) of the film r_c' and t_c' are the reflection coefficient and transmission coefficient in the denser medium and from denser medium to rarer medium respectively. Therefore, when a thin film of certain length and width is taken, the light rays undergo multiple reflections inside the film as shown in the Fig. 16.7. The amplitude of the wave is partly lost at each reflection. The rays that emerge out of the film through reflections, or the rays that emerge out of the films by transmission are all capable of showing interference effects. We will discuss the interference phenomenon in detail now.

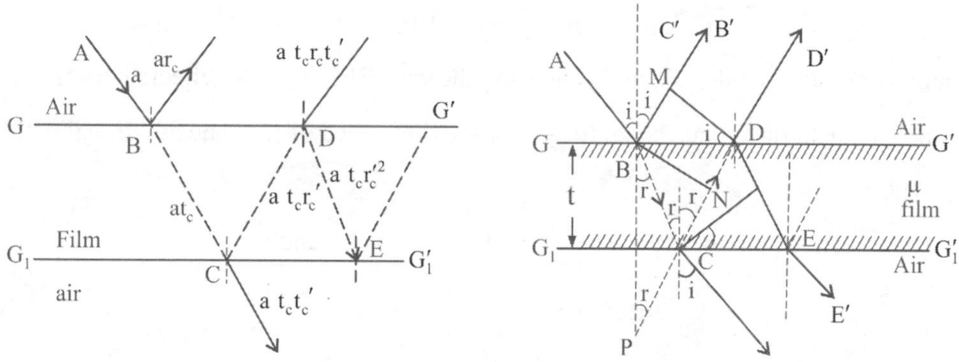

Fig. 16.7 Light ray AB on a thin film

Let us consider a uniform thin film of a transparent material like a very thin cover glass. Suppose that the film has a thickness 't' and the refractive index μ as depicted in the Fig. 16.8(a). Let a monochromatic wave AB fall on the thin film's surface GG′. At B, it is partially reflected as ray $BB'(r_1)$. It is also refracted into the film at an angle 'r' with the normal to the surface. The ray travels through the medium to C, the bottom surface, and at C, it is partially reflected towards D, and partially transmitted into air as CC′. At D, once again, the ray is partially refracted into air as ray DD′ (r_2) and partially reflected towards E. We shall consider the two rays r_1 and r_2 as they are capable of interfering. Drop a perpendicular DM on BB′ and a perpendicular BN on CD.

At point B, the light ray is reflected as BB′ after being incident on the denser medium. As the difference of refractive indices between air, and the film is such that $\mu - \mu_{air} > 1$, the wave BB′ does suffer a phase change of π at point B. However, the ray reflected at C into the film medium, does not suffer any phase change. At C, it is incident on the interface G_1G_1' and the medium beyond G_1G_1' is a rarer medium. The path difference between the two rays r_1 and r_2 can be arrived at by considering the given figure. Let the incident angle at B be 'i', and the refractive angle be 'r'. Two normals are drawn (1) from B onto CD to cut it at N, (2) from D onto BB′ to cut it at M. Then $\angle BDM = i$; $\angle NBD = r = \angle PBC$.

The optical path difference between BB′, and DD′ is $(BC + CD)\mu - (BM)$

where $(BC + CD)\,\mu$ is the path in the film, and BM is the path in air.

We have from Snell's law

$$\sin i = \mu \sin r$$

i.e.,
$$\frac{BM}{BD} = \mu \frac{ND}{BD}$$

$$\therefore \quad BM = \mu ND$$

$$\therefore \quad \Delta = \mu(BC + CN + ND) - \mu ND$$

$$= \mu(BC + CN)$$

$$\Delta = 2\mu t \cos r \quad \text{(from } \Delta\text{ BPD)} \quad \because BC = PC, \text{ we have}$$

There is an additional path difference for the ray BB′ as it is reflected from a denser medium. That path difference equal to $\frac{\lambda}{2}$, (phase difference of π). [The additional path from the ray BB′].

$\therefore$ The effective path difference between the two rays, r_1 and r_2, is

$$\Delta = 2\mu t \cos r \pm \frac{\lambda}{2} \qquad \text{.....(16.15)}$$

Then, the constructive interference occurs when

$$\Delta = 2\mu t \cos r \pm \frac{\lambda}{2} = n\lambda \qquad \text{.....(16.16)}$$

or
$$2\mu t \cos r = (2n \pm 1)\frac{\lambda}{2} \qquad \text{.....(16.17)}$$

The film appears bright in the reflected light. It should be remembered that it is the thickness and the reflective index of the film that play the role when λ is fixed.

The minima occurs when this path difference is $(2n \pm 1)\dfrac{\lambda}{2}$ or

$$2\mu t \cos r \pm \frac{\lambda}{2} = (2n \pm 1)\frac{\lambda}{2}$$

or
$$2\mu t \cos r = (2n \pm 1)\frac{\lambda}{2} \mp \frac{\lambda}{2}$$

$$= n\lambda \qquad .....(16.18)$$

i.e.,
$$2\mu t \cos r = n\lambda$$

because (n + 1) or (n – 1) can also be taken as integer, when n is integer.

The value of n = 0, 1, 2, 3 …. etc. The film can appear dark due tp the two rays r_1 and r_2 when $2\mu\, t \cos r = 0, \lambda, 2\lambda, ...$ etc.

16.5 Interference due to the Transmitted Light

We shall see the rays that are transmitted into air form the lower surface of the film as shown in Fig. 16.8(b).

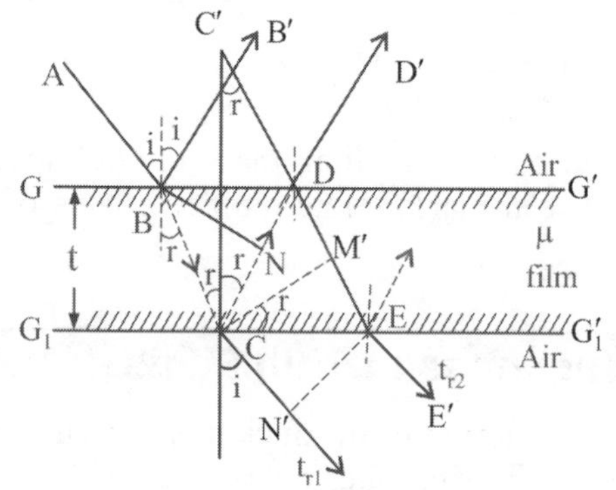

Fig. 16.8

The perpendicular on to GG′ from C is extended to C′. where CC′ = 2t. CC′ and EE′ are the two transmitted rays, t_{r1} and t_{r2}. EN′ is normal to CC′ and CM′ is normal to DE. From the geometry of the figure, it is clear that

$$\angle M'CE = r; \quad \angle CEN' = i$$

The effective path difference between the two transmitted rays is given as

$$\Delta = \mu\,(CD + DE) - CN'$$

$$\mu = \frac{\sin i}{\sin r} = \frac{CN'/CE}{EM'/CE} \quad \text{(from } \Delta CEN' \text{ and } \Delta CEM'\text{)}$$

$$= \frac{CN'}{EM'}$$

$$\therefore \quad CN' = \mu EM'$$

Then

$$\Delta = \mu(CD + DM' + M'E) - \mu EM'$$

$$= \mu(CD + DM') = \mu(2t\cos r)$$

In this case, the reflection of rays is within the thin film at the interface backed by rarer medium. Hence, there is no additional phase change for the rays EE' or CC'.

The maximum intensity, or constructive interference takes place, when

$$2\,\mu t\cos r = n\lambda \qquad \text{.....(16.19)}$$

In this case, the film appears bright.

The condition for minima or destructive interference is

$$2\mu t\cos r = (2n \pm 1)\frac{\lambda}{2} \qquad \text{.....(16.20)}$$

where n = 1, 2, 3 .... etc. Under this condition, the film does appear dark. Comparing with eq. 16.17 and eq. 16.18, it is seen that the interference effects of reflected and transmitted light are complementary.

16.6 Non Uniform Thickness in Thin Films – Wedge Shaped Films

In the case of thin films having non-uniform thickness, constructive interference occurs in certain parts and destructive interference takes place in some other parts. Bands of maximum and minimum intensity are seen. Let us look this aspect more closely.

Let us consider to surfaces GG', G_1 and $G^1{}_1$ as shown in the Fig. 16.9. A wedge is formed between the two surfaces with an angle α. Between the two surfaces, the thickness of the film varies and increases from G to G'. Let the refractive index be μ. Let a monochromatic light like sodium light fall on the surface GG'. The interference takes place between the two rays, (1) reflected beam from the front surface (suffering a phase change of π) and (2) the other beam obtained by internal reflection at the back surface and get transmitted through GG'.

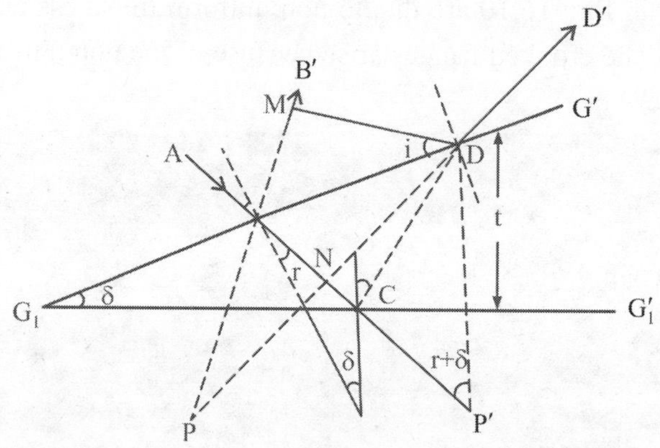

Fig. 16.9 Interference in the wedge of air (The wedge is greatly exaggerated)

From the Fig. 16.9, it can be seen that BB′ and DD′ are not parallel and they appear to diverge from a point P. Let us find the path difference Δ, between BB′ and DD′,

$$\Delta = \mu(BC + CD) - BM$$

$$= \mu\,(BN + NC + CD) - \mu BN \qquad (BM = \mu BN)$$

$$= \mu(NC + CD) = \mu(NP') \qquad \text{(from the triangle DNP}')$$

$$\therefore \quad \Delta = 2\,\mu t \cos\,(r + \delta) \qquad \text{.....(16.21)}$$

Because of the reflection at B, we have

$$\Delta = 2\,\mu\,t\cos\,(r + \delta) \pm \lambda/2 \qquad \text{.....(16.22)}$$

Maximum intensity occurs, when

$$2\,\mu t\cos\,(r + \delta) + \lambda/2 = n\lambda$$

or $$2\,\mu t\cos\,(r + \delta) = (2n \pm 1)\,\lambda/2 \qquad \text{.....(16.23)}$$

Minimum Intensity occurs when

$$2\,\mu t\cos(r + \delta) \pm \frac{\lambda}{2} = (2n + 1)\,\frac{\lambda}{2}$$

or $$2\,\mu t\cos\,(r + \delta) = n\lambda \qquad \text{.....(16.24)}$$

the integer ‘n’ can take values n = 0, 1, 2, 3. … etc.

For a fixed monochromatic light and the film, the interference pattern depends on the thickness ‘t’ only. The thickness t, remain constant, in the direction parallel to the edge, the fringes appear straight. The fringes are just parallel strips of bright and dark fringes.

They are shown in the Fig. 16.10.

The fringes shown in Fig. 16.10 are due to non uniform thickness and irregular air film. They crooked fringes. The crooked fringes arise because of the non uniformity of the surface of the lower glass plate.

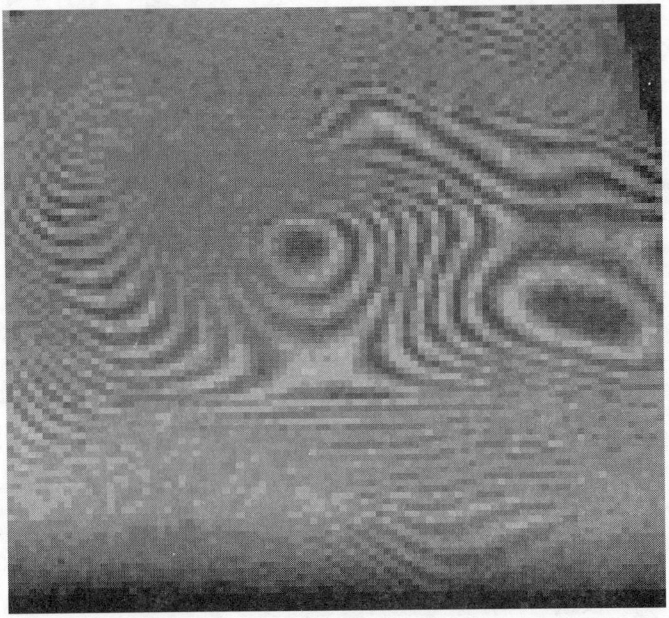

Fig. 16.10 Interference fringes – case of non uniform film or uneven glass plates in a wedge.

We can create a wedge air film shaped between two thick glass plates by placing a thin wire in between the plates at one end and by fixing the plates at the other end as shown in the Fig. 16.11. This experimental arrangement is highly useful in finding the diameter of thin wires, thickness of thin sheet etc. A monochromatic light from an extended source is shone on the arrangement and the interference fringes are observed by an observer from the top. However, a microscope with a micrometer eye piece is used to measure the fringe separation. A series of alternate bright and dark fringes are seen. The fringes are uniformly spaced and are parallel to the edge if the glass plates are smooth to the extent of $\frac{\lambda}{10}$.

The separation of the fringes depends on the angle δ or the diameter of the wire. The point of contact at G is a dark fringe. The bright fringes satisfy by the equation

$$2\mu t\cos\delta = (2n+1)\frac{\lambda}{2} \quad .....(16.25)$$

For air, $\mu = 1$

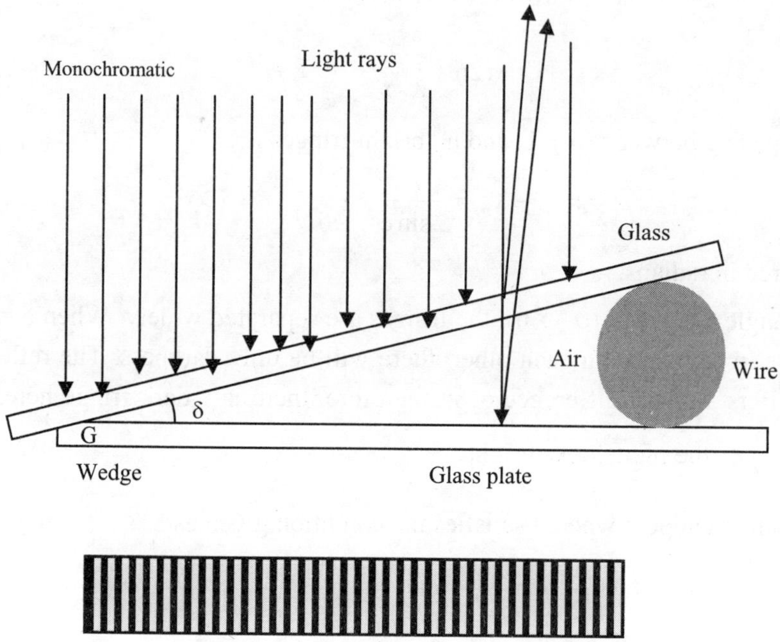

Fig. 16.11 Wedge and the fringes formed.

Suppose the nth bright fringe is at a distance x_n and $(n + 1)^{th}$ bright fringe is at a distance x_{n+1}.

We have from Fig. 16.12.

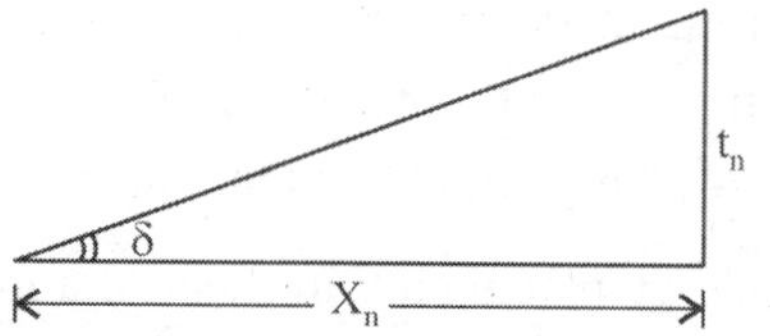

Fig. 16.12

$$\frac{t_n}{x_n} = \tan\delta$$

i.e.,
$$t_n = x_n \tan\delta$$

∴
$$2x_n \tan\delta \cos\delta = (2n+1)\frac{\lambda}{2}$$

$$2x_n \sin\delta = (2n+1)\frac{\lambda}{2} \quad \text{.....(16.26)}$$

Similarly for $(n + 1)^{th}$ fringe

$$2x_{n+1} \sin \delta = (2(n+1)+1)\frac{\lambda}{2}$$

$$= (2n+3)\frac{\lambda}{2} \qquad \text{....(16.27)}$$

Then, the spacing between (n + 1) and n^{th} bright fringes is

$$\Delta x = x_{n+1} - x_n = \frac{\lambda}{2 \sin \delta} \approx \frac{\lambda}{2\delta} \qquad \text{.....(16.28)}$$

α is measured in radians.

When the angle a is small ($\alpha \to 0$), the fringes are separated widely. When $\delta = 0$, the two glass plates are in contact with each other, there will be only darkness. The reflected ray at the contact suffers a phase difference of π, therefore, there is a dark fringe here. When the separation, $t = \frac{1}{4}\lambda$, the plate looks bright.

The dark minima appear when, t satisfies the condition given as

$$t_x = \frac{\lambda}{2}, \lambda, \frac{3\lambda}{2} \text{.... and so on}$$

The bright maxima appear at locations where 't' satisfies the condition given as

$$t_x = \frac{\lambda}{4}, \frac{3\lambda}{4}, \frac{5\lambda}{4} \text{ and so on}$$

If white light is used, bands of different colours are observed at different points, corresponding to different wavelengths. These bands get overlapped giving a beautiful colour pattern characteristic of the medium. Observation of different colors in soap bubbles, on the throat and belly of humming bird, butterflies, or feathers of peacock etc., fall under this category. One can notice the change of colours and intensity with the angle of viewing.

Example: An air wedge is made by placing a thin sheet of a metal foil between the edges of two glass plates about 80 mm from their point of contact. Suppose that the wedge is illuminated with a light of wavelength 589 nm and the fringes are 1.20 mm apart. What is the thickness of the foil?

Ans: Number of bright fringes $= \frac{80}{1.2}$

The change in vertical height from one fringe to the next $= \frac{\lambda}{2}$

$\therefore$ vertical height for $\frac{80}{1.2}$ fringes

$$= \frac{589 \times 10^{-9} \times 80}{2 \times 1.2}$$

$$= 1.47 \times 10^{-6} \text{m}$$

$\therefore$ The thickness of the foil = 1.47 microns.

16.7 An Oil Film on Water

Let us consider an oil film on the surface of another medium like water as shown in the Fig. 16.13. The refractive indices of air (μ_1), oil (μ_2), and water (μ_3) are mentioned for comparison.

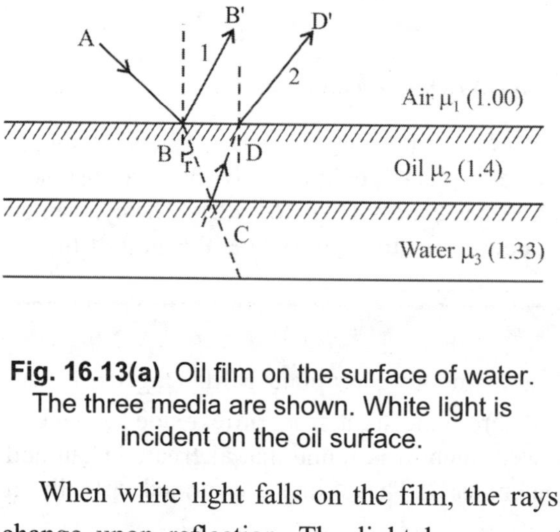

Fig. 16.13(a) Oil film on the surface of water. The three media are shown. White light is incident on the oil surface.

Fig. 16.13(b) The interference pattern from an oil slick over water.

When white light falls on the film, the rays get reflected partially suffering 180° phase change upon reflection. The light beams go to the bottom surface of the film and get reflected without suffering any phase change. The rays 1 and 2 are capable of interference. The rays interfere constructively for each wavelength in white light. For example, for wavelength λ, the condition is

$$2\mu t \cos r = \left(n - \frac{1}{2}\right)\lambda_1 \qquad ....(16.29)$$

The pattern looks like the Fig. 16.13(b). It is common to see this type of pattern on the roads on rainy days. The colours we see depend on the angle of viewing, the spread and thickness of oil.

16.8 Extended Source

The thin films (including the Newton's rings arrangement) are illuminated by an extended source of light as opposed to a point source in double slit experiment. An extended source is a necessity here. Light from a point source is made to fall on a film at different angles of incidence. As a result, different pairs of rays are formed as shown in Fig. 16.14(a).

An observer can view the film by keeping his eye at an angle with respect to the film. We should remember that for a particular angle of observation, he cannot view the whole film. He can view the film totally only if he shifts his position of eye in an arc.

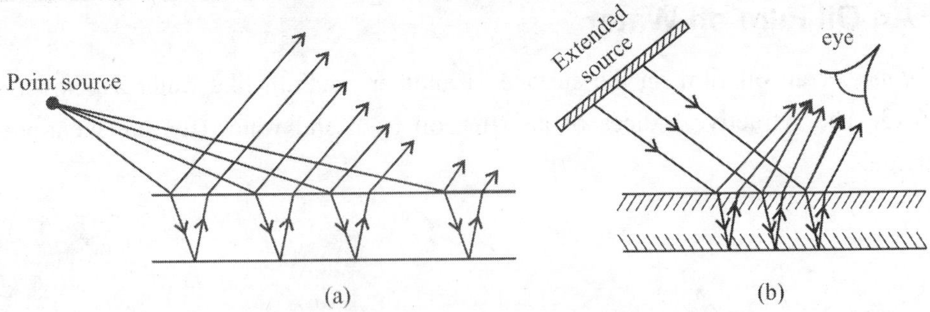

Fig. 16.14(a) Point source **(b)** extended source

Suppose, light is made to fall on the film from an extended source. An extended source which is made up of innumerable point sources placed continuously over a distance. The light reflected from every point on the film reaches the eye (Fig. 16.14(b)). The eye can be placed at a suitable angle so to observe the best contrast of the fringes over the entire film.

16.9 Newton's Rings

When a plano convex lens (AOA') is placed on a smooth glass plate MM' (Fig. 16.15(a)). There is an air film formed around O. A monochromatic light from an extended source is made to fall on the lens and the air film is viewed, then, it is found that alternate bright and dark concentric rings are formed with their centers as O. These rings are formed virtually in the air film and can be seen with naked eye.

The experimental set up is shown in the Fig. 16.15(c). In the laboratory, light is made to fall vertically on the lens.

G is a glass plate that is placed firmly fitted making an angle 45° with the vertical. The monochromatic light falls on the glass plate G. It is partially reflected on to the plano-convex lens normally. We shall consider two rays r_1 and r_2. The ray r_1 falls on the lens, enters into the lens. It is reflected back partially at Q and comes out as ray r_1. Apart of the ray at Q is transmitted into the air film Fig. 16.15(c). This ray gets reflected at the point Q' on the glass plate and it reenters the lens and emerges out as ray 2. These two rays possess phase difference that leads to interference.

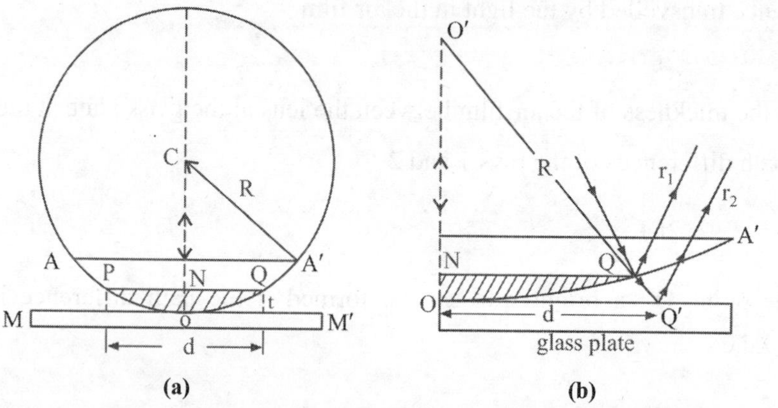

Fig. 16.15 (a) Light falls on Plano-convex lens AOA' **(b)** Experimental arrangement for observation and measurement of diameters of rings

Let us look into the analysis of the interference rings. This can be made by considering Fig. 16.15(a).

The light ray that enters the lens along CO is reflected back into the microscope. O is the point of contact of the lens and glass plate. Let R be the radius of curvature of the lens.

The reflected ray from 0 suffers a phase difference of π or a path differences of $\lambda/2$ (this is because of the fact that it is reflected at an interface beyond which there is a medium of higher refractive index as compared to air). Hence, a dark spot is formed. We consider another ray that falls on the lens at Q.

It is partially transmitted to Q_1, gets reflected suffering a phase difference of π at the glass plate. It emerges out of the lens as ray 2. At Q, the ray is partially reflected back in the lens without any phase difference. The two rays 1 and 2 can produce interference fringes as they are basically obtained from the same ray initially.

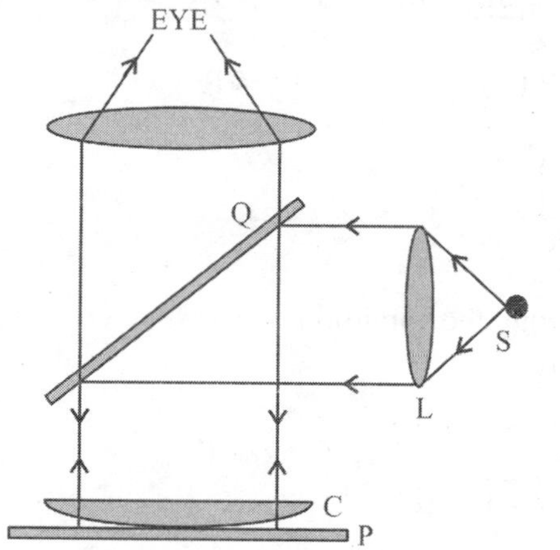

Fig. 16.15 (c) Newton's rings

The distance transvelled by the light in the air film

$$= 2t$$

Where 't' is the thickness of the air film between the lens at the glass plate at the point N.

The path differences of the rays 1 and 2

$$= 2t + \frac{\lambda}{2} \quad \text{......(16.30)}$$

At the point Q', a bright ring can be formed if the path difference is an integral multiple of λ i.e.,

$$2t + \frac{\lambda}{2} = n\lambda \quad \text{.....(16.31)}$$

$$\therefore \quad 2t = (2n-1)\frac{\lambda}{2}$$

where n = 1, 2, 3, ...etc

The dark ring can be formed at Q' if

$$2t = n\lambda \quad \text{.....(16.32)}$$

where n = 0, 1, 2, 3, ...

From the Fig. 16.15(a, b) and from the property of a circle, it can be seen

$$NP \times NQ = NO \times NO' \text{ as } CO' = R\,;\ NO = t,\ NO' = (2R - t)$$

i.e.,

$$\left(\frac{d}{2} \times \frac{d}{2}\right) = t \times (2R - t)$$

$$= 2Rt - t^2$$

$$\frac{d^2}{4} \approx 2Rt \qquad \because t << R$$

or

$$t = \frac{\left(d/2\right)^2}{2R} \quad \text{.....(16.33)}$$

Thus, for the bright ring, the condition is

$$2 \times \frac{\left(d/2\right)^2}{2R} = (2n-1)\frac{\lambda}{2}$$

i.e.,

$$d = \sqrt{(2n-1)}\sqrt{2\lambda R} \quad \text{.....(16.34)}$$

As R and λ possess fixed values for a given set of arrangement, we have

$$d \propto \sqrt{(2n-1)}$$

As n is an integer, the diameters of the bright rings in Newton ring's pattern are proportional to the square root of odd natural numbers, for a fixed system of R and λ.

In the case of dark rings, we have

$$2 \times \frac{\left(d/2\right)^2}{2R} = n\lambda$$

$$\therefore \quad d = \sqrt{4n\lambda R} \quad .....(16.35)$$

$$\text{i.e.,} \quad d \propto \sqrt{n} \quad .....(16.36)$$

Therefore, the diameters of dark rings in Newton ring's pattern are proportional to the square root of natural numbers, for a fixed a system of R and λ.

The central spot is dark for Newton rings seen under reflected monochromatic light. The lens has to be placed in perfect contact with the glass plate at the center (O). The Newton's rings formed with a monochromatic light are shown in Fig. 16.16. The rings are observed due to the reflected rays.

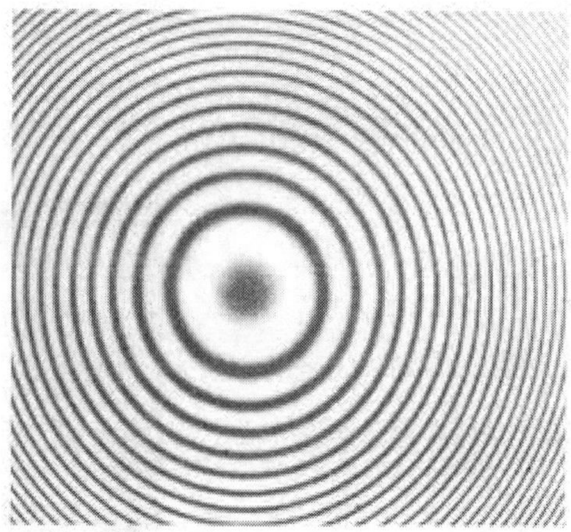

Fig. 16.16 Newton's rings observed in the reflected light. Note the central dark spot.

The rings become closely spaced i.e., as the ring number from the center increases.

The spacing between two dark rings is given as

$$\delta_R = d_n - d_{n-1}$$

$$= 2\sqrt{\lambda R}\left[\sqrt{n} - \sqrt{n-1}\right] \quad .....(16.37)$$

16.10 Newton's Rings by Transmitted Light

Newton's rings can be formed even with the light waves passing or transmitted through the lens and glass plate M M' as shown in the Fig. 16.17(a).

As the light ray at the central spot or the center of the lens does not suffer any phase reversal, the central spot of the Newton's rings is a bright spot in the case of rings due to transmitted light.

The bright ring is formed, when

$$2t = n\lambda \qquad .....(16.38)$$

The dark ring is formed at

$$2t = (2n-1)\frac{\lambda}{2} \qquad .....(16.39)$$

For bright rings, the condition is

$$\frac{2\left(\frac{d}{2}\right)^2}{2R} = n\lambda$$

or $$d^2 = 4\,n\,\lambda\,R$$

$$d = 2\sqrt{n\lambda R} \qquad .....(16.40)$$

$$\therefore \qquad d_{bright} \propto \sqrt{n}$$

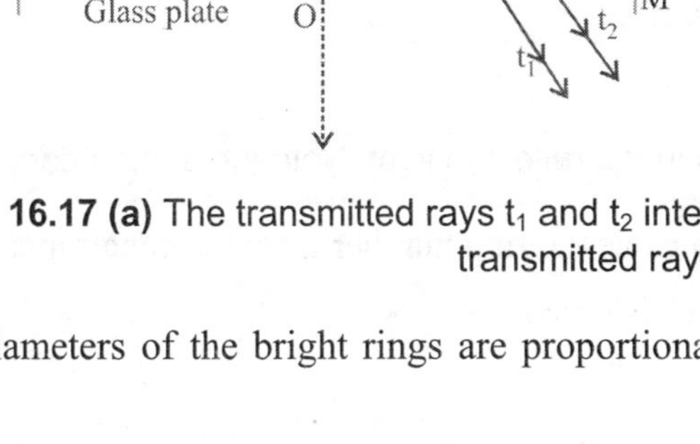

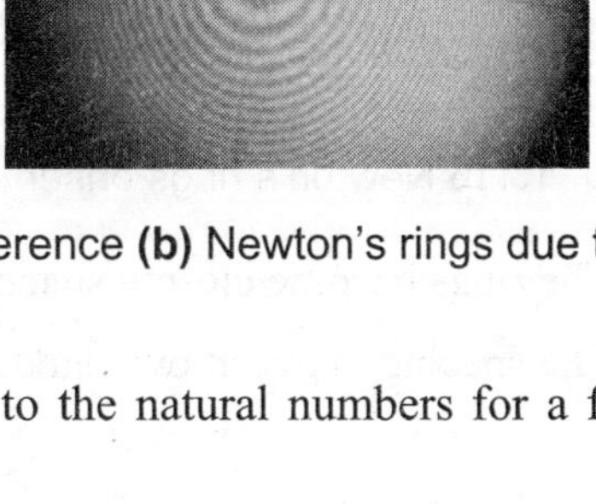

Fig. 16.17 (a) The transmitted rays t_1 and t_2 interference **(b)** Newton's rings due to transmitted rays.

∴ The diameters of the bright rings are proportional to the natural numbers for a fixed R and λ.

Similarly for dark rings

$$2\frac{\left(\frac{d}{2}\right)^2}{2R} = (2n-1)\frac{\lambda}{2}$$

$$\therefore \quad d_{dark} \propto \sqrt{2n-1} \qquad \text{.....(16.41)}$$

∴ The diameters of the dark rings in the transmitted pattern of Newton rings is proportional to odd natural numbers for a fixed R and λ.

Thus, the Newton rings by reflected light are complementory to the Newton rings formed by transmitted light. In the case of transmitted light, the central spot is bright. The rings formed by transmitted light beams are shown in Fig. 16.17(b).

The method of Newton rings is highly useful in the accurate determination of wave length of monochromatic light, or the refractive index of liquids.

When a liquid of unknown refractive index is poured in between the glass plate and the lens, a change in the diameters of the rings will occur.

Suppose we observe the diameters of nth ring and mth ring that are measured once with the air film and then with a transparent liquid, then we have

$$\mu_L = \frac{D^2_{m+n} - D^2_n}{D'^2_{m+n} - D'^2_n} \qquad \text{.....(16.42)}$$

where D' refers to the measurements made when the liquid film is present. D_{m+n} is the diameter of (m + n)th ring and D_n is the diameters of nth ring.

As the diameter is a function of the number of the ring, a plot between the (diameter)^2 vs the number of the ring is a linear plot as shown in the Fig. 16.18. The plot helps in the determination of wave length of monochromatic light accurately. The wavelength λ of the source of light is

$$\lambda = (D^2_{m+2} - D^2_n)/4mR$$

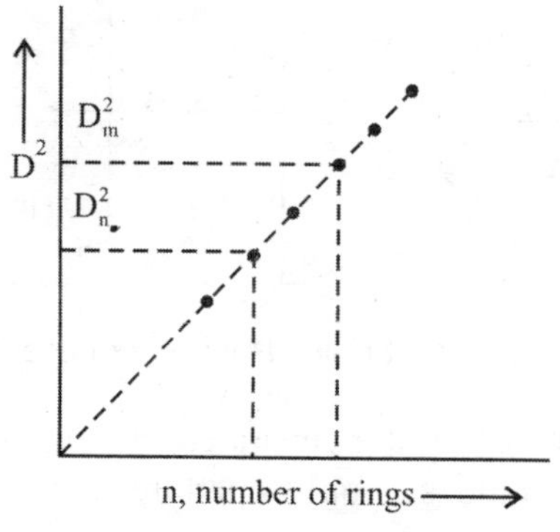

Fig. 16.18 A linear graph between D^2_n V s n.

The occurrence of distinct dark and bright rings takes place when a monochromatic light is used. The diameter is a function of the wavelength of light used. Higher wavelength of light yields rings of higher diameter as compared to the diameters of rings with lower wavelengths. When white light from incandescent bulb is used, the rings are coloured, but totally overlapped except very flow rings from the center.

Diffraction

Diffraction is a characteristic of wave phenomenon that occurs whenever a portion of a wave front is obstructed by some means. Diffraction is also described as the bending of waves that occurs around the edges of an opening or obstacle. If a region of wave front is altered by an obstacle in amplitude or phase, then the diffraction of waves takes place.

According to Huygen's principle, when a wave moves from a source, each point of the wave front becomes a source of secondary wavelets. However, Huygen's principle is independent of wavelength consideration and cannot explain diffraction or bending of visible light. It was Fresnel, who combined Huygen's principle with interference. Huygen–Fresnel principle states that when light beam spreads in space, every unobstructed point of a wave front, at a given instant of time, serves as a source of spherical secondary wavelets, with the same frequency as that of the primary wave. The optical waves generated as secondary wavelets would have amplitude at any point beyond obstruction.

The superposition of all these secondary wavelets results in relative amplitudes and phases in the geometrical shadow of the obstacle. This is called 'Diffraction'.

We suppose a small aperture (A) is present in a card board as shown in the Fig. 16.19.

When monochromatic waves of light fall on the aperture of random shape. It is possible to see light intensity pattern on the screen(s) at points like P. Each elementary area dA in the aperture becomes a source of expanding Huygens wavelet. All such wavelets interfere to give an intensity pattern at P.

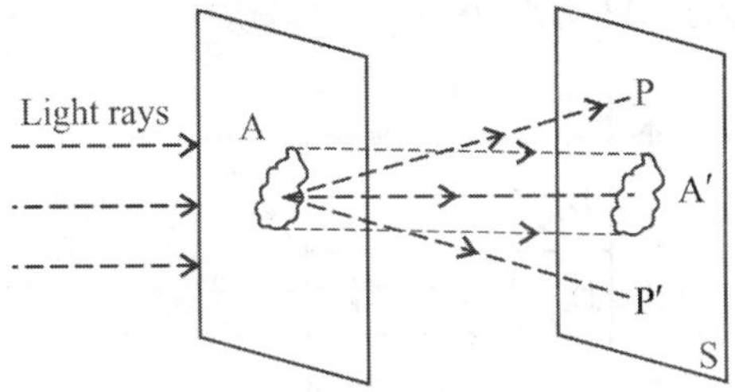

Fig. 16.19 Diffraction takes place at point P′ when coherent light falls on aperture A.

When the aperture and the screen are separated by very small distance, the diffraction pattern is not seen and are would see only aperture and its geometrical shadow of the aperture.

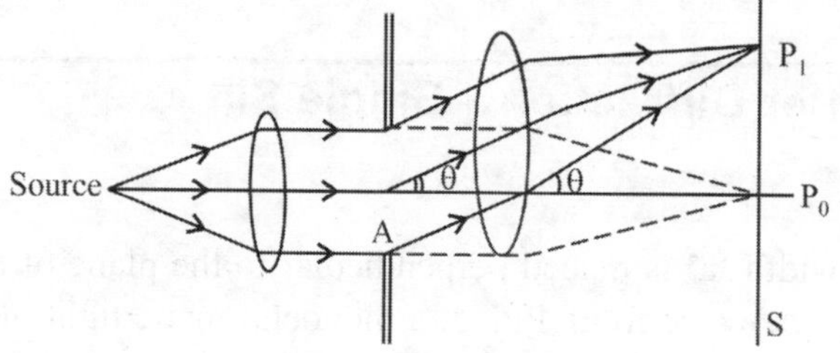

(a)

Fig. 16.20 Diffraction of light (a) Fraunhoffer diffraction

When the distance between the aperture and the screen is very large, the rays can be considered as parallel, that means, the wave fronts become plane wave fronts. One can see the diffraction effects by converging the waves by a lens on to the screen S. This is the called Fraunhoffer diffraction. Lens is kept before the aperture for diverging the rays from a source as plane waves (Fig. 16.20(a)). A converging lens is kept beyond the aperture to focus the waves to reach the points (P_1) on the screen.

If the distance between the aperture and the screen (Fig. 16.20(b)) is finite, but more than the first case, then, the screen can be at any distance beyond aperture and diffraction intensity can be seen at P_1. There are no parallel rays. This general case is called Fresnel diffraction. No lenses are needed for observation. Both cylindrical as well as spherical wave fronts can give diffraction effects.

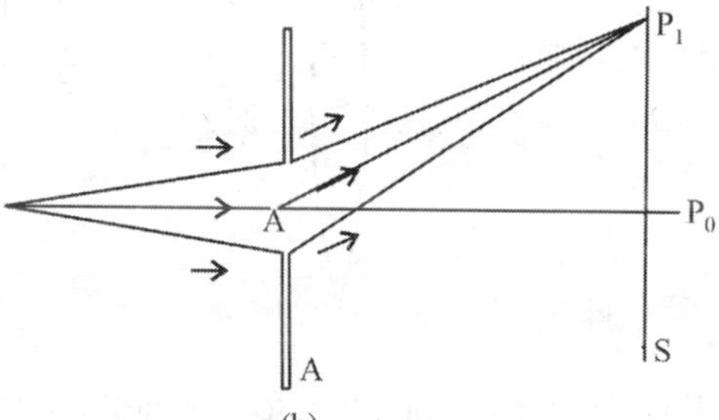

(b)

Fig. 16.20 (b) Fresnel diffraction

Fresnel diffraction: In the case of Fresnel diffraction, the source of light and the plane of observation are placed at finite distances from the aperture or obstacles. No lenses are needed for the optic waves for making the rays parallel or convergent. One wave fronts of light that are incident on the object can be spherical or cylindrical. The best example of Fresnel diffraction is the diffraction by a narrow and sharp straight edge, or by a circular aperture.

Fraunhoffer: In the type of diffraction the light source, the plane of observation are at infinity. Lenses are used both for focusing as well as for observation. Plane wave fronts are used.

We shall now consider Fraunhoffer diffraction at a single slit.

16.11 Fraunhoffer Diffraction – Single Slit

16.11.1 Analysis

A single slit, AB, of width 'a' is placed perpendicular to the plane of the paper as shown in the Fig. 16.21(a). A plane wave front PP' of a monochromatic light of wave length 'λ' falls on the slit normally as shown in Fig. 16.21(a).

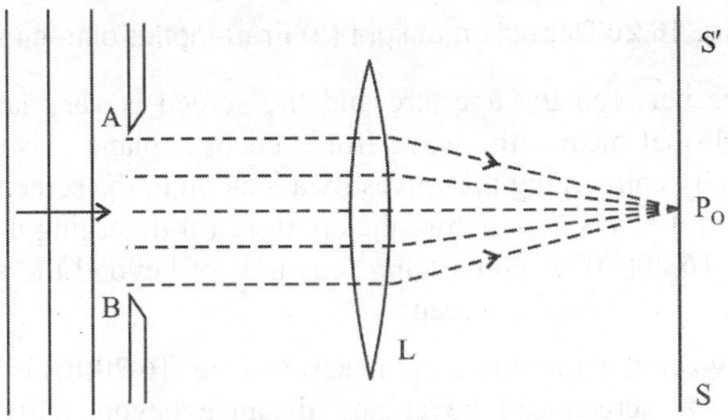

Fig. 16.21(a)

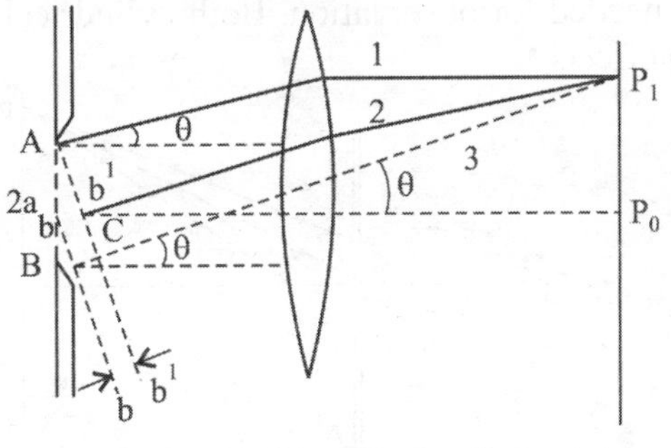

Fig. 16.21(b) Condition for the first minimum.

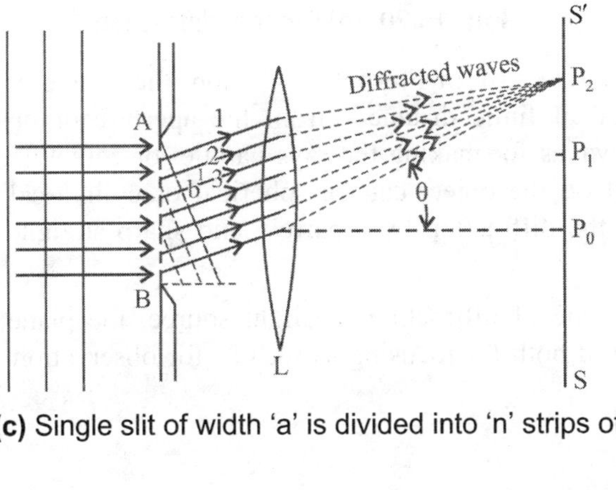

Fig. 16.21(c) Single slit of width 'a' is divided into 'n' strips of width δx.

When the light falls on the slit AB, every point on the slit becomes a source of secondary wavelets that are spherical. These wavelets spread into space beyond the slit. They are brought to focus on to the screen SS' by a covergent lens (L). All the rays parallel to the horizontal get focused at the central point P_0 as shown in Fig. 16.21(a).

Thus, P_0 is a central bright spot as the waves coming to P_0 are all in phase and interfere constructively.

We consider a point P_1 on the screen (Fig. 16.21(b)). All the wavelets travelling at an angle θ with respect to the horizontal axis are focused at P_1. If the two rays 1 and 2 are out of phase at P_1, they interfere destructively i.e., bb' is one half wavelength. AC represents the perpendicular drawn on the ray 3. Every ray coming from the upper half of the slit (width, a/2) interferes destructively with one coming from the bottom of the slit. In such a case, the point P_1, represents the first minimum of the diffraction pattern on the screen.

The path difference between the secondary wavelets traversing at an angle θ is bb' or BC,

then, $$bb' = \frac{a}{2}\sin\theta.$$

Path difference should be $\frac{\lambda}{2}$ for first minimum.

i.e. $$\frac{a}{2}\sin\theta = \frac{\lambda}{2}$$

or $$a \sin\theta = \lambda \qquad .....(16.43)$$

The central maximum becomes wider as the slit width, a_1 is decreased. One can divide the entire slit into n equal parts and consider several waves (rays) arriving at the point P_1 as in Fig. 16.21(b). Similarly there exists another point P_2 Fig. 16.21(c), the second point above P_1 that shows minimum intensity. The condition for P_2 to be of minimum intensity is

$$\frac{a}{4}\sin\theta = \frac{\lambda}{2}$$

or $$a\sin\theta = 2\lambda \qquad .....(16.44)$$

As we divide that slit into n parts, we find several points just like P_1, P_2 on either side of P_0, at which minimum intensity is found. The condition for minimum intensity is written in general as

(minima) $a\sin\theta = n\lambda$

$$\text{where } n = \pm 1, \pm 2, \pm 3 \ldots\ldots\ldots\ldots \qquad .....(16.45)$$

In between two adjacent pair of minima, there is a maximum. The screen SS' consists a series of minima and maxima alternatively on the screen as shown in the Fig. 16.21(d).

Let us now consider the intensity distribution in the diffraction pattern on the screen.

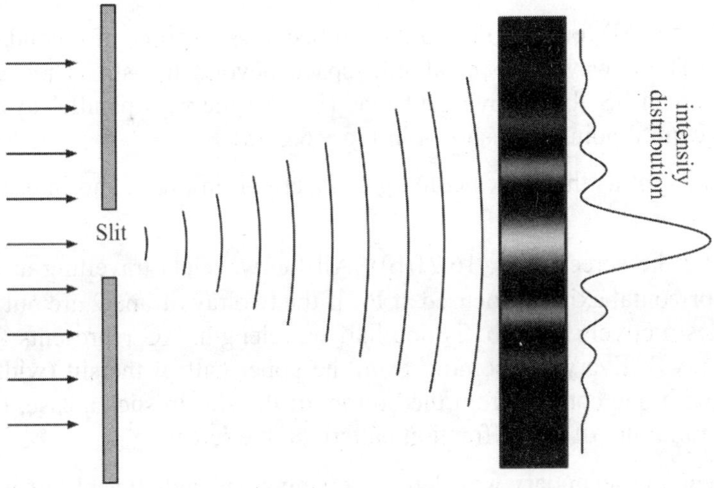

Fig. 16.21(d)

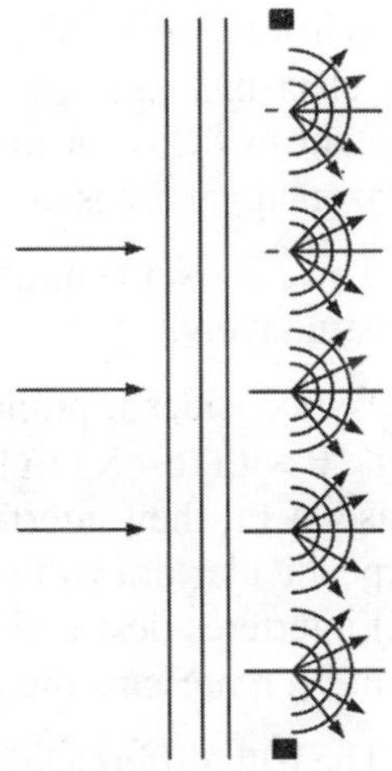

Fig. 16.21(e) Secondary wavelets from a single silt

16.11.2 Intensity distribution in Single Slit Diffraction

The slit AB is divided into n equal parts (parallel strips) as shown in Fig. 16.21(c). These strips have extremely narrow widths and each can be regarded as a source of secondary wavelets (Fig. 16.21(e)).

The phase difference between any two consecutive wavelets is given as

$$\frac{1}{n}(\text{total}) = \frac{1}{n}\left(\frac{2\pi}{\lambda} a \sin\theta\right)$$

$$= \frac{\phi}{n} \text{ (say)} \qquad .....(16.66)$$

The total phase connected with all the wave fronts (n) is dependent on number of wavelets. We can follow the method of vector addition of amplitudes.

Suppose that there are 'n' wavelets of the same wavelength, same amplitude and same phase difference between successive wavelets. Let the phase difference between successive wavelets be 'δ'

Firstly, it is necessary to find the resultant amplitude 'A_R'. A polygon of amplitudes is constructed as shown in the Fig. 16.22. Each side of the polygon has the same amplitude, 'A'. then, the component along ox_1 is

$$OP\cos\theta = A\left[1 + \cos\delta + \cos 2\delta + \cos 3\delta + ... + \cos(n-1)\delta\right] \qquad .....(16.67)$$

and the perpendicular component is

$$OP\sin\theta = A\left[\sin\delta + \sin 2\delta + ... + \sin(n-1)\delta\right]$$

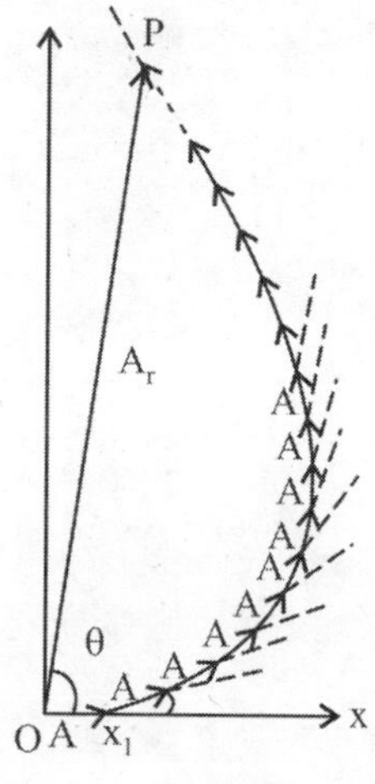

Fig. 16.22 Vector addition amplitudes

Multiplying eq. (16.67) by $2\sin\frac{\delta}{2}$, we have

$$2A_r\cos\theta\sin\frac{\delta}{2} = A\left[2\sin\frac{\delta}{2} + 2\cos\delta.\sin\frac{\delta}{2} + ... + 2\cos(n-1)\delta.\sin\frac{\delta}{2}\right]$$

$$= A\left[2\sin\frac{\delta}{2} + \left\{\sin\frac{3\delta}{2} - \sin\frac{\delta}{2}\right\} + ...\left\{\sin\left(n-\frac{1}{2}\right)\delta - \sin\left(n-\frac{3}{2}\right)\delta\right\}\right]$$

$$= A\left[\sin\frac{\delta}{2} + \sin\left(n-\frac{1}{2}\right)\delta\right]$$

$$= 2A\sin\frac{n\delta}{2}\cos\frac{(n-1)\delta}{2}$$

$$\therefore \quad A_r\cos\theta = A\frac{\sin n\delta/2}{\sin\delta/2}.\cos\frac{(n-1)\delta}{2} \qquad .....(16.68)$$

Similarly

$$A_r\sin\theta = A\frac{\sin n\delta/2}{\sin\delta/2}.\sin\frac{(n-1)\delta}{2} \qquad .....(16.69)$$

$$A_r^2(\cos^2\theta + \sin^2\theta) = A^2\left[\frac{\sin^2\frac{n\delta}{2}}{\sin^2\frac{\delta}{2}}\left\{\cos^2\frac{(n-1)\delta}{2} + \sin^2\frac{(n-1)\delta}{2}\right\}\right]$$

$$\therefore \qquad A_r^2 = A^2 \frac{\sin^2 \frac{n\delta}{2}}{\sin^2 \frac{\delta}{2}} \qquad .....(16.70)$$

But $$\frac{\sin\theta}{\cos\theta} = \frac{\sin\frac{(n-1)}{2}\delta}{\cos\frac{(n-1)}{2}\delta}$$

i.e. $$\tan\theta = \tan\frac{(n-1)}{2}\delta \qquad .....(16.71(a))$$

as θ is small, $\theta \approx \frac{(n-1)\delta}{2} \approx \frac{n\delta}{2}$ .....(16.71(b))

as n >> 1, the wavelets involved in the diffraction would be infinitely large.

The expression in eq. 16.70 can be written as

$$A_r = A\frac{\sin\frac{n\delta}{2}}{\sin\frac{\delta}{2}} = \frac{A\sin\left(\frac{\pi a\sin\theta}{\lambda}\right)}{\sin\left(\frac{\pi a\sin\theta}{n\lambda}\right)}$$

$$= A\frac{\sin(\phi)}{\sin\left(\frac{\phi}{n}\right)} \qquad \frac{\phi}{n} << 1. \qquad \therefore \sin\frac{\phi}{n} \approx \frac{\phi}{n}$$

$$\therefore \qquad A_r = nA\frac{\sin\phi}{\phi} = A_n\frac{\sin\phi}{\phi} \qquad .....(16.72)$$

A_n (= nA) is very large, nA is more or less constant,

Then, the intensity is given as

$$I = A_r^2 = A_n^2 \frac{\sin^2\phi}{\phi^2} \qquad .....(16.73)$$

Intensity of Principal maximum

The amplitude, A_r can be expressed in the ascending powers of ϕ as,

$$A_r = \frac{A_n}{\phi}\left[\phi - \frac{\phi^3}{3!} + \frac{\phi^5}{5!} - \frac{\phi^7}{7!} + ...\right]$$

$$= A_n\left[1 - \frac{\phi^2}{3!} + \frac{\phi^4}{5!} - \frac{\phi^6}{7!} + ...\right] \qquad .....(16.74)$$

when $\phi = 0$, we have the maximum value for A_r.

i.e. $\phi = 0$ means

$$\phi = \frac{2\pi}{\lambda} a \sin\theta = 0 \qquad .....(16.75)$$

or $\qquad \sin\theta = 0$, or $\theta = 0$

This is the central maximum and has the maximum intensity. If all negative terms vanish, we get maximum value of A_R.

Minimum Intensity Positions

The minimum intensity positions should be at the points where $\sin\phi = 0$.

That means that ϕ should satisfy the condition

$$\phi = \pm\pi, \pm 2\pi, + 3\pi \quad \ldots\ldots$$

$$= \pm m\pi$$

or

$$\frac{\pi a \sin\theta}{\lambda} = \pm m\pi$$

or

$$a\sin\theta = \pm m\lambda \qquad \text{(minima)} \qquad .....(16.76)$$

where m = $\pm 1, \pm 2, \pm 3$..............etc.

These minimum intensity positions are present on either side of the maximum intensity points, and at locations dictated by eq. 16.76.

Secondary maxima: In the above text, we have only discussed the information about the central principal maximum and the positions of diffraction minima. In addition to the central maximum, weak secondary maxima on either side of central maxima. They are placed between equally spaced minima. Let us now discuss about the diffraction maxima called 'the secondary maxima' We know that the intensity is given as

$$\therefore \qquad I = I_0\left[\frac{\sin^2\phi}{\phi^2}\right]$$

differentiating, we have the above expression, we get

$$\frac{dI}{d\phi} = \frac{2I_0 \sin\phi}{\phi}\left[\frac{\phi\cos\phi - \sin\phi}{\phi^2}\right] \qquad .....(16.77)$$

$\frac{dI}{d\phi}$ should be zero for maxima. Hence, either $\sin\phi = 0$ or $\phi\cos\phi - \sin\phi = 0$.

$$\left.\begin{aligned} &\text{i.e.,} \quad \sin\phi = 0 \\ &\text{or} \quad \tan\phi = 0 \end{aligned}\right\} \qquad .....(16.78)$$

One can plot y = ϕ and y = tan ϕ on the same graph. The points of inter section of two curves give the valves of ϕ that satisfy the equation. The plot of y = ϕ and y = tan ϕ are shown in Fig. 16.23.

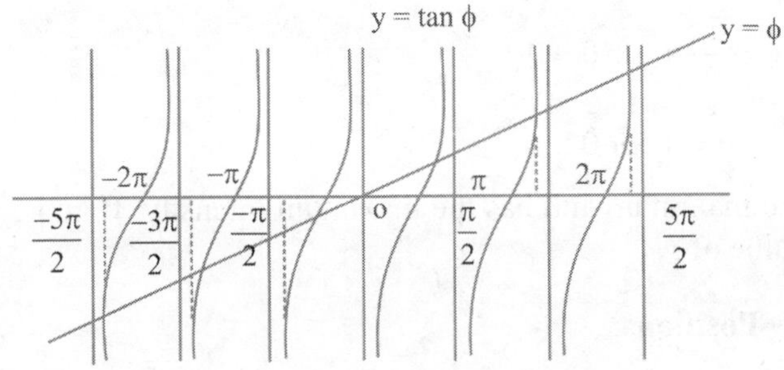

Fig. 16.23 The points of interaction of y = ϕ and y = tan ϕ are $\phi = 0, \pm\frac{3\pi}{2}, \pm\frac{5\pi}{2}$... etc.

The actual relative distribution of intensity in single-slit diffraction is dependent on the ratio $\frac{a}{\lambda}$. When $a \approx \lambda$, the central maximum would be filling the entire diffraction pattern as shown in Fig. 16.24(a). But, as the slit width is increased and made wider, the central intensity peak becomes narrower as shown in Fig. 16.24.(b). '$\Delta\theta$' gives a measure of the width of the central maximum. The Fig. 16.24(c) shows the intensity pattern of single slit diffraction for a particular slit width. The maxima should be at $\phi = 0, \pm\frac{3}{2}\pi, \pm\frac{5\pi}{2}$ ....etc.

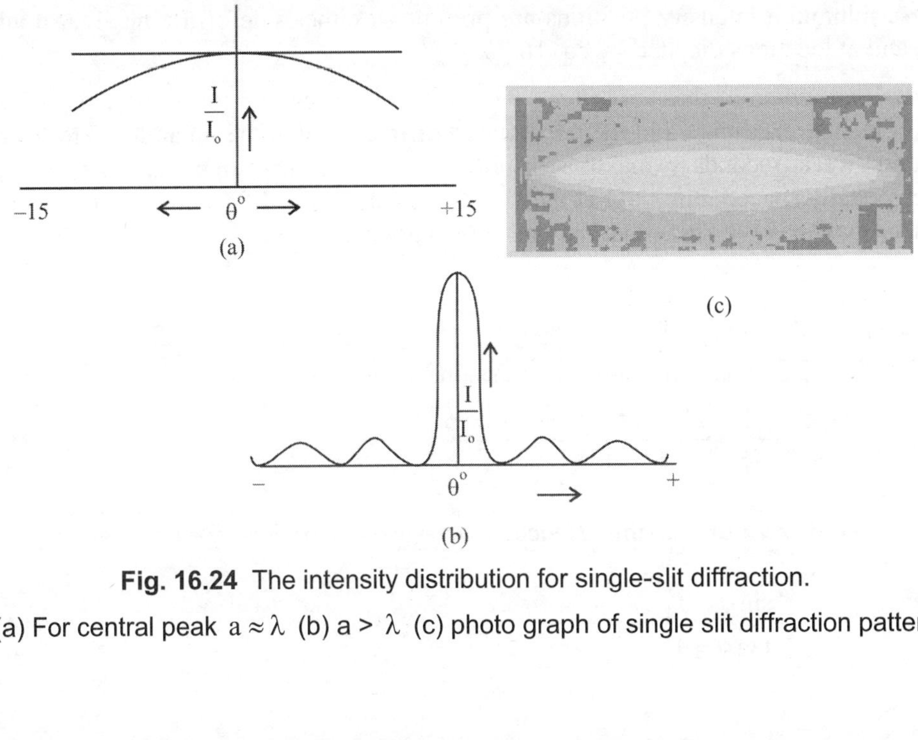

Fig. 16.24 The intensity distribution for single-slit diffraction.

(a) For central peak $a \approx \lambda$ (b) $a > \lambda$ (c) photo graph of single slit diffraction pattern

Fraunhoffer Diffraction – Double Slit

In the interference phenomenon, we have considered double-slit interference phenomenon under which the condition that the slit width is of the order of λ or less. The characteristic of such condition is that the central part of the plane of observation is covered with uniformly intense bright and dark fringes when monochromatic light is incident on it. However, if the slits are separated by a small distance 'b' and slit width is not negligible in comparison with wavelength, we get a very distinguishable diffraction pattern under Fraunhoffer diffraction condition. The focal plane of observation consists of interference fringes within an envelope of diffraction due the single-slit diffraction of the same width.

16.12 Analysis-Normal Incidence of Monochromatic Light

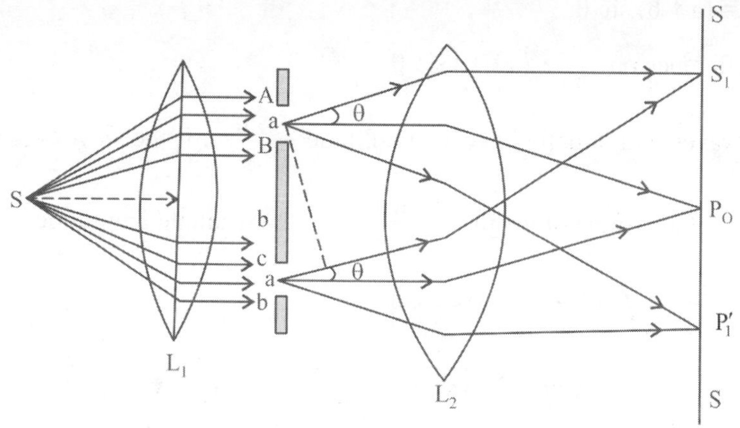

Fig. 16.25 Schematic diagram double-slit arrangement for diffraction.

Consider a double slit, having two slits, S_1 and S_2 arranged as shown in the Fig. 7. Let the slit width of S_1 and S_2 be 'a'. Let these two slits (S_1 and S_2) be separated by 'b' (BC). Suppose a parallel beam of monochromate light falls on the two slits normally. Beyond the slits, the lens L_2 conveys the detracted beams to fall on screen S.S' placed in the focal plane of the lens L_2.

A double slit is a compilation of a double slit-interference arrangement and two single slits. The two slits give diffraction pattern that is superposed over an interference fringe pattern. Both these patterns are a result of the secondary wavelets that are generated at all points in the slits. P_o is the central portion of the pattern that results from the direct light, and P_1, P_2 … l represent the focused beam of secondary wavelet pattern.

We have from the simple slit diffraction model, the resultant amplitude of A_R of all the secondary wavelets directed towards P_1 at angle θ is given as

$$A_R = \frac{A \sin \phi}{\phi}$$

where A is a constant, the amplitude of the direct bean due to each individual slit, and

$$\phi = \frac{\pi \text{ a} \sin \theta}{\lambda}$$

We make a simplification by considering the slits as equivalent to two coherent sources S_1 and S_2 arranged at mid points of the slits. Both S_1 and S_2 generate wavelets that have an amplitude $\frac{A \sin \phi}{\phi}$ along the direction θ (with respect to the direct beam). The interference on the screen is due to the two waves that differ in phase while arriving at point P_1.

To find the resultant phase difference δ, drop a perpendicular from S, to the beam that is going from S_2. Let this perpendicular be S_1 N. Then, the S_2 N represents the path difference between the wavelets from S_1 and S_2 in the direction θ.

Then $S_2 N = (a + b) \sin \theta$.

$\therefore$ Phase difference $\delta = \frac{\pi}{\lambda} \times (a + b) \sin \theta$

These two waves individually have an amplitude $\frac{A \sin \phi}{\phi}$. If the waves are represented as the two sides of a triangle of length $\frac{A \sin \phi}{\phi}$, then the resultant amplitude is given as

$$A_R^2 = \left(\frac{A \sin \phi}{\phi}\right)^2 + \left(\frac{A \sin \phi}{\phi}\right)^2$$

$$+2\left(\frac{A \sin \phi}{\phi}\right)\left(\frac{A \sin \phi}{\phi}\right) \cos \phi$$

$$= 2\frac{A^2 \sin \phi}{\phi^2}[1 + \cos \phi]$$

$$= 4\frac{A^2 \sin^2 \phi}{\phi^2} \cos^2 \phi/2$$

$$= 4 \cos^2 \beta \left(\frac{A^2 \sin^2 \phi}{\phi^2}\right) \quad \text{.....(16.80)}$$

where $\beta = \frac{\phi}{2}$. The intensity I_R is represented by A_R^2. The intensity distribution factor given above represents two factors. They are

1. The factor $\frac{A^2 \sin^2 \phi}{\phi^2}$, called the diffraction factor that represents single slit pattern of diffraction.

2. The factor $\cos^2 \beta$ represents the interference being formed by two single slits. What we observe on the focal plane of observation is the overlapped pattern of the two. The diffraction pattern is shown in the Fig. 16.26 (a) & (b).

For $\theta = 0$, we see wide central maximum with distinct interference fringes.

On either side of the central maximum, we can see the secondary maximum as in the case of single slit with a set of a dark and bright interference fringes.

The secondary maximum occur at

For $\sin\phi = 0$, i.e., $\phi = \pm m\pi$ .....(16.81)

or $a \sin\theta = \pm m\lambda$

A pattern similar to single slit pattern, discussed in single slit diffraction.

The interference pattern is represented by $\cos^2\beta$ and these fringes are seen in the diffraction maxima as depicted in the Fig.

For maxima $\cos^2\beta = 1$

i.e. $\beta = \frac{\pi}{\lambda}(a+b)\sin\theta = \pm n\pi$

i.e. $(a+b)\sin\theta = \pm n\lambda$ .....(16.82)

where n = 0. 1, 2, 3…

The Fig. 16.26(a). depicts the intensity distribution in a double-slit fraunhoffer diffraction pattern. For $\theta = 0$, we get a wide central maximum in which the interference fringes can be seen clearly.

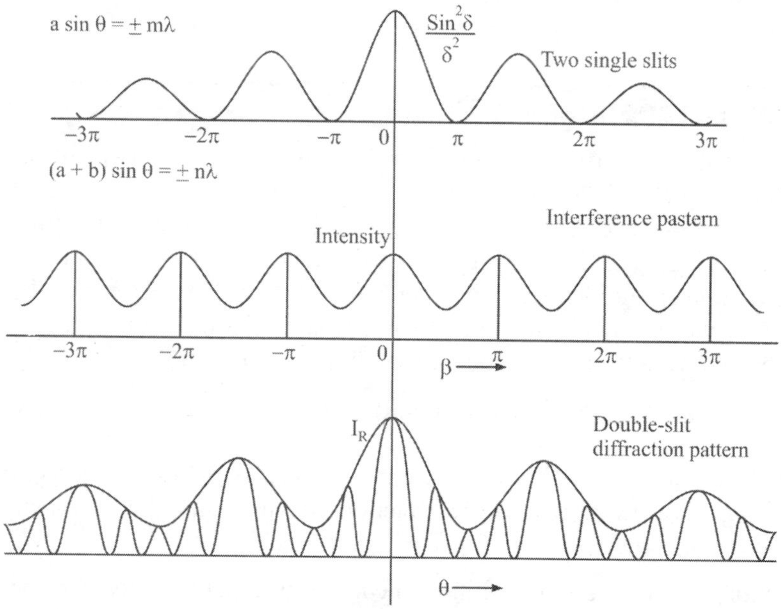

Fig. 16.26(a)

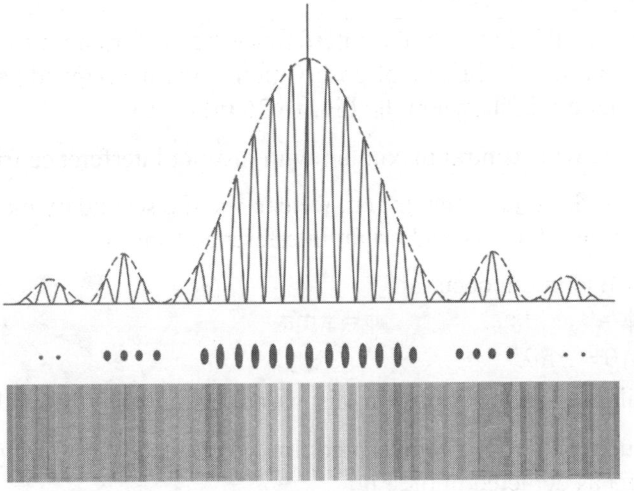

Fig. 16.26(b)

For very narrow slits $(a \approx \lambda)$, the entire diffraction pattern is a broad central peak spread out when the slit width becomes equal to several wavelengths, the envelope of the fringe pattern becomes narrower. This is shown in Fig. 16.26(a). The Fig. 16.26(b) shows double-slit diffraction pattern observed in the laboratory.

As a result, the central peak becomes sharper.

The ratio of slit separation b to λ decides the separation of fringes.

At greater wavelengths like yellow or red light, the diffraction as well as interference pattern broadens.

16.13 Plane Transmission Grating

In the case of double-slit diffraction, the diffraction patterns of the two slits coincide and we get the diffraction pattern similar to that is obtained from a single slit. Over this pattern, the interference fringes formed by the two slits is over lapped. As the slit width is increased, the maxima become sharper and narrower. Each slit is a source of diffraction. As the number of slits is increased, keeping identical slits and same spacings between them, the bright fringes become sharper and narrower. However, their positions are at the same angles. The figure in Fig. 16.27 shows diffraction pattern under Fraunhofer condition for different set of slits.

The maxima obtained are called 'diffraction orders' and are present on either side of a central maximum. The arrangement of slits and the opaque separations between the slits is called the diffraction grating. When the number of slits is made very large, of the order of 10^4, the grating gives well resolved diffraction maxima and the secondary maxima are suppressed to the maximum extent.

A diffraction grating is an optical instrument that is capable to resolve spectral lines to a high degree. A grating is also capable of dispersing white light into colours.

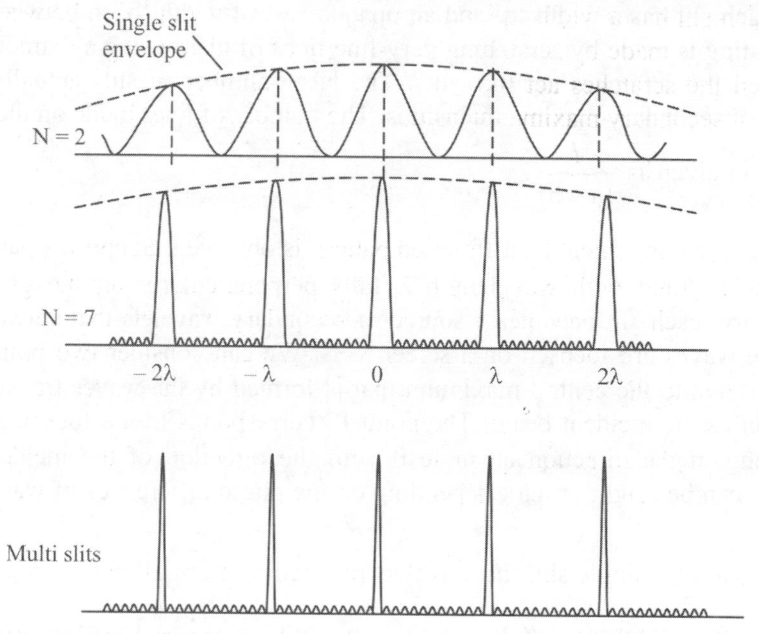

Fig 16.27 diffraction pattern under Fraunhofer condition for different set of slits

We consider only a plane transmission grating which is used to observe diffraction by means of transmitted light. The Fig. 16.28 shows how diffraction pattern is formed due to the secondary wavelets from the slits.

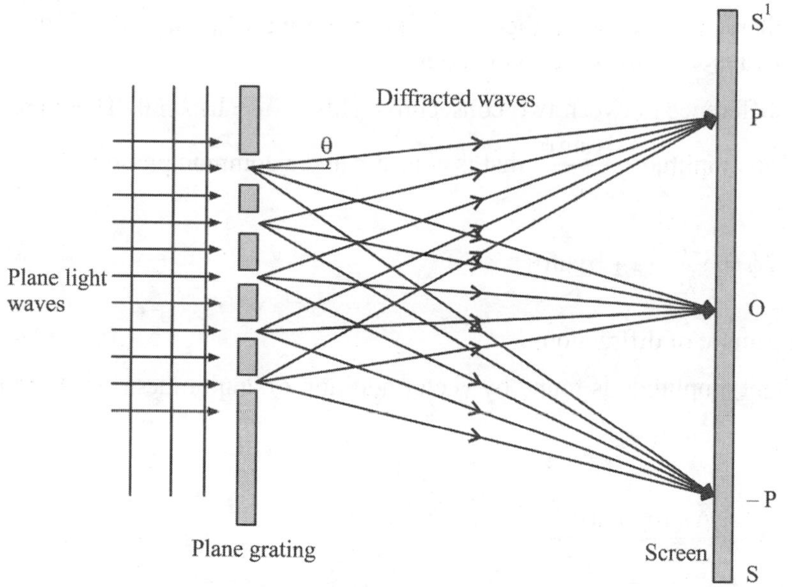

Fig. 16.28 A plane transmission grating gives Diffraction pattern

Suppose each slit has a width 'q' and an opaque part of width 'b' in between the slits. A diffraction grating is made by scratching very fine lines of glass using a diamond point. The spaces between the scratches act like slits. The large number of slits actually help in the minimization of secondary maxima intensities. The number of lines made on the grating over a length of 'l' in given as $\frac{l}{(a+b)}$.

SS' is the screen on which the diffraction pattern is observed. Suppose a parallel beam of monochromate of light with wavelength λ falls perpendicularly on the grating. As per Huygen's theory, each slit becomes a source of secondary wavelets that spread beyond the grating. These waves are focused on a screen SS'. We can consider two points P_0 and P_1. The point P_0 refers to the central maximum that is formed by the waves transversing in the same direction as the incident beam. The point P_1 corresponds to the focusing point of the waves traveling in the direction at angle θ with the direction of the incident light. The intensity at P_1 can be bright or dark depending on the phase differences of waves arriving at P_1.

As in the case of a single slit, the wavelets proceeding from all points in a slit along the direction θ are equivalent to a single wave of amplitude $\left(\frac{A\sin\phi}{\phi}\right)$ starting from the middle of the slit.

where $$\phi = \left(\frac{\pi a \sin\theta}{\lambda}\right)$$

Suppose there are 'N' slits, then from all the middle points of all the N slits, the secondary wavelets are diffracted to the point P_1.

The path difference between two consecutive slits = (a + b) sin θ. There are N wavelets, and each has an amplitude $\frac{A\sin\phi}{\phi}$ that is constant and a common phase difference

$$2\phi = \left(\frac{2\pi}{\lambda}\right)(a+b)\sin\theta$$

where θ is the angle of diffraction.

The resultant amplitude is found by vector addition of amplitudes. Let A_r be the resultant amplitude.

Then,

$$A_r = \frac{A\sin\phi}{\phi}.\frac{\sin N\phi}{\sin\phi}$$

The intensity is given as

$$I = A_r^2 = \left(\frac{A\sin\phi}{\phi}\right)^2 \left(\frac{\sin N\phi}{\phi}\right)^2 \quad .....(16.83)$$

The factor $\left(\frac{A\sin\phi}{\phi}\right)^2$ gives the distribution of intensity due to a single slit, and the $\left(\frac{\sin N\phi}{\sin\phi}\right)^2$ represents the intensity due to combined effect of all the slits.

The central or principal maxima: The central maxima is formed at $\theta = 0$; and $\sin\phi = 0$.

or $\phi = \pm m\pi$ where m = 0, 1, 2, 3, … .....(16.84)

The condition m = 0 represents the central maximum.

If $N\phi = 0$, $\frac{\sin N\phi}{\sin\phi}$ becomes indeterminate. But in the limit, ϕ is very small, we get

$$\frac{\sin^2 N\phi}{\sin^2\phi} \approx N^2 \quad .....(16.85)$$

Then, the resultant intensity is

$$I = A_R^2 = \left(\frac{A\sin\phi}{\phi}\right)^2 N^2$$

The most intense maxima is the principal maxima at the center.

The maxima are obtained at positions where θ is given as

$$\theta = \pm n\pi \quad .....(16.86)$$

or $$\frac{\pi}{\lambda}(a+b)\sin\theta = \pm n\pi$$

i.e., $$(a+b)\sin\theta = \pm n\lambda \quad .....(16.87)$$

where n = 0, 1, 2, 3,

where n is called order number. The above equation is same as that maxima of the double slit. If light passing through any two adjacent slits is in phase at a point on the screen, then light passing through any pair of slits in the array of slits (not necessarily adjacent ones) is also in phase at that point. The locations of all principal maxima is independent of the number of slits, N. The diffraction pattern shown in the Fig. 16.27 emphasises this aspect.

What is observed is the sharpening of the principal maxima as the number of slits is increased. The diffraction condition is written as

$$d\sin\theta = \pm n\lambda \qquad .....(16.88)$$

where d = (a + b). The phase difference between the two adjacent slits is $\Delta\phi$. The angle of diffraction θ is different for principal maxima of different orders. Bright lines parallel to the rulings on the gratings appear on the focal plane of observation.

We shall write the above equation for principal maximum of nth order as

$$d\sin\theta_n = n\lambda$$

or

$$Nd\sin\theta_n = Nn\lambda \qquad .....(16.89)$$

where N is the total number of slits in the grating.

Consider first minimum that is adjacent to n^{th} maximum. Its direction is given by $\theta_n + d\theta_n$.

Then, the direction of the minima is given as

$$N(a+b)\sin\theta = m\lambda \qquad .....(16.90)$$

where, $m \neq 0$, N, 2N, ...nN., but can take values given by $m = Nn + 1$.

Thus,

$$Nd\sin(\theta_n + d\theta_n) = (nN+1)\lambda$$

$$Nd[\sin\theta_n \cos d\theta_n + \cos\theta_n \sin d\theta_n] = (nN+1)\lambda$$

when $d\theta_n$ is very small, we have $-\cos d\theta_n \approx 1;\ \sin d\theta_n \approx d\theta_n$

$$Nd[\sin\theta_n + \cos\theta_n d\theta_n] = (nN+1)\lambda$$

i.e.,

$$Nn\lambda + Nd\cos\theta_n d\theta_n = nN\lambda + \lambda$$

or

$$d\theta_n = \frac{\lambda}{Nd\cos\theta_n} \qquad .....(16.91)$$

For central principal maximum, n = 0, then $d\theta_n = d\theta_0; \theta_n = \theta_0$

$$\therefore \quad d\theta_o = \frac{\lambda}{Nd\cos\theta_o} \approx \frac{\lambda}{Nd} \qquad .....(16.92)$$

$d\theta_o$ is the angular half width of the central maximum.

It is obvious that central maximum becomes narrower and narrower as N is increased. However, as θ increases, the widths of other principal maxima increase.

Dispersive Power of Grating: A diffraction grating is an optical component with a periodic structure of slits that diffract a light beam into several narrow bright beams. A diffraction grating is also a dispersive element like prism, but has more effective dispersive power than a prism. It is like a 'super prism'. A simple example in which we observe dispersion of light by diffraction is DVD. A DVD possesses finely spaced data tracks that run spirally around the center of the disc. These data tracks work as multiple grating elements. When white light falls on it, we see dispersed coloured spectra on them. We have for a grating,

$$d\sin\theta = n\lambda$$

Differentiating with respect to λ, we get

$$d\cos\theta\, d\theta = n d\lambda$$

$$\therefore \quad \frac{d\theta}{d\lambda} = \frac{n}{d\cos\theta} \qquad .....(16.93)$$

$\left(\frac{d\theta}{d\lambda}\right)$ represents dispersive power. There is always a change in diffraction angle with the change in wavelength. The dispersive power is dependent on the order of diffraction. Higher dispersion can be obtained for higher wavelength maxima. The dispersive power is inversely proportional to cos θ. For large values of θ, there exists higher dispersion as $\cos\theta \to 0$.

Differences between Prism spectra and grating spectra:

1. In prisms, the dispersion is due to refraction of white light. Different deviation angles exist for different wavelengths. Longer wavelength light beams are refracted at smaller angles as compared to shorter wavelength beams.

   In the case of diffraction grating, longer wavelength beam are diffracted at greater angles. The depicts the difference between the grating spectra and prism spectra is given in the figure below.

2. The prism spectra is studied keeping the prism at minimum deviation angle and only one spectrum is produced through refraction.

   In the case of grating, the spectra are present on either side of the central maximum. Even higher orders are produced. This enables the measurement of wave length of a particular spectral line much more accurately. The spectra can also be obtained even for oblique incidence of white light on the grating. Even minimum deviation condition exists for oblique incidence of light on grating.

   Diffraction gratings are used in mono-chromators of optical spectrophotometers. These monochrometers and spectral analysers are useful in the study of Raman effect, spectra of gases or elements and compounds. They are useful in the study of spectra of stars, in wavelength division multiplexing devices, optical pulse compressing devices etc.

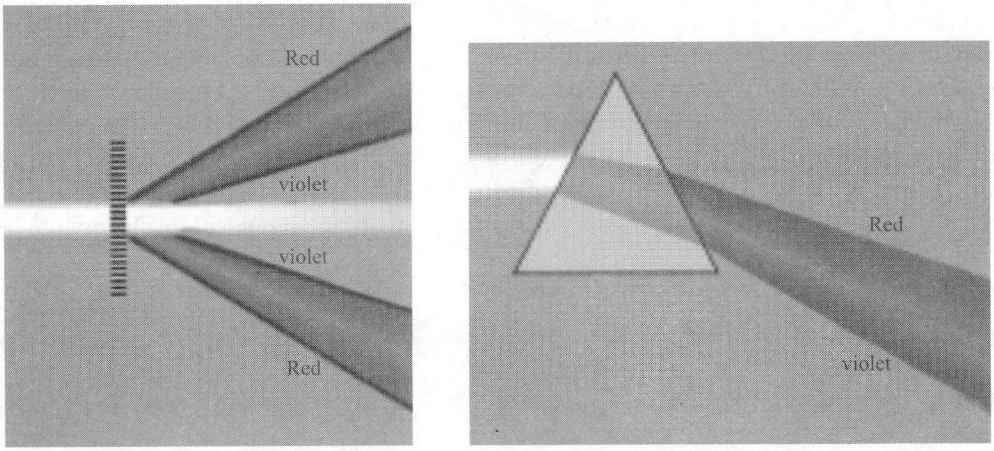

Plane grating Vs prism

Resolving Power of a Grating: The beauty of grating lies in its ability to resolve two closely spaced spectral wavelengths. The resolving power of a grating is associated with its capacity to form separate and distinguishable diffraction maxima of two wavelengths that are closely spaced. A good example is the existence of D_1 and D_2 wavelengths of sodium light and they are beautifully resolved in grating spectra in the higher orders.

Suppose a beam of light of two wavelengths, λ, and $\lambda + d\,\lambda$ is incident normally on the grating. When the diffraction maxima are viewed, two maxima corresponding to the two wavelengths are seen in the diffraction order n. According to Rayleigh Criterion, it is possible to resolve them, when the first minimum of nth order of λ falls over the principal maximum in the nth order of $(\lambda + d\,\lambda)$ in the direction $(\theta_n + d\,\theta_n)$.

The principal maximum of λ in the nth order in the direction θ_n is given as

$$(a + b) \sin \theta_n = n\,\lambda$$

The direction of minima is given as

$$N\,(a + b) \sin \theta = m\,\lambda$$

where m can take all values except 0, N, 2N; – nN

Hence, the first minimum in the direction $(\theta_n + d\,\theta_n)$ is

$$N\,(a + b) \sin (\theta_n + d\,\theta_n) = (n\,N + 1)\,\lambda \qquad \text{.....(16.94)}$$

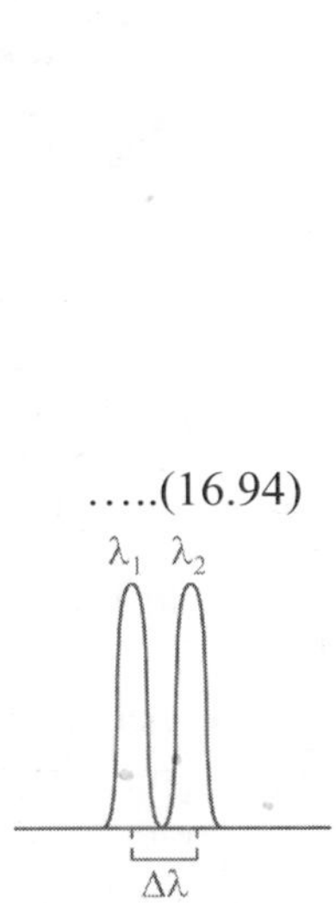

The principal maximum of $(\lambda + d\,\lambda)$ in the direction $(\theta_n + d\,\theta_n)$ is given as per the condition.

$$(a + b) \sin (\theta_n + d\,\theta_n) = n\,(\lambda + d\,\lambda)$$

It means that

$$N(a+b)\sin(\theta_n + d\,\theta_n) = n\,N(\lambda + d\,\lambda)$$

i.e.,

$$(n\,N+1)\,\lambda = n\,N(\lambda + d\,\lambda)$$

or

$$\lambda = n\,N\,d\,\lambda$$

or

$$\left(\frac{\lambda}{d\lambda}\right) = n\,N \qquad .....(16.95)$$

The above expression is the expression for resolving power of the grating

$$\therefore \qquad \frac{\lambda}{d\lambda} = \frac{N(a+b)\sin\theta_n}{\lambda} \qquad .....(16.96)$$

The resolving power can be increased with N i.e., the number of lines on the grating.

Polarization

A light wave is an electromagnetic wave. An electromagnetic wave is transverse and is associated with a direction of propagation represented by a vector, K-vector. Trasverse to the direction of propagation, the wave is associated with electric (E) and magnetic (H) field vectors. The electric and magnetic field vectors and the direction of propagation are mutually perpendicular to each other. The orientation of the electric field (E) vector represents the direction of polarization. The signals from T.V antennas (for TV signals or communication signals) are always polarized, and the receiving antennas are oriented in such a direction and orientation that they receive the signals as per their polarisation. For instance, the radio waves from a CB radio on a car are polarised in a direction parallel to the antenna.

The transverse nature of light waves makes the 'polarisation' as an important aspect of its propagation as well as its utility. The plane determined by the direction of propagation and the direction of electric field vector ($\bar{E}$) is called plane of polarisation as shown in Fig. 16.29(a). The unpolarised light is shown as a star (Fig. 16.29(b)). The natural light is unpolarised and it means that all planes of propagation are equally probable.

In the case of light waves, we have three types of polarization states namely:

1. Plane or linearly polarised waves
2. Circularly polarised waves, and
3. Elliptically polarised waves.

The light that comes to us from Sun is unpolarised light. The light from the lamps used for lighting purposes is unpolarised. It means that the emission of light takes place randomly without any preference to any single plane. The randomly coming beams of light from above said sources are transverse in nature, but there is no preferred plane of polarization. Light in the form of a plane wave is said to be linearly polarized as shown in Fig. 16.29 (a).

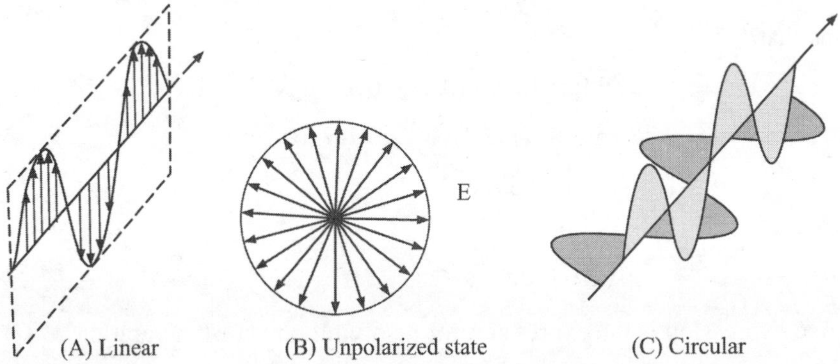

Fig. 16.29(a) Linearly polarized state (b) Unpolarised state (c) Circular polarization state.

The circularly polarized light consists of two perpendicular components of waves of same amplitude and 90° phase difference. The elliptically polarized light is thought to be made up of two perpendicular waves of unequal amplitudes and 90° phase difference.

16.14 Double Refraction

Many crystalline substances like quartz, calcite, tormaline etc are optically anisotropic. It means that the optical waves travel with different velocities in different directions in those crystals. Cubic crystals like NaCl, KCl, etc., are optically isotropic.

Suppose a polished crystal of calcite is laid over a small stick, one would see two images as shown in Fig. 16.30(a). These two images exhibit perpendicular polarizations. When a beam of light (unpolarised) falls on a calcite crystal surface normally, it splits into two beams as shown in Fig. 16.30(b).

This sort of splitting while refracting is 'double refraction' or 'birefringence'.

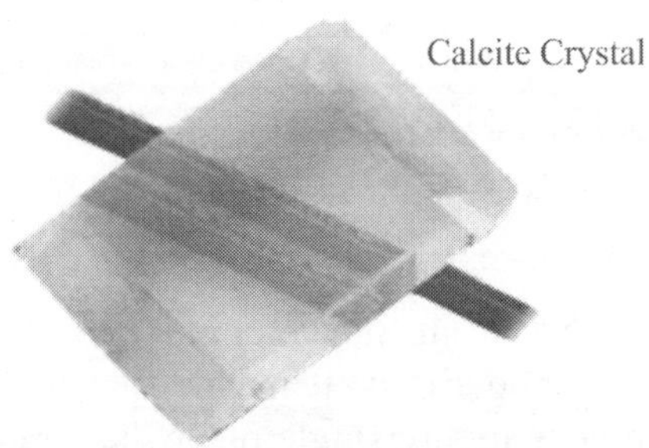

Fig. 16.30 (a) View of small stick through a birefringent crystal, like calcite

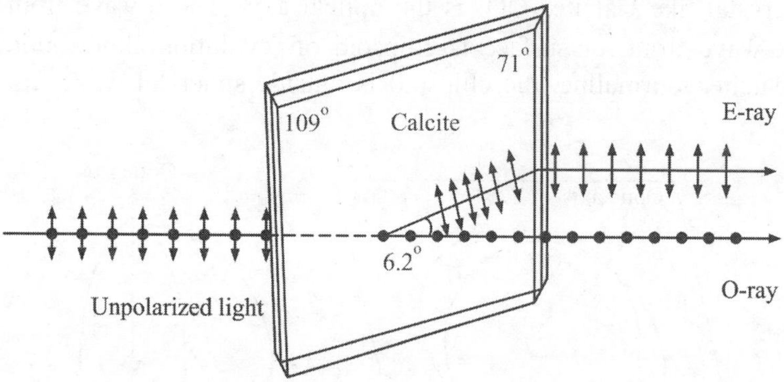

Fig. 16.30(b) Unpolarised light falling on a brifingent crystal gets split into two rays

The two beams are found to be linearly polarised with their planes of polarisation perpendicular to each other. The two beams are called ordinary ray (o-ray) and extra ordinary ray (e-ray).

The difference between o-ray and e-ray can be explained on the basis of Snell's law.

For ordinary ray, $\frac{\sin i_o}{\sin r_o} = \mu_o$ where μ_o denotes the refractive index of ordinary wave.

The ordinary ray travels as if the crystal is isotropic to it. It ovey's snell's law. The refractive index, μ_o, is same in all direction of the crystal. But the extraordinary ray travels in the crystal with a speed that varies with direction from V_o to V_e. The refractive index for the extraordinary ray is denoted by μ_e, and the velocity is noted by V_e. The angle of refraction for the extraordinary ray is dependent on the crystal. The physical quantities, μ_o and μ_e are called the principal indices of refraction. The phenomenon of the splitting of a light ray incident on a crystal into ordinary and extraordinary rays is called 'Birefrigence' or 'Double refraction'. There are many crystals that exhibit double refraction. Basically, the crystal should be anisotropic in nature.

The cubic crystals like sodium chloride, do not show double refraction. Some of the crystals that show 'double refraction' are quartz (SiO_2), zinc sulphite calcium carbonate calcite, tourmaline, potassium dihydrogen phosphate, dolomite, topaz, mica etc.

The velocity values (V_e) for e-ray can be greater or less than those (V_o) for o-ray in anisotropic crystals. It is a property that depends on crystalline nature of the crystal.

Some crystals like calcite and Dolomite crystals have $\mu_o > \mu_e$. For example, calcite has $\mu_o = 1.658$ and $\mu_e = 1.486$. Such crystals are called negative crystals. For crystals like quartz which is SiO_2 in crystalline form has $\mu_o = 1.553$ and $\mu_e = 1.544$. Such crystals are called positive crystals. It should be remembered that SiO_2 in glassy form does not exhibit double refraction.

When any wave front strike a doubly refracting crystal, every point of the crystal becomes a source of two wave fronts. The Fig. 16.31 illustrates Huygen's wave surfaces produced at a point S in a crystal like Calcite, OO' is the optical axis. The o-wave front is a spherical surface, and e-wave front, or surface is ellipsoid of revolution about optic axis OO'. In crystals like calcite, tourmaline, the ellipsoid is outside spherical wave front as shown in Fig. 16.31 (a).

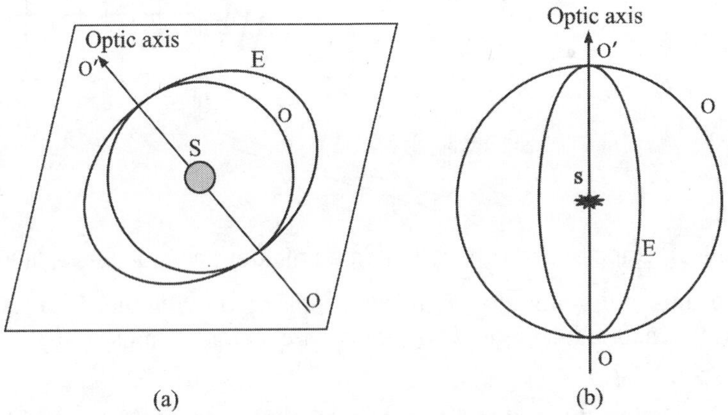

Fig. 16.31 Wave front in doubly refracting crystals.

The ellipsoid spherical surfaces touch each other at points on the optic axis. In crystals like quartz the spherical wave front lies outside the ellipsoid as shown in Fig. 16.31 (b).

16.15 Circular Polarisation

(a) Quarter wave plate: Suppose a plane polarized wave of light falls on quartz wave plate with its plane of polarization at 45° to the axis of the plate as shown in the Fig. 16.32.

The polarization of the wave is oriented at an angle 45° with respect to the optic axis. The polarisation can be resolved along X and Y-axis as two perpendicular electric field components with equal amplitudes.

The incident wave is resolved into two waves. One polarized in Z Y plane and the second in X Z plane. These components represent the o-wave and e-wave and they travel with different velocities in the slab. The phase difference between the o-wave and e-wave is dependent on the thickness of the slab and the two refractive indices.

It is possible to make the phase difference between the o-wave and e-wave equal to 90° by suitably selecting the thickness of the slab. In such a case, the slab or the plate is said to be quarter-wave plate, because 90° represents $\frac{1}{4}$ of a full cycle of oscillation. The two waves are shown in the schematic diagram in Fig. 16.32.

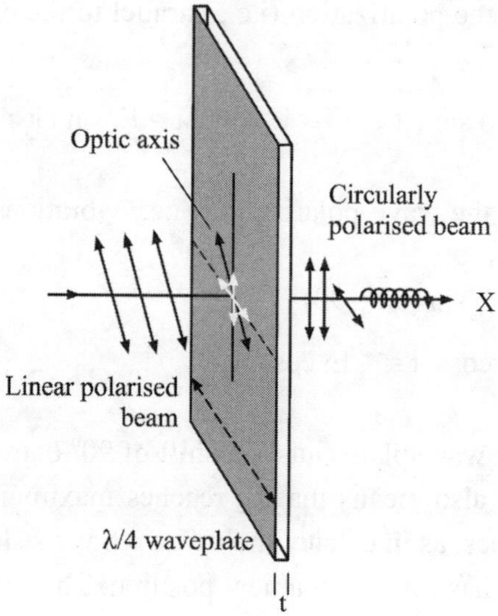

Fig. 16.32 The emergent wave from the plate possesses two components as shown resulting in circular polarisation.

The waves with these perpendicular polarization states are 90° out of phase. At any point, the net electric field is the vector sum of the two components. If one looks into the wave, the e-vector appears to rotate clock wise with time at a point i.e., on the X-axis that means that the direction of polarization changes continuously as the wave propagates. The tip of e-vector traces a circle about X-axis and this type of wave is called circularly polarised wave. The above one is a case of right-circular polarisation.

(b) Circular polarization: A quarter wave plates are useful for giving circularly polarized light from a linearly polarized ray, in ellipsometry, optical pumping, optical isolators, and unwanted reflection suppressors in optical instruments. Suppose a linearly polarised is incident on a crystal which is cut so that the optic axis is parallel to the face of the slab as shown in the Fig. 16.33. The wave is incident along the X-direction. Let the amplitude of the wave be E_o.

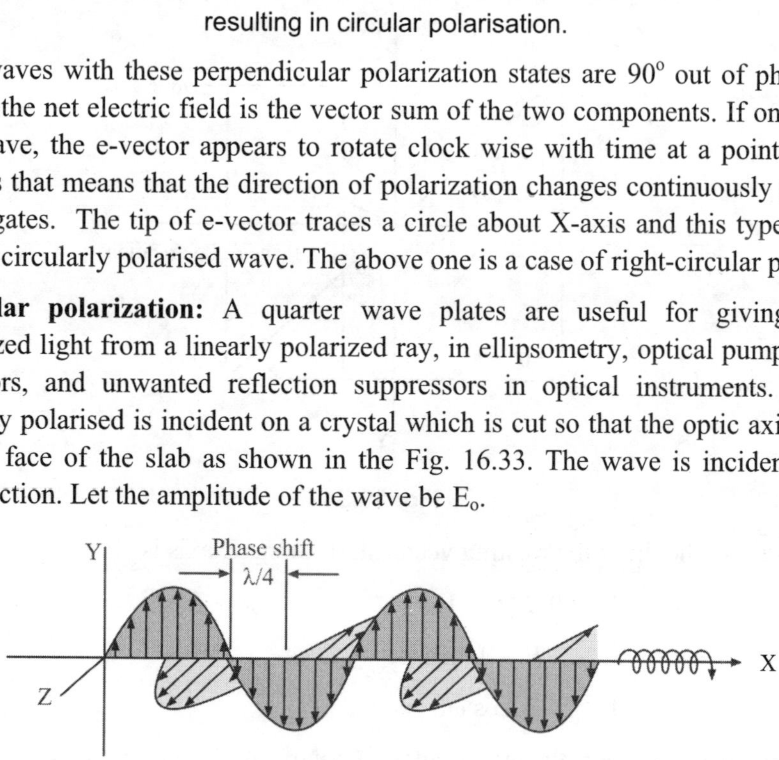

Fig. 16.33 Circularly polarised

The y-component of the polarization (i.e., parallel to the optic axis) i.e., the emerging wave is

$$E_y = (E_0 \cos 45) \sin \omega t = \frac{1}{\sqrt{2}} E_o \sin \omega t = E_r \sin \omega t$$

The z-component of the wave polarization (i.e., vibrations perpendicular to the optic axis i.e., is o-wave) is.

$$E_z = (E_0 \sin 45^o) \sin (\omega t - 90)$$

$$= -\frac{1}{\sqrt{2}} E_0 \cos \omega t = - E_r \cos \omega t$$

Therefore, the quartz wave plate causes a shift of 90° between the planes of vibrations of o and e-waves. It also means that E_z reaches maximum value for one-fourth of a cycle later than E_y does, as in calcite, the o-wave travels slower than the e-wave. After a time 't', the vector has rotated to a new position. The rotation of E vector is counter clockwise, as we look into the YZ plane in which the wave is arriving. This is called left circular polarisation.

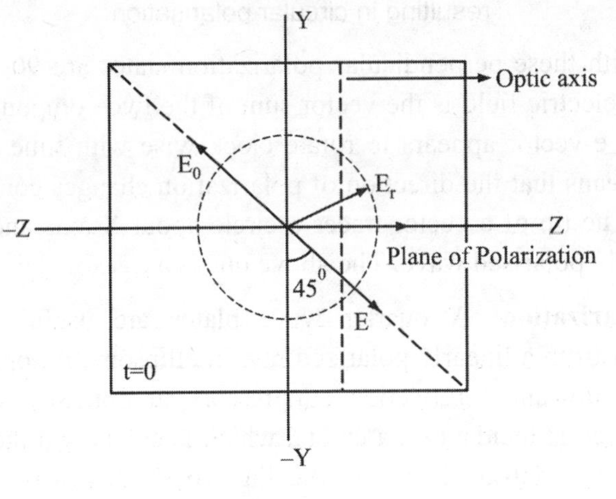

Fig. 16.34

At t = 0, the tip of the rotating vector, E_r is along Z-axis is

$$E_y = 0; E_z = - E_r$$

But at $t = t_1$, $E_y = E_r \sin \omega t_1 \approx E_r \omega t_1$

$$E_z = - E_r \cos \omega t_1 \approx - E_r$$

This case represents the left circulating polarised beam (Fig. 16.34).

16.16 Polarisation by Reflection

An unpolarised beam falls on a glass plates surface as shown in Fig.16.35.

The unpolarised wave can be resolved into two components, linearly polarised at right angles to one another with a random phase difference between them,

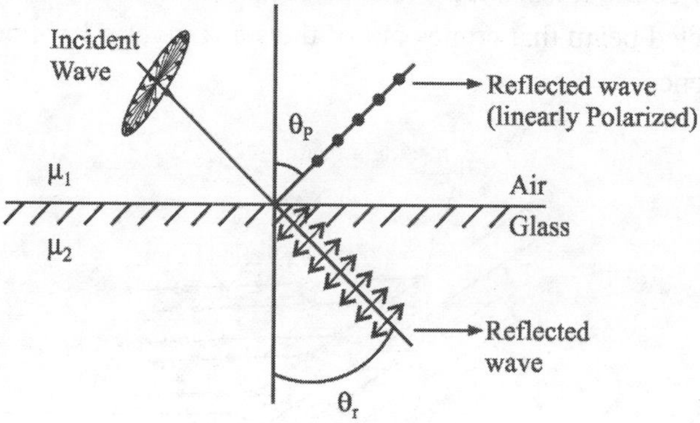

Fig. 16.35

The two components are parallel and perpendicular to the plane of incidence. There is a particular angle of incidence, θ_p, at which the reflected wave is a linearly polarised wave completely. The angle θ_p at which it takes place, is the polarizing angle. The direction of polarization is parallel to the plane or surface of the glass plate. At the polarising angle, θ_p, the reflected and the refracted beams are at right angles to each other i.e.,

$$\theta_p + \theta_r = 90^\circ$$

From Snell's law, we have

$$\mu_1 \sin \theta_p = \mu_2 \sin \theta_r = \mu_2 \cos \theta_p$$

$$\therefore \quad \tan \theta_p = \frac{\mu_2}{\mu_1}$$

The above relation is called 'Brewster's law'. For glass with refractive index 1.5, the Brewster's angle is 56°.

Most light sources emit polarised light. One important way of getting plane polarised wave is by reflection using Brewster's principle.

There are other methods of obtaining polarised beams. Polarisation of electromagnetic waves is an important aspect of spreading electromagnetic waves by transmission antennas.

At $\theta_p = 56^\circ$, the light reflected by the glass is completely polarised at right angles to the plane of incidence. However, the refracted light is not completely polarised parallel to the plane of incidence. A glass plate reflects light partially and transmits more intensity.

The intensity of the reflected beam is low and maximum intensity (about 80-85%) is refracted. This light is polarised at right angles to the plane of incidence.

If the refracted light is made to be reflected at Brewster's angle, the refracted light beam tends to be polarised more into the plane of incidence. A sack of glass plates as shown in Fig. 16.36 is used to obtain a reflected beam that is polarised at right angles to the plane of incidence. The refracted beam that comes out of the stack gives a light beam that is polarised in the plane of incidence.

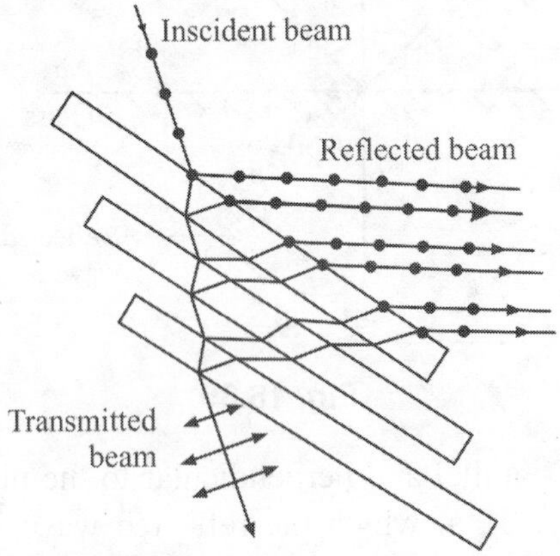

Fig. 16.36 Polarisation by reflection. Multiple glass plates are used for polarisation. Dot represents polarisation normal to the page.

16.17 Law of Malus

Double refraction is also used for getting polarised beams of light. The polarisers are used for this purpose. Two polarizers are arranged as shown in the Fig. 16.37. Suppose unpolarised beam of light falls on the first polarizer. It selects light beams that have vertical polarisation and allows them to pass through. It is then allowed to be incident on the second polarizer (called analyser) that has orientation 'θ' with respect to the fist one.

The polarized light falls on the second polariser. The linearly polarised has two components E_1 and E_2. E_1-component is polarised along the vertical direction. E_2-component is polarised perpendicular to E_1-component. The component E_1 is allowed to go out while the component E_2 is not allowed. If the amplitude of the wave incident on the analyse is E_0, the amplitude of the parallel wave is

$$E_2^1 = E_0 \cos\theta$$

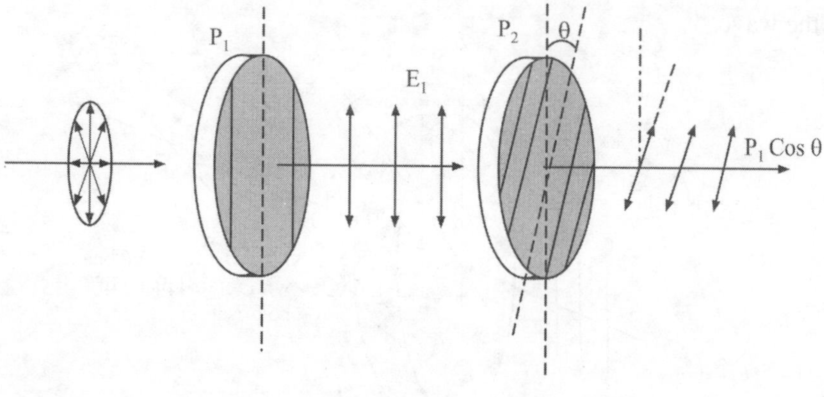

Fig. 16.37(a) A P_1 and P_2 are oriented at an angle θ with respect to each other (Law of Malus).

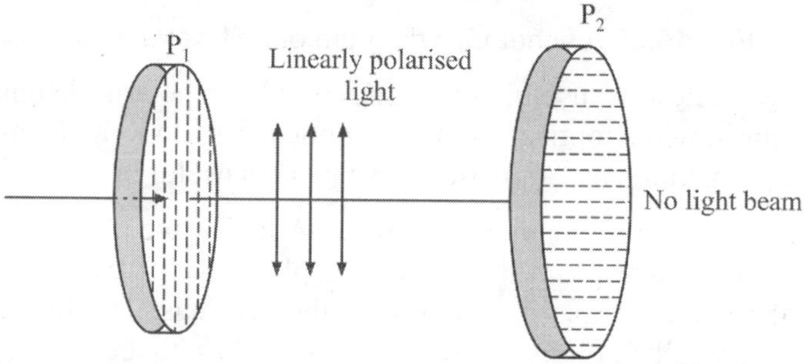

Fig. 16.37 (b) P_1 and P_2 crossed polarisers. No light beam passes through.

Then, the intensity of light that is transmitted given as

$$I_t = (E_0)^2 \cos^2 \phi$$

the above relation is called the 'law of Malus'.

If $\phi = 90^\circ$, $I_t = 0$. The second polariser completely blocks the light that is incident on first polarizer. This is the principle that is used in Polaroid sun glass, and polarization microscopes.

16.18 Half-wave plate

If the thickness of birefringent crystal is such that the incident linearly polarised light beam emerges as a linearly polarised beam but its polarisation is rotated by twice the angle of the incident beam with respect to the optic axis. This is shown in Fig. 16.38. The thickness, t needed is given as $(\mu_\sigma \mu_e)t = \frac{\lambda}{2}$. Half-wave plates are useful in rotating the plane of polarization, in the electro-optic modulation, and as a beam splitter. Half-wave plate is a retarder of the wave.

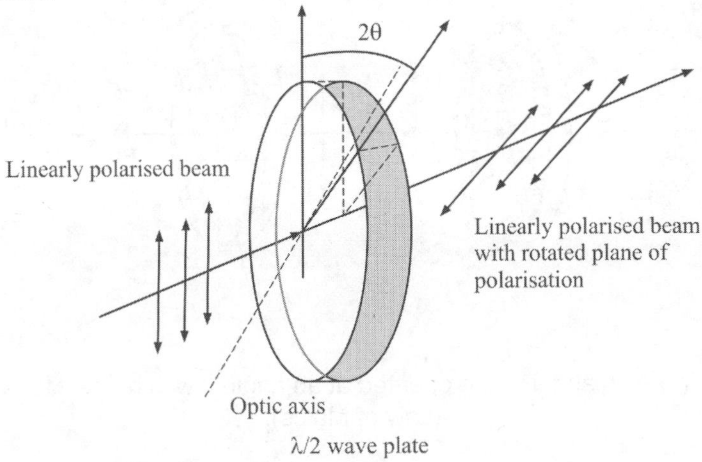

Fig. 16.38 A schematic diagram of half wave plate.

The quarter wave plates are useful for giving circularly polarised light from a linearly polarised ray, in ellipsometry, optical pumping, optical isolators, and unwanted reflection suppressors in optical instruments. A quarter-wave plate is useful in compact disc detection.

The wave plates are useful in many applications. A half wave plate handy in rotating the plane of polarisation of a light beam to any desired plane. Half wave plates are achromatic retarders flipping the plane of polarisation around the fast axis. The light beams from Ion-Lasers are mostly vertically polarised. Some times, it has to be flipped into horizontal polarisation and this is done by placing half wave plate in the beam with its fast axis or slow axis 45° to the vertical. The axis along which the wave component travels faster and experiences low refractive index. The axis along which the refractive index is high is called slow axis.

The achromatic wave plates are used in special Instruments for observation of body fluids in biomedical field, semiconductor devices, aero-space communication and in astronomical observations. Quarz plates are widely because of their strength and availability. Quarz is highly useful in the wave length range 200 nm to 2300 nm. Mica is also useful as retarder in the wavelength range 400-200 nm. Polarisation microscopes are widely used for observation body fluids, microbes, etc in biomedical studies, microstructures of metals and their alloy systems, fractured and fatigued portions of materials etc.

16.19 Elliptically Polarised light

The elliptically polarized light is thought to be made up of two perpendicular linearly polarized waves of unequal amplitudes which differ in phase by 90°. If the light wave approaches an observer, the electric vector appears to rotate following the right-hand rule. The elliptical polarisation can be obtained even if the two waves differ in phase by not necessarily 90° but by a value other than 90°. The elliptically polarised beam of light is shown in the Fig.16.39

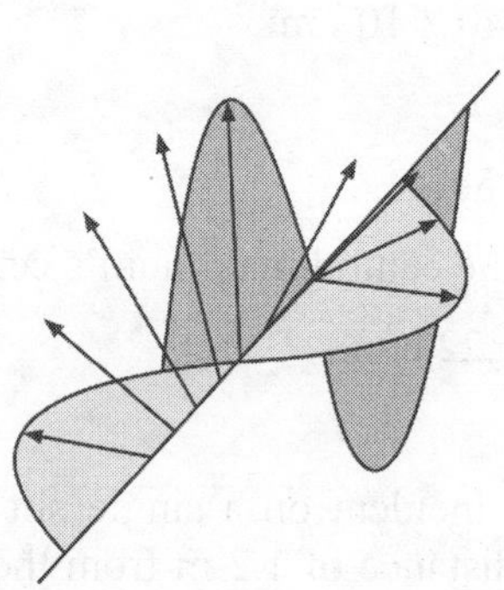

Fig. 16.39 Schematic diagram of elliptically- polarised Light beam. The tip of the electric vector rotates in an elliptical way as the wave Progresses.

Examples

Example 1

The first order bright line appears 0.25 cm from the center bright line when a double slit grating is used. The distance between the slits is 0.5 mm and the screen is 2.8 m from the grating. Find the wavelength.

Sol: x = distancc of 1st bright line from the central bright line

$$0.25 \text{ cm} = 2.5 \times 10^{-3} \text{ m}$$

$$D = 2.7 \text{ m}$$

$$d = 0.5 \text{ mm} = 5 \times 10^{-4} \text{m}$$

$$D >> x, \quad \text{then,} \quad \lambda = xd/(mD)$$

$$x = \lambda mD/d, \quad m=1$$

$$x = (589 \times 2.8)/4.9 \times 10^{-4}$$

$$x = 3.36 \text{ mm.}$$

Example 2

A diffraction grating has 450 lines per mm. The grating is illuminated normally with a light of wavelength of 440 nm to observe diffraction. The grating is placed 1.4 from a screen. Where does the first order bright line from the center of central maximum appear on the screen?

Sol:

$$d \sin\theta = \lambda$$

$$d = (1/450)\ \text{mm} = 2.02 \times 10^{-3}\ \text{mm} = 2.02 \times 10^{-6}\ \text{m}$$

$$\lambda = 440\ \text{nm} = 440 \times 10^{-9}\ \text{m}$$

$$D = 1.4\ \text{m}$$

$$\sin\theta = \lambda/d = 12.58^{\circ}.$$

Distance from the central maximum's center is

$$x = D \tan\theta = 31.22\ \text{cm}.$$

Example 3

Light of wave length 589 nm is incident on a single slit normally .The diffraction pattern appears on a screen at a distance of 1.2 m from the slit. What is the width of the central maximum if the slit width is 0.6 mm.

Sol: Let the width of the cental maximum be $2x$

$$a\text{Sin}\theta_1 = \frac{x}{1.2}$$

$$\vartheta_1 \approx \frac{x}{1.2}$$

$$\text{but } \vartheta_1 = \frac{\lambda}{a} \quad \text{from diffractin condition}$$

$$\therefore x = \frac{1.2 \times 589 \times 10^{-9}}{6 \times 10^{-4}} = 1178 \times 10^{-6}\ \text{m}$$

$$\text{width of the central max imum} = 2 \times 1178 \times 10^{-6} = 0.236\ \text{cm}.$$

Example 4

A grating has 2000 lines over a width of 2.5 cm. Suppose that a monochromatic laser light of 632.8 nm is made to fall on it normally. The diffraction pattern is formed on a screen placed at a distance. In the first three orders from the center. Suggest which order is actually seen on the screen.

Sol:

$$d\text{Sin}\vartheta = n\lambda$$

$$N = \frac{2000}{2.5}\ \text{lines per cm.}$$

1. for n=1

$$\sin\theta_1 = \frac{2\times 328\times 10^{-8}\times 2000}{2.5} = 0.5062.$$

2. for n=2

$$\sin\theta_1 = \frac{2\times 6328\times 10^{-8}\times 2000}{2.5} = 1.0124. \quad \text{greater than } 1 .$$

3. n=3

$$\sin\theta_1 = \frac{3\times 6328\times 10^{-8}\times 2000}{2.5} = 1.5186. \quad \text{greater vthan } 1.$$

∴ **Only the first order is possible.**

Example 5

A plane transmission grating is used for observing diffraction spectra with the incident white light normally falling on it. Suppose that the opaque portion is twice the slit width, what are the orders that would be absent?

Sol: In a diffraction pattern of a grating, it is possible that certain orders would be missing. The principal maximum is obtained at

$$(a+b)\text{Sin}\theta = n\lambda$$

The slits give a the diffraction pattern. The condition for minima for the slit is

$$a\ \text{Sin}\theta = m\lambda$$

The question of missing order comes when the minima due to slits falls on the principal diffraction maxima due to diffraction from the grating.

$$\therefore \quad \frac{(a+b)}{a} = \frac{n}{m}$$

In the present case b = 2a

$$\therefore \quad n = 3\text{ m} \quad \text{where} \quad m = 1,2,3\ldots$$

The 3rd principal maximum is absent, there after, 6th,9th order… will be missing.

Example 6

A plane transmission that is used in a laboratory posses a width, w, When a particular light beam having two wavelengths, λ_1= 589 nm and λ_2 = 589.596 nm is incident on the grating normally how wide the grating element should be to resolve the two wavelengths distinctly in the second order? The grating has 150 lines per mm.

Sol: The resolving power, $R = \frac{\lambda}{\nabla\lambda} = \frac{589.3}{0.59} = 1000$ Order number = 2

The number lines in width, w = 150 × w mm.

Bur we have

$$R = \text{ordernumber} \times w \times \text{number of lines / mm}$$

$$\therefore \quad w = \frac{R}{2\times 150} = \frac{1000}{2\times 150} = 3.33\text{mm}$$

Example 7

In a double slit, the separation between the slits is equal to 'd'. How many maxima can exist between the first diffraction minima on the left side and the first minima on the right side of the central maxima. Given that d/a = an integer.

Sol: Angular fringe width = λ/d

The angular position θ of the first maximum is

$$\text{Sin}\,\theta \approx \theta = \lambda/a,$$

linear separation of the first diffraction minima from the center

$$\Delta x' = \theta D = \frac{\lambda D}{d}$$

The fringe width or fringe separation = $\Delta x = \lambda D/d$

$$\text{Number of fringes} = \frac{2\Delta x'}{\Delta x} = \frac{2D.\lambda}{a\frac{\lambda D}{d}} = \frac{2d}{a} = 2n$$

The number of bright fringes = (2n – 2)/2 = (n – 1).

Example 8

In the case of quarter wave plate, obtain the relation for the thickness't' needed to produce a circularly polarised beam of light and the wave length of light.

Sol: When a wave of light strikes a double refractive crystal, two rays are created. If the original wave travels along x-axis, the o-ray travels along x-axis polarised along z-axis and experiences a refractive index, μ_0. The second ray is polarised along y-axis and experiences a refractive index, μ_e. After traversing a distance, t , it experiences a phase shift ,ϕ_0 given as

$$\phi_o = \frac{2\pi}{\lambda}\mu_0 t$$

Similarly, the e-ray experiences a phase shift

$$\phi_e = \frac{2\pi}{\lambda}\mu_e t$$

We know that initially the beam of light is polarised at 45° to coordinate axes. The electric field entering the crystal is given as

$$E_i = \frac{E_0}{\sqrt{2}}(i + j)\exp j(kx - wt)$$

The out coming wave is given as

$$E_i = \frac{E_0}{\sqrt{2}}\left(i\, e^{j\phi_o} + je^{j\phi_e}\right)\exp j(kx - wt)$$

As it travels a distance, t. The phase difference between the waves is given as

$$\Delta\phi = (\phi_e - \phi_o) = \frac{2\pi}{\lambda}(\mu_e - \mu_o)\, t$$

In the case of quarter wave plate, $\Delta\phi = \frac{\pi}{2}$. One of the waves runs $\lambda/4$ longer than the other. The out coming wave is circularly polarised.

Example 9

When a light beam enters a birefrengent crystal, the light waves are divided as ordinary and extraordinary waves. Under the condition of a quarter wave plate, the emergent wave forms a circularly polarised wave. Discuss the analysis for this occurrence.

Sol: The polarisation is due to orientation of the electric field vector with respect to the propagation direction. We have two components in that are perpendicular to the direction of propagation. Let x–axis be the direction of propagation. They are described as

$$E_y = E_{oy}\, e^{i(kx - \omega t + \phi_y)}$$

$$E_z = E_{oz}\, e^{i(kx - \omega t + \phi_z)}$$

The terms, k, ω, ϕ have usual meanings. The combined wave can be represented as

$$E = \hat{y}E_{oy}\, e^{i(kx - \omega t + \phi_y)} + \hat{z}E_{oz}\, e^{i(kx - \omega t + \phi_z)} = E_o^{'}\, e^{(kx - \omega t)}$$

Where $E_0' = \hat{y}E_{oy}e^{i(\phi_y)} + \hat{z}E_{oz}e^{i(\phi_z)}$

Let us suppose $\phi_y = 0,\ \phi_z = \phi$

We shall look into the real parts of the above expression only.

Then the two components fluctuate with a phase difference of ϕ, taking the real parts we get

$$E_y = E_{oy}\cos(kx - \omega t) \quad , \quad E_z = E_{oz}\cos(kx - \omega t + \phi)$$

Suppose $\phi = \frac{\pi}{2}$, then the vibrations pertaining to z-component becomes

$$E_z = E_{oz}\cos(kx - \omega t + \pi/2) = -E_{oz}\sin(kx - \omega t)$$

Therefore, we have $\frac{E_x^2}{E_0^2} + \frac{E_y^2}{E_0^2} = 1.$

This is an equation of a circle that represents a circularly polarised light beam.

Example 10

Discuss the origin of elliptically polarised light with necessary mathematical steps.

Sol: The elliptically polarised emergent beam of light is a result of the superposition of two perpendicular polarised components with unequal amplitudes.

$$E_y = \hat{j}E_{oy}e^{i(kx - \omega t)}$$

$$E_z = \hat{k}E_{oz}e^{i(kx - \omega t + \phi)}$$

The superposed wave is $E_y = (\hat{j}E_{oy} + \hat{k}E_{oz})e^{i(kx - \omega t)}$

The y component of the resultant is

The real part of $E_y = E_{oy}e^{i(kx - \omega t)}$

i.e., $E_y = E_{oy}\cos(kx - \omega t)$

Similarly the real part of the z-component is $E_z = E_{oz}\cos(kx - \omega t + \phi)$

$$E_z = E_{oz}\left[\cos\phi\cos(kx - \omega t\right] - \sin\phi\sin(kx - \omega t)$$

$$\frac{E_z}{E_{oz}} = \left(\frac{E_y}{E_{oy}} \cos\phi - \sqrt{1 - \frac{E_y^2}{E_{oy}^2}} \sin\phi \right)$$

Squaring the above equation

$$\frac{E_z^2}{E_{oz}^2} + \frac{E_y^2}{E_{oy}^2} \cos^2\phi - \frac{2E_zE_y}{E_{oz}E_{oy}} \cos\phi = \left(1 - \frac{E_y^2}{E_{oy}^2} \right) \sin^2\phi$$

$$\frac{E_z^2}{E_{oz}^2} + \frac{E_y^2}{E_{oy}^2} - \frac{2E_zE_y}{E_{oz}E_{oy}} = \sin^2\phi$$

This is an equation of ellipse ϕ, can take values $\phi = (2n+1)\frac{\pi}{2}$, where n is an integer.

n = 0,1,2,3,…. The simplest case is $\frac{E_z^2}{E_{oz}^2} + \frac{E_y^2}{E_{oy}^2} = 1$, The axes coincide with y and z directions.

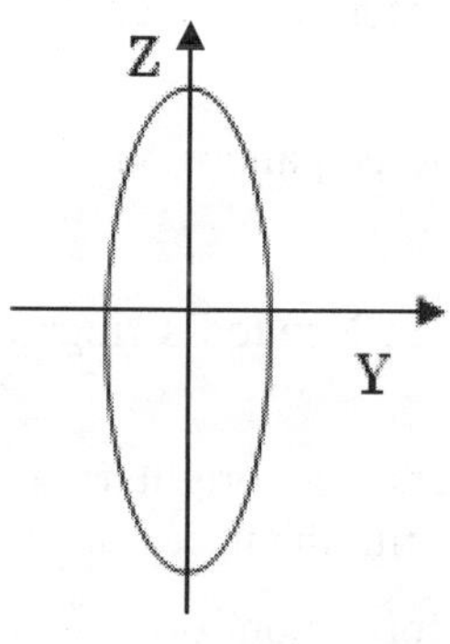

If a linearly polarised light falls at angles 0°,45°, 90° with respect to the optic axis on a calcite plate that has its optic axis parallel to its face. The emerging light is elliptically polarised. If the thickness of the crystal is such that a phase difference $\pi/2$, $5\pi/2$, $9\pi/2$…exists between o- and e- rays, then the emergent light is left handed elliptical light with its axes to parallel and normal to optic ais..

Questions

1. What is super position of waves? State the principle of super position of waves.
2. What is the significance of double-slit experiment for observation of interference of light?

3. Describe how interference fringes are formed in Young's double-slit experiment?
4. Obtain the conditions for occurrence of bright fringes and dark fringes on a screen by means of Young's double-slit experiment.
5. Mention important conditions for interference of light.
6. A monochromatic is used to obtain interference pattern on a screen in a double slit experiment. Obtain the expression for the spacing between two consecutive maxima and minima.
7. What happens when white light is used as source of light in double-slit experiments. What is the significance of using a monochromatic light source for obtaining interference pattern.
8. What is 'coherence' of light sources? Write briefly about (a) temporal coherence and (b) spatial coherence.
9. Discuss the importance of coherence in interference phenomena.
10. Discuss the conditions for brightness and darkness when a monochromatic light falls on a thin film and observation is made from the top.
11. Prove that in a thin transparent film, the interference pattern obtained with transmitted light is complementary to that in the reflected light.
12. Discuss the formation of interference pattern by a wedge formed between two glass plates.
13. Explain the formation of coloured pattern on oil slick on the road on a rainy day.
14. What are Newton's rings?
15. Why the central spot is dark in Newton's rings observed in reflected monochromatic light?
16. Discuss the formation of dark and bright rings in reflected monocromatie light in Newton's rings arrangement with the necessary mathematical equations.
17. Obtain the conditions for formation of bright and dark rings in Newton's ring experimental set up. What is the relation between the diameter of the rings and the number of the ring from the center.
18. Explain why the interference fringes formed in air wedge are straight and parallel.
19. Does the diameter of Newton's rings depend on the radius of curvature of the convex lens? Give reasons for your answer.
20. Distinguish between diffraction and interference
21. What are the differences between Fresnel diffraction and Frauhoffer diffraction.
22. Discuss in detail the Fraunhoffer diffraction due to a single slit illuminated by a monochromatic light.

23. Discuss the relation between the slit width and Fraunhoffer the diffraction pattern obtained when monochromatic light illuminates a single slit.

24. Discuss the intensity distribution in a single slit – Fraunhoffer diffraction.

25. Discuss the relation between diffraction pattern produced by a single slit and
    (a) the slit width
    (b) the change in the wave length of light

26. Discuss the Fraunhoffer diffraction pattern produced by a double-slit.

27. Explain the intensity distribution of light in the Fraunhoffer diffraction pattern formed by a double-slit.

28. Compare the Fraunhoffer diffraction pattern formed by a single slit and a double-slit having the same slit width.

29. What is the significance of the separation of the distance between two slit in Fraunhoffer diffraction with double-slit arrangement.

30. Discuss the conditions for the formation diffraction pattern in a double-slit arrangement.

31. Keeping the same slit width and distance of separation, discuss the formation of diffraction pattern with increased number of slits.

32. Discuss the relation between the increase in the number of slits per cm and the bright fringe appearance from plane transmission grating.

33. Discuss the formation of diffraction orders by a plane transmission grating when illuminated by parallel beam of monochromatic light.

34. Discuss the theory of a plane diffraction grating (transmission) giving diffraction pattern.

35. Obtain the expression for dispersive power of a grating. How is a grating different from a prism as far as dispersion is concerned.

36. Obtain the relation for resolving power of a diffraction grating. Discuss the relation between
    (a) The number of lines per cm in the grating
    (b) Order number and the resolving power of a grating

37. Discuss the number of lines per cm and dispersive power (Give the formula for dispersive power of grating).

38. Discuss the effect of increase of rulings per cm on a grating on
    (a) principal maxima (b) secondary maxima
    (c) dispersive power (c) width of principal maxima

39. What is polarisation? What are the three states of polarization.
40. Explain the phenomenon of 'Double refraction' – Distinguish between e – ray and o – ray. Mention some examples.
41. Describe what is 'circular polarisation'. What is a quarter wave plate?
42. How do you produce a circularly polarised beam of light using a quarter-wave plate?
43. What is elliptically polarised light? What are its features.
44. Explain the functioning of a half-wave plate and quarter-wave plate.
45. What is 'Brewster's law'. What is its use.
46. Explain the 'Law of Malus'.
47. Two narrow slits separated by a distance of 0.017 mm are illuminated with light of wave length 589 nm. What is the angular position of the first lateral maximum from the centre if the distance between the screen and the slits is 2.0 m. What is the distance of separation between the maxima?
48. Explain the formation of Newton's rings when a plane-convex lens is kept on a smooth glass plate and illuminated by sodium light normally.
49. What is the relation between the radius of curvature of the plane-convex lens and the diameter of the bright ring in Newton's rings – set up.
50. A monochromatic light passes through a single slit of width of 0.21 mm and falls on a screen 3.51 mm away. The first dark fringe appears 9.82 mm away from the center of the bright hand. What is the wave length of light?
51. In a double-slit arrangement, thc first order is formed at a distance x from the central bright line. The distance between the slits is 0.28 mm and the screen is about 1.5 m away from the double-slit. The wavelength of light is 632.8 nm. What is the value of x?
52. In a double-slit arrangement, the slits are illuminated by a light of wave length in 589 nm. The first maximum has an intensity of 75% of the central maximum. If the slits are separated by a distance 0.0024 cm, find the portion of the first maximum's centre from the center of the central maximum.
53. Two glass plates form a wedge when a wire of diameter 0.05 mm is kept at one end. When light from sodium lamp (589 nm) falls on the wedge normally, a number of interference fringes are formed over a wedge length of 12 cm. What is the number of fringes formed?
54. The distance between the slits and the biprism and between the biprism and the screen are 50 cm each. The obtuse angle of biprism is 179° and its refractive index is 1.5. If the width of the fringes is 0.0135 cm, calculate the wavelength of light.

55. Given the light of wavelength 5100 A° from a narrow slit is incident on a double slit. If the separation of 10 fringes on a screen 200 cm away is 2.0 cm, find the double slit separation?

56. In a Young's double slit experiment, the angular width of a fringe formed on a distant screen is 0.1°. The wavelength of light used is 6000 A°. What is the spacing between the slits.

57. Two closely spaced spectral lines in a spectrum of sodium have a wavelength difference of 6 A. If these two spectral lines have to be separated (the wavelengths of the lines, 589 nm and 589.6 nm) clearly by a grating in the first order, what is the minimum number of slits that should be there in a plane grating?

58. How many interference maxima should be in the central maximum envelope for a double slit with d = 0.025 mm and a = 0.004 mm. How many interference maxima are present actually in the central envelope.

59. Explain the functioning of half-wave plate.

60. What are the applications of wave plates.

61. Discuss the difference between a quarter-wave plate and a half-wave plate.

62. A pair of slits are separated by distance of one micron. What is the angle for the interference maximum with m=1 for light with wavelength 400 nm and for red light with wave length 700 nm?

63. A slit of width 0.11mm is illuminated with a light of wave length 577 nm. A screen is placed 4 m beyond the slit and the diffraction pattern is made to focus on a screen. What is the width of the central maximum in the diffraction pattern?

64. A slit of width 0.25mm is illuminated with plane waves of wavelength 66 nm to give a Fraunhoffer diffraction pattern focused on a screen by a lens of focal length 35 cm at what place the first dark band is located from the center of the pattern? What is location of first secondary maximum from the center?

Objective Questions

I. Choose the correct choice

1. A monochromatic light of unknown wave length λ shines normally on a plano-convex lens and glass plate combination in Newton's ring set up. If D_n and D_m represent the diameters of n th and m th rings (m>n) respectively, then the wave length, λ, is given by relation (R=radius of curvature of lens)

(a) $\frac{D_n^2 - D_m^2}{4R}$ (b) $\frac{D_m^2 - D_n^2}{4(m-n)}$ (c) $\frac{D_m^2 - D_n^2}{4mR}$ (d) $\frac{D_m^2 - D_n^2}{4(m-n)R}$

2. The radius of dark Newton's rings formed by transmitted light in Newton's ring's experimental set up is given by (n –fringe number, R -the radius of curvature, λ- wave length)

(a) $\sqrt{n\lambda}$ (b) $\sqrt{nR}$

(c) $\sqrt{n\lambda R}$ (d) $\sqrt{\frac{n}{(2n-1)}\lambda R}$

3. The fringe width, β of fringes produced by a wedge formed by two glass plates inclined at an angle θ (refractive index of medium is μ, and the wedge is illuminated by λ) is

(a) $\frac{\lambda}{\mu\theta}$ (b) $\frac{\lambda}{2\mu}$

(c) $\frac{\lambda}{2\mu\theta}$ (d) $\frac{\lambda\mu}{\theta}$

4. In a double- slit diffraction pattern, the distance between two consecutive maxima of interference depends on

(a) Width of central maxima

(b) Slit width only

(c) The total width the opaque portion of slits

(d) Width of the opaque portion only.

5. If we look into a soap film on the vertical hoop, we see coloured horizontal fringes. As time progresses, the film increases at the bottom while decreasing at the top. Then the spacing between the fringes

(a) increases (b) remains the same

(c) becomes irregular (d) decreases.

6. A wave of sodium light is propagating along Z-direction, then the directions of electric field (**E**) and the magnetic vectors (**B**) are

(a) **E** in x-direction
B in z-direction

(b) **E** in z-direction
B in x-direction

(c) **E** in y-direction
B in z-direction or vice versa

(d) **E** in x-direction
B in y-direction or vice versa

7. An electromagnetic plane wave propagating in free space has the electric field given by

$\mathbf{E}(z, t) = (\mathbf{i}\, E_{ox} + \mathbf{j}\, E_{oy}) \sin (kz - wt)$

then, the wave is propagating

(a) + x-direction (b) – y-direction

(c) + z-direction (d) – z-direction

8. Light passes through a polariser whose transmission axis is aligned horizontally. It then passes through a second polariser whose axis is aligned vertically. Then, the light that comes out of second polariser is

(a) polarised at 45° to the original polarisation

(b) unpolarised

(c) 3 times less in intensity

(d) returned to original polarization

9. If a_1 and a_2 are the amplitudes of the waves coming from two closely spaced sources of light, then, the maximum intensity of bright interference fringes formed on a non-absorbing screen at a distance 1 m away from sources is

(a) $a_1^2 + a_2^2$ (b) $(a_1 + a_2)^2$

(c) $4(a_1 + a_2)^2$ (d) $(a_1 - a_2)^2$

10. A birefringent crystal has refractive indices μ_0 and μ_e for the rays. If the crystal is to function as a quarter-wave plate, the number of wavelengths each component as the waves travel within the crystal thickness must differ by

(a) 4 (b) 2

(c) $\frac{1}{4}$ (d) $\frac{1}{8}$

11. A light beam of wavelength λ_o falls on a quarter-wave plate, then, the wave plate should have a minimum thickness 't' to cause a change of 0.25 fraction of wavelength for each component of the two rays in it then 't' should be

(a) $\frac{\lambda_0}{4(\mu_0 - \mu_e)}$ (b) $\frac{\lambda_0^2}{4}$

(c) $\frac{4\lambda_0^2}{(\mu_0 - \mu_e)}$ (d) $2\frac{\lambda_0 \mu_e}{(\mu_0 - \mu_e)}$

II. Fill in the blanks

1. When linearly polarised light with its plane at 45° to the optic axis through a quarter-wave plate, ______________ polarized light may be produced.
2. If light is composed of two plane waves of equal amplitude but differing in phase by 90°, then the resulting light wave is said to be ___________.

3. If two plane waves of differing amplitude are related in phase by 90°, or if the relative phase is other than 90° then the resulting wave of light is said to be ________ polarized.
4. Half-wave plates can be used as continuously adjustable polarization ________.
5. When a monochromatic light of wavelength, 520 nm passes through a 0.2 mm wide slit onto a screen 3.5 m away, the first dark band in the pattern appears at X mm from the center of the bright band. What is the value of X?
6. The wide bright fringe in the middle of the Fraunhoffer diffraction pattern of a double slit is caused by diffraction of the light due to ____________ interference of light from ________ traveling the same distance to the screen.
7. A un polarized beam of light is incident on a glass plate at its 'Brewster's angle. Then the reflected beam of light is polarised at ____________ angle to the plane of incidence.
8. The colourful pattern on soap bubbles we see under day light are due to __________ of light.
9. When light encounters an entire array of identical, equally-spaced slits, called a diffraction grating, the bright fringes arise due to ______________ of the light waves from different slits, are found at the same angles at which bright fringes are found in the __________ of double slit diffraction pattern.
10. The transmitted intensity through two ideal polarisers oriented at an angle θ is given as proportional to ____________.
11. A grating has 4000 lines per cm. For light of wavelength, 589 nm, the maximum number of orders that can be seen is __________.
12. The intensity of central maximum of a single –slit diffraction pattern is ______________ times less than the intensity of central maximum of double –slit pattern.
13. In a double –slit interference experiment, the first minimum is found at an angle of 1.3×10^{-4} rad. When illuminated by light of 589 nm wavelength. When the light source is replaced by a source giving light of 632.8 nm, the second maximum now is placed at an angle of ____________ radians.
14. The resolving power of a grating is directly proportional to the product of __________ and __________.
15. There are two closely spaced wavelengths in an elemental spectrum are $\lambda_1 = 500.57$ nm and $\lambda_2 = 500.61$ nm. If these two wave lengths are to be separated clearly in the first order spectrum formed by a plane grating, the minimum number of lines needed in the diffraction grating is _______.

16. A quarter wave plate introduces _______ radians of phase difference between the two linearly polarised beams of light.
17. A quarter wave plate is used to produce _________ polarised beam of light.
18. From the point of view of double refraction, quartz is _______ crystal.
19. A diffraction grating having 600 lines per mm is used to give a diffraction pattern with a laser of wavelength 633 nm. If the laser beam falls on the grating normally, then the distance of first order on a screen placed parallel at a distance of 10 m to the grating is________.

CHAPTER 17

Photoconductors and Solar Cells

17.0 Photoconductors

17.1 Introduction

Photoconductivity is the electrical conductivity produced in semiconducting crystal caused by the light radiation incident on the crystal. It arises when the energy of the incident photon is higher than the energy gap, E_g, of the crystal; under this condition, free electron-hole pairs are produced in the crystal by the absorption of incident photon, and the electrons and holes serve as the carriers of electrical conductivity. The electrons are in the conduction band and holes in the valence band of the crystal. That is, the incident radiation excites the electrons in the valence band into conduction band, where they become mobile; and the holes thus created in the valence band also become mobile as a natural consequence. Photoconductors are good detectors of radiation.

17.2 Operation of Photo Conductor

A Photoconductor as shown in Fig. 17.1 is fabricated by placing ohmic contacts on to a semiconductor material like CdS, PbS, InSb, nickel doped germanium, phosphorous doped silicon, HdCdTe, InGaAs, InGaAs P, etc.

When such a material is illuminated, electron-hole pairs are generated and the electrical conductivity of the material increases. More is the photon flux incident, greater will be the increase in conductivity What the photon incidence on the crystal is to create excess carriers of charge in addition to the thermally created change carrier concentration. The incident photon with energy sufficient to excite an electron from valence band, or from the metastable levels in the bandgap. When a voltage is applied to the contacts, electrons and holes move in opposite direction, resulting in a photocurrent. Since the electron velocity is greater than that of the hole, the electron reaches the side of the photoconductor before the hole does. Continuity of current requires that another electron enter the device through the external circuit. It means that one photon can cause several electrons to move through the device until recombination occurs. Photoconductive detectors have the advantages of conductive gain and ease of fabrication, but lack the speed capabilities. The processes of intrinsic and extrinsic photoexcitation of carriers are shown in Fig. 17.2.

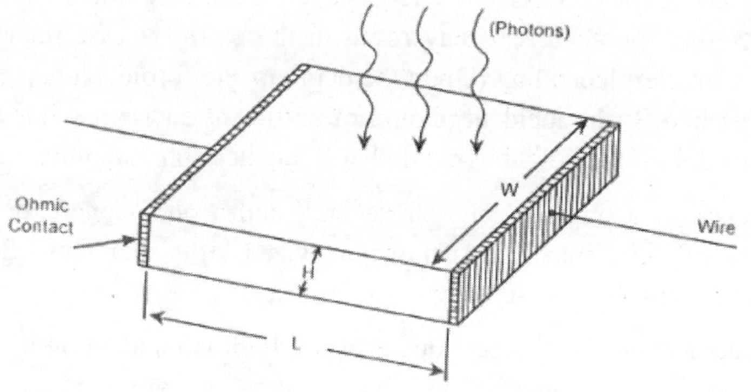

Fig. 17.1 Structure of a photoconductor with ohmic contacts

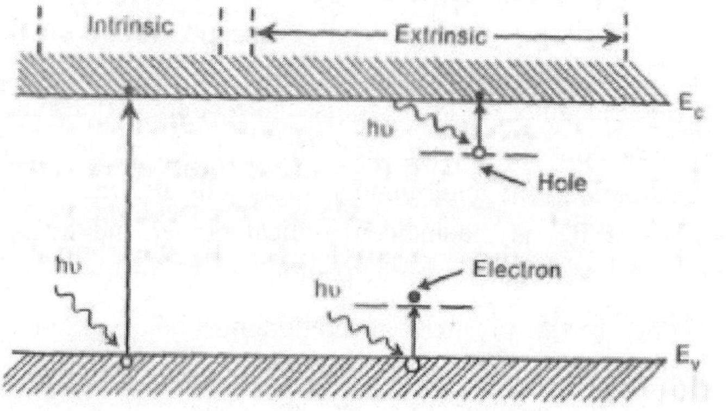

Fig. 17.2 Intrinsic (band-to-band) and extrinsic photo excitation processes.

For an intrinsic photo conductor, the conductivity is given by

$$\sigma = q(\mu_n n + \mu_p \rho) \quad .....(17.1)$$

where q → charge, μ_n, μ_p,→ mobility of electron and hole,

n, p → Concentration of electrons and holes

and the increase of conductivity under illumination is mainly due to the increase in the number of carriers. The long wavelength cut-off for this case is given by

$$\lambda_c = \frac{hc}{E_g} = \frac{1.24}{E_g(eV)}(\mu m) \qquad .....(17.2)$$

where λ_c is the wavelength corresponding to the semiconductor bandgap E_g. For wavelengths smaller than λ_c, the incident photon generates electron-hole pairs. For the extrinsic case, photoexcitation may occur between the energy level in the forbidden gap and either of the band edges. Therefore, photocurrent may result in the extrinsic case for energies equal to band gap or even smaller depending (Upon the depth of the forbidden-gap energy level. For example, at levels like B, the incident photon of sufficient energy excites the electron into the conduction band. Into the hole thus created at B, an electron may jump into it.

The performance of a photodetector in general, and a photoconductor in particular, is estimated interms of three important parameters viz, (i) the quantum efficiency or gain, (ii) the response time and (iii) the sensitivity (detectivity).

Consider the operation of a photoconductor when light is incident on it. Suppose at time zero, the number of carriers generated in a unit volume by a given photon flux is n_o. At a later time t, the number of carriers n(t) in the same volume decays by recombination as

$$n = n_o \exp\left(-\frac{t}{\tau}\right) \qquad .....(17.3)$$

where τ is the carrier life time. In other words, the recombination rate is $1/\tau$. Assuming a steady flow of photon flux impinging uniformly on the surface of a photoconductor (Fig. 9.1) with area A = WL, the total number of photons arriving at the surface is $(P_{opt}/h\nu)$ per unit volume, where P_{opt} is the incident optical power and $h\nu$ is the photon energy (h - Planck's constant, $\nu \rightarrow$ frequency of the photon).

In equilibrium state, the rate of carrier generation must be equal to the recombination rate. If the detector height H is much larger than the light penetration depth l/α, the total equilibrium state carrier generation rate per unit volume is given as (G is the carrier generation rate)

$$G = \frac{n}{\tau} = \frac{\eta(P_{opt}/h\nu)}{WLH} \qquad .....(17.4)$$

where η is the quantum efficiency and n is the number of carriers per unit volume (carrier density). Then the photocurrent between the electrodes is

$$I_p = (\sigma E)WH = (q\mu_n nE)WH = qnv_d WH \qquad .....(17.5)$$

where E is the electric field inside the photoconductor and v_d, is the carrier drift velocity (since $v_d = \mu_n E$). Substituting (17.3) in (17.5) one gets

$$I_p = q\left[\eta\frac{P_{opt}}{h\nu}\right]\left[\frac{\mu_n \tau E}{L}\right] \quad .....(17.6)$$

Now, we define the primary photocurrent as

$$I_{ph} \equiv q\left[\eta\frac{P_{opt}}{h\nu}\right] \quad .....(17.7)$$

then the photocurrent gain g, is

$$g \equiv \frac{I_p}{I_{ph}} = \frac{\mu_n \tau E}{L} = \frac{\tau}{t_r} \quad .....(17.8)$$

where $t_r \equiv L / v_d$, is the carrier transit time. The gain depends upon the ratio of carrier life time to the transit time. The carrier life time is the time that the charge either hole or electron spends before recombination. The recombination is a process in which an electron and hole recombine during the transit. The carrier transit time is the time which the change takes to reach the other electrode.

For a sample with long-life time and short electrode spacing, the gain can be substantially greater than unity. Gains as high on 10^6 have been obtained in some photoconductors. The response time of a photoconductor depends on transit time t_r. For obtaining shorter transit time, small electrode spacing and high electric fields are used. Typical response time of photoconductors lie in the range of 10^{-3} to 10^{-10}s. They are extensively used for detection of infrared radiation greater than a few microns.

17.3 Photodiode

Photodiode is a photosensitive semiconductor diode, relying on the photo voltaic effect. It has a depleted semiconductor region with high electric field, that serves to separate photo generated electron - hole pairs. The depletion region must be kept thin, to reduce the transit time for high speed operation. On the other hand, to increase the quantum efficiency (the number of electron-hole pairs generated per incident photon), the depletion layer must be sufficiently thick to allow a large fraction of the incident light to be absorbed. Thus, the speed of response and quantum efficiency require exactly opposite conditions to be satisfied.

The photo-diode can be operated in a photovoltaic mode, that is, the photodiode is unbiased or reverse biased and connected to a load impedance, similar to a solar cell. However, the device designs are fundamentally different. For a photodiode, only a narow wavelength range centred at the optical signal wavelength is important, where as for a solar cell, high spectral response over a broad solar wavelength range is required. To minimize the junction capacitance, photodiode junction areas are made small, but in solar cells, the surface is made large to absorb more light flux. In photodiode, the most important parameter is quantum efficiency, whereas in solar cells, it is power conversion efficiency. More details of solar cells would be dealt in the next chapter.

Internal mechanism of operation of a photodiode can be understood as follows:

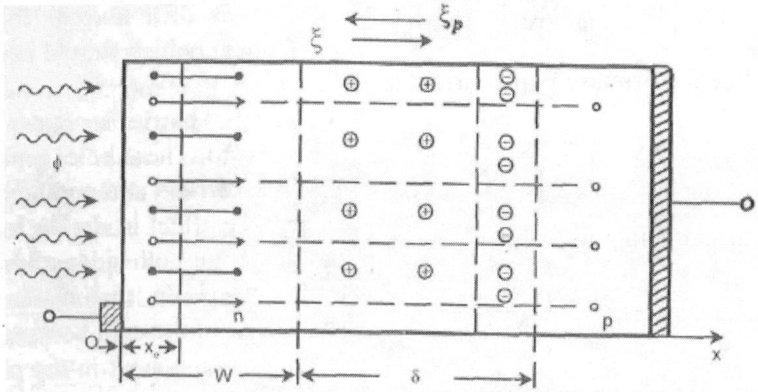

Fig. 17.3 Photodiode structure on exposure to radiation.

Let the radiation act in the direction normal to pn junction plane (Fig. 17.3). As the material absorbs photons of energy higher than the bandgap energy, electron-hole pairs (photo cariers) appear in the n-type base, at a depth of about x_o from the semiconductor surface. The photo carriers diffuse into the n-region. The width W, of the n-region, is such that a major fraction of photo carriers have no time to recombine and reach the pn junction boundary (x = w), where the electric field x_o seperates the electrons and holes. The holes move into the o-region, while the electrons cannot negotiate the junction field ξ_o and pile up at the pn junction boundary in the n-region. The photo carries through the pn junction thus results from the drift of minority carriers namely, holes.

The magnitude of photo current through photodiode depends on the level of illumination i.e., intensity of high. It is the optical control of the current through a photodiode that determines the specific operating conditions of this photo detector. To illustrate the point, we will compare the principle of normal pn junction diode with that of a photodiode.

In the equilibrium state when there is no illumination of the diode element, the photodiode should act as a normal diode, with a bandgap diagram as shown in Fig. 17.4(a). The Fermi level E_f is the same in both p and n regions. The diffusion current will be equal to drift current due to minority carriers across the junction.

In case of normal diode, the forward bias (voltage), i.e., positive potential to p and negative potential to n region, disturbs the equilibrium. Depending upon the voltage, diffusion current sets in due to lowering of internal barrier and increases with the increases in voltage, where as drift current through the junction is usually small and is a parasitic component, which should be avoided.

In a pn junction of a photodiode, the potential barrier separates the photo carriers generated by the incident radiation (Fig. 17.4(b)). These holes and electrons thus generated, move in the direction of minority carriers and combine with the ion cores reducing the internal potential barrier (the effect is similar to applying forward bias to normal pn diode). The Fermi levels on both sides of pn junction no longer coincide, and one becomes higher than the other. It is important to note that radiation flux disturbs the equilibrium in favour of the drift component. This is indicative of the need, to suppress the diffusion component in the photodiode to enable a more effective optical control of the photo current. The diffusion current is parasitic in the photo diode and should be avoided. Therefore photodiode is operated without bias or in reverse bias.

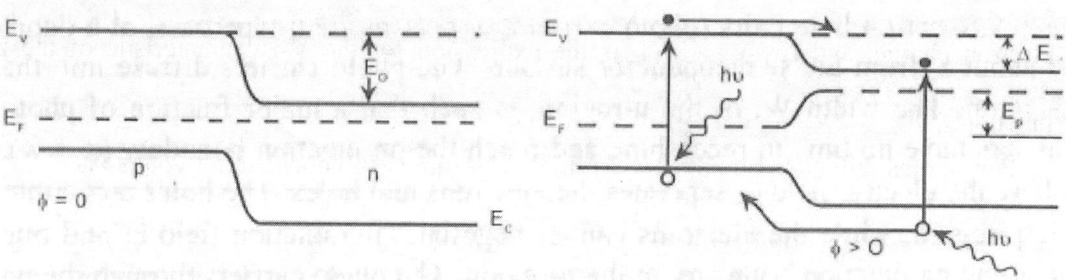

Fig. 17.4 Internal photoeffect in a pn junction
(a) pn junction in equilibrium state (b) pn junction on exposure to radiation

The drift flow of photo carriers produces a current called the photo current I_p. The holes charge the p-region positively with respect to n-region and the electrons charge the n-region negatively with respect to p-region. The resultant potential difference referred to as the photo electromotive force E_p, lowers the internal potential barrier ΔE, Fig. 17.4.

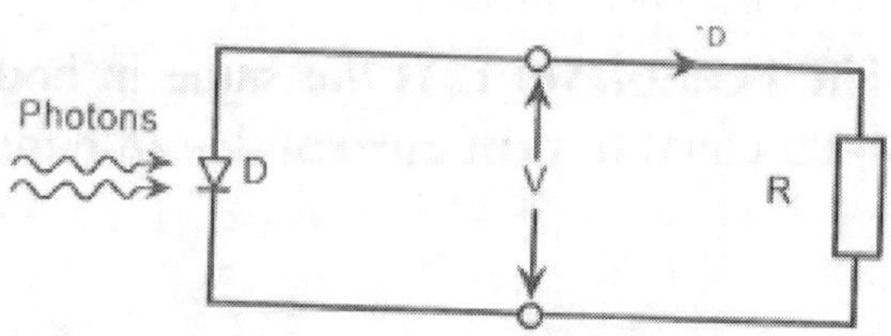

Fig. 17.5 Photodiode operating in PV mode.

17.3.1 Modes of Operations

For the photodiode to be made more responsive to the incident radiation, the diffusion component of current must be kept at a minimum. To achieve this, the photodiode is made to operate either without a bias at all (in the photovoltaic mode) or with a reverse bias (in the photoconductive mode).

In the photovoltaic (PV) mode of operation, the photodiode thus needs no external voltage source; the diode acts as a photocurrent generator Fig. 17.5. The expression for I_D, through the photodiode D can be written as

$$I_D = \frac{V}{R} - I_p - I_{pn} = I_p(\phi) - I_o\left[\exp\left(\frac{V}{\phi_T} - 1\right)\right] \qquad \text{.....(17.9)}$$

where I_p, is the current of photo carriers (photocurrent); R is the load resistance; I_{pm} is the current through the pn junction; V is the voltage across the diode; I_o is the thermal current through the pn junction; $\phi_T = \frac{kT}{e}$ is the thermal potential; k is the Boltzmann constant; e the electron charge; and $I_o\exp\left(\frac{V}{\phi_T} - 1\right)$ is the expression for the V-I characteristic of the pn junction.

For on open-circuited output (R = ∞), I_p, = I_{pn}. From (17.9) open circuit voltage is

$$V_{oc} = \phi_T \log_e\left(1 + \frac{I_p}{I_o}\right)$$

For R = 0, short circuit, $I_{sc} = I_p$.

In the photoconductive (PC) mode, the photodiode operates from a reverse bias source V_{SR} placed in series with the diode (Fig. 17.6).

If the mode potential barrier gets higher, so the pn junction current I_{pm} will be governed by the thermal current I_o which flows through the diode in the absence of illumination. The current through the illuminated photodiode is

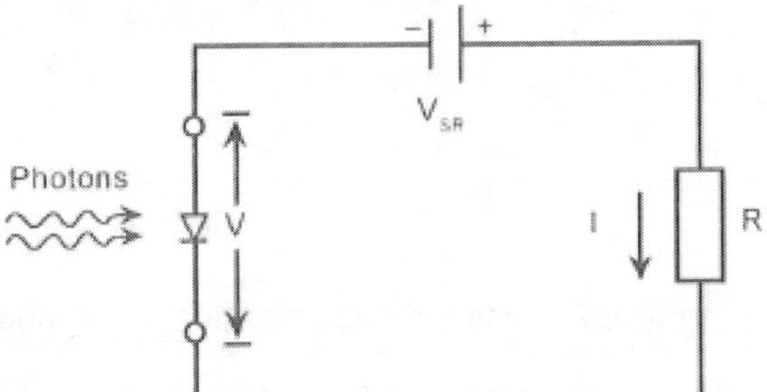

Fig. 17.6 Photodiode operating in PC mode.

17.3.2 Voltage – current Characteristics

In the general case, the current of a photodiode is defined by the expression (17.9). This expression gives the current I_D, as a function of voltage V across the diode with radiant flux

ϕ as a parameter, i.e., it is the equation for the family of V-I characteristics illustrated in Fig. 17.7.

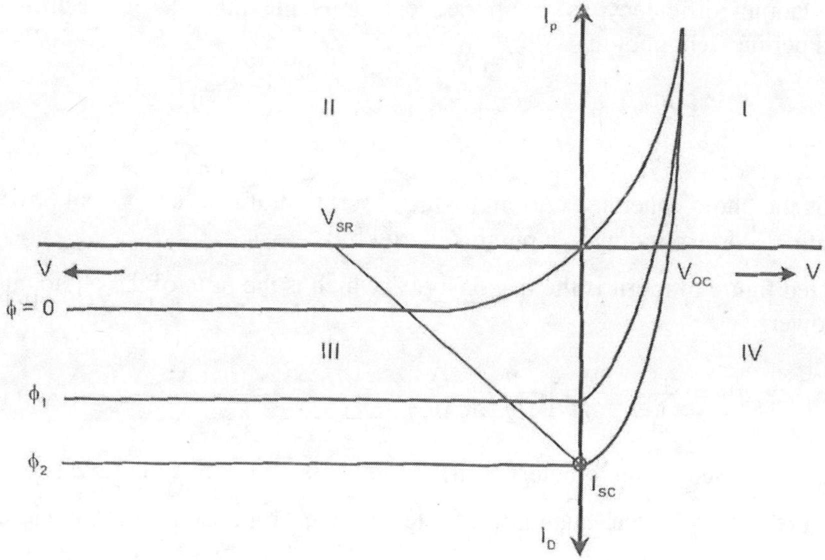

Fig. 17.7 V-I characteristics of photodiode.

The V-I characteristics for a photodiode lie in quadrants I, III, and IV. Quadrant I is not an operating region for the diode; the forward bias on the pn junction gives rise to the diffusion current, which totally swamps the photo current, since I_{pn}>> I_p. Therefore, optical control of the current through the diode is thus impossible.

Quadrant III is the photoconductive (PC) region of the diode operating with reverse bias. It should be borne in mind that in the working range of reverse voltages, the photocurrent is practically independent of the reverse voltage and

load resistance. The V-I curve for the load resister R becomes a straight line, whose equation has the from

$$V_{SR} - I_p R = V \qquad .....(17.12)$$

where V_{SR} is the reverse voltage of the source; V is the reverse voltage across the diode; and I_p is the photocurrent. The value of I_p is practically independent of the external circuit parameters such as V_{SR} and R(1oad). Quadrant IV for the family of V-I curves describes the operation of the photodiode in the PV mode. The points of intersection of V-I curves with the voltage axis correspond to the values of the photo-emf or open-circuit voltage at different levels of illumination. In silicon photodiodes, a typical value of the photo-emf is 0.50-0.55v. Given the value of current I, the V-I characteristics can be used to select the optimum operating conditions for the diode as a photo voltaic device. The optimum operation here implies, that the diode delivers the highest electric power to the load of the adequately selected resistance.

We shall now briefly consider the general characteristics of a photodiode: its quantum efficiency, response speed and device noise.

The quantum efficiency as mentioned earlier is the number of electron - hole pairs generated per incident photon:

$$\eta = \left(\frac{I_p / q}{P_{opt/h\nu}} \right) \quad .....(17.13)$$

where I_p is the photo generated current by the absorption of incident optical power P_{opt} at a wavelength λ (corresponding to a photon energy $h\nu$).

A related figure of merit is the responsivity, which is the ratio of the photo current to the optical power:

$$R = \frac{I_p}{P_{opt}} = \frac{\eta q}{h\nu} = \left(\frac{\eta\lambda}{1.24} \right) \left(\frac{A}{W} \right) \quad .....(17.14)$$

where λ is expressed in micrometers (μm). R is expressed in units of (AJW).

Therefore, for a given quantum efficiency, the responsivity increases linearly with wavelength. For an ideal photodiode $(\eta = 1), R = \frac{\lambda}{1.24} \cdot \left(\frac{A}{W} \right)$, where λ is expressed in microns. One of the key factors that determines the quantum efficiency is the absorption coefficient (α), which is again a strong function of wavelength.

The response speed is limited by a combination of three factors: diffusion of carriers, drift time in the depletion region, and capacitance of the depletion region. Carriers generated outside the depletion region must diffuse to the junction resulting in considerable time delay. To minimize the diffusion effect, the junction should be formed very close to the surface. Most light will be absorbed when the depletion region is sufficiently wide (approximately $1/\alpha$); with sufficient reverse bias, the carriers will drift at their saturation velocities. The depletion layer must not be too wide, otherwise transit-time effects will limit the frequency response. It also should not be too thin, or excessive capacitance C will result in large RC time constant, where R is the load resistance. The optimum compromise occurs when the depletion layer is chosen, so that, the transit time is of the order of one-half the modulation period. In Si for modulation frequency of 10GHZ, the optimum thickness is 5μm. Another important factor that effects the performance of a photodiode is the noise that is associated with incident optical signal.

17.4 Solar Cells

Introduction: Solar cells or photo voltaic cells convert light energy into electrical energy. Solar cells are used to power calculators, watches, irrigation systems, artificial satellites etc., Solar cells offer clean, pollution free and simple alternative sources of energy. Over 95% of solar cells produced are silicon based. Single crystal Si, polysilicon, CdTe and GaAs etc., are some of the materials used in solar cells. The recent technological innovations make solar cells economically feasible for large-scale use. Various solar cell geometries used are flat-planed systems, thin film devices, concentration systems etc.

Radiation from Sun: Solar radiation is a result of thermonuclear fusion reaction inside Sun. The spectrum of solar radiation resembles black body radiation at 5800°K. The energy emitted by Sun is about 5.26×10^{26} Joules/year. Every second about 6×10^{11} kg of H_2, is converted into the He with a net loss of 4×10^3 kg mass which is converted into light (Einstein's $E = mc^2$ relation). It is projected that at this rate Sun can provide radiation for over 10^{10} years.

The intensity of solar radiation in free space at the average distance of the earth from Sun is 1353 w/mt^2 and is called solar constant. Due to various absorption mechanisms in the atmosphere of the earth, the sun light received on the earth is less. Water vapour absorbs infrared radiation where as Ozone layer of the atmosphere cuts off ultra violet radiation of solar spectrum. Air borne dust and air particles scatter the radiation. The degree to which atmosphere affects the sun light received by earth surface (I) is measured by the air mass. The secant of the angle between Sun and zenith is called air mass and measures the path length of the Sun light in atmosphere. Table 17.1 summaries various air masses.

The energy output from the Sun is primarily electromagnetic radiation that has wavelength range 200-3000nm. The radiation energy decreases as it reaches earth's surface depending on angle of incidence and atmospheric conditions. AMO refers to solar radiation received in outer atmosphere, AM1 refers to radiation received when zenith is zero, AM 1.5 is when Sun is at 45° above horizon.

Table 17.1

Air mass	Zenith (angle)	Intensity w/m^2
AM0	Outer atmosphere	1,353
AM1	0	925
AM1.5	45	844
AM2	60	691

Intensity of radiation received can be written as

$$I = 1353(0.7)^{\sec\alpha}\ \mathrm{w/mt^s} \qquad \text{.....(17.15)}$$

where α is the angle.

17.5 Working Principle of Solar Cells

Broadly there are four types of solar cells, they are

1. p-n homo junction
2. p-n hetero junction
3. Schottky barrier

4. Semiconductor electrolyte (Photoelectron chemical)

p-n junction solar cell has attracted substantial effort, hence we discuss this type in detail.

A photon of wavelength λ when incident on a semiconductor material with band gap Eg will be absorbed if E = hc/λ > Eg. This absorbed energy excites valence electrons to conduction band. (E – Eg) energy will be lost as heat (phonons). Photons of energy E < Eg are not absorbed, they are transmitted. Thus the rate of carrier generation is a function of wave length and intensity of incident radiation.

The generation charge carries recombine after a short interval of time, called life time rate of recombination r and is given by the relation

$$r = \frac{n - n_o}{\tau} = \frac{p - p_o}{\tau} (cm^{-3}\ sec^{-1}) \quad .....(17.16)$$

where n_0 and p_0 are equilibrium concentration of electrons and holes, n and p are concentrations at any given instant of time. The recombination processes can be any one among the processes already discussed in chapter 6.

A fraction of generated minority charges diffuse to the edges of the junction depletion region and are swept across the junction due to junction potential and consequent electric field there. Thus holes are collected on the p-side and electrons on n side.

This collection gives rise to light induced current (Photocurrent) through load resistance connected across the cell. (Fig. 10.1(a)). Let F_{ph} be the photon flux (number of photons incident per unit area per sec). If 100% quantum efficiency is assumed, then each absorbed photon generates one electron-hole pair, the pair generation rate G is given by

$$G = \alpha F_{Ph}(1 - R)\exp(-\alpha x) \quad .....(17.17)$$

where $\alpha\ cm^{-1}$ is absorption coefficient.

R is the reflection coefficient.

x is the distance where G is measured

Under steady state condition, continuity equation for holes on n side can be written as

$$\frac{1}{e}\frac{dj_p}{dx} + \frac{p_e}{\tau_p} - G = 0 \quad .....(17.18)$$

$J = \text{current density} = -eD_p(dp_e / dx)$

D_p = hole diffusion coefficient

P_c = hole concentration on n side

Hence, eqn. (17.18) can be written as

$$D_p \frac{d^2 p_e}{dx^2} + \frac{p_e}{\tau_p} - G = 0$$

Similar expression can be arrived at for electrons on p side. Solving these expressions and adding them together gives current passing through cell and is given by:

$$I = I_s(\exp(V / V_T) - 1) - I_L \qquad \text{.....(17.19)}$$

V is the bias voltage applied for collection of charges generated due to light, I_s is the reverse saturation current, I_L is the light generated current that passes through load. The current I_L is given by:

$$I_L = eAG(L_p + L_n) \qquad \text{.....(17.20)}$$

In equation (17.20), A is the cross-sectional area of the junction. L_p and L_n are the diffusion lengths of minority carries pn p-side and n-side respectively. If the equation for current I is plotted against V, we get the graph shown in the Fig. 17.8. The different curves obtained are for different G values. The total current (I) is the difference between the injection current and photon induced current.

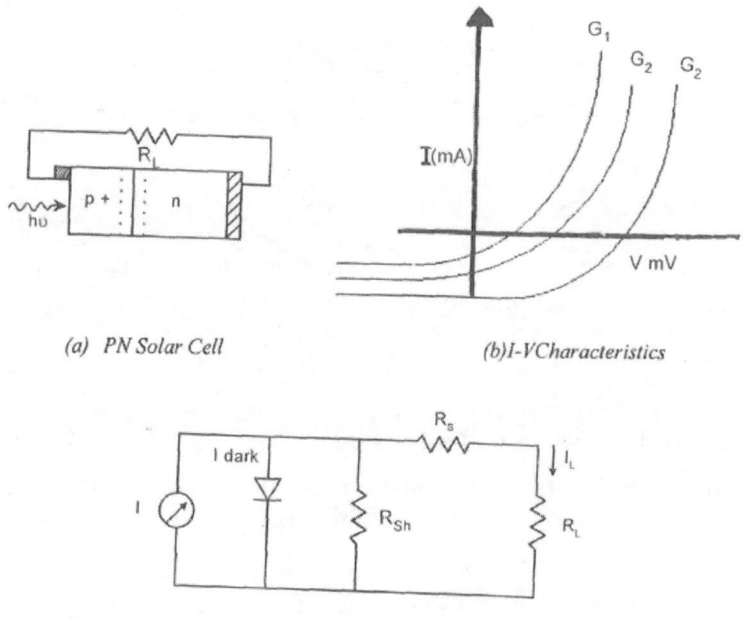

(a) PN Solar Cell

(b)I-VCharacteristics

(c) Equivalent Circuit

Fig. 17.8 Solar cell. Its I-V characteristics, and equivalent circuit.

The equivalent circuit of the solar cell is shown in Fig. 17.8(c) under illuminated condition. R_s, the series resistance is due to the bulk resistance of n and p regions and sue to the contacts. R_{sh}, the shunt resistance can be attributed to the leakage current. Photo current is the current flowing through the load R_L. Heavy doping of the top layer of solar cell and deep junction Fig. 17.8 can reduce R_s.

17.6 Solar Cell Characteristics

Short circuit current (I_{SC}), open circuit voltage (V_{OC}) are the important solar cell parameters. Short circuit current is the maximum current that exists under short circuit condition i.e., when the load resistance $R_L = 0$.

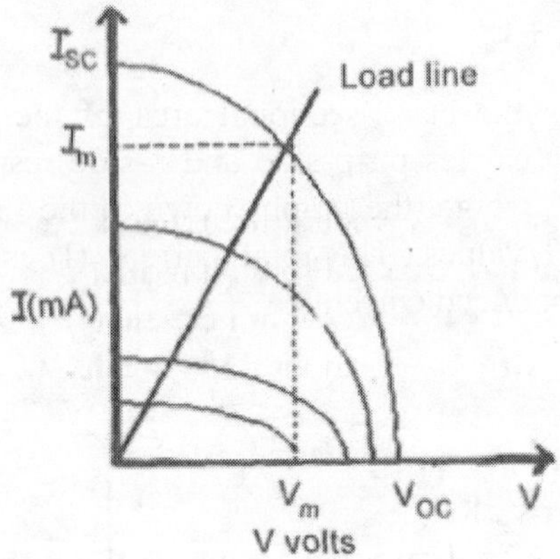

Fig. 17.9

The voltage developed when $R_L = \infty$ is the open circuit voltage. If I_o and V_o are the current through the solar cell and voltage across the cell respectively then the power (P) delivered by the cell is

$$P = I_o V_o \qquad \text{.....(17.21)}$$

The ratio of P with I_{SC} V_{OC} is defined as fill factor.

When the load is shorted ($R_L = 0$), voltage across the load resistance V_L doesn't exist, that means it can be taken as zero. The short circuit current I_{SC} is related to open circuit voltage as

$$I_{sc} = I_s(\exp(V_{oc} / V_T) - 1) \qquad \text{.....(17.22)}$$

where VT represented thermal energy equivalent voltage. Solving for V_{oc}, we get:

$$V_{oc} = V_T \ln[(I_{sc} / I_s) + 1] \qquad \text{.....(17.23)}$$

The above equation gives the relation between open circuit voltage and short circuit current. If reverse saturation current I_s is small, V_{oc} is large and vice-versa. V_{oc}, and I_{sc}, are determined for a given light level or intensity by the cell properties. Fig.17.9 illustrates these characteristics. The power is given by the product VI, where V is the voltage and I is the current flowing. The maximum power delivery is possible when V and I attain their maximum values, namely, V_m & I_m which are shown in the Fig.17.9 by the dashed line. This maximum power output is less than V_{sc} I_{oc}. Conversion efficiency of solar cell is defined as

$$\eta = \frac{\text{Power delivered}}{\text{Solar power incident}} = \frac{VI}{\text{Solar power incident}}$$

$$= \frac{V_{oc} I_{sc}.FF}{\text{Solar power}} \quad .....(17.24)$$

The fill factor (FF) or the figure of merit of the solar cell is the ratio of the maximum power that can be delivered to the product I_{sc} E_{oc}. This is an important characteristic of a solar cell. The maximum efficiency is attained as on today in any semi conducting solar cell is about 18% while the maximum theoretical efficiency is 25%.

17.7 pn Junction Solar Cell

The efficiency depends on the doping of the p-region and it should be such that electrons generated in this region can diffuse to the junction before the recombination takes place. This implies a proper match between the diffusion length of the electrons L_n, and thickness of the p-region, and the mean optical penetration depth. It is desirable to have a large contact potential V_o, to obtain a large photo voltage. This can be achieved by heavy doping. On the other hand, longer life times are needed for the charge carriers, but heavy doping reduces lifetime of electrons and holes generated. It is also important that the series resistance (R_s) of the solar cell should be small so that the power is not lost as heat due to Joule heat in the device itself. This is the major problem in solar cells, especially in the thin region at the surface. Therefore the surface is partially contacted. A series resistance of few ohms can reduce the output voltage to appreciable levels. The area of the p-side of the diode is kept large so that the resistance of p-body of the diode of solar cell is kept low. The n-region is normally thin. This region is contacted at the edge and current should be along the thin n-region to the contact. This results in large series resistance. To prevent this, contact fingers are distributed over the surface of the n-region. There are many compromises that have to be made in a solar cell design. Much of the absorbed. The high-energy photons are absorbed, but the electron-hole pairs may recombine at the surface itself. Therefore, the output from a solar cell is limited. A well-prepared Si solar cell is supposed to give an efficiency of 10% for solar energy conversion and this provides an output $100W/m^2$ of electrical power due to full illumination. From the point of view of the cost this output is extremely small. Beside the solar energy absorbed can increase the temperature of the cell thereby generating non-equilibrium concentration of charge carriers. GaAs and related compound with direct band gaps can be used up to 100 °C. The hetro junctions are found to be very useful for good conversion efficiency and good for operating temperatures with reasonable efficiency.

The configuration of solar cell is shown in the Fig. 17.10. The surface of the solar cell is coated with anti reflection coating to reduce the surface recombination. The maximum output power is obtained by keeping large area junction located near the surface of the device. The thickness 'd' of the n-region must be less than the diffusion length L_p in the n material.

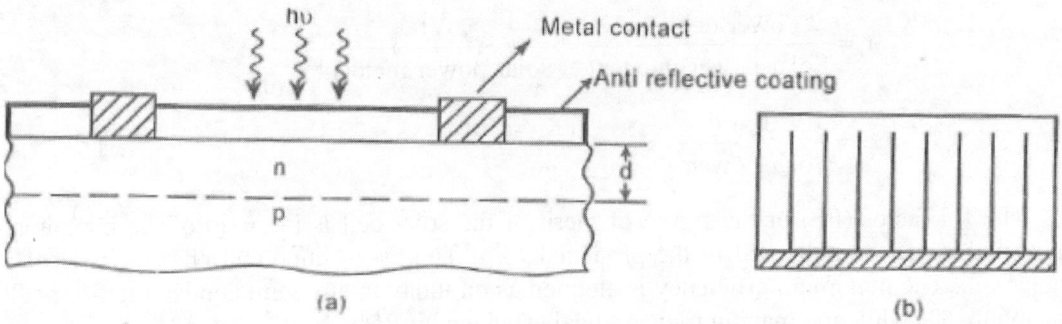

Fig. 17.10 Construction of a solar cell (a) Side view (b) top view of the contact "fingers".

Examples

Example 1

Example 1

V_{oc}, I_{sc} and fill factor of a solar cell are 600 mV, 160 mA and 81.4%. Find out the power generated by the solar cell and if incident solar power is 840 W find its conversion efficiency, the area of the solar cell is one unit.

Sol: Power = open circuit voltage V_{oc} × short circuit current I_{sc} × fill factor FF

$= 0.600 \times 160 \times 0.814 = 78.14$ mW

Percentage of efficiency = (Power/Power incident) × 100

$= (78.14 \text{ mW}/840 \text{ W}) \times 100 = 9.3\%.$

Example 2

A solar cell has a reverse saturation current I_s of 2 nA. Its short circuit current is 0.32 mA. Find the open circuit voltage.

Sol: From equation, $V_{oc} = V_T \ln((I_{sc}/I_s) + 1)$

Where V_T = temperature equivalent of voltage = kT = 0.026volts.

$V_{oc} = 0.026 \ln((0.32 \times 10^{-3}/2 \times 10^{-9}) + 1) = 0.31\text{V}$

Example 3

Injection (dark) current of a solar cell with unit cross-section area is 0.5 nA and short circuit current is 100 mA. Find its efficiency if AM0 radiation is incident and the fill factor is unity.

Sol: $V_{oc} = V_T \ln((I_{sc}/I_s) + 1) = 0.026 \ln((100 \times 10^{-3}/0.5 \times 10^{-9}) + 1)$

$= 496$ mV

Efficiency = I_{sc} V_{oc} × fill factor/power incident

= 496 mV × 100 mA × 1/0.135 W = 0.3674

Hence efficiency = 36.74%.

Questions

1. For a uniformly doped pn-junction solar cell, draw the energy band diagram under dark and illuminated conditions.
2. Discuss various aspects of solar radiation. Define air mass. How is it affected due to earth's atmosphere?
3. Obtain an expression for efficiency of pn-junction solar cell.
4. Describe the equivalent circuit of a solar cell.
5. What are the characteristics of solar cells.
6. Explain the working of a solar cells.
7. Define fill factor and efficiency of solar cell and explain the factors that affect them.
8. What is dark current? How does this depend on the solar cell material characteristics?
9. A solar cell has a saturation dark current of 5 nA. When illuminated, it has short circuit current of 200 mA. Draw 1-V characteristics for the solar cell.
10. A solar cell 3 cm × 3 cm area has I_{th} as 32Ma. When illuminated, concentration of electron hole pairs generated are $10^{20}/cm^3S$ with in $L_p = L_n = 2$ μm of the junction. The depletion layer thickness is 1 μm. Compute I_{sc} and V_{oc}, (the area of the cell is 2.5 cm × 2.5 cm) short circuit current and open circuit voltage.

Objective Questions

I. Choose the correct choice

1. The value of Solar constant is ________

(a) 1.4 M Wm^{-2} (b) 0.14 KWm^{-2}

(c) 1.4 Wm^{-2} (d) 140 Wm^{-2}

2. In a solar panel the metal that is used is
   (a) steel (b) nickel
   (c) copper (d) silver
3. The device that is coated on rollers of photo copiers and laser printers is
   (a) piezo-photo resistors
   (b) photo diode
   (c) photo conductors
   (d) photo emitting powder
4. When a solar panel is rotated so that light falls on the solar cells at an angle θ to the panel, the power delivered by the solar cells
   (a) increases (b) decreases
   (c) decreases as square of cosine θ (d) remains unaltered.

II. Fill in the blanks

1. When a light beam is incident on a photodiode, the current that increases is __________.
2. The accumulation of electrons and holes on the two sides of the junction in a solar cell give rise to __________ voltage.
3. The maximum output from a solar cell is obtained when the cell is operated at the __________ of the output current-out put voltage curve.
4. A photo conductor has its resistance very____ when it is not exposed to light. (High) Compared to a photoconductor, a solar photo diode or solar cell possesses ___response time.
5. Compared to a photo conductor, a photo diode or solar cell possesses a __________ response time.
6. Photo voltaic cells are useful as _______ energy trappers.
7. A solar cell is a _________ terminal optoelectronic device.
8. A silicon solar cell with an exposed area of 75 cm^2 provides nearly ________volts.
9. Solar cells are basically Photo-_________ cells.
10. Selenium solar cells supply ________ output power than silicon cells.

CHAPTER 18

Superconductivity

18.1 Introduction

The discovery of superconductivity was an off shoot of the results of the study of the properties of gases at low temperatures. During the second half of the nineteenth century, many scientists were developing methods to condense gases with lower and lower boiling points. One of thc outstanding methods was Joule-Thomson porous plug experiment which showed that gases when pumped through a porous plug and made to expand, they suffer a fall in temperature (cooling). All real gases suffer this coo!ing. Carbondioxide was first frozen by M.Faraday in 1845. Raoul Pictet and Paul Cailletet were successful in liquefying air in 1877. James Dewar liquefied hydrogen in 1898 and he invented Dewar flask, to keep the liquids safe and cool. All the gases have to be cooled first below their inversion temperatures before subjecting them to further cooling in porous plug technique. For helium and hydrogen, the inversion temperatures are much below room temperature (He, 23.6 K; H_2, 196 K). Hydrogen can be precooled by liquid nitrogen (boiling point 77.3 K at one atmospheric pressure) and helium can be precooled by liquid hydrogen (boiling point 20.4 K at one atmospheric pressure). For a long time the liquefaction of helium posed a serious problem till Kammerlingh Onnes in 1908 could liquefy it applying thc method of Joule-Thomson expansion. Liquid helium below 4.22 K found a number of applications. The thermal, electrical and magnetic properties of a number of substances were determined at low temperatures and they exhibited property–parameters which could be explained only on the basis of quantum physics thereby paving the way for strong roots of quantum mechanics in physics. One of them was the discovery of superconductivity in a number of metals. The discovery of superconductivity was one of the outstanding discoveries made in the twentieth century and it heralded the dawn of a new era in low temperature physics and engineering.

18.2 The Properties of Superconductors

18.2.1 Electrical Resistivity

Matthiesen predicted that resistivity of pure metal would level off as absolute temperature approached zero. At higher temperature the resistivity of metal would be determined by lattice vibrations. The vibrations would be responsible for scattering of electrons resulting in resistance. Kammerlingh Onnes investigated the electrical resistance of metals at low temperatures. This was a hot subject in physics in those days. Dewar predicted that the resistivity of a metal would approach zero at absolute temperature. The experiments conducted by Onnes on metals like gold, showed that the resistivity of pure metals would decrease with the decrease of temperature (below 10 K) and at very low temperatures, would tend to a constant value determined by residual impurities or other defects in the metallic sample. However Kammerlingh Onnes who achieved the liquefaction of helium and investigated the electrical resistivity of pure mercury below 10 K, found that the electrical resistivity of pure mercury dropped abruptly to zero at around 4.2 K. The mercury was incorporated into a 'U' shaped glass capillary tube with electrodes on both ends so that the current could be measured through it while still in liquid. Finally mercury was cooled into a solid wire. Onnes found a regular decrease in resistance with decrease of temperature and at liquid helium temperature, the resistance appeared to have completely vanished. (Fig. 18.1). The result was checked and rechecked, and the discovery was finally announced.

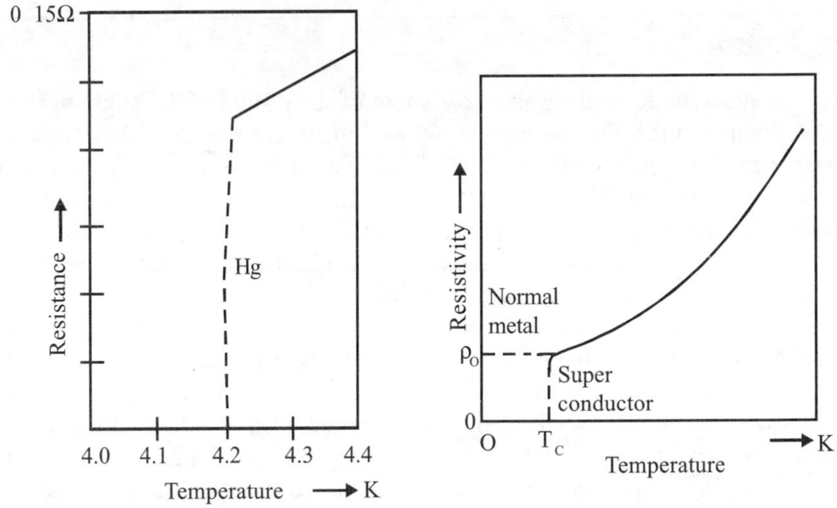

Fig. 18.1 (a) and (b) The variation of resistivity (a) for mercury (b) comparison between pure metal and a superconductor.

From his observation, Onnes concluded, "mercury has passed into a new state which on account of its extraordinary electrical properties may be called the superconducting state." The temperature at which the metal had passed into the superconductivity is known as the superconducting transition temperature (T_c). H.Kammerlingh Onnes received the 1913 Nobel prize in physics for this great discovery. Onnes in 1914 performed a similar experiment on a coil of lead and established a 'persistent current' in it below 7.2 K. This discovery of superconductivity opened up tremendous curiosity about the scope and nature of the phenomenon. It was subsequently found by several scientists that twenty-one elements and a large number of alloys become superconducting, the transition temperature being characteristic of the particular metal or alloy. In the case of pure metals it was found to vary as low as 0.35 K for hafnium to 9.3 K for niobium. Table 18.1 gives the list of superconducting elements along with their transition temperatures.

Table 18.1 Superconducting metals and their transition temperatures

S.No.	Metal	T_cK	S.No.	Metal	T_cK
1.	^{13}Al	1.20	11.	^{44}Ru	0.47
2.	^{40}Cd	0.56	12.	^{50}Sn	3.74
3.	^{31}Ga	1.10	13.	^{43}Te	11.20
4.	^{78}Hf	0.37	14.	^{81}Tl	2.39
5.	^{80}Hg	4.12	15.	^{73}Ta	4.38
6.	^{49}In	3.40	16.	^{90}Th	1.37
7.	^{57}La	4.28-5.8	17.	^{92}U	1.10
8.	^{41}Nb	8-9	18.	^{23}V	4.89
9.	^{82}Pb	7.22	19.	^{30}Zn	0.93
10.	^{45}Rh	0.90	20.	^{40}Zr	0.55

Later Onnes also discovered that even a minor magnetic field as a result of the currents in the superconductor extinguished the superconductivity. All elements do not show superconductivity phenomenon. Normally, the superconducting elements have been found to have the valence electrons between 2 and 8. For example, alkali metals, noble metals like copper, gold, silver etc., ferromagnetic and antiferromagnetic metals do not exhibit superconductivity . It was also found that the purity of the metal determines the sharpness of the transition from normal state to the superconducting state. For example, in Onnes' experiment, he observed that the resistance of mercury decreased from 0.08 W at 4K to about $2.8 \times 10^{-6}\Omega$ over just 0.01 temperature interval. So the transition may not be sharp, but may exhibit a width near the transition temperature. The Fig. 18.2 depicts the effect of the nature of the metal on the sharpness of the transition.

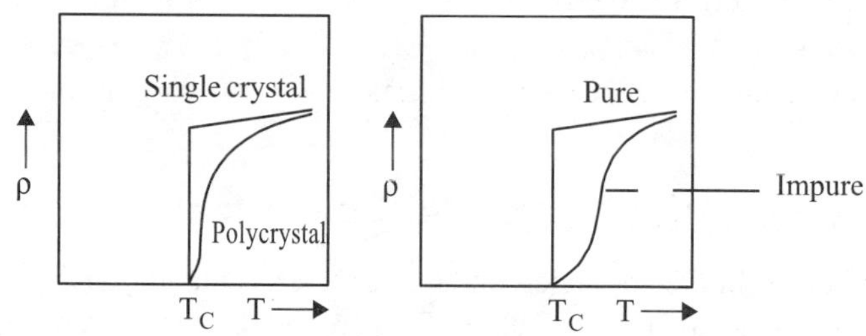

Fig. 18.2 Superconducting transition depends on the purity and form of superconducting Specimen.

Effects of Alloying

From the observations carried out on many elements and alloys, it became evident that an important characteristic of superconductivity is the zero resistance ($\rho = 0$) or infinite conductivity. It was also discovered that non-magnetic impurities had little effect on superconducting properties of metals. However the addition of impurities like the magnetic ions to a pure superconductor lowers the transition temperature. For example one atomic percent of gadolinium lowers the transition temperature of lanthanum from 5.6 K to 0.6 K. Pure molybdenum (Mo) has a transition temperature of 0.92 K, but the addition of a few parts per million of Fe destroys the superconductivity in it. Many alloys are known to become superconducting at low temperatures. The most important alloy is Nb_3Ge which exhibits a T_c of 23 K.Table 18.2 gives the list of some superconducting alloys along with their transition temperatures. Onnes observed that the increase of superconducting current in the specimen of a superconductor resulted in the quenching of superconductivity .

Table 18.2 Some Superconducting alloys

System	T_cK	System	T_cK
Pb-As	8.4	NbB	6
Pb-Bi	7.3-8.8	NbN	15-16
Pb-Cd	7	Nb_3Sn	18.05
Pb-Cu	7.3-7.8	Nb_3Ga	20.3
Pb-Hg	4.1-7.3	Nb_3Ge	23.2
Pb-In	3.4-7.3	NiBi	4.25
$PbSb_2$	1.25	V_3Ga	16.5
La_3In	10.4	Sn-Cu	3.6-3.7
Woods metal	8.2	Robes metal	8.5

Properties that do not Change :

1. The X-ray diffraction pattern of the lattice was same both above and below transition. It means that there was no crystal lattice change upon transition.
2. No change in reflectivity of visible or infrared radiation for a metal in normal and superconducting state.
3. There was no change in latent heat or volume upon transition in the absence of magnetic field.

18.2.2 The Magnetic Properties : Meissner Effect

The magnetic properties exhibited by superconductors were as dramatic as their electrical properties. The observation of H. Onnes that the increase of superconductivity current resulted in destroying the superconductivity lead to the conclusion that the magnetic fields should have a definite effect on their property . It was found by Tuny and Onnes that keeping a super-conductor in a magnetic field of a definite intensity restored the resistance of the normal conducting state. The magnetic field intensity at which this restoration takes place is called critical field (H_c). The critical field needed to quench superconductivity depends on the temperature. Its relation to temperature is expressed empirically by

$$H_c = H_o[1 - (T/T_c)^2] \qquad .....(18.1)$$

where H_o is the critical field at T = 0 K and T, the temperature.

A super conductor is supposed to be a perfect conductor. A perfect conductor possesses infinite conductivity. No electric field can exist inside it. However a surface current would be induced in a super conductor when placed in a changing magnetic field. This surface current should keep the magnetic field inside the super conductor, constant. However the experimental observations were different. Meissner and Ochsenfeld (1933) performed experiment to show that a superconductor is a perfect dia-magnetic material in nature.

Their experiments on a narrow metallic cylinder showed that the magnetic field strength outside the metallic sample changed suddenly when the sample was cooled through the transition temperature. A coil was wound on a thin and long cylinder of metal, which was to be made superconducting. A magnetic field was switched on parallel to the axis of the cylinder while the metal was in normal state. A galvanometer connected in series to the coil showed a ballistic deflection proportional to the magnetic flux present in the metal. Now the specimen was cooled, a ballistic deflection opposite in sense, but equal in magnitude to the original one, was produced, at the instant the transition temperature appropriate to the field was reached. When the field was switched off, there was no deflection. This clearly demonstrated that the superconductor expelled the magnetic flux lines from it. The superconducting, sample showed no magnetic flux inside as shown in Fig. 18.3.

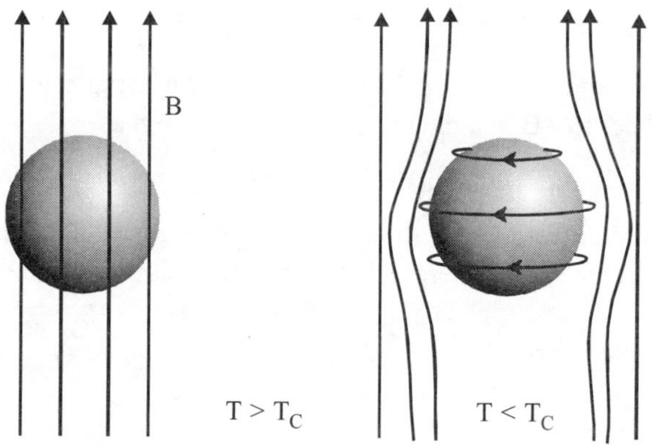

Fig. 18.3 Meissner effect in a superconductor.

The above observation known as Meissner effect was regarded as an additional fundamental property apart from zero resistivity. Therefore, a superconductor does not only behave like a perfect conductor, but behaves like a perfect diamagnetic in which B = 0. A perfect conductor resists a change in magnetic flux, while a superconductor expels it.

Critical field Vs temperature :

The resistance of a superconductor depends on the intensity of the external magnetic field. The current in the superconductor has the same effect as a magnetic field if it is sufficient to generate a critical field B_c. At T = 0, this field must be B_o and zero at $T = T_c$. Expression 18.1 can be written in terms of magnetic induction as

$$B_c = [B_0(1 - (T/T_c)^2] \quad .....(18.2)$$

The above parabolic relation or equation (18.2) is often called Tuny's law and it provides a convenient representation of the relation between temperature and the critical field. The Fig. 18.4 depicts the temperature dependence of the magnetic field necessary to quench superconductivity. The transition between the normal and superconducting states is thermodynamically reversible, just as the transition between liquid → vapour is reversible.

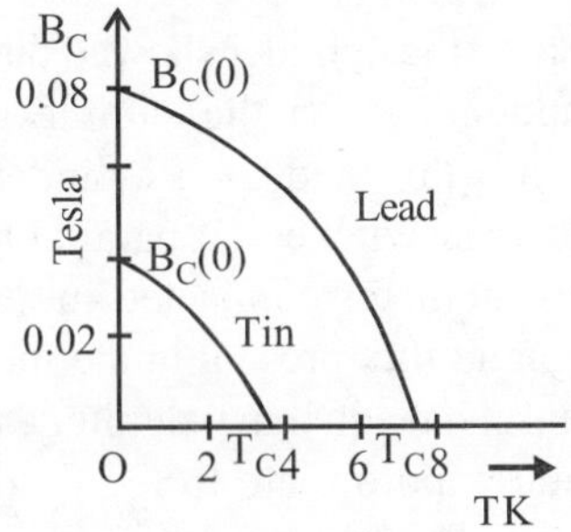

Fig. 18.4 Variation of critical induction with temperature for a superconductor. The specimen is superconducting below the curve and normal above the curve.

Type I Superconductors :

One can get useful representation by considering a long thin sample in a magnetic field. The magnetic induction field density B in a magnetic material is given as

$$\mathbf{B} = \mu_o(\mathbf{H} + \mathbf{M}) = \mathbf{B_a} + \mu_o\mathbf{M} \quad .....(18.3)$$

But, B = 0 for a superconductor. H is the applied field intensity, B_a is the applied field. The magnetization curve expected for a superconductor as per Meissner-Oschenfeld's observation is shown in Fig. 18.5.

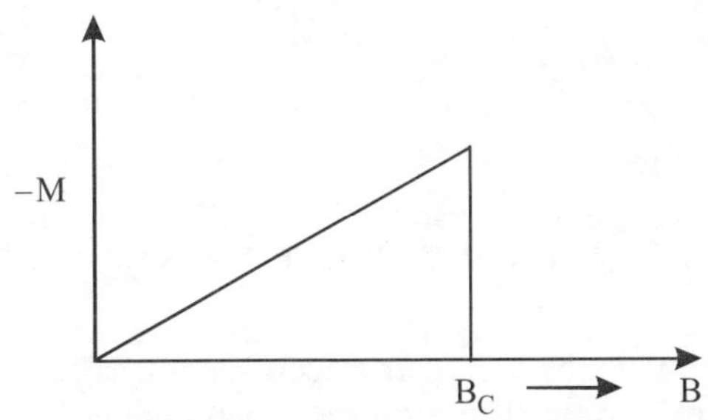

Fig. 18.5 Magnetization vs the applied magnetic field induction for a bulk superconductor.

$$-\frac{1}{\mu_0} = \frac{M}{B_a} \qquad \text{......(18.4)}$$

There is a sudden change of the sign of magnetisation at the critical field. Such conductors, which exhibit the above behaviour, are known as Type I superconductors. Above the critical field, the specimen is in the normal state and the magnetization is extremely small. The negative value of M corresponds to diamagnetism. The original elemental superconductors–like mercury, aluminium, zinc etc., are all classified as type I superconductors. They possess a critical magnetic field for any given temperature. They exhibit perfect diamagnetism.

Type II Superconductors :

The magnetization curve for type II superconductors is shown in Fig.18.6. Normally the type II superconductors are alloys or transition metals with high values of the electrical resistivity in the normal state. These metals as compared to the good conducting metals like copper have lower values of electronic mean free path in the normal state. It is the mean free path which determines their magnetic properties. When one considers the example of lead in pure form, it exhibits type I superconducting behaviour. The critical field is around 0.8 Tesla. However in the case of indium-doped lead, the flux lines start penetrating around a field less than 0.8 Tesla (Bc_1), but the superconductivity is not destroyed. The specimen shows the superconducting property upto a critical induction field Bc_2. But above Bc_2, the flux density is expelled from the sample. The specimen is said to be in mixed or vortex state between Bc_1 and Bc_2. In the vortex state, the material specimen has both normal and superconducting regions and can possess zero resistance and partial flux penetration (Type II superconductor permits some penetration of magnetic flux above a first critical flux Bc_1). It is threaded through by flux lines at the normal regions. The vortex regions are essentially the filaments of normal regions that run through the sample when the applied field density is greater than Bc_1. As the applied field is increased in strength, the number of filaments (Fig. 18.6 (b)) increases until the field density reaches Bc_2 and the entire sample becomes normal.

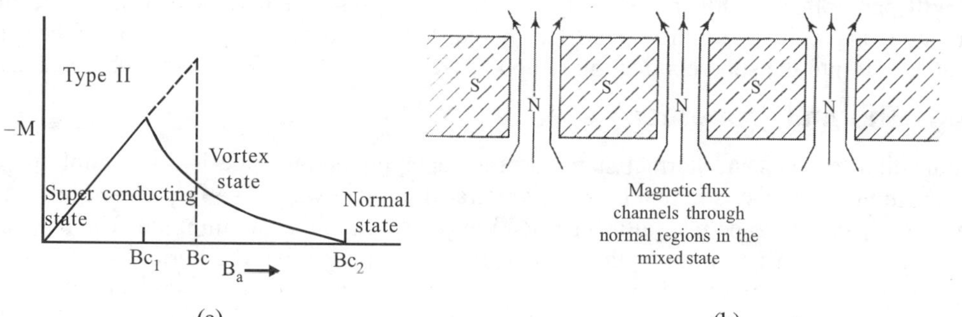

Fig. 18.6 (a) Magnetization curve for type II superconductor (b) Mixed state.

The vortex state can be viewed as a cylindrical swirl of super currents surrounding a cylindrical normal metal core which allows some flux to penetrate the interior of the type II superconductor. Associated with each vortex filament is a magnetic field which is maximum at the core center and falls off exponentially out side the core with the characteristic penetration depth λ. The super currents are the 'source' of B for each vortex state. The radius of normal core is smaller than the penetration depth. The value of Bc_2 generally is far greater than Bc_1. Many alloys of Nb, AI and Ge etc., are type II superconductors and the Table 18.3 gives a list of some useful alloys with their Bc_2 values.

Table 18.3 Superconducting alloys along with their Bc_2 values

Alloy	**T_cK**	**Bc_2 Tesla**
Nb_3AI	18.7	32.4
Nb_3AI	18.0	24.5
Nb_3Ge	23.0	38.0
NbTi	9.3	15.0
V_3Si	16.9	23.5
V_3Ga	14.8	20.8
PbMoS	14.4	60.0

The type II superconductors can carry large super currents even in fields approaching 100 KG. This has led to the development of strong field-super conducting magnets. The interface between a normal region and the superconductor region should have a surface energy. If the surface energy is positive, then the superconductor is type I superconductor. In type II superconductor, the surface energy becomes negative as the magnetic field is increased. The penetration of the applied field into the superconducting specimen causes the surface energy to become negative. It should be remembered that $(Bc_1\ Bc_2)^{1/2} \approx B_c$, the critical field for the pure specimen, a type I superconductor. They are some superconductors in which Bc_2 is as large as 35 Tesla at absolute zero.

The type II superconductors have some resistivity in mixed or intermediate state. The normal material takes the form of laminae or filaments in thin films of superconductors. (Fig. 18.6 b). Each normal region is surrounded by a superconducting region. In the mixed state, the external field will penetrate the thin normal regions uniformly. The field will also penetrate in the superconducting regions to some extent. The mixed state is stable as the penetration of the applied magnetic field into the superconducting region causes the surface energy to become negative.

18.2.3 Thermal Properties

The transition between the normal state and the superconducting state is a reversible process. This transition is analogous to that of any phase transition like vapour → liquid → solid and vice versa. For any system to be in equilibrium, its free energy should be a minimum. The free energy is decided by the configuration of the system. If it is supposed that the normal state has a free energy G_n, and the superconducting state, G_s, and the destruction of superconductivity occurs at a critical field H_c, then this transition can be described as

$$G_s(H_c) = G_n \qquad ......(18.5)$$

In the absence of the magnetic field, the specific heat of the superconducting phase is greater than that of the normal phase. The specific heat is entirely electronic in a superconductor and the electronic specific heat increases with temperature exponentially in the superconducting state. There is a sudden jump at the transition temperature. (Fig. 18.7, the dotted curve).

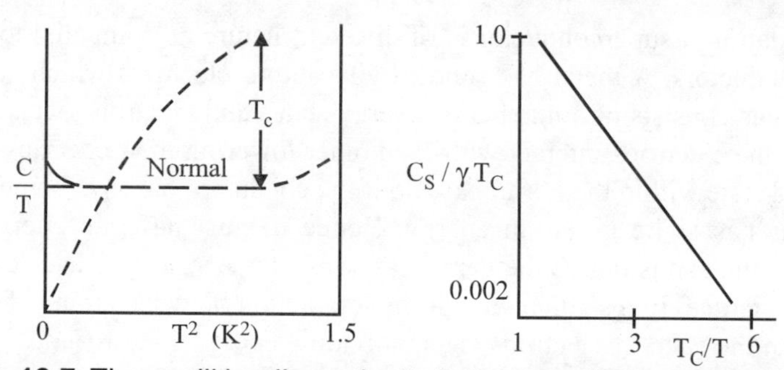

Fig. 18.7 There will be discontinuity in the specific heat for gallium.

Consider a typical isothermal-isobaric process, at a temperature TK and pressure P, causing a transition of a system from initial state (i) to final state (f). Then, the work done by it during transformation is $\Delta W = pdV$, where dV is the change in volume. $\Delta V = V_f - V_i$. The transition is isothermal, $pdV = p(V_f - V_i) \leq [U_i - TS_i] - [U_f - TS_f]$, where $U_i - U_f$ refers to change in internal energy, and $T(S_f - S_i)$ refers to the quantity of heat absorbed or evolved. $(S_f - S_i)$ is called change in entropy. Hence $[pV_f - TS_f + U_f] \leq [pV_i - TS_i + U_i]$. We call this quantity $U_i - TS + pV$ as G, the Gibb's free energy.

This sort of variation at the transition is called a second order phase transition. A phase transition is called the second order transition when there is no latent heat. The variation of specific heat in a superconductor is expressed as

$$C_v = AT^3 + \gamma_s \exp\left(\frac{-E}{kT}\right) \qquad \text{........(18.6)}$$

In normal state, the specific heat at constant volume C_v is expressed as

$$C_v = AT^3 + \gamma_s T \qquad \text{........(18.7)}$$

A and γ_s are constants that depend on the nature of the superconductor. The type of variation that is shown in specific heat with temperature also suggests that there is an energy gap (Fig. 18.8) between the normal state and the superconducting state. This is taken care of in Bardeen-Cooper-Shrieffer (BCS) theory of superconductivity which will be discussed later.

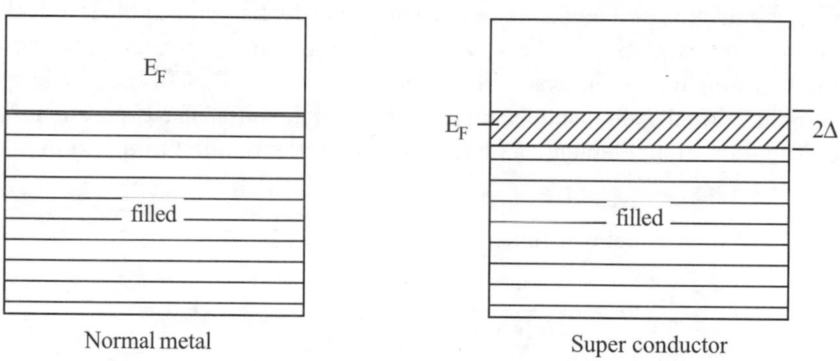

(a) Conduction band in normal state

(b) Energy gap at the Fermi level in the super conducting state.

Fig. 18.8 Energy gap in a superconductor.

The energy gap in a superconductor is of different nature as compared to the energy gap in intrinsic semiconductors. A metal has conductivity due to electrons which are treated as Fermi gas. The Fermi gas consists of a number of energy states and electrons occupy these states. In a superconductor, the electrons interact with each other lowering their energies. Therefore they go into a condensed state within the Fermi gas and if the conductor has to become normal conductor, the electron gas has to be given an energy. Hence it must have an energy gap near E_f. In superconductors, the gap is tied to the Fermi gas, while the energy gap in a semiconductor is tied to the crystalline lattice. In a semiconductor the interaction between atoms of the solid gives rise to the changes in the state of electrons. The individual energy states of the free atoms combine, overlap and give rise to energy bands of the crystal (Fig. 18.8).

The argument in the exponential in eq. (18.6) is equal to one-half of the gap energy. Table 18.4 gives the values of energy gap in some superconductors. The energy gap decreases continuously to zero as the temperature is increased to T_c from 0 K.

Table 18.4

Element	Energy gap in 10^{-3} ev at T = 0
V	1.60
Nb	3.05
Ta	1.40
Hg	1.65
Ga	0.30
In	1.05
Tl	0.73
Pb	2.73

18.2.4 Isotope Effect

The effect of isotopic constitution was first studied by Onnes and Tuyn (1922). They found that ordinary lead (atomic weight 207) and uranium lead (atomic weight 206.06) differ in transition temperature by more than 0.025 K. The isotopic constitution has been found to have an effect on the transition temperatures of elements since then. It is now known that the transition temperature varies to a good approximation as the square root of atomic mass. Similarly, the critical field also varies with the isotopic mass in exactly the same way as does T_c. The electronic specific heat of an element in its normal state is independent of its isotopic mass. The experimental results on the isotopic constitution of an element can be expressed by the relation of the form.

$$M^{+\alpha} T_c = \text{constant} \quad \text{........(18.8)}$$

α is called the isotopic effect coefficient and is given as

$$\alpha = -\frac{\partial \ln T_c}{\partial(\ln M)}$$

The value of α is normally of the order of 0.5, but there are deviations for many elements. There is no theory to explain the variations of α values for different elements.

The above observation indicates that the phenomenon of superconductivity is determined in some way by the interaction between electrons and atoms in the superconductor. Electrons are responsible foı the conductivity of metals. The electrons interact with lattice vibrations and their interaction affects the conductivity. In superconductors also there should be interactions between electrons and lattice vibrations.

The values of α are given in Table 18.5 for various elements.

Table 18.5

Substance	α
Hg	0.50
Pb	0.49
TI	0.60
Os	0.15
Mo	0.33
Mo_3Ir	0.33
Zr	0.00
Cd	0.32

18.2.5 London Penetration Depth

The characteristic property of a superconductor is that B = 0 inside, and therefore, that any current existing in it must be superficial. The magnetic susceptibility of a superconductor is -1 (SI units). However, this can not be generalised and the current can not entirely be superficial. There must be a certain depth, however small, of penetration of the magnetic field into the superconductor. London (1935, 1939) observed the penetration effects became important when the samples of superconductors wcrc takcn as thin films. In bulk sample, the depth is very small. Let it be supposed that the field inside the surface of an infinitely thick sample falls from its value H_0. The field decays exponentially from the surface into the material. The fall in H in x-direction is expressed as

$$-dH = \frac{H(x)dx}{\lambda} \qquad .....(18.9)$$

i.e,

$$H = -\int_{H_0}^{0} \frac{H(x)dx}{\lambda} = H_0 e^{-x/\lambda}$$

In terms of magnetic induction or magnetic field density, we can write the above equation as

$$B = B_0\, e^{-x/\lambda} \qquad .....(18.10)$$

λ is known as London penetration depth. B_0 is induction field at the surface of the super conductor and it falls off with distance as shown in (Fig 18.9).The exponential decay is the characteristic of Meissner effect. The London penetration depth is the depth from the surface through which the field decays 1/e of its value at the surface.

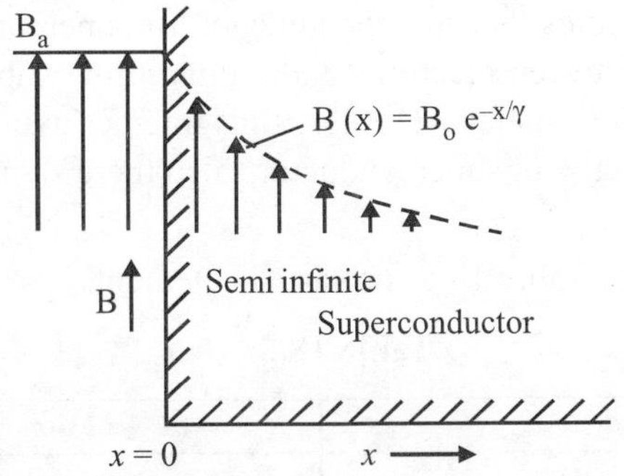

Fig. 18.9 Shows the penetration of an applied magnetic field into semi infinite superconductor.

The macroscopic description of a superconductor is given as a body with B = 0 and E = 0 inside it. London considered a superconductor as a condensed state consisting of superelectrons and normal electrons. The total number of electrons in the superconductors N is equal to the sum of the concentration of normal electrons n, and the concentration of superelectrons n_s.

In a superconductor, λ, the London penetration depth is a function of temperature and the relation between temperature and λ is

$$\lambda/\lambda(o) = [\ 1 - (T/T_c)^4]^{-1/2} \qquad .....(18.11)$$

where λ(o) is the London penetration depth at T = 0 K. It is of the order of 10^{-6} to 10^{-5} cm for many materials.

London Penetration depths at absolute zero for various metals are given in the Table 18.6. The equation (18.11) also implies that the density of superconducting electrons increases from zero at T_c to maximum at 0 K.

Table 18.6

Element	$\lambda 10^{-8}$m
Sn	3.4
Al	1.6
Pb	3.7
Nb	3.9
Hg	5.0

Thus, in a superconductor, Meissner effect is not complete, London penetration depth λ is a fundamental length that characterizes a superconductor. λ is of the order of 10^{-8} m. The penetration depth λ decreases as temperature is reduced.

The diamagnetic nature of a superconductor can be demonstrated by floating magnet experiment. This was one of the earliest expriments done. If a small permanent magnet is lowered over a superconducting surface, the lines of force of the magnet can not enter the superconductor and this results in a repulsive force large enough to over come the weight of the magnet. The

Fig. 18.10 shows a floating magnet above a superconductor. This sort of levitation is important in applications.

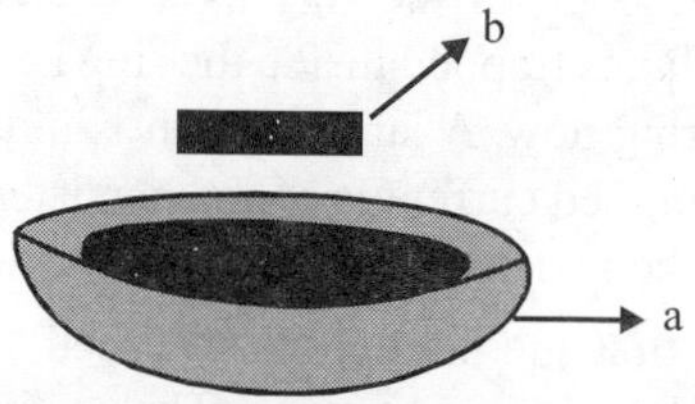

a bowl of superconductor
b floating magnet

Fig. 18.10 A magnet floats over a superconducting bowl.

18.3 Flux Quantization

The magnetic flux is ejected out at the transition temperature due to the induction current in the specimen which produces a flux equal and opposite to the original flux due to the magnetic field. As there is no resistance, the current due to induction in the specimen persists as long as the field is on. As a result of the persistent current, the specimen acquires a negative magnetic moment producing a field in a direction opposite to the direction of the applied field. Due to this negative moment, a superconductor is treated as a diamagnetic substance. When the field is removed, the persistent current goes to zero.

It appears from above that the persistent current exists as long as there is external field. This opinion can not fully support the occurrence of superconducting state when a conductor is cooled in the absence of the field as was experimentally verified. Meissner observed that persistent currents could be set up in any sample of suitable shape. Consider that the sample is taken in the form of an anchor ring as shown in the Fig.18.11.

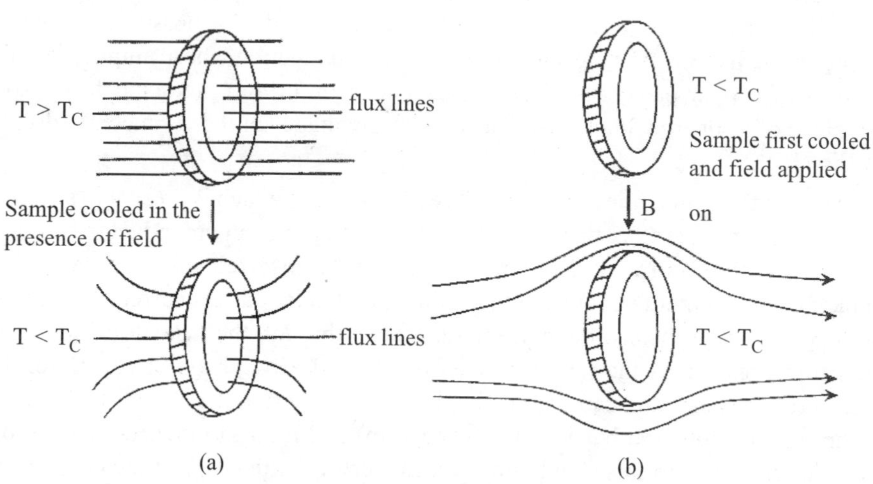

Fig. 18.11 A ring in magnetic field.

Suppose the magnetic field is put on while the ring is in normal state. Then the magnetic flux penetrates through the ring, When the ring is cooled to the transition temperature while the field is on, the flux is excluded from the body of the ring as a result of the induction current on the surface of the ring. However the flux is trapped inside the ring i.e. in the hole of the ring as the flux lines can not thread through the ring now. A large current is induced by the collapsing field that maintains the trapped flux. This trapped current is called persistent current. It will be present even after the field is switched off as the resistivity of the superconductor is zero.

However, if the ring is cooled first and the field is switched on after the critical temperature is reached, the magnetic flux is excluded from the ring. There will not be any flux trapping in the hole of the ring. The current induced in the ring keeps the flux out of the ring. This current disappears when the field is switched off.

A superconductor allows the current through. There is a minimum quantity of current that can exist in a superconductor without destroying the superconductivity. This minimum current is called critical current I_c. Suppose the superconductor is in the form of a wire of radius r. The current I_c in it will set up a transverse magnetic field H_t and $H_t = I_c/2\pi r$. For any other current value I, $H'_t = I/2\pi r$.

The magnetic flux that passes through the ring is the sum of the flux from the applied field and the magnetic flux from the superconducting currents which flow in the ring. The flux lines from the applied field are not quantized. However, the flux from superconductor adjusts itself such that the total flux is quantized.

The magnetic flux is quantized in integral multiples of $\phi_0 = h/q$, and q is equal to '2e', the charge of an electron pair. The value of h/q was found experimentally to be 2.07×10^{-15} weber. It confirmed the existence of an electron pair in the superconductor. The unit of flux is called 'fluxoid'.

18.4 Josephson Effect

Josephson junctions are useful in creating fast memory elements of electronic computers. A brief discussion will be given here. It has been said that a superconductor possesses cooper pairs. All the cooper pairs are in the same state and possess one wave function whose amplitude is dependent on the number of cooper pairs. Based on these, Josephson (1962) predicted Josephson effects. Suppose a piece of a superconductor is connected to a voltage source, ammeter and voltmeter as shown in Fig. 18.12.

When the circuit is closed, the ammeter will read a current I which is present in superconductor. The voltage across it is zero, as there is no resistance. Let this superconductor be cut, into two halves (Fig. 18.12(b)) and a thin (1 mm) insulator be inserted into the gap between the superconductors and let them be joined. On closing the circuit, a direct current still flows through the superconductor. That means, the current also flows through the insulator. This is due to the tunnelling of charges across the junction. This is called dc Josephson effect. Here, the cooper pairs tunnel through (tunneling can only be explained by quantum theory) from one conductor to the other through the potential barrier (insulating film) and return to the first conductor through the external circuit. The currents should not exceed a critical value (I_o). If the currents exceed I_o, a potential difference develops across the junction indicating a resistance in it (Fig.18.12(c)). This

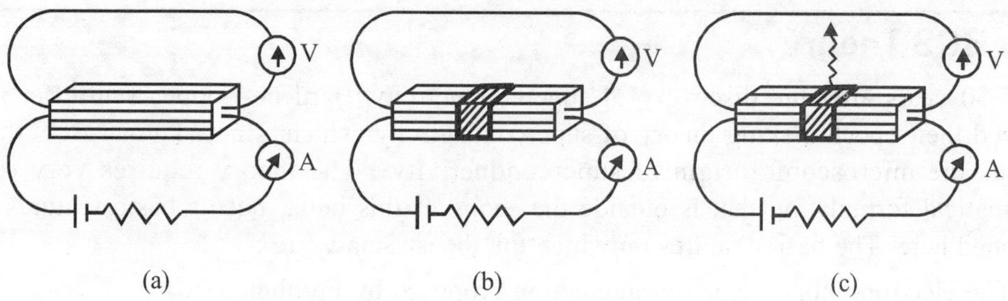

Fig. 18.12 Superconductor Junction.

principle is used to develop Josephson tunnel junction cryotrons. Since the transition from the zero to a finite resistance is not related to the elimination of the superconductivity, the speed of the transition is much higher. The speeds are of the order of 10^{-11} sec and the dissipated power is of the order of only 10^{-7} Watts. The voltage characteristic of a Josephson's junction is given in Fig. 18.13.

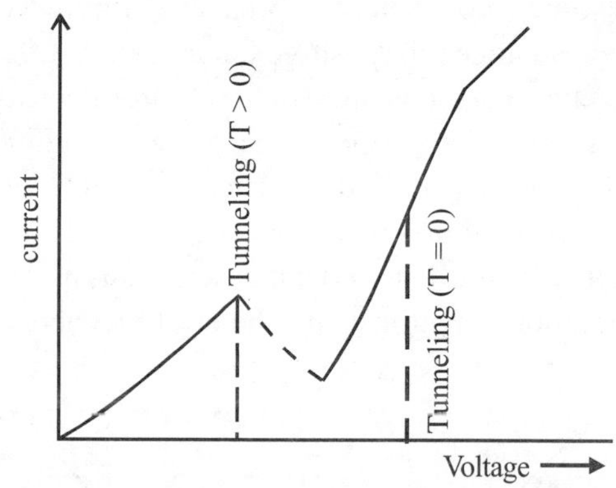

Fig. 18.13 The current-voltage characteristic of Josephson junction.

The tunneling current can be given as $I = I_0 \sin \varphi$, where φ is the phase difference in the state of the waves describing the cooper pairs on both sides of the insulator.

When current exceeds I_o, a resistance appears and the energies of cooper pairs on either side of the barrier (insulators) differ by 2qV, but the energy is zero, then the wave frequency difference $\Delta\nu = 2qV/h$ and this will appear on both sides of the junction.

The phase difference is $\varphi = 2\pi t\,(2qV/h)$ and the superconducting current through the junction becomes $I = I_o \sin\left(\frac{4\pi qV}{h}t\right)$. This is a sinusoidal current whose oscillation frequency is equal to 2qV/h. When V = 1mV, the frequency is of the order of 480 GHZ. This is in Microwave region. This type of phenomenon enables one to construct microwave resonators (with a quality factor of 10^{11}). The theoretical limit of the quality factor for conventional resonators lies near $1–8 \times 10^3$.

18.5 BCS Theory

Nearly 50 years after the discovery of superconductivity, Bardeen-Cooper-Schrieffer in 1957 produced their epoch making theory of superconductivity, which is now known BCS theory. It explains the microscopic origin of superconductivity . The theory requires very rigorous mathematical formulation that is outside the scope of this book, but its basic features can be mentioned here. The basic features on which the theory stands are :

1. The electron-phonon-electron interaction proposed by Frohlich.
2. The formation of Cooper pairs of electrons.
3. The existence of energy gap.

1. Electron-Phonon-Electron interaction and formation of Cooper pairs :

The two observations that the critical transition temperature T_c varies as $M^{-1/2}$ where M is the atomic weight of the metal that is superconducting and the sudden discontinuity in specific heat at transition temperature did make Bardeen to think that the superconducting state should involve the dynamics of ionic motions, lattice vibrations and associated phonons. Further T_c, attains value zero when M approaches infinity. When an electron with a wave vector K_1 encounters a lattice ion (positive ion cores), it suffers a Coulomb interaction. As a result of this interaction, the momentum of the electron decreases by an amount proportional to $K_1 - q$ and the lattice is set into motion gaining a momentum from the electron proportional to q. The lattice vibration is quantised and characterised by energy packets called phonons (virtual). Another electron with a wave vector K_2 (Fig. 18.14) may enter this volume and this electron interacts with the distorted lattice and gains momentum, the gain depending on the phase of the vibration at that instant. The electron energy is lowered.

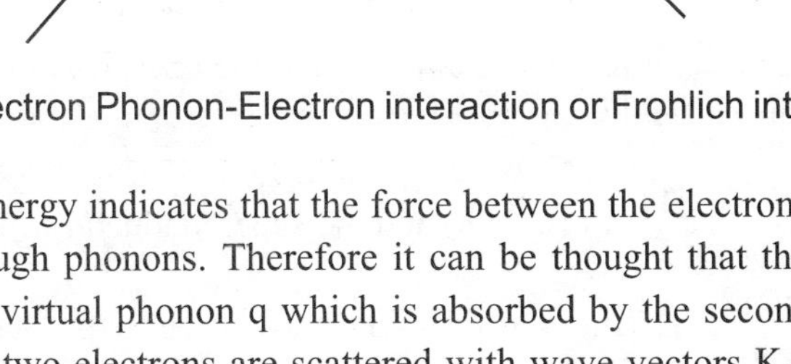

Fig. 18.14 Electron Phonon-Electron interaction or Frohlich interaction.

The lowering of energy indicates that the force between the electrons is attractive and the interaction is through phonons. Therefore it can be thought that the electron with wave vector K_1 emits a virtual phonon q which is absorbed by the second electron with wave vector K_2 and the two electrons are scattered with wave vectors $K_I - q$ and $K_2 + q$. The process is a virtual one and the energy is not conserved. The process is very rapid with the emission and absorption of the phonon. The short lived phonon is called virtual phonon. The two electrons should have equal and opposite spins and their energies are the same. Then the interaction between them is attractive. As a result of such attractive interaction, there exists pairs of electrons called Cooper pairs. They are called the Cooper pairs in honor of L. N. Cooper who was the first to show that binding of electrons in pairs is energetically advantageous. A Cooper pair is denoted by (K↑, K↓). Each electron participating in the pair formation is called mate.

Therefore, the electrons in a superconductor are in the form of Cooper pairs. A Cooper pair is a system of two electrons having equal and opposite momenta and spin (K↑, K↓). The energy of the pair is lowered when compared to the free separate electrons. All electrons in the metal can not participate in this process and only a fraction of the total electrons in the specimen can do so. The energy difference between the free state of the electron (energy of individual electrons) and the pairs appears as the energy gap at the Fermi surface (Fig. 18.15). Therefore, the energy gap is said to be tied to the Fermi gas. The attraction in between Fermi electrons tends to reduce their total energy (this total energy must be removed from the electron gas to transfer it to the superconducting state). The energy spectrum of electrons must then exhibit an energy gap.

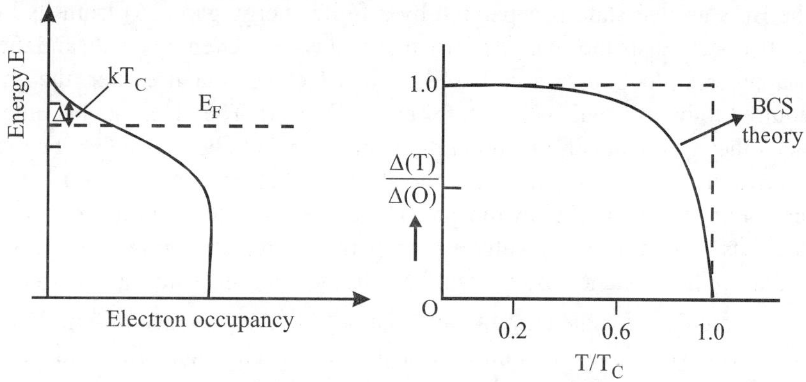

Fig. 18.15 Energy gap in a superconductor.

2. **The energy gap :**

The existence of energy gap can be proved experimentally. One such experiment is the study of the spectrum of absorption of electromagnetic waves by a superconductor in the microwave band. At temperatures close to 0 K, a superconductor does not absorb energy of incident radiation until the energy of incident radiation exceeds the width of the gap (2Δ). After absorption of energy, the electrons become free or normal.

Although BCS model is a brilliant model describing the mechanism for superconductivity, it can not explain which materials are superconductors and which materials are not. For example gold, copper and silver which are excellent conductors at room temperatures are not superconductors even at temperatures very close to 0 K. Mercury, tin, lead etc. which have higher resistivity at room temperature (compared to gold, silver) are excellent superconductors at low temperature. We have new materials which are insulators at room temperature and are now found to be superconducting at liquid nitrogen temperature (These superconductors are called high T_c superconductors).

The energy gap is a function of temperature. At absolute zero, a superconductor does not absorb any energy when microwave radiation falls on it until the energy supplied by the radiation is greater than the width of the energy gap 2Δ. When the absorption takes place, the electrons go to the energy states that lie above the gap. The Cooper pairs are broken and the superconductor becomes a normal conductor. The energy gap represented the energy needed to break up one of the cooper pairs. The width depends on the number of unpaired electrons. The presence of energy gap also explains the occurrence of persistent currents. In a Cooper pair, the momentum of individual electrons can not be changed. The pair can not be scattered as a whole unlike the individual electrons by lattice ions. Each Cooper pair is associated with a wave function and this wave function spreads over a large volume and overlaps with wave functions of other Cooper pairs. The volume of a given pair may extend to over one million other pairs. Therefore, the superconducting state is a collective state. The conduction electrons in a superconductor are condensed into a single entity some thing like a big macro molecule extending over entire volume of the superconducting system. They move as a whole.

The BCS ground state is separated by a finite energy gap (2Δ) from its lowest state. The Cooper pair resists the change of its state as it is a lower energy state and the zero resistivity of a superconducting state is based on this fact. As stated earlier, the amount of energy required to break a pair is 2Δ (Δ for each electron). The electrons become independent of each other at the transition temperature and are said to be 'normal'. The thermal vibrations of the lattice break the pair. Then thc electrons will appear in the energy states above the gap leaving vacancies below the gap. Thus, a certain equilibrium concentration of 'normal' electrons sets in at every value of temperature from 0 K onwards. This phenomenon is similar to the existence of holes and electrons in an intrinsic semiconductor.

In a superconductor the current can exist despite the presence of gap. In a semiconductor, electrons have to be excited into conduction band. The interaction and pausing of electrons extends over a considerable distance in the specimen. The maximum distance unto which the states of pairs of electrons are correlated to produce superconductivity is known as coherence length L. It can be many thousands of atomic spacing ($\sim 10^{-6}$m) apart. The properties of a superconductor depend on the correlating electrons within a volume ξ_L^3, called coherence volume. The ratio of London penetration depth to the coherence length is $K = \lambda / \xi_L$. For type I superconductors $K < 1/\sqrt{2}$, and for type II superconductors $K > 1/\sqrt{2}$. The difference in K values is related to the surface energy. Therefore the interaction of electrons follows the long-range order. The electron states responsible for superconductivity exist within $k\,T_c$ (k is Boltzman constant) of the Fermi surface. Their life time t is obtained from uncertainty principle as $t \approx h/2\pi\, kT_c$. The wave function extends over a distance ξ_L and $\xi_L = 2kv_F/\pi\Delta$ where v_F is the velocity of electron at the Fermi surface. This is shown in Fig.(18.15).

BCS theory explains the thermal properties, critical fields and most of the electromagnetic properties of superconductors in terms of the energy gap. The penetration depth, coherent length, flux quantization, isotope effect, etc., emerge as natural consequence of BCS theory.

London equations, Meissner effect become a part of BCS explanation of superconductivity. The rigidity of macromolecule to a change of its state also explains Meissner effect. When a superconductor is placed in a magnetic field, the wave function of the macromolecule does not change i.e. the orbital configuration of the electrons does not change. However, the electrons or the Cooper pairs, suffer a change in velocity while remaining in the same orbital configuration. The change of velocity causes an induced current, which expels the magnetic field from the bulk of the superconductor.

18.6 Application of Superconductors

The applications of superconductors are diverse and many. Some of them are mentioned below.

1. ***Generation of Electromagnetic Field and Electromagnets :*** The type II superconducting wires are wound in the form of solenoids to generate strong magnetic fields of an induction about 20 tesla. The size of these magnets will be small compared to the ordinary metal wire wound electromagnets. The superconducting wire has to be maintained at liquid-He temperature and the cost of the power will be 1000 times less than the cost of power required in an ordinary electromagnet of the same capacity. The energy stored in the field will be substantial about 4 KJ/m^3 at 10 Tesla. The superconducting wires are also used in electric machines and transformers. The transformers do not require any magnetic cores and the superconducting wires generate strong fields.The high T_c super conducting wries are now prepared for the electromagnets.
2. ***Bearings :*** The Meissner effect is made use of in bearings. The mutual repulsion between two superconductors due to the repulsion between the magnetic fields they generate can be embodied in bearings and the bearings would operate without power loss and friction.
3. ***High power transmission :*** The superconducting cables permit high power transmission without power loss. The superconducting wires used in the above applications are produced from solid solution alloys such as niobium-zirconium, niobium-titanium, or niobium-titanium-zirconium. High temperature superconducting wires may soon replace the wires operating at liquid-He temperatures.
4. ***Cryotronics :*** The new branch of electronics based on superconductors low temperature electronics or cryogenic electronics or Cryotronics has become a very important branch of electronics. This includes creating 1. Supersensitive miniature receivers capable of detecting extremely weak radio signals. 2. Manufacture of large-scale integrated and very large-scale integrated circuits (VLSI) for the development of new generation of electronic computers, 3. An increase in the frequency stability and frequency selectivity of microwave apparatus, and 4. The extension of radio wave bands into the microwave region up to the infrared, earlier considered inaccessible for long range reception. There are many cryogenic devices like memories, logic units, cryotron switches, oscillators, amplifiers, resonators, modulators etc. A cryotron switch and a memory cell will be described here. It relies on the effect of inducing a persistent current in a closed superconducting circuit and destroying the same state by magnetic field.

   (a) ***Cryotron Switch :*** It consists of (Fig 18.16) a controlled gate 1 a strip of thin film tin ($T_c = 3.7$ K) and a control strip 2, a thin film of lead ($T_c = 7.2$ K) deposited on a substrate and separated by a dielectric layer.

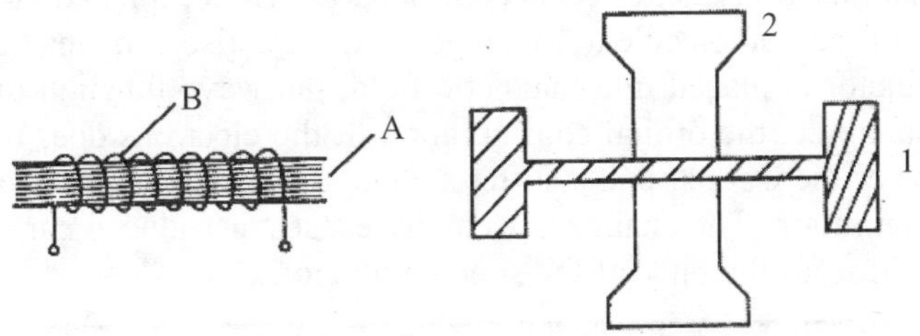

(a) (b)

Fig. 18.16 Cryotron switch.

Fig. 18.16 (a) illustrates the schematic diagram of a simple cryotron. It consists of a controlled (gate) wire A commonly made of tantalum with $T_c = 4.4$ K and a control coil B from niobium ($T_c = 9.3$ K) or lead ($T_c = 7.2$ K). At a working temperature slightly below 4.4 K, the gate wire is kept in the superconducting state and offers no resistance to the flow of current I, through it. When a current I_2 passes through control coil, the superconducting state of the gate breaks down and it is cut into normal state, possessing a finite resistance R. This is something like a relay.

The cross film cryotron offer high speed operation Fig. 18.16(b), consists of two crossed strips deposited on the substate and seperated by a thin dielectric layer. The control strip 1 is the gate (normally tin) and control strip 2 is made of lead.

At working temperature T slightly less than 3.7K, the gate is kept in superconducting state. A current through it passes without resistance. When current is passed through 2, the superconducting state of the gate 1 is broken and it goes into the normal state. Therefore the current in the circuit consisting the gate (1) is broken. By varying the current in 2 alternately one can get opening and closing of the gate in the circuit having the gate. The thickness of the strips is about 0.1 microns and by suitable means the switching speeds can be made 10^{-7}sec. Cryotrons can perform all logic and storage operations required in computing with essentially zero power dissipation except when switching. The low power consumption is an attraction to use the superconducting devices, but the reproducibility, maintenance of the equipment are the formidable obstacles for actual use.

(b) ***Superconducting Memory Cell :*** A closed superconducting circuit, in which persistent currents can be induced, can work as a memory cell. When the superconductor is in the normal state, there will not be any current in the circuit. It is called '0' state. In the superconducting state, the circuit will have a current and it is called '1' state. A superconducting memory cell is in the binary system depicted in the Fig. 18.17.

A lead substrate S is coated with a thin dielectric layer on it. A thin tin strip T is deposited on the substrate in the form of loops T and connected into groups by a bit line L. The tin strip is coated with a dielectric layer on top of it. Over the tin strip, lead strips A and B are over layered. To write the information, a current is passed through the bit line.

Simultaneously current is fed into the strips A and B. They produce a magnetic field H capable of destroying the superconducting state in the tin loop located below the lead

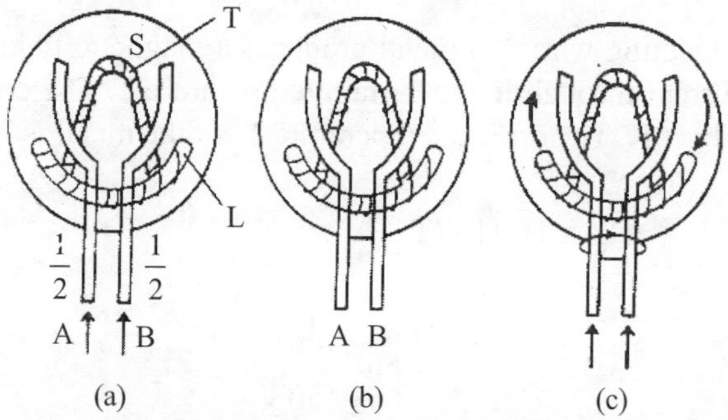

Fig. 18.17 A superconducting memory cell.

strips. The current is made to flow only in the upper section of the loop. This state is maintained stable even after cutting off the current in the lines A and B, although the tin strip becomes fully superconducting. Now when a current is allowed to go through the bit line, a persistent current will circulate in the loop, which will then store the delivered information. (Fig. 18. 17(b))

Reading out the information, requires a total current in A and B that destroys the superconducting state in the tin loop where this state was broken earlier. This eliminates the current in the loop and induces the information pulse in the bit line (Fig. 18.17(c)). The capacity of such a memory cell is up to billion bits and the speed is in the order of $10^{-6} - 10^{-7}$sec.

Cryotronics are supposed to become very popular in future as the low temperature devices operate with low noise levels and raise the performance level of the electronic devices to a high degree.

The outer space is vacuum, extremely cold and radiation full. Rockets nowadays are based on cryogenic liquids (liquid oxygen, hydrogen, helium, Nitrogen). Therefore the satellite borne or rocket-borne apparatus should not only be mini in size but also should be able to use the low temperatures for their operation. In such cases compact electronic devices based on Cryotronics are useful.

5. ***Superconducting magneto meters :*** They are capable of measuring magnetic field whose induction B is smaller than 10^{-11} Tesla. SQUID (Superconducting Quantum Interference Device) is useful for measurement of very low magnetic moments. If a Josephson junction is placed in a magnetic field, the current density in the junction varies just like the single slit diffraction pattern in optics. If two junctions are placed side by side, then the super current density exhibits a fringe pattern like in the double slit arrangement in optics. SQUID magnetometer is useful for the precision determination of h/e; SQUID are fundamentally superconducting rings that act as storage houses for magnetic flux. SQUID is extremely sensitive to very low magnetic fields and it is able to detect the magnetic field of earth very accurately. SQUID is used to detect the mineral deposits or oil deposits inside the earth. Superconducting magnetic maps are useful for this purpose. SQUIDs are useful in medical diagnostics of heart and brain. Nuclear magnetic resonance imaging is useful in medical diagnostics.

Example 1

A long thin superconducting wire of a metal produces a magnetic field 105×10^3 amp/m on its surface due to the current through it at a certain temperature T. The critical field of the metal is 150×10^3 amp/m at T = 0K. The critical temperature T_c of the metal is 9.2 K. What is the value of T ?

$$H_c = H_o [1 - (T/T_c)^2)$$

$\therefore$ $T = 4.3$ K.

$$\therefore \quad T^2 = 9.2 \times 9.2 \left[1 - \frac{105}{150}\right]$$

$$= 5.03 \text{ K}$$

Example 2

A lead wire has a critical magnetic field of 6.5 x 10^3 Am^{-1} at 0K. The critical temperature is 7.18 K. At what temperature the critical field would drop to 4.5×10^3 Am^{-1}. The diameter of the wire is 2 mm. What is the critical current density at that temperature?

$$\frac{[H_c - H_o]}{H_o} = -\left(\frac{T}{T_c}\right)^2$$

$$\frac{[H_o - H_c]}{H_o} = \frac{T^2}{T_c^2}$$

$$\frac{[H_o - H_c]}{H_o} T_c^2 = T^2$$

i.e.,

$$T^2 = \frac{6.5\times10^3 - 4.5\times10^3}{6.5\times10^3} \; 7.18 \times 7.18 = 15.86 \text{ K}^2$$

i.e., $T = 3.98$ K

The critical current density

$$J_C = \frac{H_C \times 2\pi r}{(\pi r^2)} = \frac{4.5\times10^3\times2}{1\times10^{-3}} = 9.0 \times 10^6 Am^{-2} \text{ at } 3.98 \text{ K.}$$

Example 3

The transition temperature of Nb is 9.5 K. The critical magnetic field induction density is 0.2 Tesla at 0 K. What is its stabilization energy? What is the magnetic field induction at 3.5 K.

$$\Delta U = \frac{(B_c)^2}{2\mu_o} = \frac{(0.2)^2 \times 7}{2 \times 4 \times 22 \times 10^{-7}} = 159 \times 10^2 \text{ Joules.}$$

$$B = B_o[1 - (T/T_c)^2] = 0.2\left[\frac{(9.5)^2 - (3.5)^2}{(9.5)^2}\right] = 0.173 \text{ Tesla}$$

Questions

1. What is Superconductivity?
2. What is Meissner effect?
3. Explain the terms
   (a) Critical transition temperature, (b) Critical field, (c) Persistent Current
4. What is the relation between temperature and the critical magnetic field in a superconductor?
5. Define the term 'London Penetration depth'.
6. Does the transition temperature depend on the applied magnetic field in a superconductor ? If so, how ? A superconductor has T_c = 3. 72 K in a zero magnetic field. The critical induction field at 0 K is 3.2 × 10 –2 Tesla. Find the critical field at 2.5 K.
7. What is the important feature of Type I Superconductors. Mention a few type I superconductors.
8. How does the magnetization M vary with applied magnetic field in type II superconductors?
9. What is isotope effect in superconductors?
10. What are Cooper pairs? How are they formed in a superconductor?
11. What is the difference between the energy gap in a superconductor and energy gap in an intrinsic semiconductor?
12. What is magnetic levitation? A superconducting lead wire has a radius of 1 mm. Is it possible to float it in the magnetic field of earth. The horizontal component of earth's magnetic field is 0.5 gauss.
13. What is Josephson effect?
14. What are the properties that do not change in the superconducting state and normal state?
15. What are Superconductors ? Discuss the earlier experiment on superconductivity and how zero resistivity was thought to be the characteristic of superconductivity?
16. Can the superconductor be thought of only a zero resistance material ? If not why ?
17. Explain how Meissner effect proves the superconductor to be a perfect diamagnetic material? What is the difference in critical fields for lead at 7.1 K and 1K.
18. Obtain the relation between London penetration depth and the applied magnetic field. How does the London Penetration depth change with temperature ?

19. Distinguish between Type I and Type II Superconductors ?
20. Discuss the effect of magnetic field on critical transition temperature. What is the energy stored in a superconductor in the presence of magnetic field ? Determine the amount of energy in Joules that is required to destroy superconductivity in 1 cm^3 of lead at 0K.
21. Arrive at London equations and hence discuss Meissner effect. Compute the stabilization energy density of the superconducting state at 0K for aluminum. The critical induction field density is 1.1 Tesla.
22. What is Superconductivity ? Discuss the some properties, which distinguish them from normal metals? When placed over a superconducting lead plate at what height will a tiny magnet levitate ? (its magnetic moment is 0.25 Am^2 and its mass is 4 gm).
23. Discuss the formation of Cooper pairs and the energy gap in superconductors on the basis of BCS theory .
24. Discuss the applications of superconductors.
25. What is Josephson effect ? Discuss the applications based on Josephson effect.
26. Explain : (a) Cryotron Switch (b) Memory Cell (c) Energy gap in a superconductor.

I. Choose the correct choice

1. A super conductor exhibits
   (a) only zero resistance
   (b) only diamagnetism
   (c) zero resistance and paramagnetism
   (d) zero resistance and diamagnetism
2. The property that does not exhibit any appreciable change in superconducting state as compared to the normal state is
   (a) specific heat
   (b) magnetization
   (c) internal energy
   (d) volume
3. Meissner-ochben field effect consists of
   (a) freezing in' of the magnetic field in a super conductor
   (b) occurrence of zero resistance at the transition temperature
   (c) exhibition of perfect diamagnetism by super conductor
   (d) showing ideal conduction
4. Choose the correct statement
   (a) A perfect crystal expels all the magnetic flux when placed in a magnetic field.
   (b) The metal in superconducting state exhibits a different crystal structure as compared to that in the normal state.
   (c) A perfct conductor when placed in a magnetic field would be left with "frozen in" paramagnetic moments
   (d) A type I superconductor can not have a finite critical magnetic field at any given temperature

5. The addition of magnetic ions to a pure superconductor
   (a) reduces the thermal expansion
   (b) reduces the transition temperature
   (c) increases the volume of the superconductor
   (d) does not affect the superconducting properties

6. In a perfect conductor
   (a) $\frac{\partial \mathbf{B}}{\partial t} = 0$
   (b) $\nabla . \mathbf{B}$ = constant
   (c) curl $\mathbf{J} = -\frac{\partial \mathbf{B}}{\partial t}$
   (d) curl $\mathbf{B} = 0$

7. The critical field needed to quench super conductivity at a temperature T for superconductor with a critical temperature T_c is
   (a) $H_o(T/T_c)^2$
   (b) $H_o(T_c/T)$
   (c) $H_o[1-(T/T_c)^2]$
   (d) $H_o[1-(T_c/T)^2]$

8. Choose the correct statement
   (a) At any critical field value B_c, the superconductivity disappears at any temperature.
   (b) At a critical field B_c, the current in superconductor becomes infinite.
   (c) If the temperature of a superconductor is increased, the critical field value is increased.
   (d) If the temperature of a superconductor is increased, the critical field value becomes infinite.

9. Two statements are given below. Verify their correctness
   (A) At temperatures close to the critical temperatures, a small magnetic field is sufficient to destroy super conductivity.
   (B) A super conductor has a transition temperature 7^0 K, At 1^0 K, the critical field density is 0.08 Tesla and at 5^0K, the critical field density is 3.0 Tesla.
   (a) (A) and (B) are correct
   (b) (A) is correct (B) is wrong
   (c) (A) is wrong, (B) is correct
   (d) (A) and (B) are wrong

10. The London perretiation depth λ
    (a) depends on the critical temperature of the super conductor
    (b) varies inverrely as the external magnetic field intensity
    (c) is dirictly proportional to the square of the isotopic mass of the superconductor
    (d) depends on the temperature and the nature of material of the superconductor

11. Verify the correctness of the statements
    (A) In a type II superconductor, the critical field density can be far higher than the critical field density in a type I superconductor
    (B) Alloying of superconductor with magnetic ions may destroy super conductivity
    (a) (A) and (B) are wrong
    (b) (A) is wrong, B is correct
    (c) (A) is correct, (B) is wrong
    (d) (A) and (B) are correct

12. The energy gap in a superconductor

(a) is minimum at a temperature $T = T_c /2$, Where T_c is the critical temperature

(b) is maximum at a temperature $T = T_c/2$,

(c) is zero at $T= T_c$

(d) is zero at $T = 0^o K$

13. Select the proper combination

(A) Type II - superconductor (p) Cooper pairs

(B) Electron-phonon-Electron interaction (q) Meissner effect

(C) Operation of bearings without (r) Vortex state power loss or friction

(a) A-q B-r C-p (b) A-r B-p C-q

(c) A-p B-r C-q (d) A-r B-q C-p

14. High T_c super-conductor in the following is

(a) $N d_3 Ge$ (b) $Ba Ti O_3$

(c) Lithumtitanium oxide (d) ythrium-Barium-copperoxide

15. High T_c super conductors are found to exhibit transition temperatures

(a) close to liquid hydrozen temperature

(b) above liquid Nitrogen temperatures

(c) room temperature and above

(d) above liquid helium temperature and below liquid N_2 temperature

16. The electron pairs in a super conductor are called

(a) Bardeen pairs (b) BCS pairs

(c) Electron-hole pairs (d) Cooper pairs.

17. In a cooper pair

(a) The momenta $\hbar$ k must be equal to in magnitude and opposite in driection

(b) The moment $\hbar$ k must be equal in magnitude and parallel in direction

(c) One electron revolves round the paired electron

(d) Energy of formation is zero,

18. The energy required to break a cooper pair is

(a) equal to fermi energy of the electron

(b) equal to the magnetic energy associated with the electron

(c) is equal to the energy gap in a super conductor

(d) equal to vibrational energy of the atoms in superconductor

19. The cooper pairs are

(a) Bore-Einstein particles or Bosons
(b) fermi-Dirac particles or Fermions
(c) classical particles
(d) electron positron pairs

20. The magnetic susceptibility of a super conductor is (in SI units)

(a) zero
(b) +1
(c) – 1
(d) infinity

21. The type-II superconductor in the following is

(a) Nb_3Al
(b) 82 Pb
(c) ^{41} Nb
(d) ^{31} Ga

22. The type -I superconductors in the following are

(a) Nb Ti, Pb MOS
(b) Nb_3Ge, V_3 Si
(c) Nb_3 Sn, Nb_3 Ga
(d) ^{31} Ga, ^{41}Nb

23. Read the following statements and answer

(A) The resistance of a superconductor depends on the intensity of external magneticfield in which it is placed

(B) The specific heat of a superconductor in super conducting state is greater than that in normal state.

(a) (A) is wrong, (B) is correct
(b) (A) and (B) are wrong
(c) (A) is correct, (B) is wrong
(d) A and B are correct

24. The critical magnetic field for a superconductor at T = 0 K is 0.2 Tesla. Then the critical magnetic field at 3.5 K is

(a) 0.2 Tesla
(b) 0.1 Tesla
(c) 0 Tesla
(d) 0.173 Tesla

25. The lead wire has a critical magnetic field intensity is 6.5×10^3 A m^{-1} at 0K. The critical temperature is 7.18 K. At what temperature the critical field would drop to 4.5×10^3 A m^{-1}

(a) 6 K
(b) 4.9 K
(c) 3.98 K
(d) 7.18 K

26. The thin layer of insulator between two parts of a superconductor

(a) permits current without any net loss of energy
(b) forms a strong link and prevents cooper pairs to cross
(c) permits flow of current that is inversely proportional to the voltage
(d) works like a pn junction in reverse bias

27. A ring of superconducting material is placed in a magnetic field acting in its axial drection, and its temperature is reduced below the critical temperature. Then critical temperature

(a) the ring stays in normal state only
(b) the cooper pairs get separated
(c) the field will get trapped through the hole in the ring
(d) the magnetic flux through the ring material is quantized

28. The basic unit of quantized flux is

(a) 4 webers (b) 2.1webers

(c) 0.5×10^{-15} weber (d) 2.1×10^{-15} weber

29. Choose the currect statement

(a) Super conductors are employed in switching devices

(b) Super conducting wries can never be used to produce high magnetic fields

(c) High T_c super conductors are metallic alloys

(d) Quantized lines of flux can never be trapped within the interiors of any form of superconductor

30. The wave functions of cooper pairs on the two sides of the josephson junction differ in frequency by

(a) $\hbar$ (b) $ev/\hbar$

(c) $2\,ev/\hbar$ (d) 2 ev/h

31. When a dc voltage V is applied across a Josephson junction, Then

(a) an alternating current with frequency 2 ev/h is produced

(b) an alternating current with frequency $ev/\hbar$ is produced

(c) a dc current is produced

(d) a dc current that is proportional to the voltage is produced

32. One of the following is a high T_c super conductor

(a) $Y_2\,O_3.\,B_a\,O$ (b) $B_a\,TiO_3$

(c) $Y\,Ba_2\,cu_3\,O_7$ (d) $L\,a_{1.8}\,Sr_{0.2}\,O_{2.7}$

33. The minimum value of quantized flux is

(a) $\frac{2\pi\hbar}{e}$ (b) $\frac{\pi\hbar}{e}$

(c) h /e (d) $\frac{2h}{e}$

34. The dependence of T_c on isotopic mass is given as

(a) $T_c = M^{-1}$ (b) $T_c \alpha M^{1/2}$

(c) $T_c \;\alpha\; M$ (d) $T_c \alpha M^{-1/2}$

II. Fill up the Blanks

1. In Type-I superconductors, the coherence length is __________ as compared to type II superconductors.
2. The relation between the isotopic mass and the critical temperature of a super conductor is __________, where α is the isotopic effect coefficient.

3. A super conductor is placed in the magnectic field of intensity H_0. The fall in the value of the intensity H_0 in the super conductor is expressed as __________, λ is the London penetication depth.
4. The magnetic field density __________ exponentially with distance inside the specimen of a super conductor that is placed in a magnetic field density B.
5. The existence of London penetration depth (λ) indicates that __________ effect is not complete in the super conductor.
6. If a tiny magnet placed is kept over a superconducting bowl, the magnet __________ over the super conductor.
7. The energy spectrum of electrons in a superconductor exhibits __________.
8. The unit of flux quantized in a superconductor is called __________.
9. The quantum of magnetic flux that is trapped in the hole of superconducting ring in a magnetic field is equal to __________.
10. A type II super conductor exhibits __________ critical magnetic field strengths
11. The high T_c Superconductors possess the critical temperatures above the boiling temperature of __________.
12. The London penetiation depth in a superconductor is a function of __________.
13. The London penetiation depth __________ as the temperature of a superconductor is reduced.
14. If λ is the London penetration depth, the magnetic flux density B, and the distance inside from the surface of the superconductor is x, then, B is related to the distance as __________.
15. Superconducting bearings operate without __________.
16. A high T_c super conductor piece is placed on a permanent magnet placed in a boule. When liquid nitrogen is poured into the boule to imerse the superconductor, the superconducting pliece __________ the magnet.
17. Nuclear magnetic resonance imaging used in medical diagnostics of heart and brain is based on __________.
18. The energy gap in a superconductor has maximum width at __________ K.
19. The cooper pair in a superconductor is formed as a result of strong __________ interaction.
20. The cooper pairs are broken at __________ temperature of the superconductor.

APPENDIX 1

The divergence of a vector field:

∇ is a vector operator and it must operate on an operand. If it operates on a scalar function, we get the gradient of a function that is scalar. We will see what will happen when it operates on a vector field. Consider a vector $A = A_x i + A_y j + A_z k$

The divergence of vector A is written as

$$\nabla \bullet A = \left(\mathbf{i}\frac{\partial}{\partial x} + \mathbf{j}\frac{\partial}{\partial y} + \mathbf{k}\frac{\partial}{\partial z} \right) \bullet \left(A_x i + A_y j + A_z k \right)$$

$$= \frac{\partial A_x}{\partial x} + \frac{\partial A_y}{\partial y} + \frac{\partial A_z}{\partial z}$$

Therefore, the divergence of a vector function or field is a scalar.

In order to understand the concept clearly, we take the example of a flow of fluid in a medium. Consider a parallel piped section in the mass of the fluid.

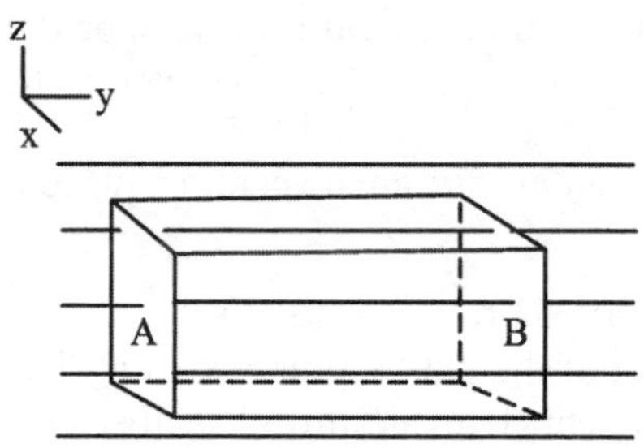

Along y direction, the velocity of the fluid is v_y. With this velocity the fluid enters the surface A. Then, the rate at which the mass of the fluid that enters the face A is

$\rho V_y x = \Delta x \ \Delta z$. The rate at which the fluid mass leaves the surface B is

$$\rho\left[V_y + \frac{\partial V_y}{\partial y}\Delta y\right]\Delta x \ \Delta z$$

Therefore, the net outward flow in direction is

$$-\rho\frac{\partial V_z}{\partial y}\Delta x \ \Delta y \ \Delta z$$

Similarly for x and z direction, the net outward flows are

$$\rho\frac{\partial V_x}{\partial y}\Delta x \ \Delta y \ \Delta z \text{ and } \rho\frac{\partial V_z}{\partial y}\Delta x \ \Delta y \ \Delta z$$

The total net outward flow from the volume $\Delta x \ \Delta y \ \Delta z$ of the parallel piped section is

$$F_z + \frac{\partial F_z}{\partial y}\Delta y \text{ and } F_y + \frac{\partial H_y}{\partial z}\Delta z$$

$$\nabla \bullet (\rho V) = \text{div}(\rho V)$$

This is true for heat flow, or for the electronic flux lines or magnetic flux lines.

If the divergence of a field is positive at a point, it means that there is a source of the field at that point. If it is negative, there is a sink at that point. If there are no sources or sinks, the divergence of a vector field **F** is zero i.e. div **F** = 0 such a field is called solenoidal field. The best example is the magnetic flux density **B**. The divergence of **B** is zero i.e. **B** is solenoidal.

There are there kinds of integrals in vector integration. They are 1. Line integral, 2. Surface integral, 3. Volume integral. The integrand can be vector or scalar field or a function.

Line integral:

Consider a space. Let a vector function or field **F** (x, y, z) be defined at any point in this space. Consider two points A and B in this space. The line integral from A to B along the path C is denoted by $\int_A^B \mathbf{F}.\text{d}$ where **dl** is an infinitesimal vector displacement along the curve C.

We divide the path into very short segments. Each segment is represented by a vector dl and we take the scalar product of the path segment vector dl with the vector **F**. Now, we add these products over whole length. If dl is inifiniliseally small, we can replace the sum by integration.

Line integral $\int \mathbf{F.dl} = \int \mathbf{F.dr}$.

Where $\mathbf{dr} = \text{dr}\mathbf{i} + \text{dr}\mathbf{j} + \text{dr}\mathbf{k}$, i.e. $\int \mathbf{F.dr} = \int (F_x dx + F_y dy + F_z dz)$

The line integral may not depend on the path. There can be infinite number of paths from A to B. Consider two path C_1 and C_2, then for a conservative field

$$\int_{c_1} \mathbf{F.dr} = \int_{c_2} \mathbf{F.dr}$$

If we travel along path C_1, and return along the path C_2, then

$$\oint \mathrm{F.d} = \int_A^B \mathrm{F.d} + \int_B^C \mathbf{F.d} + \int_C^D \mathbf{F.d} + \int_D^A \mathbf{F.d} = \int_A^B \mathbf{F.d} - \int_C^D \mathbf{F.d}$$

$\oint$ is called closed integral.

If $F = \nabla\phi$ then

$$\int_A^B \mathbf{F.dl} = \int_A^B \nabla\phi.\mathrm{dl} = \int_A^B d\phi = \phi_B = \phi_A$$

This is independent of the path but only depends on the end points. If the closed integral is equal to zero, then the field is called irrotational field.

Surface Integral:

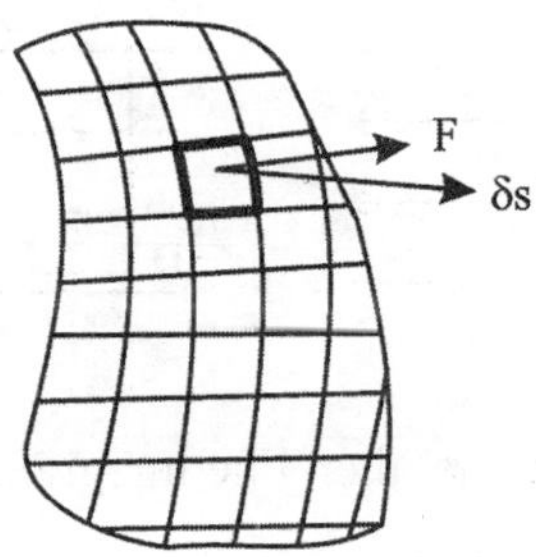

Consider **F** as a vector field.

The dot product **F.** δS represents the flux through the path of surface area S. if we add all these products for all patches of surface we get the flux through the entire surface.

$$\phi = \Sigma \mathbf{F.\delta s} = \int\int_S \mathrm{F.ds}$$

$\mathbf{Ds} = \mathbf{I}\ ds_x + \mathbf{j}\ ds_y + \mathbf{k}\ ds_z$. The double integral indicates the two dimensional nature of the surface.

$$\phi = \int_S (F_x ds_x + F_y ds_y + F_z ds_z)$$

A closed surface encloses a volume e.g., a sphere. Then the surface integral of the normal component of a vector point function **F** taken over a closed surface **S** enclosing a volume V is equal to the volume integral of the divergence of a vector F taken through the volume V.

$$\iint_S \mathbf{F}.\mathbf{Ds} = \iiint_V \text{div } \mathbf{F} \text{ dV}$$

or $$\iint_S (F_x ds_x + F_y ds_y + F_z ds_z) = \iiint_V \left(\frac{\partial F_x}{\partial x} + \frac{\partial F_y}{\partial y} + \frac{\partial F_z}{\partial z} \right) dxdydz$$

The above theorem is called Gauss's divergence theorem. The proof of this theorem can be found in any standard text books of vector analysis.

Curl of a Vector: The line integral around a closed path is zero. Then the field is called irrotational field. There are several other vector fields like the magnetic field, whose line integral around a finite closed path is not zero. These fields are known as rotational fields.

Consider a small or an incremental plane area Δs in space. The region Δs is a very small region of rotational vector field.

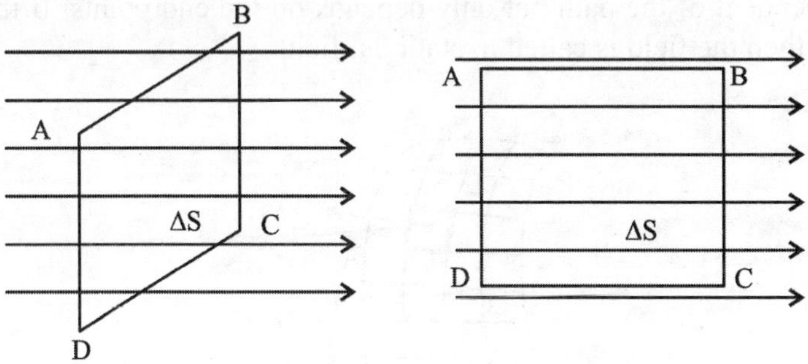

If the vector field is normal to plane, then line integral along its sides is

$$\oint \mathbf{F}.\mathbf{dl} = \int_A^B \mathbf{F}.\mathbf{dl} + \int_B^C \mathbf{F}.\mathbf{dl} + \int_C^D \mathbf{F}.\mathbf{dl} + \int_P^A \mathbf{F}.\mathbf{dl} = 0$$

If the plane is rotation through 90° such that the vector field is parallel to it, then

$$\oint \mathbf{F}.\mathbf{dl} = \int_A^B \mathbf{F}.\mathbf{dl} + \int_B^C \mathbf{F}.\mathbf{dl} + \int_C^D \mathbf{F}.\mathbf{dl} + \int_D^A \mathbf{F}.\mathbf{dl} = \int_A^B \mathbf{F}.\mathbf{dl} - \int_C^D \mathbf{F}.\mathbf{dl}$$

This gives a finite value. Therefore, the line integral along the edges of the surface Δs depends on its orientation with respect to the vector field. At particular value of the area through which the vector field penetrates, the line integral takes maximum value. This maximum line integral at any point per unit area in such a way the vector field forms a closed path or loop, is called Curl of a vector field.

$$\text{Curl } \mathbf{F} = \text{Limit}_{\Delta s \to 0} \frac{\oint \mathbf{F.dl}}{\Delta \mathbf{S}}$$

$$\text{Curl } \mathbf{F} = \mathbf{J}$$

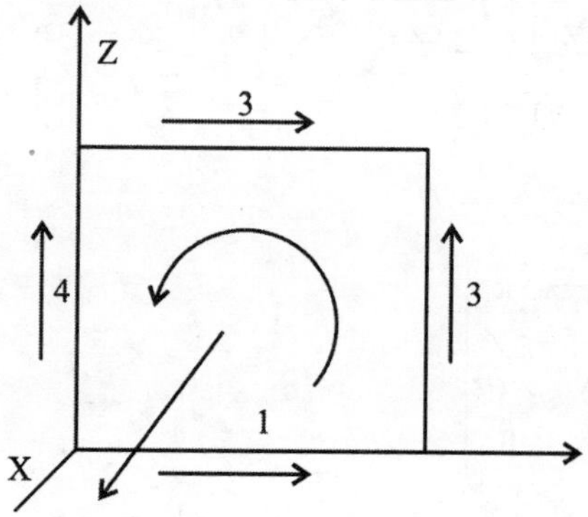

The Curl of a vector field is a vector quantity directed along the normal to the unit area towards which a right handed screw moves when turned in the sense in which the line integral along the boundary of the test area is maximum.

Referring to the Figure let us calculate the value of the component of F in the x direction by taking the integral of F around the periphery of the area $\Delta y \Delta z(\Delta s)$ in four incremental steps. Let F along side 1 equal Fy and along side 4 equal Hz. if the field is not uniform, the values of **F** along sides 2 and 3 are given by:

$$F_z + \frac{\partial F_z}{\partial y} \Delta y \text{ and } F_y + \frac{\partial F_y}{\partial y} \Delta z$$

$$\oint \mathbf{F.dl} = F_y \Delta y + F_z \Delta z + \frac{\partial H_z}{\partial y} \Delta y \Delta z - F_y \Delta y$$

$$- \frac{\partial F_y}{\partial z} \Delta z \Delta y - F_z \Delta z = J_x \Delta y \Delta z$$

Dividing by the area Δy Δz, we obtained the component of Curl **F** in the x-direction as

$$\text{Curl}_x \mathbf{F} = \text{Lim}_{\Delta y \Delta z \to 0} \frac{\oint \mathbf{F.dl}}{\Delta y \Delta z} = \frac{\partial F_z}{\partial y} - \frac{\partial F_y}{\partial z} = J_x$$

If the field is in all direction, then Curl **F** also has components in the y and z direction.

∴ Generally

$$\text{Curl}\,\mathbf{F} = \mathbf{i}\left[\frac{\partial F_z}{\partial y} - \frac{\partial F_y}{\partial z}\right] + \mathbf{j}\left[\frac{\partial F_x}{\partial z} - \frac{\partial H_z}{\partial x}\right] + \mathbf{k}\left[\frac{\partial F_y}{\partial x} - \frac{\partial F_z}{\partial y}\right]$$

$$= \mathbf{i}J_x + \mathbf{j}J_y + \mathbf{k}J_z = \mathbf{J}$$

Curl **F** = **J**

$$\nabla \times F = \begin{vmatrix} \mathbf{i} & \mathbf{j} & \mathbf{k} \\ \partial/\partial x & \partial/\partial y & \partial/\partial z \\ F_x & F_y & F_z \end{vmatrix}$$

If F is replaced by magnetic field intensity then $\nabla \times \mathbf{H} = \mathbf{J}$

J is the current density. This is differential form of ampere's law.

Stokes theorem: Consider a infinitesimal area ∇s is the xy plane.

The vector field E have components Ey and Ex as shown.

$$\text{Curl}\,E = \text{Limit}_{\Delta s \to 0} \frac{\oint E \cdot dl}{\Delta s}$$

Let an area be x_1, y_1 as shown in Fig. Let this area be divided into infinitesimal areas as shown. The Curl of **E** at & point in the small area is equal to the line integral around it. If the Curl of E is integrated over entire area x_1, y_1, all contributions to the line integral cancel except those along the periphery of the area x_1, y_1. The integral of the normal component of the curl over the area x_1 y_1 must equal the line integral of **E** around the periphery of the area. That is

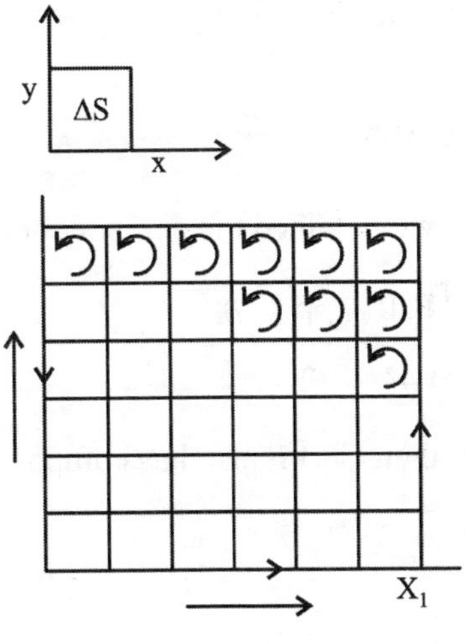

$$\oint \mathrm{E} \cdot \mathbf{dl} = \int_{s} \nabla \times \mathbf{E.ds}$$

This is called Stoke's theorem. It states that the line integral of a vector function around a closed contour C is equal to the integral of the normal component of the having the curl of that vector function over any surface having the contour C as bounding edge. If E is the electric field at every point, then

$$\iint_{s} \nabla \times \mathrm{E} = \int \mathbf{E} \cdot \mathrm{dl} = -\int \frac{\partial \mathbf{B}}{\partial \mathrm{t}} .\mathbf{ds}$$

This later point of the equation is from Faradays law of electromagnetic induction.

$$\therefore \qquad \nabla \times \mathbf{E} = \frac{\partial \mathbf{B}}{\partial \mathrm{t}}$$

APPENDIX 2

London penetration depth: The characteristic property of a superconductor is that B = 0 inside and therefore, that any current existing must be superficial. The magnetic susceptibility of a superconductor is –1 (SI units). However, this cannot be generalized and the current can not entirely be superficial. There must be a certain depth, however small, of penetration of the magnetic field into the superconductor. London (1935, 1939) observed the penetration effects as important when the samples of superconductors were taken an thin films. In bulk sample the depth is very small. Let it be supposed that the field inside the surface of an infinitely thick sample falls from its value Ho. If the field varies as exponentially from the surface into the material, then it is expressed as the value at the surface to a value H(x) inside along the x direction. The fall in H is expressed as

$$-dH = \frac{H(x)dx}{\lambda} \quad .....(A.1)$$

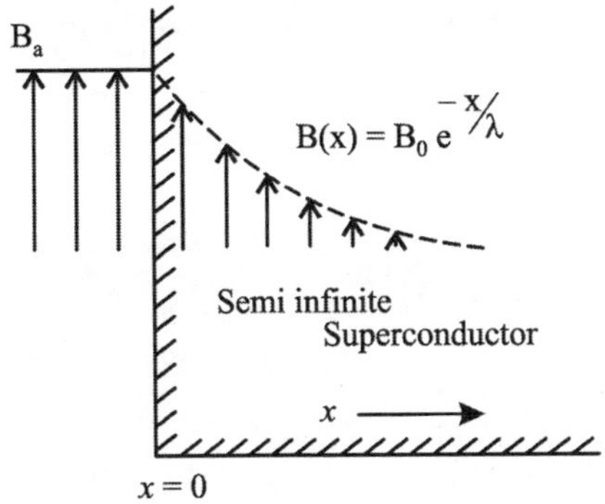

Fig. A.1 Shows the penetration of an applied magnetic field into semi infinite superconductor.

$$H = -\int_{H_0}^{0} \frac{H(x)dx}{\lambda} = H_0 e^{-x/\lambda}$$

In terms of magnetic induction

$$B = B_0 e^{-x/\lambda} \qquad .....(A.2)$$

λ is known as London penetration depth.

The macroscopic description of a superconductor is given as a body with B = 0 and E = 0 inside it. London considered a superconductor as a condensed state consisting of super electrons and normal electrons. The total number of electrons in the superconductors N is equal to the sum of the concentration of normal electrons n, and the concentration of super electrons n_s. The concentrations are taken as per unit volume.

If the super electrons move with a velocity v_s, then,

$$m\frac{dv_s}{dt} = e\mathbf{E} \qquad .....(A.3)$$

Let J_s represent the super current density, then it is given by

$$\mathbf{J}_s = n_s e v_s \qquad .....(A.4)$$

$$\therefore \quad \frac{d\mathbf{J}_s}{dt} = \left(\frac{e^2 n_s}{m}\right)\mathbf{E} \qquad .....(A.5)$$

Taking curl on both sides of (11.19)

$$\text{Curl}\frac{d\mathbf{J}_s}{dt} = \text{Curl}\left(\frac{e^2 n_s}{m}\right)\mathbf{E} = \left(\frac{e^2 n_s}{m}\right)\text{Curl}\,\mathbf{E} \qquad .....(A.6)$$

But Curl $\mathbf{E} = -\delta\mathbf{B}/\delta t$, (equation 11.3)

$$\therefore \text{Curl}\frac{d\mathbf{J}_s}{dt} = -\left(\frac{e^2 n_s}{m}\right)\frac{\partial \mathbf{B}}{\partial t} \qquad .....(A.7)$$

or

$$\text{Curl }\mathbf{J}_s = -\frac{e^2 n_s}{m}\mathbf{B} \qquad .....(A.8)$$

The electric field can not penetrate, but the magnetic field penetrates the superconductor, But according to Maxwell's equation based on Ampere's law. Curl $\mathbf{B} = \mu_o \mathbf{J}_s$ (n_s is equal to the superconducting electrons), it can be seen that

$$\text{Curl}\,\text{Curl}\,\mathbf{B} = \mu_o\,\text{Curl}\,\mathbf{J}_s \qquad .....(A.9)$$

But Curl Curl $\mathbf{B}$ = gradient div $\mathbf{B} = -\nabla^2\mathbf{B}$

$$\text{i.e., } -\nabla^2\mathbf{B} = \mu_0\text{Curl }\mathbf{J}_s = -\frac{e^2 n_2 \mu_o}{m}\mathbf{B} \qquad .....(A.10)$$

$$\text{i.e. } \nabla^2 \mathbf{B} = \left(\frac{1}{\lambda^2}\right)\mathbf{B} \quad \text{.....(A.11)}$$

$$\lambda^2 = \left(\frac{m}{\mu_o n_s e^2}\right) \quad \text{.....(A.12)}$$

i.e. B decays exponentially with distance inside the specimen from surface.

Equation (A.5) and (A.8) are known as London equations.

From above, it can be seen that $J_s = 0$ whereever $B = 0$, i.e. well inside the superconductor. λ, the London penetration depth is a function of temperature and the relation between temperature and λ is

$$\lambda/\lambda(o) = [1-(T/T_c)^4]^{-1/2} \quad \text{.....(A.13)}$$

where λ(o) is the London penetration depth at T = 0K.

London Penetration depths at absolute zero for various metals is given in the following Table A.1. The equation (A.13) also implies that the density of superconducting electrons increases from zero at T_c to maximum at 0°K

Table A.1

Element	$\lambda 10^{-8}$ m
Sn	3.4
Al	1.6
Pb	3.7
Nb	3.9
Hg	5.0

Thus in a superconductor, Meissner effect is not complete, London penetration depth λ is a fundamental length that characterizes a superconductor. λ is of the order of 10^{-8} m. The penetration depth λ decreases as temperature is reduced.

The diamagnetic nature of a superconductor can be demonstrated by floating magnet experiment. If a small permanent magnet is lowered over a superconducting surface, the lines of force of the magnet can not enter the superconductor and this results in a repulsive force large enough to over come the weight of the magnet. The Fig. A.2 shows a floating magnet above a superconductor. This sort of levitation is important in applications.

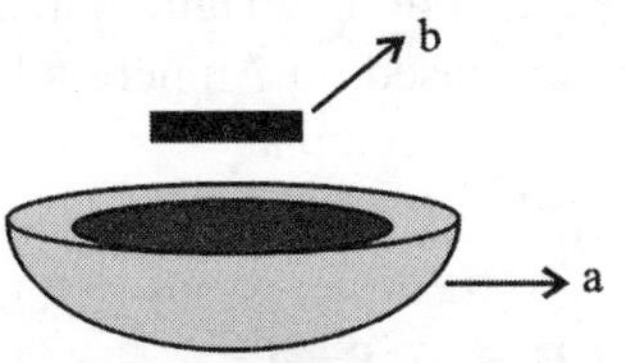

a is bowl of superconductor
b is floating magnet

8.tif

Fig. A.2 A magnet floats over a superconducting bowl.